STUDENT SOLUTIONS MANUAL
for Stewart's
SINGLE VARIABLE CALCULUS
CONCEPTS AND CONTEXTS
FOURTH EDITION

JEFFERY A. COLE
Anoka Ramsey Community College

TIMOTHY J. FLAHERTY
Carnegie Mellon University

BROOKS/COLE
CENGAGE Learning™

Australia • Brazil • Japan • Korea • Mexico • Singapore • Spain • United Kingdom • United States

For product information and technology assistance, contact us at **Cengage Learning Customer & Sales Support, 1-800-354-9706**

For permission to use material from this text or product, submit all requests online at **www.cengage.com/permissions.** Further permissions questions can be emailed to **permissionrequest@cengage.com.**

ISBN-13: 978-0-495-56061-6
ISBN-10: 0-495-56061-8

Brooks/Cole Cengage Learning
20 Davis Drive
Belmont, CA 94002-3098
USA

Cengage Learning is a leading provider of customized learning solutions with office locations around the globe, including Singapore, the United Kingdom, Australia, Mexico, Brazil, and Japan. Locate your local office at **www.cengage.com/international.**

Cengage Learning products are represented in Canada by Nelson Education, Ltd.

To learn more about Brooks/Cole, visit **www.cengage.com/brookscole**

Purchase any of our products at your local college store or at our preferred online store **www.cengagebrain.com.**

Printed in the United States of America
2 3 4 5 6   20 19 18 17 16

□ PREFACE

This Student Solutions Manual contains problem-solving strategies and solutions to selected exercises in the texts *Single Variable Calculus: Concepts and Contexts,* Fourth Edition, and Chapters 1–8 of *Calculus: Concepts and Contexts,* Fourth Edition, by James Stewart. It contains solutions to the odd-numbered exercises in each chapter section, the review sections, the True-False Quizzes, and the Focus on Problem Solving sections, as well as solutions to all the exercises in the Concept Checks.

This manual is a text supplement and should be read along with the text. You should read all exercise solutions in this manual because many concept explanations are given and then used in subsequent solutions. All concepts necessary to solve a particular problem are not reviewed for every exercise. If you are having difficulty with a previously covered concept, refer back to the section where it was covered for more complete help.

A significant number of today's students are involved in various outside activities, and find it difficult, if not impossible, to attend all class sessions; this manual should help meet the needs of these students. In addition, it is our hope that this manual will enhance the understanding of all readers and provide insights that will help them solve other exercises.

Some nonstandard notation is used in order to save space. If you see a symbol that you don't recognize, refer to the table of Abbreviations and Symbols on page v.

We appreciate feedback concerning errors, solution correctness or style, and manual style. Any comments may be sent directly to us at jeff.cole@anokaramsey.edu or tim@andrew.cmu.edu, or in care of the publisher: Cengage Learning Brooks/Cole, 10 Davis Drive, Belmont, CA 94002.

We would like to thank Jim Stewart, for his guidance; Brian Betsill, Kathi Townes, and Rebekah Million, of TECH-arts, for their production services; and Richard Stratton and Jeannine Lawless, of Cengage Learning Brooks/Cole, for entrusting us with this project as well as for their patience and support.

Jeffery A. Cole
Anoka-Ramsey Community College

Timothy J. Flaherty
Carnegie Mellon University

□ ABBREVIATIONS AND SYMBOLS

CD	concave downward
CU	concave upward
D	the domain of f
FDT	First Derivative Test
HA	horizontal asymptote(s)
I	interval of convergence
I/D	Increasing/Decreasing Test
IP	inflection point(s)
R	radius of convergence
VA	vertical asymptote(s)
$\overset{CAS}{=}$	indicates the use of a computer algebra system.
$\overset{H}{=}$	indicates the use of l'Hospital's Rule.
$\overset{j}{=}$	indicates the use of Formula j in the Table of Integrals in the back endpapers.
$\overset{s}{=}$	indicates the use of the substitution $\{u = \sin x, du = \cos x\, dx\}$.
$\overset{c}{=}$	indicates the use of the substitution $\{u = \cos x, du = -\sin x\, dx\}$.

CONTENTS

6 □ APPLICATIONS OF INTEGRATION 259

7 □ DIFFERENTIAL EQUATIONS 301

8 □ INFINITE SEQUENCES AND SERIES 331

☐ APPENDIXES   379

☐ DIAGNOSTIC TESTS

Test A   Algebra

1. (a) $(-3)^4 = (-3)(-3)(-3)(-3) = 81$

   (b) $-3^4 = -(3)(3)(3)(3) = -81$

   (c) $3^{-4} = \dfrac{1}{3^4} = \dfrac{1}{81}$

   (d) $\dfrac{5^{23}}{5^{21}} = 5^{23-21} = 5^2 = 25$

   (e) $\left(\dfrac{2}{3}\right)^{-2} = \left(\dfrac{3}{2}\right)^2 = \dfrac{9}{4}$

   (f) $16^{-3/4} = \dfrac{1}{16^{3/4}} = \dfrac{1}{\left(\sqrt[4]{16}\right)^3} = \dfrac{1}{2^3} = \dfrac{1}{8}$

2. (a) Note that $\sqrt{200} = \sqrt{100 \cdot 2} = 10\sqrt{2}$ and $\sqrt{32} = \sqrt{16 \cdot 2} = 4\sqrt{2}$. Thus $\sqrt{200} - \sqrt{32} = 10\sqrt{2} - 4\sqrt{2} = 6\sqrt{2}$.

   (b) $(3a^3b^3)(4ab^2)^2 = 3a^3b^3 16a^2b^4 = 48a^5b^7$

   (c) $\left(\dfrac{3x^{3/2}y^3}{x^2y^{-1/2}}\right)^{-2} = \left(\dfrac{x^2y^{-1/2}}{3x^{3/2}y^3}\right)^2 = \dfrac{(x^2y^{-1/2})^2}{(3x^{3/2}y^3)^2} = \dfrac{x^4y^{-1}}{9x^3y^6} = \dfrac{x^4}{9x^3y^6y} = \dfrac{x}{9y^7}$

3. (a) $3(x+6) + 4(2x-5) = 3x + 18 + 8x - 20 = 11x - 2$

   (b) $(x+3)(4x-5) = 4x^2 - 5x + 12x - 15 = 4x^2 + 7x - 15$

   (c) $\left(\sqrt{a}+\sqrt{b}\right)\left(\sqrt{a}-\sqrt{b}\right) = \left(\sqrt{a}\right)^2 - \sqrt{a}\sqrt{b} + \sqrt{a}\sqrt{b} - \left(\sqrt{b}\right)^2 = a - b$

   Or: Use the formula for the difference of two squares to see that $\left(\sqrt{a}+\sqrt{b}\right)\left(\sqrt{a}-\sqrt{b}\right) = \left(\sqrt{a}\right)^2 - \left(\sqrt{b}\right)^2 = a - b.$

   (d) $(2x+3)^2 = (2x+3)(2x+3) = 4x^2 + 6x + 6x + 9 = 4x^2 + 12x + 9.$

   Note: A quicker way to expand this binomial is to use the formula $(a+b)^2 = a^2 + 2ab + b^2$ with $a = 2x$ and $b = 3$:
   $$(2x+3)^2 = (2x)^2 + 2(2x)(3) + 3^2 = 4x^2 + 12x + 9$$

   (e) See Reference Page 1 for the binomial formula $(a+b)^3 = a^3 + 3a^2b + 3ab^2 + b^3$. Using it, we get
   $$(x+2)^3 = x^3 + 3x^2(2) + 3x(2^2) + 2^3 = x^3 + 6x^2 + 12x + 8.$$

4. (a) Using the difference of two squares formula, $a^2 - b^2 = (a+b)(a-b)$, we have
   $$4x^2 - 25 = (2x)^2 - 5^2 = (2x+5)(2x-5).$$

   (b) Factoring by trial and error, we get $2x^2 + 5x - 12 = (2x-3)(x+4)$.

   (c) Using factoring by grouping and the difference of two squares formula, we have
   $$x^3 - 3x^2 - 4x + 12 = x^2(x-3) - 4(x-3) = (x^2-4)(x-3) = (x-2)(x+2)(x-3).$$

   (d) $x^4 + 27x = x(x^3 + 27) = x(x+3)(x^2 - 3x + 9)$

   This last expression was obtained using the sum of two cubes formula, $a^3 + b^3 = (a+b)(a^2 - ab + b^2)$ with $a = x$ and $b = 3$. [See Reference Page 1 in the textbook.]

   (e) The smallest exponent on x is $-\frac{1}{2}$, so we will factor out $x^{-1/2}$.
   $$3x^{3/2} - 9x^{1/2} + 6x^{-1/2} = 3x^{-1/2}(x^2 - 3x + 2) = 3x^{-1/2}(x-1)(x-2)$$

   (f) $x^3y - 4xy = xy(x^2 - 4) = xy(x-2)(x+2)$

5. (a) $\dfrac{x^2 + 3x + 2}{x^2 - x - 2} = \dfrac{(x+1)(x+2)}{(x+1)(x-2)} = \dfrac{x+2}{x-2}$

(b) $\dfrac{2x^2 - x - 1}{x^2 - 9} \cdot \dfrac{x+3}{2x+1} = \dfrac{(2x+1)(x-1)}{(x-3)(x+3)} \cdot \dfrac{x+3}{2x+1} = \dfrac{x-1}{x-3}$

(c) $\dfrac{x^2}{x^2 - 4} - \dfrac{x+1}{x+2} = \dfrac{x^2}{(x-2)(x+2)} - \dfrac{x+1}{x+2} = \dfrac{x^2}{(x-2)(x+2)} - \dfrac{x+1}{x+2} \cdot \dfrac{x-2}{x-2} = \dfrac{x^2 - (x+1)(x-2)}{(x-2)(x+2)}$

$\quad = \dfrac{x^2 - (x^2 - x - 2)}{(x+2)(x-2)} = \dfrac{x+2}{(x+2)(x-2)} = \dfrac{1}{x-2}$

(d) $\dfrac{\dfrac{y}{x} - \dfrac{x}{y}}{\dfrac{1}{y} - \dfrac{1}{x}} = \dfrac{\dfrac{y}{x} - \dfrac{x}{y}}{\dfrac{1}{y} - \dfrac{1}{x}} \cdot \dfrac{xy}{xy} = \dfrac{y^2 - x^2}{x - y} = \dfrac{(y-x)(y+x)}{-(y-x)} = \dfrac{y+x}{-1} = -(x+y)$

6. (a) $\dfrac{\sqrt{10}}{\sqrt{5} - 2} = \dfrac{\sqrt{10}}{\sqrt{5} - 2} \cdot \dfrac{\sqrt{5} + 2}{\sqrt{5} + 2} = \dfrac{\sqrt{50} + 2\sqrt{10}}{\left(\sqrt{5}\right)^2 - 2^2} = \dfrac{5\sqrt{2} + 2\sqrt{10}}{5 - 4} = 5\sqrt{2} + 2\sqrt{10}$

(b) $\dfrac{\sqrt{4+h} - 2}{h} = \dfrac{\sqrt{4+h} - 2}{h} \cdot \dfrac{\sqrt{4+h} + 2}{\sqrt{4+h} + 2} = \dfrac{4 + h - 4}{h(\sqrt{4+h} + 2)} = \dfrac{h}{h(\sqrt{4+h} + 2)} = \dfrac{1}{\sqrt{4+h} + 2}$

7. (a) $x^2 + x + 1 = \left(x^2 + x + \frac{1}{4}\right) + 1 - \frac{1}{4} = \left(x + \frac{1}{2}\right)^2 + \frac{3}{4}$

(b) $2x^2 - 12x + 11 = 2(x^2 - 6x) + 11 = 2(x^2 - 6x + 9 - 9) + 11 = 2(x^2 - 6x + 9) - 18 + 11 = 2(x - 3)^2 - 7$

8. (a) $x + 5 = 14 - \frac{1}{2}x \iff x + \frac{1}{2}x = 14 - 5 \iff \frac{3}{2}x = 9 \iff x = \frac{2}{3} \cdot 9 \iff x = 6$

(b) $\dfrac{2x}{x+1} = \dfrac{2x-1}{x} \implies 2x^2 = (2x-1)(x+1) \iff 2x^2 = 2x^2 + x - 1 \iff x = 1$

(c) $x^2 - x - 12 = 0 \iff (x+3)(x-4) = 0 \iff x + 3 = 0 \text{ or } x - 4 = 0 \iff x = -3 \text{ or } x = 4$

(d) By the quadratic formula, $2x^2 + 4x + 1 = 0 \iff$

$\quad x = \dfrac{-4 \pm \sqrt{4^2 - 4(2)(1)}}{2(2)} = \dfrac{-4 \pm \sqrt{8}}{4} = \dfrac{-4 \pm 2\sqrt{2}}{4} = \dfrac{2\left(-2 \pm \sqrt{2}\right)}{4} = \dfrac{-2 \pm \sqrt{2}}{2} = -1 \pm \frac{1}{2}\sqrt{2}.$

(e) $x^4 - 3x^2 + 2 = 0 \iff (x^2 - 1)(x^2 - 2) = 0 \iff x^2 - 1 = 0 \text{ or } x^2 - 2 = 0 \iff x^2 = 1 \text{ or } x^2 = 2 \iff$

$\quad x = \pm 1 \text{ or } x = \pm\sqrt{2}$

(f) $3\,|x - 4| = 10 \iff |x - 4| = \frac{10}{3} \iff x - 4 = -\frac{10}{3} \text{ or } x - 4 = \frac{10}{3} \iff x = \frac{2}{3} \text{ or } x = \frac{22}{3}$

(g) Multiplying through $2x(4 - x)^{-1/2} - 3\sqrt{4 - x} = 0$ by $(4 - x)^{1/2}$ gives $2x - 3(4 - x) = 0 \iff$

$\quad 2x - 12 + 3x = 0 \iff 5x - 12 = 0 \iff 5x = 12 \iff x = \frac{12}{5}.$

9. (a) $-4 < 5 - 3x \le 17 \iff -9 < -3x \le 12 \iff 3 > x \ge -4 \text{ or } -4 \le x < 3.$

In interval notation, the answer is $[-4, 3)$.

(b) $x^2 < 2x + 8 \iff x^2 - 2x - 8 < 0 \iff (x + 2)(x - 4) < 0$. Now, $(x + 2)(x - 4)$ will change sign at the critical values $x = -2$ and $x = 4$. Thus the possible intervals of solution are $(-\infty, -2)$, $(-2, 4)$, and $(4, \infty)$. By choosing a single test value from each interval, we see that $(-2, 4)$ is the only interval that satisfies the inequality.

(c) The inequality $x(x-1)(x+2) > 0$ has critical values of $-2, 0,$ and 1. The corresponding possible intervals of solution are $(-\infty, -2)$, $(-2, 0)$, $(0, 1)$ and $(1, \infty)$. By choosing a single test value from each interval, we see that both intervals $(-2, 0)$ and $(1, \infty)$ satisfy the inequality. Thus, the solution is the union of these two intervals: $(-2, 0) \cup (1, \infty)$.

(d) $|x-4| < 3 \iff -3 < x - 4 < 3 \iff 1 < x < 7$. In interval notation, the answer is $(1, 7)$.

(e) $\dfrac{2x-3}{x+1} \le 1 \iff \dfrac{2x-3}{x+1} - 1 \le 0 \iff \dfrac{2x-3}{x+1} - \dfrac{x+1}{x+1} \le 0 \iff \dfrac{2x-3-x-1}{x+1} \le 0 \iff \dfrac{x-4}{x+1} \le 0.$

Now, the expression $\dfrac{x-4}{x+1}$ may change signs at the critical values $x = -1$ and $x = 4$, so the possible intervals of solution are $(-\infty, -1)$, $(-1, 4]$, and $[4, \infty)$. By choosing a single test value from each interval, we see that $(-1, 4]$ is the only interval that satisfies the inequality.

10. (a) False. In order for the statement to be true, it must hold for all real numbers, so, to show that the statement is false, pick $p = 1$ and $q = 2$ and observe that $(1+2)^2 \ne 1^2 + 2^2$. In general, $(p+q)^2 = p^2 + 2pq + q^2$.

(b) True as long as a and b are nonnegative real numbers. To see this, think in terms of the laws of exponents:
$$\sqrt{ab} = (ab)^{1/2} = a^{1/2}b^{1/2} = \sqrt{a}\,\sqrt{b}.$$

(c) False. To see this, let $p = 1$ and $q = 2$, then $\sqrt{1^2 + 2^2} \ne 1 + 2$.

(d) False. To see this, let $T = 1$ and $C = 2$, then $\dfrac{1 + 1(2)}{2} \ne 1 + 1$.

(e) False. To see this, let $x = 2$ and $y = 3$, then $\dfrac{1}{2-3} \ne \dfrac{1}{2} - \dfrac{1}{3}$.

(f) True since $\dfrac{1/x}{a/x - b/x} \cdot \dfrac{x}{x} = \dfrac{1}{a-b}$, as long as $x \ne 0$ and $a - b \ne 0$.

Test B   Analytic Geometry

1. (a) Using the point $(2, -5)$ and $m = -3$ in the point-slope equation of a line, $y - y_1 = m(x - x_1)$, we get
$$y - (-5) = -3(x - 2) \implies y + 5 = -3x + 6 \implies y = -3x + 1.$$

(b) A line parallel to the x-axis must be horizontal and thus have a slope of 0. Since the line passes through the point $(2, -5)$, the y-coordinate of every point on the line is -5, so the equation is $y = -5$.

(c) A line parallel to the y-axis is vertical with undefined slope. So the x-coordinate of every point on the line is 2 and so the equation is $x = 2$.

(d) Note that $2x - 4y = 3 \implies -4y = -2x + 3 \implies y = \frac{1}{2}x - \frac{3}{4}$. Thus the slope of the given line is $m = \frac{1}{2}$. Hence, the slope of the line we're looking for is also $\frac{1}{2}$ (since the line we're looking for is required to be parallel to the given line). So the equation of the line is $y - (-5) = \frac{1}{2}(x - 2) \implies y + 5 = \frac{1}{2}x - 1 \implies y = \frac{1}{2}x - 6.$

2. First we'll find the distance between the two given points in order to obtain the radius, r, of the circle:
$$r = \sqrt{[3 - (-1)]^2 + (-2 - 4)^2} = \sqrt{4^2 + (-6)^2} = \sqrt{52}.$$ Next use the standard equation of a circle, $(x-h)^2 + (y-k)^2 = r^2$, where (h, k) is the center, to get $(x+1)^2 + (y-4)^2 = 52$.

3. We must rewrite the equation in standard form in order to identify the center and radius. Note that

$x^2 + y^2 - 6x + 10y + 9 = 0 \Rightarrow x^2 - 6x + 9 + y^2 + 10y = 0$. For the left-hand side of the latter equation, we

factor the first three terms and complete the square on the last two terms as follows: $x^2 - 6x + 9 + y^2 + 10y = 0 \Rightarrow$

$(x - 3)^2 + y^2 + 10y + 25 = 25 \Rightarrow (x - 3)^2 + (y + 5)^2 = 25$. Thus, the center of the circle is $(3, -5)$ and the radius is 5.

4. (a) $A(-7, 4)$ and $B(5, -12) \Rightarrow m_{AB} = \dfrac{-12 - 4}{5 - (-7)} = \dfrac{-16}{12} = -\dfrac{4}{3}$

(b) $y - 4 = -\frac{4}{3}[x - (-7)] \Rightarrow y - 4 = -\frac{4}{3}x - \frac{28}{3} \Rightarrow 3y - 12 = -4x - 28 \Rightarrow 4x + 3y + 16 = 0$. Putting $y = 0$,

we get $4x + 16 = 0$, so the x-intercept is -4, and substituting 0 for x results in a y-intercept of $-\frac{16}{3}$.

(c) The midpoint is obtained by averaging the corresponding coordinates of both points: $\left(\frac{-7+5}{2}, \frac{4+(-12)}{2}\right) = (-1, -4)$.

(d) $d = \sqrt{[5 - (-7)]^2 + (-12 - 4)^2} = \sqrt{12^2 + (-16)^2} = \sqrt{144 + 256} = \sqrt{400} = 20$

(e) The perpendicular bisector is the line that intersects the line segment $\overline{AB}$ at a right angle through its midpoint. Thus the

perpendicular bisector passes through $(-1, -4)$ and has slope $\frac{3}{4}$ [the slope is obtained by taking the negative reciprocal of

the answer from part (a)]. So the perpendicular bisector is given by $y + 4 = \frac{3}{4}[x - (-1)]$ or $3x - 4y = 13$.

(f) The center of the required circle is the midpoint of $\overline{AB}$, and the radius is half the length of $\overline{AB}$, which is 10. Thus, the

equation is $(x + 1)^2 + (y + 4)^2 = 100$.

5. (a) Graph the corresponding horizontal lines (given by the equations $y = -1$ and

$y = 3$) as solid lines. The inequality $y \geq -1$ describes the points (x, y) that lie

on or *above* the line $y = -1$. The inequality $y \leq 3$ describes the points (x, y)

that lie on or *below* the line $y = 3$. So the pair of inequalities $-1 \leq y \leq 3$

describes the points that lie on or *between* the lines $y = -1$ and $y = 3$.

(b) Note that the given inequalities can be written as $-4 < x < 4$ and $-2 < y < 2$,

respectively. So the region lies between the vertical lines $x = -4$ and $x = 4$ and

between the horizontal lines $y = -2$ and $y = 2$. As shown in the graph, the

region common to both graphs is a rectangle (minus its edges) centered at the

origin.

(c) We first graph $y = 1 - \frac{1}{2}x$ as a dotted line. Since $y < 1 - \frac{1}{2}x$, the points in the

region lie *below* this line.

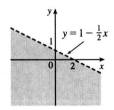

(d) We first graph the parabola $y = x^2 - 1$ using a solid curve. Since $y \geq x^2 - 1$, the points in the region lie on or *above* the parabola.

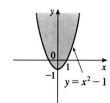

(e) We graph the circle $x^2 + y^2 = 4$ using a dotted curve. Since $\sqrt{x^2 + y^2} < 2$, the region consists of points whose distance from the origin is less than 2, that is, the points that lie *inside* the circle.

(f) The equation $9x^2 + 16y^2 = 144$ is an ellipse centered at $(0, 0)$. We put it in standard form by dividing by 144 and get $\dfrac{x^2}{16} + \dfrac{y^2}{9} = 1$. The x-intercepts are located at a distance of $\sqrt{16} = 4$ from the center while the y-intercepts are a distance of $\sqrt{9} = 3$ from the center (see the graph).

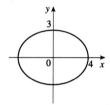

Test C   Functions

1. (a) Locate -1 on the x-axis and then go down to the point on the graph with an x-coordinate of -1. The corresponding y-coordinate is the value of the function at $x = -1$, which is -2. So, $f(-1) = -2$.

   (b) Using the same technique as in part (a), we get $f(2) \approx 2.8$.

   (c) Locate 2 on the y-axis and then go left and right to find all points on the graph with a y-coordinate of 2. The corresponding x-coordinates are the x-values we are searching for. So $x = -3$ and $x = 1$.

   (d) Using the same technique as in part (c), we get $x \approx -2.5$ and $x \approx 0.3$.

   (e) The domain is all the x-values for which the graph exists, and the range is all the y-values for which the graph exists. Thus, the domain is $[-3, 3]$, and the range is $[-2, 3]$.

2. Note that $f(2 + h) = (2 + h)^3$ and $f(2) = 2^3 = 8$. So the difference quotient becomes

$$\frac{f(2+h) - f(2)}{h} = \frac{(2+h)^3 - 8}{h} = \frac{8 + 12h + 6h^2 + h^3 - 8}{h} = \frac{12h + 6h^2 + h^3}{h} = \frac{h(12 + 6h + h^2)}{h} = 12 + 6h + h^2.$$

3. (a) Set the denominator equal to 0 and solve to find restrictions on the domain: $x^2 + x - 2 = 0 \;\Rightarrow\;$ $(x - 1)(x + 2) = 0 \;\Rightarrow\; x = 1$ or $x = -2$. Thus, the domain is all real numbers except 1 or -2 or, in interval notation, $(-\infty, -2) \cup (-2, 1) \cup (1, \infty)$.

   (b) Note that the denominator is always greater than or equal to 1, and the numerator is defined for all real numbers. Thus, the domain is $(-\infty, \infty)$.

   (c) Note that the function h is the sum of two root functions. So h is defined on the intersection of the domains of these two root functions. The domain of a square root function is found by setting its radicand greater than or equal to 0. Now,

$4 - x \geq 0 \;\Rightarrow\; x \leq 4$ and $x^2 - 1 \geq 0 \;\Rightarrow\; (x-1)(x+1) \geq 0 \;\Rightarrow\; x \leq -1$ or $x \geq 1$. Thus, the domain of h is $(-\infty, -1] \cup [1, 4]$.

4. (a) Reflect the graph of f about the x-axis.

(b) Stretch the graph of f vertically by a factor of 2, then shift 1 unit downward.

(c) Shift the graph of f right 3 units, then up 2 units.

5. (a) Make a table and then connect the points with a smooth curve:

x	-2	-1	0	1	2
y	-8	-1	0	1	8

(b) Shift the graph from part (a) left 1 unit.

(c) Shift the graph from part (a) right 2 units and up 3 units.

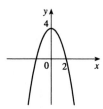

(d) First plot $y = x^2$. Next, to get the graph of $f(x) = 4 - x^2$, reflect f about the x-axis and then shift it upward 4 units.

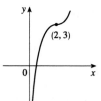

(e) Make a table and then connect the points with a smooth curve:

x	0	1	4	9
y	0	1	2	3

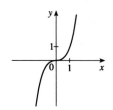

(f) Stretch the graph from part (e) vertically by a factor of two.

(g) First plot $y = 2^x$. Next, get the graph of $y = -2^x$ by reflecting the graph of $y = 2^x$ about the x-axis.

(h) Note that $y = 1 + x^{-1} = 1 + 1/x$. So first plot $y = 1/x$ and then shift it upward 1 unit.

6. (a) $f(-2) = 1 - (-2)^2 = -3$ and $f(1) = 2(1) + 1 = 3$

(b) For $x \le 0$ plot $f(x) = 1 - x^2$ and, on the same plane, for $x > 0$ plot the graph of $f(x) = 2x + 1$.

7. (a) $(f \circ g)(x) = f(g(x)) = f(2x - 3) = (2x - 3)^2 + 2(2x - 3) - 1 = 4x^2 - 12x + 9 + 4x - 6 - 1 = 4x^2 - 8x + 2$

(b) $(g \circ f)(x) = g(f(x)) = g(x^2 + 2x - 1) = 2(x^2 + 2x - 1) - 3 = 2x^2 + 4x - 2 - 3 = 2x^2 + 4x - 5$

(c) $(g \circ g \circ g)(x) = g(g(g(x))) = g(g(2x - 3)) = g(2(2x - 3) - 3) = g(4x - 9) = 2(4x - 9) - 3$
$= 8x - 18 - 3 = 8x - 21$

Test D   Trigonometry

1. (a) $300° = 300° \left(\dfrac{\pi}{180°} \right) = \dfrac{300\pi}{180} = \dfrac{5\pi}{3}$ (b) $-18° = -18° \left(\dfrac{\pi}{180°} \right) = -\dfrac{18\pi}{180} = -\dfrac{\pi}{10}$

2. (a) $\dfrac{5\pi}{6} = \dfrac{5\pi}{6} \left(\dfrac{180°}{\pi} \right) = 150°$ (b) $2 = 2 \left(\dfrac{180°}{\pi} \right) = \dfrac{360°}{\pi} \approx 114.6°$

3. We will use the arc length formula, $s = r\theta$, where s is arc length, r is the radius of the circle, and θ is the measure of the central angle in radians. First, note that $30° = 30° \left(\dfrac{\pi}{180°} \right) = \dfrac{\pi}{6}$. So $s = (12) \left(\dfrac{\pi}{6} \right) = 2\pi$ cm.

4. (a) $\tan(\pi/3) = \sqrt{3}$ [You can read the value from a right triangle with sides 1, 2, and $\sqrt{3}$.]

(b) Note that $7\pi/6$ can be thought of as an angle in the third quadrant with reference angle $\pi/6$. Thus, $\sin(7\pi/6) = -\frac{1}{2}$, since the sine function is negative in the third quadrant.

(c) Note that $5\pi/3$ can be thought of as an angle in the fourth quadrant with reference angle $\pi/3$. Thus,

$\sec(5\pi/3) = \dfrac{1}{\cos(5\pi/3)} = \dfrac{1}{1/2} = 2$, since the cosine function is positive in the fourth quadrant.

5. $\sin\theta = a/24 \;\Rightarrow\; a = 24\sin\theta$ and $\cos\theta = b/24 \;\Rightarrow\; b = 24\cos\theta$

6. $\sin x = \frac{1}{3}$ and $\sin^2 x + \cos^2 x = 1$ $\Rightarrow$ $\cos x = \sqrt{1 - \frac{1}{9}} = \dfrac{2\sqrt{2}}{3}$. Also, $\cos y = \frac{4}{5}$ $\Rightarrow$ $\sin y = \sqrt{1 - \frac{16}{25}} = \frac{3}{5}$.

So, using the sum identity for the sine, we have

$$\sin(x+y) = \sin x \, \cos y + \cos x \, \sin y = \frac{1}{3} \cdot \frac{4}{5} + \frac{2\sqrt{2}}{3} \cdot \frac{3}{5} = \frac{4 + 6\sqrt{2}}{15} = \frac{1}{15}\left(4 + 6\sqrt{2}\right)$$

7. (a) $\tan\theta \, \sin\theta + \cos\theta = \dfrac{\sin\theta}{\cos\theta}\sin\theta + \cos\theta = \dfrac{\sin^2\theta}{\cos\theta} + \dfrac{\cos^2\theta}{\cos\theta} = \dfrac{1}{\cos\theta} = \sec\theta$

(b) $\dfrac{2\tan x}{1 + \tan^2 x} = \dfrac{2\sin x/(\cos x)}{\sec^2 x} = 2\,\dfrac{\sin x}{\cos x}\cos^2 x = 2\sin x \, \cos x = \sin 2x$

8. $\sin 2x = \sin x$ $\Leftrightarrow$ $2\sin x \, \cos x = \sin x$ $\Leftrightarrow$ $2\sin x \, \cos x - \sin x = 0$ $\Leftrightarrow$ $\sin x \, (2\cos x - 1) = 0$ $\Leftrightarrow$

$\sin x = 0$ or $\cos x = \frac{1}{2}$ $\Rightarrow$ $x = 0, \frac{\pi}{3}, \pi, \frac{5\pi}{3}, 2\pi$.

9. We first graph $y = \sin 2x$ (by compressing the graph of $\sin x$ by a factor of 2) and then shift it upward 1 unit.

1 □ FUNCTIONS AND MODELS

1.1 Four Ways To Represent a Function

In exercises requiring estimations or approximations, your answers may vary slightly from the answers given here.

1. (a) The point $(1, 3)$ is on the graph of f, so $f(1) = 3$.

(b) When $x = -1$, y is about -0.2, so $f(-1) \approx -0.2$.

(c) $f(x) = 1$ is equivalent to $y = 1$. When $y = 1$, we have $x = 0$ and $x = 3$.

(d) A reasonable estimate for x when $y = 0$ is $x = -0.8$.

(e) The domain of f consists of all x-values on the graph of f. For this function, the domain is $-2 \le x \le 4$, or $[-2, 4]$.

The range of f consists of all y-values on the graph of f. For this function, the range is $-1 \le y \le 3$, or $[-1, 3]$.

(f) As x increases from -2 to 1, y increases from -1 to 3. Thus, f is increasing on the interval $[-2, 1]$.

3. From Figure 1 in the text, the lowest point occurs at about $(t, a) = (12, -85)$. The highest point occurs at about $(17, 115)$.

Thus, the range of the vertical ground acceleration is $-85 \le a \le 115$. Written in interval notation, we get $[-85, 115]$.

5. No, the curve is not the graph of a function because a vertical line intersects the curve more than once. Hence, the curve fails the Vertical Line Test.

7. Yes, the curve is the graph of a function because it passes the Vertical Line Test. The domain is $[-3, 2]$ and the range is $[-3, -2) \cup [-1, 3]$.

9. The person's weight increased to about 160 pounds at age 20 and stayed fairly steady for 10 years. The person's weight dropped to about 120 pounds for the next 5 years, then increased rapidly to about 170 pounds. The next 30 years saw a gradual increase to 190 pounds. Possible reasons for the drop in weight at 30 years of age: diet, exercise, health problems.

11. The water will cool down almost to freezing as the ice melts. Then, when the ice has melted, the water will slowly warm up to room temperature.

13. (a) The power consumption at 6 AM is 500 MW, which is obtained by reading the value of power P when $t = 6$ from the graph. At 6 PM we read the value of P when $t = 18$, obtaining approximately 730 MW.

(b) The minimum power consumption is determined by finding the time for the lowest point on the graph, $t = 4$, or 4 AM. The maximum power consumption corresponds to the highest point on the graph, which occurs just before $t = 12$, or right before noon. These times are reasonable, considering the power consumption schedules of most individuals and businesses.

15. Of course, this graph depends strongly on the geographical location!

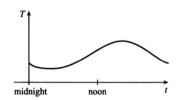

17. As the price increases, the amount sold decreases.

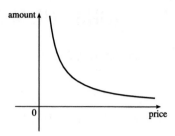

19.

Height of grass graph with axis labeled Wed. Wed. Wed. Wed. Wed. t

21. (a)

N versus t graph labeled 1996 1998 2000 2002 2004 2006 (midyear)

(b) From the graph, we estimate the number of US cell-phone subscribers to be about 126 million in 2001 and 207 million in 2005.

23. $f(x) = 3x^2 - x + 2$.

$f(2) = 3(2)^2 - 2 + 2 = 12 - 2 + 2 = 12$.

$f(-2) = 3(-2)^2 - (-2) + 2 = 12 + 2 + 2 = 16$.

$f(a) = 3a^2 - a + 2$.

$f(-a) = 3(-a)^2 - (-a) + 2 = 3a^2 + a + 2$.

$f(a + 1) = 3(a + 1)^2 - (a + 1) + 2 = 3(a^2 + 2a + 1) - a - 1 + 2 = 3a^2 + 6a + 3 - a + 1 = 3a^2 + 5a + 4$.

$2f(a) = 2 \cdot f(a) = 2(3a^2 - a + 2) = 6a^2 - 2a + 4$.

$f(2a) = 3(2a)^2 - (2a) + 2 = 3(4a^2) - 2a + 2 = 12a^2 - 2a + 2$.

$f(a^2) = 3(a^2)^2 - (a^2) + 2 = 3(a^4) - a^2 + 2 = 3a^4 - a^2 + 2$.

$[f(a)]^2 = [3a^2 - a + 2]^2 = (3a^2 - a + 2)(3a^2 - a + 2)$
$\quad = 9a^4 - 3a^3 + 6a^2 - 3a^3 + a^2 - 2a + 6a^2 - 2a + 4 = 9a^4 - 6a^3 + 13a^2 - 4a + 4$.

$f(a + h) = 3(a + h)^2 - (a + h) + 2 = 3(a^2 + 2ah + h^2) - a - h + 2 = 3a^2 + 6ah + 3h^2 - a - h + 2$.

25. $f(x) = 4 + 3x - x^2$, so $f(3 + h) = 4 + 3(3 + h) - (3 + h)^2 = 4 + 9 + 3h - (9 + 6h + h^2) = 4 - 3h - h^2$,

and $\dfrac{f(3 + h) - f(3)}{h} = \dfrac{(4 - 3h - h^2) - 4}{h} = \dfrac{h(-3 - h)}{h} = -3 - h$.

27. $\dfrac{f(x) - f(a)}{x - a} = \dfrac{\dfrac{1}{x} - \dfrac{1}{a}}{x - a} = \dfrac{\dfrac{a - x}{xa}}{x - a} = \dfrac{a - x}{xa(x - a)} = \dfrac{-1(x - a)}{xa(x - a)} = -\dfrac{1}{ax}$

29. $f(x) = (x + 4)/(x^2 - 9)$ is defined for all x except when $0 = x^2 - 9$ ⟺ $0 = (x + 3)(x - 3)$ ⟺ $x = -3$ or 3, so the domain is $\{x \in \mathbb{R} \mid x \neq -3, 3\} = (-\infty, -3) \cup (-3, 3) \cup (3, \infty)$.

31. $f(t) = \sqrt[3]{2t - 1}$ is defined for all real numbres. In fact $\sqrt[3]{p(t)}$, where $p(t)$ is a polynomial, is defined for all real numbers. Thus, the domain is $\mathbb{R}$, or $(-\infty, \infty)$.

33. $h(x) = 1 / \sqrt[4]{x^2 - 5x}$ is defined when $x^2 - 5x > 0$ ⟺ $x(x - 5) > 0$. Note that $x^2 - 5x \neq 0$ since that would result in division by zero. The expression $x(x - 5)$ is positive if $x < 0$ or $x > 5$. (See Appendix A for methods for solving inequalities.) Thus, the domain is $(-\infty, 0) \cup (5, \infty)$.

35. $f(x) = 2 - 0.4x$ is defined for all real numbers, so the domain is $\mathbb{R}$, or $(-\infty, \infty)$. The graph of f is a line with slope -0.4 and y-intercept 2.

37. $f(t) = 2t + t^2$ is defined for all real numbers, so the domain is $\mathbb{R}$, or $(-\infty, \infty)$. The graph of f is a parabola opening upward since the coefficient of t^2 is positive. To find the t-intercepts, let $y = 0$ and solve for t. $0 = 2t + t^2 = t(2 + t)$ ⟹ $t = 0$ or $t = -2$. The t-coordinate of the vertex is halfway between the t-intercepts, that is, at $t = -1$. Since $f(-1) = 2(-1) + (-1)^2 = -2 + 1 = -1$, the vertex is $(-1, -1)$.

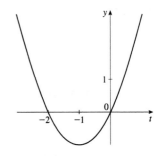

39. $g(x) = \sqrt{x - 5}$ is defined when $x - 5 \geq 0$ or $x \geq 5$, so the domain is $[5, \infty)$. Since $y = \sqrt{x - 5}$ ⟹ $y^2 = x - 5$ ⟹ $x = y^2 + 5$, we see that g is the top half of a parabola.

41. $G(x) = \dfrac{3x + |x|}{x}$. Since $|x| = \begin{cases} x & \text{if } x \geq 0 \\ -x & \text{if } x < 0 \end{cases}$, we have

$$G(x) = \begin{cases} \dfrac{3x + x}{x} & \text{if } x > 0 \\ \dfrac{3x - x}{x} & \text{if } x < 0 \end{cases} = \begin{cases} \dfrac{4x}{x} & \text{if } x > 0 \\ \dfrac{2x}{x} & \text{if } x < 0 \end{cases} = \begin{cases} 4 & \text{if } x > 0 \\ 2 & \text{if } x < 0 \end{cases}$$

Note that G is not defined for $x = 0$. The domain is $(-\infty, 0) \cup (0, \infty)$.

43. $f(x) = \begin{cases} x + 2 & \text{if } x < 0 \\ 1 - x & \text{if } x \geq 0 \end{cases}$

The domain is $\mathbb{R}$.

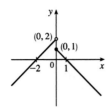

45. $f(x) = \begin{cases} x + 2 & \text{if } x \leq -1 \\ x^2 & \text{if } x > -1 \end{cases}$

Note that for $x = -1$, both $x + 2$ and x^2 are equal to 1.

The domain is $\mathbb{R}$.

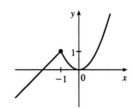

47. Recall that the slope m of a line between the two points (x_1, y_1) and (x_2, y_2) is $m = \dfrac{y_2 - y_1}{x_2 - x_1}$ and an equation of the line

connecting those two points is $y - y_1 = m(x - x_1)$. The slope of the line segment joining the points $(1, -3)$ and $(5, 7)$ is

$\dfrac{7 - (-3)}{5 - 1} = \dfrac{5}{2}$, so an equation is $y - (-3) = \frac{5}{2}(x - 1)$. The function is $f(x) = \frac{5}{2}x - \frac{11}{2}, 1 \leq x \leq 5$.

49. We need to solve the given equation for y. $x + (y - 1)^2 = 0 \iff (y - 1)^2 = -x \iff y - 1 = \pm\sqrt{-x} \iff$

$y = 1 \pm \sqrt{-x}$. The expression with the positive radical represents the top half of the parabola, and the one with the negative

radical represents the bottom half. Hence, we want $f(x) = 1 - \sqrt{-x}$. Note that the domain is $x \leq 0$.

51. For $0 \leq x \leq 3$, the graph is the line with slope -1 and y-intercept 3, that is, $y = -x + 3$. For $3 < x \leq 5$, the graph is the line

with slope 2 passing through $(3, 0)$; that is, $y - 0 = 2(x - 3)$, or $y = 2x - 6$. So the function is

$$f(x) = \begin{cases} -x + 3 & \text{if } 0 \leq x \leq 3 \\ 2x - 6 & \text{if } 3 < x \leq 5 \end{cases}$$

53. Let the length and width of the rectangle be L and W. Then the perimeter is $2L + 2W = 20$ and the area is $A = LW$.

Solving the first equation for W in terms of L gives $W = \dfrac{20 - 2L}{2} = 10 - L$. Thus, $A(L) = L(10 - L) = 10L - L^2$. Since

lengths are positive, the domain of A is $0 < L < 10$. If we further restrict L to be larger than W, then $5 < L < 10$ would be

the domain.

55. Let the length of a side of the equilateral triangle be x. Then by the Pythagorean Theorem, the height y of the triangle satisfies

$y^2 + \left(\frac{1}{2}x\right)^2 = x^2$, so that $y^2 = x^2 - \frac{1}{4}x^2 = \frac{3}{4}x^2$ and $y = \frac{\sqrt{3}}{2}x$. Using the formula for the area A of a triangle,

$A = \frac{1}{2}(\text{base})(\text{height})$, we obtain $A(x) = \frac{1}{2}(x)\left(\frac{\sqrt{3}}{2}x\right) = \frac{\sqrt{3}}{4}x^2$, with domain $x > 0$.

57. Let each side of the base of the box have length x, and let the height of the box be h. Since the volume is 2, we know that

$2 = hx^2$, so that $h = 2/x^2$, and the surface area is $S = x^2 + 4xh$. Thus, $S(x) = x^2 + 4x(2/x^2) = x^2 + (8/x)$, with

domain $x > 0$.

59. The height of the box is x and the length and width are $L = 20 - 2x$, $W = 12 - 2x$. Then $V = LWx$ and so

$$V(x) = (20 - 2x)(12 - 2x)(x) = 4(10 - x)(6 - x)(x) = 4x(60 - 16x + x^2) = 4x^3 - 64x^2 + 240x.$$

The sides L, W, and x must be positive. Thus, $L > 0$ $\Leftrightarrow$ $20 - 2x > 0$ $\Leftrightarrow$ $x < 10$;

$W > 0$ $\Leftrightarrow$ $12 - 2x > 0$ $\Leftrightarrow$ $x < 6$; and $x > 0$. Combining these restrictions gives us the domain $0 < x < 6$.

61. (a)

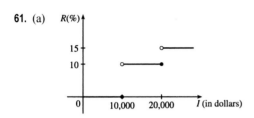

(b) On \$14,000, tax is assessed on \$4000, and $10\%(\$4000) = \$400$.

On \$26,000, tax is assessed on \$16,000, and

$$10\%(\$10,000) + 15\%(\$6000) = \$1000 + \$900 = \$1900.$$

(c) As in part (b), there is \$1000 tax assessed on \$20,000 of income, so

the graph of T is a line segment from $(10{,}000, 0)$ to $(20{,}000, 1000)$.

The tax on \$30,000 is \$2500, so the graph of T for $x > 20{,}000$ is

the ray with initial point $(20{,}000, 1000)$ that passes through

$(30{,}000, 2500)$.

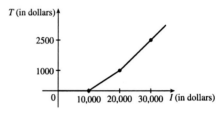

63. f is an odd function because its graph is symmetric about the origin. g is an even function because its graph is symmetric with respect to the y-axis.

65. (a) Because an even function is symmetric with respect to the y-axis, and the point $(5, 3)$ is on the graph of this even function, the point $(-5, 3)$ must also be on its graph.

(b) Because an odd function is symmetric with respect to the origin, and the point $(5, 3)$ is on the graph of this odd function, the point $(-5, -3)$ must also be on its graph.

67. $f(x) = \dfrac{x}{x^2 + 1}$.

$$f(-x) = \frac{-x}{(-x)^2 + 1} = \frac{-x}{x^2 + 1} = -\frac{x}{x^2 + 1} = -f(x).$$

So f is an odd function.

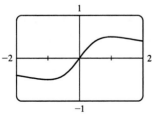

69. $f(x) = \dfrac{x}{x + 1}$, so $f(-x) = \dfrac{-x}{-x + 1} = \dfrac{x}{x - 1}$.

Since this is neither $f(x)$ nor $-f(x)$, the function f is

neither even nor odd.

71. $f(x) = 1 + 3x^2 - x^4$.

$f(-x) = 1 + 3(-x)^2 - (-x)^4 = 1 + 3x^2 - x^4 = f(x)$.

So f is an even function.

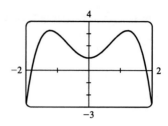

73. (i) If f and g are both even functions, then $f(-x) = f(x)$ and $g(-x) = g(x)$. Now

$(f + g)(-x) = f(-x) + g(-x) = f(x) + g(x) = (f + g)(x)$, so $f + g$ is an *even* function.

(ii) If f and g are both odd functions, then $f(-x) = -f(x)$ and $g(-x) = -g(x)$. Now

$(f + g)(-x) = f(-x) + g(-x) = -f(x) + [-g(x)] = -[f(x) + g(x)] = -(f + g)(x)$, so $f + g$ is an *odd* function.

(iii) If f is an even function and g is an odd function, then $(f + g)(-x) = f(-x) + g(-x) = f(x) + [-g(x)] = f(x) - g(x)$,

which is not $(f + g)(x)$ nor $-(f + g)(x)$, so $f + g$ is *neither* even nor odd. (Exception: if f is the zero function, then

$f + g$ will be *odd*. If g is the zero function, then $f + g$ will be *even*.)

1.2 Mathematical Models: A Catalog of Essential Functions

1. (a) $f(x) = \log_2 x$ is a logarithmic function.

(b) $g(x) = \sqrt[4]{x}$ is a root function with $n = 4$.

(c) $h(x) = \dfrac{2x^3}{1 - x^2}$ is a rational function because it is a ratio of polynomials.

(d) $u(t) = 1 - 1.1t + 2.54t^2$ is a polynomial of degree 2 (also called a *quadratic function*).

(e) $v(t) = 5^t$ is an exponential function.

(f) $w(\theta) = \sin \theta \cos^2\theta$ is a trigonometric function.

3. We notice from the figure that g and h are even functions (symmetric with respect to the y-axis) and that f is an odd function

(symmetric with respect to the origin). So (b) $\left[y = x^5\right]$ must be f. Since g is flatter than h near the origin, we must have

(c) $\left[y = x^8\right]$ matched with g and (a) $\left[y = x^2\right]$ matched with h.

5. (a) An equation for the family of linear functions with slope 2

is $y = f(x) = 2x + b$, where b is the y-intercept.

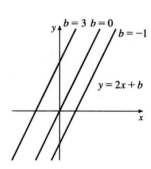

(b) $f(2) = 1$ means that the point $(2, 1)$ is on the graph of f. We can use the point-slope form of a line to obtain an equation for the family of linear functions through the point $(2, 1)$. $y - 1 = m(x - 2)$, which is equivalent to $y = mx + (1 - 2m)$ in slope-intercept form.

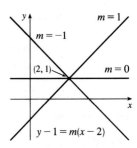

(c) To belong to both families, an equation must have slope $m = 2$, so the equation in part (b), $y = mx + (1 - 2m)$, becomes $y = 2x - 3$. It is the *only* function that belongs to both families.

7. All members of the family of linear functions $f(x) = c - x$ have graphs that are lines with slope -1. The y-intercept is c.

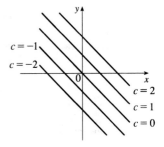

9. Since $f(-1) = f(0) = f(2) = 0$. f has zeros of -1, 0, and 2, so an equation for f is $f(x) = a[x - (-1)](x - 0)(x - 2)$, or $f(x) = ax(x + 1)(x - 2)$. Because $f(1) = 6$, we'll substitute 1 for x and 6 for $f(x)$.
$6 = a(1)(2)(-1) \Rightarrow -2a = 6 \Rightarrow a = -3$, so an equation for f is $f(x) = -3x(x + 1)(x - 2)$.

11. (a) $D = 200$, so $c = 0.0417D(a + 1) = 0.0417(200)(a + 1) = 8.34a + 8.34$. The slope is 8.34, which represents the change in mg of the dosage for a child for each change of 1 year in age.

(b) For a newborn, $a = 0$, so $c = 8.34$ mg.

13. (a)

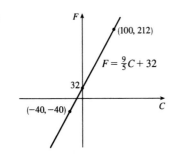

(b) The slope of $\frac{9}{5}$ means that F increases $\frac{9}{5}$ degrees for each increase of $1°C$. (Equivalently, F increases by 9 when C increases by 5 and F decreases by 9 when C decreases by 5.) The F-intercept of 32 is the Fahrenheit temperature corresponding to a Celsius temperature of 0.

15. (a) Using N in place of x and T in place of y, we find the slope to be $\dfrac{T_2 - T_1}{N_2 - N_1} = \dfrac{80 - 70}{173 - 113} = \dfrac{10}{60} = \dfrac{1}{6}$. So a linear equation is $T - 80 = \frac{1}{6}(N - 173) \Leftrightarrow T - 80 = \frac{1}{6}N - \frac{173}{6} \Leftrightarrow T = \frac{1}{6}N + \frac{307}{6}$ $\left[\frac{307}{6} = 51.1\overline{6}\right]$.

(b) The slope of $\frac{1}{6}$ means that the temperature in Fahrenheit degrees increases one-sixth as rapidly as the number of cricket chirps per minute. Said differently, each increase of 6 cricket chirps per minute corresponds to an increase of $1°F$.

(c) When $N = 150$, the temperature is given approximately by $T = \frac{1}{6}(150) + \frac{307}{6} = 76.1\overline{6}\,°F \approx 76\,°F$.

17. (a) We are given $\dfrac{\text{change in pressure}}{10 \text{ feet change in depth}} = \dfrac{4.34}{10} = 0.434$. Using P for pressure and d for depth with the point

$(d, P) = (0, 15)$, we have the slope-intercept form of the line, $P = 0.434d + 15$.

(b) When $P = 100$, then $100 = 0.434d + 15 \iff 0.434d = 85 \iff d = \frac{85}{0.434} \approx 195.85$ feet. Thus, the pressure is

100 lb/in^2 at a depth of approximately 196 feet.

19. (a) The data appear to be periodic and a sine or cosine function would make the best model. A model of the form

$f(x) = a\cos(bx) + c$ seems appropriate.

(b) The data appear to be decreasing in a linear fashion. A model of the form $f(x) = mx + b$ seems appropriate.

Exercises 21 – 24: Some values are given to many decimal places. These are the results given by several computer algebra systems — rounding is left to the reader.

21. (a)

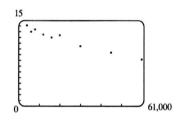

A linear model does seem appropriate.

(b) Using the points $(4000, 14.1)$ and $(60{,}000, 8.2)$, we obtain

$$y - 14.1 = \frac{8.2 - 14.1}{60{,}000 - 4000}(x - 4000) \text{ or, equivalently,}$$

$$y \approx -0.000105357x + 14.521429.$$

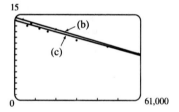

(c) Using a computing device, we obtain the least squares regression line $y = -0.0000997855x + 13.950764$.

The following commands and screens illustrate how to find the least squares regression line on a TI-84 Plus.

Enter the data into list one (L1) and list two (L2). Press $\boxed{\text{STAT}}\,\boxed{1}$ to enter the editor.

Find the regession line and store it in Y_1. Press $\boxed{\text{2nd}}\,\boxed{\text{QUIT}}\,\boxed{\text{STAT}}\,\boxed{▶}\,\boxed{4}\,\boxed{\text{VARS}}\,\boxed{▶}\,\boxed{1}\,\boxed{1}\,\boxed{\text{ENTER}}$.

Note from the last figure that the regression line has been stored in Y_1 and that Plot1 has been turned on (Plot1 is highlighted). You can turn on Plot1 from the Y= menu by placing the cursor on Plot1 and pressing ENTER or by pressing 2nd STAT PLOT 1 ENTER .

Now press ZOOM 9 to produce a graph of the data and the regression line. Note that choice 9 of the ZOOM menu automatically selects a window that displays all of the data.

(d) When $x = 25{,}000$, $y \approx 11.456$; or about 11.5 per 100 population.

(e) When $x = 80{,}000$, $y \approx 5.968$; or about a 6% chance.

(f) When $x = 200{,}000$, y is negative, so the model does not apply.

23. (a) A linear model seems appropriate over the time interval considered.

(b) Using a computing device, we obtain the least squares regression line $y \approx 0.027t - 47.758$.

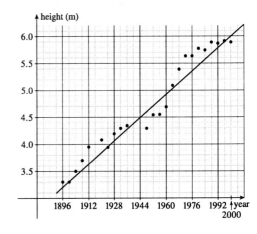

(c) When $t = 2004$, $y = 6.35$, which is higher than the actual winning height of 5.95 m.

(d) No, since the times appear to be leveling off and getting further away from the model.

25.

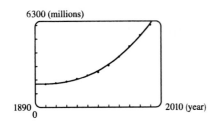

Using a computing device, we obtain the cubic function
$y = ax^3 + bx^2 + cx + d$ with $a = 0.0012937$,
$b = -7.06142$, $c = 12{,}823$, and $d = -7{,}743{,}770$. When
$x = 1925$, $y \approx 1914$ (million).

1.3 New Functions from Old Functions

1. (a) If the graph of f is shifted 3 units upward, its equation becomes $y = f(x) + 3$.

(b) If the graph of f is shifted 3 units downward, its equation becomes $y = f(x) - 3$.

(c) If the graph of f is shifted 3 units to the right, its equation becomes $y = f(x - 3)$.

(d) If the graph of f is shifted 3 units to the left, its equation becomes $y = f(x + 3)$.

(e) If the graph of f is reflected about the x-axis, its equation becomes $y = -f(x)$.

(f) If the graph of f is reflected about the y-axis, its equation becomes $y = f(-x)$.

(g) If the graph of f is stretched vertically by a factor of 3, its equation becomes $y = 3f(x)$.

(h) If the graph of f is shrunk vertically by a factor of 3, its equation becomes $y = \frac{1}{3}f(x)$.

3. (a) (graph 3) The graph of f is shifted 4 units to the right and has equation $y = f(x - 4)$.

(b) (graph 1) The graph of f is shifted 3 units upward and has equation $y = f(x) + 3$.

(c) (graph 4) The graph of f is shrunk vertically by a factor of 3 and has equation $y = \frac{1}{3}f(x)$.

(d) (graph 5) The graph of f is shifted 4 units to the left and reflected about the x-axis. Its equation is $y = -f(x + 4)$.

(e) (graph 2) The graph of f is shifted 6 units to the left and stretched vertically by a factor of 2. Its equation is $y = 2f(x + 6)$.

5. (a) To graph $y = f(2x)$ we shrink the graph of f horizontally by a factor of 2.

The point $(4, -1)$ on the graph of f corresponds to the point $\left(\frac{1}{2} \cdot 4, -1\right) = (2, -1)$.

(b) To graph $y = f\left(\frac{1}{2}x\right)$ we stretch the graph of f horizontally by a factor of 2.

The point $(4, -1)$ on the graph of f corresponds to the point $(2 \cdot 4, -1) = (8, -1)$.

(c) To graph $y = f(-x)$ we reflect the graph of f about the y-axis.

The point $(4, -1)$ on the graph of f corresponds to the point $(-1 \cdot 4, -1) = (-4, -1)$.

(d) To graph $y = -f(-x)$ we reflect the graph of f about the y-axis, then about the x-axis.

The point $(4, -1)$ on the graph of f corresponds to the point $(-1 \cdot 4, -1 \cdot -1) = (-4, 1)$.

7. The graph of $y = f(x) = \sqrt{3x - x^2}$ has been shifted 4 units to the left, reflected about the x-axis, and shifted downward 1 unit. Thus, a function describing the graph is

$$y = \underbrace{-1 \cdot}_{\substack{\text{reflect} \\ \text{about } x\text{-axis}}} \underbrace{f\ (x+4)}_{\substack{\text{shift} \\ \text{4 units left}}} \underbrace{-\ 1}_{\substack{\text{shift} \\ \text{1 unit left}}}$$

This function can be written as

$$y = -f(x+4) - 1 = -\sqrt{3(x+4) - (x+4)^2} - 1 = -\sqrt{3x + 12 - (x^2 + 8x + 16)} - 1 = -\sqrt{-x^2 - 5x - 4} - 1$$

9. $y = -x^3$: Start with the graph of $y = x^3$ and reflect about the x-axis. Note: Reflecting about the y-axis gives the same result since substituting $-x$ for x gives us $y = (-x)^3 = -x^3$.

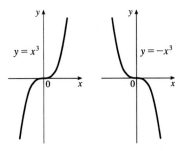

11. $y = (x+1)^2$: Start with the graph of $y = x^2$ and shift 1 unit to the left.

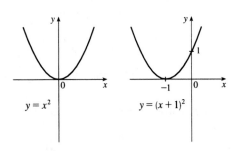

13. $y = 1 + 2\cos x$: Start with the graph of $y = \cos x$, stretch vertically by a factor of 2, and then shift 1 unit upward.

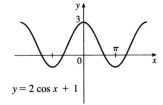

15. $y = \sin(x/2)$: Start with the graph of $y = \sin x$ and stretch horizontally by a factor of 2.

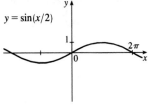

17. $y = \sqrt{x+3}$: Start with the graph of
$y = \sqrt{x}$ and shift 3 units to the left.

19. $y = \frac{1}{2}(x^2 + 8x) = \frac{1}{2}(x^2 + 8x + 16) - 8 = \frac{1}{2}(x+4)^2 - 8$: Start with the graph of $y = x^2$, compress vertically by a
factor of 2, shift 4 units to the left, and then shift 8 units downward.

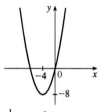

$y = x^2$ $\qquad$ $y = \frac{1}{2}x^2$ $\qquad$ $y = \frac{1}{2}(x+4)^2$ $\qquad$ $y = \frac{1}{2}(x+4)^2 - 8$

21. $y = |x - 2|$: Start with the graph of $y = |x|$ and shift 2 units to the right.

23. $y = |\sqrt{x} - 1|$: Start with the graph of $y = \sqrt{x}$, shift it 1 unit downward, and then reflect the portion of the graph below the
x-axis about the x-axis.

25. This is just like the solution to Example 4 except the amplitude of the curve (the 30°N curve in Figure 9 on June 21) is
$14 - 12 = 2$. So the function is $L(t) = 12 + 2\sin\left[\frac{2\pi}{365}(t - 80)\right]$. March 31 is the 90th day of the year, so the model gives
$L(90) \approx 12.34$ h. The daylight time (5:51 AM to 6:18 PM) is 12 hours and 27 minutes, or 12.45 h. The model value differs
from the actual value by $\frac{12.45 - 12.34}{12.45} \approx 0.009$, less than 1%.

27. (a) To obtain $y = f(|x|)$, the portion of the graph of $y = f(x)$ to the right of the y-axis is reflected about the y-axis.

(b) $y = \sin|x|$ $\qquad\qquad\qquad\qquad$ (c) $y = \sqrt{|x|}$

29. $f(x) = x^3 + 2x^2$; $g(x) = 3x^2 - 1$. $D = \mathbb{R}$ for both f and g.

(a) $(f + g)(x) = (x^3 + 2x^2) + (3x^2 - 1) = x^3 + 5x^2 - 1$, $D = \mathbb{R}$.

(b) $(f - g)(x) = (x^3 + 2x^2) - (3x^2 - 1) = x^3 - x^2 + 1$, $D = \mathbb{R}$.

(c) $(fg)(x) = (x^3 + 2x^2)(3x^2 - 1) = 3x^5 + 6x^4 - x^3 - 2x^2$, $D = \mathbb{R}$.

(d) $\left(\dfrac{f}{g}\right)(x) = \dfrac{x^3 + 2x^2}{3x^2 - 1}$, $D = \left\{x \mid x \neq \pm\dfrac{1}{\sqrt{3}}\right\}$ since $3x^2 - 1 \neq 0$.

31. $f(x) = x^2 - 1$, $D = \mathbb{R}$; $g(x) = 2x + 1$, $D = \mathbb{R}$.

(a) $(f \circ g)(x) = f(g(x)) = f(2x + 1) = (2x + 1)^2 - 1 = (4x^2 + 4x + 1) - 1 = 4x^2 + 4x$, $D = \mathbb{R}$.

(b) $(g \circ f)(x) = g(f(x)) = g(x^2 - 1) = 2(x^2 - 1) + 1 = (2x^2 - 2) + 1 = 2x^2 - 1$, $D = \mathbb{R}$.

(c) $(f \circ f)(x) = f(f(x)) = f(x^2 - 1) = (x^2 - 1)^2 - 1 = (x^4 - 2x^2 + 1) - 1 = x^4 - 2x^2$, $D = \mathbb{R}$.

(d) $(g \circ g)(x) = g(g(x)) = g(2x + 1) = 2(2x + 1) + 1 = (4x + 2) + 1 = 4x + 3$, $D = \mathbb{R}$.

33. $f(x) = 1 - 3x$; $g(x) = \cos x$. $D = \mathbb{R}$ for both f and g, and hence for their composites.

(a) $(f \circ g)(x) = f(g(x)) = f(\cos x) = 1 - 3\cos x$.

(b) $(g \circ f)(x) = g(f(x)) = g(1 - 3x) = \cos(1 - 3x)$.

(c) $(f \circ f)(x) = f(f(x)) = f(1 - 3x) = 1 - 3(1 - 3x) = 1 - 3 + 9x = 9x - 2$.

(d) $(g \circ g)(x) = g(g(x)) = g(\cos x) = \cos(\cos x)$ [Note that this is *not* $\cos x \cdot \cos x$.]

35. $f(x) = x + \dfrac{1}{x}$, $D = \{x \mid x \neq 0\}$; $g(x) = \dfrac{x + 1}{x + 2}$, $D = \{x \mid x \neq -2\}$

(a) $(f \circ g)(x) = f(g(x)) = f\left(\dfrac{x + 1}{x + 2}\right) = \dfrac{x + 1}{x + 2} + \dfrac{1}{\dfrac{x + 1}{x + 2}} = \dfrac{x + 1}{x + 2} + \dfrac{x + 2}{x + 1}$

$= \dfrac{(x + 1)(x + 1) + (x + 2)(x + 2)}{(x + 2)(x + 1)} = \dfrac{(x^2 + 2x + 1) + (x^2 + 4x + 4)}{(x + 2)(x + 1)} = \dfrac{2x^2 + 6x + 5}{(x + 2)(x + 1)}$

Since $g(x)$ is not defined for $x = -2$ and $f(g(x))$ is not defined for $x = -2$ and $x = -1$,

the domain of $(f \circ g)(x)$ is $D = \{x \mid x \neq -2, -1\}$.

(b) $(g \circ f)(x) = g(f(x)) = g\left(x + \dfrac{1}{x}\right) = \dfrac{\left(x + \dfrac{1}{x}\right) + 1}{\left(x + \dfrac{1}{x}\right) + 2} = \dfrac{\dfrac{x^2 + 1 + x}{x}}{\dfrac{x^2 + 1 + 2x}{x}} = \dfrac{x^2 + x + 1}{x^2 + 2x + 1} = \dfrac{x^2 + x + 1}{(x + 1)^2}$

Since $f(x)$ is not defined for $x = 0$ and $g(f(x))$ is not defined for $x = -1$,

the domain of $(g \circ f)(x)$ is $D = \{x \mid x \neq -1, 0\}$.

(c) $(f \circ f)(x) = f(f(x)) = f\left(x + \dfrac{1}{x}\right) = \left(x + \dfrac{1}{x}\right) + \dfrac{1}{x + \dfrac{1}{x}} = x + \dfrac{1}{x} + \dfrac{1}{\dfrac{x^2 + 1}{x}} = x + \dfrac{1}{x} + \dfrac{x}{x^2 + 1}$

$= \dfrac{x(x)\left(x^2 + 1\right) + 1\left(x^2 + 1\right) + x(x)}{x(x^2 + 1)} = \dfrac{x^4 + x^2 + x^2 + 1 + x^2}{x(x^2 + 1)}$

$= \dfrac{x^4 + 3x^2 + 1}{x(x^2 + 1)}$, $D = \{x \mid x \neq 0\}$

(d) $(g \circ g)(x) = g(g(x)) = g\left(\dfrac{x+1}{x+2}\right) = \dfrac{\dfrac{x+1}{x+2}+1}{\dfrac{x+1}{x+2}+2} = \dfrac{\dfrac{x+1+1(x+2)}{x+2}}{\dfrac{x+1+2(x+2)}{x+2}} = \dfrac{x+1+x+2}{x+1+2x+4} = \dfrac{2x+3}{3x+5}$

Since $g(x)$ is not defined for $x = -2$ and $g(g(x))$ is not defined for $x = -\frac{5}{3}$,

the domain of $(g \circ g)(x)$ is $D = \left\{x \mid x \neq -2, -\frac{5}{3}\right\}$.

37. $(f \circ g \circ h)(x) = f(g(h(x))) = f(g(x-1)) = f(2(x-1)) = 2(x-1)+1 = 2x-1$

39. $(f \circ g \circ h)(x) = f(g(h(x))) = f(g(x^3+2)) = f[(x^3+2)^2]$

$\qquad = f(x^6 + 4x^3 + 4) = \sqrt{(x^6 + 4x^3 + 4) - 3} = \sqrt{x^6 + 4x^3 + 1}$

41. Let $g(x) = 2x + x^2$ and $f(x) = x^4$. Then $(f \circ g)(x) = f(g(x)) = f(2x + x^2) = (2x + x^2)^4 = F(x)$.

43. Let $g(x) = \sqrt[3]{x}$ and $f(x) = \dfrac{x}{1+x}$. Then $(f \circ g)(x) = f(g(x)) = f(\sqrt[3]{x}) = \dfrac{\sqrt[3]{x}}{1+\sqrt[3]{x}} = F(x)$.

45. Let $g(t) = \cos t$ and $f(t) = \sqrt{t}$. Then $(f \circ g)(t) = f(g(t)) = f(\cos t) = \sqrt{\cos t} = u(t)$.

47. Let $h(x) = x^2$, $g(x) = 3^x$, and $f(x) = 1 - x$. Then

$(f \circ g \circ h)(x) = f(g(h(x))) = f(g(x^2)) = f\left(3^{x^2}\right) = 1 - 3^{x^2} = H(x)$.

49. Let $h(x) = \sqrt{x}$, $g(x) = \sec x$, and $f(x) = x^4$. Then

$(f \circ g \circ h)(x) = f(g(h(x))) = f(g(\sqrt{x})) = f(\sec \sqrt{x}) = (\sec \sqrt{x})^4 = \sec^4 (\sqrt{x}) = H(x)$.

51. (a) $g(2) = 5$, because the point $(2, 5)$ is on the graph of g. Thus, $f(g(2)) = f(5) = 4$, because the point $(5, 4)$ is on the graph of f.

(b) $g(f(0)) = g(0) = 3$

(c) $(f \circ g)(0) = f(g(0)) = f(3) = 0$

(d) $(g \circ f)(6) = g(f(6)) = g(6)$. This value is not defined, because there is no point on the graph of g that has x-coordinate 6.

(e) $(g \circ g)(-2) = g(g(-2)) = g(1) = 4$

(f) $(f \circ f)(4) = f(f(4)) = f(2) = -2$

53. (a) Using the relationship *distance* $=$ *rate* $\cdot$ *time* with the radius r as the distance, we have $r(t) = 60t$.

(b) $A = \pi r^2 \;\Rightarrow\; (A \circ r)(t) = A(r(t)) = \pi(60t)^2 = 3600\pi t^2$. This formula gives us the extent of the rippled area (in cm^2) at any time t.

55. (a) From the figure, we have a right triangle with legs 6 and d, and hypotenuse s.

By the Pythagorean Theorem, $d^2 + 6^2 = s^2 \;\Rightarrow\; s = f(d) = \sqrt{d^2 + 36}$.

(b) Using $d = rt$, we get $d = (30 \text{ km/h})(t \text{ hours}) = 30t$ (in km). Thus,

$d = g(t) = 30t$.

(c) $(f \circ g)(t) = f(g(t)) = f(30t) = \sqrt{(30t)^2 + 36} = \sqrt{900t^2 + 36}$. This function represents the distance between the lighthouse and the ship as a function of the time elapsed since noon.

57. (a)

$$H(t) = \begin{cases} 0 & \text{if } t < 0 \\ 1 & \text{if } t \geq 0 \end{cases}$$

(b)

$$V(t) = \begin{cases} 0 & \text{if } t < 0 \\ 120 & \text{if } t \geq 0 \end{cases} \quad \text{so } V(t) = 120H(t).$$

(c)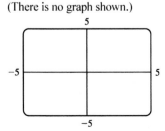

Starting with the formula in part (b), we replace 120 with 240 to reflect the different voltage. Also, because we are starting 5 units to the right of $t = 0$, we replace t with $t - 5$. Thus, the formula is $V(t) = 240H(t - 5)$.

59. If $f(x) = m_1 x + b_1$ and $g(x) = m_2 x + b_2$, then

$$(f \circ g)(x) = f(g(x)) = f(m_2 x + b_2) = m_1(m_2 x + b_2) + b_1 = m_1 m_2 x + m_1 b_2 + b_1.$$

So $f \circ g$ is a linear function with slope $m_1 m_2$.

61. (a) By examining the variable terms in g and h, we deduce that we must square g to get the terms $4x^2$ and $4x$ in h. If we let $f(x) = x^2 + c$, then $(f \circ g)(x) = f(g(x)) = f(2x + 1) = (2x + 1)^2 + c = 4x^2 + 4x + (1 + c)$. Since $h(x) = 4x^2 + 4x + 7$, we must have $1 + c = 7$. So $c = 6$ and $f(x) = x^2 + 6$.

(b) We need a function g so that $f(g(x)) = 3(g(x)) + 5 = h(x)$. But

$h(x) = 3x^2 + 3x + 2 = 3(x^2 + x) + 2 = 3(x^2 + x - 1) + 5$, so we see that $g(x) = x^2 + x - 1$.

63. We need to examine $h(-x)$.

$$h(-x) = (f \circ g)(-x) = f(g(-x)) = f(g(x)) \quad \text{[because } g \text{ is even]} \quad = h(x)$$

Because $h(-x) = h(x)$, h is an even function.

1.4 Graphing Calculators and Computers

1. $f(x) = \sqrt{x^3 - 5x^2}$

(a) $[-5, 5]$ by $[-5, 5]$

(There is no graph shown.)

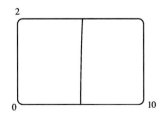

(b) $[0, 10]$ by $[0, 2]$

(c) $[0, 10]$ by $[0, 10]$

The most appropriate graph is produced in viewing rectangle (c).

3. Since the graph of $f(x) = x^2 - 36x + 32$ is a parabola opening upward, an appropriate viewing rectangle should include the minimum point. Completing the square, we get $f(x) = (x - 18)^2 - 292$, and so the minimum point is $(18, -292)$.

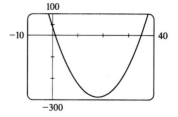

5. $f(x) = \sqrt[4]{81 - x^4}$ is defined when $81 - x^4 \geq 0 \Leftrightarrow x^4 \leq 81 \Leftrightarrow |x| \leq 3$, so the domain of f is $[-3, 3]$. Also $0 \leq \sqrt[4]{81 - x^4} \leq \sqrt[4]{81} = 3$, so the range is $[0, 3]$.

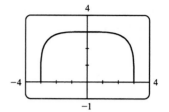

7. The graph of $f(x) = x^3 - 225x$ is symmetric with respect to the origin. Since $f(x) = x^3 - 225x = x(x^2 - 225) = x(x + 15)(x - 15)$, there are x-intercepts at 0, -15, and 15. $f(20) = 3500$.

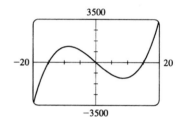

9. The period of $g(x) = \sin(1000x)$ is $\frac{2\pi}{1000} \approx 0.0063$ and its range is $[-1, 1]$. Since $f(x) = \sin^2(1000x)$ is the square of g, its range is $[0, 1]$ and a viewing rectangle of $[-0.01, 0.01]$ by $[0, 1.1]$ seems appropriate.

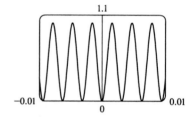

11. The domain of $y = \sqrt{x}$ is $x \geq 0$, so the domain of $f(x) = \sin\sqrt{x}$ is $[0, \infty)$ and the range is $[-1, 1]$. With a little trial-and-error experimentation, we find that an Xmax of 100 illustrates the general shape of f, so an appropriate viewing rectangle is $[0, 100]$ by $[-1.5, 1.5]$.

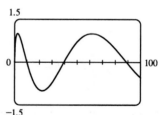

13. The first term, $10 \sin x$, has period 2π and range $[-10, 10]$. It will be the dominant term in any "large" graph of $y = 10 \sin x + \sin 100x$, as shown in the first figure. The second term, $\sin 100x$, has period $\frac{2\pi}{100} = \frac{\pi}{50}$ and range $[-1, 1]$. It causes the bumps in the first figure and will be the dominant term in any "small" graph, as shown in the view near the origin in the second figure.

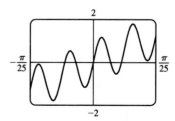

15. (a) The first figure shows the "big picture" for $f(x) = (x - 10)^3 2^{-x}$. The second figure shows a maximum near $x = 10$.

(b) You need more than one window because no single window can show what the function looks like globally and the detail of the function near $x = 10$.

17. We must solve the given equation for y to obtain equations for the upper and lower halves of the ellipse.

$$4x^2 + 2y^2 = 1 \quad \Leftrightarrow \quad 2y^2 = 1 - 4x^2 \quad \Leftrightarrow \quad y^2 = \frac{1 - 4x^2}{2} \quad \Leftrightarrow$$

$$y = \pm\sqrt{\frac{1 - 4x^2}{2}}$$

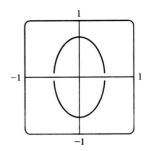

19. From the graph of $y = 3x^2 - 6x + 1$ and $y = 0.23x - 2.25$ in the viewing rectangle $[-1, 3]$ by $[-2.5, 1.5]$, it is difficult to see if the graphs intersect. If we zoom in on the fourth quadrant, we see the graphs do not intersect.

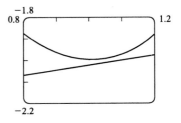

21. We see that the graphs of $f(x) = x^4 - x$ and $g(x) = 1$ intersect twice. The x-coordinates of these points (which are the solutions of the equations) are approximately -0.72 and 1.22. Alternatively, we could find these values by finding the zeros of $h(x) = x^4 - x - 1$.

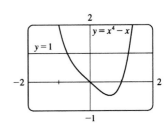

23. We see that the graphs of $f(x) = \tan x$ and $g(x) = \sqrt{1 - x^2}$ intersect once. Using an intersect feature or zooming in, we find this value to be approximately 0.65. Alternatively, we could find this value by finding the positive zero of $h(x) = \tan x - \sqrt{1 - x^2}$.

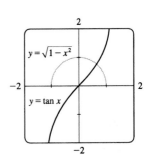

Note: After producing the graph on a TI-84 Plus, we can find the approximate value 0.65 by using the following keystrokes:

2nd CALC 5 ENTER ENTER .6 ENTER . The ".6" is just a guess for 0.65.

25. $g(x) = x^3/10$ is larger than $f(x) = 10x^2$ whenever $x > 100$.

27.

We see from the graphs of $y = |\sin x - x|$ and $y = 0.1$ that there are two solutions to the equation $|\sin x - x| = 0.1$: $x \approx -0.85$ and $x \approx 0.85$. The condition $|\sin x - x| < 0.1$ holds for any x lying between these two values, that is, $-0.85 < x < 0.85$.

29. (a) The root functions $y = \sqrt{x}$, $y = \sqrt[4]{x}$ and $y = \sqrt[6]{x}$

(b) The root functions $y = x$, $y = \sqrt[3]{x}$ and $y = \sqrt[5]{x}$

(c) The root functions $y = \sqrt{x}$, $y = \sqrt[3]{x}$, $y = \sqrt[4]{x}$ and $y = \sqrt[5]{x}$

(d) • For any n, the nth root of 0 is 0 and the nth root of 1 is 1; that is, all nth root functions pass through the points $(0, 0)$ and $(1, 1)$.
   • For odd n, the domain of the nth root function is $\mathbb{R}$, while for even n, it is $\{x \in \mathbb{R} \mid x \geq 0\}$.
   • Graphs of even root functions look similar to that of $\sqrt{x}$, while those of odd root functions resemble that of $\sqrt[3]{x}$.
   • As n increases, the graph of $\sqrt[n]{x}$ becomes steeper near 0 and flatter for $x > 1$.

31. $f(x) = x^4 + cx^2 + x$. If $c < -1.5$, there are three humps: two minimum points and a maximum point. These humps get flatter as c increases, until at $c = -1.5$ two of the humps disappear and there is only one minimum point. This single hump then moves to the right and approaches the origin as c increases.

33. $y = x^n 2^{-x}$. As n increases, the maximum of the function moves further from the origin, and gets larger. Note, however, that regardless of n, the function approaches 0 as $x \to \infty$.

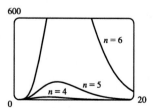

35. $y^2 = cx^3 + x^2$. If $c < 0$, the loop is to the right of the origin, and if c is positive, it is to the left. In both cases, the closer c is to 0, the larger the loop is. (In the limiting case, $c = 0$, the loop is "infinite," that is, it doesn't close.) Also, the larger $|c|$ is, the steeper the slope is on the loopless side of the origin.

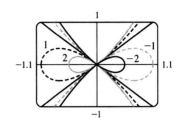

37. The graphing window is 95 pixels wide and we want to start with $x = 0$ and end with $x = 2\pi$. Since there are 94 "gaps" between pixels, the distance between pixels is $\frac{2\pi - 0}{94}$. Thus, the x-values that the calculator actually plots are $x = 0 + \frac{2\pi}{94} \cdot n$, where $n = 0, 1, 2, \ldots, 93, 94$. For $y = \sin 2x$, the actual points plotted by the calculator are $\left(\frac{2\pi}{94} \cdot n, \sin\left(2 \cdot \frac{2\pi}{94} \cdot n\right)\right)$ for $n = 0, 1, \ldots, 94$. For $y = \sin 96x$, the points plotted are $\left(\frac{2\pi}{94} \cdot n, \sin\left(96 \cdot \frac{2\pi}{94} \cdot n\right)\right)$ for $n = 0, 1, \ldots, 94$. But

$$\sin\left(96 \cdot \tfrac{2\pi}{94} \cdot n\right) = \sin\left(94 \cdot \tfrac{2\pi}{94} \cdot n + 2 \cdot \tfrac{2\pi}{94} \cdot n\right) = \sin\left(2\pi n + 2 \cdot \tfrac{2\pi}{94} \cdot n\right)$$

$$= \sin\left(2 \cdot \tfrac{2\pi}{94} \cdot n\right) \quad \text{[by periodicity of sine]}, \quad n = 0, 1, \ldots, 94$$

So the y-values, and hence the points, plotted for $y = \sin 96x$ are identical to those plotted for $y = \sin 2x$.

Note: Try graphing $y = \sin 94x$. Can you see why all the y-values are zero?

1.5 Exponential Functions

1. (a) $\dfrac{4^{-3}}{2^{-8}} = \dfrac{2^8}{4^3} = \dfrac{2^8}{(2^2)^3} = \dfrac{2^8}{2^6} = 2^{8-6} = 2^2 = 4$ **(b)** $\dfrac{1}{\sqrt[3]{x^4}} = \dfrac{1}{x^{4/3}} = x^{-4/3}$

3. (a) $b^8(2b)^4 = b^8 \cdot 2^4 b^4 = 16b^{12}$ **(b)** $\dfrac{(6y^3)^4}{2y^5} = \dfrac{6^4(y^3)^4}{2y^5} = \dfrac{1296y^{12}}{2y^5} = 648y^7$

5. (a) $f(x) = a^x$, $a > 0$ **(b)** $\mathbb{R}$ **(c)** $(0, \infty)$ **(d)** See Figures 4(c), 4(b), and 4(a), respectively.

7. All of these graphs approach 0 as $x \to -\infty$, all of them pass through the point $(0, 1)$, and all of them are increasing and approach ∞ as $x \to \infty$. The larger the base, the faster the function increases for $x > 0$, and the faster it approaches 0 as $x \to -\infty$.

9. The functions with bases greater than 1 (3^x and 10^x) are increasing, while those with bases less than 1 $\left[\left(\frac{1}{3}\right)^x \text{ and } \left(\frac{1}{10}\right)^x\right]$ are decreasing. The graph of $\left(\frac{1}{3}\right)^x$ is the reflection of that of 3^x about the y-axis, and the graph of $\left(\frac{1}{10}\right)^x$ is the reflection of that of 10^x about the y-axis. The graph of 10^x increases more quickly than that of 3^x for $x > 0$, and approaches 0 faster as $x \to -\infty$.

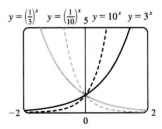

11. We start with the graph of $y = 10^x$ (Figure 3) and shift it 2 units to the left to obtain the graph of $y = 10^{x+2}$.

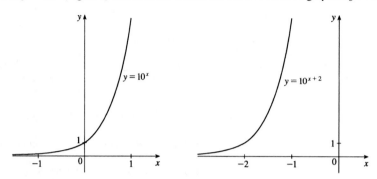

13. We start with the graph of $y = 2^x$ (Figure 3), reflect it about the y-axis, and then about the x-axis (or just rotate $180°$ to handle both reflections) to obtain the graph of $y = -2^{-x}$. In each graph, $y = 0$ is the horizontal asymptote.

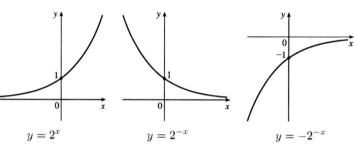

$$y = 2^x \qquad y = 2^{-x} \qquad y = -2^{-x}$$

15. We start with the graph of $y = e^x$ (Figure 13) and reflect about the y-axis to get the graph of $y = e^{-x}$. Then we compress the graph vertically by a factor of 2 to obtain the graph of $y = \frac{1}{2}e^{-x}$ and then reflect about the x-axis to get the graph of $y = -\frac{1}{2}e^{-x}$. Finally, we shift the graph upward one unit to get the graph of $y = 1 - \frac{1}{2}e^{-x}$.

17. (a) To find the equation of the graph that results from shifting the graph of $y = e^x$ 2 units downward, we subtract 2 from the original function to get $y = e^x - 2$.

(b) To find the equation of the graph that results from shifting the graph of $y = e^x$ 2 units to the right, we replace x with $x - 2$ in the original function to get $y = e^{(x-2)}$.

(c) To find the equation of the graph that results from reflecting the graph of $y = e^x$ about the x-axis, we multiply the original function by -1 to get $y = -e^x$.

(d) To find the equation of the graph that results from reflecting the graph of $y = e^x$ about the y-axis, we replace x with $-x$ in the original function to get $y = e^{-x}$.

(e) To find the equation of the graph that results from reflecting the graph of $y = e^x$ about the x-axis and then about the y-axis, we first multiply the original function by -1 (to get $y = -e^x$) and then replace x with $-x$ in this equation to get $y = -e^{-x}$.

19. (a) The denominator is zero when $1 - e^{1-x^2} = 0 \Leftrightarrow e^{1-x^2} = 1 \Leftrightarrow 1 - x^2 = 0 \Leftrightarrow x = \pm 1$. Thus,

the function $f(x) = \dfrac{1 - e^{x^2}}{1 - e^{1-x^2}}$ has domain $\{x \mid x \neq \pm 1\} = (-\infty, -1) \cup (-1, 1) \cup (1, \infty)$.

(b) The denominator is never equal to zero, so the function $f(x) = \dfrac{1 + x}{e^{\cos x}}$ has domain $\mathbb{R}$, or $(-\infty, \infty)$.

21. Use $y = Ca^x$ with the points $(1, 6)$ and $(3, 24)$. $\quad 6 = Ca^1 \quad \left[C = \frac{6}{a}\right] \quad$ and $24 = Ca^3 \Rightarrow 24 = \left(\dfrac{6}{a}\right)a^3 \Rightarrow$

$4 = a^2 \Rightarrow a = 2$ [since $a > 0$] and $C = \frac{6}{2} = 3$. The function is $f(x) = 3 \cdot 2^x$.

23. If $f(x) = 5^x$, then $\dfrac{f(x+h) - f(x)}{h} = \dfrac{5^{x+h} - 5^x}{h} = \dfrac{5^x 5^h - 5^x}{h} = \dfrac{5^x \left(5^h - 1\right)}{h} = 5^x \left(\dfrac{5^h - 1}{h}\right)$.

25. 2 ft $= 24$ in, $f(24) = 24^2$ in $= 576$ in $= 48$ ft. $\quad g(24) = 2^{24}$ in $= 2^{24}/(12 \cdot 5280)$ mi ≈ 265 mi

27. The graph of g finally surpasses that of f at $x \approx 35.8$.

 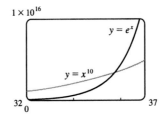

29. (a) Fifteen hours represents 5 doubling periods (one doubling period is three hours). $100 \cdot 2^5 = 3200$

(b) In t hours, there will be $t/3$ doubling periods. The initial population is 100,

so the population y at time t is $y = 100 \cdot 2^{t/3}$.

(c) $t = 20 \Rightarrow y = 100 \cdot 2^{20/3} \approx 10{,}159$

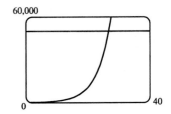

(d) We graph $y_1 = 100 \cdot 2^{x/3}$ and $y_2 = 50{,}000$. The two curves intersect at
$x \approx 26.9$, so the population reaches 50,000 in about 26.9 hours.

31. (a) Fifteen days represents 3 half-life periods (one half-life period is 5 days). $\quad 200 \left(\frac{1}{2}\right)^3 = 25$ mg

(b) In t hours, there will be $t/5$ half-life periods. The initial amount is 200 mg,

so the amount remaining after t days is $y = 200 \left(\frac{1}{2}\right)^{t/5}$, or equivalently,

$y = 200 \cdot 2^{-t/5}$.

(c) $t = 3$ weeks $= 21$ days $\Rightarrow y = 200 \cdot 2^{-21/5} \approx 10.9$ mg

(d) We graph $y_1 = 200 \cdot 2^{-t/5}$ and $y_2 = 1$. The two curves intersect at
$t \approx 38.2$, so the mass will be reduced to 1 mg in about 38.2 days.

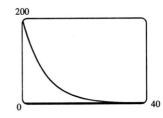

33. An exponential model is $y = ab^t$, where $a = 3.154832569 \times 10^{-12}$ and $b = 1.017764706$. This model gives $y(1993) \approx 5498$ million and $y(2010) \approx 7417$ million.

35.

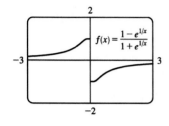

From the graph, it appears that f is an odd function (f is undefined for $x = 0$). To prove this, we must show that $f(-x) = -f(x)$.

$$f(-x) = \frac{1 - e^{1/(-x)}}{1 + e^{1/(-x)}} = \frac{1 - e^{(-1/x)}}{1 + e^{(-1/x)}} = \frac{1 - \dfrac{1}{e^{1/x}}}{1 + \dfrac{1}{e^{1/x}}} \cdot \frac{e^{1/x}}{e^{1/x}} = \frac{e^{1/x} - 1}{e^{1/x} + 1}$$

$$= -\frac{1 - e^{1/x}}{1 + e^{1/x}} = -f(x)$$

so f is an odd function.

1.6 Inverse Functions and Logarithms

1. (a) See Definition 1.

(b) It must pass the Horizontal Line Test.

3. f is not one-to-one because $2 \neq 6$, but $f(2) = 2.0 = f(6)$.

5. We could draw a horizontal line that intersects the graph in more than one point. Thus, by the Horizontal Line Test, the function is not one-to-one.

7. No horizontal line intersects the graph more than once. Thus, by the Horizontal Line Test, the function is one-to-one.

9. The graph of $f(x) = x^2 - 2x$ is a parabola with axis of symmetry $x = -\dfrac{b}{2a} = -\dfrac{-2}{2(1)} = 1$. Pick any x-values equidistant from 1 to find two equal function values. For example, $f(0) = 0$ and $f(2) = 0$, so f is not one-to-one.

11. $g(x) = 1/x$. $x_1 \neq x_2 \;\Rightarrow\; 1/x_1 \neq 1/x_2 \;\Rightarrow\; g(x_1) \neq g(x_2)$, so g is one-to-one.
Geometric solution: The graph of g is the hyperbola shown in Figure 14 in Section 1.2. It passes the Horizontal Line Test, so g is one-to-one.

13. A football will attain every height h up to its maximum height twice: once on the way up, and again on the way down. Thus, even if t_1 does not equal t_2, $f(t_1)$ may equal $f(t_2)$, so f is not 1-1.

15. Since $f(2) = 9$ and f is 1-1, we know that $f^{-1}(9) = 2$. Remember, if the point $(2, 9)$ is on the graph of f, then the point $(9, 2)$ is on the graph of f^{-1}.

17. First, we must determine x such that $g(x) = 4$. By inspection, we see that if $x = 0$, then $g(x) = 4$. Since g is 1-1 (g is an increasing function), it has an inverse, and $g^{-1}(4) = 0$.

19. We solve $C = \frac{5}{9}(F - 32)$ for F: $\frac{9}{5}C = F - 32$ $\Rightarrow$ $F = \frac{9}{5}C + 32$. This gives us a formula for the inverse function, that

is, the Fahrenheit temperature F as a function of the Celsius temperature C. $F \geq -459.67$ $\Rightarrow$ $\frac{9}{5}C + 32 \geq -459.67$ $\Rightarrow$

$\frac{9}{5}C \geq -491.67$ $\Rightarrow$ $C \geq -273.15$, the domain of the inverse function.

21. $y = f(x) = 1 + \sqrt{2 + 3x}$ $(y \geq 1)$ $\Rightarrow$ $y - 1 = \sqrt{2 + 3x}$ $\Rightarrow$ $(y - 1)^2 = 2 + 3x$ $\Rightarrow$ $(y - 1)^2 - 2 = 3x$ $\Rightarrow$

$x = \frac{1}{3}(y - 1)^2 - \frac{2}{3}$. Interchange x and y: $y = \frac{1}{3}(x - 1)^2 - \frac{2}{3}$. So $f^{-1}(x) = \frac{1}{3}(x - 1)^2 - \frac{2}{3}$. Note that the domain of f^{-1}

is $x \geq 1$.

23. $y = f(x) = e^{2x-1}$ $\Rightarrow$ $\ln y = 2x - 1$ $\Rightarrow$ $1 + \ln y = 2x$ $\Rightarrow$ $x = \frac{1}{2}(1 + \ln y)$.

Interchange x and y: $y = \frac{1}{2}(1 + \ln x)$. So $f^{-1}(x) = \frac{1}{2}(1 + \ln x)$.

25. $y = f(x) = \ln(x + 3)$ $\Rightarrow$ $x + 3 = e^y$ $\Rightarrow$ $x = e^y - 3$. Interchange x and y: $y = e^x - 3$. So $f^{-1}(x) = e^x - 3$.

27. $y = f(x) = x^4 + 1$ $\Rightarrow$ $y - 1 = x^4$ $\Rightarrow$ $x = \sqrt[4]{y - 1}$ [not $\pm$ since

$x \geq 0$]. Interchange x and y: $y = \sqrt[4]{x - 1}$. So $f^{-1}(x) = \sqrt[4]{x - 1}$. The

graph of $y = \sqrt[4]{x - 1}$ is just the graph of $y = \sqrt[4]{x}$ shifted right one unit.

From the graph, we see that f and f^{-1} are reflections about the line $y = x$.

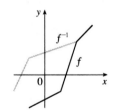

29. Reflect the graph of f about the line $y = x$. The points $(-1, -2)$, $(1, -1)$,

$(2, 2)$, and $(3, 3)$ on f are reflected to $(-2, -1)$, $(-1, 1)$, $(2, 2)$, and $(3, 3)$

on f^{-1}.

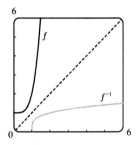

31. (a) $y = f(x) = \sqrt{1 - x^2}$ $(0 \leq x \leq 1$ and note that $y \geq 0)$ $\Rightarrow$

$y^2 = 1 - x^2$ $\Rightarrow$ $x^2 = 1 - y^2$ $\Rightarrow$ $x = \sqrt{1 - y^2}$. So

$f^{-1}(x) = \sqrt{1 - x^2}$, $0 \leq x \leq 1$. We see that f^{-1} and f are the same

function.

(b) The graph of f is the portion of the circle $x^2 + y^2 = 1$ with $0 \leq x \leq 1$ and

$0 \leq y \leq 1$ (quarter-circle in the first quadrant). The graph of f is symmetric

with respect to the line $y = x$, so its reflection about $y = x$ is itself, that is,

$f^{-1} = f$.

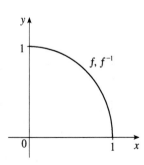

33. (a) It is defined as the inverse of the exponential function with base a, that is, $\log_a x = y$ $\Leftrightarrow$ $a^y = x$.

(b) $(0, \infty)$ (c) $\mathbb{R}$ (d) See Figure 11.

35. (a) $\log_5 125 = 3$ since $5^3 = 125$.   (b) $\log_3 \dfrac{1}{27} = -3$ since $3^{-3} = \dfrac{1}{3^3} = \dfrac{1}{27}$.

37. (a) $\log_2 6 - \log_2 15 + \log_2 20 = \log_2\left(\frac{6}{15}\right) + \log_2 20$ [by Law 2]

$$= \log_2\left(\frac{6}{15} \cdot 20\right) \qquad \text{[by Law 1]}$$

$$= \log_2 8, \text{ and } \log_2 8 = 3 \text{ since } 2^3 = 8.$$

(b) $\log_3 100 - \log_3 18 - \log_3 50 = \log_3\left(\frac{100}{18}\right) - \log_3 50 = \log_3\left(\frac{100}{18 \cdot 50}\right)$

$$= \log_3\left(\frac{1}{9}\right), \text{ and } \log_3\left(\frac{1}{9}\right) = -2 \text{ since } 3^{-2} = \frac{1}{9}.$$

39. $\ln 5 + 5\ln 3 = \ln 5 + \ln 3^5$ [by Law 3]

$$= \ln(5 \cdot 3^5) \qquad \text{[by Law 1]}$$

$$= \ln 1215$$

41. $\ln(1 + x^2) + \frac{1}{2}\ln x - \ln\sin x = \ln(1 + x^2) + \ln x^{1/2} - \ln\sin x = \ln[(1 + x^2)\sqrt{x}] - \ln\sin x = \ln\dfrac{(1 + x^2)\sqrt{x}}{\sin x}$

43. To graph these functions, we use $\log_{1.5} x = \dfrac{\ln x}{\ln 1.5}$ and $\log_{50} x = \dfrac{\ln x}{\ln 50}$.

These graphs all approach $-\infty$ as $x \to 0^+$, and they all pass through the

point $(1, 0)$. Also, they are all increasing, and all approach ∞ as $x \to \infty$.

The functions with larger bases increase extremely slowly, and the ones with

smaller bases do so somewhat more quickly. The functions with large bases

approach the y-axis more closely as $x \to 0^+$.

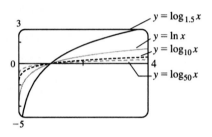

45. 3 ft $=$ 36 in, so we need x such that $\log_2 x = 36$ $\Leftrightarrow$ $x = 2^{36} = 68{,}719{,}476{,}736$. In miles, this is

$$68{,}719{,}476{,}736 \text{ in} \cdot \frac{1 \text{ ft}}{12 \text{ in}} \cdot \frac{1 \text{ mi}}{5280 \text{ ft}} \approx 1{,}084{,}587.7 \text{ mi}.$$

47. (a) Shift the graph of $y = \log_{10} x$ five units to the left to obtain the graph of $y = \log_{10}(x + 5)$. Note the vertical asymptote of $x = -5$.

(b) Reflect the graph of $y = \ln x$ about the x-axis to obtain the graph of $y = -\ln x$.

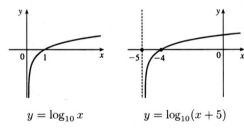

$$y = \log_{10} x \qquad\qquad y = \log_{10}(x + 5)$$

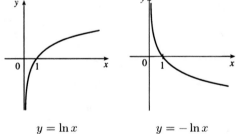

$$y = \ln x \qquad\qquad y = -\ln x$$

49. (a) $e^{7-4x} = 6$ $\Leftrightarrow$ $7 - 4x = \ln 6$ $\Leftrightarrow$ $7 - \ln 6 = 4x$ $\Leftrightarrow$ $x = \frac{1}{4}(7 - \ln 6)$

(b) $\ln(3x - 10) = 2$ $\Leftrightarrow$ $3x - 10 = e^2$ $\Leftrightarrow$ $3x = e^2 + 10$ $\Leftrightarrow$ $x = \frac{1}{3}(e^2 + 10)$

51. (a) $2^{x-5} = 3$ $\Leftrightarrow$ $\log_2 3 = x - 5$ $\Leftrightarrow$ $x = 5 + \log_2 3$.

Or: $2^{x-5} = 3$ $\Leftrightarrow$ $\ln(2^{x-5}) = \ln 3$ $\Leftrightarrow$ $(x - 5)\ln 2 = \ln 3$ $\Leftrightarrow$ $x - 5 = \dfrac{\ln 3}{\ln 2}$ $\Leftrightarrow$ $x = 5 + \dfrac{\ln 3}{\ln 2}$

(b) $\ln x + \ln(x-1) = \ln(x(x-1)) = 1 \iff x(x-1) = e^1 \iff x^2 - x - e = 0$. The quadratic formula (with $a = 1$,

$b = -1$, and $c = -e$) gives $x = \frac{1}{2}\left(1 \pm \sqrt{1+4e}\right)$, but we reject the negative root since the natural logarithm is not

defined for $x < 0$. So $x = \frac{1}{2}\left(1 + \sqrt{1+4e}\right)$.

53. (a) $e^x < 10 \implies \ln e^x < \ln 10 \implies x < \ln 10 \implies x \in (-\infty, \ln 10)$

(b) $\ln x > -1 \implies e^{\ln x} > e^{-1} \implies x > e^{-1} \implies x \in (1/e, \infty)$

55. (a) For $f(x) = \sqrt{3 - e^{2x}}$, we must have $3 - e^{2x} \geq 0 \implies e^{2x} \leq 3 \implies 2x \leq \ln 3 \implies x \leq \frac{1}{2}\ln 3$.

Thus, the domain of f is $(-\infty, \frac{1}{2}\ln 3]$.

(b) $y = f(x) = \sqrt{3 - e^{2x}}$ [note that $y \geq 0$] $\implies y^2 = 3 - e^{2x} \implies e^{2x} = 3 - y^2 \implies 2x = \ln(3 - y^2) \implies$

$x = \frac{1}{2}\ln(3 - y^2)$. Interchange x and y: $y = \frac{1}{2}\ln(3 - x^2)$. So $f^{-1}(x) = \frac{1}{2}\ln(3 - x^2)$. For the domain of f^{-1},

we must have $3 - x^2 > 0 \implies x^2 < 3 \implies |x| < \sqrt{3} \implies -\sqrt{3} < x < \sqrt{3} \implies 0 \leq x < \sqrt{3}$ since $x \geq 0$. Note

that the domain of f^{-1}, $[0, \sqrt{3})$, equals the range of f.

57. We see that the graph of $y = f(x) = \sqrt{x^3 + x^2 + x + 1}$ is increasing, so f is 1-1.

Enter $x = \sqrt{y^3 + y^2 + y + 1}$ and use your CAS to solve the equation for y.

Using Derive, we get two (irrelevant) solutions involving imaginary expressions,

as well as one which can be simplified to the following:

$y = f^{-1}(x) = -\frac{\sqrt[3]{4}}{6}\left(\sqrt[3]{D - 27x^2 + 20} - \sqrt[3]{D + 27x^2 - 20} + \sqrt[3]{2}\right)$

where $D = 3\sqrt{3}\sqrt{27x^4 - 40x^2 + 16}$.

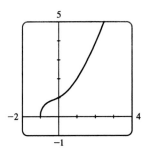

Maple and Mathematica each give two complex expressions and one real expression, and the real expression is equivalent

to that given by Derive. For example, Maple's expression simplifies to $\frac{1}{6}\dfrac{M^{2/3} - 8 - 2M^{1/3}}{2M^{1/3}}$, where

$M = 108x^2 + 12\sqrt{48 - 120x^2 + 81x^4} - 80$.

59. (a) $n = 100 \cdot 2^{t/3} \implies \dfrac{n}{100} = 2^{t/3} \implies \log_2\left(\dfrac{n}{100}\right) = \dfrac{t}{3} \implies t = 3\log_2\left(\dfrac{n}{100}\right)$. Using formula (10), we can write

this as $t = f^{-1}(n) = 3 \cdot \dfrac{\ln(n/100)}{\ln 2}$. This function tells us how long it will take to obtain n bacteria (given the number n).

(b) $n = 50{,}000 \implies t = f^{-1}(50{,}000) = 3 \cdot \dfrac{\ln\left(\frac{50{,}000}{100}\right)}{\ln 2} = 3\left(\dfrac{\ln 500}{\ln 2}\right) \approx 26.9$ hours

61. (a) To find the equation of the graph that results from shifting the graph of $y = \ln x$ 3 units upward, we add 3 to the original

function to get $y = \ln x + 3$.

(b) To find the equation of the graph that results from shifting the graph of $y = \ln x$ 3 units to the left, we replace x with $x + 3$

in the original function to get $y = \ln(x + 3)$.

(c) To find the equation of the graph that results from reflecting the graph of $y = \ln x$ about the x-axis, we multiply the

original equation by -1 to get $y = -\ln x$.

(d) To find the equation of the graph that results from reflecting the graph of $y = \ln x$ about the y-axis, we replace x with $-x$ in the original equation to get $y = \ln(-x)$.

(e) To find the equation of the graph that results from reflecting the graph of $y = \ln x$ about the line $y = x$, we interchange x and y in the original equation to get $x = \ln y \iff y = e^x$.

(f) To find the equation of the graph that results from reflecting the graph of $y = \ln x$ about the x-axis and then about the line $y = x$, we first multiply the original equation by -1 [to get $y = -\ln x$] and then interchange x and y in this equation to get $x = -\ln y \iff \ln y = -x \iff y = e^{-x}$.

(g) To find the equation of the graph that results from reflecting the graph of $y = \ln x$ about the y-axis and then about the line $y = x$, we first replace x with $-x$ in the original equation [to get $y = \ln(-x)$] and then interchange x and y to get $x = \ln(-y) \iff -y = e^x \iff y = -e^x$.

(h) To find the equation of the graph that results from shifting the graph of $y = \ln x$ 3 units to the left and then reflecting it about the line $y = x$, we first replace x with $x + 3$ in the original equation [to get $y = \ln(x + 3)$] and then interchange x and y in this equation to get $x = \ln(y + 3) \iff y + 3 = e^x \iff y = e^x - 3$.

1.7 Parametric Curves

1. $x = t^2 + t$, $\quad y = t^2 - t$, $\quad -2 \le t \le 2$

t	-2	-1	0	1	2
x	2	0	0	2	6
y	6	2	0	0	2

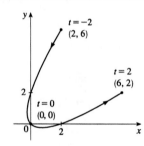

3. $x = \cos^2 t$, $\quad y = 1 - \sin t$, $\quad 0 \le t \le \pi/2$

t	0	$\pi/6$	$\pi/3$	$\pi/2$
x	1	3/4	1/4	0
y	1	1/2	$1 - \frac{\sqrt{3}}{2} \approx 0.13$	0

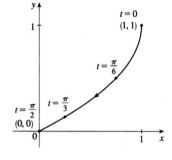

5. $x = 3t - 5$, $\quad y = 2t + 1$

(a)

t	-2	-1	0	1	2	3	4
x	-11	-8	-5	-2	1	4	7
y	-3	-1	1	3	5	7	9

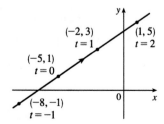

(b) $x = 3t - 5 \implies 3t = x + 5 \implies t = \frac{1}{3}(x + 5) \implies$

$y = 2 \cdot \frac{1}{3}(x + 5) + 1$, so $y = \frac{2}{3}x + \frac{13}{3}$.

7. $x = \sqrt{t}$, $y = 1 - t$

(a)

t	0	1	2	3	4
x	0	1	1.414	1.732	2
y	1	0	−1	−2	−3

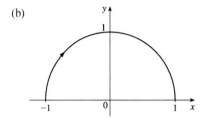

(b) $x = \sqrt{t}$ $\Rightarrow$ $t = x^2$ $\Rightarrow$ $y = 1 - t = 1 - x^2$. Since $t \geq 0$, $x \geq 0$.

So the curve is the right half of the parabola $y = 1 - x^2$.

9. (a) $x = \sin \frac{1}{2}\theta$, $y = \cos \frac{1}{2}\theta$, $-\pi \leq \theta \leq \pi$.

$x^2 + y^2 = \sin^2 \frac{1}{2}\theta + \cos^2 \frac{1}{2}\theta = 1$. For $-\pi \leq \theta \leq 0$, we have

$-1 \leq x \leq 0$ and $0 \leq y \leq 1$. For $0 < \theta \leq \pi$, we have $0 < x \leq 1$

and $1 > y \geq 0$. The graph is a semicircle.

(b)

11. (a) $x = \sin t$, $y = \csc t$, $0 < t < \frac{\pi}{2}$. $y = \csc t = \dfrac{1}{\sin t} = \dfrac{1}{x}$.

For $0 < t < \frac{\pi}{2}$, we have $0 < x < 1$ and $y > 1$. Thus, the curve is the

portion of the hyperbola $y = 1/x$ with $y > 1$.

(b)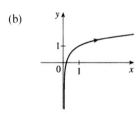

13. (a) $x = e^{2t}$ $\Rightarrow$ $2t = \ln x$ $\Rightarrow$ $t = \frac{1}{2}\ln x$.

$y = t + 1 = \frac{1}{2}\ln x + 1$.

(b)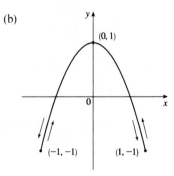

15. (a) $x = \sin \theta$, $y = \cos 2\theta$.

$y = \cos^2 \theta - \sin^2 \theta = 1 - \sin^2 \theta - \sin^2 \theta$

$\quad = 1 - 2\sin^2 \theta = 1 - 2x^2$.

Since $-1 \leq \sin \theta \leq 1$ and $-1 \leq \cos 2\theta \leq 1$, $-1 \leq x \leq 1$, and

$-1 \leq y \leq 1$. The point (x, y) moves back and forth infinitely often along

the parabola $y = 1 - 2x^2$ from $(1, -1)$ to $(-1, -1)$.

(b)

17. $x = 3 + 2\cos t$, $y = 1 + 2\sin t$, $\pi/2 \leq t \leq 3\pi/2$.   By Example 4 with $r = 2$, $h = 3$, and $k = 1$, the motion of the particle

takes place on a circle centered at $(3, 1)$ with a radius of 2. As t goes from $\frac{\pi}{2}$ to $\frac{3\pi}{2}$, the particle starts at the point $(3, 3)$ and

moves counterclockwise to $(3, -1)$ [one-half of a circle].

19. $x = 5 \sin t$, $y = 2 \cos t$ $\Rightarrow$ $\sin t = \dfrac{x}{5}$, $\cos t = \dfrac{y}{2}$. $\sin^2 t + \cos^2 t = 1$ $\Rightarrow$ $\left(\dfrac{x}{5}\right)^2 + \left(\dfrac{y}{2}\right)^2 = 1$. The motion of the

particle takes place on an ellipse centered at $(0, 0)$. As t goes from $-\pi$ to 5π, the particle starts at the point $(0, -2)$ and moves clockwise around the ellipse 3 times.

21. We must have $1 \leq x \leq 4$ and $2 \leq y \leq 3$. So the graph of the curve must be contained in the rectangle $[1, 4]$ by $[2, 3]$.

23. When $t = -1$, $(x, y) = (0, -1)$. As t increases to 0, x decreases to -1 and y

increases to 0. As t increases from 0 to 1, x increases to 0 and y increases to 1.

As t increases beyond 1, both x and y increase. For $t < -1$, x is positive and

decreasing and y is negative and increasing. We could achieve greater accuracy

by estimating x- and y-values for selected values of t from the given graphs and

plotting the corresponding points.

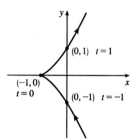

25. When $t = 0$ we see that $x = 0$ and $y = 0$, so the curve starts at the origin. As t

increases from 0 to $\frac{1}{2}$, the graphs show that y increases from 0 to 1 while x

increases from 0 to 1, decreases to 0 and to -1, then increases back to 0, so we

arrive at the point $(0, 1)$. Similarly, as t increases from $\frac{1}{2}$ to 1, y decreases from 1

to 0 while x repeats its pattern, and we arrive back at the origin. We could achieve greater accuracy by estimating x- and

y-values for selected values of t from the given graphs and plotting the corresponding points.

27. Use $y = t$ and $x = t - 2 \sin \pi t$ with a t-interval of $[-\pi, \pi]$.

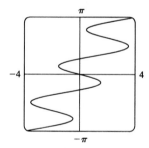

29. (a) $x = x_1 + (x_2 - x_1)t$, $y = y_1 + (y_2 - y_1)t$, $0 \leq t \leq 1$. Clearly the curve passes through $P_1(x_1, y_1)$ when $t = 0$ and

through $P_2(x_2, y_2)$ when $t = 1$. For $0 < t < 1$, x is strictly between x_1 and x_2 and y is strictly between y_1 and y_2. For

every value of t, x and y satisfy the relation $y - y_1 = \dfrac{y_2 - y_1}{x_2 - x_1}(x - x_1)$, which is the equation of the line through

$P_1(x_1, y_1)$ and $P_2(x_2, y_2)$.

Finally, any point (x, y) on that line satisfies $\dfrac{y - y_1}{y_2 - y_1} = \dfrac{x - x_1}{x_2 - x_1}$; if we call that common value t, then the given

parametric equations yield the point (x, y); and any (x, y) on the line between $P_1(x_1, y_1)$ and $P_2(x_2, y_2)$ yields a value of

t in $[0, 1]$. So the given parametric equations exactly specify the line segment from $P_1(x_1, y_1)$ to $P_2(x_2, y_2)$.

(b) $x = -2 + [3 - (-2)]t = -2 + 5t$ and $y = 7 + (-1 - 7)t = 7 - 8t$ for $0 \leq t \leq 1$.

31. The circle $x^2 + (y-1)^2 = 4$ has center $(0,1)$ and radius 2, so by Example 4 it can be represented by $x = 2\cos t$, $y = 1 + 2\sin t$, $0 \le t \le 2\pi$. This representation gives us the circle with a counterclockwise orientation starting at $(2,1)$.

(a) To get a clockwise orientation, we could change the equations to $x = 2\cos t$, $y = 1 - 2\sin t$, $0 \le t \le 2\pi$.

(b) To get three times around in the counterclockwise direction, we use the original equations $x = 2\cos t$, $y = 1 + 2\sin t$ with the domain expanded to $0 \le t \le 6\pi$.

(c) To start at $(0,3)$ using the original equations, we must have $x_1 = 0$; that is, $2\cos t = 0$. Hence, $t = \frac{\pi}{2}$. So we use

$$x = 2\cos t, \quad y = 1 + 2\sin t, \quad \tfrac{\pi}{2} \le t \le \tfrac{3\pi}{2}.$$

Alternatively, if we want t to start at 0, we could change the equations of the curve. For example, we could use

$$x = -2\sin t, \quad y = 1 + 2\cos t, \quad 0 \le t \le \pi.$$

33. *Big circle:* It's centered at $(2,2)$ with a radius of 2, so by Example 4, parametric equations are

$$x = 2 + 2\cos t, \qquad y = 2 + 2\sin t, \qquad 0 \le t \le 2\pi$$

Small circles: They are centered at $(1,3)$ and $(3,3)$ with a radius of 0.1. By Example 4, parametric equations are

$$\text{(left)} \qquad x = 1 + 0.1\cos t, \qquad y = 3 + 0.1\sin t, \qquad 0 \le t \le 2\pi$$

and $\qquad$ (right) $\qquad x = 3 + 0.1\cos t, \qquad y = 3 + 0.1\sin t, \qquad 0 \le t \le 2\pi$

Semicircle: It's the lower half of a circle centered at $(2,2)$ with radius 1. By Example 4, parametric equations are

$$x = 2 + 1\cos t, \qquad y = 2 + 1\sin t, \qquad \pi \le t \le 2\pi$$

To get all four graphs on the same screen with a typical graphing calculator, we need to change the last t-interval to $[0, 2\pi]$ in order to match the others. We can do this by changing t to $0.5t$. This change gives us the upper half. There are several ways to get the lower half—one is to change the "$+$" to a "$-$" in the y-assignment, giving us

$$x = 2 + 1\cos(0.5t), \qquad y = 2 - 1\sin(0.5t), \qquad 0 \le t \le 2\pi$$

35. (a) $x = t^3 \;\Rightarrow\; t = x^{1/3}$, so $y = t^2 = x^{2/3}$.

We get the entire curve $y = x^{2/3}$ traversed in a left to right direction.

(b) $x = t^6 \;\Rightarrow\; t = x^{1/6}$, so $y = t^4 = x^{4/6} = x^{2/3}$.

Since $x = t^6 \ge 0$, we only get the right half of the curve $y = x^{2/3}$.

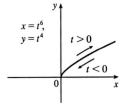

(c) $x = e^{-3t} = (e^{-t})^3$ [so $e^{-t} = x^{1/3}$],

$y = e^{-2t} = (e^{-t})^2 = (x^{1/3})^2 = x^{2/3}$.

If $t < 0$, then x and y are both larger than 1. If $t > 0$, then x and y are between 0 and 1. Since $x > 0$ and $y > 0$, the curve never quite reaches the origin.

37. The case $\frac{\pi}{2} < \theta < \pi$ is illustrated. C has coordinates $(r\theta, r)$ as in Example 7,

and Q has coordinates $(r\theta, r + r\cos(\pi - \theta)) = (r\theta, r(1 - \cos\theta))$

[since $\cos(\pi - \alpha) = \cos\pi\cos\alpha + \sin\pi\sin\alpha = -\cos\alpha$], so P has

coordinates $(r\theta - r\sin(\pi - \theta), r(1 - \cos\theta)) = (r(\theta - \sin\theta), r(1 - \cos\theta))$

[since $\sin(\pi - \alpha) = \sin\pi\cos\alpha - \cos\pi\sin\alpha = \sin\alpha$]. Again we have the

parametric equations $x = r(\theta - \sin\theta)$, $y = r(1 - \cos\theta)$.

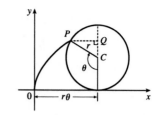

39. It is apparent that $x = |OQ|$ and $y = |QP| = |ST|$. From the diagram,

$x = |OQ| = a\cos\theta$ and $y = |ST| = b\sin\theta$. Thus, the parametric equations are

$x = a\cos\theta$ and $y = b\sin\theta$. To eliminate θ we rearrange: $\sin\theta = y/b$ $\Rightarrow$

$\sin^2\theta = (y/b)^2$ and $\cos\theta = x/a$ $\Rightarrow$ $\cos^2\theta = (x/a)^2$. Adding the two

equations: $\sin^2\theta + \cos^2\theta = 1 = x^2/a^2 + y^2/b^2$. Thus, we have an ellipse.

41. (a)

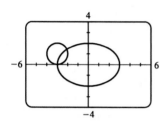

There are 2 points of intersection:

$(-3, 0)$ and approximately $(-2.1, 1.4)$.

(b) A collision point occurs when $x_1 = x_2$ and $y_1 = y_2$ for the same t. So solve the equations:

$$3\sin t = -3 + \cos t \quad \textbf{(1)}$$
$$2\cos t = 1 + \sin t \quad \textbf{(2)}$$

From **(2)**, $\sin t = 2\cos t - 1$. Substituting into **(1)**, we get $3(2\cos t - 1) = -3 + \cos t$ $\Rightarrow$ $5\cos t = 0$ $(\star)$ $\Rightarrow$

$\cos t = 0$ $\Rightarrow$ $t = \frac{\pi}{2}$ or $\frac{3\pi}{2}$. We check that $t = \frac{3\pi}{2}$ satisfies **(1)** and **(2)** but $t = \frac{\pi}{2}$ does not. So the only collision point

occurs when $t = \frac{3\pi}{2}$, and this gives the point $(-3, 0)$. [We could check our work by graphing x_1 and x_2 together as

functions of t and, on another plot, y_1 and y_2 as functions of t. If we do so, we see that the only value of t for which *both*

pairs of graphs intersect is $t = \frac{3\pi}{2}$.]

(c) The circle is centered at $(3, 1)$ instead of $(-3, 1)$. There are still 2 intersection points: $(3, 0)$ and $(2.1, 1.4)$, but there are

no collision points, since $(\star)$ in part (b) becomes $5\cos t = 6$ $\Rightarrow$ $\cos t = \frac{6}{5} > 1$.

43. $x = t^2$, $y = t^3 - ct$. We use a graphing device to produce the graphs for various values of c with $-\pi \le t \le \pi$. Note that all

the members of the family are symmetric about the x-axis. For $c < 0$, the graph does not cross itself, but for $c = 0$ it has a

cusp at $(0, 0)$ and for $c > 0$ the graph crosses itself at $x = c$, so the loop grows larger as c increases.

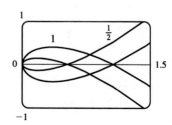

45. Note that all the Lissajous figures are symmetric about the x-axis. The parameters a and b simply stretch the graph in the x- and y-directions respectively. For $a = b = n = 1$ the graph is simply a circle with radius 1. For $n = 2$ the graph crosses itself at the origin and there are loops above and below the x-axis. In general, the figures have $n - 1$ points of intersection, all of which are on the y-axis, and a total of n closed loops.

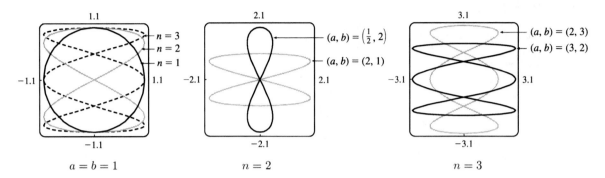

1   Review

CONCEPT CHECK

1. (a) A **function** f is a rule that assigns to each element x in a set A exactly one element, called $f(x)$, in a set B. The set A is called the **domain** of the function. The **range** of f is the set of all possible values of $f(x)$ as x varies throughout the domain.

(b) If f is a function with domain A, then its **graph** is the set of ordered pairs $\{(x, f(x)) \mid x \in A\}$.

(c) Use the Vertical Line Test on page 17.

2. The four ways to represent a function are: verbally, numerically, visually, and algebraically. An example of each is given below.

Verbally: An assignment of students to chairs in a classroom (a description in words)

Numerically: A tax table that assigns an amount of tax to an income (a table of values)

Visually: A graphical history of the Dow Jones average (a graph)

Algebraically: A relationship between distance, rate, and time: $d = rt$ (an explicit formula)

3. (a) An **even function** f satisfies $f(-x) = f(x)$ for every number x in its domain. It is symmetric with respect to the y-axis.

(b) An **odd function** g satisfies $g(-x) = -g(x)$ for every number x in its domain. It is symmetric with respect to the origin.

4. A function f is called **increasing** on an interval I if $f(x_1) < f(x_2)$ whenever $x_1 < x_2$ in I.

5. A **mathematical model** is a mathematical description (often by means of a function or an equation) of a real-world phenomenon.

6. (a) Linear function: $f(x) = 2x + 1$, $f(x) = ax + b$

(b) Power function: $f(x) = x^2$, $f(x) = x^a$

(c) Exponential function: $f(x) = 2^x$, $f(x) = a^x$

(d) Quadratic function: $f(x) = x^2 + x + 1$, $f(x) = ax^2 + bx + c$

(e) Polynomial of degree 5: $f(x) = x^5 + 2$

(f) Rational function: $f(x) = \dfrac{x}{x+2}$, $f(x) = \dfrac{P(x)}{Q(x)}$ where $P(x)$ and $Q(x)$ are polynomials

7.

8. (a)

(b)

(c)

(d)

(e)

(f)

(g)

9. (a) The domain of $f + g$ is the intersection of the domain of f and the domain of g; that is, $A \cap B$.

(b) The domain of fg is also $A \cap B$.

(c) The domain of f/g must exclude values of x that make g equal to 0; that is, $\{x \in A \cap B \mid g(x) \neq 0\}$.

10. Given two functions f and g, the **composite** function $f \circ g$ is defined by $(f \circ g)(x) = f(g(x))$. The domain of $f \circ g$ is the set of all x in the domain of g such that $g(x)$ is in the domain of f.

11. (a) If the graph of f is shifted 2 units upward, its equation becomes $y = f(x) + 2$.

(b) If the graph of f is shifted 2 units downward, its equation becomes $y = f(x) - 2$.

(c) If the graph of f is shifted 2 units to the right, its equation becomes $y = f(x - 2)$.

(d) If the graph of f is shifted 2 units to the left, its equation becomes $y = f(x + 2)$.

(e) If the graph of f is reflected about the x-axis, its equation becomes $y = -f(x)$.

(f) If the graph of f is reflected about the y-axis, its equation becomes $y = f(-x)$.

(g) If the graph of f is stretched vertically by a factor of 2, its equation becomes $y = 2f(x)$.

(h) If the graph of f is shrunk vertically by a factor of 2, its equation becomes $y = \frac{1}{2}f(x)$.

(i) If the graph of f is stretched horizontally by a factor of 2, its equation becomes $y = f\left(\frac{1}{2}x\right)$.

(j) If the graph of f is shrunk horizontally by a factor of 2, its equation becomes $y = f(2x)$.

12. (a) A function f is called a *one-to-one function* if it never takes on the same value twice; that is, if $f(x_1) \neq f(x_2)$ whenever $x_1 \neq x_2$. (Or, f is 1-1 if each output corresponds to only one input.)

Use the Horizontal Line Test: A function is one-to-one if and only if no horizontal line intersects its graph more than once.

(b) If f is a one-to-one function with domain A and range B, then its *inverse function* f^{-1} has domain B and range A and is defined by

$$f^{-1}(y) = x \quad \Leftrightarrow \quad f(x) = y$$

for any y in B. The graph of f^{-1} is obtained by reflecting the graph of f about the line $y = x$.

13. (a) A parametric curve is a set of points of the form $(x, y) = (f(t), g(t))$, where f and g are continuous functions of a variable t.

(b) Sketching a parametric curve, like sketching the graph of a function, is difficult to do in general. We can plot points on the curve by finding $f(t)$ and $g(t)$ for various values of t, either by hand or with a calculator or computer. Sometimes, when f and g are given by formulas, we can eliminate t from the equations $x = f(t)$ and $y = g(t)$ to get a Cartesian equation relating x and y. It may be easier to graph that equation than to work with the original formulas for x and y in terms of t.

(c) See the margin note on page 72.

TRUE-FALSE QUIZ

1. False. Let $f(x) = x^2$, $s = -1$, and $t = 1$. Then $f(s + t) = (-1 + 1)^2 = 0^2 = 0$, but $f(s) + f(t) = (-1)^2 + 1^2 = 2 \neq 0 = f(s + t)$.

3. False. Let $f(x) = x^2$. Then $f(3x) = (3x)^2 = 9x^2$ and $3f(x) = 3x^2$. So $f(3x) \neq 3f(x)$.

5. True. See the Vertical Line Test.

7. False. Let $f(x) = x^3$. Then f is one-to-one and $f^{-1}(x) = \sqrt[3]{x}$. But $1/f(x) = 1/x^3$, which is not equal to $f^{-1}(x)$.

9. True. The function $\ln x$ is an increasing function on $(0, \infty)$.

11. False. Let $x = e^2$ and $a = e$. Then $\dfrac{\ln x}{\ln a} = \dfrac{\ln e^2}{\ln e} = \dfrac{2 \ln e}{\ln e} = 2$ and $\ln \dfrac{x}{a} = \ln \dfrac{e^2}{e} = \ln e = 1$, so in general the statement is false. What *is* true, however, is that $\ln \dfrac{x}{a} = \ln x - \ln a$.

EXERCISES

1. (a) When $x = 2$, $y \approx 2.7$. Thus, $f(2) \approx 2.7$.

(b) $f(x) = 3 \Rightarrow x \approx 2.3, 5.6$

(c) The domain of f is $-6 \le x \le 6$, or $[-6, 6]$.

(d) The range of f is $-4 \le y \le 4$, or $[-4, 4]$.

(e) f is increasing on $[-4, 4]$, that is, on $-4 \le x \le 4$.

(f) f is not one-to-one since it fails the Horizontal Line Test.

(g) f is odd since its graph is symmetric about the origin.

3. $f(x) = x^2 - 2x + 3$, so $f(a + h) = (a + h)^2 - 2(a + h) + 3 = a^2 + 2ah + h^2 - 2a - 2h + 3$, and

$$\frac{f(a + h) - f(a)}{h} = \frac{(a^2 + 2ah + h^2 - 2a - 2h + 3) - (a^2 - 2a + 3)}{h} = \frac{h(2a + h - 2)}{h} = 2a + h - 2.$$

5. $f(x) = 2/(3x - 1)$.   Domain: $3x - 1 \ne 0 \Rightarrow 3x \ne 1 \Rightarrow x \ne \frac{1}{3}$. $D = \left(-\infty, \frac{1}{3}\right) \cup \left(\frac{1}{3}, \infty\right)$

                     Range: all reals except 0 ($y = 0$ is the horizontal asymptote for f.) $R = (-\infty, 0) \cup (0, \infty)$

7. $h(x) = \ln(x + 6)$.   Domain: $x + 6 > 0 \Rightarrow x > -6$. $D = (-6, \infty)$

                       Range: $x + 6 > 0$, so $\ln(x + 6)$ takes on all real numbers and, hence, the range is $\mathbb{R}$.

                       $R = (-\infty, \infty)$

9. (a) To obtain the graph of $y = f(x) + 8$, we shift the graph of $y = f(x)$ up 8 units.

(b) To obtain the graph of $y = f(x + 8)$, we shift the graph of $y = f(x)$ left 8 units.

(c) To obtain the graph of $y = 1 + 2f(x)$, we stretch the graph of $y = f(x)$ vertically by a factor of 2, and then shift the resulting graph 1 unit upward.

(d) To obtain the graph of $y = f(x - 2) - 2$, we shift the graph of $y = f(x)$ right 2 units (for the "-2" inside the parentheses), and then shift the resulting graph 2 units downward.

(e) To obtain the graph of $y = -f(x)$, we reflect the graph of $y = f(x)$ about the x-axis.

(f) To obtain the graph of $y = f^{-1}(x)$, we reflect the graph of $y = f(x)$ about the line $y = x$ (assuming f is one-to-one).

11. $y = -\sin 2x$: Start with the graph of $y = \sin x$, compress horizontally by a factor of 2, and reflect about the x-axis.

13. $y = \frac{1}{2}(1 + e^x)$:

Start with the graph of $y = e^x$,

shift 1 unit upward, and compress

vertically by a factor of 2.

15. $f(x) = \dfrac{1}{x+2}$:

Start with the graph of $f(x) = 1/x$

and shift 2 units to the left.

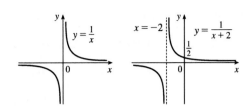

17. (a) The terms of f are a mixture of odd and even powers of x, so f is neither even nor odd.

(b) The terms of f are all odd powers of x, so f is odd.

(c) $f(-x) = e^{-(-x)^2} = e^{-x^2} = f(x)$, so f is even.

(d) $f(-x) = 1 + \sin(-x) = 1 - \sin x$. Now $f(-x) \neq f(x)$ and $f(-x) \neq -f(x)$, so f is neither even nor odd.

19. $f(x) = \ln x$, $D = (0, \infty)$; $g(x) = x^2 - 9$, $D = \mathbb{R}$.

(a) $(f \circ g)(x) = f(g(x)) = f(x^2 - 9) = \ln(x^2 - 9)$.

   Domain: $x^2 - 9 > 0 \Rightarrow x^2 > 9 \Rightarrow |x| > 3 \Rightarrow x \in (-\infty, -3) \cup (3, \infty)$

(b) $(g \circ f)(x) = g(f(x)) = g(\ln x) = (\ln x)^2 - 9$.   Domain: $x > 0$, or $(0, \infty)$

(c) $(f \circ f)(x) = f(f(x)) = f(\ln x) = \ln(\ln x)$.   Domain: $\ln x > 0 \Rightarrow x > e^0 = 1$, or $(1, \infty)$

(d) $(g \circ g)(x) = g(g(x)) = g(x^2 - 9) = (x^2 - 9)^2 - 9$.   Domain: $x \in \mathbb{R}$, or $(-\infty, \infty)$

21.

Many models appear to be plausible. Your choice depends on whether you think medical advances will keep increasing life expectancy, or if there is bound to be a natural leveling-off of life expectancy. A linear model, $y = 0.2493x - 423.4818$, gives us an estimate of 77.6 years for the year 2010.

23. We need to know the value of x such that $f(x) = 2x + \ln x = 2$. Since $x = 1$ gives us $y = 2$, $f^{-1}(2) = 1$.

25. (a) $e^{2 \ln 3} = (e^{\ln 3})^2 = 3^2 = 9$

(b) $\log_{10} 25 + \log_{10} 4 = \log_{10}(25 \cdot 4) = \log_{10} 100 = \log_{10} 10^2 = 2$

27. (a) After 4 days, $\frac{1}{2}$ gram remains; after 8 days, $\frac{1}{4}$ g; after 12 days, $\frac{1}{8}$ g; after 16 days, $\frac{1}{16}$ g.

(b) $m(4) = \dfrac{1}{2}$, $m(8) = \dfrac{1}{2^2}$, $m(12) = \dfrac{1}{2^3}$, $m(16) = \dfrac{1}{2^4}$. From the pattern, we see that $m(t) = \dfrac{1}{2^{t/4}}$, or $2^{-t/4}$.

(c) $m = 2^{-t/4} \Rightarrow \log_2 m = -t/4 \Rightarrow t = -4 \log_2 m$; this is the time elapsed when there are m grams of ^{100}Pd.

(d) $m = 0.01 \Rightarrow t = -4 \log_2 0.01 = -4 \left(\dfrac{\ln 0.01}{\ln 2} \right) \approx 26.6$ days

29. $f(x) = \ln(x^2 - c)$. If $c < 0$, the domain of f is $\mathbb{R}$. If $c = 0$, the

domain of f is $(-\infty, 0) \cup (0, \infty)$. If $c > 0$, the domain of f is

$(-\infty, -\sqrt{c}) \cup (\sqrt{c}, \infty)$. As c increases, the dip at $x = 0$ becomes

deeper. For $c \geq 0$, the graph has asymptotes at $x = \pm\sqrt{c}$.

31. (a)

(e, 1)

(b) $x = e^t \;\Rightarrow\; t = \ln x; y = \sqrt{t}$ so $y = \sqrt{\ln x}$.

$0 \leq t \leq 1 \;\Rightarrow\; 0 \leq y \leq 1$ and $1 \leq x \leq e$.

33. We sketch $x = t$, $y = 2t + \ln t$ (the function) and $x = 2t + \ln t$, $y = t$ (its inverse) for $t > 0$.

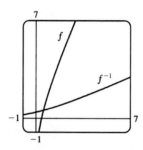

PRINCIPLES OF PROBLEM SOLVING

1.

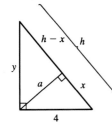

By using the area formula for a triangle, $\frac{1}{2}$ (base) (height), in two ways, we see that

$$\tfrac{1}{2}\,(4)\,(y) = \tfrac{1}{2}\,(h)\,(a), \text{ so } a = \frac{4y}{h}. \text{ Since } 4^2 + y^2 = h^2, y = \sqrt{h^2 - 16}, \text{ and}$$

$$a = \frac{4\sqrt{h^2 - 16}}{h}.$$

3. $|2x - 1| = \begin{cases} 2x - 1 & \text{if } x \geq \frac{1}{2} \\ 1 - 2x & \text{if } x < \frac{1}{2} \end{cases}$ and $|x + 5| = \begin{cases} x + 5 & \text{if } x \geq -5 \\ -x - 5 & \text{if } x < -5 \end{cases}$

Therefore, we consider the three cases $x < -5$, $-5 \leq x < \frac{1}{2}$, and $x \geq \frac{1}{2}$.

If $x < -5$, we must have $1 - 2x - (-x - 5) = 3 \iff x = 3$, which is false, since we are considering $x < -5$.

If $-5 \leq x < \frac{1}{2}$, we must have $1 - 2x - (x + 5) = 3 \iff x = -\frac{7}{3}$.

If $x \geq \frac{1}{2}$, we must have $2x - 1 - (x + 5) = 3 \iff x = 9$.

So the two solutions of the equation are $x = -\frac{7}{3}$ and $x = 9$.

5. $f(x) = \left|x^2 - 4\,|x| + 3\right|$. If $x \geq 0$, then $f(x) = \left|x^2 - 4x + 3\right| = |(x - 1)(x - 3)|$.

  Case (i): If $0 < x \leq 1$, then $f(x) = x^2 - 4x + 3$.

  Case (ii): If $1 < x \leq 3$, then $f(x) = -(x^2 - 4x + 3) = -x^2 + 4x - 3$.

  Case (iii): If $x > 3$, then $f(x) = x^2 - 4x + 3$.

This enables us to sketch the graph for $x \geq 0$. Then we use the fact that f is an even

function to reflect this part of the graph about the y-axis to obtain the entire graph. Or, we could consider also the cases $x < -3$, $-3 \leq x < -1$, and $-1 \leq x < 0$.

7. Remember that $|a| = a$ if $a \geq 0$ and that $|a| = -a$ if $a < 0$. Thus,

$$x + |x| = \begin{cases} 2x & \text{if } x \geq 0 \\ 0 & \text{if } x < 0 \end{cases} \quad \text{and} \quad y + |y| = \begin{cases} 2y & \text{if } y \geq 0 \\ 0 & \text{if } y < 0 \end{cases}$$

We will consider the equation $x + |x| = y + |y|$ in four cases.

(1) $x \geq 0, y \geq 0$	(2) $x \geq 0, y < 0$	(3) $x < 0, y \geq 0$	(4) $x < 0, y < 0$
$2x = 2y$	$2x = 0$	$0 = 2y$	$0 = 0$
$x = y$	$x = 0$	$0 = y$	

Case 1 gives us the line $y = x$ with nonnegative x and y.

Case 2 gives us the portion of the y-axis with y negative.

Case 3 gives us the portion of the x-axis with x negative.

Case 4 gives us the entire third quadrant.

9. $|x| + |y| \le 1$. The boundary of the region has equation $|x| + |y| = 1$. In quadrants

I, II, III, and IV, this becomes the lines $x + y = 1$, $-x + y = 1$, $-x - y = 1$, and

$x - y = 1$ respectively.

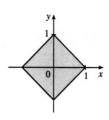

11. $(\log_2 3)(\log_3 4)(\log_4 5) \cdots (\log_{31} 32) = \left(\dfrac{\ln 3}{\ln 2}\right)\left(\dfrac{\ln 4}{\ln 3}\right)\left(\dfrac{\ln 5}{\ln 4}\right) \cdots \left(\dfrac{\ln 32}{\ln 31}\right) = \dfrac{\ln 32}{\ln 2} = \dfrac{\ln 2^5}{\ln 2} = \dfrac{5\ln 2}{\ln 2} = 5$

13. $\ln(x^2 - 2x - 2) \le 0 \;\Rightarrow\; x^2 - 2x - 2 \le e^0 = 1 \;\Rightarrow\; x^2 - 2x - 3 \le 0 \;\Rightarrow\; (x-3)(x+1) \le 0 \;\Rightarrow\; x \in [-1, 3]$.

Since the argument must be positive, $x^2 - 2x - 2 > 0 \;\Rightarrow\; \left[x - (1 - \sqrt{3})\right]\left[x - (1 + \sqrt{3})\right] > 0 \;\Rightarrow$

$x \in (-\infty, 1 - \sqrt{3}) \cup (1 + \sqrt{3}, \infty)$. The intersection of these intervals is $\left[-1, 1 - \sqrt{3}\right) \cup \left(1 + \sqrt{3}, 3\right]$.

15. Let d be the distance traveled on each half of the trip. Let t_1 and t_2 be the times taken for the first and second halves of the trip.

For the first half of the trip we have $t_1 = d/30$ and for the second half we have $t_2 = d/60$. Thus, the average speed for the

entire trip is $\dfrac{\text{total distance}}{\text{total time}} = \dfrac{2d}{t_1 + t_2} = \dfrac{2d}{\dfrac{d}{30} + \dfrac{d}{60}} \cdot \dfrac{60}{60} = \dfrac{120d}{2d + d} = \dfrac{120d}{3d} = 40$. The average speed for the entire trip

is 40 mi/h.

17. Let S_n be the statement that $7^n - 1$ is divisible by 6.

- S_1 is true because $7^1 - 1 = 6$ is divisible by 6.

- Assume S_k is true, that is, $7^k - 1$ is divisible by 6. In other words, $7^k - 1 = 6m$ for some positive integer m. Then

  $7^{k+1} - 1 = 7^k \cdot 7 - 1 = (6m + 1) \cdot 7 - 1 = 42m + 6 = 6(7m + 1)$, which is divisible by 6, so S_{k+1} is true.

- Therefore, by mathematical induction, $7^n - 1$ is divisible by 6 for every positive integer n.

19. $f_0(x) = x^2$ and $f_{n+1}(x) = f_0(f_n(x))$ for $n = 0, 1, 2, \ldots$.

$f_1(x) = f_0(f_0(x)) = f_0(x^2) = (x^2)^2 = x^4$, $f_2(x) = f_0(f_1(x)) = f_0(x^4) = (x^4)^2 = x^8$,

$f_3(x) = f_0(f_2(x)) = f_0(x^8) = (x^8)^2 = x^{16}, \ldots$. Thus, a general formula is $f_n(x) = x^{2^{n+1}}$.

2 □ LIMITS AND DERIVATIVES

2.1 The Tangent and Velocity Problems

1. (a) Using $P(15, 250)$, we construct the following table:

t	Q	slope $= m_{PQ}$
5	$(5, 694)$	$\frac{694-250}{5-15} = -\frac{444}{10} = -44.4$
10	$(10, 444)$	$\frac{444-250}{10-15} = -\frac{194}{5} = -38.8$
20	$(20, 111)$	$\frac{111-250}{20-15} = -\frac{139}{5} = -27.8$
25	$(25, 28)$	$\frac{28-250}{25-15} = -\frac{222}{10} = -22.2$
30	$(30, 0)$	$\frac{0-250}{30-15} = -\frac{250}{15} = -16.\overline{6}$

(b) Using the values of t that correspond to the points closest to P ($t = 10$ and $t = 20$), we have

$$\frac{-38.8 + (-27.8)}{2} = -33.3$$

(c) From the graph, we can estimate the slope of the tangent line at P to be $\frac{-300}{9} = -33.\overline{3}$.

3. (a) $y = \dfrac{x}{1 + x}$, $P\left(1, \frac{1}{2}\right)$

(b) The slope appears to be $\frac{1}{4}$.

(c) $y - \frac{1}{2} = \frac{1}{4}(x - 1)$ or $y = \frac{1}{4}x + \frac{1}{4}$.

	x	Q	m_{PQ}
(i)	0.5	$(0.5, 0.333333)$	0.333333
(ii)	0.9	$(0.9, 0.473684)$	0.263158
(iii)	0.99	$(0.99, 0.497487)$	0.251256
(iv)	0.999	$(0.999, 0.499750)$	0.250125
(v)	1.5	$(1.5, 0.6)$	0.2
(vi)	1.1	$(1.1, 0.523810)$	0.238095
(vii)	1.01	$(1.01, 0.502488)$	0.248756
(viii)	1.001	$(1.001, 0.500250)$	0.249875

5. (a) $y = y(t) = 40t - 16t^2$. At $t = 2$, $y = 40(2) - 16(2)^2 = 16$. The average velocity between times 2 and $2 + h$ is

$$v_{\text{ave}} = \frac{y(2 + h) - y(2)}{(2 + h) - 2} = \frac{[40(2 + h) - 16(2 + h)^2] - 16}{h} = \frac{-24h - 16h^2}{h} = -24 - 16h, \text{ if } h \neq 0.$$

(i) $[2, 2.5]$: $h = 0.5$, $v_{\text{ave}} = -32$ ft/s

(ii) $[2, 2.1]$: $h = 0.1$, $v_{\text{ave}} = -25.6$ ft/s

(iii) $[2, 2.05]$: $h = 0.05$, $v_{\text{ave}} = -24.8$ ft/s

(iv) $[2, 2.01]$: $h = 0.01$, $v_{\text{ave}} = -24.16$ ft/s

(b) The instantaneous velocity when $t = 2$ (h approaches 0) is -24 ft/s.

7. (a) (i) On the interval $[1, 3]$, $v_{\text{ave}} = \dfrac{s(3) - s(1)}{3 - 1} = \dfrac{10.7 - 1.4}{2} = \dfrac{9.3}{2} = 4.65$ m/s.

(ii) On the interval $[2, 3]$, $v_{\text{ave}} = \dfrac{s(3) - s(2)}{3 - 2} = \dfrac{10.7 - 5.1}{1} = 5.6$ m/s.

(iii) On the interval $[3, 5]$, $v_{\text{ave}} = \dfrac{s(5) - s(3)}{5 - 3} = \dfrac{25.8 - 10.7}{2} = \dfrac{15.1}{2} = 7.55$ m/s.

(iv) On the interval $[3, 4]$, $v_{\text{ave}} = \dfrac{s(4) - s(3)}{4 - 3} = \dfrac{17.7 - 10.7}{1} = 7$ m/s.

(b)

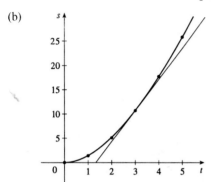

Using the points $(2, 4)$ and $(5, 23)$ from the approximate tangent line, the instantaneous velocity at $t = 3$ is about $\dfrac{23 - 4}{5 - 2} \approx 6.3$ m/s.

9. (a) For the curve $y = \sin(10\pi/x)$ and the point $P(1, 0)$:

x	Q	m_{PQ}
2	$(2, 0)$	0
1.5	$(1.5, 0.8660)$	1.7321
1.4	$(1.4, -0.4339)$	-1.0847
1.3	$(1.3, -0.8230)$	-2.7433
1.2	$(1.2, 0.8660)$	4.3301
1.1	$(1.1, -0.2817)$	-2.8173

x	Q	m_{PQ}
0.5	$(0.5, 0)$	0
0.6	$(0.6, 0.8660)$	-2.1651
0.7	$(0.7, 0.7818)$	-2.6061
0.8	$(0.8, 1)$	-5
0.9	$(0.9, -0.3420)$	3.4202

As x approaches 1, the slopes do not appear to be approaching any particular value.

(b)

We see that problems with estimation are caused by the frequent oscillations of the graph. The tangent is so steep at P that we need to take x-values much closer to 1 in order to get accurate estimates of its slope.

(c) If we choose $x = 1.001$, then the point Q is $(1.001, -0.0314)$ and $m_{PQ} \approx -31.3794$. If $x = 0.999$, then Q is $(0.999, 0.0314)$ and $m_{PQ} = -31.4422$. The average of these slopes is -31.4108. So we estimate that the slope of the tangent line at P is about -31.4.

2.2   The Limit of a Function

1. As x approaches 2, $f(x)$ approaches 5. [Or, the values of $f(x)$ can be made as close to 5 as we like by taking x sufficiently close to 2 (but $x \neq 2$).] Yes, the graph could have a hole at $(2, 5)$ and be defined such that $f(2) = 3$.

3. (a) $f(x)$ approaches 2 as x approaches 1 from the left, so $\lim\limits_{x \to 1^-} f(x) = 2$.

   (b) $f(x)$ approaches 3 as x approaches 1 from the right, so $\lim\limits_{x \to 1^+} f(x) = 3$.

   (c) $\lim\limits_{x \to 1} f(x)$ does not exist because the limits in part (a) and part (b) are not equal.

   (d) $f(x)$ approaches 4 as x approaches 5 from the left and from the right, so $\lim\limits_{x \to 5} f(x) = 4$.

   (e) $f(5)$ is not defined, so it doesn't exist.

5. (a) $\lim\limits_{t \to 0^-} g(t) = -1$ (b) $\lim\limits_{t \to 0^+} g(t) = -2$

   (c) $\lim\limits_{t \to 0} g(t)$ does not exist because the limits in part (a) and part (b) are not equal.

   (d) $\lim\limits_{t \to 2^-} g(t) = 2$ (e) $\lim\limits_{t \to 2^+} g(t) = 0$

   (f) $\lim\limits_{t \to 2} g(t)$ does not exist because the limits in part (d) and part (e) are not equal.

   (g) $g(2) = 1$ (h) $\lim\limits_{t \to 4} g(t) = 3$

7. From the graph of

$$f(x) = \begin{cases} 1 + x & \text{if } x < -1 \\ x^2 & \text{if } -1 \le x < 1, \\ 2 - x & \text{if } x \ge 1 \end{cases}$$

we see that $\lim\limits_{x \to a} f(x)$ exists for all a except $a = -1$. Notice that the right and left limits are different at $a = -1$.

9. (a) $\lim\limits_{x \to 0^-} f(x) = 1$

   (b) $\lim\limits_{x \to 0^+} f(x) = 0$

   (c) $\lim\limits_{x \to 0} f(x)$ does not exist because the limits

       in part (a) and part (b) are not equal.

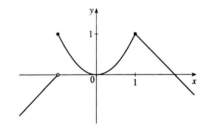

11. (a) $\lim\limits_{x \to 0^-} f(x) = -2$

   (b) $\lim\limits_{x \to 0^+} f(x) = 2$

   (c) $\lim\limits_{x \to 0} f(x)$ does not exist because the limits

       in part (a) and part (b) are not equal.

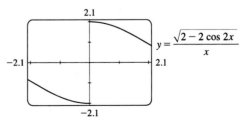

13. $\lim\limits_{x\to0^-} f(x) = -1,$ $\lim\limits_{x\to0^+} f(x) = 2,$ $f(0) = 1$

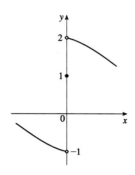

15. $\lim\limits_{x\to3^+} f(x) = 4,$ $\lim\limits_{x\to3^-} f(x) = 2,$ $\lim\limits_{x\to-2} f(x) = 2,$

$f(3) = 3,$ $f(-2) = 1$

17. For $f(x) = \dfrac{x^2 - 2x}{x^2 - x - 2}$:

x	$f(x)$
2.5	0.714286
2.1	0.677419
2.05	0.672131
2.01	0.667774
2.005	0.667221
2.001	0.666778

x	$f(x)$
1.9	0.655172
1.95	0.661017
1.99	0.665552
1.995	0.666110
1.999	0.666556

It appears that $\lim\limits_{x\to2} \dfrac{x^2 - 2x}{x^2 - x - 2} = 0.\overline{6} = \frac{2}{3}$.

19. For $f(t) = \dfrac{e^{5t} - 1}{t}$:

t	$f(t)$
0.5	22.364988
0.1	6.487213
0.01	5.127110
0.001	5.012521
0.0001	5.001250

t	$f(t)$
-0.5	1.835830
-0.1	3.934693
-0.01	4.877058
-0.001	4.987521
-0.0001	4.998750

It appears that $\lim\limits_{t\to0} \dfrac{e^{5t} - 1}{t} = 5$.

21. For $f(x) = \dfrac{\sqrt{x + 4} - 2}{x}$:

x	$f(x)$
1	0.236068
0.5	0.242641
0.1	0.248457
0.05	0.249224
0.01	0.249844

x	$f(x)$
-1	0.267949
-0.5	0.258343
-0.1	0.251582
-0.05	0.250786
-0.01	0.250156

It appears that $\lim\limits_{x\to0} \dfrac{\sqrt{x + 4} - 2}{x} = 0.25 = \frac{1}{4}$.

23. For $f(x) = \dfrac{x^6 - 1}{x^{10} - 1}$:

x	$f(x)$
0.5	0.985337
0.9	0.719397
0.95	0.660186
0.99	0.612018
0.999	0.601200

x	$f(x)$
1.5	0.183369
1.1	0.484119
1.05	0.540783
1.01	0.588022
1.001	0.598800

It appears that $\lim\limits_{x\to1} \dfrac{x^6 - 1}{x^{10} - 1} = 0.6 = \frac{3}{5}$.

25. (a) From the graphs, it seems that $\lim\limits_{x\to0} \dfrac{\cos 2x - \cos x}{x^2} = -1.5$.

(b)

x	$f(x)$
±0.1	-1.493759
±0.01	-1.499938
±0.001	-1.499999
±0.0001	-1.500000

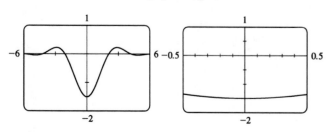

27. (a) Let $h(x) = (1 + x)^{1/x}$.

(b)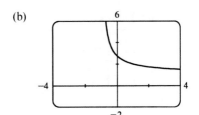

x	$h(x)$
-0.001	2.71964
-0.0001	2.71842
-0.00001	2.71830
-0.000001	2.71828
0.000001	2.71828
0.00001	2.71827
0.0001	2.71815
0.001	2.71692

It appears that $\lim\limits_{x \to 0} (1 + x)^{1/x} \approx 2.71828$, which is approximately e.

In Section 3.7 we will see that the value of the limit is exactly e.

29. For $f(x) = x^2 - (2^x/1000)$:

(a)

x	$f(x)$
1	0.998000
0.8	0.638259
0.6	0.358484
0.4	0.158680
0.2	0.038851
0.1	0.008928
0.05	0.001465

It appears that $\lim\limits_{x \to 0} f(x) = 0$.

(b)

x	$f(x)$
0.04	0.000572
0.02	-0.000614
0.01	-0.000907
0.005	-0.000978
0.003	-0.000993
0.001	-0.001000

It appears that $\lim\limits_{x \to 0} f(x) = -0.001$.

31. We need to have $5.8 < x^3 - 3x + 4 < 6.2$. From the graph we obtain the approximate points of intersection $P(1.9774, 5.8)$ and $Q(2.0219, 6.2)$. So if x is within 0.021 of 2, then y will be within 0.2 of 6. If we must have $x^3 - 3x + 4$ within 0.1 of 6, we get $P(1.9888, 5.9)$ and $Q(2.0110, 6.1)$. We would then need x to be within 0.011 of 2.

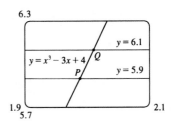

2.3 Calculating Limits Using the Limit Laws

1. (a) $\lim\limits_{x \to 2} [f(x) + 5g(x)] = \lim\limits_{x \to 2} f(x) + \lim\limits_{x \to 2} [5g(x)]$ [Limit Law 1]

$\qquad\qquad\qquad = \lim\limits_{x \to 2} f(x) + 5 \lim\limits_{x \to 2} g(x)$ [Limit Law 3]

$\qquad\qquad\qquad = 4 + 5(-2) = -6$

(b) $\lim\limits_{x \to 2} [g(x)]^3 = \left[\lim\limits_{x \to 2} g(x) \right]^3$ [Limit Law 6]

$\qquad\qquad = (-2)^3 = -8$

(c) $\lim\limits_{x \to 2} \sqrt{f(x)} = \sqrt{\lim\limits_{x \to 2} f(x)}$ [Limit Law 11]

$\qquad\qquad = \sqrt{4} = 2$

(d) $\lim\limits_{x \to 2} \dfrac{3f(x)}{g(x)} = \dfrac{\lim\limits_{x \to 2} [3f(x)]}{\lim\limits_{x \to 2} g(x)}$ [Limit Law 5]

$\qquad\qquad = \dfrac{3 \lim\limits_{x \to 2} f(x)}{\lim\limits_{x \to 2} g(x)}$ [Limit Law 3]

$\qquad\qquad = \dfrac{3(4)}{-2} = -6$

(e) Because the limit of the denominator is 0, we can't use Limit Law 5. The given limit, $\lim\limits_{x \to 2} \dfrac{g(x)}{h(x)}$, does not exist because the denominator approaches 0 while the numerator approaches a nonzero number.

(f) $\lim\limits_{x \to 2} \dfrac{g(x)\,h(x)}{f(x)} = \dfrac{\lim\limits_{x \to 2} [g(x)\,h(x)]}{\lim\limits_{x \to 2} f(x)}$ [Limit Law 5]

$\qquad\qquad = \dfrac{\lim\limits_{x \to 2} g(x) \cdot \lim\limits_{x \to 2} h(x)}{\lim\limits_{x \to 2} f(x)}$ [Limit Law 4]

$\qquad\qquad = \dfrac{-2 \cdot 0}{4} = 0$

3. $\lim\limits_{x \to -2} (3x^4 + 2x^2 - x + 1) = \lim\limits_{x \to -2} 3x^4 + \lim\limits_{x \to -2} 2x^2 - \lim\limits_{x \to -2} x + \lim\limits_{x \to -2} 1$ [Limit Laws 1 and 2]

$\qquad\qquad\qquad\qquad = 3 \lim\limits_{x \to -2} x^4 + 2 \lim\limits_{x \to -2} x^2 - \lim\limits_{x \to -2} x + \lim\limits_{x \to -2} 1$ [3]

$\qquad\qquad\qquad\qquad = 3(-2)^4 + 2(-2)^2 - (-2) + (1)$ [9, 8, and 7]

$\qquad\qquad\qquad\qquad = 48 + 8 + 2 + 1 = 59$

5. $\lim\limits_{x \to 8} (1 + \sqrt[3]{x})(2 - 6x^2 + x^3) = \lim\limits_{x \to 8} (1 + \sqrt[3]{x}) \cdot \lim\limits_{x \to 8} (2 - 6x^2 + x^3)$ [Limit Law 4]

$\qquad\qquad\qquad = \left(\lim\limits_{x \to 8} 1 + \lim\limits_{x \to 8} \sqrt[3]{x} \right) \cdot \left(\lim\limits_{x \to 8} 2 - 6 \lim\limits_{x \to 8} x^2 + \lim\limits_{x \to 8} x^3 \right)$ [1, 2, and 3]

$\qquad\qquad\qquad = \left(1 + \sqrt[3]{8} \right) \cdot \left(2 - 6 \cdot 8^2 + 8^3 \right)$ [7, 10, 9]

$\qquad\qquad\qquad = (3)(130) = 390$

7. $\lim\limits_{x\to2}\sqrt{\dfrac{2x^2+1}{3x-2}}=\sqrt{\lim\limits_{x\to2}\dfrac{2x^2+1}{3x-2}}$ [Limit Law 11]

$\qquad=\sqrt{\dfrac{\lim\limits_{x\to2}(2x^2+1)}{\lim\limits_{x\to2}(3x-2)}}$ [5]

$\qquad=\sqrt{\dfrac{2\lim\limits_{x\to2}x^2+\lim\limits_{x\to2}1}{3\lim\limits_{x\to2}x-\lim\limits_{x\to2}2}}$ [1, 2, and 3]

$\qquad=\sqrt{\dfrac{2(2)^2+1}{3(2)-2}}=\sqrt{\dfrac{9}{4}}=\dfrac{3}{2}$ [9, 8, and 7]

9. $\lim\limits_{x\to5}\dfrac{x^2-6x+5}{x-5}=\lim\limits_{x\to5}\dfrac{(x-5)(x-1)}{x-5}=\lim\limits_{x\to5}(x-1)=5-1=4$

11. $\lim\limits_{x\to5}\dfrac{x^2-5x+6}{x-5}$ does not exist since $x-5\to0$, but $x^2-5x+6\to6$ as $x\to5$.

13. $\lim\limits_{t\to-3}\dfrac{t^2-9}{2t^2+7t+3}=\lim\limits_{t\to-3}\dfrac{(t+3)(t-3)}{(2t+1)(t+3)}=\lim\limits_{t\to-3}\dfrac{t-3}{2t+1}=\dfrac{-3-3}{2(-3)+1}=\dfrac{-6}{-5}=\dfrac{6}{5}$

15. $\lim\limits_{h\to0}\dfrac{(4+h)^2-16}{h}=\lim\limits_{h\to0}\dfrac{(16+8h+h^2)-16}{h}=\lim\limits_{h\to0}\dfrac{8h+h^2}{h}=\lim\limits_{h\to0}\dfrac{h(8+h)}{h}=\lim\limits_{h\to0}(8+h)=8+0=8$

17. By the formula for the sum of cubes, we have

$\qquad\lim\limits_{x\to-2}\dfrac{x+2}{x^3+8}=\lim\limits_{x\to-2}\dfrac{x+2}{(x+2)(x^2-2x+4)}=\lim\limits_{x\to-2}\dfrac{1}{x^2-2x+4}=\dfrac{1}{4+4+4}=\dfrac{1}{12}.$

19. $\lim\limits_{x\to-4}\dfrac{\frac{1}{4}+\frac{1}{x}}{4+x}=\lim\limits_{x\to-4}\dfrac{\frac{x+4}{4x}}{4+x}=\lim\limits_{x\to-4}\dfrac{x+4}{4x(4+x)}=\lim\limits_{x\to-4}\dfrac{1}{4x}=\dfrac{1}{4(-4)}=-\dfrac{1}{16}$

21. $\lim\limits_{x\to16}\dfrac{4-\sqrt{x}}{16x-x^2}=\lim\limits_{x\to16}\dfrac{(4-\sqrt{x})(4+\sqrt{x})}{(16x-x^2)(4+\sqrt{x})}=\lim\limits_{x\to16}\dfrac{16-x}{x(16-x)(4+\sqrt{x})}$

$\qquad=\lim\limits_{x\to16}\dfrac{1}{x(4+\sqrt{x})}=\dfrac{1}{16\left(4+\sqrt{16}\right)}=\dfrac{1}{16(8)}=\dfrac{1}{128}$

23. $\lim\limits_{t\to0}\left(\dfrac{1}{t\sqrt{1+t}}-\dfrac{1}{t}\right)=\lim\limits_{t\to0}\dfrac{1-\sqrt{1+t}}{t\sqrt{1+t}}=\lim\limits_{t\to0}\dfrac{\left(1-\sqrt{1+t}\right)\left(1+\sqrt{1+t}\right)}{t\sqrt{t+1}\left(1+\sqrt{1+t}\right)}=\lim\limits_{t\to0}\dfrac{-t}{t\sqrt{1+t}\left(1+\sqrt{1+t}\right)}$

$\qquad=\lim\limits_{t\to0}\dfrac{-1}{\sqrt{1+t}\left(1+\sqrt{1+t}\right)}=\dfrac{-1}{\sqrt{1+0}\left(1+\sqrt{1+0}\right)}=-\dfrac{1}{2}$

25. (a)

$$\lim_{x \to 0} \frac{x}{\sqrt{1+3x}-1} \approx \frac{2}{3}$$

(b)

x	$f(x)$
-0.001	0.6661663
-0.0001	0.6666167
-0.00001	0.6666617
-0.000001	0.6666662
0.000001	0.6666672
0.00001	0.6666717
0.0001	0.6667167
0.001	0.6671663

The limit appears to be $\frac{2}{3}$.

(c) $\displaystyle\lim_{x \to 0} \left(\frac{x}{\sqrt{1+3x}-1} \cdot \frac{\sqrt{1+3x}+1}{\sqrt{1+3x}+1} \right) = \lim_{x \to 0} \frac{x(\sqrt{1+3x}+1)}{(1+3x)-1} = \lim_{x \to 0} \frac{x(\sqrt{1+3x}+1)}{3x}$

$\qquad = \dfrac{1}{3} \displaystyle\lim_{x \to 0} (\sqrt{1+3x}+1)$ [Limit Law 3]

$\qquad = \dfrac{1}{3} \left[\sqrt{\displaystyle\lim_{x \to 0}(1+3x)} + \displaystyle\lim_{x \to 0} 1 \right]$ [1 and 11]

$\qquad = \dfrac{1}{3} \left(\sqrt{\displaystyle\lim_{x \to 0} 1 + 3\lim_{x \to 0} x} + 1 \right)$ [1, 3, and 7]

$\qquad = \dfrac{1}{3} \left(\sqrt{1+3 \cdot 0} + 1 \right)$ [7 and 8]

$\qquad = \dfrac{1}{3}(1+1) = \dfrac{2}{3}$

27. Let $f(x) = -x^2$, $g(x) = x^2 \cos 20\pi x$ and $h(x) = x^2$. Then

$-1 \le \cos 20\pi x \le 1 \quad \Rightarrow \quad -x^2 \le x^2 \cos 20\pi x \le x^2 \quad \Rightarrow \quad f(x) \le g(x) \le h(x).$

So since $\displaystyle\lim_{x \to 0} f(x) = \lim_{x \to 0} h(x) = 0$, by the Squeeze Theorem we have

$\displaystyle\lim_{x \to 0} g(x) = 0.$

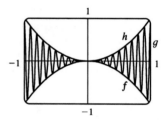

29. We have $\displaystyle\lim_{x \to 4} (4x-9) = 4(4) - 9 = 7$ and $\displaystyle\lim_{x \to 4} (x^2 - 4x + 7) = 4^2 - 4(4) + 7 = 7$. Since $4x - 9 \le f(x) \le x^2 - 4x + 7$

for $x \ge 0$, $\displaystyle\lim_{x \to 4} f(x) = 7$ by the Squeeze Theorem.

31. $-1 \le \cos(2/x) \le 1 \quad \Rightarrow \quad -x^4 \le x^4 \cos(2/x) \le x^4$. Since $\displaystyle\lim_{x \to 0} (-x^4) = 0$ and $\displaystyle\lim_{x \to 0} x^4 = 0$, we have

$\displaystyle\lim_{x \to 0} \left[x^4 \cos(2/x) \right] = 0$ by the Squeeze Theorem.

33. $|x-3| = \begin{cases} x - 3 & \text{if } x - 3 \ge 0 \\ -(x-3) & \text{if } x - 3 < 0 \end{cases} = \begin{cases} x - 3 & \text{if } x \ge 3 \\ 3 - x & \text{if } x < 3 \end{cases}$

Thus, $\displaystyle\lim_{x \to 3^+} (2x + |x-3|) = \lim_{x \to 3^+} (2x + x - 3) = \lim_{x \to 3^+} (3x - 3) = 3(3) - 3 = 6$ and

$\displaystyle\lim_{x \to 3^-} (2x + |x-3|) = \lim_{x \to 3^-} (2x + 3 - x) = \lim_{x \to 3^-} (x + 3) = 3 + 3 = 6$. Since the left and right limits are equal,

$\displaystyle\lim_{x \to 3} (2x + |x-3|) = 6.$

35. Since $|x| = -x$ for $x < 0$, we have $\lim\limits_{x \to 0^-} \left(\dfrac{1}{x} - \dfrac{1}{|x|} \right) = \lim\limits_{x \to 0^-} \left(\dfrac{1}{x} - \dfrac{1}{-x} \right) = \lim\limits_{x \to 0^-} \dfrac{2}{x}$, which does not exist since the

denominator approaches 0 and the numerator does not.

37. (a) (i) $\lim\limits_{x \to 1^-} g(x) = \lim\limits_{x \to 1^-} x = 1$

(ii) $\lim\limits_{x \to 1^+} g(x) = \lim\limits_{x \to 1^+} (2 - x^2) = 2 - 1^2 = 1$. Since $\lim\limits_{x \to 1^-} g(x) = 1$ and $\lim\limits_{x \to 1^+} g(x) = 1$, we have $\lim\limits_{x \to 1} g(x) = 1$.

Note that the fact $g(1) = 3$ does not affect the value of the limit.

(iii) When $x = 1$, $g(x) = 3$, so $g(1) = 3$.

(iv) $\lim\limits_{x \to 2^-} g(x) = \lim\limits_{x \to 2^-} (2 - x^2) = 2 - 2^2 = 2 - 4 = -2$

(v) $\lim\limits_{x \to 2^+} g(x) = \lim\limits_{x \to 2^+} (x - 3) = 2 - 3 = -1$

(vi) $\lim\limits_{x \to 2} g(x)$ does not exist since $\lim\limits_{x \to 2^-} g(x) \neq \lim\limits_{x \to 2^+} g(x)$.

(b)

$$g(x) = \begin{cases} x & \text{if } x < 1 \\ 3 & \text{if } x = 1 \\ 2 - x^2 & \text{if } 1 < x \leq 2 \\ x - 3 & \text{if } x > 2 \end{cases}$$

39. (a) (i) $[\![x]\!] = -2$ for $-2 \leq x < -1$, so $\lim\limits_{x \to -2^+} [\![x]\!] = \lim\limits_{x \to -2^+} (-2) = -2$

(ii) $[\![x]\!] = -3$ for $-3 \leq x < -2$, so $\lim\limits_{x \to -2^-} [\![x]\!] = \lim\limits_{x \to -2^-} (-3) = -3$.

The right and left limits are different, so $\lim\limits_{x \to -2} [\![x]\!]$ does not exist.

(iii) $[\![x]\!] = -3$ for $-3 \leq x < -2$, so $\lim\limits_{x \to -2.4} [\![x]\!] = \lim\limits_{x \to -2.4} (-3) = -3$.

(b) (i) $[\![x]\!] = n - 1$ for $n - 1 \leq x < n$, so $\lim\limits_{x \to n^-} [\![x]\!] = \lim\limits_{x \to n^-} (n - 1) = n - 1$.

(ii) $[\![x]\!] = n$ for $n \leq x < n + 1$, so $\lim\limits_{x \to n^+} [\![x]\!] = \lim\limits_{x \to n^+} n = n$.

(c) $\lim\limits_{x \to a} [\![x]\!]$ exists $\Leftrightarrow$ a is not an integer.

41. The graph of $f(x) = [\![x]\!] + [\![-x]\!]$ is the same as the graph of $g(x) = -1$ with holes at each integer, since $f(a) = 0$ for any

integer a. Thus, $\lim\limits_{x \to 2^-} f(x) = -1$ and $\lim\limits_{x \to 2^+} f(x) = -1$, so $\lim\limits_{x \to 2} f(x) = -1$. However,

$f(2) = [\![2]\!] + [\![-2]\!] = 2 + (-2) = 0$, so $\lim\limits_{x \to 2} f(x) \neq f(2)$.

43. Since $p(x)$ is a polynomial, $p(x) = a_0 + a_1 x + a_2 x^2 + \cdots + a_n x^n$. Thus, by the Limit Laws,

$$\lim\limits_{x \to a} p(x) = \lim\limits_{x \to a} \left(a_0 + a_1 x + a_2 x^2 + \cdots + a_n x^n \right) = a_0 + a_1 \lim\limits_{x \to a} x + a_2 \lim\limits_{x \to a} x^2 + \cdots + a_n \lim\limits_{x \to a} x^n$$

$$= a_0 + a_1 a + a_2 a^2 + \cdots + a_n a^n = p(a)$$

Thus, for any polynomial p, $\lim\limits_{x \to a} p(x) = p(a)$.

45. $\lim\limits_{x\to 1} [f(x) - 8] = \lim\limits_{x\to 1} \left[\dfrac{f(x) - 8}{x - 1} \cdot (x - 1) \right] = \lim\limits_{x\to 1} \dfrac{f(x) - 8}{x - 1} \cdot \lim\limits_{x\to 1} (x - 1) = 10 \cdot 0 = 0.$

Thus, $\lim\limits_{x\to 1} f(x) = \lim\limits_{x\to 1} \{[f(x) - 8] + 8\} = \lim\limits_{x\to 1} [f(x) - 8] + \lim\limits_{x\to 1} 8 = 0 + 8 = 8.$

Note: The value of $\lim\limits_{x\to 1} \dfrac{f(x) - 8}{x - 1}$ does not affect the answer since it's multiplied by 0. What's important is that

$\lim\limits_{x\to 1} \dfrac{f(x) - 8}{x - 1}$ exists.

47. Let $f(x) = [\![x]\!]$ and $g(x) = -[\![x]\!]$. Then $\lim\limits_{x\to 3} f(x)$ and $\lim\limits_{x\to 3} g(x)$ do not exist   [Example 9]

but $\lim\limits_{x\to 3} [f(x) + g(x)] = \lim\limits_{x\to 3} ([\![x]\!] - [\![x]\!]) = \lim\limits_{x\to 3} 0 = 0.$

49. Since the denominator approaches 0 as $x \to -2$, the limit will exist only if the numerator also approaches

0 as $x \to -2$. In order for this to happen, we need $\lim\limits_{x\to -2} (3x^2 + ax + a + 3) = 0$ ⇔

$3(-2)^2 + a(-2) + a + 3 = 0$ ⇔ $12 - 2a + a + 3 = 0$ ⇔ $a = 15$. With $a = 15$, the limit becomes

$\lim\limits_{x\to -2} \dfrac{3x^2 + 15x + 18}{x^2 + x - 2} = \lim\limits_{x\to -2} \dfrac{3(x + 2)(x + 3)}{(x - 1)(x + 2)} = \lim\limits_{x\to -2} \dfrac{3(x + 3)}{x - 1} = \dfrac{3(-2 + 3)}{-2 - 1} = \dfrac{3}{-3} = -1.$

2.4   Continuity

1. From Definition 1, $\lim\limits_{x\to 4} f(x) = f(4)$.

3. (a) f is discontinuous at -4 since $f(-4)$ is not defined and at -2, 2, and 4 since the limit does not exist (the left and right

limits are not the same).

(b) f is continuous from the left at -2 since $\lim\limits_{x\to -2^-} f(x) = f(-2)$. f is continuous from the right at 2 and 4 since

$\lim\limits_{x\to 2^+} f(x) = f(2)$ and $\lim\limits_{x\to 4^+} f(x) = f(4)$. It is continuous from neither side at -4 since $f(-4)$ is undefined.

5. The graph of $y = f(x)$ must have a discontinuity at

$x = 2$ and must show that $\lim\limits_{x\to 2^+} f(x) = f(2)$.

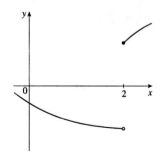

7. The graph of $y = f(x)$ must have a removable

discontinuity (a hole) at $x = 3$ and a jump discontinuity

at $x = 5$.

9. (a)

(b) There are discontinuities at times $t = 1, 2, 3$, and 4. A person parking in the lot would want to keep in mind that the charge will jump at the beginning of each hour.

11. Since f and g are continuous functions,

$$\lim_{x \to 3} [2f(x) - g(x)] = 2 \lim_{x \to 3} f(x) - \lim_{x \to 3} g(x) \qquad \text{[by Limit Laws 2 and 3]}$$

$$= 2f(3) - g(3) \qquad \text{[by continuity of } f \text{ and } g \text{ at } x = 3]$$

$$= 2 \cdot 5 - g(3) = 10 - g(3)$$

Since it is given that $\lim_{x \to 3} [2f(x) - g(x)] = 4$, we have $10 - g(3) = 4$, so $g(3) = 6$.

13. $\lim_{x \to -1} f(x) = \lim_{x \to -1} (x + 2x^3)^4 = \left(\lim_{x \to -1} x + 2 \lim_{x \to -1} x^3 \right)^4 = [-1 + 2(-1)^3]^4 = (-3)^4 = 81 = f(-1).$

By the definition of continuity, f is continuous at $a = -1$.

15. $f(x) = \begin{cases} e^x & \text{if } x < 0 \\ x^2 & \text{if } x \geq 0 \end{cases}$

The left-hand limit of f at $a = 0$ is $\lim_{x \to 0^-} f(x) = \lim_{x \to 0^-} e^x = 1$. The right-hand limit of f at $a = 0$ is $\lim_{x \to 0^+} f(x) = \lim_{x \to 0^+} x^2 = 0$. Since these limits are not equal, $\lim_{x \to 0} f(x)$ does not exist and f is discontinuous at 0.

17. $f(x) = \begin{cases} \cos x & \text{if } x < 0 \\ 0 & \text{if } x = 0 \\ 1 - x^2 & \text{if } x > 0 \end{cases}$

$\lim_{x \to 0} f(x) = 1$, but $f(0) = 0 \neq 1$, so f is discontinuous at 0.

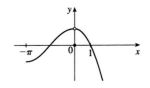

19. By Theorem 5, the polynomials x^2 and $2x - 1$ are continuous on $(-\infty, \infty)$. By Theorem 7, the root function $\sqrt{x}$ is continuous on $[0, \infty)$. By Theorem 9, the composite function $\sqrt{2x - 1}$ is continuous on its domain, $[\frac{1}{2}, \infty)$.

By part 1 of Theorem 4, the sum $R(x) = x^2 + \sqrt{2x - 1}$ is continuous on $[\frac{1}{2}, \infty)$.

21. By Theorem 7, the exponential function e^{-5t} and the trigonometric function $\cos 2\pi t$ are continuous on $(-\infty, \infty)$. By part 4 of Theorem 4, $L(t) = e^{-5t} \cos 2\pi t$ is continuous on $(-\infty, \infty)$.

23. By Theorem 5, the polynomial $t^4 - 1$ is continuous on $(-\infty, \infty)$. By Theorem 7, $\ln x$ is continuous on its domain, $(0, \infty)$. By Theorem 9, $\ln(t^4 - 1)$ is continuous on its domain, which is

$$\{t \mid t^4 - 1 > 0\} = \{t \mid t^4 > 1\} = \{t \mid |t| > 1\} = (-\infty, -1) \cup (1, \infty)$$

25. The function $y = \dfrac{1}{1 + e^{1/x}}$ is discontinuous at $x = 0$ because the

left- and right-hand limits at $x = 0$ are different.

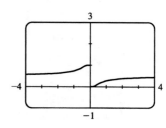

27. Because we are dealing with root functions, $5 + \sqrt{x}$ is continuous on $[0, \infty)$, $\sqrt{x + 5}$ is continuous on $[-5, \infty)$, so the

quotient $f(x) = \dfrac{5 + \sqrt{x}}{\sqrt{5 + x}}$ is continuous on $[0, \infty)$. Since f is continuous at $x = 4$, $\lim\limits_{x \to 4} f(x) = f(4) = \frac{7}{3}$.

29. Because $x^2 - x$ is continuous on $\mathbb{R}$, the composite function $f(x) = e^{x^2 - x}$ is continuous on $\mathbb{R}$, so

$\lim\limits_{x \to 1} f(x) = f(1) = e^{1 - 1} = e^0 = 1$.

31. $f(x) = \begin{cases} x^2 & \text{if } x < 1 \\ \sqrt{x} & \text{if } x \geq 1 \end{cases}$

By Theorem 5, since $f(x)$ equals the polynomial x^2 on $(-\infty, 1)$, f is continuous on $(-\infty, 1)$. By Theorem 7, since $f(x)$

equals the root function $\sqrt{x}$ on $(1, \infty)$, f is continuous on $(1, \infty)$. At $x = 1$, $\lim\limits_{x \to 1^-} f(x) = \lim\limits_{x \to 1^-} x^2 = 1$ and

$\lim\limits_{x \to 1^+} f(x) = \lim\limits_{x \to 1^+} \sqrt{x} = 1$. Thus, $\lim\limits_{x \to 1} f(x)$ exists and equals 1. Also, $f(1) = \sqrt{1} = 1$. Thus, f is continuous at $x = 1$.

We conclude that f is continuous on $(-\infty, \infty)$.

33. $f(x) = \begin{cases} x + 2 & \text{if } x < 0 \\ e^x & \text{if } 0 \leq x \leq 1 \\ 2 - x & \text{if } x > 1 \end{cases}$

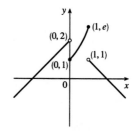

f is continuous on $(-\infty, 0)$ and $(1, \infty)$ since on each of these intervals

it is a polynomial; it is continuous on $(0, 1)$ since it is an exponential.

Now $\lim\limits_{x \to 0^-} f(x) = \lim\limits_{x \to 0^-} (x + 2) = 2$ and $\lim\limits_{x \to 0^+} f(x) = \lim\limits_{x \to 0^+} e^x = 1$, so f is discontinuous at 0. Since $f(0) = 1$, f is

continuous from the right at 0. Also $\lim\limits_{x \to 1^-} f(x) = \lim\limits_{x \to 1^-} e^x = e$ and $\lim\limits_{x \to 1^+} f(x) = \lim\limits_{x \to 1^+} (2 - x) = 1$, so f is discontinuous

at 1. Since $f(1) = e$, f is continuous from the left at 1.

35. $f(x) = \begin{cases} cx^2 + 2x & \text{if } x < 2 \\ x^3 - cx & \text{if } x \geq 2 \end{cases}$

f is continuous on $(-\infty, 2)$ and $(2, \infty)$. Now $\lim\limits_{x \to 2^-} f(x) = \lim\limits_{x \to 2^-} (cx^2 + 2x) = 4c + 4$ and

$\lim\limits_{x \to 2^+} f(x) = \lim\limits_{x \to 2^+} (x^3 - cx) = 8 - 2c$. So f is continuous $\Leftrightarrow$ $4c + 4 = 8 - 2c$ $\Leftrightarrow$ $6c = 4$ $\Leftrightarrow$ $c = \frac{2}{3}$. Thus, for f

to be continuous on $(-\infty, \infty)$, $c = \frac{2}{3}$.

37. (a) $f(x) = \dfrac{x^4 - 1}{x - 1} = \dfrac{(x^2 + 1)(x^2 - 1)}{x - 1} = \dfrac{(x^2 + 1)(x + 1)(x - 1)}{x - 1} = (x^2 + 1)(x + 1)$ [or $x^3 + x^2 + x + 1$]

for $x \neq 1$. The discontinuity is removable and $g(x) = x^3 + x^2 + x + 1$ agrees with f for $x \neq 1$ and is continuous on $\mathbb{R}$.

(b) $f(x) = \dfrac{x^3 - x^2 - 2x}{x - 2} = \dfrac{x(x^2 - x - 2)}{x - 2} = \dfrac{x(x - 2)(x + 1)}{x - 2} = x(x + 1)$ [or $x^2 + x$] for $x \neq 2$. The discontinuity

is removable and $g(x) = x^2 + x$ agrees with f for $x \neq 2$ and is continuous on $\mathbb{R}$.

(c) $\lim\limits_{x \to \pi^-} f(x) = \lim\limits_{x \to \pi^-} [\![\sin x]\!] = \lim\limits_{x \to \pi^-} 0 = 0$ and $\lim\limits_{x \to \pi^+} f(x) = \lim\limits_{x \to \pi^+} [\![\sin x]\!] = \lim\limits_{x \to \pi^+} (-1) = -1$, so $\lim\limits_{x \to \pi} f(x)$ does not

exist. The discontinuity at $x = \pi$ is a jump discontinuity.

39. $f(x) = x^2 + 10 \sin x$ is continuous on the interval $[31, 32]$. $f(31) \approx 957$, and $f(32) \approx 1030$. Since $957 < 1000 < 1030$,

there is a number c in $(31, 32)$ such that $f(c) = 1000$ by the Intermediate Value Theorem. *Note:* There is also a number c in

$(-32, -31)$ such that $f(c) = 1000$.

41. $f(x) = x^4 + x - 3$ is continuous on the interval $[1, 2]$, $f(1) = -1$, and $f(2) = 15$. Since $-1 < 0 < 15$, there is a number c

in $(1, 2)$ such that $f(c) = 0$ by the Intermediate Value Theorem. Thus, there is a root of the equation $x^4 + x - 3 = 0$ in the

interval $(1, 2)$.

43. The equation $e^x = 3 - 2x$ is equivalent to the equation $e^x + 2x - 3 = 0$. $f(x) = e^x + 2x - 3$ is continuous on the interval

$[0, 1]$, $f(0) = -2$, and $f(1) = e - 1 \approx 1.72$. Since $-2 < 0 < e - 1$, there is a number c in $(0, 1)$ such that $f(c) = 0$ by the

Intermediate Value Theorem. Thus, there is a root of the equation $e^x + 2x - 3 = 0$, or $e^x = 3 - 2x$, in the interval $(0, 1)$.

45. (a) $f(x) = \cos x - x^3$ is continuous on the interval $[0, 1]$, $f(0) = 1 > 0$, and $f(1) = \cos 1 - 1 \approx -0.46 < 0$. Since

$1 > 0 > -0.46$, there is a number c in $(0, 1)$ such that $f(c) = 0$ by the Intermediate Value Theorem. Thus, there is a root

of the equation $\cos x - x^3 = 0$, or $\cos x = x^3$, in the interval $(0, 1)$.

(b) $f(0.86) \approx 0.016 > 0$ and $f(0.87) \approx -0.014 < 0$, so there is a root between 0.86 and 0.87, that is, in the interval

$(0.86, 0.87)$.

47. (a) Let $f(x) = 100e^{-x/100} - 0.01x^2$. Then $f(0) = 100 > 0$ and

$f(100) = 100e^{-1} - 100 \approx -63.2 < 0$. So by the Intermediate

Value Theorem, there is a number c in $(0, 100)$ such that $f(c) = 0$.

This implies that $100e^{-c/100} = 0.01c^2$.

(b) Using the intersect feature of the graphing device, we find that the

root of the equation is $x = 70.347$, correct to three decimal places.

49. $\lim\limits_{h \to 0} \sin(a + h) = \lim\limits_{h \to 0} (\sin a \cos h + \cos a \sin h) = \lim\limits_{h \to 0} (\sin a \cos h) + \lim\limits_{h \to 0} (\cos a \sin h)$

$= \left(\lim\limits_{h \to 0} \sin a\right)\left(\lim\limits_{h \to 0} \cos h\right) + \left(\lim\limits_{h \to 0} \cos a\right)\left(\lim\limits_{h \to 0} \sin h\right) = (\sin a)(1) + (\cos a)(0) = \sin a$

51. If there is such a number, it satisfies the equation $x^3 + 1 = x \iff x^3 - x + 1 = 0$. Let the left-hand side of this equation be

called $f(x)$. Now $f(-2) = -5 < 0$, and $f(-1) = 1 > 0$. Note also that $f(x)$ is a polynomial, and thus continuous. So by the

Intermediate Value Theorem, there is a number c between -2 and -1 such that $f(c) = 0$, so that $c = c^3 + 1$.

53. $f(x) = x^4 \sin(1/x)$ is continuous on $(-\infty, 0) \cup (0, \infty)$ since it is the product of a polynomial and a composite of a trigonometric function and a rational function. Now since $-1 \le \sin(1/x) \le 1$, we have $-x^4 \le x^4 \sin(1/x) \le x^4$. Because $\lim_{x \to 0}(-x^4) = 0$ and $\lim_{x \to 0} x^4 = 0$, the Squeeze Theorem gives us $\lim_{x \to 0}(x^4 \sin(1/x)) = 0$, which equals $f(0)$. Thus, f is continuous at 0 and, hence, on $(-\infty, \infty)$.

55. Define $u(t)$ to be the monk's distance from the monastery, as a function of time, on the first day, and define $d(t)$ to be his distance from the monastery, as a function of time, on the second day. Let D be the distance from the monastery to the top of the mountain. From the given information we know that $u(0) = 0$, $u(12) = D$, $d(0) = D$ and $d(12) = 0$. Now consider the function $u - d$, which is clearly continuous. We calculate that $(u - d)(0) = -D$ and $(u - d)(12) = D$. So by the Intermediate Value Theorem, there must be some time t_0 between 0 and 12 such that $(u - d)(t_0) = 0 \iff u(t_0) = d(t_0)$. So at time t_0 after 7:00 AM, the monk will be at the same place on both days.

2.5 Limits Involving Infinity

1. (a) As x approaches 2 (from the right or the left), the values of $f(x)$ become large.

(b) As x approaches 1 from the right, the values of $f(x)$ become large negative.

(c) As x becomes large, the values of $f(x)$ approach 5.

(d) As x becomes large negative, the values of $f(x)$ approach 3.

3. (a) $\lim_{x \to 2} f(x) = \infty$ (b) $\lim_{x \to -1^-} f(x) = \infty$ (c) $\lim_{x \to -1^+} f(x) = -\infty$ (d) $\lim_{x \to \infty} f(x) = 1$

(e) $\lim_{x \to -\infty} f(x) = 2$ (f) Vertical: $x = -1$, $x = 2$; Horizontal: $y = 1$, $y = 2$

5. $\lim_{x \to 0} f(x) = -\infty$,

$\lim_{x \to -\infty} f(x) = 5$,

$\lim_{x \to \infty} f(x) = -5$

7. $\lim_{x \to 2} f(x) = -\infty$, $\lim_{x \to \infty} f(x) = \infty$,

$\lim_{x \to -\infty} f(x) = 0$, $\lim_{x \to 0^+} f(x) = \infty$,

$\lim_{x \to 0^-} f(x) = -\infty$

9. $f(0) = 3$, $\lim_{x \to 0^-} f(x) = 4$,

$\lim_{x \to 0^+} f(x) = 2$,

$\lim_{x \to -\infty} f(x) = -\infty$, $\lim_{x \to 4^-} f(x) = -\infty$,

$\lim_{x \to 4^+} f(x) = \infty$, $\lim_{x \to \infty} f(x) = 3$

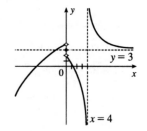

11. If $f(x) = x^2/2^x$, then a calculator gives $f(0) = 0$, $f(1) = 0.5$, $f(2) = 1$, $f(3) = 1.125$, $f(4) = 1$, $f(5) = 0.78125$, $f(6) = 0.5625$, $f(7) = 0.3828125$, $f(8) = 0.25$, $f(9) = 0.158203125$, $f(10) = 0.09765625$, $f(20) \approx 0.00038147$, $f(50) \approx 2.2204 \times 10^{-12}$, $f(100) \approx 7.8886 \times 10^{-27}$.

It appears that $\lim_{x \to \infty} (x^2/2^x) = 0$.

13. Vertical: $x \approx -1.62$, $x \approx 0.62$, $x = 1$;

   Horizontal: $y = 1$

15. $\displaystyle\lim_{x \to 1} \frac{2-x}{(x-1)^2} = \infty$ since the numerator is positive and the denominator approaches 0 through positive values as $x \to 1$.

17. Let $t = 3/(2-x)$. As $x \to 2^+$, $t \to -\infty$. So $\displaystyle\lim_{x \to 2^+} e^{3/(2-x)} = \lim_{t \to -\infty} e^t = 0$ by (7).

19. Let $t = x^2 - 9$. Then as $x \to 3^+$, $t \to 0^+$, and $\displaystyle\lim_{x \to 3^+} \ln(x^2 - 9) = \lim_{t \to 0^+} \ln t = -\infty$ by (3).

21. $\displaystyle\lim_{x \to 2\pi^-} x \csc x = \lim_{x \to 2\pi^-} \frac{x}{\sin x} = -\infty$ since the numerator is positive and the denominator approaches 0 through negative

   values as $x \to 2\pi^-$.

23. Divide both the numerator and denominator by x^3 (the highest power of x that occurs in the denominator).

$$\lim_{x \to \infty} \frac{x^3 + 5x}{2x^3 - x^2 + 4} = \lim_{x \to \infty} \frac{\dfrac{x^3 + 5x}{x^3}}{\dfrac{2x^3 - x^2 + 4}{x^3}} = \lim_{x \to \infty} \frac{1 + \dfrac{5}{x^2}}{2 - \dfrac{1}{x} + \dfrac{4}{x^3}} = \frac{\displaystyle\lim_{x \to \infty} \left(1 + \dfrac{5}{x^2}\right)}{\displaystyle\lim_{x \to \infty} \left(2 - \dfrac{1}{x} + \dfrac{4}{x^3}\right)}$$

$$= \frac{\displaystyle\lim_{x \to \infty} 1 + 5 \lim_{x \to \infty} \dfrac{1}{x^2}}{\displaystyle\lim_{x \to \infty} 2 - \lim_{x \to \infty} \dfrac{1}{x} + 4 \lim_{x \to \infty} \dfrac{1}{x^3}} = \frac{1 + 5(0)}{2 - 0 + 4(0)} = \frac{1}{2}$$

25. First, multiply the factors in the denominator. Then divide both the numerator and denominator by u^4.

$$\lim_{u \to \infty} \frac{4u^4 + 5}{(u^2 - 2)(2u^2 - 1)} = \lim_{u \to \infty} \frac{4u^4 + 5}{2u^4 - 5u^2 + 2} = \lim_{u \to \infty} \frac{\dfrac{4u^4 + 5}{u^4}}{\dfrac{2u^4 - 5u^2 + 2}{u^4}} = \lim_{u \to \infty} \frac{4 + \dfrac{5}{u^4}}{2 - \dfrac{5}{u^2} + \dfrac{2}{u^4}}$$

$$= \frac{\displaystyle\lim_{u \to \infty} \left(4 + \dfrac{5}{u^4}\right)}{\displaystyle\lim_{u \to \infty} \left(2 - \dfrac{5}{u^2} + \dfrac{2}{u^4}\right)} = \frac{\displaystyle\lim_{u \to \infty} 4 + 5 \lim_{u \to \infty} \dfrac{1}{u^4}}{\displaystyle\lim_{u \to \infty} 2 - 5 \lim_{u \to \infty} \dfrac{1}{u^2} + 2 \lim_{u \to \infty} \dfrac{1}{u^4}} = \frac{4 + 5(0)}{2 - 5(0) + 2(0)} = \frac{4}{2} = 2$$

27. $\displaystyle\lim_{x \to \infty} \left(\sqrt{9x^2 + x} - 3x\right) = \lim_{x \to \infty} \frac{\left(\sqrt{9x^2 + x} - 3x\right)\left(\sqrt{9x^2 + x} + 3x\right)}{\sqrt{9x^2 + x} + 3x} = \lim_{x \to \infty} \frac{\left(\sqrt{9x^2 + x}\right)^2 - (3x)^2}{\sqrt{9x^2 + x} + 3x}$

$\displaystyle = \lim_{x \to \infty} \frac{(9x^2 + x) - 9x^2}{\sqrt{9x^2 + x} + 3x} = \lim_{x \to \infty} \frac{x}{\sqrt{9x^2 + x} + 3x} \cdot \frac{1/x}{1/x}$

$\displaystyle = \lim_{x \to \infty} \frac{x/x}{\sqrt{9x^2/x^2 + x/x^2} + 3x/x} = \lim_{x \to \infty} \frac{1}{\sqrt{9 + 1/x} + 3} = \frac{1}{\sqrt{9} + 3} = \frac{1}{3 + 3} = \frac{1}{6}$

29. Let $t = -x^2$. As $x \to \infty$, $t \to -\infty$. So $\lim\limits_{x \to \infty} e^{-x^2} = \lim\limits_{t \to -\infty} e^t = 0$ by (7).

31. $\lim\limits_{x \to \infty} \cos x$ does not exist because as x increases $\cos x$ does not approach any one value, but oscillates between 1 and -1.

33. Since $-1 \le \cos x \le 1$ and $e^{-2x} > 0$, we have $-e^{-2x} \le e^{-2x} \cos x \le e^{-2x}$. We know that $\lim\limits_{x \to \infty} (-e^{-2x}) = 0$ and

$\lim\limits_{x \to \infty} (e^{-2x}) = 0$, so by the Squeeze Theorem, $\lim\limits_{x \to \infty} (e^{-2x} \cos x) = 0$.

35. $\lim\limits_{x \to -\infty} (x^4 + x^5) = \lim\limits_{x \to -\infty} x^5(\frac{1}{x} + 1)$ [factor out the largest power of x] $= -\infty$ because $x^5 \to -\infty$ and $1/x + 1 \to 1$

as $x \to -\infty$.

Or: $\lim\limits_{x \to -\infty} (x^4 + x^5) = \lim\limits_{x \to -\infty} x^4 (1 + x) = -\infty$.

37. $\lim\limits_{x \to \infty} \dfrac{x + x^3 + x^5}{1 - x^2 + x^4} = \lim\limits_{x \to \infty} \dfrac{(x + x^3 + x^5)/x^4}{(1 - x^2 + x^4)/x^4}$ [divide by the highest power of x in the denominator]

$= \lim\limits_{x \to \infty} \dfrac{1/x^3 + 1/x + x}{1/x^4 - 1/x^2 + 1} = \infty$

because $(1/x^3 + 1/x + x) \to \infty$ and $(1/x^4 - 1/x^2 + 1) \to 1$ as $x \to \infty$.

39. $\lim\limits_{x \to \infty} \dfrac{2x^2 + x - 1}{x^2 + x - 2} = \lim\limits_{x \to \infty} \dfrac{\dfrac{2x^2 + x - 1}{x^2}}{\dfrac{x^2 + x - 2}{x^2}} = \lim\limits_{x \to \infty} \dfrac{2 + \dfrac{1}{x} - \dfrac{1}{x^2}}{1 + \dfrac{1}{x} - \dfrac{2}{x^2}} = \dfrac{\lim\limits_{x \to \infty} \left(2 + \dfrac{1}{x} - \dfrac{1}{x^2}\right)}{\lim\limits_{x \to \infty} \left(1 + \dfrac{1}{x} - \dfrac{2}{x^2}\right)}$

$= \dfrac{\lim\limits_{x \to \infty} 2 + \lim\limits_{x \to \infty} \dfrac{1}{x} - \lim\limits_{x \to \infty} \dfrac{1}{x^2}}{\lim\limits_{x \to \infty} 1 + \lim\limits_{x \to \infty} \dfrac{1}{x} - 2\lim\limits_{x \to \infty} \dfrac{1}{x^2}} = \dfrac{2 + 0 - 0}{1 + 0 - 2(0)} = 2$,   so $y = 2$ is a horizontal asymptote.

$y = f(x) = \dfrac{2x^2 + x - 1}{x^2 + x - 2} = \dfrac{(2x - 1)(x + 1)}{(x + 2)(x - 1)}$, so $\lim\limits_{x \to -2^-} f(x) = \infty$,

$\lim\limits_{x \to -2^+} f(x) = -\infty$, $\lim\limits_{x \to 1^-} f(x) = -\infty$, and $\lim\limits_{x \to 1^+} f(x) = \infty$. Thus, $x = -2$

and $x = 1$ are vertical asymptotes. The graph confirms our work.

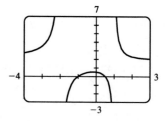

41. $y = f(x) = \dfrac{x^3 - x}{x^2 - 6x + 5} = \dfrac{x(x^2 - 1)}{(x - 1)(x - 5)} = \dfrac{x(x + 1)(x - 1)}{(x - 1)(x - 5)} = \dfrac{x(x + 1)}{x - 5} = g(x)$ for $x \ne 1$.

The graph of g is the same as the graph of f with the exception of a hole in the

graph of f at $x = 1$. By long division, $g(x) = \dfrac{x^2 + x}{x - 5} = x + 6 + \dfrac{30}{x - 5}$.

As $x \to \pm\infty$, $g(x) \to \pm\infty$, so there is no horizontal asymptote. The denominator

of g is zero when $x = 5$. $\lim\limits_{x \to 5^-} g(x) = -\infty$ and $\lim\limits_{x \to 5^+} g(x) = \infty$, so $x = 5$ is a

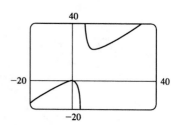

vertical asymptote. The graph confirms our work.

43. (a)

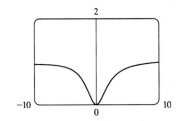

From the graph of $f(x) = \sqrt{x^2 + x + 1} + x$, we estimate the value of $\lim\limits_{x \to -\infty} f(x)$ to be -0.5.

(b)

x	$f(x)$
$-10{,}000$	-0.4999625
$-100{,}000$	-0.4999962
$-1{,}000{,}000$	-0.4999996

From the table, we estimate the limit to be -0.5.

(c) $\lim\limits_{x \to -\infty} \left(\sqrt{x^2 + x + 1} + x \right) = \lim\limits_{x \to -\infty} \left(\sqrt{x^2 + x + 1} + x \right) \left[\dfrac{\sqrt{x^2 + x + 1} - x}{\sqrt{x^2 + x + 1} - x} \right] = \lim\limits_{x \to -\infty} \dfrac{(x^2 + x + 1) - x^2}{\sqrt{x^2 + x + 1} - x}$

$= \lim\limits_{x \to -\infty} \dfrac{(x + 1)(1/x)}{\left(\sqrt{x^2 + x + 1} - x \right)(1/x)} = \lim\limits_{x \to -\infty} \dfrac{1 + (1/x)}{-\sqrt{1 + (1/x) + (1/x^2)} - 1}$

$= \dfrac{1 + 0}{-\sqrt{1 + 0 + 0} - 1} = -\dfrac{1}{2}$

Note that for $x < 0$, we have $\sqrt{x^2} = |x| = -x$, so when we divide the radical by x, with $x < 0$, we get

$\dfrac{1}{x} \sqrt{x^2 + x + 1} = -\dfrac{1}{\sqrt{x^2}} \sqrt{x^2 + x + 1} = -\sqrt{1 + (1/x) + (1/x^2)}$.

45. From the graph, it appears $y = 1$ is a horizontal asymptote.

$\lim\limits_{x \to \infty} \dfrac{3x^3 + 500x^2}{x^3 + 500x^2 + 100x + 2000} = \lim\limits_{x \to \infty} \dfrac{\dfrac{3x^3 + 500x^2}{x^3}}{\dfrac{x^3 + 500x^2 + 100x + 2000}{x^3}} = \lim\limits_{x \to \infty} \dfrac{3 + (500/x)}{1 + (500/x) + (100/x^2) + (2000/x^3)}$

$= \dfrac{3 + 0}{1 + 0 + 0 + 0} = 3$,   so $y = 3$ is a horizontal asymptote.

The discrepancy can be explained by the choice of the viewing window. Try $[-100{,}000, 100{,}000]$ by $[-1, 4]$ to get a graph that lends credibility to our calculation that $y = 3$ is a horizontal asymptote.

47. Let's look for a rational function.

(1) $\lim\limits_{x \to \pm\infty} f(x) = 0 \ \Rightarrow$ degree of numerator < degree of denominator

(2) $\lim\limits_{x \to 0} f(x) = -\infty \ \Rightarrow$ there is a factor of x^2 in the denominator (not just x, since that would produce a sign change at $x = 0$), and the function is negative near $x = 0$.

(3) $\lim\limits_{x \to 3^-} f(x) = \infty$ and $\lim\limits_{x \to 3^+} f(x) = -\infty \ \Rightarrow$ vertical asymptote at $x = 3$; there is a factor of $(x - 3)$ in the denominator.

(4) $f(2) = 0 \ \Rightarrow$ 2 is an x-intercept; there is at least one factor of $(x - 2)$ in the numerator.

Combining all of this information and putting in a negative sign to give us the desired left- and right-hand limits gives us

$f(x) = \dfrac{2 - x}{x^2(x - 3)}$ as one possibility.

49. (a) We must first find the function f. Since f has a vertical asymptote $x = 4$ and x-intercept $x = 1$, $x - 4$ is a factor of the denominator and $x - 1$ is a factor of the numerator. There is a removable discontinuity at $x = -1$, so $x - (-1) = x + 1$ is a factor of both the numerator and denominator. Thus, f now looks like this: $f(x) = \dfrac{a(x-1)(x+1)}{(x-4)(x+1)}$, where a is still to be determined. Then $\lim\limits_{x \to -1} f(x) = \lim\limits_{x \to -1} \dfrac{a(x-1)(x+1)}{(x-4)(x+1)} = \lim\limits_{x \to -1} \dfrac{a(x-1)}{x-4} = \dfrac{a(-1-1)}{(-1-4)} = \dfrac{2}{5}a$, so $\dfrac{2}{5}a = 2$, and $a = 5$. Thus $f(x) = \dfrac{5(x-1)(x+1)}{(x-4)(x+1)}$ is a ratio of quadratic functions satisfying all the given conditions and

$$f(0) = \frac{5(-1)(1)}{(-4)(1)} = \frac{5}{4}.$$

(b) $\lim\limits_{x \to \infty} f(x) = 5 \lim\limits_{x \to \infty} \dfrac{x^2 - 1}{x^2 - 3x - 4} = 5 \lim\limits_{x \to \infty} \dfrac{(x^2/x^2) - (1/x^2)}{(x^2/x^2) - (3x/x^2) - (4/x^2)} = 5 \dfrac{1 - 0}{1 - 0 - 0} = 5(1) = 5$

51. (a) Divide the numerator and the denominator by the highest power of x in $Q(x)$.

(a) If $\deg P < \deg Q$, then the numerator $\to 0$ but the denominator doesn't. So $\lim\limits_{x \to \infty} [P(x)/Q(x)] = 0$.

(b) If $\deg P > \deg Q$, then the numerator $\to \pm\infty$ but the denominator doesn't, so $\lim\limits_{x \to \infty} [P(x)/Q(x)] = \pm\infty$ (depending on the ratio of the leading coefficients of P and Q).

53. $\lim\limits_{x \to \infty} \dfrac{5\sqrt{x}}{\sqrt{x-1}} \cdot \dfrac{1/\sqrt{x}}{1/\sqrt{x}} = \lim\limits_{x \to \infty} \dfrac{5}{\sqrt{1 - (1/x)}} = \dfrac{5}{\sqrt{1-0}} = 5$ and

$\lim\limits_{x \to \infty} \dfrac{10e^x - 21}{2e^x} \cdot \dfrac{1/e^x}{1/e^x} = \lim\limits_{x \to \infty} \dfrac{10 - (21/e^x)}{2} = \dfrac{10 - 0}{2} = 5$. Since $\dfrac{10e^x - 21}{2e^x} < f(x) < \dfrac{5\sqrt{x}}{\sqrt{x-1}}$,

we have $\lim\limits_{x \to \infty} f(x) = 5$ by the Squeeze Theorem.

55. (a) After t minutes, $25t$ liters of brine with 30 g of salt per liter has been pumped into the tank, so it contains $(5000 + 25t)$ liters of water and $25t \cdot 30 = 750t$ grams of salt. Therefore, the salt concentration at time t will be

$$C(t) = \frac{750t}{5000 + 25t} = \frac{30t}{200 + t} \frac{\text{g}}{\text{L}}.$$

(b) $\lim\limits_{t \to \infty} C(t) = \lim\limits_{t \to \infty} \dfrac{30t}{200 + t} = \lim\limits_{t \to \infty} \dfrac{30t/t}{200/t + t/t} = \dfrac{30}{0 + 1} = 30$. So the salt concentration approaches that of the brine being pumped into the tank.

57. (a) If $t = -x/10$, then $x = -10t$ and as $x \to \infty$, $t \to -\infty$. Thus, $\lim\limits_{x \to \infty} e^{-x/10} = \lim\limits_{t \to -\infty} e^t = 0$ by Equation 7.

(b) $y = e^{-x/10}$ and $y = 0.1$ intersect at $x_1 \approx 23.03$.

If $x > x_1$, then $e^{-x/10} < 0.1$.

(c) $e^{-x/10} < 0.1 \;\Rightarrow\; -x/10 < \ln 0.1 \;\Rightarrow$

$x > -10 \ln \frac{1}{10} = -10 \ln 10^{-1} = 10 \ln 10 \approx 23.03$

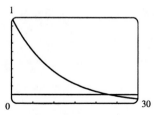

2.6 Derivatives and Rates of Change

1. (a) This is just the slope of the line through two points: $m_{PQ} = \dfrac{\Delta y}{\Delta x} = \dfrac{f(x) - f(3)}{x - 3}$.

(b) This is the limit of the slope of the secant line PQ as Q approaches P: $m = \lim\limits_{x \to 3} \dfrac{f(x) - f(3)}{x - 3}$.

3. (a) (i) Using Definition 1 with $f(x) = 4x - x^2$ and $P(1, 3)$,

$$m = \lim_{x \to a} \frac{f(x) - f(a)}{x - a} = \lim_{x \to 1} \frac{(4x - x^2) - 3}{x - 1} = \lim_{x \to 1} \frac{-(x^2 - 4x + 3)}{x - 1} = \lim_{x \to 1} \frac{-(x - 1)(x - 3)}{x - 1}$$

$$= \lim_{x \to 1}(3 - x) = 3 - 1 = 2$$

(ii) Using Equation 2 with $f(x) = 4x - x^2$ and $P(1, 3)$,

$$m = \lim_{h \to 0} \frac{f(a + h) - f(a)}{h} = \lim_{h \to 0} \frac{f(1 + h) - f(1)}{h} = \lim_{h \to 0} \frac{\left[4(1 + h) - (1 + h)^2\right] - 3}{h}$$

$$= \lim_{h \to 0} \frac{4 + 4h - 1 - 2h - h^2 - 3}{h} = \lim_{h \to 0} \frac{-h^2 + 2h}{h} = \lim_{h \to 0} \frac{h(-h + 2)}{h} = \lim_{h \to 0}(-h + 2) = 2$$

(b) An equation of the tangent line is $y - f(a) = f'(a)(x - a)$ $\Rightarrow$ $y - f(1) = f'(1)(x - 1)$ $\Rightarrow$ $y - 3 = 2(x - 1)$,

or $y = 2x + 1$.

(c)

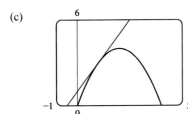

The graph of $y = 2x + 1$ is tangent to the graph of $y = 4x - x^2$ at the point $(1, 3)$. Now zoom in toward the point $(1, 3)$ until the parabola and the tangent line are indistiguishable.

5. Using (1) with $f(x) = 4x - 3x^2$ and $P(2, -4)$ [we could also use (2)],

$$m = \lim_{x \to a} \frac{f(x) - f(a)}{x - a} = \lim_{x \to 2} \frac{(4x - 3x^2) - (-4)}{x - 2} = \lim_{x \to 2} \frac{-3x^2 + 4x + 4}{x - 2}$$

$$= \lim_{x \to 2} \frac{(-3x - 2)(x - 2)}{x - 2} = \lim_{x \to 2}(-3x - 2) = -3(2) - 2 = -8$$

Tangent line: $y - (-4) = -8(x - 2)$ $\Leftrightarrow$ $y + 4 = -8x + 16$ $\Leftrightarrow$ $y = -8x + 12$.

7. Using (1), $m = \lim\limits_{x \to 1} \dfrac{\sqrt{x} - \sqrt{1}}{x - 1} = \lim\limits_{x \to 1} \dfrac{(\sqrt{x} - 1)(\sqrt{x} + 1)}{(x - 1)(\sqrt{x} + 1)} = \lim\limits_{x \to 1} \dfrac{x - 1}{(x - 1)(\sqrt{x} + 1)} = \lim\limits_{x \to 1} \dfrac{1}{\sqrt{x} + 1} = \dfrac{1}{2}$.

Tangent line: $y - 1 = \frac{1}{2}(x - 1)$ $\Leftrightarrow$ $y = \frac{1}{2}x + \frac{1}{2}$

9. (a) Using (2) with $y = f(x) = 3 + 4x^2 - 2x^3$,

$$m = \lim_{h \to 0} \frac{f(a + h) - f(a)}{h} = \lim_{h \to 0} \frac{3 + 4(a + h)^2 - 2(a + h)^3 - (3 + 4a^2 - 2a^3)}{h}$$

$$= \lim_{h \to 0} \frac{3 + 4(a^2 + 2ah + h^2) - 2(a^3 + 3a^2h + 3ah^2 + h^3) - 3 - 4a^2 + 2a^3}{h}$$

$$= \lim_{h \to 0} \frac{3 + 4a^2 + 8ah + 4h^2 - 2a^3 - 6a^2h - 6ah^2 - 2h^3 - 3 - 4a^2 + 2a^3}{h}$$

$$= \lim_{h \to 0} \frac{8ah + 4h^2 - 6a^2h - 6ah^2 - 2h^3}{h} = \lim_{h \to 0} \frac{h(8a + 4h - 6a^2 - 6ah - 2h^2)}{h}$$

$$= \lim_{h \to 0}(8a + 4h - 6a^2 - 6ah - 2h^2) = 8a - 6a^2$$

(b) At $(1,5)$: $m = 8(1) - 6(1)^2 = 2$, so an equation of the tangent line

is $y - 5 = 2(x - 1)$ ⟺ $y = 2x + 3$.

At $(2,3)$: $m = 8(2) - 6(2)^2 = -8$, so an equation of the tangent

line is $y - 3 = -8(x - 2)$ ⟺ $y = -8x + 19$.

(c)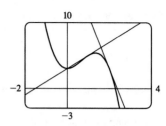

11. (a) The particle is moving to the right when s is increasing; that is, on the intervals $(0, 1)$ and $(4, 6)$. The particle is moving to

the left when s is decreasing; that is, on the interval $(2, 3)$. The particle is standing still when s is constant; that is, on the

intervals $(1, 2)$ and $(3, 4)$.

(b) The velocity of the particle is equal to the slope of the tangent line of the

graph. Note that there is no slope at the corner points on the graph. On the

interval $(0, 1)$, the slope is $\dfrac{3 - 0}{1 - 0} = 3$. On the interval $(2, 3)$, the slope is

$\dfrac{1 - 3}{3 - 2} = -2$. On the interval $(4, 6)$, the slope is $\dfrac{3 - 1}{6 - 4} = 1$.

13. Let $s(t) = 40t - 16t^2$.

$$v(2) = \lim_{t \to 2} \frac{s(t) - s(2)}{t - 2} = \lim_{t \to 2} \frac{(40t - 16t^2) - 16}{t - 2} = \lim_{t \to 2} \frac{-16t^2 + 40t - 16}{t - 2} = \lim_{t \to 2} \frac{-8(2t^2 - 5t + 2)}{t - 2}$$

$$= \lim_{t \to 2} \frac{-8(t - 2)(2t - 1)}{t - 2} = -8 \lim_{t \to 2} (2t - 1) = -8(3) = -24$$

Thus, the instantaneous velocity when $t = 2$ is -24 ft/s.

15. $v(a) = \lim_{h \to 0} \dfrac{s(a + h) - s(a)}{h} = \lim_{h \to 0} \dfrac{\dfrac{1}{(a + h)^2} - \dfrac{1}{a^2}}{h} = \lim_{h \to 0} \dfrac{\dfrac{a^2 - (a + h)^2}{a^2(a + h)^2}}{h} = \lim_{h \to 0} \dfrac{a^2 - (a^2 + 2ah + h^2)}{ha^2(a + h)^2}$

$= \lim_{h \to 0} \dfrac{-(2ah + h^2)}{ha^2(a + h)^2} = \lim_{h \to 0} \dfrac{-h(2a + h)}{ha^2(a + h)^2} = \lim_{h \to 0} \dfrac{-(2a + h)}{a^2(a + h)^2} = \dfrac{-2a}{a^2 \cdot a^2} = \dfrac{-2}{a^3}$ m/s

So $v(1) = \dfrac{-2}{1^3} = -2$ m/s, $v(2) = \dfrac{-2}{2^3} = -\dfrac{1}{4}$ m/s, and $v(3) = \dfrac{-2}{3^3} = -\dfrac{2}{27}$ m/s.

17. $g'(0)$ is the only negative value. The slope at $x = 4$ is smaller than the slope at $x = 2$ and both are smaller than the slope

at $x = -2$. Thus, $g'(0) < 0 < g'(4) < g'(2) < g'(-2)$.

19. For the tangent line $y = 4x - 5$: when $x = 2$, $y = 4(2) - 5 = 3$ and its slope is 4 (the coefficient of x). At the point of

tangency, these values are shared with the curve $y = f(x)$; that is, $f(2) = 3$ and $f'(2) = 4$.

21. We begin by drawing a curve through the origin with a slope of 3 to satisfy $f(0) = 0$ and $f'(0) = 3$. Since $f'(1) = 0$, we will round off our figure so that there is a horizontal tangent directly over $x = 1$. Last, we make sure that the curve has a slope of -1 as we pass over $x = 2$. Two of the many possibilities are shown.

 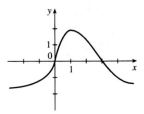

23. Using (4) with $f(x) = 3x^2 - x^3$ and $a = 1$,

$$f'(1) = \lim_{h \to 0} \frac{f(1+h) - f(1)}{h} = \lim_{h \to 0} \frac{[3(1+h)^2 - (1+h)^3] - 2}{h}$$

$$= \lim_{h \to 0} \frac{(3 + 6h + 3h^2) - (1 + 3h + 3h^2 + h^3) - 2}{h} = \lim_{h \to 0} \frac{3h - h^3}{h} = \lim_{h \to 0} \frac{h(3 - h^2)}{h}$$

$$= \lim_{h \to 0} (3 - h^2) = 3 - 0 = 3$$

Tangent line: $y - 2 = 3(x - 1) \quad \Leftrightarrow \quad y - 2 = 3x - 3 \quad \Leftrightarrow \quad y = 3x - 1$

25. (a) Using (4) with $F(x) = 5x/(1 + x^2)$ and the point $(2, 2)$, we have

$$F'(2) = \lim_{h \to 0} \frac{F(2+h) - F(2)}{h} = \lim_{h \to 0} \frac{\dfrac{5(2+h)}{1 + (2+h)^2} - 2}{h}$$

$$= \lim_{h \to 0} \frac{\dfrac{5h + 10}{h^2 + 4h + 5} - 2}{h} = \lim_{h \to 0} \frac{\dfrac{5h + 10 - 2(h^2 + 4h + 5)}{h^2 + 4h + 5}}{h}$$

$$= \lim_{h \to 0} \frac{-2h^2 - 3h}{h(h^2 + 4h + 5)} = \lim_{h \to 0} \frac{h(-2h - 3)}{h(h^2 + 4h + 5)} = \lim_{h \to 0} \frac{-2h - 3}{h^2 + 4h + 5} = \frac{-3}{5}$$

(b)

So an equation of the tangent line at $(2, 2)$ is $y - 2 = -\frac{3}{5}(x - 2)$ or $y = -\frac{3}{5}x + \frac{16}{5}$.

27. Use (4) with $f(x) = 3x^2 - 4x + 1$.

$$f'(a) = \lim_{h \to 0} \frac{f(a+h) - f(a)}{h} = \lim_{h \to 0} \frac{[3(a+h)^2 - 4(a+h) + 1] - (3a^2 - 4a + 1)}{h}$$

$$= \lim_{h \to 0} \frac{3a^2 + 6ah + 3h^2 - 4a - 4h + 1 - 3a^2 + 4a - 1}{h} = \lim_{h \to 0} \frac{6ah + 3h^2 - 4h}{h}$$

$$= \lim_{h \to 0} \frac{h(6a + 3h - 4)}{h} = \lim_{h \to 0} (6a + 3h - 4) = 6a - 4$$

29. Use (4) with $f(t) = (2t + 1)/(t + 3)$.

$$f'(a) = \lim_{h \to 0} \frac{f(a+h) - f(a)}{h} = \lim_{h \to 0} \frac{\dfrac{2(a+h) + 1}{(a+h) + 3} - \dfrac{2a + 1}{a + 3}}{h} = \lim_{h \to 0} \frac{(2a + 2h + 1)(a + 3) - (2a + 1)(a + h + 3)}{h(a + h + 3)(a + 3)}$$

$$= \lim_{h \to 0} \frac{(2a^2 + 6a + 2ah + 6h + a + 3) - (2a^2 + 2ah + 6a + a + h + 3)}{h(a + h + 3)(a + 3)}$$

$$= \lim_{h \to 0} \frac{5h}{h(a + h + 3)(a + 3)} = \lim_{h \to 0} \frac{5}{(a + h + 3)(a + 3)} = \frac{5}{(a + 3)^2}$$

31. Use (4) with $f(x) = \sqrt{1 - 2x}$.

$$f'(a) = \lim_{h \to 0} \frac{f(a+h) - f(a)}{h} = \lim_{h \to 0} \frac{\sqrt{1 - 2(a+h)} - \sqrt{1 - 2a}}{h}$$

$$= \lim_{h \to 0} \frac{\sqrt{1 - 2(a+h)} - \sqrt{1 - 2a}}{h} \cdot \frac{\sqrt{1 - 2(a+h)} + \sqrt{1 - 2a}}{\sqrt{1 - 2(a+h)} + \sqrt{1 - 2a}} = \lim_{h \to 0} \frac{\left(\sqrt{1 - 2(a+h)}\right)^2 - \left(\sqrt{1 - 2a}\right)^2}{h\left(\sqrt{1 - 2(a+h)} + \sqrt{1 - 2a}\right)}$$

$$= \lim_{h \to 0} \frac{(1 - 2a - 2h) - (1 - 2a)}{h\left(\sqrt{1 - 2(a+h)} + \sqrt{1 - 2a}\right)} = \lim_{h \to 0} \frac{-2h}{h\left(\sqrt{1 - 2(a+h)} + \sqrt{1 - 2a}\right)}$$

$$= \lim_{h \to 0} \frac{-2}{\sqrt{1 - 2(a+h)} + \sqrt{1 - 2a}} = \frac{-2}{\sqrt{1 - 2a} + \sqrt{1 - 2a}} = \frac{-2}{2\sqrt{1 - 2a}} = \frac{-1}{\sqrt{1 - 2a}}$$

Note that the answers to Exercises 33–38 are not unique.

33. By (4), $\displaystyle\lim_{h \to 0} \frac{(1+h)^{10} - 1}{h} = f'(1)$, where $f(x) = x^{10}$ and $a = 1$.

Or: By (4), $\displaystyle\lim_{h \to 0} \frac{(1+h)^{10} - 1}{h} = f'(0)$, where $f(x) = (1+x)^{10}$ and $a = 0$.

35. By Equation 5, $\displaystyle\lim_{x \to 5} \frac{2^x - 32}{x - 5} = f'(5)$, where $f(x) = 2^x$ and $a = 5$.

37. By (4), $\displaystyle\lim_{h \to 0} \frac{\cos(\pi + h) + 1}{h} = f'(\pi)$, where $f(x) = \cos x$ and $a = \pi$.

Or: By (4), $\displaystyle\lim_{h \to 0} \frac{\cos(\pi + h) + 1}{h} = f'(0)$, where $f(x) = \cos(\pi + x)$ and $a = 0$.

39. $\displaystyle v(5) = f'(5) = \lim_{h \to 0} \frac{f(5+h) - f(5)}{h} = \lim_{h \to 0} \frac{[100 + 50(5+h) - 4.9(5+h)^2] - [100 + 50(5) - 4.9(5)^2]}{h}$

$$= \lim_{h \to 0} \frac{(100 + 250 + 50h - 4.9h^2 - 49h - 122.5) - (100 + 250 - 122.5)}{h} = \lim_{h \to 0} \frac{-4.9h^2 + h}{h}$$

$$= \lim_{h \to 0} \frac{h(-4.9h + 1)}{h} = \lim_{h \to 0} (-4.9h + 1) = 1 \text{ m/s}$$

The speed when $t = 5$ is $|1| = 1$ m/s.

41. The sketch shows the graph for a room temperature of $72°$ and a refrigerator temperature of $38°$. The initial rate of change is greater in magnitude than the rate of change after an hour.

43. (a) (i) [2002, 2006]: $\displaystyle\frac{N(2006) - N(2002)}{2006 - 2002} = \frac{233 - 141}{4} = \frac{92}{4} = 23$ millions of cell phone subscribers per year

(ii) [2002, 2004]: $\displaystyle\frac{N(2004) - N(2002)}{2004 - 2002} = \frac{182 - 141}{2} = \frac{41}{2} = 20.5$ millions of cell phone subscribers per year

(iii) [2000, 2002]: $\displaystyle\frac{N(2002) - N(2000)}{2002 - 2000} = \frac{141 - 109}{2} = \frac{32}{2} = 16$ millions of cell phone subscribers per year

(b) Using the values from (ii) and (iii), we have $\dfrac{20.5 + 16}{2} = 18.25$ millions of cell phone subscribers per year.

(c) Estimating A as $(2000, 107)$ and B as $(2004, 175)$, the slope at 2002

is $\dfrac{175 - 107}{2004 - 2000} = \dfrac{68}{4} = 17$ millions of cell phone subscribers per

year.

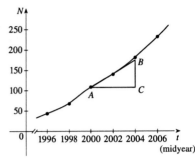

45. (a) (i) $\dfrac{\Delta C}{\Delta x} = \dfrac{C(105) - C(100)}{105 - 100} = \dfrac{6601.25 - 6500}{5} = \$20.25/\text{unit}.$

(ii) $\dfrac{\Delta C}{\Delta x} = \dfrac{C(101) - C(100)}{101 - 100} = \dfrac{6520.05 - 6500}{1} = \$20.05/\text{unit}.$

(b) $\dfrac{C(100 + h) - C(100)}{h} = \dfrac{\left[5000 + 10(100 + h) + 0.05(100 + h)^2\right] - 6500}{h} = \dfrac{20h + 0.05h^2}{h}$

$= 20 + 0.05h,\ h \neq 0$

So the instantaneous rate of change is $\displaystyle\lim_{h \to 0} \dfrac{C(100 + h) - C(100)}{h} = \lim_{h \to 0}(20 + 0.05h) = \$20/\text{unit}.$

47. (a) $f'(x)$ is the rate of change of the production cost with respect to the number of ounces of gold produced. Its units are dollars per ounce.

(b) After 800 ounces of gold have been produced, the rate at which the production cost is increasing is $\$17/\text{ounce}$. So the cost of producing the 800th (or 801st) ounce is about $\$17$.

(c) In the short term, the values of $f'(x)$ will decrease because more efficient use is made of start-up costs as x increases. But eventually $f'(x)$ might increase due to large-scale operations.

49. $T'(10)$ is the rate at which the temperature is changing at 10:00 A.M. To estimate the value of $T'(10)$, we will average the

difference quotients obtained using the times $t = 8$ and $t = 12$. Let $A = \dfrac{T(8) - T(10)}{8 - 10} = \dfrac{65 - 76}{-2} = 5.5$ and

$B = \dfrac{T(12) - T(10)}{12 - 10} = \dfrac{85 - 76}{2} = 4.5.$ Then $T'(10) = \displaystyle\lim_{t \to 10} \dfrac{T(t) - T(10)}{t - 10} \approx \dfrac{A + B}{2} = \dfrac{5.5 + 4.5}{2} = 5°\text{F}/\text{h}.$

51. (a) $S'(T)$ is the rate at which the oxygen solubility changes with respect to the water temperature. Its units are $(\text{mg/L})/°\text{C}$.

(b) For $T = 16°\text{C}$, it appears that the tangent line to the curve goes through the points $(0, 14)$ and $(32, 6)$. So

$S'(16) \approx \dfrac{6 - 14}{32 - 0} = -\dfrac{8}{32} = -0.25\ (\text{mg/L})/°\text{C}.$ This means that as the temperature increases past $16°\text{C}$, the oxygen

solubility is decreasing at a rate of $0.25\ (\text{mg/L})/°\text{C}.$

53. Since $f(x) = x \sin(1/x)$ when $x \neq 0$ and $f(0) = 0$, we have

$f'(0) = \displaystyle\lim_{h \to 0} \dfrac{f(0 + h) - f(0)}{h} = \lim_{h \to 0} \dfrac{h \sin(1/h) - 0}{h} = \lim_{h \to 0} \sin(1/h).$ This limit does not exist since $\sin(1/h)$ takes the

values -1 and 1 on any interval containing 0. (Compare with Example 4 in Section 2.2.)

2.7 The Derivative as a Function

1. It appears that f is an odd function, so f' will be an even function—that is, $f'(-a) = f'(a)$.

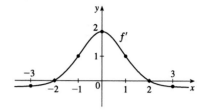

(a) $f'(-3) \approx -0.2$

(b) $f'(-2) \approx 0$

(c) $f'(-1) \approx 1$

(d) $f'(0) \approx 2$

(e) $f'(1) \approx 1$

(f) $f'(2) \approx 0$

(g) $f'(3) \approx -0.2$

3. $(a)' = $ II, since from left to right, the slopes of the tangents to graph (a) start out negative, become 0, then positive, then 0, then negative again. The actual function values in graph II follow the same pattern.

$(b)' = $ IV, since from left to right, the slopes of the tangents to graph (b) start out at a fixed positive quantity, then suddenly become negative, then positive again. The discontinuities in graph IV indicate sudden changes in the slopes of the tangents.

$(c)' = $ I, since the slopes of the tangents to graph (c) are negative for $x < 0$ and positive for $x > 0$, as are the function values of graph I.

$(d)' = $ III, since from left to right, the slopes of the tangents to graph (d) are positive, then 0, then negative, then 0, then positive, then 0, then negative again, and the function values in graph III follow the same pattern.

Hints for Exercises 4–11: First plot x-intercepts on the graph of f' for any horizontal tangents on the graph of f. Look for any corners on the graph of f—there will be a discontinuity on the graph of f'. On any interval where f has a tangent with positive (or negative) slope, the graph of f' will be positive (or negative). If the graph of the function is linear, the graph of f' will be a horizontal line.

5.

7.

9.

11.

13. It appears that there are horizontal tangents on the graph of M for $t = 1963$ and $t = 1971$. Thus, there are zeros for those values of t on the graph of M'. The derivative is negative for the years 1963 to 1971.

15.

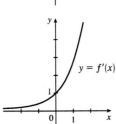

The slope at 0 appears to be 1 and the slope at 1 appears to be 2.7. As x decreases, the slope gets closer to 0. Since the graphs are so similar, we might guess that $f'(x) = e^x$.

17. (a) By zooming in, we estimate that $f'(0) = 0$, $f'\left(\frac{1}{2}\right) = 1$, $f'(1) = 2$, and $f'(2) = 4$.

(b) By symmetry, $f'(-x) = -f'(x)$. So $f'\left(-\frac{1}{2}\right) = -1$, $f'(-1) = -2$, and $f'(-2) = -4$.

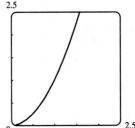

(c) It appears that $f'(x)$ is twice the value of x, so we guess that $f'(x) = 2x$.

(d)
$$\begin{aligned}
f'(x) &= \lim_{h \to 0} \frac{f(x+h) - f(x)}{h} = \lim_{h \to 0} \frac{(x+h)^2 - x^2}{h} \\
&= \lim_{h \to 0} \frac{(x^2 + 2hx + h^2) - x^2}{h} = \lim_{h \to 0} \frac{2hx + h^2}{h} = \lim_{h \to 0} \frac{h(2x+h)}{h} = \lim_{h \to 0}(2x+h) = 2x
\end{aligned}$$

19. $f'(x) = \lim\limits_{h \to 0} \dfrac{f(x+h) - f(x)}{h} = \lim\limits_{h \to 0} \dfrac{\left[\frac{1}{2}(x+h) - \frac{1}{3}\right] - \left(\frac{1}{2}x - \frac{1}{3}\right)}{h} = \lim\limits_{h \to 0} \dfrac{\frac{1}{2}x + \frac{1}{2}h - \frac{1}{3} - \frac{1}{2}x + \frac{1}{3}}{h}$

$= \lim\limits_{h \to 0} \dfrac{\frac{1}{2}h}{h} = \lim\limits_{h \to 0} \frac{1}{2} = \frac{1}{2}$

Domain of f = domain of $f' = \mathbb{R}$.

21. $f'(t) = \lim\limits_{h \to 0} \dfrac{f(t+h) - f(t)}{h} = \lim\limits_{h \to 0} \dfrac{\left[5(t+h) - 9(t+h)^2\right] - (5t - 9t^2)}{h}$

$= \lim\limits_{h \to 0} \dfrac{5t + 5h - 9(t^2 + 2th + h^2) - 5t + 9t^2}{h} = \lim\limits_{h \to 0} \dfrac{5t + 5h - 9t^2 - 18th - 9h^2 - 5t + 9t^2}{h}$

$= \lim\limits_{h \to 0} \dfrac{5h - 18th - 9h^2}{h} = \lim\limits_{h \to 0} \dfrac{h(5 - 18t - 9h)}{h} = \lim\limits_{h \to 0} (5 - 18t - 9h) = 5 - 18t$

Domain of f = domain of $f' = \mathbb{R}$.

23. $f'(x) = \lim\limits_{h \to 0} \dfrac{f(x+h) - f(x)}{h} = \lim\limits_{h \to 0} \dfrac{\left[(x+h)^2 - 2(x+h)^3\right] - (x^2 - 2x^3)}{h}$

$= \lim\limits_{h \to 0} \dfrac{x^2 + 2xh + h^2 - 2x^3 - 6x^2h - 6xh^2 - 2h^3 - x^2 + 2x^3}{h}$

$= \lim\limits_{h \to 0} \dfrac{2xh + h^2 - 6x^2h - 6xh^2 - 2h^3}{h} = \lim\limits_{h \to 0} \dfrac{h(2x + h - 6x^2 - 6xh - 2h^2)}{h}$

$= \lim\limits_{h \to 0} (2x + h - 6x^2 - 6xh - 2h^2) = 2x - 6x^2$

Domain of f = domain of $f' = \mathbb{R}$.

25. $g'(x) = \lim\limits_{h \to 0} \dfrac{g(x+h) - g(x)}{h} = \lim\limits_{h \to 0} \dfrac{\sqrt{1 + 2(x+h)} - \sqrt{1 + 2x}}{h} \left[\dfrac{\sqrt{1 + 2(x+h)} + \sqrt{1 + 2x}}{\sqrt{1 + 2(x+h)} + \sqrt{1 + 2x}}\right]$

$= \lim\limits_{h \to 0} \dfrac{(1 + 2x + 2h) - (1 + 2x)}{h\left[\sqrt{1 + 2(x+h)} + \sqrt{1 + 2x}\right]} = \lim\limits_{h \to 0} \dfrac{2}{\sqrt{1 + 2x + 2h} + \sqrt{1 + 2x}} = \dfrac{2}{2\sqrt{1 + 2x}} = \dfrac{1}{\sqrt{1 + 2x}}$

Domain of $g = \left[-\frac{1}{2}, \infty\right)$, domain of $g' = \left(-\frac{1}{2}, \infty\right)$.

27. $G'(t) = \lim\limits_{h \to 0} \dfrac{G(t+h) - G(t)}{h} = \lim\limits_{h \to 0} \dfrac{\dfrac{4(t+h)}{(t+h) + 1} - \dfrac{4t}{t+1}}{h} = \lim\limits_{h \to 0} \dfrac{\dfrac{4(t+h)(t+1) - 4t(t+h+1)}{(t+h+1)(t+1)}}{h}$

$= \lim\limits_{h \to 0} \dfrac{\left(4t^2 + 4ht + 4t + 4h\right) - \left(4t^2 + 4ht + 4t\right)}{h(t+h+1)(t+1)} = \lim\limits_{h \to 0} \dfrac{4h}{h(t+h+1)(t+1)}$

$= \lim\limits_{h \to 0} \dfrac{4}{(t+h+1)(t+1)} = \dfrac{4}{(t+1)^2}$

Domain of G = domain of $G' = (-\infty, -1) \cup (-1, \infty)$.

29. $f'(x) = \lim\limits_{h \to 0} \dfrac{f(x+h) - f(x)}{h} = \lim\limits_{h \to 0} \dfrac{(x+h)^4 - x^4}{h} = \lim\limits_{h \to 0} \dfrac{\left(x^4 + 4x^3h + 6x^2h^2 + 4xh^3 + h^4\right) - x^4}{h}$

$= \lim\limits_{h \to 0} \dfrac{4x^3h + 6x^2h^2 + 4xh^3 + h^4}{h} = \lim\limits_{h \to 0} \left(4x^3 + 6x^2h + 4xh^2 + h^3\right) = 4x^3$

Domain of f = domain of $f' = \mathbb{R}$.

31. (a) $f'(x) = \lim\limits_{h \to 0} \dfrac{f(x+h) - f(x)}{h} = \lim\limits_{h \to 0} \dfrac{[(x+h)^4 + 2(x+h)] - (x^4 + 2x)}{h}$

$= \lim\limits_{h \to 0} \dfrac{x^4 + 4x^3h + 6x^2h^2 + 4xh^3 + h^4 + 2x + 2h - x^4 - 2x}{h}$

$= \lim\limits_{h \to 0} \dfrac{4x^3h + 6x^2h^2 + 4xh^3 + h^4 + 2h}{h} = \lim\limits_{h \to 0} \dfrac{h(4x^3 + 6x^2h + 4xh^2 + h^3 + 2)}{h}$

$= \lim\limits_{h \to 0} (4x^3 + 6x^2h + 4xh^2 + h^3 + 2) = 4x^3 + 2$

(b) Notice that $f'(x) = 0$ when f has a horizontal tangent, $f'(x)$ is

positive when the tangents have positive slope, and $f'(x)$ is

negative when the tangents have negative slope.

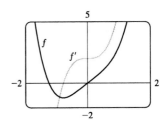

33. (a) $U'(t)$ is the rate at which the unemployment rate is changing with respect to time. Its units are percent per year.

(b) To find $U'(t)$, we use $\lim\limits_{h \to 0} \dfrac{U(t+h) - U(t)}{h} \approx \dfrac{U(t+h) - U(t)}{h}$ for small values of h.

For 1998: $U'(1998) \approx \dfrac{U(1999) - U(1998)}{1999 - 1998} = \dfrac{4.2 - 4.5}{1} = -0.30$

For 1999: We estimate $U'(1999)$ by using $h = -1$ and $h = 1$, and then average the two results to obtain a final estimate.

$h = -1 \quad \Rightarrow \quad U'(1999) \approx \dfrac{U(1998) - U(1999)}{1998 - 1999} = \dfrac{4.5 - 4.2}{-1} = -0.30;$

$h = 1 \quad \Rightarrow \quad U'(1999) \approx \dfrac{U(2000) - U(1999)}{2000 - 1999} = \dfrac{4.0 - 4.2}{1} = -0.20.$

So we estimate that $U'(1999) \approx \frac{1}{2}[(-0.30) + (-0.20)] = -0.25.$

t	1998	1999	2000	2001	2002	2003	2004	2005	2006	2007
$U'(t)$	−0.30	−0.25	0.25	0.90	0.65	−0.15	−0.45	−0.45	−0.25	0.00

35. f is not differentiable at $x = -4$, because the graph has a corner there, and at $x = 0$, because there is a discontinuity there.

37. f is not differentiable at $x = -1$, because the graph has a vertical tangent there, and at $x = 4$, because the graph has a corner there.

39. As we zoom in toward $(-1, 0)$, the curve appears more and more like a straight

line, so $f(x) = x + \sqrt{|x|}$ is differentiable at $x = -1$. But no matter how much

we zoom in toward the origin, the curve doesn't straighten out—we can't

eliminate the sharp point (a cusp). So f is not differentiable at $x = 0$.

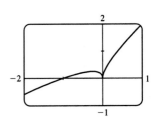

41. $a = f, b = f', c = f''$. We can see this because where a has a horizontal tangent, $b = 0$, and where b has a horizontal tangent, $c = 0$. We can immediately see that c can be neither f nor f', since at the points where c has a horizontal tangent, neither a nor b is equal to 0.

43. We can immediately see that a is the graph of the acceleration function, since at the points where a has a horizontal tangent, neither c nor b is equal to 0. Next, we note that $a = 0$ at the point where b has a horizontal tangent, so b must be the graph of the velocity function, and hence, $b' = a$. We conclude that c is the graph of the position function.

45. $f'(x) = \lim_{h \to 0} \dfrac{f(x+h) - f(x)}{h} = \lim_{h \to 0} \dfrac{[3(x+h)^2 + 2(x+h) + 1] - (3x^2 + 2x + 1)}{h}$

$= \lim_{h \to 0} \dfrac{(3x^2 + 6xh + 3h^2 + 2x + 2h + 1) - (3x^2 + 2x + 1)}{h} = \lim_{h \to 0} \dfrac{6xh + 3h^2 + 2h}{h}$

$= \lim_{h \to 0} \dfrac{h(6x + 3h + 2)}{h} = \lim_{h \to 0} (6x + 3h + 2) = 6x + 2$

$f''(x) = \lim_{h \to 0} \dfrac{f'(x+h) - f'(x)}{h} = \lim_{h \to 0} \dfrac{[6(x+h) + 2] - (6x + 2)}{h} = \lim_{h \to 0} \dfrac{(6x + 6h + 2) - (6x + 2)}{h}$

$= \lim_{h \to 0} \dfrac{6h}{h} = \lim_{h \to 0} 6 = 6$

We see from the graph that our answers are reasonable because the graph of f' is that of a linear function and the graph of f'' is that of a constant function.

47. $f'(x) = \lim_{h \to 0} \dfrac{f(x+h) - f(x)}{h} = \lim_{h \to 0} \dfrac{[2(x+h)^2 - (x+h)^3] - (2x^2 - x^3)}{h}$

$= \lim_{h \to 0} \dfrac{h(4x + 2h - 3x^2 - 3xh - h^2)}{h} = \lim_{h \to 0} (4x + 2h - 3x^2 - 3xh - h^2) = 4x - 3x^2$

$f''(x) = \lim_{h \to 0} \dfrac{f'(x+h) - f'(x)}{h} = \lim_{h \to 0} \dfrac{[4(x+h) - 3(x+h)^2] - (4x - 3x^2)}{h} = \lim_{h \to 0} \dfrac{h(4 - 6x - 3h)}{h}$

$= \lim_{h \to 0} (4 - 6x - 3h) = 4 - 6x$

$f'''(x) = \lim_{h \to 0} \dfrac{f''(x+h) - f''(x)}{h} = \lim_{h \to 0} \dfrac{[4 - 6(x+h)] - (4 - 6x)}{h} = \lim_{h \to 0} \dfrac{-6h}{h} = \lim_{h \to 0} (-6) = -6$

$f^{(4)}(x) = \lim_{h \to 0} \dfrac{f'''(x+h) - f'''(x)}{h} = \lim_{h \to 0} \dfrac{-6 - (-6)}{h} = \lim_{h \to 0} \dfrac{0}{h} = \lim_{h \to 0} (0) = 0$

The graphs are consistent with the geometric interpretations of the derivatives because f' has zeros where f has a local minimum and a local maximum, f'' has a zero where f' has a local maximum, and f''' is a constant function equal to the slope of f''.

49. (a) Note that we have factored $x - a$ as the difference of two cubes in the third step.

$f'(a) = \lim_{x \to a} \dfrac{f(x) - f(a)}{x - a} = \lim_{x \to a} \dfrac{x^{1/3} - a^{1/3}}{x - a} = \lim_{x \to a} \dfrac{x^{1/3} - a^{1/3}}{(x^{1/3} - a^{1/3})(x^{2/3} + x^{1/3}a^{1/3} + a^{2/3})}$

$= \lim_{x \to a} \dfrac{1}{x^{2/3} + x^{1/3}a^{1/3} + a^{2/3}} = \dfrac{1}{3a^{2/3}}$ or $\frac{1}{3}a^{-2/3}$

(b) $f'(0) = \lim\limits_{h \to 0} \dfrac{f(0 + h) - f(0)}{h} = \lim\limits_{h \to 0} \dfrac{\sqrt[3]{h} - 0}{h} = \lim\limits_{h \to 0} \dfrac{1}{h^{2/3}}$. This function increases without bound, so the limit does not

exist, and therefore $f'(0)$ does not exist.

(c) $\lim\limits_{x \to 0} |f'(x)| = \lim\limits_{x \to 0} \dfrac{1}{3x^{2/3}} = \infty$ and f is continuous at $x = 0$ (root function), so f has a vertical tangent at $x = 0$.

51. $f(x) = |x - 6| = \begin{cases} x - 6 & \text{if } x - 6 \geq 6 \\ -(x - 6) & \text{if } x - 6 < 0 \end{cases} = \begin{cases} x - 6 & \text{if } x \geq 6 \\ 6 - x & \text{if } x < 6 \end{cases}$

So the right-hand limit is $\lim\limits_{x \to 6^+} \dfrac{f(x) - f(6)}{x - 6} = \lim\limits_{x \to 6^+} \dfrac{|x - 6| - 0}{x - 6} = \lim\limits_{x \to 6^+} \dfrac{x - 6}{x - 6} = \lim\limits_{x \to 6^+} 1 = 1$, and the left-hand limit

is $\lim\limits_{x \to 6^-} \dfrac{f(x) - f(6)}{x - 6} = \lim\limits_{x \to 6^-} \dfrac{|x - 6| - 0}{x - 6} = \lim\limits_{x \to 6^-} \dfrac{6 - x}{x - 6} = \lim\limits_{x \to 6^-} (-1) = -1$. Since these limits are not equal,

$f'(6) = \lim\limits_{x \to 6} \dfrac{f(x) - f(6)}{x - 6}$ does not exist and f is not differentiable at 6.

However, a formula for f' is $f'(x) = \begin{cases} 1 & \text{if } x > 6 \\ -1 & \text{if } x < 6 \end{cases}$

Another way of writing the formula is $f'(x) = \dfrac{x - 6}{|x - 6|}$.

53. (a) If f is even, then

$$f'(-x) = \lim\limits_{h \to 0} \dfrac{f(-x + h) - f(-x)}{h} = \lim\limits_{h \to 0} \dfrac{f[-(x - h)] - f(-x)}{h}$$

$$= \lim\limits_{h \to 0} \dfrac{f(x - h) - f(x)}{h} = -\lim\limits_{h \to 0} \dfrac{f(x - h) - f(x)}{-h} \qquad \text{[let } \Delta x = -h\text{]}$$

$$= -\lim\limits_{\Delta x \to 0} \dfrac{f(x + \Delta x) - f(x)}{\Delta x} = -f'(x)$$

Therefore, f' is odd.

(b) If f is odd, then

$$f'(-x) = \lim\limits_{h \to 0} \dfrac{f(-x + h) - f(-x)}{h} = \lim\limits_{h \to 0} \dfrac{f[-(x - h)] - f(-x)}{h}$$

$$= \lim\limits_{h \to 0} \dfrac{-f(x - h) + f(x)}{h} = \lim\limits_{h \to 0} \dfrac{f(x - h) - f(x)}{-h} \qquad \text{[let } \Delta x = -h\text{]}$$

$$= \lim\limits_{\Delta x \to 0} \dfrac{f(x + \Delta x) - f(x)}{\Delta x} = f'(x)$$

Therefore, f' is even.

55.

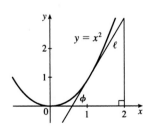

In the right triangle in the diagram, let Δy be the side opposite angle ϕ and Δx the side adjacent angle ϕ. Then the slope of the tangent line ℓ is $m = \Delta y / \Delta x = \tan \phi$. Note that $0 < \phi < \frac{\pi}{2}$. We know (see Exercise 17) that the derivative of $f(x) = x^2$ is $f'(x) = 2x$. So the slope of the tangent to the curve at the point $(1, 1)$ is 2. Thus, ϕ is the angle between 0 and $\frac{\pi}{2}$ whose tangent is 2; that is, $\phi = \tan^{-1} 2 \approx 63°$.

2.8 What Does f' Say about f?

1. (a) Since $f'(x) < 0$ on $(1, 4)$, f is decreasing on this interval. Since $f'(x) > 0$ on $(0, 1)$ and $(4, 5)$, f is increasing on these

   intervals.

(b) At $x = 1$, $f'(x) = 0$ and f' changes from positive to negative there, f changes from increasing to decreasing and has a

   local maximum at $x = 1$. At $x = 4$, $f'(x) = 0$ and f' changes from negative to positive there, f changes from decreasing

   to increasing and has local minimum at $x = 4$.

(c) Since $f(0) = 0$, start at the origin. Draw an increasing function on $(0, 1)$

   with a local maximum at $x = 1$. Now draw a decreasing function on $(1, 4)$

   and the steepest slope should occur at $x = 2.5$ since that's where the

   smallest value of f' occurs. Last, draw an increasing function on $(4, 5)$

   making sure you have a local minimum at $x = 4$.

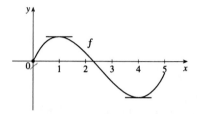

3. (a) $f'(x) > 0$ and f is increasing on $(-2, -1)$, $(0, 1)$, and $(2, 3)$. $f'(x) < 0$

   and f is decreasing on $(-1, 0)$ and $(1, 2)$.

(b) At $x = -1$ and $x = 1$, $f'(x) = 0$ and f' changes from positive to negative,

   so f has local maxima at $x = -1$ and $x = 1$. At $x = 0$ and $x = 2$,

   $f'(x) = 0$ and f' changes from negative to positive, so f has local minima

   at $x = 0$ and $x = 2$.

(c)

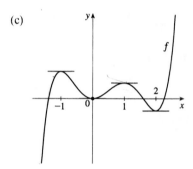

5. The derivative f' is increasing when the slopes of the tangent lines of f are becoming larger as x increases. This seems to be

the case on the interval $(2, 5)$. The derivative is decreasing when the slopes of the tangent lines of f are becoming smaller as x

increases, and this seems to be the case on $(-\infty, 2)$ and $(5, \infty)$. So f' is increasing on $(2, 5)$ and decreasing on $(-\infty, 2)$

and $(5, \infty)$.

7. Call the curve with the positive y-intercept g and the other curve h. Notice that g has a maximum (horizontal tangent) at

$x = 0$, but $h \neq 0$, so h cannot be the derivative of g. Also notice that where g is positive, h is increasing. Thus, $h = f$ and

$g = f'$. Now $f'(-1)$ is negative since f' is below the x-axis there and $f''(1)$ is positive since f is concave upward at $x = 1$.

Therefore, $f''(1)$ is greater than $f'(-1)$.

9. If $D(t)$ is the size of the deficit as a function of time, then at the time of the speech $D'(t) > 0$, but $D''(t) < 0$ because

$D''(t) = (D')'(t)$ is the rate of change of $D'(t)$.

11. (a) The rate of increase of the population is initially very small, then increases rapidly until about 1932 when it starts decreasing. The rate becomes negative by 1936, peaks in magnitude in 1937, and approaches 0 in 1940.

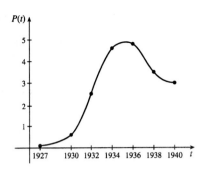

(b) Inflection points (IP) appear to be at $(1932, 2.5)$ and $(1937, 4.3)$. The rate of change of population density starts to decrease in 1932 and starts to increase in 1937. The rates of population increase and decrease have their maximum values at those points.

13. Most students learn more in the third hour of studying than in the eighth hour, so $K(3) - K(2)$ is larger than $K(8) - K(7)$. In other words, as you begin studying for a test, the rate of knowledge gain is large and then starts to taper off, so $K'(t)$ decreases and the graph of K is concave downward.

15. (a) f is increasing where f' is positive, that is, on $(0, 2)$, $(4, 6)$, and $(8, \infty)$; and decreasing where f' is negative, that is, on $(2, 4)$ and $(6, 8)$.

(b) f has local maxima where f' changes from positive to negative, at $x = 2$ and at $x = 6$, and local minima where f' changes from negative to positive, at $x = 4$ and at $x = 8$.

(c) f is concave upward (CU) where f' is increasing, that is, on $(3, 6)$ and $(6, \infty)$, and concave downward (CD) where f' is decreasing, that is, on $(0, 3)$.

(d) There is a point of inflection where f changes from being CD to being CU, that is, at $x = 3$.

(e)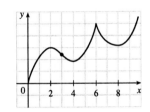

17. The function must be always decreasing (since the first derivative is always negative) and concave downward (since the second derivative is always negative).

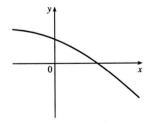

19. $f'(0) = f'(4) = 0 \Rightarrow$ horizontal tangents at $x = 0, 4$.

$f'(x) > 0$ if $x < 0 \Rightarrow$ f is increasing on $(-\infty, 0)$.

$f'(x) < 0$ if $0 < x < 4$ or if $x > 4 \Rightarrow$ f is decreasing on $(0, 4)$ and $(4, \infty)$.

$f''(x) > 0$ if $2 < x < 4 \Rightarrow$ f is concave upward on $(2, 4)$.

$f''(x) < 0$ if $x < 2$ or $x > 4 \Rightarrow$ f is concave downward on $(-\infty, 2)$

and $(4, \infty)$. There are inflection points when $x = 2$ and 4.

21. $f'(0) = f'(2) = f'(4) = 0 \Rightarrow$ horizontal tangents at $x = 0$, 2, 4.

$f'(x) > 0$ if $x < 0$ or $2 < x < 4 \Rightarrow$ f is increasing on $(-\infty, 0)$ and $(2, 4)$.

$f'(x) < 0$ if $0 < x < 2$ or $x > 4 \Rightarrow$ f is decreasing on $(0, 2)$ and $(4, \infty)$.

$f''(x) > 0$ if $1 < x < 3 \Rightarrow$ f is concave upward on $(1, 3)$.

$f''(x) < 0$ if $x < 1$ or $x > 3 \Rightarrow$ f is concave downward on $(-\infty, 1)$

and $(3, \infty)$. There are inflection points when $x = 1$ and 3.

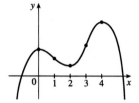

23. $f'(x) > 0$ if $|x| < 2 \Rightarrow$ f is increasing on $(-2, 2)$.

$f'(x) < 0$ if $|x| > 2 \Rightarrow$ f is decreasing on $(-\infty, -2)$

and $(2, \infty)$. $f'(-2) = 0 \Rightarrow$ horizontal tangent at

$x = -2$. $\lim_{x \to 2} |f'(x)| = \infty \Rightarrow$ there is a vertical

 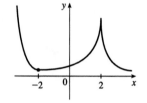

asymptote or vertical tangent (cusp) at $x = 2$. $f''(x) > 0$ if $x \ne 2 \Rightarrow$ f is concave upward on $(-\infty, 2)$ and $(2, \infty)$.

25. (a) Since e^{-x^2} is positive for all x, $f'(x) = xe^{-x^2}$ is positive where $x > 0$ and negative where $x < 0$. Thus, f is increasing

on $(0, \infty)$ and decreasing on $(-\infty, 0)$.

(b) Since f changes from decreasing to increasing at $x = 0$, f has a minimum value there.

27. (a) To find the intervals on which f is increasing, we need to find the intervals on which $f'(x) = 3x^2 - 1$ is positive.

$3x^2 - 1 > 0 \Leftrightarrow 3x^2 > 1 \Leftrightarrow x^2 > \frac{1}{3} \Leftrightarrow |x| > \sqrt{\frac{1}{3}}$, so $x \in \left(-\infty, -\sqrt{\frac{1}{3}}\right) \cup \left(\sqrt{\frac{1}{3}}, \infty\right)$. Thus, f is

increasing on $\left(-\infty, -\sqrt{\frac{1}{3}}\right)$ and on $\left(\sqrt{\frac{1}{3}}, \infty\right)$. In a similar fashion, f is decreasing on $\left(-\sqrt{\frac{1}{3}}, \sqrt{\frac{1}{3}}\right)$.

(b) To find the intervals on which f is concave upward, we need to find the intervals on which $f''(x) = 6x$ is positive.

$6x > 0 \Leftrightarrow x > 0$. So f is concave upward on $(0, \infty)$ and f is concave downward on $(-\infty, 0)$.

(c) There is an inflection point at $(0, 0)$ since f changes its direction of concavity at $x = 0$.

29. b is the antiderivative of f. For small x, f is negative, so the graph of its antiderivative must be decreasing. But both a and c

are increasing for small x, so only b can be f's antiderivative. Also, f is positive where b is increasing, which supports our

conclusion.

31.

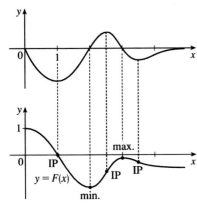

The graph of F must start at $(0, 1)$. Where the given graph, $y = f(x)$, has a local minimum or maximum, the graph of F will have an inflection point.

Where f is negative (positive), F is decreasing (increasing).

Where f changes from negative to positive, F will have a minimum.

Where f changes from positive to negative, F will have a maximum.

Where f is decreasing (increasing), F is concave downward (upward).

33. $f(x) = \dfrac{\sin x}{1 + x^2}, \quad -2\pi \le x \le 2\pi$

Note that the graph of f is one of an odd function, so the graph of F will be one of an even function.

2 Review

CONCEPT CHECK

1. (a) $\lim\limits_{x \to a} f(x) = L$: See Definition 2.2.1 and Figures 1 and 2 in Section 2.2.

 (b) $\lim\limits_{x \to a^+} f(x) = L$: See the paragraph after Definition 2.2.2 and Figure 9(b) in Section 2.2.

 (c) $\lim\limits_{x \to a^-} f(x) = L$: See Definition 2.2.2 and Figure 9(a) in Section 2.2.

 (d) $\lim\limits_{x \to a} f(x) = \infty$: See Definition 2.5.1 and Figure 2 in Section 2.5.

 (e) $\lim\limits_{x \to \infty} f(x) = L$: See Definition 2.5.4 and Figure 9 in Section 2.5.

2. In general, the limit of a function fails to exist when the function does not approach a fixed number. For each of the following functions, the limit fails to exist at $x = 2$.

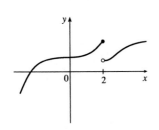

The left- and right-hand limits are not equal.

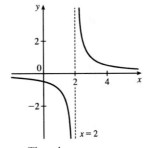

There is an infinite discontinuity.

There are an infinite number of oscillations.

3. (a)–(g) See the statements of Limit Laws 1–6 and 11 in Section 2.3.

4. See Theorem 3 in Section 2.3.

5. (a) See Definition 2.5.2 and Figures 2–4 in Section 2.5.

(b) See Definition 2.5.5 and Figures 9 and 10 in Section 2.5.

6. (a) $y = x^4$: No asymptote

(b) $y = \sin x$: No asymptote

(c) $y = \tan x$: Vertical asymptotes $x = \frac{\pi}{2} + \pi n$, n an integer

(d) $y = e^x$: Horizontal asymptote $y = 0$ $\left(\lim\limits_{x \to -\infty} e^x = 0 \right)$

(e) $y = \ln x$: Vertical asymptote $x = 0$ $\left(\lim\limits_{x \to 0^+} \ln x = -\infty \right)$

(f) $y = 1/x$: Vertical asymptote $x = 0$, horizontal asymptote $y = 0$

(g) $y = \sqrt{x}$: No asymptote

7. (a) A function f is continuous at a number a if $f(x)$ approaches $f(a)$ as x approaches a; that is, $\lim\limits_{x \to a} f(x) = f(a)$.

(b) A function f is continuous on the interval $(-\infty, \infty)$ if f is continuous at every real number a. The graph of such a function has no breaks and every vertical line crosses it.

8. See Theorem 2.4.10.

9. See Definition 2.6.1.

10. See the paragraph containing Formula 3 in Section 2.6.

11. (a) The average rate of change of y with respect to x over the interval $[x_1, x_2]$ is $\dfrac{f(x_2) - f(x_1)}{x_2 - x_1}$.

(b) The instantaneous rate of change of y with respect to x at $x = x_1$ is $\lim\limits_{x_2 \to x_1} \dfrac{f(x_2) - f(x_1)}{x_2 - x_1}$.

12. See Definition 2.7.2. The pages following the definition discuss interpretations of $f'(a)$ as the slope of a tangent line to the graph of f at $x = a$ and as an instantaneous rate of change of $f(x)$ with respect to x when $x = a$.

13. See the paragraphs before and after Example 7 in Section 2.7.

14. (a) A function f is differentiable at a number a if its derivative f' exists at $x = a$; that is, if $f'(a)$ exists.

(b) See Theorem 2.7.4. This theorem also tells us that if f is *not* continuous at a, then f is *not* differentiable at a.

(c)

15. See the discussion and Figure 8 on page 152.

16. (a) See the first box in Section 2.8.

(b) See the second box in Section 2.8.

17. (a) An antiderivative of a function f is a function F such that $F' = f$.

(b) The antiderivative of a velocity function is a position function (the derivative of a position function is a velocity function). The antiderivative of an acceleration function is a velocity function (the derivative of a velocity function is an acceleration function).

TRUE-FALSE QUIZ

1. False. Limit Law 2 applies only if the individual limits exist (these don't).

3. True. Limit Law 5 applies.

5. False. Consider $\lim\limits_{x \to 5} \dfrac{x(x-5)}{x-5}$ or $\lim\limits_{x \to 5} \dfrac{\sin(x-5)}{x-5}$. The first limit exists and is equal to 5. By Example 3 in Section 2.2, we know that the latter limit exists (and it is equal to 1).

7. True. A polynomial is continuous everywhere, so $\lim\limits_{x \to b} p(x)$ exists and is equal to $p(b)$.

9. True. See Figure 11 in Section 2.5.

11. False. Consider $f(x) = \begin{cases} 1/(x-1) & \text{if } x \neq 1 \\ 2 & \text{if } x = 1 \end{cases}$

13. True. Use Theorem 2.4.8 with $a = 2$, $b = 5$, and $g(x) = 4x^2 - 11$. Note that $f(4) = 3$ is not needed.

15. False. See the note after Theorem 4 in Section 2.7.

17. False. $\dfrac{d^2y}{dx^2}$ is the second derivative while $\left(\dfrac{dy}{dx}\right)^2$ is the first derivative squared. For example, if $y = x$, then $\dfrac{d^2y}{dx^2} = 0$, but $\left(\dfrac{dy}{dx}\right)^2 = 1$.

EXERCISES

1. (a) (i) $\lim\limits_{x \to 2^+} f(x) = 3$ (ii) $\lim\limits_{x \to -3^+} f(x) = 0$

(iii) $\lim\limits_{x \to -3} f(x)$ does not exist since the left and right limits are not equal. (The left limit is -2.)

(iv) $\lim\limits_{x \to 4} f(x) = 2$

(v) $\lim\limits_{x \to 0} f(x) = \infty$ (vi) $\lim\limits_{x \to 2^-} f(x) = -\infty$

(vii) $\lim\limits_{x \to \infty} f(x) = 4$ (viii) $\lim\limits_{x \to -\infty} f(x) = -1$

(b) The equations of the horizontal asymptotes are $y = -1$ and $y = 4$.

(c) The equations of the vertical asymptotes are $x = 0$ and $x = 2$.

(d) f is discontinuous at $x = -3, 0, 2$, and 4. The discontinuities are jump, infinite, infinite, and removable, respectively.

3. Since the exponential function is continuous, $\lim\limits_{x \to 1} e^{x^3 - x} = e^{1-1} = e^0 = 1$.

5. $\lim\limits_{x \to -3} \dfrac{x^2 - 9}{x^2 + 2x - 3} = \lim\limits_{x \to -3} \dfrac{(x+3)(x-3)}{(x+3)(x-1)} = \lim\limits_{x \to -3} \dfrac{x-3}{x-1} = \dfrac{-3-3}{-3-1} = \dfrac{-6}{-4} = \dfrac{3}{2}$

7. $\lim\limits_{h \to 0} \dfrac{(h-1)^3 + 1}{h} = \lim\limits_{h \to 0} \dfrac{(h^3 - 3h^2 + 3h - 1) + 1}{h} = \lim\limits_{h \to 0} \dfrac{h^3 - 3h^2 + 3h}{h} = \lim\limits_{h \to 0} (h^2 - 3h + 3) = 3$

Another solution: Factor the numerator as a sum of two cubes and then simplify.

$$\lim\limits_{h \to 0} \dfrac{(h-1)^3 + 1}{h} = \lim\limits_{h \to 0} \dfrac{(h-1)^3 + 1^3}{h} = \lim\limits_{h \to 0} \dfrac{[(h-1) + 1]\left[(h-1)^2 - 1(h-1) + 1^2\right]}{h}$$
$$= \lim\limits_{h \to 0} \left[(h-1)^2 - h + 2\right] = 1 - 0 + 2 = 3$$

9. $\lim\limits_{r \to 9} \dfrac{\sqrt{r}}{(r-9)^4} = \infty$ since $(r-9)^4 \to 0$ as $r \to 9$ and $\dfrac{\sqrt{r}}{(r-9)^4} > 0$ for $r \neq 9$.

11. $\lim\limits_{u \to 1} \dfrac{u^4 - 1}{u^3 + 5u^2 - 6u} = \lim\limits_{u \to 1} \dfrac{(u^2 + 1)(u^2 - 1)}{u(u^2 + 5u - 6)} = \lim\limits_{u \to 1} \dfrac{(u^2 + 1)(u+1)(u-1)}{u(u+6)(u-1)} = \lim\limits_{u \to 1} \dfrac{(u^2 + 1)(u+1)}{u(u+6)} = \dfrac{2(2)}{1(7)} = \dfrac{4}{7}$

13. Let $t = \sin x$. Then as $x \to \pi^-$, $\sin x \to 0^+$, so $t \to 0^+$. Thus, $\lim\limits_{x \to \pi^-} \ln(\sin x) = \lim\limits_{t \to 0^+} \ln t = -\infty$.

15. Since x is positive, $\sqrt{x^2} = |x| = x$. Thus,

$$\lim\limits_{x \to \infty} \dfrac{\sqrt{x^2 - 9}}{2x - 6} = \lim\limits_{x \to \infty} \dfrac{\sqrt{x^2 - 9}/\sqrt{x^2}}{(2x - 6)/x} = \lim\limits_{x \to \infty} \dfrac{\sqrt{1 - 9/x^2}}{2 - 6/x} = \dfrac{\sqrt{1-0}}{2-0} = \dfrac{1}{2}$$

17. $\lim\limits_{x \to \infty} \left(\sqrt{x^2 + 4x + 1} - x\right) = \lim\limits_{x \to \infty} \left[\dfrac{\sqrt{x^2 + 4x + 1} - x}{1} \cdot \dfrac{\sqrt{x^2 + 4x + 1} + x}{\sqrt{x^2 + 4x + 1} + x}\right] = \lim\limits_{x \to \infty} \dfrac{(x^2 + 4x + 1) - x^2}{\sqrt{x^2 + 4x + 1} + x}$

$\qquad = \lim\limits_{x \to \infty} \dfrac{(4x + 1)/x}{\left(\sqrt{x^2 + 4x + 1} + x\right)/x} \qquad \left[\text{divide by } x = \sqrt{x^2} \text{ for } x > 0\right]$

$\qquad = \lim\limits_{x \to \infty} \dfrac{4 + 1/x}{\sqrt{1 + 4/x + 1/x^2} + 1} = \dfrac{4 + 0}{\sqrt{1 + 0 + 0} + 1} = \dfrac{4}{2} = 2$

19. From the graph of $y = (\cos^2 x)/x^2$, it appears that $y = 0$ is the horizontal

asymptote and $x = 0$ is the vertical asymptote. Now $0 \leq (\cos x)^2 \leq 1 \quad \Rightarrow$

$\dfrac{0}{x^2} \leq \dfrac{\cos^2 x}{x^2} \leq \dfrac{1}{x^2} \quad \Rightarrow \quad 0 \leq \dfrac{\cos^2 x}{x^2} \leq \dfrac{1}{x^2}$. But $\lim\limits_{x \to \pm\infty} 0 = 0$ and

$\lim\limits_{x \to \pm\infty} \dfrac{1}{x^2} = 0$, so by the Squeeze Theorem, $\lim\limits_{x \to \pm\infty} \dfrac{\cos^2 x}{x^2} = 0$.

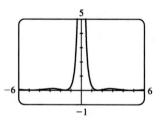

Thus, $y = 0$ is the horizontal asymptote. $\lim\limits_{x \to 0} \dfrac{\cos^2 x}{x^2} = \infty$ because $\cos^2 x \to 1$ and $x^2 \to 0$ as $x \to 0$, so $x = 0$ is the

vertical asymptote.

21. Since $2x - 1 \leq f(x) \leq x^2$ for $0 < x < 3$ and $\lim\limits_{x \to 1} (2x - 1) = 1 = \lim\limits_{x \to 1} x^2$, we have $\lim\limits_{x \to 1} f(x) = 1$ by the Squeeze Theorem.

23. (a) $f(x) = \sqrt{-x}$ if $x < 0$, $f(x) = 3 - x$ if $0 \le x < 3$, $f(x) = (x-3)^2$ if $x > 3$.

(i) $\lim\limits_{x \to 0^+} f(x) = \lim\limits_{x \to 0^+} (3 - x) = 3$

(ii) $\lim\limits_{x \to 0^-} f(x) = \lim\limits_{x \to 0^-} \sqrt{-x} = 0$

(iii) Because of (i) and (ii), $\lim\limits_{x \to 0} f(x)$ does not exist.

(iv) $\lim\limits_{x \to 3^-} f(x) = \lim\limits_{x \to 3^-} (3 - x) = 0$

(v) $\lim\limits_{x \to 3^+} f(x) = \lim\limits_{x \to 3^+} (x-3)^2 = 0$

(vi) Because of (iv) and (v), $\lim\limits_{x \to 3} f(x) = 0$.

(b) f is discontinuous at 0 since $\lim\limits_{x \to 0} f(x)$ does not exist.

f is discontinuous at 3 since $f(3)$ does not exist.

(c)

25. $f(x) = 2x^3 + x^2 + 2$ is a polynomial, so it is continuous on $[-2, -1]$ and $f(-2) = -10 < 0 < 1 = f(-1)$. So by the Intermediate Value Theorem there is a number c in $(-2, -1)$ such that $f(c) = 0$, that is, the equation $2x^3 + x^2 + 2 = 0$ has a root in $(-2, -1)$.

27. (a) $s = s(t) = 1 + 2t + t^2/4$. The average velocity over the time interval $[1, 1+h]$ is

$$v_{\text{ave}} = \frac{s(1+h) - s(1)}{(1+h) - 1} = \frac{1 + 2(1+h) + (1+h)^2/4 - 13/4}{h} = \frac{10h + h^2}{4h} = \frac{10 + h}{4}$$

So for the following intervals the average velocities are:

(i) $[1, 3]$: $h = 2$, $v_{\text{ave}} = (10 + 2)/4 = 3$ m/s

(ii) $[1, 2]$: $h = 1$, $v_{\text{ave}} = (10 + 1)/4 = 2.75$ m/s

(iii) $[1, 1.5]$: $h = 0.5$, $v_{\text{ave}} = (10 + 0.5)/4 = 2.625$ m/s

(iv) $[1, 1.1]$: $h = 0.1$, $v_{\text{ave}} = (10 + 0.1)/4 = 2.525$ m/s

(b) When $t = 1$, the instantaneous velocity is $\lim\limits_{h \to 0} \dfrac{s(1+h) - s(1)}{h} = \lim\limits_{h \to 0} \dfrac{10 + h}{4} = \dfrac{10}{4} = 2.5$ m/s.

29. Estimating the slopes of the tangent lines at $x = 2$, 3, and 5, we obtain approximate values 0.4, 2, and 0.1. Since the graph is concave downward at $x = 5$, $f''(5)$ is negative. Arranging the numbers in increasing order, we have:

$$f''(5) < 0 < f'(5) < f'(2) < 1 < f'(3).$$

31. (a) Estimating $f'(1)$ from the triangle in the graph,

we get $\dfrac{\Delta y}{\Delta x} \approx \dfrac{-0.37}{0.50} = -0.74$.

To estimate $f'(1)$ numerically, we have

$$f'(1) = \lim_{h \to 0} \frac{f(1+h) - f(1)}{h} = \lim_{h \to 0} \frac{e^{-(1+h)^2} - e^{-1}}{h} = y$$

From the table, we have $f'(1) \approx -0.736$.

(b) $y - e^{-1} \approx -0.736(x-1)$ or $y \approx -0.736x + 1.104$

(c) See the graph in part (a).

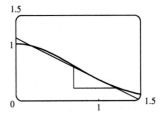

h	y
0.01	-0.732
0.001	-0.735
0.0001	-0.736
-0.01	-0.739
-0.001	-0.736
-0.0001	-0.736

33. (a) $f'(r)$ is the rate at which the total cost changes with respect to the interest rate. Its units are dollars/(percent per year).

(b) The total cost of paying off the loan is increasing by $1200/(percent per year) as the interest rate reaches 10%. So if the interest rate goes up from 10% to 11%, the cost goes up approximately $1200.

(c) As r increases, C increases. So $f'(r)$ will always be positive.

35.

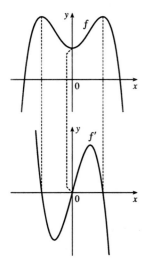

37. (a) $f'(x) = \lim\limits_{h \to 0} \dfrac{f(x+h) - f(x)}{h} = \lim\limits_{h \to 0} \dfrac{\sqrt{3 - 5(x+h)} - \sqrt{3 - 5x}}{h} \cdot \dfrac{\sqrt{3 - 5(x+h)} + \sqrt{3 - 5x}}{\sqrt{3 - 5(x+h)} + \sqrt{3 - 5x}}$

$= \lim\limits_{h \to 0} \dfrac{[3 - 5(x+h)] - (3 - 5x)}{h\left(\sqrt{3 - 5(x+h)} + \sqrt{3 - 5x}\right)} = \lim\limits_{h \to 0} \dfrac{-5}{\sqrt{3 - 5(x+h)} + \sqrt{3 - 5x}} = \dfrac{-5}{2\sqrt{3 - 5x}}$

(b) Domain of f: (the radicand must be nonnegative) $3 - 5x \geq 0 \quad \Rightarrow$

$5x \leq 3 \quad \Rightarrow \quad x \in \left(-\infty, \frac{3}{5}\right]$

Domain of f': exclude $\frac{3}{5}$ because it makes the denominator zero;

$x \in \left(-\infty, \frac{3}{5}\right)$

(c) Our answer to part (a) is reasonable because $f'(x)$ is always negative and f is always decreasing.

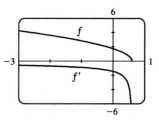

39. f is not differentiable: at $x = -4$ because f is not continuous, at $x = -1$ because f has a corner, at $x = 2$ because f is not continuous, and at $x = 5$ because f has a vertical tangent.

41. $C'(1990)$ is the rate at which the total value of US currency in circulation is changing in billions of dollars per year. To estimate the value of $C'(1990)$, we will average the difference quotients obtained using the times $t = 1985$ and $t = 1995$.

Let $A = \dfrac{C(1985) - C(1990)}{1985 - 1990} = \dfrac{187.3 - 271.9}{-5} = \dfrac{-84.6}{-5} = 16.92$ and

$B = \dfrac{C(1995) - C(1990)}{1995 - 1990} = \dfrac{409.3 - 271.9}{5} = \dfrac{137.4}{5} = 27.48$. Then

$C'(1990) = \lim\limits_{t \to 1990} \dfrac{C(t) - C(1990)}{t - 1990} \approx \dfrac{A + B}{2} = \dfrac{16.92 + 27.48}{2} = \dfrac{44.4}{2} = 22.2$ billion dollars/year.

43. (a) $f'(x) > 0$ on $(-2, 0)$ and $(2, \infty)$ $\Rightarrow$ f is increasing on those intervals. $f'(x) < 0$ on $(-\infty, -2)$ and $(0, 2)$ $\Rightarrow$ f is decreasing on those intervals.

(b) $f'(x) = 0$ at $x = -2, 0$, and 2, so these are where local maxima or minima will occur. At $x = \pm 2$, f' changes from negative to positive, so f has local minima at those values. At $x = 0$, f' changes from positive to negative, so f has a local maximum there.

(c) f' is increasing on $(-\infty, -1)$ and $(1, \infty)$ $\Rightarrow$ (d)

$f'' > 0$ and f is concave upward on those intervals.

f' is decreasing on $(-1, 1)$ $\Rightarrow$ $f'' < 0$ and

f is concave downward on this interval.

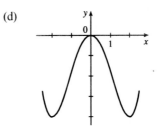

45. $f(0) = 0$, $f'(-2) = f'(1) = f'(9) = 0$, $\lim\limits_{x \to \infty} f(x) = 0$, $\lim\limits_{x \to 6} f(x) = -\infty$,

$f'(x) < 0$ on $(-\infty, -2)$, $(1, 6)$, and $(9, \infty)$, $f'(x) > 0$ on $(-2, 1)$ and $(6, 9)$,

$f''(x) > 0$ on $(-\infty, 0)$ and $(12, \infty)$, $f''(x) < 0$ on $(0, 6)$ and $(6, 12)$

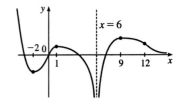

47. (a) Using the data closest to $t = 6$, we have $\dfrac{s(8) - s(6)}{8 - 6} = \dfrac{180 - 95}{2} = 42.5$

and $\dfrac{s(4) - s(6)}{4 - 6} = \dfrac{40 - 95}{-2} = 27.5$. Averaging these two values gives us

$\dfrac{42.5 + 27.5}{2} = 35$ ft/s as an estimate for the speed of the car after

6 seconds.

(b) From the graph, it appears that the inflection point is at $(8, 180)$.

(c) The velocity of the car is at a maximum at the inflection point.

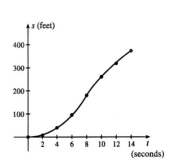

☐ FOCUS ON PROBLEM SOLVING

1. Let $t = \sqrt[6]{x}$, so $x = t^6$. Then $t \to 1$ as $x \to 1$, so

$$\lim_{x \to 1} \frac{\sqrt[3]{x} - 1}{\sqrt{x} - 1} = \lim_{t \to 1} \frac{t^2 - 1}{t^3 - 1} = \lim_{t \to 1} \frac{(t-1)(t+1)}{(t-1)\left(t^2 + t + 1\right)} = \lim_{t \to 1} \frac{t+1}{t^2 + t + 1} = \frac{1+1}{1^2 + 1 + 1} = \frac{2}{3}.$$

Another method: Multiply both the numerator and the denominator by $(\sqrt{x} + 1)\left(\sqrt[3]{x^2} + \sqrt[3]{x} + 1\right)$.

3. For $-\frac{1}{2} < x < \frac{1}{2}$, we have $2x - 1 < 0$ and $2x + 1 > 0$, so $|2x - 1| = -(2x - 1)$ and $|2x + 1| = 2x + 1$.

Therefore, $\lim_{x \to 0} \dfrac{|2x - 1| - |2x + 1|}{x} = \lim_{x \to 0} \dfrac{-(2x - 1) - (2x + 1)}{x} = \lim_{x \to 0} \dfrac{-4x}{x} = \lim_{x \to 0} (-4) = -4.$

5. Since $[\![x]\!] \leq x < [\![x]\!] + 1$, we have $\dfrac{[\![x]\!]}{[\![x]\!]} \leq \dfrac{x}{[\![x]\!]} < \dfrac{[\![x]\!] + 1}{[\![x]\!]} \quad \Rightarrow \quad 1 \leq \dfrac{x}{[\![x]\!]} < 1 + \dfrac{1}{[\![x]\!]}$ for $x \geq 1$. As $x \to \infty$, $[\![x]\!] \to \infty$,

so $\dfrac{1}{[\![x]\!]} \to 0$ and $1 + \dfrac{1}{[\![x]\!]} \to 1$. Thus, $\lim_{x \to \infty} \dfrac{x}{[\![x]\!]} = 1$ by the Squeeze Theorem.

7. f is continuous on $(-\infty, a)$ and (a, ∞). To make f continuous on $\mathbb{R}$, we must have continuity at a. Thus,

$$\lim_{x \to a^+} f(x) = \lim_{x \to a^-} f(x) \quad \Rightarrow \quad \lim_{x \to a^+} x^2 = \lim_{x \to a^-} (x + 1) \quad \Rightarrow \quad a^2 = a + 1 \quad \Rightarrow \quad a^2 - a - 1 = 0 \quad \Rightarrow$$

[by the quadratic formula] $a = \left(1 \pm \sqrt{5}\right)/2 \approx 1.618$ or -0.618.

9. (a) Consider $G(x) = T(x + 180°) - T(x)$. Fix any number a. If $G(a) = 0$, we are done: Temperature at $a =$ Temperature

at $a + 180°$. If $G(a) > 0$, then $G(a + 180°) = T(a + 360°) - T(a + 180°) = T(a) - T(a + 180°) = -G(a) < 0$.

Also, G is continuous since temperature varies continuously. So, by the Intermediate Value Theorem, G has a zero on the

interval $[a, a + 180°]$. If $G(a) < 0$, then a similar argument applies.

(b) Yes. The same argument applies.

(c) The same argument applies for quantities that vary continuously, such as barometric pressure. But one could argue that

altitude above sea level is sometimes discontinuous, so the result might not always hold for that quantity.

11. Let a be the x-coordinate of Q. Since the derivative of $y = 1 - x^2$ is $y' = -2x$, the slope at Q is $-2a$. But since the triangle

is equilateral, $\overline{AO}/\overline{OC} = \sqrt{3}/1$, so the slope at Q is $-\sqrt{3}$. Therefore, we must have that $-2a = -\sqrt{3} \quad \Rightarrow \quad a = \frac{\sqrt{3}}{2}$.

Thus, the point Q has coordinates $\left(\frac{\sqrt{3}}{2}, 1 - \left(\frac{\sqrt{3}}{2}\right)^2\right) = \left(\frac{\sqrt{3}}{2}, \frac{1}{4}\right)$ and by symmetry, P has coordinates $\left(-\frac{\sqrt{3}}{2}, \frac{1}{4}\right)$.

13. (a) Put $x = 0$ and $y = 0$ in the equation: $f(0 + 0) = f(0) + f(0) + 0^2 \cdot 0 + 0 \cdot 0^2 \quad \Rightarrow \quad f(0) = 2f(0)$.

Subtracting $f(0)$ from each side of this equation gives $f(0) = 0$.

(b) $f'(0) = \lim\limits_{h \to 0} \dfrac{f(0+h) - f(0)}{h} = \lim\limits_{h \to 0} \dfrac{\left[f(0) + f(h) + 0^2 h + 0h^2\right] - f(0)}{h} = \lim\limits_{h \to 0} \dfrac{f(h)}{h} = \lim\limits_{x \to 0} \dfrac{f(x)}{x} = 1$

(c) $f'(x) = \lim\limits_{h \to 0} \dfrac{f(x+h) - f(x)}{h} = \lim\limits_{h \to 0} \dfrac{\left[f(x) + f(h) + x^2 h + xh^2\right] - f(x)}{h} = \lim\limits_{h \to 0} \dfrac{f(h) + x^2 h + xh^2}{h}$

$\qquad = \lim\limits_{h \to 0} \left[\dfrac{f(h)}{h} + x^2 + xh\right] = 1 + x^2$

15. $\lim\limits_{x \to a} f(x) = \lim\limits_{x \to a} \left(\tfrac{1}{2}\left[f(x) + g(x)\right] + \tfrac{1}{2}\left[f(x) - g(x)\right]\right) = \tfrac{1}{2} \lim\limits_{x \to a} \left[f(x) + g(x)\right] + \tfrac{1}{2} \lim\limits_{x \to a} \left[f(x) - g(x)\right]$

$\qquad = \tfrac{1}{2} \cdot 2 + \tfrac{1}{2} \cdot 1 = \tfrac{3}{2}$,

and $\lim\limits_{x \to a} g(x) = \lim\limits_{x \to a} \left(\left[f(x) + g(x)\right] - f(x)\right) = \lim\limits_{x \to a}\left[f(x) + g(x)\right] - \lim\limits_{x \to a} f(x) = 2 - \tfrac{3}{2} = \tfrac{1}{2}$.

So $\lim\limits_{x \to a} \left[f(x)g(x)\right] = \left[\lim\limits_{x \to a} f(x)\right]\left[\lim\limits_{x \to a} g(x)\right] = \tfrac{3}{2} \cdot \tfrac{1}{2} = \tfrac{3}{4}$.

Another solution: Since $\lim\limits_{x \to a} \left[f(x) + g(x)\right]$ and $\lim\limits_{x \to a} \left[f(x) - g(x)\right]$ exist, we must have

$\lim\limits_{x \to a} \left[f(x) + g(x)\right]^2 = \left(\lim\limits_{x \to a} \left[f(x) + g(x)\right]\right)^2$ and $\lim\limits_{x \to a} \left[f(x) - g(x)\right]^2 = \left(\lim\limits_{x \to a} \left[f(x) - g(x)\right]\right)^2$, so

$\lim\limits_{x \to a} \left[f(x) g(x)\right] = \lim\limits_{x \to a} \tfrac{1}{4}\left(\left[f(x) + g(x)\right]^2 - \left[f(x) - g(x)\right]^2\right) \qquad$ [because all of the f^2 and g^2 cancel]

$\qquad = \tfrac{1}{4}\left(\lim\limits_{x \to a} \left[f(x) + g(x)\right]^2 - \lim\limits_{x \to a} \left[f(x) - g(x)\right]^2\right) = \tfrac{1}{4}\left(2^2 - 1^2\right) = \tfrac{3}{4}$.

17. We are given that $|f(x)| \leq x^2$ for all x. In particular, $|f(0)| \leq 0$, but $|a| \geq 0$ for all a. The only conclusion is

that $f(0) = 0$. Now $\left|\dfrac{f(x) - f(0)}{x - 0}\right| = \left|\dfrac{f(x)}{x}\right| = \dfrac{|f(x)|}{|x|} \leq \dfrac{x^2}{|x|} = \dfrac{|x^2|}{|x|} = |x| \quad \Rightarrow \quad -|x| \leq \dfrac{f(x) - f(0)}{x - 0} \leq |x|$.

But $\lim\limits_{x \to 0} (-|x|) = 0 = \lim\limits_{x \to 0} |x|$, so by the Squeeze Theorem, $\lim\limits_{x \to 0} \dfrac{f(x) - f(0)}{x - 0} = 0$. So by the definition of a derivative,

f is differentiable at 0 and, furthermore, $f'(0) = 0$.

3 □ DIFFERENTIATION RULES

3.1 Derivatives of Polynomials and Exponential Functions

1. (a) e is the number such that $\lim\limits_{h \to 0} \dfrac{e^h - 1}{h} = 1$.

(b)

x	$\dfrac{2.7^x - 1}{x}$
-0.001	0.9928
-0.0001	0.9932
0.001	0.9937
0.0001	0.9933

x	$\dfrac{2.8^x - 1}{x}$
-0.001	1.0291
-0.0001	1.0296
0.001	1.0301
0.0001	1.0297

From the tables (to two decimal places),

$\lim\limits_{h \to 0} \dfrac{2.7^h - 1}{h} = 0.99$ and $\lim\limits_{h \to 0} \dfrac{2.8^h - 1}{h} = 1.03$.

Since $0.99 < 1 < 1.03$, $2.7 < e < 2.8$.

3. $f(x) = 186.5$ is a constant function, so its derivative is 0, that is, $f'(x) = 0$.

5. $f(t) = 2 - \frac{2}{3}t \quad \Rightarrow \quad f'(t) = 0 - \frac{2}{3} = -\frac{2}{3}$

7. $f(x) = x^3 - 4x + 6 \quad \Rightarrow \quad f'(x) = 3x^2 - 4(1) + 0 = 3x^2 - 4$

9. $f(t) = \frac{1}{4}(t^4 + 8) \quad \Rightarrow \quad f'(t) = \frac{1}{4}(t^4 + 8)' = \frac{1}{4}(4t^{4-1} + 0) = t^3$

11. $A(s) = -\dfrac{12}{s^5} = -12s^{-5} \quad \Rightarrow \quad A'(s) = -12(-5s^{-6}) = 60s^{-6} \quad$ or $\quad 60/s^6$

13. $g(t) = 2t^{-3/4} \quad \Rightarrow \quad g'(t) = 2\left(-\frac{3}{4}t^{-7/4}\right) = -\frac{3}{2}t^{-7/4}$

15. $y = 3e^x + \dfrac{4}{\sqrt[3]{x}} = 3e^x + 4x^{-1/3} \quad \Rightarrow \quad y' = 3(e^x) + 4\left(-\frac{1}{3}\right)x^{-4/3} = 3e^x - \frac{4}{3}x^{-4/3}$

17. $F(x) = \left(\frac{1}{2}x\right)^5 = \left(\frac{1}{2}\right)^5 x^5 = \frac{1}{32}x^5 \quad \Rightarrow \quad F'(x) = \frac{1}{32}(5x^4) = \frac{5}{32}x^4$

19. $y = \dfrac{x^2 + 4x + 3}{\sqrt{x}} = x^{3/2} + 4x^{1/2} + 3x^{-1/2} \quad \Rightarrow$

$y' = \frac{3}{2}x^{1/2} + 4\left(\frac{1}{2}\right)x^{-1/2} + 3\left(-\frac{1}{2}\right)x^{-3/2} = \frac{3}{2}\sqrt{x} + \dfrac{2}{\sqrt{x}} - \dfrac{3}{2x\sqrt{x}} \quad \left[\text{note that } x^{3/2} = x^{2/2} \cdot x^{1/2} = x\sqrt{x}\right]$

The last expression can be written as $\dfrac{3x^2}{2x\sqrt{x}} + \dfrac{4x}{2x\sqrt{x}} - \dfrac{3}{2x\sqrt{x}} = \dfrac{3x^2 + 4x - 3}{2x\sqrt{x}}$.

21. $y = 4\pi^2 \quad \Rightarrow \quad y' = 0$ since $4\pi^2$ is a constant.

23. $u = \sqrt[5]{t} + 4\sqrt{t^5} = t^{1/5} + 4t^{5/2} \quad \Rightarrow \quad u' = \frac{1}{5}t^{-4/5} + 4\left(\frac{5}{2}t^{3/2}\right) = \frac{1}{5}t^{-4/5} + 10t^{3/2} \quad$ or $\quad 1/\left(5\sqrt[5]{t^4}\right) + 10\sqrt{t^3}$

25. $z = \dfrac{A}{y^{10}} + Be^y = Ay^{-10} + Be^y \quad \Rightarrow \quad z' = -10Ay^{-11} + Be^y = -\dfrac{10A}{y^{11}} + Be^y$

27. $y = \sqrt[4]{x} = x^{1/4} \Rightarrow y' = \frac{1}{4}x^{-3/4} = \dfrac{1}{4\sqrt[4]{x^3}}$. At $(1,1)$, $y' = \frac{1}{4}$ and an equation of the tangent line is

$y - 1 = \frac{1}{4}(x - 1)$ or $y = \frac{1}{4}x + \frac{3}{4}$.

29. $y = x^4 + 2e^x \Rightarrow y' = 4x^3 + 2e^x$. At $(0, 2)$, $y' = 2$ and an equation of the tangent line is $y - 2 = 2(x - 0)$

or $y = 2x + 2$. The slope of the normal line is $-\frac{1}{2}$ (the negative reciprocal of 2) and an equation of the normal line is

$y - 2 = -\frac{1}{2}(x - 0)$ or $y = -\frac{1}{2}x + 2$.

31. $y = 3x^2 - x^3 \Rightarrow y' = 6x - 3x^2$.

At $(1, 2)$, $y' = 6 - 3 = 3$, so an equation of the tangent line is

$y - 2 = 3(x - 1)$ or $y = 3x - 1$.

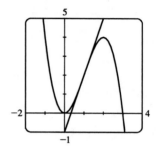

33. $f(x) = e^x - 5x \Rightarrow f'(x) = e^x - 5$.

Notice that $f'(x) = 0$ when f has a horizontal tangent, f' is positive

when f is increasing, and f' is negative when f is decreasing.

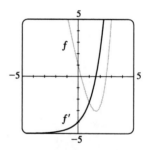

35. $f(x) = 3x^{15} - 5x^3 + 3 \Rightarrow f'(x) = 45x^{14} - 15x^2$.

Notice that $f'(x) = 0$ when f has a horizontal tangent, f' is positive

when f is increasing, and f' is negative when f is decreasing.

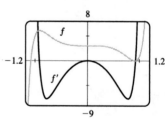

37. To graphically estimate the value of $f'(1)$ for $f(x) = 3x^2 - x^3$, we'll graph f in the viewing rectangle $[1 - 0.1, 1 + 0.1]$ by

$[f(0.9), f(1.1)]$, as shown in the figure. [When assigning values to the window variables, it is convenient to use $Y_1(0.9)$ for

$Y_{\min}$ and $Y_1(1.1)$ for $Y_{\max}$.] If we have sufficiently zoomed in on the graph of f, we should obtain a graph that looks like a

diagonal line; if not, graph again with $1 - 0.01$ and $1 + 0.01$, etc.

Estimated value:

$f'(1) \approx \dfrac{2.299 - 1.701}{1.1 - 0.9} = \dfrac{0.589}{0.2} = 2.99$.

Exact value: $f(x) = 3x^2 - x^3 \Rightarrow f'(x) = 6x - 3x^2$,

so $f'(1) = 6 - 3 = 3$.

39. (a)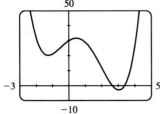

(b) From the graph in part (a), it appears that f' is zero at $x_1 \approx -1.25$, $x_2 \approx 0.5$, and $x_3 \approx 3$. The slopes are negative (so f' is negative) on $(-\infty, x_1)$ and (x_2, x_3). The slopes are positive (so f' is positive) on (x_1, x_2) and (x_3, ∞).

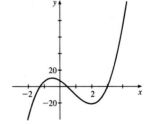

(c) $f(x) = x^4 - 3x^3 - 6x^2 + 7x + 30 \Rightarrow$

$f'(x) = 4x^3 - 9x^2 - 12x + 7$

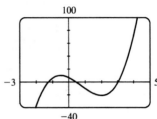

41. $f(x) = 10x^{10} + 5x^5 - x \Rightarrow f'(x) = 100x^9 + 25x^4 - 1 \Rightarrow f''(x) = 900x^8 + 100x^3$

43. $f(x) = 2x - 5x^{3/4} \Rightarrow f'(x) = 2 - \frac{15}{4}x^{-1/4} \Rightarrow f''(x) = \frac{15}{16}x^{-5/4}$

Note that f' is negative when f is decreasing and positive when f is increasing. f'' is always positive since f' is always increasing.

45. (a) $s = t^3 - 3t \Rightarrow v(t) = s'(t) = 3t^2 - 3 \Rightarrow a(t) = v'(t) = 6t$

(b) $a(2) = 6(2) = 12 \text{ m/s}^2$

(c) $v(t) = 3t^2 - 3 = 0$ when $t^2 = 1$, that is, $t = 1$ and $a(1) = 6 \text{ m/s}^2$.

47. $f(x) = 5x - e^x \Rightarrow f'(x) = 5 - e^x$. $f'(x) > 0 \Rightarrow 5 - e^x > 0 \Rightarrow e^x < 5 \Rightarrow x < \ln 5 \approx 1.61$.
f is increasing when f' is positive; that is, on $(-\infty, \ln 5)$.

49. The curve $y = 2x^3 + 3x^2 - 12x + 1$ has a horizontal tangent when $y' = 6x^2 + 6x - 12 = 0 \Leftrightarrow 6(x^2 + x - 2) = 0 \Leftrightarrow$
$6(x + 2)(x - 1) = 0 \Leftrightarrow x = -2$ or $x = 1$. The points on the curve are $(-2, 21)$ and $(1, -6)$.

51. $y = 6x^3 + 5x - 3 \Rightarrow m = y' = 18x^2 + 5$, but $x^2 \geq 0$ for all x, so $m \geq 5$ for all x.

53. The slope of the line $12x - y = 1$ (or $y = 12x - 1$) is 12, so the slope of both lines tangent to the curve is 12.
$y = 1 + x^3 \Rightarrow y' = 3x^2$. Thus, $3x^2 = 12 \Rightarrow x^2 = 4 \Rightarrow x = \pm 2$, which are the x-coordinates at which the tangent
lines have slope 12. The points on the curve are $(2, 9)$ and $(-2, -7)$, so the tangent line equations are $y - 9 = 12(x - 2)$
or $y = 12x - 15$ and $y + 7 = 12(x + 2)$ or $y = 12x + 17$.

55. The slope of $y = x^2 - 5x + 4$ is given by $m = y' = 2x - 5$. The slope of $x - 3y = 5$ $\Leftrightarrow$ $y = \frac{1}{3}x - \frac{5}{3}$ is $\frac{1}{3}$,

so the desired normal line must have slope $\frac{1}{3}$, and hence, the tangent line to the parabola must have slope -3. This occurs if

$2x - 5 = -3$ $\Rightarrow$ $2x = 2$ $\Rightarrow$ $x = 1$. When $x = 1$, $y = 1^2 - 5(1) + 4 = 0$, and an equation of the normal line is

$y - 0 = \frac{1}{3}(x - 1)$ or $y = \frac{1}{3}x - \frac{1}{3}$.

57.

Let (a, a^2) be a point on the parabola at which the tangent line passes

through the point $(0, -4)$. The tangent line has slope $2a$ and equation

$y - (-4) = 2a(x - 0)$ $\Leftrightarrow$ $y = 2ax - 4$. Since (a, a^2) also lies on the

line, $a^2 = 2a(a) - 4$, or $a^2 = 4$. So $a = \pm 2$ and the points are $(2, 4)$

and $(-2, 4)$.

59. $f'(x) = \lim_{h \to 0} \dfrac{f(x+h) - f(x)}{h} = \lim_{h \to 0} \dfrac{\frac{1}{x+h} - \frac{1}{x}}{h} = \lim_{h \to 0} \dfrac{x - (x+h)}{hx(x+h)} = \lim_{h \to 0} \dfrac{-h}{hx(x+h)} = \lim_{h \to 0} \dfrac{-1}{x(x+h)} = -\dfrac{1}{x^2}$

61. Let $P(x) = ax^2 + bx + c$. Then $P'(x) = 2ax + b$ and $P''(x) = 2a$. $P''(2) = 2$ $\Rightarrow$ $2a = 2$ $\Rightarrow$ $a = 1$.

$P'(2) = 3$ $\Rightarrow$ $2(1)(2) + b = 3$ $\Rightarrow$ $4 + b = 3$ $\Rightarrow$ $b = -1$.

$P(2) = 5$ $\Rightarrow$ $1(2)^2 + (-1)(2) + c = 5$ $\Rightarrow$ $2 + c = 5$ $\Rightarrow$ $c = 3$. So $P(x) = x^2 - x + 3$.

63. (a) At this stage, we would guess that an antiderivative of x^2 must have x^3 in it. Differentiating x^3 gives us $3x^2$, so we know

that we must divide x^3 by 3. That gives us $F(x) = \frac{1}{3}x^3$. Checking, we have $F'(x) = \frac{1}{3}(3x^2) = x^2 = f(x)$. Because we

can add an arbitrary constant C to F without changing its derivative, we have an infinite number of antiderivatives of the

form $F(x) = \frac{1}{3}x^3 + C$.

(b) As in part (a), antiderivatives of $f(x) = x^3$ and $f(x) = x^4$ are $F(x) = \frac{1}{4}x^4 + C$ and $F(x) = \frac{1}{5}x^5 + C$.

(c) Similarly, an antiderivative for $f(x) = x^n$ is $F(x) = \dfrac{1}{n+1}x^{n+1} + C$, since then

$F'(x) = \dfrac{1}{n+1}[(n+1)x^n] = x^n = f(x)$ for $n \neq -1$.

65. Substituting $x = 1$ and $y = 1$ into $y = ax^2 + bx$ gives us $a + b = 1$ **(1)**. The slope of the tangent line $y = 3x - 2$ is 3 and the

slope of the tangent to the parabola at (x, y) is $y' = 2ax + b$. At $x = 1$, $y' = 3$ $\Rightarrow$ $3 = 2a + b$ **(2)**. Subtracting **(1)** from

(2) gives us $2 = a$ and it follows that $b = -1$. The parabola has equation $y = 2x^2 - x$.

67. $y = f(x) = ax^3 + bx^2 + cx + d$ $\Rightarrow$ $f'(x) = 3ax^2 + 2bx + c$. The point $(-2, 6)$ is on f, so $f(-2) = 6$ $\Rightarrow$

$-8a + 4b - 2c + d = 6$ **(1)**. The point $(2, 0)$ is on f, so $f(2) = 0$ $\Rightarrow$ $8a + 4b + 2c + d = 0$ **(2)**. Since there are

horizontal tangents at $(-2, 6)$ and $(2, 0)$, $f'(\pm 2) = 0$. $f'(-2) = 0$ $\Rightarrow$ $12a - 4b + c = 0$ **(3)** and $f'(2) = 0$ $\Rightarrow$

$12a + 4b + c = 0$ **(4)**. Subtracting equation **(3)** from **(4)** gives $8b = 0$ $\Rightarrow$ $b = 0$. Adding **(1)** and **(2)** gives $8b + 2d = 6$, so $d = 3$ since $b = 0$. From **(3)** we have $c = -12a$, so **(2)** becomes $8a + 4(0) + 2(-12a) + 3 = 0$ $\Rightarrow$ $3 = 16a$ $\Rightarrow$

$a = \frac{3}{16}$. Now $c = -12a = -12\left(\frac{3}{16}\right) = -\frac{9}{4}$ and the desired cubic function is $y = \frac{3}{16}x^3 - \frac{9}{4}x + 3$.

69. $y = f(x) = ax^2$ $\Rightarrow$ $f'(x) = 2ax$.   So the slope of the tangent to the parabola at $x = 2$ is $m = 2a(2) = 4a$. The slope

of the given line, $2x + y = b$ $\Leftrightarrow$ $y = -2x + b$, is seen to be -2, so we must have $4a = -2$ $\Leftrightarrow$ $a = -\frac{1}{2}$. So when

$x = 2$, the point in question has y-coordinate $-\frac{1}{2} \cdot 2^2 = -2$. Now we simply require that the given line, whose equation is

$2x + y = b$, pass through the point $(2, -2)$: $2(2) + (-2) = b$ $\Leftrightarrow$ $b = 2$. So we must have $a = -\frac{1}{2}$ and $b = 2$.

71. *Solution 1:*   Let $f(x) = x^{1000}$. Then, by the definition of a derivative, $f'(1) = \lim\limits_{x \to 1} \dfrac{f(x) - f(1)}{x - 1} = \lim\limits_{x \to 1} \dfrac{x^{1000} - 1}{x - 1}$.

But this is just the limit we want to find, and we know (from the Power Rule) that $f'(x) = 1000x^{999}$, so

$f'(1) = 1000(1)^{999} = 1000$. So $\lim\limits_{x \to 1} \dfrac{x^{1000} - 1}{x - 1} = 1000$.

Solution 2:   Note that $(x^{1000} - 1) = (x - 1)(x^{999} + x^{998} + x^{997} + \cdots + x^2 + x + 1)$. So

$$\lim_{x \to 1} \frac{x^{1000} - 1}{x - 1} = \lim_{x \to 1} \frac{(x - 1)(x^{999} + x^{998} + x^{997} + \cdots + x^2 + x + 1)}{x - 1} = \lim_{x \to 1}(x^{999} + x^{998} + x^{997} + \cdots + x^2 + x + 1)$$

$$= \underbrace{1 + 1 + 1 + \cdots + 1 + 1 + 1}_{1000 \text{ ones}} = 1000, \text{ as above.}$$

73. $y = x^2$ $\Rightarrow$ $y' = 2x$, so the slope of a tangent line at the point (a, a^2) is $y' = 2a$ and the slope of a normal line is $-1/(2a)$,

for $a \neq 0$. The slope of the normal line through the points (a, a^2) and $(0, c)$ is $\dfrac{a^2 - c}{a - 0}$, so $\dfrac{a^2 - c}{a} = -\dfrac{1}{2a}$ $\Rightarrow$

$a^2 - c = -\frac{1}{2}$ $\Rightarrow$ $a^2 = c - \frac{1}{2}$. The last equation has two solutions if $c > \frac{1}{2}$, one solution if $c = \frac{1}{2}$, and no solution if

$c < \frac{1}{2}$. Since the y-axis is normal to $y = x^2$ regardless of the value of c (this is the case for $a = 0$), we have three normal lines

if $c > \frac{1}{2}$ and one normal line if $c \leq \frac{1}{2}$.

3.2   The Product and Quotient Rules

1. Product Rule: $f(x) = (1 + 2x^2)(x - x^2)$ $\Rightarrow$

$$f'(x) = (1 + 2x^2)(1 - 2x) + (x - x^2)(4x) = 1 - 2x + 2x^2 - 4x^3 + 4x^2 - 4x^3 = 1 - 2x + 6x^2 - 8x^3.$$

Multiplying first: $f(x) = (1 + 2x^2)(x - x^2) = x - x^2 + 2x^3 - 2x^4$ $\Rightarrow$ $f'(x) = 1 - 2x + 6x^2 - 8x^3$ (equivalent).

3. By the Product Rule, $f(x) = (x^3 + 2x)e^x$ $\Rightarrow$

$$f'(x) = (x^3 + 2x)(e^x)' + e^x(x^3 + 2x)' = (x^3 + 2x)e^x + e^x(3x^2 + 2)$$

$$= e^x[(x^3 + 2x) + (3x^2 + 2)] = e^x(x^3 + 3x^2 + 2x + 2)$$

5. By the Quotient Rule, $y = \dfrac{e^x}{x^2}$ $\Rightarrow$ $y' = \dfrac{x^2 \dfrac{d}{dx}(e^x) - e^x \dfrac{d}{dx}(x^2)}{(x^2)^2} = \dfrac{x^2(e^x) - e^x(2x)}{x^4} = \dfrac{xe^x(x - 2)}{x^4} = \dfrac{e^x(x - 2)}{x^3}$.

The notations $\overset{PR}{\Rightarrow}$ and $\overset{QR}{\Rightarrow}$ indicate the use of the Product and Quotient Rules, respectively.

7. $g(x) = \dfrac{3x-1}{2x+1} \quad \overset{QR}{\Rightarrow} \quad g'(x) = \dfrac{(2x+1)(3) - (3x-1)(2)}{(2x+1)^2} = \dfrac{6x+3-6x+2}{(2x+1)^2} = \dfrac{5}{(2x+1)^2}$

9. $F(y) = \left(\dfrac{1}{y^2} - \dfrac{3}{y^4}\right)(y + 5y^3) = (y^{-2} - 3y^{-4})(y + 5y^3) \quad \overset{PR}{\Rightarrow}$

$\quad F'(y) = (y^{-2} - 3y^{-4})(1 + 15y^2) + (y + 5y^3)(-2y^{-3} + 12y^{-5})$

$\qquad = (y^{-2} + 15 - 3y^{-4} - 45y^{-2}) + (-2y^{-2} + 12y^{-4} - 10 + 60y^{-2})$

$\qquad = 5 + 14y^{-2} + 9y^{-4} \text{ or } 5 + 14/y^2 + 9/y^4$

11. $y = \dfrac{x^3}{1-x^2} \quad \overset{QR}{\Rightarrow} \quad y' = \dfrac{(1-x^2)(3x^2) - x^3(-2x)}{(1-x^2)^2} = \dfrac{x^2(3 - 3x^2 + 2x^2)}{(1-x^2)^2} = \dfrac{x^2(3 - x^2)}{(1-x^2)^2}$

13. $y = \dfrac{t^2+2}{t^4 - 3t^2 + 1} \quad \overset{QR}{\Rightarrow}$

$\quad y' = \dfrac{(t^4 - 3t^2 + 1)(2t) - (t^2+2)(4t^3 - 6t)}{(t^4 - 3t^2 + 1)^2} = \dfrac{2t[(t^4 - 3t^2 + 1) - (t^2+2)(2t^2 - 3)]}{(t^4 - 3t^2 + 1)^2}$

$\qquad = \dfrac{2t(t^4 - 3t^2 + 1 - 2t^4 - 4t^2 + 3t^2 + 6)}{(t^4 - 3t^2 + 1)^2} = \dfrac{2t(-t^4 - 4t^2 + 7)}{(t^4 - 3t^2 + 1)^2}$

15. $y = (r^2 - 2r)e^r \quad \overset{PR}{\Rightarrow} \quad y' = (r^2 - 2r)(e^r) + e^r(2r - 2) = e^r(r^2 - 2r + 2r - 2) = e^r(r^2 - 2)$

17. $y = \dfrac{v^3 - 2v\sqrt{v}}{v} = v^2 - 2\sqrt{v} = v^2 - 2v^{1/2} \quad \Rightarrow \quad y' = 2v - 2\left(\tfrac{1}{2}\right)v^{-1/2} = 2v - v^{-1/2}.$

We can change the form of the answer as follows: $2v - v^{-1/2} = 2v - \dfrac{1}{\sqrt{v}} = \dfrac{2v\sqrt{v} - 1}{\sqrt{v}} = \dfrac{2v^{3/2} - 1}{\sqrt{v}}$

19. $f(t) = \dfrac{2t}{2+\sqrt{t}} \quad \overset{QR}{\Rightarrow} \quad f'(t) = \dfrac{(2+t^{1/2})(2) - 2t\left(\frac{1}{2}t^{-1/2}\right)}{(2+\sqrt{t})^2} = \dfrac{4 + 2t^{1/2} - t^{1/2}}{(2+\sqrt{t})^2} = \dfrac{4 + t^{1/2}}{(2+\sqrt{t})^2} \text{ or } \dfrac{4 + \sqrt{t}}{(2+\sqrt{t})^2}$

21. $f(x) = \dfrac{A}{B + Ce^x} \quad \overset{QR}{\Rightarrow} \quad f'(x) = \dfrac{(B + Ce^x) \cdot 0 - A(Ce^x)}{(B + Ce^x)^2} = -\dfrac{ACe^x}{(B + Ce^x)^2}$

23. $f(x) = \dfrac{x}{x + c/x} \quad \Rightarrow \quad f'(x) = \dfrac{(x + c/x)(1) - x(1 - c/x^2)}{\left(x + \dfrac{c}{x}\right)^2} = \dfrac{x + c/x - x + c/x}{\left(\dfrac{x^2+c}{x}\right)^2} = \dfrac{2c/x}{\dfrac{(x^2+c)^2}{x^2}} \cdot \dfrac{x^2}{x^2} = \dfrac{2cx}{(x^2+c)^2}$

25. $f(x) = x^4 e^x \quad \Rightarrow \quad f'(x) = x^4 e^x + e^x \cdot 4x^3 = (x^4 + 4x^3)e^x \quad [\text{or } x^3 e^x(x + 4)] \quad \Rightarrow$

$\quad f''(x) = (x^4 + 4x^3)e^x + e^x(4x^3 + 12x^2) = (x^4 + 4x^3 + 4x^3 + 12x^2)e^x$

$\qquad = (x^4 + 8x^3 + 12x^2)e^x \quad [\text{or } x^2 e^x(x+2)(x+6)]$

27. $f(x) = \dfrac{x^2}{1 + 2x}$ $\Rightarrow$ $f'(x) = \dfrac{(1 + 2x)(2x) - x^2(2)}{(1 + 2x)^2} = \dfrac{2x + 4x^2 - 2x^2}{(1 + 2x)^2} = \dfrac{2x^2 + 2x}{(1 + 2x)^2}$ $\Rightarrow$

$f''(x) = \dfrac{(1 + 2x)^2(4x + 2) - (2x^2 + 2x)(1 + 4x + 4x^2)'}{[(1 + 2x)^2]^2} = \dfrac{2(1 + 2x)^2(2x + 1) - 2x(x + 1)(4 + 8x)}{(1 + 2x)^4}$

$\quad = \dfrac{2(1 + 2x)[(1 + 2x)^2 - 4x(x + 1)]}{(1 + 2x)^4} = \dfrac{2(1 + 4x + 4x^2 - 4x^2 - 4x)}{(1 + 2x)^3} = \dfrac{2}{(1 + 2x)^3}$

29. $y = \dfrac{2x}{x + 1}$ $\Rightarrow$ $y' = \dfrac{(x + 1)(2) - (2x)(1)}{(x + 1)^2} = \dfrac{2}{(x + 1)^2}$.

At $(1, 1)$, $y' = \frac{1}{2}$, and an equation of the tangent line is $y - 1 = \frac{1}{2}(x - 1)$, or $y = \frac{1}{2}x + \frac{1}{2}$.

31. $y = 2xe^x$ $\Rightarrow$ $y' = 2(x \cdot e^x + e^x \cdot 1) = 2e^x(x + 1)$.

At $(0, 0)$, $y' = 2e^0(0 + 1) = 2 \cdot 1 \cdot 1 = 2$, and an equation of the tangent line is $y - 0 = 2(x - 0)$, or $y = 2x$. The slope of

the normal line is $-\frac{1}{2}$, so an equation of the normal line is $y - 0 = -\frac{1}{2}(x - 0)$, or $y = -\frac{1}{2}x$.

33. (a) $y = f(x) = \dfrac{1}{1 + x^2}$ $\Rightarrow$

$f'(x) = \dfrac{(1 + x^2)(0) - 1(2x)}{(1 + x^2)^2} = \dfrac{-2x}{(1 + x^2)^2}$. So the slope of the

tangent line at the point $\left(-1, \frac{1}{2}\right)$ is $f'(-1) = \dfrac{2}{2^2} = \frac{1}{2}$ and its

equation is $y - \frac{1}{2} = \frac{1}{2}(x + 1)$ or $y = \frac{1}{2}x + 1$.

(b)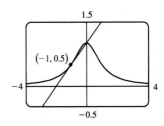

35. (a) $f(x) = (x^3 - x)e^x$ $\Rightarrow$ $f'(x) = (x^3 - x)e^x + e^x(3x^2 - 1) = e^x(x^3 + 3x^2 - x - 1)$

(b)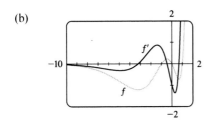

$f' = 0$ when f has a horizontal tangent line. f' is negative when f is
decreasing, and f' is positive when f is increasing.

37. (a) $f(x) = \dfrac{x^2 - 1}{x^2 + 1}$ $\Rightarrow$

$f'(x) = \dfrac{(x^2 + 1)(2x) - (x^2 - 1)(2x)}{(x^2 + 1)^2} = \dfrac{(2x)[(x^2 + 1) - (x^2 - 1)]}{(x^2 + 1)^2} = \dfrac{(2x)(2)}{(x^2 + 1)^2} = \dfrac{4x}{(x^2 + 1)^2}$ $\Rightarrow$

$f''(x) = \dfrac{(x^2 + 1)^2(4) - 4x(x^4 + 2x^2 + 1)'}{[(x^2 + 1)^2]^2} = \dfrac{4(x^2 + 1)^2 - 4x(4x^3 + 4x)}{(x^2 + 1)^4}$

$\quad = \dfrac{4(x^2 + 1)^2 - 16x^2(x^2 + 1)}{(x^2 + 1)^4} = \dfrac{4(x^2 + 1)[(x^2 + 1) - 4x^2]}{(x^2 + 1)^4} = \dfrac{4(1 - 3x^2)}{(x^2 + 1)^3}$

(b)

$f' = 0$ when f has a horizontal tangent and $f'' = 0$ when f' has a
horizontal tangent. f' is negative when f is decreasing and positive when f
is increasing. f'' is negative when f' is decreasing and positive when f' is
increasing. f'' is negative when f is concave down and positive when f is
concave up.

39. $f(x) = \dfrac{x^2}{1+x}$ $\Rightarrow$ $f'(x) = \dfrac{(1+x)(2x) - x^2(1)}{(1+x)^2} = \dfrac{2x + 2x^2 - x^2}{(1+x)^2} = \dfrac{x^2 + 2x}{x^2 + 2x + 1}$ $\Rightarrow$

$$f''(x) = \frac{(x^2 + 2x + 1)(2x + 2) - (x^2 + 2x)(2x + 2)}{(x^2 + 2x + 1)^2} = \frac{(2x + 2)(x^2 + 2x + 1 - x^2 - 2x)}{[(x+1)^2]^2}$$

$$= \frac{2(x+1)(1)}{(x+1)^4} = \frac{2}{(x+1)^3},$$

so $f''(1) = \dfrac{2}{(1+1)^3} = \dfrac{2}{8} = \dfrac{1}{4}$.

41. We are given that $f(5) = 1$, $f'(5) = 6$, $g(5) = -3$, and $g'(5) = 2$.

(a) $(fg)'(5) = f(5)g'(5) + g(5)f'(5) = (1)(2) + (-3)(6) = 2 - 18 = -16$

(b) $\left(\dfrac{f}{g}\right)'(5) = \dfrac{g(5)f'(5) - f(5)g'(5)}{[g(5)]^2} = \dfrac{(-3)(6) - (1)(2)}{(-3)^2} = -\dfrac{20}{9}$

(c) $\left(\dfrac{g}{f}\right)'(5) = \dfrac{f(5)g'(5) - g(5)f'(5)}{[f(5)]^2} = \dfrac{(1)(2) - (-3)(6)}{(1)^2} = 20$

43. $f(x) = e^x g(x)$ $\Rightarrow$ $f'(x) = e^x g'(x) + g(x)e^x = e^x[g'(x) + g(x)]$. $f'(0) = e^0[g'(0) + g(0)] = 1(5 + 2) = 7$

45. (a) From the graphs of f and g, we obtain the following values: $f(1) = 2$ since the point $(1, 2)$ is on the graph of f;

$g(1) = 1$ since the point $(1, 1)$ is on the graph of g; $f'(1) = 2$ since the slope of the line segment between $(0, 0)$ and

$(2, 4)$ is $\dfrac{4 - 0}{2 - 0} = 2$; $g'(1) = -1$ since the slope of the line segment between $(-2, 4)$ and $(2, 0)$ is $\dfrac{0 - 4}{2 - (-2)} = -1$.

Now $u(x) = f(x)g(x)$, so $u'(1) = f(1)g'(1) + g(1)\,f'(1) = 2 \cdot (-1) + 1 \cdot 2 = 0$.

(b) $v(x) = f(x)/g(x)$, so $v'(5) = \dfrac{g(5)f'(5) - f(5)g'(5)}{[g(5)]^2} = \dfrac{2\left(-\frac{1}{3}\right) - 3 \cdot \frac{2}{3}}{2^2} = \dfrac{-\frac{8}{3}}{4} = -\dfrac{2}{3}$

47. (a) $y = xg(x)$ $\Rightarrow$ $y' = xg'(x) + g(x) \cdot 1 = xg'(x) + g(x)$

(b) $y = \dfrac{x}{g(x)}$ $\Rightarrow$ $y' = \dfrac{g(x) \cdot 1 - xg'(x)}{[g(x)]^2} = \dfrac{g(x) - xg'(x)}{[g(x)]^2}$

(c) $y = \dfrac{g(x)}{x}$ $\Rightarrow$ $y' = \dfrac{xg'(x) - g(x) \cdot 1}{(x)^2} = \dfrac{xg'(x) - g(x)}{x^2}$

49. If $P(t)$ denotes the population at time t and $A(t)$ the average annual income, then $T(t) = P(t)A(t)$ is the total personal income. The rate at which $T(t)$ is rising is given by $T'(t) = P(t)A'(t) + A(t)P'(t)$ $\Rightarrow$

$$T'(1999) = P(1999)A'(1999) + A(1999)P'(1999) = (961{,}400)(\$1400/\text{yr}) + (\$30{,}593)(9200/\text{yr})$$

$$= \$1{,}345{,}960{,}000/\text{yr} + \$281{,}455{,}600/\text{yr} = \$1{,}627{,}415{,}600/\text{yr}$$

So the total personal income was rising by about $1.627 billion per year in 1999.

The term $P(t)A'(t) \approx \$1.346$ billion represents the portion of the rate of change of total income due to the existing population's increasing income. The term $A(t)P'(t) \approx \$281$ million represents the portion of the rate of change of total income due to increasing population.

51. f is increasing when f' is positive. $f(x) = x^3 e^x \ \Rightarrow \ f'(x) = x^3 e^x + e^x(3x^2) = x^2 e^x(x+3)$. Now $x^2 \geq 0$ and $e^x > 0$

for all x, so $f'(x) > 0$ when $x + 3 > 0$ and $x \neq 0$; that is, when $x \in (-3, 0) \cup (0, \infty)$. So f is increasing on $(-3, \infty)$.

53. If $y = f(x) = \dfrac{x}{x+1}$, then $f'(x) = \dfrac{(x+1)(1) - x(1)}{(x+1)^2} = \dfrac{1}{(x+1)^2}$. When $x = a$, the equation of the tangent line is

$y - \dfrac{a}{a+1} = \dfrac{1}{(a+1)^2}(x-a)$. This line passes through $(1, 2)$ when $2 - \dfrac{a}{a+1} = \dfrac{1}{(a+1)^2}(1-a) \ \Leftrightarrow$

$2(a+1)^2 - a(a+1) = 1 - a \ \Leftrightarrow \ 2a^2 + 4a + 2 - a^2 - a - 1 + a = 0 \ \Leftrightarrow \ a^2 + 4a + 1 = 0.$

The quadratic formula gives the roots of this equation as $a = \dfrac{-4 \pm \sqrt{4^2 - 4(1)(1)}}{2(1)} = \dfrac{-4 \pm \sqrt{12}}{2} = -2 \pm \sqrt{3}$,

so there are two such tangent lines. Since

$$f\left(-2 \pm \sqrt{3}\right) = \frac{-2 \pm \sqrt{3}}{-2 \pm \sqrt{3} + 1} = \frac{-2 \pm \sqrt{3}}{-1 \pm \sqrt{3}} \cdot \frac{-1 \mp \sqrt{3}}{-1 \mp \sqrt{3}}$$

$$= \frac{2 \pm 2\sqrt{3} \mp \sqrt{3} - 3}{1 - 3} = \frac{-1 \pm \sqrt{3}}{-2} = \frac{1 \mp \sqrt{3}}{2},$$

the lines touch the curve at $A\left(-2 + \sqrt{3}, \frac{1 - \sqrt{3}}{2}\right) \approx (-0.27, -0.37)$

and $B\left(-2 - \sqrt{3}, \frac{1 + \sqrt{3}}{2}\right) \approx (-3.73, 1.37)$.

55. $R = \dfrac{f}{g} \ \Rightarrow \ R' = \dfrac{gf' - fg'}{g^2}$. For $f(x) = x - 3x^3 + 5x^5$, $f'(x) = 1 - 9x^2 + 25x^4$,

and for $g(x) = 1 + 3x^3 + 6x^6 + 9x^9$, $g'(x) = 9x^2 + 36x^5 + 81x^8$.

Thus, $R'(0) = \dfrac{g(0)f'(0) - f(0)g'(0)}{[g(0)]^2} = \dfrac{1 \cdot 1 - 0 \cdot 0}{1^2} = \dfrac{1}{1} = 1.$

57. (a) $(fgh)' = [(fg)h]' = (fg)'h + (fg)h' = (f'g + fg')h + (fg)h' = f'gh + fg'h + fgh'$

(b) Putting $f = g = h$ in part (a), we have $\dfrac{d}{dx}[f(x)]^3 = (fff)' = f'ff + ff'f + fff' = 3fff' = 3[f(x)]^2 f'(x)$.

(c) $\dfrac{d}{dx}(e^{3x}) = \dfrac{d}{dx}(e^x)^3 = 3(e^x)^2 e^x = 3e^{2x} e^x = 3e^{3x}$

59. For $f(x) = x^2 e^x$, $f'(x) = x^2 e^x + e^x(2x) = e^x(x^2 + 2x)$. Similarly, we have

$$f''(x) = e^x(x^2 + 4x + 2)$$

$$f'''(x) = e^x(x^2 + 6x + 6)$$

$$f^{(4)}(x) = e^x(x^2 + 8x + 12)$$

$$f^{(5)}(x) = e^x(x^2 + 10x + 20)$$

It appears that the coefficient of x in the quadratic term increases by 2 with each differentiation. The pattern for the

constant terms seems to be $0 = 1 \cdot 0$, $2 = 2 \cdot 1$, $6 = 3 \cdot 2$, $12 = 4 \cdot 3$, $20 = 5 \cdot 4$. So a reasonable guess is that

$f^{(n)}(x) = e^x[x^2 + 2nx + n(n-1)]$.

[continued]

Proof: Let S_n be the statement that $f^{(n)}(x) = e^x[x^2 + 2nx + n(n-1)]$.

    1. S_1 is true because $f'(x) = e^x(x^2 + 2x)$.

    2. Assume that S_k is true; that is, $f^{(k)}(x) = e^x[x^2 + 2kx + k(k-1)]$. Then

$$f^{(k+1)}(x) = \frac{d}{dx}\left[f^{(k)}(x)\right] = e^x(2x + 2k) + [x^2 + 2kx + k(k-1)]e^x$$
$$= e^x[x^2 + (2k+2)x + (k^2 + k)] = e^x[x^2 + 2(k+1)x + (k+1)k]$$

    This shows that S_{k+1} is true.

    3. Therefore, by mathematical induction, S_n is true for all n; that is, $f^{(n)}(x) = e^x[x^2 + 2nx + n(n-1)]$ for every positive integer n.

3.3 Derivatives of Trigonometric Functions

1. $f(x) = 3x^2 - 2\cos x \implies f'(x) = 6x - 2(-\sin x) = 6x + 2\sin x$

3. $f(x) = \sin x + \frac{1}{2}\cot x \implies f'(x) = \cos x - \frac{1}{2}\csc^2 x$

5. $y = \sec\theta\tan\theta \implies y' = \sec\theta\,(\sec^2\theta) + \tan\theta\,(\sec\theta\tan\theta) = \sec\theta\,(\sec^2\theta + \tan^2\theta)$. Using the identity

    $1 + \tan^2\theta = \sec^2\theta$, we can write alternative forms of the answer as $\sec\theta\,(1 + 2\tan^2\theta)$ or $\sec\theta\,(2\sec^2\theta - 1)$.

7. $y = c\cos t + t^2\sin t \implies y' = c(-\sin t) + t^2(\cos t) + \sin t\,(2t) = -c\sin t + t(t\cos t + 2\sin t)$

9. $y = \dfrac{x}{2 - \tan x} \implies y' = \dfrac{(2 - \tan x)(1) - x(-\sec^2 x)}{(2 - \tan x)^2} = \dfrac{2 - \tan x + x\sec^2 x}{(2 - \tan x)^2}$

11. $f(\theta) = \dfrac{\sec\theta}{1 + \sec\theta} \implies$

    $f'(\theta) = \dfrac{(1 + \sec\theta)(\sec\theta\tan\theta) - (\sec\theta)(\sec\theta\tan\theta)}{(1 + \sec\theta)^2} = \dfrac{(\sec\theta\tan\theta)\,[(1 + \sec\theta) - \sec\theta]}{(1 + \sec\theta)^2} = \dfrac{\sec\theta\tan\theta}{(1 + \sec\theta)^2}$

13. Using Exercise 3.2.57(a), $f(x) = xe^x\csc x \implies$

    $f'(x) = (x)'e^x\csc x + x(e^x)'\csc x + xe^x(\csc x)' = 1e^x\csc x + xe^x\csc x + xe^x(-\cot x\,\csc x)$

    $= e^x\csc x\,(1 + x - x\cot x)$

15. $\dfrac{d}{dx}(\csc x) = \dfrac{d}{dx}\left(\dfrac{1}{\sin x}\right) = \dfrac{(\sin x)(0) - 1(\cos x)}{\sin^2 x} = \dfrac{-\cos x}{\sin^2 x} = -\dfrac{1}{\sin x}\cdot\dfrac{\cos x}{\sin x} = -\csc x\cot x$

17. $\dfrac{d}{dx}(\cot x) = \dfrac{d}{dx}\left(\dfrac{\cos x}{\sin x}\right) = \dfrac{(\sin x)(-\sin x) - (\cos x)(\cos x)}{\sin^2 x} = -\dfrac{\sin^2 x + \cos^2 x}{\sin^2 x} = -\dfrac{1}{\sin^2 x} = -\csc^2 x$

19. $y = \sec x \implies y' = \sec x\tan x$, so $y'\left(\frac{\pi}{3}\right) = \sec\frac{\pi}{3}\tan\frac{\pi}{3} = 2\sqrt{3}$. An equation of the tangent line to the curve $y = \sec x$

    at the point $\left(\frac{\pi}{3}, 2\right)$ is $y - 2 = 2\sqrt{3}\left(x - \frac{\pi}{3}\right)$ or $y = 2\sqrt{3}\,x + 2 - \frac{2}{3}\sqrt{3}\,\pi$.

21. $y = x + \cos x \Rightarrow y' = 1 - \sin x$. At $(0,1)$, $y' = 1$, and an equation of the tangent line is $y - 1 = 1(x - 0)$, or $y = x + 1$.

23. (a) $y = 2x \sin x \Rightarrow y' = 2(x \cos x + \sin x \cdot 1)$.   At $\left(\frac{\pi}{2}, \pi\right)$,

(b)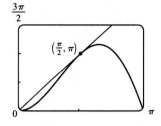

$y' = 2\left(\frac{\pi}{2} \cos \frac{\pi}{2} + \sin \frac{\pi}{2}\right) = 2(0 + 1) = 2$, and an equation of the

tangent line is $y - \pi = 2\left(x - \frac{\pi}{2}\right)$, or $y = 2x$.

25. (a) $f(x) = \sec x - x \Rightarrow f'(x) = \sec x \tan x - 1$

(b)

Note that $f' = 0$ where f has a minimum. Also note that f' is negative when f is decreasing and f' is positive when f is increasing.

27. $H(\theta) = \theta \sin \theta \Rightarrow H'(\theta) = \theta (\cos \theta) + (\sin \theta) \cdot 1 = \theta \cos \theta + \sin \theta \Rightarrow$

$H''(\theta) = \theta(-\sin \theta) + (\cos \theta) \cdot 1 + \cos \theta = -\theta \sin \theta + 2 \cos \theta$

29. (a) $f(x) = \dfrac{\tan x - 1}{\sec x} \Rightarrow$

$$f'(x) = \frac{\sec x(\sec^2 x) - (\tan x - 1)(\sec x \tan x)}{(\sec x)^2} = \frac{\sec x(\sec^2 x - \tan^2 x + \tan x)}{\sec^2 x} = \frac{1 + \tan x}{\sec x}$$

(b) $f(x) = \dfrac{\tan x - 1}{\sec x} = \dfrac{\dfrac{\sin x}{\cos x} - 1}{\dfrac{1}{\cos x}} = \dfrac{\dfrac{\sin x - \cos x}{\cos x}}{\dfrac{1}{\cos x}} = \sin x - \cos x \Rightarrow f'(x) = \cos x - (-\sin x) = \cos x + \sin x$

(c) From part (a), $f'(x) = \dfrac{1 + \tan x}{\sec x} = \dfrac{1}{\sec x} + \dfrac{\tan x}{\sec x} = \cos x + \sin x$, which is the expression for $f'(x)$ in part (b).

31. $f(x) = x + 2 \sin x$ has a horizontal tangent when $f'(x) = 0 \Leftrightarrow 1 + 2 \cos x = 0 \Leftrightarrow \cos x = -\frac{1}{2} \Leftrightarrow$

$x = \frac{2\pi}{3} + 2\pi n$ or $\frac{4\pi}{3} + 2\pi n$, where n is an integer. Note that $\frac{4\pi}{3}$ and $\frac{2\pi}{3}$ are $\pm \frac{\pi}{3}$ units from π. This allows us to write the

solutions in the more compact equivalent form $(2n + 1)\pi \pm \frac{\pi}{3}$, n an integer.

33. $f(x) = x - 2 \sin x$, $0 \le x \le 2\pi$. $f'(x) = 1 - 2 \cos x$. So $f'(x) > 0 \Leftrightarrow 1 - 2 \cos x > 0 \Leftrightarrow -2 \cos x > -1 \Leftrightarrow$

$\cos x < \frac{1}{2} \Leftrightarrow \frac{\pi}{3} < x < \frac{5\pi}{3} \Rightarrow f$ is increasing on $\left(\frac{\pi}{3}, \frac{5\pi}{3}\right)$.

35. (a) $x(t) = 8 \sin t \Rightarrow v(t) = x'(t) = 8 \cos t \Rightarrow a(t) = x''(t) = -8 \sin t$

(b) The mass at time $t = \frac{2\pi}{3}$ has position $x\left(\frac{2\pi}{3}\right) = 8 \sin \frac{2\pi}{3} = 8\left(\frac{\sqrt{3}}{2}\right) = 4\sqrt{3}$, velocity $v\left(\frac{2\pi}{3}\right) = 8 \cos \frac{2\pi}{3} = 8\left(-\frac{1}{2}\right) = -4$,

and acceleration $a\left(\frac{2\pi}{3}\right) = -8 \sin \frac{2\pi}{3} = -8\left(\frac{\sqrt{3}}{2}\right) = -4\sqrt{3}$. Since $v\left(\frac{2\pi}{3}\right) < 0$, the particle is moving to the left.

37. From the diagram we can see that $\sin\theta = x/10$ $\Leftrightarrow$ $x = 10\sin\theta$. We want to find the rate of change of x with respect to θ, that is, $dx/d\theta$. Taking the derivative of $x = 10\sin\theta$, we get

$$dx/d\theta = 10(\cos\theta). \text{ So when } \theta = \tfrac{\pi}{3}, \ \tfrac{dx}{d\theta} = 10\cos\tfrac{\pi}{3} = 10\left(\tfrac{1}{2}\right) = 5 \text{ ft/rad.}$$

39. $\dfrac{d}{dx}(\sin x) = \cos x$ $\Rightarrow$ $\dfrac{d^2}{dx^2}(\sin x) = -\sin x$ $\Rightarrow$ $\dfrac{d^3}{dx^3}(\sin x) = -\cos x$ $\Rightarrow$ $\dfrac{d^4}{dx^4}(\sin x) = \sin x$.

The derivatives of $\sin x$ occur in a cycle of four. Since $99 = 4(24) + 3$, we have $\dfrac{d^{99}}{dx^{99}}(\sin x) = \dfrac{d^3}{dx^3}(\sin x) = -\cos x$.

41. $y = A\sin x + B\cos x$ $\Rightarrow$ $y' = A\cos x - B\sin x$ $\Rightarrow$ $y'' = -A\sin x - B\cos x$. Substituting these

expressions for y, y', and y'' into the given differential equation $y'' + y' - 2y = \sin x$ gives us

$(-A\sin x - B\cos x) + (A\cos x - B\sin x) - 2(A\sin x + B\cos x) = \sin x$ $\Leftrightarrow$

$-3A\sin x - B\sin x + A\cos x - 3B\cos x = \sin x$ $\Leftrightarrow$ $(-3A - B)\sin x + (A - 3B)\cos x = 1\sin x$, so we must have

$-3A - B = 1$ and $A - 3B = 0$ (since 0 is the coefficient of $\cos x$ on the right side). Solving for A and B, we add the first

equation to three times the second to get $B = -\tfrac{1}{10}$ and $A = -\tfrac{3}{10}$.

43. $\displaystyle\lim_{t\to0}\frac{\tan 6t}{\sin 2t} = \lim_{t\to0}\left(\frac{\sin 6t}{t}\cdot\frac{1}{\cos 6t}\cdot\frac{t}{\sin 2t}\right) = \lim_{t\to0}\frac{6\sin 6t}{6t}\cdot\lim_{t\to0}\frac{1}{\cos 6t}\cdot\lim_{t\to0}\frac{2t}{2\sin 2t}$

$$= 6\lim_{t\to0}\frac{\sin 6t}{6t}\cdot\lim_{t\to0}\frac{1}{\cos 6t}\cdot\frac{1}{2}\lim_{t\to0}\frac{2t}{\sin 2t} = 6(1)\cdot\frac{1}{1}\cdot\frac{1}{2}(1) = 3$$

45. $\displaystyle\lim_{\theta\to0}\frac{\sin\theta}{\theta + \tan\theta} = \frac{\displaystyle\lim_{\theta\to0}\frac{\sin\theta}{\theta}}{\displaystyle\lim_{\theta\to0}\frac{\theta + \tan\theta}{\theta}} = \frac{1}{\displaystyle\lim_{\theta\to0}\left(1 + \frac{\sin\theta}{\theta}\cdot\frac{1}{\cos\theta}\right)} = \frac{1}{1 + 1\cdot 1} = \frac{1}{2}$

47. (a) $\dfrac{d}{dx}\tan x = \dfrac{d}{dx}\dfrac{\sin x}{\cos x}$ $\Rightarrow$ $\sec^2 x = \dfrac{\cos x\cos x - \sin x(-\sin x)}{\cos^2 x} = \dfrac{\cos^2 x + \sin^2 x}{\cos^2 x}$. So $\sec^2 x = \dfrac{1}{\cos^2 x}$.

(b) $\dfrac{d}{dx}\sec x = \dfrac{d}{dx}\dfrac{1}{\cos x}$ $\Rightarrow$ $\sec x\tan x = \dfrac{(\cos x)(0) - 1(-\sin x)}{\cos^2 x}$. So $\sec x\tan x = \dfrac{\sin x}{\cos^2 x}$.

(c) $\dfrac{d}{dx}(\sin x + \cos x) = \dfrac{d}{dx}\dfrac{1 + \cot x}{\csc x}$ $\Rightarrow$

$\cos x - \sin x = \dfrac{\csc x(-\csc^2 x) - (1 + \cot x)(-\csc x\cot x)}{\csc^2 x} = \dfrac{\csc x[-\csc^2 x + (1 + \cot x)\cot x]}{\csc^2 x} \cdot$

$$= \dfrac{-\csc^2 x + \cot^2 x + \cot x}{\csc x} = \dfrac{-1 + \cot x}{\csc x}$$

So $\cos x - \sin x = \dfrac{\cot x - 1}{\csc x}$.

49. By the definition of radian measure, $s = r\theta$, where r is the radius of the circle. By drawing the bisector of the angle θ, we can

see that $\sin\dfrac{\theta}{2} = \dfrac{d/2}{r} \;\Rightarrow\; d = 2r\sin\dfrac{\theta}{2}$. So $\displaystyle\lim_{\theta\to 0^+}\dfrac{s}{d} = \lim_{\theta\to 0^+}\dfrac{r\theta}{2r\sin(\theta/2)} = \lim_{\theta\to 0^+}\dfrac{2\cdot(\theta/2)}{2\sin(\theta/2)} = \lim_{\theta\to 0}\dfrac{\theta/2}{\sin(\theta/2)} = 1$.

[This is just the reciprocal of the limit $\displaystyle\lim_{x\to 0}\dfrac{\sin x}{x} = 1$ combined with the fact that as $\theta\to 0$, $\dfrac{\theta}{2}\to 0$ also.]

3.4 The Chain Rule

1. Let $u = g(x) = 1 + 4x$ and $y = f(u) = \sqrt[3]{u}$. Then $\dfrac{dy}{dx} = \dfrac{dy}{du}\dfrac{du}{dx} = \left(\tfrac{1}{3}u^{-2/3}\right)(4) = \dfrac{4}{3\sqrt[3]{(1+4x)^2}}$.

3. Let $u = g(x) = \pi x$ and $y = f(u) = \tan u$. Then $\dfrac{dy}{dx} = \dfrac{dy}{du}\dfrac{du}{dx} = (\sec^2 u)(\pi) = \pi\sec^2\pi x$.

5. Let $u = g(x) = \sqrt{x}$ and $y = f(u) = e^u$. Then $\dfrac{dy}{dx} = \dfrac{dy}{du}\dfrac{du}{dx} = (e^u)\left(\tfrac{1}{2}x^{-1/2}\right) = e^{\sqrt{x}}\cdot\dfrac{1}{2\sqrt{x}} = \dfrac{e^{\sqrt{x}}}{2\sqrt{x}}$.

7. $F(x) = (x^4 + 3x^2 - 2)^5 \;\Rightarrow\; F'(x) = 5(x^4 + 3x^2 - 2)^4\cdot\dfrac{d}{dx}(x^4 + 3x^2 - 2) = 5(x^4 + 3x^2 - 2)^4(4x^3 + 6x)$

$\left[\text{or } 10x(x^4 + 3x^2 - 2)^4(2x^2 + 3)\right]$

9. $F(x) = \sqrt{1 - 2x} = (1 - 2x)^{1/2} \;\Rightarrow\; F'(x) = \tfrac{1}{2}(1 - 2x)^{-1/2}(-2) = -\dfrac{1}{\sqrt{1 - 2x}}$

11. $f(z) = \dfrac{1}{z^2 + 1} = (z^2 + 1)^{-1} \;\Rightarrow\; f'(z) = -1(z^2 + 1)^{-2}(2z) = -\dfrac{2z}{(z^2 + 1)^2}$

13. $y = \cos(a^3 + x^3) \;\Rightarrow\; y' = -\sin(a^3 + x^3)\cdot 3x^2 \quad [a^3 \text{ is just a constant}] \quad = -3x^2\sin(a^3 + x^3)$

15. $h(t) = t^3 - 3^t \;\Rightarrow\; h'(t) = 3t^2 - 3^t\ln 3 \qquad$ [by Formula 5]

17. $y = xe^{-kx} \;\Rightarrow\; y' = x\left[e^{-kx}(-k)\right] + e^{-kx}\cdot 1 = e^{-kx}(-kx + 1) \quad \left[\text{or } (1 - kx)e^{-kx}\right]$

19. $y = (2x - 5)^4(8x^2 - 5)^{-3} \;\Rightarrow$

$y' = 4(2x - 5)^3(2)(8x^2 - 5)^{-3} + (2x - 5)^4(-3)(8x^2 - 5)^{-4}(16x)$

$= 8(2x - 5)^3(8x^2 - 5)^{-3} - 48x(2x - 5)^4(8x^2 - 5)^{-4}$

[This simplifies to $8(2x - 5)^3(8x^2 - 5)^{-4}(-4x^2 + 30x - 5)$.]

21. $y = e^{x\cos x} \;\Rightarrow\; y' = e^{x\cos x}\cdot\dfrac{d}{dx}(x\cos x) = e^{x\cos x}[x(-\sin x) + (\cos x)\cdot 1] = e^{x\cos x}(\cos x - x\sin x)$

23. $y = \left(\dfrac{x^2 + 1}{x^2 - 1}\right)^3 \;\Rightarrow$

$y' = 3\left(\dfrac{x^2 + 1}{x^2 - 1}\right)^2\cdot\dfrac{d}{dx}\left(\dfrac{x^2 + 1}{x^2 - 1}\right) = 3\left(\dfrac{x^2 + 1}{x^2 - 1}\right)^2\cdot\dfrac{(x^2 - 1)(2x) - (x^2 + 1)(2x)}{(x^2 - 1)^2}$

$= 3\left(\dfrac{x^2 + 1}{x^2 - 1}\right)^2\cdot\dfrac{2x[x^2 - 1 - (x^2 + 1)]}{(x^2 - 1)^2} = 3\left(\dfrac{x^2 + 1}{x^2 - 1}\right)^2\cdot\dfrac{2x(-2)}{(x^2 - 1)^2} = \dfrac{-12x(x^2 + 1)^2}{(x^2 - 1)^4}$

25. $y = \sec^2 x + \tan^2 x = (\sec x)^2 + (\tan x)^2 \quad \Rightarrow$

$y' = 2(\sec x)(\sec x \tan x) + 2(\tan x)(\sec^2 x) = 2\sec^2 x \tan x + 2\sec^2 x \tan x = 4\sec^2 x \tan x$

27. $y = \dfrac{r}{\sqrt{r^2+1}} \quad \Rightarrow$

$y' = \dfrac{\sqrt{r^2+1}\,(1) - r \cdot \frac{1}{2}(r^2+1)^{-1/2}(2r)}{\left(\sqrt{r^2+1}\right)^2} = \dfrac{\sqrt{r^2+1} - \dfrac{r^2}{\sqrt{r^2+1}}}{\left(\sqrt{r^2+1}\right)^2} = \dfrac{\dfrac{\sqrt{r^2+1}\sqrt{r^2+1} - r^2}{\sqrt{r^2+1}}}{\left(\sqrt{r^2+1}\right)^2}$

$= \dfrac{(r^2+1) - r^2}{\left(\sqrt{r^2+1}\right)^3} = \dfrac{1}{(r^2+1)^{3/2}}$ or $(r^2+1)^{-3/2}$

Another solution: Write y as a product and make use of the Product Rule. $y = r(r^2+1)^{-1/2} \quad \Rightarrow$

$y' = r \cdot -\frac{1}{2}(r^2+1)^{-3/2}(2r) + (r^2+1)^{-1/2} \cdot 1 = (r^2+1)^{-3/2}[-r^2 + (r^2+1)^1] = (r^2+1)^{-3/2}(1) = (r^2+1)^{-3/2}.$

The step that students usually have trouble with is factoring out $(r^2+1)^{-3/2}$. But this is no different than factoring out x^2 from $x^2 + x^5$; that is, we are just factoring out a factor with the *smallest* exponent that appears on it. In this case, $-\frac{3}{2}$ is smaller than $-\frac{1}{2}$.

29. $y = \sin(\tan 2x) \quad \Rightarrow \quad y' = \cos(\tan 2x) \cdot \dfrac{d}{dx}(\tan 2x) = \cos(\tan 2x) \cdot \sec^2(2x) \cdot \dfrac{d}{dx}(2x) = 2\cos(\tan 2x)\sec^2(2x)$

31. Using Formula 5 and the Chain Rule, $y = 2^{\sin \pi x} \quad \Rightarrow$

$y' = 2^{\sin \pi x}(\ln 2) \cdot \dfrac{d}{dx}(\sin \pi x) = 2^{\sin \pi x}(\ln 2) \cdot \cos \pi x \cdot \pi = 2^{\sin \pi x}(\pi \ln 2)\cos \pi x$

33. $y = \cot^2(\sin \theta) = [\cot(\sin \theta)]^2 \quad \Rightarrow$

$y' = 2[\cot(\sin \theta)] \cdot \dfrac{d}{d\theta}[\cot(\sin \theta)] = 2\cot(\sin \theta) \cdot [-\csc^2(\sin \theta) \cdot \cos \theta] = -2\cos \theta \cot(\sin \theta)\csc^2(\sin \theta)$

35. $y = \cos \sqrt{\sin(\tan \pi x)} = \cos(\sin(\tan \pi x))^{1/2} \quad \Rightarrow$

$y' = -\sin(\sin(\tan \pi x))^{1/2} \cdot \dfrac{d}{dx}(\sin(\tan \pi x))^{1/2} = -\sin(\sin(\tan \pi x))^{1/2} \cdot \frac{1}{2}(\sin(\tan \pi x))^{-1/2} \cdot \dfrac{d}{dx}(\sin(\tan \pi x))$

$= \dfrac{-\sin \sqrt{\sin(\tan \pi x)}}{2\sqrt{\sin(\tan \pi x)}} \cdot \cos(\tan \pi x) \cdot \dfrac{d}{dx}\tan \pi x = \dfrac{-\sin \sqrt{\sin(\tan \pi x)}}{2\sqrt{\sin(\tan \pi x)}} \cdot \cos(\tan \pi x) \cdot \sec^2(\pi x) \cdot \pi$

$= \dfrac{-\pi \cos(\tan \pi x)\sec^2(\pi x)\sin \sqrt{\sin(\tan \pi x)}}{2\sqrt{\sin(\tan \pi x)}}$

37. $y = \cos(x^2) \quad \Rightarrow \quad y' = -\sin(x^2) \cdot 2x = -2x\sin(x^2) \quad \Rightarrow$

$y'' = -2x\cos(x^2) \cdot 2x + \sin(x^2) \cdot (-2) = -4x^2\cos(x^2) - 2\sin(x^2)$

39. $y = e^{\alpha x}\sin \beta x \quad \Rightarrow \quad y' = e^{\alpha x} \cdot \beta \cos \beta x + \sin \beta x \cdot \alpha e^{\alpha x} = e^{\alpha x}(\beta \cos \beta x + \alpha \sin \beta x) \quad \Rightarrow$

$y'' = e^{\alpha x}(-\beta^2 \sin \beta x + \alpha \beta \cos \beta x) + (\beta \cos \beta x + \alpha \sin \beta x) \cdot \alpha e^{\alpha x}$

$= e^{\alpha x}(-\beta^2 \sin \beta x + \alpha \beta \cos \beta x + \alpha \beta \cos \beta x + \alpha^2 \sin \beta x) = e^{\alpha x}(\alpha^2 \sin \beta x - \beta^2 \sin \beta x + 2\alpha \beta \cos \beta x)$

$= e^{\alpha x}\left[(\alpha^2 - \beta^2)\sin \beta x + 2\alpha \beta \cos \beta x\right]$

41. $y = (1 + 2x)^{10}$ $\Rightarrow$ $y' = 10(1 + 2x)^9 \cdot 2 = 20(1 + 2x)^9$.

At $(0, 1)$, $y' = 20(1 + 0)^9 = 20$, and an equation of the tangent line is $y - 1 = 20(x - 0)$, or $y = 20x + 1$.

43. $y = \sin(\sin x)$ $\Rightarrow$ $y' = \cos(\sin x) \cdot \cos x$. At $(\pi, 0)$, $y' = \cos(\sin \pi) \cdot \cos \pi = \cos(0) \cdot (-1) = 1(-1) = -1$, and an equation of the tangent line is $y - 0 = -1(x - \pi)$, or $y = -x + \pi$.

45. (a) $y = \dfrac{2}{1 + e^{-x}}$ $\Rightarrow$ $y' = \dfrac{(1 + e^{-x})(0) - 2(-e^{-x})}{(1 + e^{-x})^2} = \dfrac{2e^{-x}}{(1 + e^{-x})^2}$.

(b)

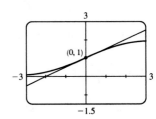

At $(0, 1)$, $y' = \dfrac{2e^0}{(1 + e^0)^2} = \dfrac{2(1)}{(1 + 1)^2} = \dfrac{2}{2^2} = \dfrac{1}{2}$. So an equation of the

tangent line is $y - 1 = \frac{1}{2}(x - 0)$ or $y = \frac{1}{2}x + 1$.

47. (a) $f(x) = x\sqrt{2 - x^2} = x(2 - x^2)^{1/2}$ $\Rightarrow$

$$f'(x) = x \cdot \tfrac{1}{2}(2 - x^2)^{-1/2}(-2x) + (2 - x^2)^{1/2} \cdot 1 = (2 - x^2)^{-1/2}\left[-x^2 + (2 - x^2)\right] = \dfrac{2 - 2x^2}{\sqrt{2 - x^2}}$$

(b)

$f' = 0$ when f has a horizontal tangent line, f' is negative when f is decreasing, and f' is positive when f is increasing.

49. For the tangent line to be horizontal, $f'(x) = 0$. $f(x) = 2\sin x + \sin^2 x$ $\Rightarrow$ $f'(x) = 2\cos x + 2\sin x \cos x = 0$ $\Leftrightarrow$ $2\cos x(1 + \sin x) = 0$ $\Leftrightarrow$ $\cos x = 0$ or $\sin x = -1$, so $x = \frac{\pi}{2} + 2n\pi$ or $\frac{3\pi}{2} + 2n\pi$, where n is any integer. Now $f\left(\frac{\pi}{2}\right) = 3$ and $f\left(\frac{3\pi}{2}\right) = -1$, so the points on the curve with a horizontal tangent are $\left(\frac{\pi}{2} + 2n\pi, 3\right)$ and $\left(\frac{3\pi}{2} + 2n\pi, -1\right)$, where n is any integer.

51. $F(x) = f(g(x))$ $\Rightarrow$ $F'(x) = f'(g(x)) \cdot g'(x)$, so $F'(5) = f'(g(5)) \cdot g'(5) = f'(-2) \cdot 6 = 4 \cdot 6 = 24$

53. (a) $h(x) = f(g(x))$ $\Rightarrow$ $h'(x) = f'(g(x)) \cdot g'(x)$, so $h'(1) = f'(g(1)) \cdot g'(1) = f'(2) \cdot 6 = 5 \cdot 6 = 30$.

(b) $H(x) = g(f(x))$ $\Rightarrow$ $H'(x) = g'(f(x)) \cdot f'(x)$, so $H'(1) = g'(f(1)) \cdot f'(1) = g'(3) \cdot 4 = 9 \cdot 4 = 36$.

55. (a) $u(x) = f(g(x))$ $\Rightarrow$ $u'(x) = f'(g(x))g'(x)$. So $u'(1) = f'(g(1))g'(1) = f'(3)g'(1)$. To find $f'(3)$, note that f is linear from $(2, 4)$ to $(6, 3)$, so its slope is $\dfrac{3 - 4}{6 - 2} = -\dfrac{1}{4}$. To find $g'(1)$, note that g is linear from $(0, 6)$ to $(2, 0)$, so its slope is $\dfrac{0 - 6}{2 - 0} = -3$. Thus, $f'(3)g'(1) = \left(-\frac{1}{4}\right)(-3) = \frac{3}{4}$.

(b) $v(x) = g(f(x))$ $\Rightarrow$ $v'(x) = g'(f(x))f'(x)$. So $v'(1) = g'(f(1))f'(1) = g'(2)f'(1)$, which does not exist since $g'(2)$ does not exist.

(c) $w(x) = g(g(x))$ $\Rightarrow$ $w'(x) = g'(g(x))g'(x)$. So $w'(1) = g'(g(1))g'(1) = g'(3)g'(1)$. To find $g'(3)$, note that g is linear from $(2, 0)$ to $(5, 2)$, so its slope is $\dfrac{2 - 0}{5 - 2} = \dfrac{2}{3}$. Thus, $g'(3)g'(1) = \left(\frac{2}{3}\right)(-3) = -2$.

57. $h(x) = f(g(x)) \Rightarrow h'(x) = f'(g(x))g'(x)$. So $h'(0.5) = f'(g(0.5))g'(0.5) = f'(0.1)g'(0.5)$.

We can estimate the derivatives by taking the average of two secant slopes.

For $f'(0.1)$: $m_1 = \dfrac{14.8 - 12.6}{0.1 - 0} = 22$, $m_2 = \dfrac{18.4 - 14.8}{0.2 - 0.1} = 36$. So $f'(0.1) \approx \dfrac{m_1 + m_2}{2} = \dfrac{22 + 36}{2} = 29$.

For $g'(0.5)$: $m_1 = \dfrac{0.10 - 0.17}{0.5 - 0.4} = -0.7$, $m_2 = \dfrac{0.05 - 0.10}{0.6 - 0.5} = -0.5$. So $g'(0.5) \approx \dfrac{m_1 + m_2}{2} = -0.6$.

Hence, $h'(0.5) = f'(0.1)g'(0.5) \approx (29)(-0.6) = -17.4$.

59. (a) $F(x) = f(e^x) \Rightarrow F'(x) = f'(e^x)\dfrac{d}{dx}(e^x) = f'(e^x)e^x$

(b) $G(x) = e^{f(x)} \Rightarrow G'(x) = e^{f(x)}\dfrac{d}{dx}f(x) = e^{f(x)}f'(x)$

61. $r(x) = f(g(h(x))) \Rightarrow r'(x) = f'(g(h(x))) \cdot g'(h(x)) \cdot h'(x)$, so

$r'(1) = f'(g(h(1))) \cdot g'(h(1)) \cdot h'(1) = f'(g(2)) \cdot g'(2) \cdot 4 = f'(3) \cdot 5 \cdot 4 = 6 \cdot 5 \cdot 4 = 120$

63. $F(x) = f(3f(4f(x))) \Rightarrow$

$F'(x) = f'(3f(4f(x))) \cdot \dfrac{d}{dx}(3f(4f(x))) = f'(3f(4f(x))) \cdot 3f'(4f(x)) \cdot \dfrac{d}{dx}(4f(x))$

$\qquad = f'(3f(4f(x))) \cdot 3f'(4f(x)) \cdot 4f'(x)$, so

$F'(0) = f'(3f(4f(0))) \cdot 3f'(4f(0)) \cdot 4f'(0) = f'(3f(4 \cdot 0)) \cdot 3f'(4 \cdot 0) \cdot 4 \cdot 2 = f'(3 \cdot 0) \cdot 3 \cdot 2 \cdot 4 \cdot 2 = 2 \cdot 3 \cdot 2 \cdot 4 \cdot 2 = 96$.

65. $y = e^{2x}(A\cos 3x + B\sin 3x) \Rightarrow$

$y' = e^{2x}(-3A\sin 3x + 3B\cos 3x) + (A\cos 3x + B\sin 3x) \cdot 2e^{2x}$

$\quad = e^{2x}(-3A\sin 3x + 3B\cos 3x + 2A\cos 3x + 2B\sin 3x)$

$\quad = e^{2x}[(2A + 3B)\cos 3x + (2B - 3A)\sin 3x] \Rightarrow$

$y'' = e^{2x}[-3(2A + 3B)\sin 3x + 3(2B - 3A)\cos 3x] + [(2A + 3B)\cos 3x + (2B - 3A)\sin 3x] \cdot 2e^{2x}$

$\quad = e^{2x}\{[-3(2A + 3B) + 2(2B - 3A)]\sin 3x + [3(2B - 3A) + 2(2A + 3B)]\cos 3x\}$

$\quad = e^{2x}[(-12A - 5B)\sin 3x + (-5A + 12B)\cos 3x]$

Substitute the expressions for y, y', and y'' in $y'' - 4y' + 13y$ to get

$y'' - 4y' + 13y = e^{2x}[(-12A - 5B)\sin 3x + (-5A + 12B)\cos 3x]$

$\qquad\qquad\qquad - 4e^{2x}[(2A + 3B)\cos 3x + (2B - 3A)\sin 3x] + 13e^{2x}(A\cos 3x + B\sin 3x)$

$\qquad\quad = e^{2x}[(-12A - 5B - 8B + 12A + 13B)\sin 3x + (-5A + 12B - 8A - 12B + 13A)\cos 3x]$

$\qquad\quad = e^{2x}[(0)\sin 3x + (0)\cos 3x] = 0$

Thus, the function y satisfies the differential equation $y'' - 4y' + 13y = 0$.

67. The use of $D, D^2, \ldots, D^n$ is just a derivative notation (see text page 150). In general, $Df(2x) = 2f'(2x)$,

$D^2f(2x) = 4f''(2x), \ldots, D^nf(2x) = 2^n f^{(n)}(2x)$. Since $f(x) = \cos x$ and $50 = 4(12) + 2$, we have

$f^{(50)}(x) = f^{(2)}(x) = -\cos x$, so $D^{50}\cos 2x = -2^{50}\cos 2x$.

69. $s(t) = 10 + \frac{1}{4}\sin(10\pi t)$ $\Rightarrow$ the velocity after t seconds is $v(t) = s'(t) = \frac{1}{4}\cos(10\pi t)(10\pi) = \frac{5\pi}{2}\cos(10\pi t)$ cm/s.

71. (a) $B(t) = 4.0 + 0.35\sin\dfrac{2\pi t}{5.4}$ $\Rightarrow$ $\dfrac{dB}{dt} = \left(0.35\cos\dfrac{2\pi t}{5.4}\right)\left(\dfrac{2\pi}{5.4}\right) = \dfrac{0.7\pi}{5.4}\cos\dfrac{2\pi t}{5.4} = \dfrac{7\pi}{54}\cos\dfrac{2\pi t}{5.4}$

　　(b) At $t = 1$, $\dfrac{dB}{dt} = \dfrac{7\pi}{54}\cos\dfrac{2\pi}{5.4} \approx 0.16$.

73. $s(t) = 2e^{-1.5t}\sin 2\pi t$ $\Rightarrow$

　　$v(t) = s'(t) = 2[e^{-1.5t}(\cos 2\pi t)(2\pi) + (\sin 2\pi t)e^{-1.5t}(-1.5)] = 2e^{-1.5t}(2\pi\cos 2\pi t - 1.5\sin 2\pi t)$

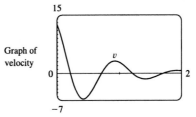

75. By the Chain Rule, $a(t) = \dfrac{dv}{dt} = \dfrac{dv}{ds}\dfrac{ds}{dt} = \dfrac{dv}{ds}v(t) = v(t)\dfrac{dv}{ds}$.   The derivative dv/dt is the rate of change of the velocity

with respect to time (in other words, the acceleration) whereas the derivative dv/ds is the rate of change of the velocity with

respect to the displacement.

77. (a) Using a calculator or CAS, we obtain the model $Q = ab^t$ with $a \approx 100.0124369$ and $b \approx 0.000045145933$.

　　(b) Use $Q'(t) = ab^t \ln b$ (from Formula 5) to get $Q'(0.04) \approx -670.63$ μA. The result of Example 2 in Section 2.1 was

　　-670 μA.

79. $x = t^4 + 1$, $y = t^3 + t$; $t = -1$. $\dfrac{dy}{dt} = 3t^2 + 1$, $\dfrac{dx}{dt} = 4t^3$, and $\dfrac{dy}{dx} = \dfrac{dy/dt}{dx/dt} = \dfrac{3t^2 + 1}{4t^3}$. When $t = -1$,

　　$(x, y) = (2, -2)$ and $dy/dx = \frac{4}{-4} = -1$, so an equation of the tangent to the curve at the point corresponding to $t = -1$

　　is $y - (-2) = (-1)(x - 2)$, or $y = -x$.

81. $x = e^{\sqrt{t}}$, $y = t - \ln t^2$; $t = 1$. $\dfrac{dy}{dt} = 1 - \dfrac{2t}{t^2} = 1 - \dfrac{2}{t}$, $\dfrac{dx}{dt} = \dfrac{e^{\sqrt{t}}}{2\sqrt{t}}$, and $\dfrac{dy}{dx} = \dfrac{dy/dt}{dx/dt} = \dfrac{1 - 2/t}{e^{\sqrt{t}}/(2\sqrt{t})} \cdot \dfrac{2t}{2t} = \dfrac{2t - 4}{\sqrt{t}e^{\sqrt{t}}}$.

　　When $t = 1$, $(x, y) = (e, 1)$ and $\dfrac{dy}{dx} = -\dfrac{2}{e}$, so an equation of the tangent line is $y - 1 = -\dfrac{2}{e}(x - e)$, or $y = -\dfrac{2}{e}x + 3$.

83. $x = 10 - t^2$, $y = t^3 - 12t$. $\dfrac{dy}{dt} = 3t^2 - 12 = 3(t + 2)(t - 2)$, so $\dfrac{dy}{dt} = 0$ $\Leftrightarrow$ $t = \pm 2$ $\Leftrightarrow$

　　$(x, y) = (6, \mp 16)$. $\dfrac{dx}{dt} = -2t$, so $\dfrac{dx}{dt} = 0$ $\Leftrightarrow$ $t = 0$ $\Leftrightarrow$ $(x, y) = (10, 0)$. The curve has horizontal tangents at $(6, \pm 16)$

　　and a vertical tangent at $(10, 0)$.

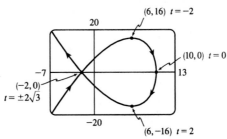

85. (a) $x = t^2$, $y = t^3 - 3t$ $\Rightarrow$ $\dfrac{dy}{dx} = \dfrac{dy/dt}{dx/dt} = \dfrac{3t^2 - 3}{2t}$. At the point $(3, 0)$, $x = 3$ $\Rightarrow$ $t^2 = 3$ $\Rightarrow$ $t = \pm\sqrt{3}$ $\Rightarrow$

$\dfrac{dy}{dx} = \dfrac{3(\pm\sqrt{3})^2 - 3}{2(\pm\sqrt{3})} = \dfrac{6}{2(\pm\sqrt{3})} = \pm\dfrac{3}{\sqrt{3}} = \pm\sqrt{3}$. When $t = \sqrt{3}$, an equation of the tangent line is

$y - 0 = \sqrt{3}(x - 3)$ or $y = \sqrt{3}\,x - 3\sqrt{3}$. When $t = -\sqrt{3}$, an equation of the tangent line is $y - 0 = -\sqrt{3}(x - 3)$

or $y = -\sqrt{3}\,x + 3\sqrt{3}$.

(b) Horizontal tangent: $dy/dx = 0$ $\Leftrightarrow$ $3t^2 - 3 = 0$ $\Leftrightarrow$

$3(t^2 - 1) = 0$ $\Leftrightarrow$ $t^2 = 1$ $\Leftrightarrow$ $t = \pm 1$. $t = 1$ corresponds to

the point $(x, y) = (t^2, t^3 - 3t) = (1^2, 1^3 - 3 \cdot 1) = (1, -2)$

and $t = -1$ to $(1, 2)$.

Vertical tangent: dy/dx is undefined $\Leftrightarrow$ $2t = 0$ $\Leftrightarrow$ $t = 0$.

The value $t = 0$ corresponds to the origin; that is, $(0, 0)$.

(c)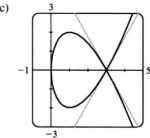

87. (a) Derive gives $g'(t) = \dfrac{45(t - 2)^8}{(2t + 1)^{10}}$ without simplifying. With either Maple or Mathematica, we first get

$g'(t) = 9\dfrac{(t - 2)^8}{(2t + 1)^9} - 18\dfrac{(t - 2)^9}{(2t + 1)^{10}}$, and the simplification command results in the expression given by Derive.

(b) Derive gives $y' = 2(x^3 - x + 1)^3(2x + 1)^4(17x^3 + 6x^2 - 9x + 3)$ without simplifying. With either Maple or

Mathematica, we first get $y' = 10(2x + 1)^4(x^3 - x + 1)^4 + 4(2x + 1)^5(x^3 - x + 1)^3(3x^2 - 1)$. If we use

Mathematica's Factor or Simplify, or Maple's factor, we get the above expression, but Maple's simplify gives

the polynomial expansion instead. For locating horizontal tangents, the factored form is the most helpful.

89. (a) $\dfrac{d}{dx}(\sin^n x \cos nx) = n\sin^{n-1} x \cos x \cos nx + \sin^n x (-n\sin nx)$ [Product Rule]

$= n\sin^{n-1} x (\cos nx \cos x - \sin nx \sin x)$ [factor out $n\sin^{n-1} x$]

$= n\sin^{n-1} x \cos(nx + x)$ [Addition Formula for cosine]

$= n\sin^{n-1} x \cos[(n + 1)x]$ [factor out x]

(b) $\dfrac{d}{dx}(\cos^n x \cos nx) = n\cos^{n-1} x (-\sin x)\cos nx + \cos^n x (-n\sin nx)$ [Product Rule]

$= -n\cos^{n-1} x (\cos nx \sin x + \sin nx \cos x)$ [factor out $-n\cos^{n-1} x$]

$= -n\cos^{n-1} x \sin(nx + x)$ [Addition Formula for sine]

$= -n\cos^{n-1} x \sin[(n + 1)x]$ [factor out x]

91. Since $\theta° = \left(\frac{\pi}{180}\right)\theta$ rad, we have $\dfrac{d}{d\theta}(\sin \theta°) = \dfrac{d}{d\theta}\left(\sin \frac{\pi}{180}\theta\right) = \frac{\pi}{180}\cos\frac{\pi}{180}\theta = \frac{\pi}{180}\cos\theta°$.

93. $\dfrac{d^2y}{dx^2} = \dfrac{d}{dx}\left(\dfrac{dy}{dx}\right)$ [Leibniz notation for the second derivative]

$\qquad = \dfrac{d}{dx}\left(\dfrac{dy}{du}\dfrac{du}{dx}\right)$ [Chain Rule]

$\qquad = \dfrac{dy}{du}\cdot\dfrac{d}{dx}\left(\dfrac{du}{dx}\right) + \dfrac{du}{dx}\cdot\dfrac{d}{dx}\left(\dfrac{dy}{du}\right)$ [Product Rule]

$\qquad = \dfrac{dy}{du}\cdot\dfrac{d^2u}{dx^2} + \dfrac{du}{dx}\cdot\dfrac{d}{du}\left(\dfrac{dy}{du}\right)\cdot\dfrac{du}{dx}$ [dy/du is a function of u]

$\qquad = \dfrac{dy}{du}\dfrac{d^2u}{dx^2} + \dfrac{d^2y}{du^2}\left(\dfrac{du}{dx}\right)^2$

Or: Using function notation for $y = f(u)$ and $u = g(x)$, we have $y = f(g(x))$, so

$y' = f'(g(x))\cdot g'(x)$ [by the Chain Rule] $\Rightarrow$

$(y')' = [f'(g(x))\cdot g'(x)]' = f'(g(x))\cdot g''(x) + g'(x)\cdot f''(g(x))\cdot g'(x) = f'(g(x))\cdot g''(x) + f''(g(x))\cdot [g'(x)]^2$.

3.5 Implicit Differentiation

1. (a) $\dfrac{d}{dx}\left(xy + 2x + 3x^2\right) = \dfrac{d}{dx}(4)$ $\Rightarrow$ $(x\cdot y' + y\cdot 1) + 2 + 6x = 0$ $\Rightarrow$ $xy' = -y - 2 - 6x$ $\Rightarrow$

$y' = \dfrac{-y - 2 - 6x}{x}$ or $y' = -6 - \dfrac{y+2}{x}$.

(b) $xy + 2x + 3x^2 = 4$ $\Rightarrow$ $xy = 4 - 2x - 3x^2$ $\Rightarrow$ $y = \dfrac{4 - 2x - 3x^2}{x} = \dfrac{4}{x} - 2 - 3x$, so $y' = -\dfrac{4}{x^2} - 3$.

(c) From part (a), $y' = \dfrac{-y - 2 - 6x}{x} = \dfrac{-(4/x - 2 - 3x) - 2 - 6x}{x} = \dfrac{-4/x - 3x}{x} = -\dfrac{4}{x^2} - 3$.

3. $\dfrac{d}{dx}\left(x^3 + y^3\right) = \dfrac{d}{dx}(1)$ $\Rightarrow$ $3x^2 + 3y^2\cdot y' = 0$ $\Rightarrow$ $3y^2\,y' = -3x^2$ $\Rightarrow$ $y' = -\dfrac{x^2}{y^2}$

5. $\dfrac{d}{dx}\left(x^2 + xy - y^2\right) = \dfrac{d}{dx}(4)$ $\Rightarrow$ $2x + x\cdot y' + y\cdot 1 - 2y\,y' = 0$ $\Rightarrow$

$xy' - 2y\,y' = -2x - y$ $\Rightarrow$ $(x - 2y)\,y' = -2x - y$ $\Rightarrow$ $y' = \dfrac{-2x - y}{x - 2y} = \dfrac{2x + y}{2y - x}$

7. $\dfrac{d}{dx}\left[x^4(x+y)\right] = \dfrac{d}{dx}\left[y^2(3x - y)\right]$ $\Rightarrow$ $x^4(1 + y') + (x + y)\cdot 4x^3 = y^2(3 - y') + (3x - y)\cdot 2y\,y'$ $\Rightarrow$

$x^4 + x^4\,y' + 4x^4 + 4x^3y = 3y^2 - y^2\,y' + 6xy\,y' - 2y^2\,y'$ $\Rightarrow$ $x^4\,y' + 3y^2\,y' - 6xy\,y' = 3y^2 - 5x^4 - 4x^3y$ $\Rightarrow$

$(x^4 + 3y^2 - 6xy)\,y' = 3y^2 - 5x^4 - 4x^3y$ $\Rightarrow$ $y' = \dfrac{3y^2 - 5x^4 - 4x^3y}{x^4 + 3y^2 - 6xy}$

9. $\dfrac{d}{dx}\left(x^2y^2 + x\sin y\right) = \dfrac{d}{dx}(4)$ $\Rightarrow$ $x^2\cdot 2y\,y' + y^2\cdot 2x + x\cos y\cdot y' + \sin y\cdot 1 = 0$ $\Rightarrow$

$2x^2y\,y' + x\cos y\cdot y' = -2xy^2 - \sin y$ $\Rightarrow$ $(2x^2y + x\cos y)y' = -2xy^2 - \sin y$ $\Rightarrow$ $y' = \dfrac{-2xy^2 - \sin y}{2x^2y + x\cos y}$

11. $\dfrac{d}{dx}(4\cos x\,\sin y) = \dfrac{d}{dx}(1) \quad\Rightarrow\quad 4\left[\cos x\cdot\cos y\cdot y' + \sin y\cdot(-\sin x)\right] = 0 \quad\Rightarrow$

$y'(4\cos x\,\cos y) = 4\sin x\,\sin y \quad\Rightarrow\quad y' = \dfrac{4\sin x\,\sin y}{4\cos x\,\cos y} = \tan x\,\tan y$

13. $\dfrac{d}{dx}(e^{x/y}) = \dfrac{d}{dx}(x-y) \quad\Rightarrow\quad e^{x/y}\cdot\dfrac{d}{dx}\left(\dfrac{x}{y}\right) = 1 - y' \quad\Rightarrow$

$e^{x/y}\cdot\dfrac{y\cdot 1 - x\cdot y'}{y^2} = 1 - y' \quad\Rightarrow\quad e^{x/y}\cdot\dfrac{1}{y} - \dfrac{xe^{x/y}}{y^2}\cdot y' = 1 - y' \quad\Rightarrow\quad y' - \dfrac{xe^{x/y}}{y^2}\cdot y' = 1 - \dfrac{e^{x/y}}{y} \quad\Rightarrow$

$y'\left(1 - \dfrac{xe^{x/y}}{y^2}\right) = \dfrac{y - e^{x/y}}{y} \quad\Rightarrow\quad y' = \dfrac{\dfrac{y - e^{x/y}}{y}}{\dfrac{y^2 - xe^{x/y}}{y^2}} = \dfrac{y(y - e^{x/y})}{y^2 - xe^{x/y}}$

15. $\dfrac{d}{dx}(e^y\cos x) = \dfrac{d}{dx}[1 + \sin(xy)] \quad\Rightarrow\quad e^y(-\sin x) + \cos x\cdot e^y\cdot y' = \cos(xy)\cdot(xy' + y\cdot 1) \quad\Rightarrow$

$-e^y\sin x + e^y\cos x\cdot y' = x\cos(xy)\cdot y' + y\cos(xy) \quad\Rightarrow\quad e^y\cos x\cdot y' - x\cos(xy)\cdot y' = e^y\sin x + y\cos(xy) \quad\Rightarrow$

$[e^y\cos x - x\cos(xy)]\,y' = e^y\sin x + y\cos(xy) \quad\Rightarrow\quad y' = \dfrac{e^y\sin x + y\cos(xy)}{e^y\cos x - x\cos(xy)}$

17. $\dfrac{d}{dx}\left\{f(x) + x^2[f(x)]^3\right\} = \dfrac{d}{dx}(10) \quad\Rightarrow\quad f'(x) + x^2\cdot 3[f(x)]^2\cdot f'(x) + [f(x)]^3\cdot 2x = 0.$ If $x = 1$, we have

$f'(1) + 1^2\cdot 3[f(1)]^2\cdot f'(1) + [f(1)]^3\cdot 2(1) = 0 \quad\Rightarrow\quad f'(1) + 1\cdot 3\cdot 2^2\cdot f'(1) + 2^3\cdot 2 = 0 \quad\Rightarrow$

$f'(1) + 12f'(1) = -16 \quad\Rightarrow\quad 13f'(1) = -16 \quad\Rightarrow\quad f'(1) = -\tfrac{16}{13}.$

19. $\dfrac{d}{dy}(x^4y^2 - x^3y + 2xy^3) = \dfrac{d}{dy}(0) \quad\Rightarrow\quad x^4\cdot 2y + y^2\cdot 4x^3\,x' - (x^3\cdot 1 + y\cdot 3x^2\,x') + 2(x\cdot 3y^2 + y^3\cdot x') = 0 \quad\Rightarrow$

$4x^3y^2\,x' - 3x^2y\,x' + 2y^3\,x' = -2x^4y + x^3 - 6xy^2 \quad\Rightarrow\quad (4x^3y^2 - 3x^2y + 2y^3)\,x' = -2x^4y + x^3 - 6xy^2 \quad\Rightarrow$

$x' = \dfrac{dx}{dy} = \dfrac{-2x^4y + x^3 - 6xy^2}{4x^3y^2 - 3x^2y + 2y^3}$

21. $y\sin 2x = x\cos 2y \quad\Rightarrow\quad y\cdot\cos 2x\cdot 2 + \sin 2x\cdot y' = x(-\sin 2y\cdot 2y') + \cos(2y)\cdot 1 \quad\Rightarrow$

$\sin 2x\cdot y' + 2x\sin 2y\cdot y' = -2y\cos 2x + \cos 2y \quad\Rightarrow$

$y'(\sin 2x + 2x\sin 2y) = -2y\cos 2x + \cos 2y \quad\Rightarrow\quad y' = \dfrac{-2y\cos 2x + \cos 2y}{\sin 2x + 2x\sin 2y}.$ When $x = \tfrac{\pi}{2}$ and $y = \tfrac{\pi}{4}$, we have

$y' = \dfrac{(-\pi/2)(-1) + 0}{0 + \pi\cdot 1} = \dfrac{\pi/2}{\pi} = \dfrac{1}{2}$, so an equation of the tangent line is $y - \tfrac{\pi}{4} = \tfrac{1}{2}(x - \tfrac{\pi}{2})$, or $y = \tfrac{1}{2}x.$

23. $x^2 + xy + y^2 = 3 \quad\Rightarrow\quad 2x + x\,y' + y\cdot 1 + 2yy' = 0 \quad\Rightarrow\quad x\,y' + 2y\,y' = -2x - y \quad\Rightarrow\quad y'(x + 2y) = -2x - y \quad\Rightarrow$

$y' = \dfrac{-2x - y}{x + 2y}.$ When $x = 1$ and $y = 1$, we have $y' = \dfrac{-2 - 1}{1 + 2\cdot 1} = \dfrac{-3}{3} = -1$, so an equation of the tangent line is

$y - 1 = -1(x - 1)$ or $y = -x + 2.$

25. $x^2 + y^2 = (2x^2 + 2y^2 - x)^2$ $\Rightarrow$ $2x + 2y\,y' = 2(2x^2 + 2y^2 - x)(4x + 4y\,y' - 1)$. When $x = 0$ and $y = \frac{1}{2}$, we have

$0 + y' = 2(\frac{1}{2})(2y' - 1)$ $\Rightarrow$ $y' = 2y' - 1$ $\Rightarrow$ $y' = 1$, so an equation of the tangent line is $y - \frac{1}{2} = 1(x - 0)$

or $y = x + \frac{1}{2}$.

27. $2(x^2 + y^2)^2 = 25(x^2 - y^2)$ $\Rightarrow$ $4(x^2 + y^2)(2x + 2y\,y') = 25(2x - 2y\,y')$ $\Rightarrow$

$4(x + y\,y')(x^2 + y^2) = 25(x - y\,y')$ $\Rightarrow$ $4y\,y'(x^2 + y^2) + 25yy' = 25x - 4x(x^2 + y^2)$ $\Rightarrow$

$y' = \dfrac{25x - 4x(x^2 + y^2)}{25y + 4y(x^2 + y^2)}$. When $x = 3$ and $y = 1$, we have $y' = \frac{75 - 120}{25 + 40} = -\frac{45}{65} = -\frac{9}{13}$,

so an equation of the tangent line is $y - 1 = -\frac{9}{13}(x - 3)$ or $y = -\frac{9}{13}x + \frac{40}{13}$.

29. (a) $y^2 = 5x^4 - x^2$ $\Rightarrow$ $2y\,y' = 5(4x^3) - 2x$ $\Rightarrow$ $y' = \dfrac{10x^3 - x}{y}$. (b)

So at the point $(1, 2)$ we have $y' = \dfrac{10(1)^3 - 1}{2} = \dfrac{9}{2}$, and an equation

of the tangent line is $y - 2 = \frac{9}{2}(x - 1)$ or $y = \frac{9}{2}x - \frac{5}{2}$.

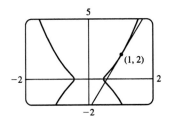

31. $9x^2 + y^2 = 9$ $\Rightarrow$ $18x + 2y\,y' = 0$ $\Rightarrow$ $2y\,y' = -18x$ $\Rightarrow$ $y' = -9x/y$ $\Rightarrow$

$y'' = -9\left(\dfrac{y \cdot 1 - x \cdot y'}{y^2}\right) = -9\left(\dfrac{y - x(-9x/y)}{y^2}\right) = -9 \cdot \dfrac{y^2 + 9x^2}{y^3} = -9 \cdot \dfrac{9}{y^3}$ [since x and y must satisfy the

original equation, $9x^2 + y^2 = 9$]. Thus, $y'' = -81/y^3$.

33. $x^3 + y^3 = 1$ $\Rightarrow$ $3x^2 + 3y^2\,y' = 0$ $\Rightarrow$ $y' = -\dfrac{x^2}{y^2}$ $\Rightarrow$

$y'' = -\dfrac{y^2(2x) - x^2 \cdot 2y\,y'}{(y^2)^2} = -\dfrac{2xy^2 - 2x^2y(-x^2/y^2)}{y^4} = -\dfrac{2xy^4 + 2x^4y}{y^6} = -\dfrac{2xy(y^3 + x^3)}{y^6} = -\dfrac{2x}{y^5}$,

since x and y must satisfy the original equation, $x^3 + y^3 = 1$.

35. If $x = 0$ in $xy + e^y = e$, then we get $0 + e^y = e$, so $y = 1$ and the point where $x = 0$ is $(0, 1)$. Differentiating implicitly

with respect to x gives us $xy' + y \cdot 1 + e^y y' = 0$. Substituting 0 for x and 1 for y gives us

$0 + 1 + ey' = 0$ $\Rightarrow$ $ey' = -1$ $\Rightarrow$ $y' = -1/e$. Differentiating $xy' + y + e^y y' = 0$ implicitly with respect to x gives

us $xy'' + y' \cdot 1 + y' + e^y y'' + y' \cdot e^y y' = 0$. Now substitute 0 for x, 1 for y, and $-1/e$ for y'.

$0 + \left(-\dfrac{1}{e}\right) + \left(-\dfrac{1}{e}\right) + ey'' + \left(-\dfrac{1}{e}\right)(e)\left(-\dfrac{1}{e}\right) = 0$ $\Rightarrow$ $-\dfrac{2}{e} + ey'' + \dfrac{1}{e} = 0$ $\Rightarrow$ $ey'' = \dfrac{1}{e}$ $\Rightarrow$ $y'' = \dfrac{1}{e^2}$.

37. (a) There are eight points with horizontal tangents: four at $x \approx 1.57735$ and

four at $x \approx 0.42265$.

(b) $y' = \dfrac{3x^2 - 6x + 2}{2(2y^3 - 3y^2 - y + 1)}$ $\Rightarrow$ $y' = -1$ at $(0, 1)$ and $y' = \frac{1}{3}$ at $(0, 2)$.

Equations of the tangent lines are $y = -x + 1$ and $y = \frac{1}{3}x + 2$.

(c) $y' = 0$ $\Rightarrow$ $3x^2 - 6x + 2 = 0$ $\Rightarrow$ $x = 1 \pm \frac{1}{3}\sqrt{3}$

(d) By multiplying the right side of the equation by $x - 3$, we obtain the first graph. By modifying the equation in other ways, we can generate the other graphs.

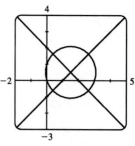

$$y(y^2 - 1)(y - 2)$$
$$= x(x - 1)(x - 2)(x - 3)$$

$$y(y^2 - 4)(y - 2)$$
$$= x(x - 1)(x - 2)$$

$$y(y + 1)(y^2 - 1)(y - 2)$$
$$= x(x - 1)(x - 2)$$

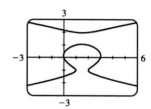

$$(y + 1)(y^2 - 1)(y - 2)$$
$$= (x - 1)(x - 2)$$

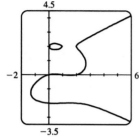

$$x(y + 1)(y^2 - 1)(y - 2)$$
$$= y(x - 1)(x - 2)$$

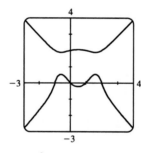

$$y(y^2 + 1)(y - 2)$$
$$= x(x^2 - 1)(x - 2)$$

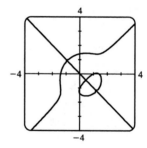

$$y(y + 1)(y^2 - 2)$$
$$= x(x - 1)(x^2 - 2)$$

39. From Exercise 27, a tangent to the lemniscate will be horizontal if $y' = 0 \Rightarrow 25x - 4x(x^2 + y^2) = 0 \Rightarrow$ $x[25 - 4(x^2 + y^2)] = 0 \Rightarrow x^2 + y^2 = \frac{25}{4}$ (**1**). (Note that when x is 0, y is also 0, and there is no horizontal tangent at the origin.) Substituting $\frac{25}{4}$ for $x^2 + y^2$ in the equation of the lemniscate, $2(x^2 + y^2)^2 = 25(x^2 - y^2)$, we get $x^2 - y^2 = \frac{25}{8}$ (**2**). Solving (**1**) and (**2**), we have $x^2 = \frac{75}{16}$ and $y^2 = \frac{25}{16}$, so the four points are $\left(\pm\frac{5\sqrt{3}}{4}, \pm\frac{5}{4}\right)$.

41. $x^2 + y^2 = r^2$ is a circle with center O and $ax + by = 0$ is a line through O [assume a and b are not both zero]. $x^2 + y^2 = r^2 \Rightarrow 2x + 2yy' = 0 \Rightarrow y' = -x/y$, so the slope of the tangent line at P_0 (x_0, y_0) is $-x_0/y_0$. The slope of the line OP_0 is y_0/x_0, which is the negative reciprocal of $-x_0/y_0$. Hence, the curves are orthogonal, and the families of curves are orthogonal trajectories of each other.

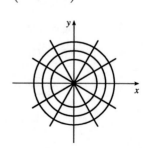

43. $y = cx^2 \Rightarrow y' = 2cx$ and $x^2 + 2y^2 = k$ [assume $k > 0$] $\Rightarrow 2x + 4yy' = 0 \Rightarrow$

$2yy' = -x \Rightarrow y' = -\dfrac{x}{2(y)} = -\dfrac{x}{2(cx^2)} = -\dfrac{1}{2cx}$, so the curves are orthogonal if

$c \neq 0$. If $c = 0$, then the horizontal line $y = cx^2 = 0$ intersects $x^2 + 2y^2 = k$ orthogonally

at $\left(\pm\sqrt{k}, 0\right)$, since the ellipse $x^2 + 2y^2 = k$ has vertical tangents at those two points.

45. Since $A^2 < a^2$, we are assured that there are four points of intersection.

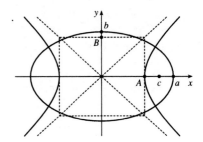

(1) $\dfrac{x^2}{a^2} + \dfrac{y^2}{b^2} = 1 \Rightarrow \dfrac{2x}{a^2} + \dfrac{2yy'}{b^2} = 0 \Rightarrow \dfrac{yy'}{b^2} = -\dfrac{x}{a^2} \Rightarrow y' = m_1 = -\dfrac{xb^2}{ya^2}$.

(2) $\dfrac{x^2}{A^2} - \dfrac{y^2}{B^2} = 1 \Rightarrow \dfrac{2x}{A^2} - \dfrac{2yy'}{B^2} = 0 \Rightarrow \dfrac{yy'}{B^2} = \dfrac{x}{A^2} \Rightarrow y' = m_2 = \dfrac{xB^2}{yA^2}$.

Now $m_1 m_2 = -\dfrac{xb^2}{ya^2} \cdot \dfrac{xB^2}{yA^2} = -\dfrac{b^2 B^2}{a^2 A^2} \cdot \dfrac{x^2}{y^2}$ **(3)**. Subtracting equations, **(1)** − **(2)**, gives us $\dfrac{x^2}{a^2} + \dfrac{y^2}{b^2} - \dfrac{x^2}{A^2} + \dfrac{y^2}{B^2} = 0 \Rightarrow$

$\dfrac{y^2}{b^2} + \dfrac{y^2}{B^2} = \dfrac{x^2}{A^2} - \dfrac{x^2}{a^2} \Rightarrow \dfrac{y^2 B^2 + y^2 b^2}{b^2 B^2} = \dfrac{x^2 a^2 - x^2 A^2}{A^2 a^2} \Rightarrow \dfrac{y^2(b^2 + B^2)}{b^2 B^2} = \dfrac{x^2(a^2 - A^2)}{a^2 A^2}$ **(4)**. Since

$a^2 - b^2 = A^2 + B^2$, we have $a^2 - A^2 = b^2 + B^2$. Thus, equation **(4)** becomes $\dfrac{y^2}{b^2 B^2} = \dfrac{x^2}{A^2 a^2} \Rightarrow \dfrac{x^2}{y^2} = \dfrac{A^2 a^2}{b^2 B^2}$, and

substituting for $\dfrac{x^2}{y^2}$ in equation **(3)** gives us $m_1 m_2 = -\dfrac{b^2 B^2}{a^2 A^2} \cdot \dfrac{a^2 A^2}{b^2 B^2} = -1$. Hence, the ellipse and hyperbola are orthogonal

trajectories.

47. (a) $\left(P + \dfrac{n^2 a}{V^2}\right)(V - nb) = nRT \Rightarrow PV - Pnb + \dfrac{n^2 a}{V} - \dfrac{n^3 ab}{V^2} = nRT \Rightarrow$

$\dfrac{d}{dP}(PV - Pnb + n^2 aV^{-1} - n^3 abV^{-2}) = \dfrac{d}{dP}(nRT) \Rightarrow$

$PV' + V \cdot 1 - nb - n^2 aV^{-2} \cdot V' + 2n^3 abV^{-3} \cdot V' = 0 \Rightarrow V'(P - n^2 aV^{-2} + 2n^3 abV^{-3}) = nb - V \Rightarrow$

$V' = \dfrac{nb - V}{P - n^2 aV^{-2} + 2n^3 abV^{-3}}$ or $\dfrac{dV}{dP} = \dfrac{V^3(nb - V)}{PV^3 - n^2 aV + 2n^3 ab}$

(b) Using the last expression for dV/dP from part (a), we get

$$\dfrac{dV}{dP} = \dfrac{(10\text{ L})^3[(1\text{ mole})(0.04267\text{ L/mole}) - 10\text{ L}]}{\left[\begin{array}{c}(2.5\text{ atm})(10\text{ L})^3 - (1\text{ mole})^2(3.592\text{ L}^2\text{-atm/mole}^2)(10\text{ L}) \\ + 2(1\text{ mole})^3(3.592\text{ L}^2\text{-atm/mole}^2)(0.04267\text{ L/mole})\end{array}\right]}$$

$$= \dfrac{-9957.33\text{ L}^4}{2464.386541\text{ L}^3\text{-atm}} \approx -4.04\text{ L/atm}.$$

49. If the circle has radius r, its equation is $x^2 + y^2 = r^2$ $\Rightarrow$ $2x + 2yy' = 0$ $\Rightarrow$ $y' = -\dfrac{x}{y}$, so the slope of the tangent line

at $P(x_0, y_0)$ is $-\dfrac{x_0}{y_0}$. The negative reciprocal of that slope is $\dfrac{-1}{-x_0/y_0} = \dfrac{y_0}{x_0}$, which is the slope of OP, so the tangent line at

P is perpendicular to the radius OP.

51. To find the points at which the ellipse $x^2 - xy + y^2 = 3$ crosses the x-axis, let $y = 0$ and solve for x.

$y = 0$ $\Rightarrow$ $x^2 - x(0) + 0^2 = 3$ $\Leftrightarrow$ $x = \pm\sqrt{3}$. So the graph of the ellipse crosses the x-axis at the points $(\pm\sqrt{3}, 0)$.

Using implicit differentiation to find y', we get $2x - xy' - y + 2yy' = 0$ $\Rightarrow$ $y'(2y - x) = y - 2x$ $\Leftrightarrow$ $y' = \dfrac{y - 2x}{2y - x}$.

So y' at $(\sqrt{3}, 0)$ is $\dfrac{0 - 2\sqrt{3}}{2(0) - \sqrt{3}} = 2$ and y' at $(-\sqrt{3}, 0)$ is $\dfrac{0 + 2\sqrt{3}}{2(0) + \sqrt{3}} = 2$. Thus, the tangent lines at these points are parallel.

53. $x^2y^2 + xy = 2$ $\Rightarrow$ $x^2 \cdot 2yy' + y^2 \cdot 2x + x \cdot y' + y \cdot 1 = 0$ $\Leftrightarrow$ $y'(2x^2y + x) = -2xy^2 - y$ $\Leftrightarrow$

$y' = -\dfrac{2xy^2 + y}{2x^2y + x}$. So $-\dfrac{2xy^2 + y}{2x^2y + x} = -1$ $\Leftrightarrow$ $2xy^2 + y = 2x^2y + x$ $\Leftrightarrow$ $y(2xy + 1) = x(2xy + 1)$ $\Leftrightarrow$

$y(2xy + 1) - x(2xy + 1) = 0$ $\Leftrightarrow$ $(2xy + 1)(y - x) = 0$ $\Leftrightarrow$ $xy = -\frac{1}{2}$ or $y = x$. But $xy = -\frac{1}{2}$ $\Rightarrow$

$x^2y^2 + xy = \frac{1}{4} - \frac{1}{2} \neq 2$, so we must have $x = y$. Then $x^2y^2 + xy = 2$ $\Rightarrow$ $x^4 + x^2 = 2$ $\Leftrightarrow$ $x^4 + x^2 - 2 = 0$ $\Leftrightarrow$

$(x^2 + 2)(x^2 - 1) = 0$. So $x^2 = -2$, which is impossible, or $x^2 = 1$ $\Leftrightarrow$ $x = \pm 1$. Since $x = y$, the points on the curve

where the tangent line has a slope of -1 are $(-1, -1)$ and $(1, 1)$.

55. (a) $y = J(x)$ and $xy'' + y' + xy = 0$ $\Rightarrow$ $xJ''(x) + J'(x) + xJ(x) = 0$. If $x = 0$, we have $0 + J'(0) + 0 = 0$,

so $J'(0) = 0$.

(b) Differentiating $xy'' + y' + xy = 0$ implicitly, we get $xy''' + y'' \cdot 1 + y'' + xy' + y \cdot 1 = 0$ $\Rightarrow$

$xy''' + 2y'' + xy' + y = 0$, so $xJ'''(x) + 2J''(x) + xJ'(x) + J(x) = 0$. If $x = 0$, we have

$0 + 2J''(0) + 0 + 1$ $[J(0) = 1$ is given$]$ $= 0$ $\Rightarrow$ $2J''(0) = -1$ $\Rightarrow$ $J''(0) = -\frac{1}{2}$.

3.6 Inverse Trigonometric Functions and Their Derivatives

1. (a) $\sin^{-1}\left(\frac{\sqrt{3}}{2}\right) = \frac{\pi}{3}$ since $\sin\frac{\pi}{3} = \frac{\sqrt{3}}{2}$ and $\frac{\pi}{3}$ is in $\left[-\frac{\pi}{2}, \frac{\pi}{2}\right]$.

(b) $\cos^{-1}(-1) = \pi$ since $\cos\pi = -1$ and π is in $[0, \pi]$.

3. (a) $\arctan 1 = \frac{\pi}{4}$ since $\tan\frac{\pi}{4} = 1$ and $\frac{\pi}{4}$ is in $\left(-\frac{\pi}{2}, \frac{\pi}{2}\right)$.

(b) $\sin^{-1}\frac{1}{\sqrt{2}} = \frac{\pi}{4}$ since $\sin\frac{\pi}{4} = \frac{1}{\sqrt{2}}$ and $\frac{\pi}{4}$ is in $\left[-\frac{\pi}{2}, \frac{\pi}{2}\right]$.

5. Let $\theta = \sin^{-1}\left(\frac{2}{3}\right)$.

Then $\tan\left(\sin^{-1}\left(\frac{2}{3}\right)\right) = \tan\theta = \dfrac{2}{\sqrt{5}}$.

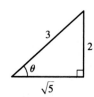

7. Let $\theta = \tan^{-1} \sqrt{2}$. Then

$$\sin\left(2 \tan^{-1} \sqrt{2}\right) = \sin(2\theta) = 2 \sin = 2\left(\frac{\sqrt{2}}{\sqrt{3}}\right)\left(\frac{1}{\sqrt{3}}\right) = \frac{2\sqrt{2}}{3}$$

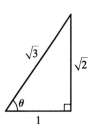

9. Let $y = \sin^{-1} x$. Then $-\frac{\pi}{2} \le y \le \frac{\pi}{2}$ $\Rightarrow$ $\cos y \ge 0$, so $\cos(\sin^{-1} x) = \cos y = \sqrt{1 - \sin^2 y} = \sqrt{1 - x^2}$.

11. Let $y = \tan^{-1} x$. Then $\tan y = x$, so from the triangle we see that

$$\sin(\tan^{-1} x) = \sin y = \frac{x}{\sqrt{1 + x^2}}.$$

13.

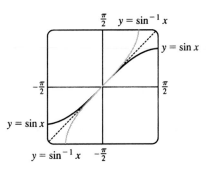

The graph of $\sin^{-1} x$ is the reflection of the graph of $\sin x$ about the line $y = x$.

15. Let $y = \cos^{-1} x$. Then $\cos y = x$ and $0 \le y \le \pi$ $\Rightarrow$ $-\sin y \dfrac{dy}{dx} = 1$ $\Rightarrow$

$$\frac{dy}{dx} = -\frac{1}{\sin y} = -\frac{1}{\sqrt{1 - \cos^2 y}} = -\frac{1}{\sqrt{1 - x^2}}. \quad \text{[Note that } \sin y \ge 0 \text{ for } 0 \le y \le \pi.]$$

17. $y = (\tan^{-1} x)^2$ $\Rightarrow$ $y' = 2(\tan^{-1} x)^1 \cdot \dfrac{d}{dx}(\tan^{-1} x) = 2 \tan^{-1} x \cdot \dfrac{1}{1 + x^2} = \dfrac{2 \tan^{-1} x}{1 + x^2}$

19. $y = \sin^{-1}(2x + 1)$ $\Rightarrow$

$$y' = \frac{1}{\sqrt{1 - (2x + 1)^2}} \cdot \frac{d}{dx}(2x + 1) = \frac{1}{\sqrt{1 - (4x^2 + 4x + 1)}} \cdot 2 = \frac{2}{\sqrt{-4x^2 - 4x}} = \frac{1}{\sqrt{-x^2 - x}}$$

21. $G(x) = \sqrt{1 - x^2} \arccos x$ $\Rightarrow$ $G'(x) = \sqrt{1 - x^2} \cdot \dfrac{-1}{\sqrt{1 - x^2}} + \arccos x \cdot \dfrac{1}{2}(1 - x^2)^{-1/2}(-2x) = -1 - \dfrac{x \arccos x}{\sqrt{1 - x^2}}$

23. $y = \cos^{-1}(e^{2x})$ $\Rightarrow$ $y' = -\dfrac{1}{\sqrt{1 - (e^{2x})^2}} \cdot \dfrac{d}{dx}(e^{2x}) = -\dfrac{2e^{2x}}{\sqrt{1 - e^{4x}}}$

25. $y = \arctan(\cos \theta)$ $\Rightarrow$ $y' = \dfrac{1}{1 + (\cos \theta)^2}(-\sin \theta) = -\dfrac{\sin \theta}{1 + \cos^2 \theta}$

27. $y = x \sin^{-1} x + \sqrt{1 - x^2}$ $\Rightarrow$

$$y' = x \cdot \frac{1}{\sqrt{1 - x^2}} + (\sin^{-1} x)(1) + \frac{1}{2}(1 - x^2)^{-1/2}(-2x) = \frac{x}{\sqrt{1 - x^2}} + \sin^{-1} x - \frac{x}{\sqrt{1 - x^2}} = \sin^{-1} x$$

29. $y = \arccos\left(\dfrac{b + a\cos x}{a + b\cos x}\right)$ $\Rightarrow$

$$y' = -\frac{1}{\sqrt{1 - \left(\dfrac{b + a\cos x}{a + b\cos x}\right)^2}} \frac{(a + b\cos x)(-a\sin x) - (b + a\cos x)(-b\sin x)}{(a + b\cos x)^2}$$

$$= \frac{1}{\sqrt{a^2 + b^2\cos^2 x - b^2 - a^2\cos^2 x}} \frac{(a^2 - b^2)\sin x}{|a + b\cos x|}$$

$$= \frac{1}{\sqrt{a^2 - b^2}\sqrt{1 - \cos^2 x}} \frac{(a^2 - b^2)\sin x}{|a + b\cos x|} = \frac{\sqrt{a^2 - b^2}}{|a + b\cos x|} \frac{\sin x}{|\sin x|}$$

But $0 \le x \le \pi$, so $|\sin x| = \sin x$. Also $a > b > 0$ $\Rightarrow$ $b\cos x \ge -b > -a$, so $a + b\cos x > 0$. Thus $y' = \dfrac{\sqrt{a^2 - b^2}}{a + b\cos x}$.

31. $g(x) = \cos^{-1}(3 - 2x)$ $\Rightarrow$ $g'(x) = -\dfrac{1}{\sqrt{1 - (3 - 2x)^2}}(-2) = \dfrac{2}{\sqrt{1 - (3 - 2x)^2}}$.

Domain$(g) = \{x \mid -1 \le 3 - 2x \le 1\} = \{x \mid -4 \le -2x \le -2\} = \{x \mid 2 \ge x \ge 1\} = [1, 2]$.

Domain$(g') = \{x \mid 1 - (3 - 2x)^2 > 0\} = \{x \mid (3 - 2x)^2 < 1\} = \{x \mid |3 - 2x| < 1\}$

$\quad = \{x \mid -1 < 3 - 2x < 1\} = \{x \mid -4 < -2x < -2\} = \{x \mid 2 > x > 1\} = (1, 2)$

33. $g(x) = x\sin^{-1}\left(\dfrac{x}{4}\right) + \sqrt{16 - x^2}$ $\Rightarrow$ $g'(x) = \sin^{-1}\left(\dfrac{x}{4}\right) + \dfrac{x}{4\sqrt{1 - (x/4)^2}} - \dfrac{x}{\sqrt{16 - x^2}} = \sin^{-1}\left(\dfrac{x}{4}\right)$ $\Rightarrow$

$g'(2) = \sin^{-1}\left(\tfrac{1}{2}\right) = \tfrac{\pi}{6}$

35. $f(x) = \sqrt{1 - x^2}\arcsin x$ $\Rightarrow$ $f'(x) = \sqrt{1 - x^2} \cdot \dfrac{1}{\sqrt{1 - x^2}} + \arcsin x \cdot \tfrac{1}{2}\left(1 - x^2\right)^{-1/2}(-2x) = 1 - \dfrac{x\arcsin x}{\sqrt{1 - x^2}}$

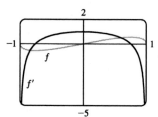

Note that $f' = 0$ where the graph of f has a horizontal tangent. Also note that f' is negative when f is decreasing and f' is positive when f is increasing.

37. $\lim\limits_{x \to -1^+} \sin^{-1} x = \sin^{-1}(-1) = -\tfrac{\pi}{2}$

39. Let $t = e^x$. As $x \to \infty$, $t \to \infty$. $\lim\limits_{x \to \infty} \arctan(e^x) = \lim\limits_{t \to \infty} \arctan t = \tfrac{\pi}{2}$ by (3).

41. (a) If $y = f^{-1}(x)$, then $f(y) = x$. Differentiating implicitly with respect to x and remembering that y is a function of x,

we get $f'(y)\dfrac{dy}{dx} = 1$, so $\dfrac{dy}{dx} = \dfrac{1}{f'(y)}$ $\Rightarrow$ $\left(f^{-1}\right)'(x) = \dfrac{1}{f'(f^{-1}(x))}$.

(b) $f(4) = 5$ $\Rightarrow$ $f^{-1}(5) = 4$. By part (a), $\left(f^{-1}\right)'(5) = \dfrac{1}{f'(f^{-1}(5))} = \dfrac{1}{f'(4)} = 1/\left(\tfrac{2}{3}\right) = \tfrac{3}{2}$.

43. (a) Let $f(x) = \sin x$, so $f^{-1}(x) = \sin^{-1} x$ and $f'(x) = \cos x$.

$$(f^{-1})'(x) = \frac{1}{f'(f^{-1}(x))} \quad \Rightarrow$$

$$\frac{d}{dx}(\sin^{-1} x) = \frac{1}{\cos(\sin^{-1} x)} = \frac{1}{\sqrt{1 - x^2}/1} = \frac{1}{\sqrt{1 - x^2}}, \text{ which is Formula 1.}$$

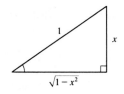

(Note that we could have used the result of Exercise 9.)

(b) Let $f(x) = \tan x$, so $f^{-1}(x) = \tan^{-1} x$ and $f'(x) = \sec^2 x$.

$$(f^{-1})'(x) = \frac{1}{f'(f^{-1}(x))} \quad \Rightarrow$$

$$\frac{d}{dx}(\tan^{-1} x) = \frac{1}{\sec^2(\tan^{-1} x)} = \frac{1}{[\sec(\tan^{-1} x)]^2} = \frac{1}{(\sqrt{1 + x^2})^2} = \frac{1}{1 + x^2}.$$

which is Formula 4.

3.7   Derivatives of Logarithmic Functions

1. The differentiation formula for logarithmic functions, $\dfrac{d}{dx}(\log_a x) = \dfrac{1}{x \ln a}$, is simplest when $a = e$ because $\ln e = 1$.

3. $f(x) = \sin(\ln x) \quad \Rightarrow \quad f'(x) = \cos(\ln x) \cdot \dfrac{d}{dx} \ln x = \cos(\ln x) \cdot \dfrac{1}{x} = \dfrac{\cos(\ln x)}{x}$

5. $f(x) = \log_2(1 - 3x) \quad \Rightarrow \quad f'(x) = \dfrac{1}{(1 - 3x)\ln 2} \dfrac{d}{dx}(1 - 3x) = \dfrac{-3}{(1 - 3x)\ln 2}$ or $\dfrac{3}{(3x - 1)\ln 2}$

7. $f(x) = \sqrt[5]{\ln x} = (\ln x)^{1/5} \quad \Rightarrow \quad f'(x) = \frac{1}{5}(\ln x)^{-4/5} \dfrac{d}{dx}(\ln x) = \dfrac{1}{5(\ln x)^{4/5}} \cdot \dfrac{1}{x} = \dfrac{1}{5x\sqrt[5]{(\ln x)^4}}$

9. $f(x) = \sin x \ln(5x) \quad \Rightarrow \quad f'(x) = \sin x \cdot \dfrac{1}{5x} \cdot \dfrac{d}{dx}(5x) + \ln(5x) \cdot \cos x = \dfrac{\sin x \cdot 5}{5x} + \cos x \ln(5x) = \dfrac{\sin x}{x} + \cos x \ln(5x)$

11. $F(t) = \ln \dfrac{(2t + 1)^3}{(3t - 1)^4} = \ln(2t + 1)^3 - \ln(3t - 1)^4 = 3\ln(2t + 1) - 4\ln(3t - 1) \quad \Rightarrow$

$$F'(t) = 3 \cdot \dfrac{1}{2t + 1} \cdot 2 - 4 \cdot \dfrac{1}{3t - 1} \cdot 3 = \dfrac{6}{2t + 1} - \dfrac{12}{3t - 1}, \text{ or combined, } \dfrac{-6(t + 3)}{(2t + 1)(3t - 1)}.$$

13. $g(x) = \ln(x\sqrt{x^2 - 1}) = \ln x + \ln(x^2 - 1)^{1/2} = \ln x + \frac{1}{2}\ln(x^2 - 1) \quad \Rightarrow$

$$g'(x) = \dfrac{1}{x} + \dfrac{1}{2} \cdot \dfrac{1}{x^2 - 1} \cdot 2x = \dfrac{1}{x} + \dfrac{x}{x^2 - 1} = \dfrac{x^2 - 1 + x \cdot x}{x(x^2 - 1)} = \dfrac{2x^2 - 1}{x(x^2 - 1)}$$

15. $y = \ln|2 - x - 5x^2| \quad \Rightarrow \quad y' = \dfrac{1}{2 - x - 5x^2} \cdot (-1 - 10x) = \dfrac{-10x - 1}{2 - x - 5x^2}$ or $\dfrac{10x + 1}{5x^2 + x - 2}$

17. $y = \ln(e^{-x} + xe^{-x}) = \ln(e^{-x}(1 + x)) = \ln(e^{-x}) + \ln(1 + x) = -x + \ln(1 + x) \quad \Rightarrow$

$$y' = -1 + \dfrac{1}{1 + x} = \dfrac{-1 - x + 1}{1 + x} = -\dfrac{x}{1 + x}$$

19. $y = 2x \log_{10} \sqrt{x} = 2x \log_{10} x^{1/2} = 2x \cdot \frac{1}{2} \log_{10} x = x \log_{10} x \quad \Rightarrow \quad y' = x \cdot \frac{1}{x \ln 10} + \log_{10} x \cdot 1 = \frac{1}{\ln 10} + \log_{10} x$

Note: $\frac{1}{\ln 10} = \frac{\ln e}{\ln 10} = \log_{10} e$, so the answer could be written as $\frac{1}{\ln 10} + \log_{10} x = \log_{10} e + \log_{10} x = \log_{10} ex$.

21. $y = x^2 \ln(2x) \quad \Rightarrow \quad y' = x^2 \cdot \frac{1}{2x} \cdot 2 + \ln(2x) \cdot (2x) = x + 2x \ln(2x) \quad \Rightarrow$

$y'' = 1 + 2x \cdot \frac{1}{2x} \cdot 2 + \ln(2x) \cdot 2 = 1 + 2 + 2 \ln(2x) = 3 + 2 \ln(2x)$

23. $f(x) = \frac{x}{1 - \ln(x - 1)} \quad \Rightarrow$

$f'(x) = \frac{[1 - \ln(x - 1)] \cdot 1 - x \cdot \frac{-1}{x - 1}}{[1 - \ln(x - 1)]^2} = \frac{\dfrac{(x - 1)[1 - \ln(x - 1)] + x}{x - 1}}{[1 - \ln(x - 1)]^2} = \frac{x - 1 - (x - 1) \ln(x - 1) + x}{(x - 1)[1 - \ln(x - 1)]^2}$

$= \frac{2x - 1 - (x - 1) \ln(x - 1)}{(x - 1)[1 - \ln(x - 1)]^2}$

$\text{Dom}(f) = \{x \mid x - 1 > 0 \quad \text{and} \quad 1 - \ln(x - 1) \neq 0\} = \{x \mid x > 1 \quad \text{and} \quad \ln(x - 1) \neq 1\}$

$= \{x \mid x > 1 \quad \text{and} \quad x - 1 \neq e^1\} = \{x \mid x > 1 \quad \text{and} \quad x \neq 1 + e\} = (1, 1 + e) \cup (1 + e, \infty)$

25. $y = \ln(x^2 - 3x + 1) \quad \Rightarrow \quad y' = \frac{1}{x^2 - 3x + 1} \cdot (2x - 3) \quad \Rightarrow \quad y'(3) = \frac{1}{1} \cdot 3 = 3$, so an equation of a tangent line at

$(3, 0)$ is $y - 0 = 3(x - 3)$, or $y = 3x - 9$.

27. $y = \ln\left(xe^{x^2}\right) = \ln x + \ln e^{x^2} = \ln x + x^2 \quad \Rightarrow \quad y' = \frac{1}{x} + 2x$. At $(1, 1)$, the slope of the tangent line is

$y'(1) = 1 + 2 = 3$, and an equation of the tangent line is $y - 1 = 3(x - 1)$, or $y = 3x - 2$.

29. (a) The domain of $f(x) = x \ln x$ is $(0, \infty)$. $f'(x) = x(1/x) + (\ln x) \cdot 1 = 1 + \ln x$. So $f'(x) < 0$ when

$1 + \ln x < 0 \quad \Leftrightarrow \quad \ln x < -1 \quad \Leftrightarrow \quad x < e^{-1}$. Therefore, f is decreasing on $(0, 1/e)$.

(b) $f'(x) = 1 + \ln x \quad \Rightarrow \quad f''(x) = 1/x > 0$ for $x > 0$. So the curve is concave upward on $(0, \infty)$.

31. $f(x) = cx + \ln(\cos x) \quad \Rightarrow \quad f'(x) = c + \frac{1}{\cos x} \cdot (-\sin x) = c - \tan x$.

$f'(\frac{\pi}{4}) = 6 \quad \Rightarrow \quad c - \tan \frac{\pi}{4} = 6 \quad \Rightarrow \quad c - 1 = 6 \quad \Rightarrow \quad c = 7$.

33. $y = (2x + 1)^5 (x^4 - 3)^6 \quad \Rightarrow \quad \ln y = \ln\left((2x + 1)^5 (x^4 - 3)^6\right) \quad \Rightarrow \quad \ln y = 5 \ln(2x + 1) + 6 \ln(x^4 - 3) \quad \Rightarrow$

$\frac{1}{y} y' = 5 \cdot \frac{1}{2x + 1} \cdot 2 + 6 \cdot \frac{1}{x^4 - 3} \cdot 4x^3 \quad \Rightarrow$

$y' = y\left(\frac{10}{2x + 1} + \frac{24x^3}{x^4 - 3}\right) = (2x + 1)^5 (x^4 - 3)^6 \left(\frac{10}{2x + 1} + \frac{24x^3}{x^4 - 3}\right)$.

[The answer could be simplified to $y' = 2(2x + 1)^4 (x^4 - 3)^5 (29x^4 + 12x^3 - 15)$, but this is unnecessary.]

35. $y = \dfrac{\sin^2 x \, \tan^4 x}{(x^2+1)^2}$ $\Rightarrow$ $\ln y = \ln(\sin^2 x \, \tan^4 x) - \ln(x^2+1)^2$ $\Rightarrow$

$\ln y = \ln(\sin x)^2 + \ln(\tan x)^4 - \ln(x^2+1)^2$ $\Rightarrow$ $\ln y = 2\ln|\sin x| + 4\ln|\tan x| - 2\ln(x^2+1)$ $\Rightarrow$

$\dfrac{1}{y}\,y' = 2 \cdot \dfrac{1}{\sin x} \cdot \cos x + 4 \cdot \dfrac{1}{\tan x} \cdot \sec^2 x - 2 \cdot \dfrac{1}{x^2+1} \cdot 2x$ $\Rightarrow$ $y' = \dfrac{\sin^2 x \, \tan^4 x}{(x^2+1)^2}\left(2\cot x + \dfrac{4\sec^2 x}{\tan x} - \dfrac{4x}{x^2+1}\right)$

37. $y = x^x$ $\Rightarrow$ $\ln y = \ln x^x$ $\Rightarrow$ $\ln y = x \ln x$ $\Rightarrow$ $y'/y = x(1/x) + (\ln x) \cdot 1$ $\Rightarrow$ $y' = y(1 + \ln x)$ $\Rightarrow$

$y' = x^x(1 + \ln x)$

39. $y = (\cos x)^x$ $\Rightarrow$ $\ln y = \ln(\cos x)^x$ $\Rightarrow$ $\ln y = x \ln \cos x$ $\Rightarrow$ $\dfrac{1}{y}\,y' = x \cdot \dfrac{1}{\cos x} \cdot (-\sin x) + \ln \cos x \cdot 1$ $\Rightarrow$

$y' = y\left(\ln \cos x - \dfrac{x \sin x}{\cos x}\right)$ $\Rightarrow$ $y' = (\cos x)^x(\ln \cos x - x \tan x)$

41. $y = (\tan x)^{1/x}$ $\Rightarrow$ $\ln y = \ln(\tan x)^{1/x}$ $\Rightarrow$ $\ln y = \dfrac{1}{x}\ln \tan x$ $\Rightarrow$

$\dfrac{1}{y}\,y' = \dfrac{1}{x} \cdot \dfrac{1}{\tan x} \cdot \sec^2 x + \ln \tan x \cdot \left(-\dfrac{1}{x^2}\right)$ $\Rightarrow$ $y' = y\left(\dfrac{\sec^2 x}{x \tan x} - \dfrac{\ln \tan x}{x^2}\right)$ $\Rightarrow$

$y' = (\tan x)^{1/x}\left(\dfrac{\sec^2 x}{x \tan x} - \dfrac{\ln \tan x}{x^2}\right)$ or $y' = (\tan x)^{1/x} \cdot \dfrac{1}{x}\left(\csc x \sec x - \dfrac{\ln \tan x}{x}\right)$

43. $y = \ln(x^2 + y^2)$ $\Rightarrow$ $y' = \dfrac{1}{x^2+y^2}\dfrac{d}{dx}(x^2+y^2)$ $\Rightarrow$ $y' = \dfrac{2x + 2yy'}{x^2+y^2}$ $\Rightarrow$ $x^2 y' + y^2 y' = 2x + 2yy'$ $\Rightarrow$

$x^2 y' + y^2 y' - 2yy' = 2x$ $\Rightarrow$ $(x^2 + y^2 - 2y)y' = 2x$ $\Rightarrow$ $y' = \dfrac{2x}{x^2+y^2-2y}$

45. $f(x) = \ln(x-1)$ $\Rightarrow$ $f'(x) = \dfrac{1}{(x-1)} = (x-1)^{-1}$ $\Rightarrow$ $f''(x) = -(x-1)^{-2}$ $\Rightarrow$ $f'''(x) = 2(x-1)^{-3}$ $\Rightarrow$

$f^{(4)}(x) = -2 \cdot 3(x-1)^{-4}$ $\Rightarrow$ $\cdots$ $\Rightarrow$ $f^{(n)}(x) = (-1)^{n-1} \cdot 2 \cdot 3 \cdot 4 \cdot \cdots \cdot (n-1)(x-1)^{-n} = (-1)^{n-1}\dfrac{(n-1)!}{(x-1)^n}$

47. If $f(x) = \ln(1+x)$, then $f'(x) = \dfrac{1}{1+x}$, so $f'(0) = 1$.

Thus, $\lim\limits_{x \to 0}\dfrac{\ln(1+x)}{x} = \lim\limits_{x \to 0}\dfrac{f(x)}{x} = \lim\limits_{x \to 0}\dfrac{f(x) - f(0)}{x - 0} = f'(0) = 1$.

3.8 Rates of Change in the Natural and Social Sciences

1. (a) $s = f(t) = t^3 - 12t^2 + 36t$ $\Rightarrow$ $v(t) = f'(t) = 3t^2 - 24t + 36$

(b) $v(3) = 27 - 72 + 36 = -9$ ft/s

(c) The particle is at rest when $v(t) = 0$. $3t^2 - 24t + 36 = 0$ $\Leftrightarrow$ $3(t-2)(t-6) = 0$ $\Leftrightarrow$ $t = 2$ s or 6 s.

(d) The particle is moving in the positive direction when $v(t) > 0$. $3(t-2)(t-6) > 0$ $\Leftrightarrow$ $0 \le t < 2$ or $t > 6$.

(e) Since the particle is moving in the positive direction and in the negative direction, we need to calculate the distance traveled in the intervals $[0, 2]$, $[2, 6]$, and $[6, 8]$ separately.

$$|f(2) - f(0)| = |32 - 0| = 32.$$

$$|f(6) - f(2)| = |0 - 32| = 32.$$

$$|f(8) - f(6)| = |32 - 0| = 32.$$

The total distance is $32 + 32 + 32 = 96$ ft.

(f)

(g) $v(t) = 3t^2 - 24t + 36 \Rightarrow$

$a(t) = v'(t) = 6t - 24.$

$a(3) = 6(3) - 24 = -6$ (ft/s)/s or ft/s^2.

(h)

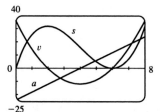

(i) The particle is speeding up when v and a have the same sign. This occurs when $2 < t < 4$ [v and a are both negative] and when $t > 6$ [v and a are both positive]. It is slowing down when v and a have opposite signs; that is, when $0 \le t < 2$ and when $4 < t < 6$.

3. (a) $s = f(t) = \cos(\pi t/4) \Rightarrow v(t) = f'(t) = -\sin(\pi t/4) \cdot (\pi/4)$

(b) $v(3) = -\frac{\pi}{4} \sin \frac{3\pi}{4} = -\frac{\pi}{4} \cdot \frac{\sqrt{2}}{2} = -\frac{\pi\sqrt{2}}{8}$ ft/s $[\approx -0.56]$

(c) The particle is at rest when $v(t) = 0$. $-\frac{\pi}{4} \sin \frac{\pi t}{4} = 0 \Rightarrow \sin \frac{\pi t}{4} = 0 \Rightarrow \frac{\pi t}{4} = \pi n \Rightarrow t = 0, 4, 8$ s.

(d) The particle is moving in the positive direction when $v(t) > 0$. $-\frac{\pi}{4} \sin \frac{\pi t}{4} > 0 \Rightarrow \sin \frac{\pi t}{4} < 0 \Rightarrow 4 < t < 8.$

(e) From part (c), $v(t) = 0$ for $t = 0, 4, 8$. As in Exercise 1, we'll find the distance traveled in the intervals $[0, 4]$ and $[4, 8]$.

$$|f(4) - f(0)| = |-1 - 1| = 2$$

$$|f(8) - f(4)| = |1 - (-1)| = 2.$$

The total distance is $2 + 2 = 4$ ft.

(f)

(g) $v(t) = -\frac{\pi}{4} \sin \frac{\pi t}{4} \Rightarrow$

$a(t) = v'(t) = -\frac{\pi}{4} \cos \frac{\pi t}{4} \cdot \frac{\pi}{4} = -\frac{\pi^2}{16} \cos \frac{\pi t}{4}.$

$a(3) = -\frac{\pi^2}{16} \cos \frac{3\pi}{4} = -\frac{\pi^2}{16} \left(-\frac{\sqrt{2}}{2} \right) = \frac{\pi^2 \sqrt{2}}{32}$ (ft/s)/s or ft/s^2.

(h)

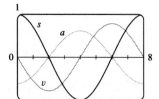

(i) The particle is speeding up when v and a have the same sign. This occurs when $0 < t < 2$ or $8 < t < 10$ [v and a are both negative] and when $4 < t < 6$ [v and a are both positive]. It is slowing down when v and a have opposite signs; that is, when $2 < t < 4$ and when $6 < t < 8$.

5. (a) From the figure, the velocity v is positive on the interval $(0, 2)$ and negative on the interval $(2, 3)$. The acceleration a is positive (negative) when the slope of the tangent line is positive (negative), so the acceleration is positive on the interval

$(0, 1)$, and negative on the interval $(1, 3)$. The particle is speeding up when v and a have the same sign, that is, on the interval $(0, 1)$ when $v > 0$ and $a > 0$, and on the interval $(2, 3)$ when $v < 0$ and $a < 0$. The particle is slowing down when v and a have opposite signs, that is, on the interval $(1, 2)$ when $v > 0$ and $a < 0$.

(b) $v > 0$ on $(0, 3)$ and $v < 0$ on $(3, 4)$. $a > 0$ on $(1, 2)$ and $a < 0$ on $(0, 1)$ and $(2, 4)$. The particle is speeding up on $(1, 2)$ $[v > 0, a > 0]$ and on $(3, 4)$ $[v < 0, a < 0]$. The particle is slowing down on $(0, 1)$ and $(2, 3)$ $[v > 0, a < 0]$.

7. (a) $s(t) = t^3 - 4.5t^2 - 7t$ $\Rightarrow$ $v(t) = s'(t) = 3t^2 - 9t - 7 = 5$ $\Leftrightarrow$ $3t^2 - 9t - 12 = 0$ $\Leftrightarrow$

$3(t - 4)(t + 1) = 0$ $\Leftrightarrow$ $t = 4$ or -1. Since $t \geq 0$, the particle reaches a velocity of 5 m/s at $t = 4$ s.

(b) $a(t) = v'(t) = 6t - 9 = 0$ $\Leftrightarrow$ $t = 1.5$. The acceleration changes from negative to positive, so the velocity changes from decreasing to increasing. Thus, at $t = 1.5$ s, the velocity has its minimum value.

9. (a) $h = 10t - 0.83t^2$ $\Rightarrow$ $v(t) = \dfrac{dh}{dt} = 10 - 1.66t$, so $v(3) = 10 - 1.66(3) = 5.02$ m/s.

(b) $h = 25$ $\Rightarrow$ $10t - 0.83t^2 = 25$ $\Rightarrow$ $0.83t^2 - 10t + 25 = 0$ $\Rightarrow$ $t = \dfrac{10 \pm \sqrt{17}}{1.66} \approx 3.54$ or 8.51.

The value $t_1 = \dfrac{10 - \sqrt{17}}{1.66}$ corresponds to the time it takes for the stone to rise 25 m and $t_2 = \dfrac{10 + \sqrt{17}}{1.66}$ corresponds to the time when the stone is 25 m high on the way down. Thus, $v(t_1) = 10 - 1.66\left(\dfrac{10 - \sqrt{17}}{1.66}\right) = \sqrt{17} \approx 4.12$ m/s.

11. (a) $A(x) = x^2$ $\Rightarrow$ $A'(x) = 2x$. $A'(15) = 30$ mm²/mm is the rate at which the area is increasing with respect to the side length as x reaches 15 mm.

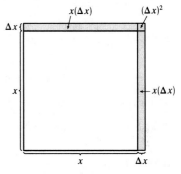

(b) The perimeter is $P(x) = 4x$, so $A'(x) = 2x = \frac{1}{2}(4x) = \frac{1}{2}P(x)$. The figure suggests that if Δx is small, then the change in the area of the square is approximately half of its perimeter (2 of the 4 sides) times Δx. From the figure, $\Delta A = 2x\,(\Delta x) + (\Delta x)^2$. If Δx is small, then $\Delta A \approx 2x\,(\Delta x)$ and so $\Delta A / \Delta x \approx 2x$.

13. (a) Using $A(r) = \pi r^2$, we find that the average rate of change is:

(i) $\dfrac{A(3) - A(2)}{3 - 2} = \dfrac{9\pi - 4\pi}{1} = 5\pi$

(ii) $\dfrac{A(2.5) - A(2)}{2.5 - 2} = \dfrac{6.25\pi - 4\pi}{0.5} = 4.5\pi$

(iii) $\dfrac{A(2.1) - A(2)}{2.1 - 2} = \dfrac{4.41\pi - 4\pi}{0.1} = 4.1\pi$

(b) $A(r) = \pi r^2$ $\Rightarrow$ $A'(r) = 2\pi r$, so $A'(2) = 4\pi$.

(c) The circumference is $C(r) = 2\pi r = A'(r)$. The figure suggests that if Δr is small, then the change in the area of the circle (a ring around the outside) is approximately equal to its circumference times Δr. Straightening out this ring gives us a shape that is approximately rectangular with length $2\pi r$ and width Δr, so $\Delta A \approx 2\pi r(\Delta r)$. Algebraically, $\Delta A = A(r + \Delta r) - A(r) = \pi(r + \Delta r)^2 - \pi r^2 = 2\pi r(\Delta r) + \pi(\Delta r)^2$. So we see that if Δr is small, then $\Delta A \approx 2\pi r(\Delta r)$ and therefore, $\Delta A / \Delta r \approx 2\pi r$.

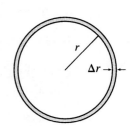

15. $S(r) = 4\pi r^2 \quad \Rightarrow \quad S'(r) = 8\pi r \quad \Rightarrow$

(a) $S'(1) = 8\pi$ ft²/ft   (b) $S'(2) = 16\pi$ ft²/ft   (c) $S'(3) = 24\pi$ ft²/ft

As the radius increases, the surface area grows at an increasing rate. In fact, the rate of change is linear with respect to the radius.

17. The mass is $f(x) = 3x^2$, so the linear density at x is $\rho(x) = f'(x) = 6x$.

(a) $\rho(1) = 6$ kg/m   (b) $\rho(2) = 12$ kg/m   (c) $\rho(3) = 18$ kg/m

Since ρ is an increasing function, the density will be the highest at the right end of the rod and lowest at the left end.

19. The quantity of charge is $Q(t) = t^3 - 2t^2 + 6t + 2$, so the current is $Q'(t) = 3t^2 - 4t + 6$.

(a) $Q'(0.5) = 3(0.5)^2 - 4(0.5) + 6 = 4.75$ A   (b) $Q'(1) = 3(1)^2 - 4(1) + 6 = 5$ A

The current is lowest when Q' has a minimum. $Q''(t) = 6t - 4 < 0$ when $t < \frac{2}{3}$. So the current decreases when $t < \frac{2}{3}$ and increases when $t > \frac{2}{3}$. Thus, the current is lowest at $t = \frac{2}{3}$ s.

21. (a) To find the rate of change of volume with respect to pressure, we first solve for V in terms of P.

$$PV = C \quad \Rightarrow \quad V = \frac{C}{P} \quad \Rightarrow \quad \frac{dV}{dP} = -\frac{C}{P^2}.$$

(b) From the formula for dV/dP in part (a), we see that as P increases, the absolute value of dV/dP decreases.

Thus, the volume is decreasing more rapidly at the beginning.

(c) $\beta = -\dfrac{1}{V}\dfrac{dV}{dP} = -\dfrac{1}{V}\left(-\dfrac{C}{P^2}\right) = \dfrac{C}{(PV)P} = \dfrac{C}{CP} = \dfrac{1}{P}$

23. In Example 6, the population function was $n = 2^t n_0$. Since we are tripling instead of doubling and the initial population is 400, the population function is $n(t) = 400 \cdot 3^t$. The rate of growth is $n'(t) = 400 \cdot 3^t \cdot \ln 3$, so the rate of growth after 2.5 hours is $n'(2.5) = 400 \cdot 3^{2.5} \cdot \ln 3 \approx 6850$ bacteria/hour.

25. (a) **1920:** $m_1 = \dfrac{1860 - 1750}{1920 - 1910} = \dfrac{110}{10} = 11$, $m_2 = \dfrac{2070 - 1860}{1930 - 1920} = \dfrac{210}{10} = 21$,

$(m_1 + m_2)/2 = (11 + 21)/2 = 16$ million/year

1980: $m_1 = \dfrac{4450 - 3710}{1980 - 1970} = \dfrac{740}{10} = 74$, $m_2 = \dfrac{5280 - 4450}{1990 - 1980} = \dfrac{830}{10} = 83$,

$(m_1 + m_2)/2 = (74 + 83)/2 = 78.5$ million/year

(b) $P(t) = at^3 + bt^2 + ct + d$ (in millions of people), where $a \approx 0.0012937063$, $b \approx -7.061421911$, $c \approx 12{,}822.97902$, and $d \approx -7{,}743{,}770.396$.

(c) $P(t) = at^3 + bt^2 + ct + d \quad \Rightarrow \quad P'(t) = 3at^2 + 2bt + c$ (in millions of people per year)

(d) $P'(1920) = 3(0.0012937063)(1920)^2 + 2(-7.061421911)(1920) + 12{,}822.97902$

≈ 14.48 million/year [smaller than the answer in part (a), but close to it]

$P'(1980) \approx 75.29$ million/year (smaller, but close)

(e) $P'(1985) \approx 81.62$ million/year, so the rate of growth in 1985 was about 81.62 million/year.

27. (a) Using $v = \dfrac{P}{4\eta l}(R^2 - r^2)$ with $R = 0.01$, $l = 3$, $P = 3000$, and $\eta = 0.027$, we have v as a function of r:

$$v(r) = \frac{3000}{4(0.027)3}(0.01^2 - r^2). \quad v(0) = 0.9\overline{25} \text{ cm/s}, \ v(0.005) = 0.69\overline{4} \text{ cm/s}, \ v(0.01) = 0.$$

(b) $v(r) = \dfrac{P}{4\eta l}(R^2 - r^2) \ \Rightarrow \ v'(r) = \dfrac{P}{4\eta l}(-2r) = -\dfrac{Pr}{2\eta l}.$ When $l = 3$, $P = 3000$, and $\eta = 0.027$, we have

$$v'(r) = -\frac{3000r}{2(0.027)3}. \quad v'(0) = 0, \ v'(0.005) = -92.5\overline{92} \text{ (cm/s)/cm, and } v'(0.01) = -185.\overline{185} \text{ (cm/s)/cm}.$$

(c) The velocity is greatest where $r = 0$ (at the center) and the velocity is changing most where $r = R = 0.01$ cm

(at the edge).

29. (a) $C(x) = 1200 + 12x - 0.1x^2 + 0.0005x^3 \ \Rightarrow \ C'(x) = 12 - 0.2x + 0.0015x^2$ \$/yard, which is the marginal cost

function.

(b) $C'(200) = 12 - 0.2(200) + 0.0015(200)^2 = \$32/\text{yard}$, and this is the rate at which costs are increasing with respect to

the production level when $x = 200$. $C'(200)$ predicts the cost of producing the 201st yard.

(c) The cost of manufacturing the 201st yard of fabric is $C(201) - C(200) = 3632.2005 - 3600 \approx \$32.20$, which is

approximately $C'(200)$.

31. (a) $A(x) = \dfrac{p(x)}{x} \ \Rightarrow \ A'(x) = \dfrac{xp'(x) - p(x) \cdot 1}{x^2} = \dfrac{xp'(x) - p(x)}{x^2}.$

$A'(x) > 0 \ \Rightarrow \ A(x)$ is increasing; that is, the average productivity increases as the size of the workforce increases.

(b) $p'(x)$ is greater than the average productivity $\ \Rightarrow \ p'(x) > A(x) \ \Rightarrow \ p'(x) > \dfrac{p(x)}{x} \ \Rightarrow \ xp'(x) > p(x) \ \Rightarrow$

$xp'(x) - p(x) > 0 \ \Rightarrow \ \dfrac{xp'(x) - p(x)}{x^2} > 0 \ \Rightarrow \ A'(x) > 0.$

33. $PV = nRT \ \Rightarrow \ T = \dfrac{PV}{nR} = \dfrac{PV}{(10)(0.0821)} = \dfrac{1}{0.821}(PV).$ Using the Product Rule, we have

$$\frac{dT}{dt} = \frac{1}{0.821}[P(t)V'(t) + V(t)P'(t)] = \frac{1}{0.821}[(8)(-0.15) + (10)(0.10)] \approx -0.2436 \text{ K/min}.$$

35. (a) If the populations are stable, then the growth rates are neither positive nor negative; that is, $\dfrac{dC}{dt} = 0$ and $\dfrac{dW}{dt} = 0.$

(b) "The caribou go extinct" means that the population is zero, or mathematically, $C = 0.$

(c) We have the equations $\dfrac{dC}{dt} = aC - bCW$ and $\dfrac{dW}{dt} = -cW + dCW.$ Let $dC/dt = dW/dt = 0$, $a = 0.05$, $b = 0.001$,

$c = 0.05$, and $d = 0.0001$ to obtain $0.05C - 0.001CW = 0$ **(1)** and $-0.05W + 0.0001CW = 0$ **(2)**. Adding 10 times

(2) to **(1)** eliminates the CW-terms and gives us $0.05C - 0.5W = 0 \ \Rightarrow \ C = 10W.$ Substituting $C = 10W$ into **(1)**

results in $0.05(10W) - 0.001(10W)W = 0 \ \Leftrightarrow \ 0.5W - 0.01W^2 = 0 \ \Leftrightarrow \ 50W - W^2 = 0 \ \Leftrightarrow$

$W(50 - W) = 0 \ \Leftrightarrow \ W = 0$ or 50. Since $C = 10W$, $C = 0$ or 500. Thus, the population pairs (C, W) that lead to

stable populations are $(0, 0)$ and $(500, 50)$. So it is possible for the two species to live in harmony.

3.9 Linear Approximations and Differentials

1. As in Example 1, $T(0) = 185$, $T(10) = 172$, $T(20) = 160$, and

$$T'(20) \approx \frac{T(10) - T(20)}{10 - 20} = \frac{172 - 160}{-10} = -1.2°\text{F/min}.$$

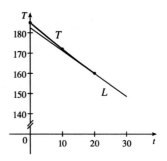

$$T(30) \approx T(20) + T'(20)(30 - 20) \approx 160 - 1.2(10) = 148°\text{F}.$$

We would expect the temperature of the turkey to get closer to $75°$F as time

increases. Since the temperature decreased $13°$F in the first 10 minutes and $12°$F

in the second 10 minutes, we can assume that the slopes of the tangent line are

increasing through negative values: $-1.3, -1.2, \ldots$. Hence, the tangent lines are

under the curve and $148°$F is an underestimate. From the figure, we estimate the slope of the tangent line at $t = 20$ to be

$\frac{184 - 147}{0 - 30} = -\frac{37}{30}$. Then the linear approximation becomes $T(30) \approx T(20) + T'(20) \cdot 10 \approx 160 - \frac{37}{30}(10) = 147\frac{2}{3} \approx 147.7$.

3. Extend the tangent line at the point $(2030, 21)$ to the t-axis. Answers

will vary based on this approximation—we'll use $t = 1900$ as our

t-intercept. The linearization is then

$$P(t) \approx P(2030) + P'(2030)(t - 2030) \approx 21 + \frac{21}{130}(t - 2030)$$

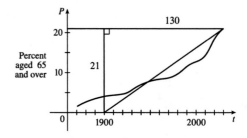

$$P(2040) = 21 + \frac{21}{130}(2040 - 2030) \approx 22.6\%$$

$$P(2050) = 21 + \frac{21}{130}(2050 - 2030) \approx 24.2\%$$

These predictions are probably too high since the tangent line lies above the graph at $t = 2030$.

5. $f(x) = x^4 + 3x^2 \Rightarrow f'(x) = 4x^3 + 6x$, so $f(-1) = 4$ and $f'(-1) = -10$.

Thus, $L(x) = f(-1) + f'(-1)(x - (-1)) = 4 + (-10)(x + 1) = -10x - 6$.

7. $f(x) = \cos x \Rightarrow f'(x) = -\sin x$, so $f\left(\frac{\pi}{2}\right) = 0$ and $f'\left(\frac{\pi}{2}\right) = -1$.

Thus, $L(x) = f\left(\frac{\pi}{2}\right) + f'\left(\frac{\pi}{2}\right)\left(x - \frac{\pi}{2}\right) = 0 - 1\left(x - \frac{\pi}{2}\right) = -x + \frac{\pi}{2}$.

9. $f(x) = \sqrt{1 - x} \Rightarrow f'(x) = \dfrac{-1}{2\sqrt{1 - x}}$, so $f(0) = 1$ and $f'(0) = -\frac{1}{2}$.

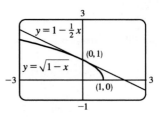

Therefore,

$$\sqrt{1 - x} = f(x) \approx f(0) + f'(0)(x - 0) = 1 + \left(-\frac{1}{2}\right)(x - 0) = 1 - \frac{1}{2}x.$$

So $\sqrt{0.9} = \sqrt{1 - 0.1} \approx 1 - \frac{1}{2}(0.1) = 0.95$

and $\sqrt{0.99} = \sqrt{1 - 0.01} \approx 1 - \frac{1}{2}(0.01) = 0.995$.

11. $f(x) = \sqrt[3]{1 - x} = (1 - x)^{1/3} \Rightarrow f'(x) = -\frac{1}{3}(1 - x)^{-2/3}$, so $f(0) = 1$

and $f'(0) = -\frac{1}{3}$. Thus, $f(x) \approx f(0) + f'(0)(x - 0) = 1 - \frac{1}{3}x$. We need

$\sqrt[3]{1 - x} - 0.1 < 1 - \frac{1}{3}x < \sqrt[3]{1 - x} + 0.1$, which is true when

$-1.204 < x < 0.706$.

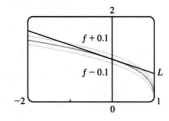

13. $f(x) = \dfrac{1}{(1+2x)^4} = (1+2x)^{-4}$ $\Rightarrow$

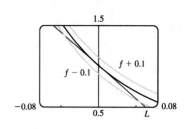

$f'(x) = -4(1+2x)^{-5}(2) = \dfrac{-8}{(1+2x)^5}$, so $f(0) = 1$ and $f'(0) = -8$.

Thus, $f(x) \approx f(0) + f'(0)(x-0) = 1 + (-8)(x-0) = 1 - 8x$.

We need $\dfrac{1}{(1+2x)^4} - 0.1 < 1 - 8x < \dfrac{1}{(1+2x)^4} + 0.1$, which is true

when $-0.045 < x < 0.055$.

15. To estimate $(2.001)^5$, we'll find the linearization of $f(x) = x^5$ at $a = 2$. Since $f'(x) = 5x^4$, $f(2) = 32$, and $f'(2) = 80$,

we have $L(x) = 32 + 80(x-2) = 80x - 128$. Thus, $x^5 \approx 80x - 128$ when x is near 2 , so

$(2.001)^5 \approx 80(2.001) - 128 = 160.08 - 128 = 32.08$.

17. To estimate $(8.06)^{2/3}$, we'll find the linearization of $f(x) = x^{2/3}$ at $a = 8$. Since $f'(x) = \frac{2}{3}x^{-1/3} = 2/\left(3\sqrt[3]{x}\right)$,

$f(8) = 4$, and $f'(8) = \frac{1}{3}$, we have $L(x) = 4 + \frac{1}{3}(x-8) = \frac{1}{3}x + \frac{4}{3}$. Thus, $x^{2/3} \approx \frac{1}{3}x + \frac{4}{3}$ when x is near 8, so

$(8.06)^{2/3} \approx \frac{1}{3}(8.06) + \frac{4}{3} = \frac{12.06}{3} = 4.02$.

19. $y = f(x) = \sec x$ $\Rightarrow$ $f'(x) = \sec x \tan x$, so $f(0) = 1$ and $f'(0) = 1 \cdot 0 = 0$. The linear approximation of f at 0 is

$f(0) + f'(0)(x-0) = 1 + 0(x) = 1$. Since 0.08 is close to 0, approximating $\sec 0.08$ with 1 is reasonable.

21. $y = f(x) = \ln x$ $\Rightarrow$ $f'(x) = 1/x$, so $f(1) = 0$ and $f'(1) = 1$. The linear approximation of f at 1 is

$f(1) + f'(1)(x-1) = 0 + 1(x-1) = x - 1$. Now $f(1.05) = \ln 1.05 \approx 1.05 - 1 = 0.05$, so the approximation

is reasonable.

23. (a) For $y = f(u) = \dfrac{u+1}{u-1}$, $f'(u) = \dfrac{(u-1)(1) - (u+1)(1)}{(u-1)^2} = \dfrac{-2}{(u-1)^2}$, so $dy = \dfrac{-2}{(u-1)^2}\,du$.

(b) For $y = f(r) = (1+r^3)^{-2}$, $f'(r) = -2(1+r^3)^{-3}(3r^2) = \dfrac{-6r^2}{(1+r^3)^3}$, so $dy = \dfrac{-6r^2}{(1+r^3)^3}\,dr$.

25. (a) $y = e^{x/10}$ $\Rightarrow$ $dy = e^{x/10} \cdot \frac{1}{10}\,dx = \frac{1}{10}e^{x/10}dx$

(b) $x = 0$ and $dx = 0.1$ $\Rightarrow$ $dy = \frac{1}{10}e^{0/10}(0.1) = 0.01$.

$\Delta y = f(x + \Delta x) - f(x) = e^{0+0.1} - e^0 = e^{0.1} - 1 \approx 0.0101$.

27. (a) If x is the edge length, then $V = x^3$ $\Rightarrow$ $dV = 3x^2\,dx$. When $x = 30$ and $dx = 0.1$, $dV = 3(30)^2(0.1) = 270$, so the

maximum possible error in computing the volume of the cube is about 270 cm^3. The relative error is calculated by dividing

the change in V, ΔV, by V. We approximate ΔV with dV.

Relative error $= \dfrac{\Delta V}{V} \approx \dfrac{dV}{V} = \dfrac{3x^2\,dx}{x^3} = 3\,\dfrac{dx}{x} = 3\left(\dfrac{0.1}{30}\right) = 0.01$.

Percentage error $=$ relative error $\times\ 100\% = 0.01 \times 100\% = 1\%$.

(b) $S = 6x^2 \Rightarrow dS = 12x\,dx$. When $x = 30$ and $dx = 0.1$, $dS = 12(30)(0.1) = 36$, so the maximum possible error in

computing the surface area of the cube is about 36 cm^2.

Relative error $= \dfrac{\Delta S}{S} \approx \dfrac{dS}{S} = \dfrac{12x\,dx}{6x^2} = 2\dfrac{dx}{x} = 2\left(\dfrac{0.1}{30}\right) = 0.00\overline{6}$.

Percentage error $=$ relative error $\times\, 100\% = 0.00\overline{6} \times 100\% = 0.\overline{6}\%$.

29. (a) For a sphere of radius r, the circumference is $C = 2\pi r$ and the surface area is $S = 4\pi r^2$, so

$r = \dfrac{C}{2\pi} \Rightarrow S = 4\pi \left(\dfrac{C}{2\pi}\right)^2 = \dfrac{C^2}{\pi} \Rightarrow dS = \dfrac{2}{\pi} C\,dC$. When $C = 84$ and $dC = 0.5$, $dS = \dfrac{2}{\pi}(84)(0.5) = \dfrac{84}{\pi}$,

so the maximum error is about $\dfrac{84}{\pi} \approx 27$ cm^2. Relative error $\approx \dfrac{dS}{S} = \dfrac{84/\pi}{84^2/\pi} = \dfrac{1}{84} \approx 0.012$

(b) $V = \dfrac{4}{3}\pi r^3 = \dfrac{4}{3}\pi\left(\dfrac{C}{2\pi}\right)^3 = \dfrac{C^3}{6\pi^2} \Rightarrow dV = \dfrac{1}{2\pi^2} C^2\,dC$. When $C = 84$ and $dC = 0.5$,

$dV = \dfrac{1}{2\pi^2}(84)^2(0.5) = \dfrac{1764}{\pi^2}$, so the maximum error is about $\dfrac{1764}{\pi^2} \approx 179$ cm^3.

The relative error is approximately $\dfrac{dV}{V} = \dfrac{1764/\pi^2}{(84)^3/(6\pi^2)} = \dfrac{1}{56} \approx 0.018$.

31. (a) $V = \pi r^2 h \Rightarrow \Delta V \approx dV = 2\pi r h\,dr = 2\pi r h\,\Delta r$

(b) The error is

$\Delta V - dV = [\pi(r + \Delta r)^2 h - \pi r^2 h] - 2\pi r h\,\Delta r = \pi r^2 h + 2\pi r h\,\Delta r + \pi(\Delta r)^2 h - \pi r^2 h - 2\pi r h\,\Delta r = \pi(\Delta r)^2 h$.

33. $F = kR^4 \Rightarrow dF = 4kR^3\,dR \Rightarrow \dfrac{dF}{F} = \dfrac{4kR^3\,dR}{kR^4} = 4\left(\dfrac{dR}{R}\right)$. Thus, the relative change in F is about 4 times the

relative change in R. So a 5% increase in the radius corresponds to a 20% increase in blood flow.

35. (a) The graph shows that $f'(1) = 2$, so $L(x) = f(1) + f'(1)(x - 1) = 5 + 2(x - 1) = 2x + 3$.

$f(0.9) \approx L(0.9) = 4.8$ and $f(1.1) \approx L(1.1) = 5.2$.

(b) From the graph, we see that $f'(x)$ is positive and decreasing. This means that the slopes of the tangent lines are positive,

but the tangents are becoming less steep. So the tangent lines lie *above* the curve. Thus, the estimates in part (a) are too

large.

3 Review

CONCEPT CHECK

1. (a) The Power Rule: If n is any real number, then $\dfrac{d}{dx}(x^n) = nx^{n-1}$. The derivative of a variable base raised to a constant

power is the power times the base raised to the power minus one.

(b) The Constant Multiple Rule: If c is a constant and f is a differentiable function, then $\dfrac{d}{dx}[cf(x)] = c\dfrac{d}{dx}f(x)$.

The derivative of a constant times a function is the constant times the derivative of the function.

(c) The Sum Rule: If f and g are both differentiable, then $\dfrac{d}{dx}\,[f(x)+g(x)] = \dfrac{d}{dx}\,f(x) + \dfrac{d}{dx}\,g(x)$. The derivative of a sum of functions is the sum of the derivatives.

(d) The Difference Rule: If f and g are both differentiable, then $\dfrac{d}{dx}\,[f(x)-g(x)] = \dfrac{d}{dx}\,f(x) - \dfrac{d}{dx}\,g(x)$. The derivative of a difference of functions is the difference of the derivatives.

(e) The Product Rule: If f and g are both differentiable, then $\dfrac{d}{dx}\,[f(x)\,g(x)] = f(x)\,\dfrac{d}{dx}\,g(x) + g(x)\,\dfrac{d}{dx}\,f(x)$. The derivative of a product of two functions is the first function times the derivative of the second function plus the second function times the derivative of the first function.

(f) The Quotient Rule: If f and g are both differentiable, then $\dfrac{d}{dx}\left[\dfrac{f(x)}{g(x)}\right] = \dfrac{g(x)\,\dfrac{d}{dx}\,f(x) - f(x)\,\dfrac{d}{dx}\,g(x)}{[g(x)]^2}$.

The derivative of a quotient of functions is the denominator times the derivative of the numerator minus the numerator times the derivative of the denominator, all divided by the square of the denominator.

(g) The Chain Rule: If f and g are both differentiable and $F = f \circ g$ is the composite function defined by $F(x) = f(g(x))$, then F is differentiable and F' is given by the product $F'(x) = f'(g(x))\,g'(x)$. The derivative of a composite function is the derivative of the outer function evaluated at the inner function times the derivative of the inner function.

2. (a) $y = x^n \ \Rightarrow \ y' = nx^{n-1}$

(b) $y = e^x \ \Rightarrow \ y' = e^x$

(c) $y = a^x \ \Rightarrow \ y' = a^x \ln a$

(d) $y = \ln x \ \Rightarrow \ y' = 1/x$

(e) $y = \log_a x \ \Rightarrow \ y' = 1/(x \ln a)$

(f) $y = \sin x \ \Rightarrow \ y' = \cos x$

(g) $y = \cos x \ \Rightarrow \ y' = -\sin x$

(h) $y = \tan x \ \Rightarrow \ y' = \sec^2 x$

(i) $y = \csc x \ \Rightarrow \ y' = -\csc x \cot x$

(j) $y = \sec x \ \Rightarrow \ y' = \sec x \tan x$

(k) $y = \cot x \ \Rightarrow \ y' = -\csc^2 x$

(l) $y = \sin^{-1} x \ \Rightarrow \ y' = 1/\sqrt{1-x^2}$

(m) $y = \cos^{-1} x \ \Rightarrow \ y' = -1/\sqrt{1-x^2}$

(n) $y = \tan^{-1} x \ \Rightarrow \ y' = 1/(1+x^2)$

3. (a) e is the number such that $\displaystyle\lim_{h\to 0} \frac{e^h - 1}{h} = 1$.

(b) $e = \displaystyle\lim_{x\to 0} (1+x)^{1/x}$

(c) The differentiation formula for $y = a^x$ $[y' = a^x \ln a]$ is simplest when $a = e$ because $\ln e = 1$.

(d) The differentiation formula for $y = \log_a x$ $[y' = 1/(x \ln a)]$ is simplest when $a = e$ because $\ln e = 1$.

4. (a) Implicit differentiation consists of differentiating both sides of an equation involving x and y with respect to x, and then solving the resulting equation for y'. Use implicit differentiation when it is difficult to solve an equation for y in terms of x.

(b) Logarithmic differentiation consists of taking natural logarithms of both sides of an equation $y = f(x)$, simplifying, differentiating implicitly with respect to x, and then solving the resulting equation for y'. Use logarithmic differentiation when the calculation of derivatives of complicated functions involving products, quotients, or powers can be simplified by taking logarithms.

5. You can find $\dfrac{dy}{dx}$ as a function of t by calculating $\dfrac{dy}{dx} = \dfrac{dy/dt}{dx/dt} = \dfrac{g'(t)}{f'(t)}$ [if $dx/dt = f'(t) \neq 0$].

6. The linearization L of f at $x = a$ is $L(x) = f(a) + f'(a)(x - a)$.

TRUE-FALSE QUIZ

1. True.   This is the Sum Rule.

3. True.   This is the Chain Rule.

5. False. $\dfrac{d}{dx} f(\sqrt{x}) = \dfrac{f'(\sqrt{x})}{2\sqrt{x}}$ by the Chain Rule.

7. False. $\dfrac{d}{dx} 10^x = 10^x \ln 10$

9. True. $\dfrac{d}{dx}(\tan^2 x) = 2 \tan x \sec^2 x$, and $\dfrac{d}{dx}(\sec^2 x) = 2 \sec x (\sec x \tan x) = 2 \tan x \sec^2 x$.

Or: $\dfrac{d}{dx}(\sec^2 x) = \dfrac{d}{dx}(1 + \tan^2 x) = \dfrac{d}{dx}(\tan^2 x)$.

11. True. $g(x) = x^5 \Rightarrow g'(x) = 5x^4 \Rightarrow g'(2) = 5(2)^4 = 80$, and by the definition of the derivative,

$\displaystyle\lim_{x \to 2} \dfrac{g(x) - g(2)}{x - 2} = g'(2) = 80$.

EXERCISES

1. $y = (x^4 - 3x^2 + 5)^3 \Rightarrow$

$y' = 3(x^4 - 3x^2 + 5)^2 \dfrac{d}{dx}(x^4 - 3x^2 + 5) = 3(x^4 - 3x^2 + 5)^2(4x^3 - 6x) = 6x(x^4 - 3x^2 + 5)^2(2x^2 - 3)$

3. $y = \sqrt{x} + \dfrac{1}{\sqrt[3]{x^4}} = x^{1/2} + x^{-4/3} \Rightarrow y' = \frac{1}{2}x^{-1/2} - \frac{4}{3}x^{-7/3} = \dfrac{1}{2\sqrt{x}} - \dfrac{4}{3\sqrt[3]{x^7}}$

5. $y = 2x\sqrt{x^2 + 1} \Rightarrow$

$y' = 2x \cdot \frac{1}{2}(x^2 + 1)^{-1/2}(2x) + \sqrt{x^2 + 1}\,(2) = \dfrac{2x^2}{\sqrt{x^2 + 1}} + 2\sqrt{x^2 + 1} = \dfrac{2x^2 + 2(x^2 + 1)}{\sqrt{x^2 + 1}} = \dfrac{2(2x^2 + 1)}{\sqrt{x^2 + 1}}$

7. $y = e^{\sin 2\theta} \Rightarrow y' = e^{\sin 2\theta} \dfrac{d}{d\theta}(\sin 2\theta) = e^{\sin 2\theta}(\cos 2\theta)(2) = 2\cos 2\theta\, e^{\sin 2\theta}$

9. $y = \dfrac{t}{1 - t^2} \Rightarrow y' = \dfrac{(1 - t^2)(1) - t(-2t)}{(1 - t^2)^2} = \dfrac{1 - t^2 + 2t^2}{(1 - t^2)^2} = \dfrac{t^2 + 1}{(1 - t^2)^2}$

11. $y = \dfrac{e^{1/x}}{x^2} \Rightarrow y' = \dfrac{x^2(e^{1/x})' - e^{1/x}\,(x^2)'}{(x^2)^2} = \dfrac{x^2(e^{1/x})(-1/x^2) - e^{1/x}(2x)}{x^4} = \dfrac{-e^{1/x}(1 + 2x)}{x^4}$

13. $\dfrac{d}{dx}(xy^4 + x^2 y) = \dfrac{d}{dx}(x + 3y) \Rightarrow x \cdot 4y^3 y' + y^4 \cdot 1 + x^2 \cdot y' + y \cdot 2x = 1 + 3y' \Rightarrow$

$y'(4xy^3 + x^2 - 3) = 1 - y^4 - 2xy \Rightarrow y' = \dfrac{1 - y^4 - 2xy}{4xy^3 + x^2 - 3}$

15. $y = \dfrac{\sec 2\theta}{1 + \tan 2\theta}$ $\Rightarrow$

$$y' = \frac{(1 + \tan 2\theta)(\sec 2\theta \, \tan 2\theta \cdot 2) - (\sec 2\theta)(\sec^2 2\theta \cdot 2)}{(1 + \tan 2\theta)^2} = \frac{2 \sec 2\theta \left[(1 + \tan 2\theta) \tan 2\theta - \sec^2 2\theta\right]}{(1 + \tan 2\theta)^2}$$

$$= \frac{2 \sec 2\theta \, (\tan 2\theta + \tan^2 2\theta - \sec^2 2\theta)}{(1 + \tan 2\theta)^2} = \frac{2 \sec 2\theta \, (\tan 2\theta - 1)}{(1 + \tan 2\theta)^2} \qquad \left[1 + \tan^2 x = \sec^2 x\right]$$

17. $y = e^{cx}(c \sin x - \cos x)$ $\Rightarrow$

$$y' = e^{cx}(c \cos x + \sin x) + ce^{cx}(c \sin x - \cos x) = e^{cx}(c^2 \sin x - c \cos x + c \cos x + \sin x)$$

$$= e^{cx}(c^2 \sin x + \sin x) = e^{cx} \sin x \, (c^2 + 1)$$

19. $y = \log_5(1 + 2x)$ $\Rightarrow$ $y' = \dfrac{1}{(1 + 2x) \ln 5} \dfrac{d}{dx}(1 + 2x) = \dfrac{2}{(1 + 2x) \ln 5}$

21. $\sin(xy) = x^2 - y$ $\Rightarrow$ $\cos(xy)(xy' + y \cdot 1) = 2x - y'$ $\Rightarrow$ $x \cos(xy)y' + y' = 2x - y \cos(xy)$ $\Rightarrow$

$$y'[x \cos(xy) + 1] = 2x - y \cos(xy) \quad \Rightarrow \quad y' = \frac{2x - y \cos(xy)}{x \cos(xy) + 1}$$

23. $y = 3^{x \ln x}$ $\Rightarrow$ $y' = 3^{x \ln x}(\ln 3) \dfrac{d}{dx}(x \ln x) = 3^{x \ln x}(\ln 3)\left(x \cdot \dfrac{1}{x} + \ln x \cdot 1\right) = 3^{x \ln x}(\ln 3)(1 + \ln x)$

25. $y = \ln \sin x - \frac{1}{2} \sin^2 x$ $\Rightarrow$ $y' = \dfrac{1}{\sin x} \cdot \cos x - \frac{1}{2} \cdot 2 \sin x \cdot \cos x = \cot x - \sin x \, \cos x$

27. $y = x \tan^{-1}(4x)$ $\Rightarrow$ $y' = x \cdot \dfrac{1}{1 + (4x)^2} \cdot 4 + \tan^{-1}(4x) \cdot 1 = \dfrac{4x}{1 + 16x^2} + \tan^{-1}(4x)$

29. $y = \ln|\sec 5x + \tan 5x|$ $\Rightarrow$

$$y' = \frac{1}{\sec 5x + \tan 5x}(\sec 5x \, \tan 5x \cdot 5 + \sec^2 5x \cdot 5) = \frac{5 \sec 5x \, (\tan 5x + \sec 5x)}{\sec 5x + \tan 5x} = 5 \sec 5x$$

31. $y = \tan^2(\sin \theta) = [\tan(\sin \theta)]^2$ $\Rightarrow$ $y' = 2[\tan(\sin \theta)] \cdot \sec^2(\sin \theta) \cdot \cos \theta$

33. $y = \sin\left(\tan \sqrt{1 + x^3}\right)$ $\Rightarrow$ $y' = \cos\left(\tan \sqrt{1 + x^3}\right)\left(\sec^2 \sqrt{1 + x^3}\right)\left[3x^2 / \left(2\sqrt{1 + x^3}\right)\right]$

35. $y = \cos\left(e^{\sqrt{\tan 3x}}\right)$ $\Rightarrow$

$$y' = -\sin\left(e^{\sqrt{\tan 3x}}\right) \cdot \left(e^{\sqrt{\tan 3x}}\right)' = -\sin\left(e^{\sqrt{\tan 3x}}\right) e^{\sqrt{\tan 3x}} \cdot \tfrac{1}{2}(\tan 3x)^{-1/2} \cdot \sec^2(3x) \cdot 3$$

$$= \frac{-3 \sin\left(e^{\sqrt{\tan 3x}}\right) e^{\sqrt{\tan 3x}} \sec^2(3x)}{2\sqrt{\tan 3x}}$$

37. $f(t) = \sqrt{4t + 1}$ $\Rightarrow$ $f'(t) = \frac{1}{2}(4t + 1)^{-1/2} \cdot 4 = 2(4t + 1)^{-1/2}$ $\Rightarrow$

$$f''(t) = 2(-\tfrac{1}{2})(4t + 1)^{-3/2} \cdot 4 = -4/(4t + 1)^{3/2}, \text{ so } f''(2) = -4/9^{3/2} = -\tfrac{4}{27}.$$

39. $f(x) = 2^x$ $\Rightarrow$ $f'(x) = 2^x \ln 2$ $\Rightarrow$ $f''(x) = (2^x \ln 2) \ln 2 = 2^x (\ln 2)^2$ $\Rightarrow$

$$f'''(x) = (2^x \ln 2)(\ln 2)^2 = 2^x (\ln 2)^3 \quad \Rightarrow \quad \cdots \quad \Rightarrow \quad f^{(n)}(x) = (2^x \ln 2)(\ln 2)^{n-1} = 2^x (\ln 2)^n$$

41. $y = 4\sin^2 x \;\Rightarrow\; y' = 4 \cdot 2\sin x \cos x$. At $\left(\frac{\pi}{6}, 1\right)$, $y' = 8 \cdot \frac{1}{2} \cdot \frac{\sqrt{3}}{2} = 2\sqrt{3}$, so an equation of the tangent line

is $y - 1 = 2\sqrt{3}\left(x - \frac{\pi}{6}\right)$, or $y = 2\sqrt{3}\,x + 1 - \pi\sqrt{3}/3$.

43. $x = \ln t$, $y = t^2 + 1$, $(0, 2)$. If $(x, y) = (0, 2)$, then $x = \ln t = 0 \;\Rightarrow\; t = 1$.

$\dfrac{dy}{dt} = 2t$, $\dfrac{dx}{dt} = \dfrac{1}{t}$, and $\dfrac{dy}{dx} = \dfrac{dy/dt}{dx/dt} = \dfrac{2t}{1/t} = 2t^2$, so $\dfrac{dy}{dx} = 2$ when $t = 1$.

An equation of the tangent to the curve at $(0, 2)$ is $y - 2 = 2(x - 0)$, or $y = 2x + 2$.

45. $y = (2 + x)e^{-x} \;\Rightarrow\; y' = (2 + x)(-e^{-x}) + e^{-x} \cdot 1 = e^{-x}[-(2 + x) + 1] = e^{-x}(-x - 1)$.

At $(0, 2)$, $y' = 1(-1) = -1$, so an equation of the tangent line is $y - 2 = -1(x - 0)$, or $y = -x + 2$.

The slope of the normal line is 1, so an equation of the normal line is $y - 2 = 1(x - 0)$, or $y = x + 2$.

47. (a) $f(x) = x\sqrt{5 - x} \;\Rightarrow$

$$f'(x) = x\left[\frac{1}{2}(5 - x)^{-1/2}(-1)\right] + \sqrt{5 - x} = \frac{-x}{2\sqrt{5 - x}} + \sqrt{5 - x} \cdot \frac{2\sqrt{5 - x}}{2\sqrt{5 - x}} = \frac{-x}{2\sqrt{5 - x}} + \frac{2(5 - x)}{2\sqrt{5 - x}}$$

$$= \frac{-x + 10 - 2x}{2\sqrt{5 - x}} = \frac{10 - 3x}{2\sqrt{5 - x}}$$

(b) At $(1, 2)$: $f'(1) = \frac{7}{4}$.

So an equation of the tangent line is $y - 2 = \frac{7}{4}(x - 1)$ or $y = \frac{7}{4}x + \frac{1}{4}$.

At $(4, 4)$: $f'(4) = -\frac{2}{2} = -1$.

So an equation of the tangent line is $y - 4 = -1(x - 4)$ or $y = -x + 8$.

(c)

(d)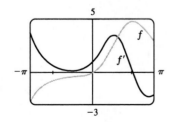

The graphs look reasonable, since f' is positive where f has tangents with positive slope, and f' is negative where f has tangents with negative slope.

49. $f(x) = xe^{\sin x} \;\Rightarrow\; f'(x) = x[e^{\sin x}(\cos x)] + e^{\sin x}(1) = e^{\sin x}(x\cos x + 1)$. As a check on our work, we notice from the graphs that $f'(x) > 0$ when f is increasing. Also, we see in the larger viewing rectangle a certain similarity in the graphs of f and f': the sizes of the oscillations of f and f' are linked.

51. (a) $h(x) = f(x) g(x) \;\Rightarrow\; h'(x) = f(x) g'(x) + g(x) f'(x) \;\Rightarrow$

$h'(2) = f(2) g'(2) + g(2) f'(2) = (3)(4) + (5)(-2) = 12 - 10 = 2$

(b) $F(x) = f(g(x)) \;\Rightarrow\; F'(x) = f'(g(x)) g'(x) \;\Rightarrow\; F'(2) = f'(g(2)) g'(2) = f'(5)(4) = 11 \cdot 4 = 44$

53. $f(x) = x^2 g(x) \;\Rightarrow\; f'(x) = x^2 g'(x) + g(x)(2x) = x[xg'(x) + 2g(x)]$

55. $f(x) = [g(x)]^2 \;\Rightarrow\; f'(x) = 2[g(x)] \cdot g'(x) = 2g(x) g'(x)$

57. $f(x) = g(e^x) \;\Rightarrow\; f'(x) = g'(e^x) e^x$

59. $f(x) = \ln|g(x)| \;\Rightarrow\; f'(x) = \dfrac{1}{g(x)} g'(x) = \dfrac{g'(x)}{g(x)}$

61. $h(x) = \dfrac{f(x) g(x)}{f(x) + g(x)} \;\Rightarrow$

$$h'(x) = \frac{[f(x) + g(x)]\,[f(x) g'(x) + g(x) f'(x)] - f(x) g(x)\,[f'(x) + g'(x)]}{[f(x) + g(x)]^2}$$

$$= \frac{[f(x)]^2 g'(x) + f(x) g(x) f'(x) + f(x) g(x) g'(x) + [g(x)]^2 f'(x) - f(x) g(x) f'(x) - f(x) g(x) g'(x)}{[f(x) + g(x)]^2}$$

$$= \frac{f'(x)\,[g(x)]^2 + g'(x)\,[f(x)]^2}{[f(x) + g(x)]^2}$$

63. $y = [\ln(x + 4)]^2 \;\Rightarrow\; y' = 2[\ln(x + 4)]^1 \cdot \dfrac{1}{x + 4} \cdot 1 = 2\dfrac{\ln(x + 4)}{x + 4}$ and $y' = 0 \;\Leftrightarrow\; \ln(x + 4) = 0 \;\Leftrightarrow$

$x + 4 = e^0 \;\Rightarrow\; x + 4 = 1 \;\Leftrightarrow\; x = -3$, so the tangent is horizontal at the point $(-3, 0)$.

65. $x^2 + 2y^2 = 1 \;\Rightarrow\; 2x + 4yy' = 0 \;\Rightarrow\; y' = -x/(2y) = 1 \;\Leftrightarrow\; x = -2y$. Since the points lie on the ellipse,

we have $(-2y)^2 + 2y^2 = 1 \;\Rightarrow\; 6y^2 = 1 \;\Rightarrow\; y = \pm\frac{1}{\sqrt{6}}$. The points are $\left(-\frac{2}{\sqrt{6}}, \frac{1}{\sqrt{6}}\right)$ and $\left(\frac{2}{\sqrt{6}}, -\frac{1}{\sqrt{6}}\right)$.

67. $y = f(x) = ax^2 + bx + c \;\Rightarrow\; f'(x) = 2ax + b$. We know that $f'(-1) = 6$ and $f'(5) = -2$, so $-2a + b = 6$ and

$10a + b = -2$. Subtracting the first equation from the second gives $12a = -8 \;\Rightarrow\; a = -\frac{2}{3}$. Substituting $-\frac{2}{3}$ for a in the

first equation gives $b = \frac{14}{3}$. Now $f(1) = 4 \;\Rightarrow\; 4 = a + b + c$, so $c = 4 + \frac{2}{3} - \frac{14}{3} = 0$ and hence, $f(x) = -\frac{2}{3}x^2 + \frac{14}{3}x$.

69. $s(t) = Ae^{-ct} \cos(\omega t + \delta) \;\Rightarrow$

$v(t) = s'(t) = A\{e^{-ct}[-\omega \sin(\omega t + \delta)] + \cos(\omega t + \delta)(-ce^{-ct})\} = -Ae^{-ct}[\omega \sin(\omega t + \delta) + c\cos(\omega t + \delta)] \;\Rightarrow$

$a(t) = v'(t) = -A\{e^{-ct}[\omega^2 \cos(\omega t + \delta) - c\omega \sin(\omega t + \delta)] + [\omega \sin(\omega t + \delta) + c\cos(\omega t + \delta)](-ce^{-ct})\}$

$= -Ae^{-ct}[\omega^2 \cos(\omega t + \delta) - c\omega \sin(\omega t + \delta) - c\omega \sin(\omega t + \delta) - c^2 \cos(\omega t + \delta)]$

$= -Ae^{-ct}[(\omega^2 - c^2) \cos(\omega t + \delta) - 2c\omega \sin(\omega t + \delta)] = Ae^{-ct}[(c^2 - \omega^2) \cos(\omega t + \delta) + 2c\omega \sin(\omega t + \delta)]$

71. The linear density ρ is the rate of change of mass m with respect to length x.

$$m = x\left(1 + \sqrt{x}\right) = x + x^{3/2} \quad \Rightarrow \quad \rho = dm/dx = 1 + \tfrac{3}{2}\sqrt{x},$$ so the linear density when $x = 4$ is $1 + \tfrac{3}{2}\sqrt{4} = 4$ kg/m.

73. (a) $C(x) = 920 + 2x - 0.02x^2 + 0.00007x^3 \quad \Rightarrow \quad C'(x) = 2 - 0.04x + 0.00021x^2$

(b) $C'(100) = 2 - 4 + 2.1 = \$0.10$/unit. This value represents the rate at which costs are increasing as the hundredth unit is produced, and is the approximate cost of producing the 101st unit.

(c) The cost of producing the 101st item is $C(101) - C(100) = 990.10107 - 990 = \$0.10107$, slightly larger than $C'(100)$.

(d) $C''(x) = -0.04 + 0.00042x = 0 \quad \Rightarrow \quad x = \frac{0.04}{0.00042} \approx 95.24$ and C'' changes from negative to positive at this value of x. This is the value of x at which the marginal cost is minimized.

75. (a) $f(x) = \sqrt[3]{1 + 3x} = (1 + 3x)^{1/3} \quad \Rightarrow \quad f'(x) = (1 + 3x)^{-2/3}$, so the linearization of f at $a = 0$ is

$$L(x) = f(0) + f'(0)(x - 0) = 1^{1/3} + 1^{-2/3}x = 1 + x.$$ Thus, $\sqrt[3]{1 + 3x} \approx 1 + x \quad \Rightarrow$

$$\sqrt[3]{1.03} = \sqrt[3]{1 + 3(0.01)} \approx 1 + (0.01) = 1.01.$$

(b) The linear approximation is $\sqrt[3]{1 + 3x} \approx 1 + x$, so for the required accuracy we want $\sqrt[3]{1 + 3x} - 0.1 < 1 + x < \sqrt[3]{1 + 3x} + 0.1$. From the graph, it appears that this is true when $-0.23 < x < 0.40$.

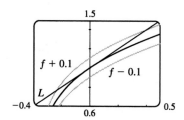

77. $\displaystyle \lim_{\theta \to \pi/3} \frac{\cos\theta - 0.5}{\theta - \pi/3} = \left[\frac{d}{d\theta}\cos\theta\right]_{\theta = \pi/3} = -\sin\frac{\pi}{3} = -\frac{\sqrt{3}}{2}$

79. $\displaystyle \lim_{x \to 0} \frac{\sqrt{1 + \tan x} - \sqrt{1 + \sin x}}{x^3} = \lim_{x \to 0} \frac{\left(\sqrt{1 + \tan x} - \sqrt{1 + \sin x}\right)\left(\sqrt{1 + \tan x} + \sqrt{1 + \sin x}\right)}{x^3\left(\sqrt{1 + \tan x} + \sqrt{1 + \sin x}\right)}$

$$= \lim_{x \to 0} \frac{(1 + \tan x) - (1 + \sin x)}{x^3\left(\sqrt{1 + \tan x} + \sqrt{1 + \sin x}\right)} = \lim_{x \to 0} \frac{\sin x\,(1/\cos x - 1)}{x^3\left(\sqrt{1 + \tan x} + \sqrt{1 + \sin x}\right)} \cdot \frac{\cos x}{\cos x}$$

$$= \lim_{x \to 0} \frac{\sin x\,(1 - \cos x)}{x^3\left(\sqrt{1 + \tan x} + \sqrt{1 + \sin x}\right)\cos x} \cdot \frac{1 + \cos x}{1 + \cos x}$$

$$= \lim_{x \to 0} \frac{\sin x \cdot \sin^2 x}{x^3\left(\sqrt{1 + \tan x} + \sqrt{1 + \sin x}\right)\cos x\,(1 + \cos x)}$$

$$= \left(\lim_{x \to 0} \frac{\sin x}{x}\right)^3 \lim_{x \to 0} \frac{1}{\left(\sqrt{1 + \tan x} + \sqrt{1 + \sin x}\right)\cos x\,(1 + \cos x)}$$

$$= 1^3 \cdot \frac{1}{\left(\sqrt{1} + \sqrt{1}\right) \cdot 1 \cdot (1 + 1)} = \frac{1}{4}$$

☐ FOCUS ON PROBLEM SOLVING

1. We must find a value x_0 such that the normal lines to the parabola $y = x^2$ at $x = \pm x_0$ intersect at a point one unit from the

points $(\pm x_0, x_0^2)$. The normals to $y = x^2$ at $x = \pm x_0$ have slopes $-\dfrac{1}{\pm 2x_0}$ and pass through $(\pm x_0, x_0^2)$ respectively, so the

normals have the equations $y - x_0^2 = -\dfrac{1}{2x_0}(x - x_0)$ and $y - x_0^2 = \dfrac{1}{2x_0}(x + x_0)$. The common y-intercept is $x_0^2 + \dfrac{1}{2}$.

We want to find the value of x_0 for which the distance from $\left(0, x_0^2 + \frac{1}{2}\right)$ to (x_0, x_0^2) equals 1. The square of the distance is

$(x_0 - 0)^2 + \left[x_0^2 - \left(x_0^2 + \frac{1}{2}\right)\right]^2 = x_0^2 + \frac{1}{4} = 1 \ \Leftrightarrow \ x_0 = \pm\frac{\sqrt{3}}{2}$. For these values of x_0, the y-intercept is $x_0^2 + \frac{1}{2} = \frac{5}{4}$, so

the center of the circle is at $\left(0, \frac{5}{4}\right)$.

Another solution: Let the center of the circle be $(0, a)$. Then the equation of the circle is $x^2 + (y - a)^2 = 1$.

Solving with the equation of the parabola, $y = x^2$, we get $x^2 + (x^2 - a)^2 = 1 \ \Leftrightarrow \ x^2 + x^4 - 2ax^2 + a^2 = 1 \ \Leftrightarrow$

$x^4 + (1 - 2a)x^2 + a^2 - 1 = 0$. The parabola and the circle will be tangent to each other when this quadratic equation in x^2

has equal roots; that is, when the discriminant is 0. Thus, $(1 - 2a)^2 - 4(a^2 - 1) = 0 \ \Leftrightarrow$

$1 - 4a + 4a^2 - 4a^2 + 4 = 0 \ \Leftrightarrow \ 4a = 5$, so $a = \frac{5}{4}$. The center of the circle is $\left(0, \frac{5}{4}\right)$.

3.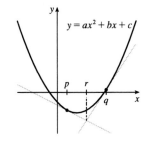

We must show that r (in the figure) is halfway between p and q, that is,

$r = (p + q)/2$. For the parabola $y = ax^2 + bx + c$, the slope of the tangent line is

given by $y' = 2ax + b$. An equation of the tangent line at $x = p$ is

$y - (ap^2 + bp + c) = (2ap + b)(x - p)$. Solving for y gives us

$$y = (2ap + b)x - 2ap^2 - bp + (ap^2 + bp + c)$$

or $$y = (2ap + b)x + c - ap^2 \quad \textbf{(1)}$$

Similarly, an equation of the tangent line at $x = q$ is $\quad y = (2aq + b)x + c - aq^2 \quad \textbf{(2)}$

We can eliminate y and solve for x by subtracting equation **(1)** from equation **(2)**.

$$[(2aq + b) - (2ap + b)]x - aq^2 + ap^2 = 0$$

$$(2aq - 2ap)x = aq^2 - ap^2$$

$$2a(q - p)x = a(q^2 - p^2)$$

$$x = \frac{a(q + p)(q - p)}{2a(q - p)} = \frac{p + q}{2}$$

Thus, the x-coordinate of the point of intersection of the two tangent lines, namely r, is $(p + q)/2$.

5. Using $f'(a) = \lim\limits_{x \to a} \dfrac{f(x) - f(a)}{x - a}$, we recognize the given expression, $f(x) = \lim\limits_{t \to x} \dfrac{\sec t - \sec x}{t - x}$, as

$g'(x)$ with $g(x) = \sec x$. Now $f'\left(\frac{\pi}{4}\right) = g''\left(\frac{\pi}{4}\right)$, so we will find $g''(x)$.

$g'(x) = \sec x \tan x \ \Rightarrow \ g''(x) = \sec x \sec^2 x + \tan x \sec x \tan x = \sec x(\sec^2 x + \tan^2 x)$, so

$g''\left(\frac{\pi}{4}\right) = \sqrt{2}\left(\sqrt{2}^2 + 1^2\right) = \sqrt{2}(2 + 1) = 3\sqrt{2}$.

7. We can assume without loss of generality that $\theta = 0$ at time $t = 0$, so that $\theta = 12\pi t$ rad. [The angular velocity of the wheel is 360 rpm $= 360 \cdot (2\pi \text{ rad})/(60 \text{ s}) = 12\pi$ rad/s.] Then the position of A as a function of time is

$$A = (40\cos\theta, 40\sin\theta) = (40\cos 12\pi t, 40\sin 12\pi t), \text{ so } \sin\alpha = \frac{y}{1.2 \text{ m}} = \frac{40\sin\theta}{120} = \frac{\sin\theta}{3} = \frac{1}{3}\sin 12\pi t.$$

(a) Differentiating the expression for $\sin\alpha$, we get $\cos\alpha \cdot \dfrac{d\alpha}{dt} = \dfrac{1}{3} \cdot 12\pi \cdot \cos 12\pi t = 4\pi\cos\theta$. When $\theta = \dfrac{\pi}{3}$, we have

$$\sin\alpha = \frac{1}{3}\sin\theta = \frac{\sqrt{3}}{6}, \text{ so } \cos\alpha = \sqrt{1 - \left(\frac{\sqrt{3}}{6}\right)^2} = \sqrt{\frac{11}{12}} \text{ and } \frac{d\alpha}{dt} = \frac{4\pi\cos\frac{\pi}{3}}{\cos\alpha} = \frac{2\pi}{\sqrt{11/12}} = \frac{4\pi\sqrt{3}}{\sqrt{11}} \approx 6.56 \text{ rad/s}.$$

(b) By the Law of Cosines, $|AP|^2 = |OA|^2 + |OP|^2 - 2|OA|\,|OP|\cos\theta \;\Rightarrow$

$$120^2 = 40^2 + |OP|^2 - 2 \cdot 40\,|OP|\cos\theta \;\Rightarrow\; |OP|^2 - (80\cos\theta)\,|OP| - 12{,}800 = 0 \;\Rightarrow$$

$$|OP| = \tfrac{1}{2}\left(80\cos\theta \pm \sqrt{6400\cos^2\theta + 51{,}200}\,\right) = 40\cos\theta \pm 40\sqrt{\cos^2\theta + 8} = 40\left(\cos\theta + \sqrt{8 + \cos^2\theta}\,\right) \text{ cm}$$

[since $|OP| > 0$].   As a check, note that $|OP| = 160$ cm when $\theta = 0$ and $|OP| = 80\sqrt{2}$ cm when $\theta = \frac{\pi}{2}$.

(c) By part (b), the x-coordinate of P is given by $x = 40\left(\cos\theta + \sqrt{8 + \cos^2\theta}\,\right)$, so

$$\frac{dx}{dt} = \frac{dx}{d\theta}\frac{d\theta}{dt} = 40\left(-\sin\theta - \frac{2\cos\theta\sin\theta}{2\sqrt{8 + \cos^2\theta}}\right) \cdot 12\pi = -480\pi\sin\theta\left(1 + \frac{\cos\theta}{\sqrt{8 + \cos^2\theta}}\right) \text{ cm/s}.$$

In particular, $dx/dt = 0$ cm/s when $\theta = 0$ and $dx/dt = -480\pi$ cm/s when $\theta = \frac{\pi}{2}$.

9. Consider the statement that $\dfrac{d^n}{dx^n}(e^{ax}\sin bx) = r^n e^{ax}\sin(bx + n\theta)$. For $n = 1$,

$$\frac{d}{dx}(e^{ax}\sin bx) = ae^{ax}\sin bx + be^{ax}\cos bx, \text{ and}$$

$$re^{ax}\sin(bx + \theta) = re^{ax}[\sin bx\cos\theta + \cos bx\sin\theta] = re^{ax}\left(\frac{a}{r}\sin bx + \frac{b}{r}\cos bx\right) = ae^{ax}\sin bx + be^{ax}\cos bx$$

since $\tan\theta = \dfrac{b}{a} \;\Rightarrow\; \sin\theta = \dfrac{b}{r}$ and $\cos\theta = \dfrac{a}{r}$. So the statement is true for $n = 1$.

Assume it is true for $n = k$. Then

$$\frac{d^{k+1}}{dx^{k+1}}(e^{ax}\sin bx) = \frac{d}{dx}\left[r^k e^{ax}\sin(bx + k\theta)\right] = r^k ae^{ax}\sin(bx + k\theta) + r^k e^{ax}b\cos(bx + k\theta)$$

$$= r^k e^{ax}[a\sin(bx + k\theta) + b\cos(bx + k\theta)]$$

But

$$\sin[bx + (k + 1)\theta] = \sin[(bx + k\theta) + \theta] = \sin(bx + k\theta)\cos\theta + \sin\theta\cos(bx + k\theta)$$

$$= \tfrac{a}{r}\sin(bx + k\theta) + \tfrac{b}{r}\cos(bx + k\theta).$$

Hence, $a\sin(bx + k\theta) + b\cos(bx + k\theta) = r\sin[bx + (k + 1)\theta]$. So

$$\frac{d^{k+1}}{dx^{k+1}}(e^{ax}\sin bx) = r^k e^{ax}[a\sin(bx + k\theta) + b\cos(bx + k\theta)] = r^k e^{ax}[r\sin(bx + (k+1)\theta)] = r^{k+1}e^{ax}[\sin(bx + (k+1)\theta)].$$

Therefore, the statement is true for all n by mathematical induction.

11. It seems from the figure that as P approaches the point $(0, 2)$ from the right, $x_T \to \infty$ and $y_T \to 2^+$. As P approaches the point $(3, 0)$ from the left, it appears that $x_T \to 3^+$ and $y_T \to \infty$. So we guess that $x_T \in (3, \infty)$ and $y_T \in (2, \infty)$. It is more difficult to estimate the range of values for x_N and y_N. We might perhaps guess that $x_N \in (0, 3)$, and $y_N \in (-\infty, 0)$ or $(-2, 0)$.

In order to actually solve the problem, we implicitly differentiate the equation of the ellipse to find the equation of the tangent line: $\dfrac{x^2}{9} + \dfrac{y^2}{4} = 1 \;\Rightarrow\; \dfrac{2x}{9} + \dfrac{2y}{4} y' = 0$, so $y' = -\dfrac{4}{9}\dfrac{x}{y}$. So at the point (x_0, y_0) on the ellipse, an equation of the tangent line is $y - y_0 = -\dfrac{4}{9}\dfrac{x_0}{y_0}(x - x_0)$ or $4x_0 x + 9y_0 y = 4x_0^2 + 9y_0^2$. This can be written as $\dfrac{x_0 x}{9} + \dfrac{y_0 y}{4} = \dfrac{x_0^2}{9} + \dfrac{y_0^2}{4} = 1$, because (x_0, y_0) lies on the ellipse. So an equation of the tangent line is $\dfrac{x_0 x}{9} + \dfrac{y_0 y}{4} = 1$.

Therefore, the x-intercept x_T for the tangent line is given by $\dfrac{x_0 x_T}{9} = 1 \;\Leftrightarrow\; x_T = \dfrac{9}{x_0}$, and the y-intercept y_T is given by $\dfrac{y_0 y_T}{4} = 1 \;\Leftrightarrow\; y_T = \dfrac{4}{y_0}$.

So as x_0 takes on all values in $(0, 3)$, x_T takes on all values in $(3, \infty)$, and as y_0 takes on all values in $(0, 2)$, y_T takes on all values in $(2, \infty)$.   At the point (x_0, y_0) on the ellipse, the slope of the normal line is $-\dfrac{1}{y'(x_0, y_0)} = \dfrac{9}{4}\dfrac{y_0}{x_0}$, and its equation is $y - y_0 = \dfrac{9}{4}\dfrac{y_0}{x_0}(x - x_0)$. So the x-intercept x_N for the normal line is given by $0 - y_0 = \dfrac{9}{4}\dfrac{y_0}{x_0}(x_N - x_0) \;\Rightarrow\; x_N = -\dfrac{4x_0}{9} + x_0 = \dfrac{5x_0}{9}$, and the y-intercept y_N is given by $y_N - y_0 = \dfrac{9}{4}\dfrac{y_0}{x_0}(0 - x_0) \;\Rightarrow\; y_N = -\dfrac{9y_0}{4} + y_0 = -\dfrac{5y_0}{4}$.

So as x_0 takes on all values in $(0, 3)$, x_N takes on all values in $\left(0, \frac{5}{3}\right)$, and as y_0 takes on all values in $(0, 2)$, y_N takes on all values in $\left(-\frac{5}{2}, 0\right)$.

13. $y = \dfrac{x}{\sqrt{a^2 - 1}} - \dfrac{2}{\sqrt{a^2 - 1}} \arctan \dfrac{\sin x}{a + \sqrt{a^2 - 1} + \cos x}$. Let $k = a + \sqrt{a^2 - 1}$. Then

$$y' = \dfrac{1}{\sqrt{a^2 - 1}} - \dfrac{2}{\sqrt{a^2 - 1}} \cdot \dfrac{1}{1 + \sin^2 x/(k + \cos x)^2} \cdot \dfrac{\cos x(k + \cos x) + \sin^2 x}{(k + \cos x)^2}$$

$$= \dfrac{1}{\sqrt{a^2 - 1}} - \dfrac{2}{\sqrt{a^2 - 1}} \cdot \dfrac{k \cos x + \cos^2 x + \sin^2 x}{(k + \cos x)^2 + \sin^2 x} = \dfrac{1}{\sqrt{a^2 - 1}} - \dfrac{2}{\sqrt{a^2 - 1}} \cdot \dfrac{k \cos x + 1}{k^2 + 2k \cos x + 1}$$

$$= \dfrac{k^2 + 2k \cos x + 1 - 2k \cos x - 2}{\sqrt{a^2 - 1}\,(k^2 + 2k \cos x + 1)} = \dfrac{k^2 - 1}{\sqrt{a^2 - 1}\,(k^2 + 2k \cos x + 1)}$$

But $k^2 = 2a^2 + 2a\sqrt{a^2 - 1} - 1 = 2a\left(a + \sqrt{a^2 - 1}\right) - 1 = 2ak - 1$, so $k^2 + 1 = 2ak$, and $k^2 - 1 = 2(ak - 1)$.

So $y' = \dfrac{2(ak - 1)}{\sqrt{a^2 - 1}\,(2ak + 2k \cos x)} = \dfrac{ak - 1}{\sqrt{a^2 - 1}\,k\,(a + \cos x)}$. But $ak - 1 = a^2 + a\sqrt{a^2 - 1} - 1 = k\sqrt{a^2 - 1}$,

so $y' = 1/(a + \cos x)$.

15.

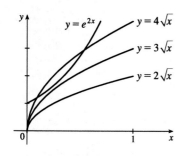

Let $f(x) = e^{2x}$ and $g(x) = k\sqrt{x}$ $[k > 0]$. From the graphs of f and g, we see that f will intersect g exactly once when f and g share a tangent line. Thus, we must have $f = g$ and $f' = g'$ at $x = a$.

$$f(a) = g(a) \quad \Rightarrow \quad e^{2a} = k\sqrt{a} \quad (\star)$$

and $\qquad f'(a) = g'(a) \quad \Rightarrow \quad 2e^{2a} = \dfrac{k}{2\sqrt{a}} \quad \Rightarrow \quad e^{2a} = \dfrac{k}{4\sqrt{a}}.$

So we must have $k\sqrt{a} = \dfrac{k}{4\sqrt{a}} \quad \Rightarrow \quad \left(\sqrt{a}\right)^2 = \dfrac{k}{4k} \quad \Rightarrow \quad a = \frac{1}{4}$. From $(\star)$, $e^{2(1/4)} = k\sqrt{1/4} \quad \Rightarrow$

$k = 2e^{1/2} = 2\sqrt{e} \approx 3.297$.

17. (a) If the two lines L_1 and L_2 have slopes m_1 and m_2 and angles of

inclination ϕ_1 and ϕ_2, then $m_1 = \tan\phi_1$ and $m_2 = \tan\phi_2$. The triangle

in the figure shows that $\phi_1 + \alpha + (180° - \phi_2) = 180°$ and so

$\alpha = \phi_2 - \phi_1$. Therefore, using the identity for $\tan(x - y)$, we have

$\tan\alpha = \tan(\phi_2 - \phi_1) = \dfrac{\tan\phi_2 - \tan\phi_1}{1 + \tan\phi_2\tan\phi_1}$ and so $\tan\alpha = \dfrac{m_2 - m_1}{1 + m_1 m_2}$.

(b) (i) The parabolas intersect when $x^2 = (x - 2)^2 \quad \Rightarrow \quad x = 1$. If $y = x^2$, then $y' = 2x$, so the slope of the tangent

to $y = x^2$ at $(1, 1)$ is $m_1 = 2(1) = 2$. If $y = (x - 2)^2$, then $y' = 2(x - 2)$, so the slope of the tangent to

$y = (x - 2)^2$ at $(1, 1)$ is $m_2 = 2(1 - 2) = -2$. Therefore, $\tan\alpha = \dfrac{m_2 - m_1}{1 + m_1 m_2} = \dfrac{-2 - 2}{1 + 2(-2)} = \dfrac{4}{3}$ and

so $\alpha = \tan^{-1}\left(\frac{4}{3}\right) \approx 53°$ [or $127°$].

(ii) $x^2 - y^2 = 3$ and $x^2 - 4x + y^2 + 3 = 0$ intersect when $x^2 - 4x + (x^2 - 3) + 3 = 0 \quad \Leftrightarrow \quad 2x(x - 2) = 0 \quad \Rightarrow$

$x = 0$ or 2, but 0 is extraneous. If $x = 2$, then $y = \pm 1$. If $x^2 - y^2 = 3$ then $2x - 2yy' = 0 \quad \Rightarrow \quad y' = x/y$ and

$x^2 - 4x + y^2 + 3 = 0 \quad \Rightarrow \quad 2x - 4 + 2yy' = 0 \quad \Rightarrow \quad y' = \dfrac{2 - x}{y}$. At $(2, 1)$ the slopes are $m_1 = 2$ and

$m_2 = 0$, so $\tan\alpha = \frac{0 - 2}{1 + 2 \cdot 0} = -2 \quad \Rightarrow \quad \alpha \approx 117°$. At $(2, -1)$ the slopes are $m_1 = -2$ and $m_2 = 0$,

so $\tan\alpha = \dfrac{0 - (-2)}{1 + (-2)(0)} = 2 \quad \Rightarrow \quad \alpha \approx 63°$ [or $117°$].

19. Since $\angle ROQ = \angle OQP = \theta$, the triangle QOR is isosceles, so

$|QR| = |RO| = x$. By the Law of Cosines, $x^2 = x^2 + r^2 - 2rx\cos\theta$. Hence,

$2rx\cos\theta = r^2$, so $x = \dfrac{r^2}{2r\cos\theta} = \dfrac{r}{2\cos\theta}$. Note that as $y \to 0^+$, $\theta \to 0^+$ (since

$\sin\theta = y/r$), and hence $x \to \dfrac{r}{2\cos 0} = \dfrac{r}{2}$. Thus, as P is taken closer and closer

to the x-axis, the point R approaches the midpoint of the radius AO.

21. $y = x^4 - 2x^2 - x$ $\Rightarrow$ $y' = 4x^3 - 4x - 1$. The equation of the tangent line at $x = a$ is

$y - (a^4 - 2a^2 - a) = (4a^3 - 4a - 1)(x - a)$ or $y = (4a^3 - 4a - 1)x + (-3a^4 + 2a^2)$ and similarly for $x = b$. So if at

$x = a$ and $x = b$ we have the same tangent line, then $4a^3 - 4a - 1 = 4b^3 - 4b - 1$ and $-3a^4 + 2a^2 = -3b^4 + 2b^2$. The first

equation gives $a^3 - b^3 = a - b$ $\Rightarrow$ $(a - b)(a^2 + ab + b^2) = (a - b)$. Assuming $a \neq b$, we have $1 = a^2 + ab + b^2$.

The second equation gives $3(a^4 - b^4) = 2(a^2 - b^2)$ $\Rightarrow$ $3(a^2 - b^2)(a^2 + b^2) = 2(a^2 - b^2)$ which is true if $a = -b$.

Substituting into $1 = a^2 + ab + b^2$ gives $1 = a^2 - a^2 + a^2$ $\Rightarrow$ $a = \pm 1$ so that $a = 1$ and $b = -1$ or vice versa. Thus,

the points $(1, -2)$ and $(-1, 0)$ have a common tangent line.

As long as there are only two such points, we are done. So we show that these are in fact the only two such points.

Suppose that $a^2 - b^2 \neq 0$. Then $3(a^2 - b^2)(a^2 + b^2) = 2(a^2 - b^2)$ gives $3(a^2 + b^2) = 2$ or $a^2 + b^2 = \frac{2}{3}$.

Thus, $ab = (a^2 + ab + b^2) - (a^2 + b^2) = 1 - \frac{2}{3} = \frac{1}{3}$, so $b = \frac{1}{3a}$. Hence, $a^2 + \frac{1}{9a^2} = \frac{2}{3}$, so $9a^4 + 1 = 6a^2$ $\Rightarrow$

$0 = 9a^4 - 6a^2 + 1 = (3a^2 - 1)^2$. So $3a^2 - 1 = 0$ $\Rightarrow$ $a^2 = \frac{1}{3}$ $\Rightarrow$ $b^2 = \frac{1}{9a^2} = \frac{1}{3} = a^2$, contradicting our assumption

that $a^2 \neq b^2$.

23. Because of the periodic nature of the lattice points, it suffices to consider the points in the 5×2 grid shown. We can see that

the minimum value of r occurs when there is a line with slope $\frac{2}{5}$ which touches the circle centered at $(3, 1)$ and the circles

centered at $(0, 0)$ and $(5, 2)$.

To find P, the point at which the line is tangent to the circle at $(0, 0)$, we simultaneously solve $x^2 + y^2 = r^2$ and

$y = -\frac{5}{2}x$ $\Rightarrow$ $x^2 + \frac{25}{4}x^2 = r^2$ $\Rightarrow$ $x^2 = \frac{4}{29}r^2$ $\Rightarrow$ $x = \frac{2}{\sqrt{29}}r, y = -\frac{5}{\sqrt{29}}r$. To find Q, we either use symmetry or

solve $(x - 3)^2 + (y - 1)^2 = r^2$ and $y - 1 = -\frac{5}{2}(x - 3)$. As above, we get $x = 3 - \frac{2}{\sqrt{29}}r, y = 1 + \frac{5}{\sqrt{29}}r$. Now the slope of

the line PQ is $\frac{2}{5}$, so $m_{PQ} = \dfrac{1 + \frac{5}{\sqrt{29}}r - \left(-\frac{5}{\sqrt{29}}r\right)}{3 - \frac{2}{\sqrt{29}}r - \frac{2}{\sqrt{29}}r} = \dfrac{1 + \frac{10}{\sqrt{29}}r}{3 - \frac{4}{\sqrt{29}}r} = \dfrac{\sqrt{29} + 10r}{3\sqrt{29} - 4r} = \dfrac{2}{5}$ $\Rightarrow$

$5\sqrt{29} + 50r = 6\sqrt{29} - 8r$ $\Leftrightarrow$ $58r = \sqrt{29}$ $\Leftrightarrow$ $r = \frac{\sqrt{29}}{58}$. So the minimum value of r for which any line with slope $\frac{2}{5}$

intersects circles with radius r centered at the lattice points on the plane is $r = \frac{\sqrt{29}}{58} \approx 0.093$.

4 □ APPLICATIONS OF DIFFERENTIATION

4.1 Related Rates

1. $V = x^3 \quad \Rightarrow \quad \dfrac{dV}{dt} = \dfrac{dV}{dx}\dfrac{dx}{dt} = 3x^2\dfrac{dx}{dt}$

3. Let s denote the side of a square. The square's area A is given by $A = s^2$. Differentiating with respect to t gives us $\dfrac{dA}{dt} = 2s\dfrac{ds}{dt}$. When $A = 16$, $s = 4$. Substituting 4 for s and 6 for $\dfrac{ds}{dt}$ gives us $\dfrac{dA}{dt} = 2(4)(6) = 48$ cm^2/s.

5. $V = \pi r^2 h = \pi(5)^2 h = 25\pi h \quad \Rightarrow \quad \dfrac{dV}{dt} = 25\pi\dfrac{dh}{dt} \quad \Rightarrow \quad 3 = 25\pi\dfrac{dh}{dt} \quad \Rightarrow \quad \dfrac{dh}{dt} = \dfrac{3}{25\pi}$ m/min.

7. (a) $y = \sqrt{2x+1}$ and $\dfrac{dx}{dt} = 3 \quad \Rightarrow \quad \dfrac{dy}{dt} = \dfrac{dy}{dx}\dfrac{dx}{dt} = \dfrac{1}{2}(2x+1)^{-1/2}\cdot 2\cdot 3 = \dfrac{3}{\sqrt{2x+1}}$. When $x = 4$, $\dfrac{dy}{dt} = \dfrac{3}{\sqrt{9}} = 1$.

  (b) $y = \sqrt{2x+1} \quad \Rightarrow \quad y^2 = 2x+1 \quad \Rightarrow \quad 2x = y^2 - 1 \quad \Rightarrow \quad x = \frac{1}{2}y^2 - \frac{1}{2}$ and $\dfrac{dy}{dt} = 5 \quad \Rightarrow$

  $\dfrac{dx}{dt} = \dfrac{dx}{dy}\dfrac{dy}{dt} = y\cdot 5 = 5y$. When $x = 12$, $y = \sqrt{25} = 5$, so $\dfrac{dx}{dt} = 5(5) = 25$.

9. $z^2 = x^2 + y^2 \quad \Rightarrow \quad 2z\dfrac{dz}{dt} = 2x\dfrac{dx}{dt} + 2y\dfrac{dy}{dt} \quad \Rightarrow \quad \dfrac{dz}{dt} = \dfrac{1}{z}\left(x\dfrac{dx}{dt} + y\dfrac{dy}{dt}\right)$. When $x = 5$ and $y = 12$,

  $z^2 = 5^2 + 12^2 \quad \Rightarrow \quad z^2 = 169 \quad \Rightarrow \quad z = \pm 13$. For $\dfrac{dx}{dt} = 2$ and $\dfrac{dy}{dt} = 3$, $\dfrac{dz}{dt} = \dfrac{1}{\pm 13}(5\cdot 2 + 12\cdot 3) = \pm\dfrac{46}{13}$.

11. (a) Given: the rate of decrease of the surface area is 1 cm^2/min. If we let t be time (in minutes) and S be the surface area (in cm^2), then we are given that $dS/dt = -1$ cm^2/s.

  (b) Unknown: the rate of decrease of the diameter when the diameter is 10 cm. If we let x be the diameter, then we want to find dx/dt when $x = 10$ cm.

  (c)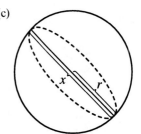

  (d) If the radius is r and the diameter $x = 2r$, then $r = \frac{1}{2}x$ and

  $S = 4\pi r^2 = 4\pi\left(\frac{1}{2}x\right)^2 = \pi x^2 \quad \Rightarrow \quad \dfrac{dS}{dt} = \dfrac{dS}{dx}\dfrac{dx}{dt} = 2\pi x\dfrac{dx}{dt}$.

  (e) $-1 = \dfrac{dS}{dt} = 2\pi x\dfrac{dx}{dt} \quad \Rightarrow \quad \dfrac{dx}{dt} = -\dfrac{1}{2\pi x}$. When $x = 10$, $\dfrac{dx}{dt} = -\dfrac{1}{20\pi}$. So the rate of decrease is $\dfrac{1}{20\pi}$ cm/min.

13. (a) Given: a plane flying horizontally at an altitude of 1 mi and a speed of 500 mi/h passes directly over a radar station. If we let t be time (in hours) and x be the horizontal distance traveled by the plane (in mi), then we are given that $dx/dt = 500$ mi/h.

  (b) Unknown: the rate at which the distance from the plane to the station is increasing when it is 2 mi from the station. If we let y be the distance from the plane to the station, then we want to find dy/dt when $y = 2$ mi.

  (c)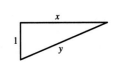

(d) By the Pythagorean Theorem, $y^2 = x^2 + 1 \implies 2y\,(dy/dt) = 2x\,(dx/dt)$.

(e) $\dfrac{dy}{dt} = \dfrac{x}{y}\dfrac{dx}{dt} = \dfrac{x}{y}(500)$. Since $y^2 = x^2 + 1$, when $y = 2$, $x = \sqrt{3}$, so $\dfrac{dy}{dt} = \dfrac{\sqrt{3}}{2}(500) = 250\sqrt{3} \approx 433$ mi/h.

15.

We are given that $\dfrac{dx}{dt} = 60$ mi/h and $\dfrac{dy}{dt} = 25$ mi/h. $z^2 = x^2 + y^2 \implies$

$2z\dfrac{dz}{dt} = 2x\dfrac{dx}{dt} + 2y\dfrac{dy}{dt} \implies z\dfrac{dz}{dt} = x\dfrac{dx}{dt} + y\dfrac{dy}{dt} \implies \dfrac{dz}{dt} = \dfrac{1}{z}\left(x\dfrac{dx}{dt} + y\dfrac{dy}{dt}\right)$.

After 2 hours, $x = 2\,(60) = 120$ and $y = 2\,(25) = 50 \implies z = \sqrt{120^2 + 50^2} = 130$,

so $\dfrac{dz}{dt} = \dfrac{1}{z}\left(x\dfrac{dx}{dt} + y\dfrac{dy}{dt}\right) = \dfrac{120(60) + 50(25)}{130} = 65$ mi/h.

17.

We are given that $\dfrac{dx}{dt} = 4$ ft/s and $\dfrac{dy}{dt} = 5$ ft/s. $z^2 = (x+y)^2 + 500^2 \implies$

$2z\dfrac{dz}{dt} = 2(x+y)\left(\dfrac{dx}{dt} + \dfrac{dy}{dt}\right)$. 15 minutes after the woman starts, we have

$x = (4\text{ ft/s})(20\text{ min})(60\text{ s/min}) = 4800$ ft and $y = 5 \cdot 15 \cdot 60 = 4500 \implies$

$z = \sqrt{(4800 + 4500)^2 + 500^2} = \sqrt{86{,}740{,}000}$, so

$\dfrac{dz}{dt} = \dfrac{x+y}{z}\left(\dfrac{dx}{dt} + \dfrac{dy}{dt}\right) = \dfrac{4800 + 4500}{\sqrt{86{,}740{,}000}}(4+5) = \dfrac{837}{\sqrt{8674}} \approx 8.99$ ft/s.

19. $A = \frac{1}{2}bh$, where b is the base and h is the altitude. We are given that $\dfrac{dh}{dt} = 1$ cm/min and $\dfrac{dA}{dt} = 2$ cm^2/min. Using the

Product Rule, we have $\dfrac{dA}{dt} = \dfrac{1}{2}\left(b\dfrac{dh}{dt} + h\dfrac{db}{dt}\right)$. When $h = 10$ and $A = 100$, we have $100 = \frac{1}{2}b(10) \implies \frac{1}{2}b = 10 \implies$

$b = 20$, so $2 = \dfrac{1}{2}\left(20 \cdot 1 + 10\dfrac{db}{dt}\right) \implies 4 = 20 + 10\dfrac{db}{dt} \implies \dfrac{db}{dt} = \dfrac{4 - 20}{10} = -1.6$ cm/min.

21.

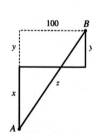

We are given that $\dfrac{dx}{dt} = 35$ km/h and $\dfrac{dy}{dt} = 25$ km/h. $z^2 = (x+y)^2 + 100^2 \implies$

$2z\dfrac{dz}{dt} = 2(x+y)\left(\dfrac{dx}{dt} + \dfrac{dy}{dt}\right)$. At 4:00 PM, $x = 4(35) = 140$ and $y = 4(25) = 100 \implies$

$z = \sqrt{(140 + 100)^2 + 100^2} = \sqrt{67{,}600} = 260$, so

$\dfrac{dz}{dt} = \dfrac{x+y}{z}\left(\dfrac{dx}{dt} + \dfrac{dy}{dt}\right) = \dfrac{140 + 100}{260}(35 + 25) = \dfrac{720}{13} \approx 55.4$ km/h.

23. From the figure and given information, we have $x^2 + y^2 = L^2$, $\dfrac{dy}{dt} = -0.15$ m/s, and

$\dfrac{dx}{dt} = 0.2$ m/s when $x = 3$ m. Differentiating implicitly with respect to t, we get

$x^2 + y^2 = L^2 \implies 2x\dfrac{dx}{dt} + 2y\dfrac{dy}{dt} = 0 \implies y\dfrac{dy}{dt} = -x\dfrac{dx}{dt}$. Substituting the given

information gives us $y(-0.15) = -3(0.2) \implies y = 4$ m. Thus, $3^2 + 4^2 = L^2 \implies$

$L^2 = 25 \implies L = 5$ m.

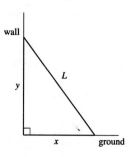

25. Using Q for the origin, we are given $\dfrac{dx}{dt} = -2$ ft/s and need to find $\dfrac{dy}{dt}$ when $x = -5$.

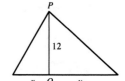

Using the Pythagorean Theorem twice, we have $\sqrt{x^2 + 12^2} + \sqrt{y^2 + 12^2} = 39$,

the total length of the rope. Differentiating with respect to t, we get

$$\frac{x}{\sqrt{x^2 + 12^2}} \frac{dx}{dt} + \frac{y}{\sqrt{y^2 + 12^2}} \frac{dy}{dt} = 0, \text{ so } \frac{dy}{dt} = -\frac{x\sqrt{y^2 + 12^2}}{y\sqrt{x^2 + 12^2}} \frac{dx}{dt}.$$

Now when $x = -5$, $39 = \sqrt{(-5)^2 + 12^2} + \sqrt{y^2 + 12^2} = 13 + \sqrt{y^2 + 12^2} \iff \sqrt{y^2 + 12^2} = 26$, and

$y = \sqrt{26^2 - 12^2} = \sqrt{532}$. So when $x = -5$, $\dfrac{dy}{dt} = -\dfrac{(-5)(26)}{\sqrt{532}\,(13)}(-2) = -\dfrac{10}{\sqrt{133}} \approx -0.87$ ft/s.

So cart B is moving towards Q at about 0.87 ft/s.

27. By similar triangles, $\dfrac{3}{1} = \dfrac{b}{h}$, so $b = 3h$. The trough has volume

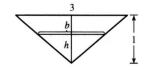

$$V = \tfrac{1}{2}bh(10) = 5(3h)h = 15h^2 \quad \Rightarrow \quad 12 = \frac{dV}{dt} = 30h\frac{dh}{dt} \quad \Rightarrow \quad \frac{dh}{dt} = \frac{2}{5h}.$$

When $h = \tfrac{1}{2}$, $\dfrac{dh}{dt} = \dfrac{2}{5 \cdot \frac{1}{2}} = \dfrac{4}{5}$ ft/min.

29. We are given that $\dfrac{dV}{dt} = 30$ ft^3/min. $V = \dfrac{1}{3}\pi r^2 h = \dfrac{1}{3}\pi\left(\dfrac{h}{2}\right)^2 h = \dfrac{\pi h^3}{12} \quad \Rightarrow$

$$\frac{dV}{dt} = \frac{dV}{dh}\frac{dh}{dt} \quad \Rightarrow \quad 30 = \frac{\pi h^2}{4}\frac{dh}{dt} \quad \Rightarrow \quad \frac{dh}{dt} = \frac{120}{\pi h^2}.$$

When $h = 10$ ft, $\dfrac{dh}{dt} = \dfrac{120}{10^2\pi} = \dfrac{6}{5\pi} \approx 0.38$ ft/min.

31. $A = \tfrac{1}{2}bh$, but $b = 5$ m and $\sin\theta = \dfrac{h}{4} \quad \Rightarrow \quad h = 4\sin\theta$, so $A = \tfrac{1}{2}(5)(4\sin\theta) = 10\sin\theta$.

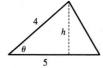

We are given $\dfrac{d\theta}{dt} = 0.06$ rad/s, so $\dfrac{dA}{dt} = \dfrac{dA}{d\theta}\dfrac{d\theta}{dt} = (10\cos\theta)(0.06) = 0.6\cos\theta$.

When $\theta = \dfrac{\pi}{3}$, $\dfrac{dA}{dt} = 0.6\left(\cos\dfrac{\pi}{3}\right) = (0.6)\left(\tfrac{1}{2}\right) = 0.3$ m^2/s.

33. Differentiating both sides of $PV = C$ with respect to t and using the Product Rule gives us $P\dfrac{dV}{dt} + V\dfrac{dP}{dt} = 0 \quad \Rightarrow$

$\dfrac{dV}{dt} = -\dfrac{V}{P}\dfrac{dP}{dt}$. When $V = 600$, $P = 150$ and $\dfrac{dP}{dt} = 20$, so we have $\dfrac{dV}{dt} = -\dfrac{600}{150}(20) = -80$. Thus, the volume is

decreasing at a rate of 80 cm^3/min.

35. With $R_1 = 80$ and $R_2 = 100$, $\dfrac{1}{R} = \dfrac{1}{R_1} + \dfrac{1}{R_2} = \dfrac{1}{80} + \dfrac{1}{100} = \dfrac{180}{8000} = \dfrac{9}{400}$, so $R = \dfrac{400}{9}$. Differentiating $\dfrac{1}{R} = \dfrac{1}{R_1} + \dfrac{1}{R_2}$

with respect to t, we have $-\dfrac{1}{R^2}\dfrac{dR}{dt} = -\dfrac{1}{R_1^2}\dfrac{dR_1}{dt} - \dfrac{1}{R_2^2}\dfrac{dR_2}{dt} \quad \Rightarrow \quad \dfrac{dR}{dt} = R^2\left(\dfrac{1}{R_1^2}\dfrac{dR_1}{dt} + \dfrac{1}{R_2^2}\dfrac{dR_2}{dt}\right)$. When $R_1 = 80$ and

$R_2 = 100$, $\dfrac{dR}{dt} = \dfrac{400^2}{9^2}\left[\dfrac{1}{80^2}(0.3) + \dfrac{1}{100^2}(0.2)\right] = \dfrac{107}{810} \approx 0.132$ Ω/s.

37. (a) By the Pythagorean Theorem, $4000^2 + y^2 = \ell^2$. Differentiating with respect to t,

we obtain $2y\,\dfrac{dy}{dt} = 2\ell\,\dfrac{d\ell}{dt}$. We know that $\dfrac{dy}{dt} = 600$ ft/s, so when $y = 3000$ ft,

$\ell = \sqrt{4000^2 + 3000^2} = \sqrt{25{,}000{,}000} = 5000$ ft

and $\dfrac{d\ell}{dt} = \dfrac{y}{\ell}\dfrac{dy}{dt} = \dfrac{3000}{5000}(600) = \dfrac{1800}{5} = 360$ ft/s.

(b) Here $\tan\theta = \dfrac{y}{4000} \ \Rightarrow\ \dfrac{d}{dt}(\tan\theta) = \dfrac{d}{dt}\left(\dfrac{y}{4000}\right) \ \Rightarrow\ \sec^2\theta\,\dfrac{d\theta}{dt} = \dfrac{1}{4000}\dfrac{dy}{dt} \ \Rightarrow\ \dfrac{d\theta}{dt} = \dfrac{\cos^2\theta}{4000}\dfrac{dy}{dt}$. When

$y = 3000$ ft, $\dfrac{dy}{dt} = 600$ ft/s, $\ell = 5000$ and $\cos\theta = \dfrac{4000}{\ell} = \dfrac{4000}{5000} = \dfrac{4}{5}$, so $\dfrac{d\theta}{dt} = \dfrac{(4/5)^2}{4000}(600) = 0.096$ rad/s.

39. $\cot\theta = \dfrac{x}{5} \ \Rightarrow\ -\csc^2\theta\,\dfrac{d\theta}{dt} = \dfrac{1}{5}\dfrac{dx}{dt} \ \Rightarrow\ -\left(\csc\dfrac{\pi}{3}\right)^2\left(-\dfrac{\pi}{6}\right) = \dfrac{1}{5}\dfrac{dx}{dt} \ \Rightarrow$

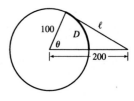

$\dfrac{dx}{dt} = \dfrac{5\pi}{6}\left(\dfrac{2}{\sqrt{3}}\right)^2 = \dfrac{10}{9}\pi$ km/min $[\approx 130$ mi/h$]$

41. We are given that $\dfrac{dx}{dt} = 300$ km/h. By the Law of Cosines,

$y^2 = x^2 + 1^2 - 2(1)(x)\cos 120° = x^2 + 1 - 2x\left(-\tfrac{1}{2}\right) = x^2 + x + 1$, so

$2y\,\dfrac{dy}{dt} = 2x\,\dfrac{dx}{dt} + \dfrac{dx}{dt} \ \Rightarrow\ \dfrac{dy}{dt} = \dfrac{2x+1}{2y}\dfrac{dx}{dt}$. After 1 minute, $x = \frac{300}{60} = 5$ km $\ \Rightarrow$

$y = \sqrt{5^2 + 5 + 1} = \sqrt{31}$ km $\ \Rightarrow\ \dfrac{dy}{dt} = \dfrac{2(5)+1}{2\sqrt{31}}(300) = \dfrac{1650}{\sqrt{31}} \approx 296$ km/h.

43. Let the distance between the runner and the friend be ℓ. Then by the Law of Cosines,

$\ell^2 = 200^2 + 100^2 - 2 \cdot 200 \cdot 100 \cdot \cos\theta = 50{,}000 - 40{,}000\cos\theta \ (\star)$. Differentiating

implicitly with respect to t, we obtain $2\ell\,\dfrac{d\ell}{dt} = -40{,}000(-\sin\theta)\dfrac{d\theta}{dt}$. Now if D is the

distance run when the angle is θ radians, then by the formula for the length of an arc

on a circle, $s = r\theta$, we have $D = 100\theta$, so $\theta = \dfrac{1}{100}D \ \Rightarrow\ \dfrac{d\theta}{dt} = \dfrac{1}{100}\dfrac{dD}{dt} = \dfrac{7}{100}$. To substitute into the expression for

$\dfrac{d\ell}{dt}$, we must know $\sin\theta$ at the time when $\ell = 200$, which we find from $(\star)$: $200^2 = 50{,}000 - 40{,}000\cos\theta \ \Leftrightarrow$

$\cos\theta = \tfrac{1}{4} \ \Rightarrow\ \sin\theta = \sqrt{1 - \left(\tfrac{1}{4}\right)^2} = \dfrac{\sqrt{15}}{4}$. Substituting, we get $2(200)\dfrac{d\ell}{dt} = 40{,}000\dfrac{\sqrt{15}}{4}\left(\tfrac{7}{100}\right) \ \Rightarrow$

$d\ell/dt = \dfrac{7\sqrt{15}}{4} \approx 6.78$ m/s. Whether the distance between them is increasing or decreasing depends on the direction in which

the runner is running.

4.2 Maximum and Minimum Values

1. A function f has an **absolute minimum** at $x = c$ if $f(c)$ is the smallest function value on the entire domain of f, whereas f has a **local minimum** at c if $f(c)$ is the smallest function value when x is near c.

3. Absolute maximum at s, absolute minimum at r, local maximum at c, local minima at b and r, neither a maximum nor a minimum at a and d.

5. Absolute maximum value is $f(4) = 5$; there is no absolute minimum value; local maximum values are $f(4) = 5$ and $f(6) = 4$; local minimum values are $f(2) = 2$ and $f(1) = f(5) = 3$.

7. Absolute minimum at 2, absolute maximum at 3, local minimum at 4

9. Absolute maximum at 5, absolute minimum at 2, local maximum at 3, local minima at 2 and 4

11. (a)

(b)

(c)

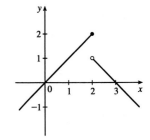

13. (a) *Note:* By the Extreme Value Theorem, f must *not* be continuous; because if it were, it would attain an absolute minimum.

(b)

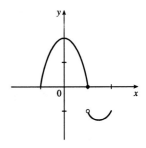

15. $f(x) = \frac{1}{2}(3x - 1)$, $x \le 3$. Absolute maximum $f(3) = 4$; no local maximum. No absolute or local minimum.

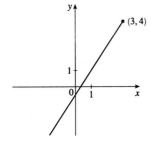

17. $f(x) = x^2$, $0 < x < 2$. No absolute or local maximum or minimum value.

19. $f(x) = \ln x, 0 < x \le 2$. Absolute maximum $f(2) = \ln 2 \approx 0.69$; no local maximum. No absolute or local minimum.

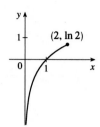

21. $f(x) = 1 - \sqrt{x}$. Absolute maximum $f(0) = 1$; no local maximum. No absolute or local minimum.

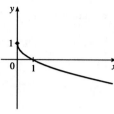

23. $f(x) = 4 + \frac{1}{3}x - \frac{1}{2}x^2 \;\Rightarrow\; f'(x) = \frac{1}{3} - x$. $f'(x) = 0 \;\Rightarrow\; x = \frac{1}{3}$. This is the only critical number.

25. $f(x) = x^3 + 3x^2 - 24x \;\Rightarrow\; f'(x) = 3x^2 + 6x - 24 = 3(x^2 + 2x - 8)$.

$f'(x) = 0 \;\Rightarrow\; 3(x+4)(x-2) = 0 \;\Rightarrow\; x = -4, 2$. These are the only critical numbers.

27. $s(t) = 3t^4 + 4t^3 - 6t^2 \;\Rightarrow\; s'(t) = 12t^3 + 12t^2 - 12t$. $s'(t) = 0 \;\Rightarrow\; 12t(t^2 + t - 1) \;\Rightarrow\;$

$t = 0$ or $t^2 + t - 1 = 0$. Using the quadratic formula to solve the latter equation gives us

$$t = \frac{-1 \pm \sqrt{1^2 - 4(1)(-1)}}{2(1)} = \frac{-1 \pm \sqrt{5}}{2} \approx 0.618, -1.618. \text{ The three critical numbers are } 0, \frac{-1 \pm \sqrt{5}}{2}.$$

29. $g(y) = \dfrac{y-1}{y^2 - y + 1} \;\Rightarrow\;$

$$g'(y) = \frac{(y^2 - y + 1)(1) - (y-1)(2y-1)}{(y^2 - y + 1)^2} = \frac{y^2 - y + 1 - (2y^2 - 3y + 1)}{(y^2 - y + 1)^2} = \frac{-y^2 + 2y}{(y^2 - y + 1)^2} = \frac{y(2-y)}{(y^2 - y + 1)^2}.$$

$g'(y) = 0 \;\Rightarrow\; y = 0, 2$. The expression $y^2 - y + 1$ is never equal to 0, so $g'(y)$ exists for all real numbers.
The critical numbers are 0 and 2.

31. $h(t) = t^{3/4} - 2t^{1/4} \;\Rightarrow\; h'(t) = \frac{3}{4}t^{-1/4} - \frac{2}{4}t^{-3/4} = \frac{1}{4}t^{-3/4}(3t^{1/2} - 2) = \dfrac{3\sqrt{t} - 2}{4\sqrt[4]{t^3}}.$

$h'(t) = 0 \;\Rightarrow\; 3\sqrt{t} = 2 \;\Rightarrow\; \sqrt{t} = \frac{2}{3} \;\Rightarrow\; t = \frac{4}{9}$. $h'(t)$ does not exist at $t = 0$, so the critical numbers are 0 and $\frac{4}{9}$.

33. $F(x) = x^{4/5}(x-4)^2 \;\Rightarrow\;$

$$F'(x) = x^{4/5} \cdot 2(x-4) + (x-4)^2 \cdot \tfrac{4}{5}x^{-1/5} = \tfrac{1}{5}x^{-1/5}(x-4)[5 \cdot x \cdot 2 + (x-4) \cdot 4]$$

$$= \frac{(x-4)(14x - 16)}{5x^{1/5}} = \frac{2(x-4)(7x-8)}{5x^{1/5}}$$

$F'(x) = 0 \;\Rightarrow\; x = 4, \frac{8}{7}$. $F'(0)$ does not exist. Thus, the three critical numbers are $0, \frac{8}{7}$, and 4.

35. $f(\theta) = 2\cos\theta + \sin^2\theta \;\Rightarrow\; f'(\theta) = -2\sin\theta + 2\sin\theta\cos\theta$. $f'(\theta) = 0 \;\Rightarrow\; 2\sin\theta(\cos\theta - 1) = 0 \;\Rightarrow\; \sin\theta = 0$

or $\cos\theta = 1 \;\Rightarrow\; \theta = n\pi$ [n an integer] or $\theta = 2n\pi$. The solutions $\theta = n\pi$ include the solutions $\theta = 2n\pi$, so the critical numbers are $\theta = n\pi$.

37. $f(x) = x^2 e^{-3x}$ ⇒ $f'(x) = x^2(-3e^{-3x}) + e^{-3x}(2x) = xe^{-3x}(-3x + 2)$. $f'(x) = 0$ ⇒ $x = 0, \frac{2}{3}$

[e^{-3x} is never equal to 0]. $f'(x)$ always exists, so the critical numbers are 0 and $\frac{2}{3}$.

39. The graph of $f'(x) = 5e^{-0.1|x|} \sin x - 1$ has 10 zeros and exists

everywhere, so f has 10 critical numbers.

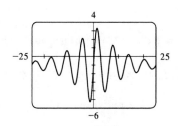

41. $f(x) = 12 + 4x - x^2$, $[0, 5]$. $f'(x) = 4 - 2x = 0$ ⇔ $x = 2$. $f(0) = 12$, $f(2) = 16$, and $f(5) = 7$.

So $f(2) = 16$ is the absolute maximum value and $f(5) = 7$ is the absolute minimum value.

43. $f(x) = 2x^3 - 3x^2 - 12x + 1$, $[-2, 3]$. $f'(x) = 6x^2 - 6x - 12 = 6(x^2 - x - 2) = 6(x - 2)(x + 1) = 0$ ⇔

$x = 2, -1$. $f(-2) = -3$, $f(-1) = 8$, $f(2) = -19$, and $f(3) = -8$. So $f(-1) = 8$ is the absolute maximum value and

$f(2) = -19$ is the absolute minimum value.

45. $f(x) = x^4 - 2x^2 + 3$, $[-2, 3]$. $f'(x) = 4x^3 - 4x = 4x(x^2 - 1) = 4x(x + 1)(x - 1) = 0$ ⇔ $x = -1, 0, 1$.

$f(-2) = 11$, $f(-1) = 2$, $f(0) = 3$, $f(1) = 2$, $f(3) = 66$. So $f(3) = 66$ is the absolute maximum value and $f(\pm 1) = 2$ is

the absolute minimum value.

47. $f(t) = t\sqrt{4 - t^2}$, $[-1, 2]$.

$f'(t) = t \cdot \frac{1}{2}(4 - t^2)^{-1/2}(-2t) + (4 - t^2)^{1/2} \cdot 1 = \frac{-t^2}{\sqrt{4 - t^2}} + \sqrt{4 - t^2} = \frac{-t^2 + (4 - t^2)}{\sqrt{4 - t^2}} = \frac{4 - 2t^2}{\sqrt{4 - t^2}}$.

$f'(t) = 0$ ⇒ $4 - 2t^2 = 0$ ⇒ $t^2 = 2$ ⇒ $t = \pm\sqrt{2}$, but $t = -\sqrt{2}$ is not in the given interval, $[-1, 2]$.

$f'(t)$ does not exist if $4 - t^2 = 0$ ⇒ $t = \pm 2$, but -2 is not in the given interval. $f(-1) = -\sqrt{3}$, $f(\sqrt{2}) = 2$, and

$f(2) = 0$. So $f(\sqrt{2}) = 2$ is the absolute maximum value and $f(-1) = -\sqrt{3}$ is the absolute minimum value.

49. $f(x) = xe^{-x^2/8}$, $[-1, 4]$. $f'(x) = x \cdot e^{-x^2/8} \cdot (-\frac{x}{4}) + e^{-x^2/8} \cdot 1 = e^{-x^2/8}(-\frac{x^2}{4} + 1)$. Since $e^{-x^2/8}$ is never 0,

$f'(x) = 0$ ⇒ $-x^2/4 + 1 = 0$ ⇒ $1 = x^2/4$ ⇒ $x^2 = 4$ ⇒ $x = \pm 2$, but -2 is not in the given interval, $[-1, 4]$.

$f(-1) = -e^{-1/8} \approx -0.88$, $f(2) = 2e^{-1/2} \approx 1.21$, and $f(4) = 4e^{-2} \approx 0.54$. So $f(2) = 2e^{-1/2} = 2/\sqrt{e}$ is the absolute

maximum value and $f(-1) = -e^{-1/8} = -1/\sqrt[8]{e}$ is the absolute minimum value.

51. $f(x) = \ln(x^2 + x + 1)$, $[-1, 1]$. $f'(x) = \frac{1}{x^2 + x + 1} \cdot (2x + 1) = 0$ ⇔ $x = -\frac{1}{2}$. Since $x^2 + x + 1 > 0$ for all x, the

domain of f and f' is $\mathbb{R}$. $f(-1) = \ln 1 = 0$, $f(-\frac{1}{2}) = \ln \frac{3}{4} \approx -0.29$, and $f(1) = \ln 3 \approx 1.10$. So $f(1) = \ln 3 \approx 1.10$ is

the absolute maximum value and $f(-\frac{1}{2}) = \ln \frac{3}{4} \approx -0.29$ is the absolute minimum value.

53. $f(t) = 2\cos t + \sin 2t$, $[0, \pi/2]$.

$f'(t) = -2\sin t + \cos 2t \cdot 2 = -2\sin t + 2(1 - 2\sin^2 t) = -2(2\sin^2 t + \sin t - 1) = -2(2\sin t - 1)(\sin t + 1)$.

$f'(t) = 0 \Rightarrow \sin t = \frac{1}{2}$ or $\sin t = -1 \Rightarrow t = \frac{\pi}{6}$. $f(0) = 2$, $f(\frac{\pi}{6}) = \sqrt{3} + \frac{1}{2}\sqrt{3} = \frac{3}{2}\sqrt{3} \approx 2.60$, and $f(\frac{\pi}{2}) = 0$.

So $f(\frac{\pi}{6}) = \frac{3}{2}\sqrt{3}$ is the absolute maximum value and $f(\frac{\pi}{2}) = 0$ is the absolute minimum value.

55. $f(x) = x^a(1-x)^b$, $0 \le x \le 1, a > 0, b > 0$.

$f'(x) = x^a \cdot b(1-x)^{b-1}(-1) + (1-x)^b \cdot ax^{a-1} = x^{a-1}(1-x)^{b-1}[x \cdot b(-1) + (1-x) \cdot a]$

$= x^{a-1}(1-x)^{b-1}(a - ax - bx)$

At the endpoints, we have $f(0) = f(1) = 0$ [the minimum value of f]. In the interval $(0, 1)$, $f'(x) = 0 \Leftrightarrow x = \dfrac{a}{a+b}$.

$$f\left(\frac{a}{a+b}\right) = \left(\frac{a}{a+b}\right)^a\left(1 - \frac{a}{a+b}\right)^b = \frac{a^a}{(a+b)^a}\left(\frac{a+b-a}{a+b}\right)^b = \frac{a^a}{(a+b)^a} \cdot \frac{b^b}{(a+b)^b} = \frac{a^a b^b}{(a+b)^{a+b}}.$$

So $f\left(\dfrac{a}{a+b}\right) = \dfrac{a^a b^b}{(a+b)^{a+b}}$ is the absolute maximum value.

57. (a)

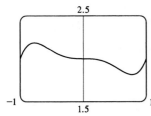

From the graph, it appears that the absolute maximum value is about $f(-0.77) = 2.19$, and the absolute minimum value is about $f(0.77) = 1.81$.

(b) $f(x) = x^5 - x^3 + 2 \Rightarrow f'(x) = 5x^4 - 3x^2 = x^2(5x^2 - 3)$. So $f'(x) = 0 \Rightarrow x = 0, \pm\sqrt{\frac{3}{5}}$.

$$f\left(-\sqrt{\tfrac{3}{5}}\right) = \left(-\sqrt{\tfrac{3}{5}}\right)^5 - \left(-\sqrt{\tfrac{3}{5}}\right)^3 + 2 = -\left(\tfrac{3}{5}\right)^2\sqrt{\tfrac{3}{5}} + \tfrac{3}{5}\sqrt{\tfrac{3}{5}} + 2$$

$$= \left(\tfrac{3}{5} - \tfrac{9}{25}\right)\sqrt{\tfrac{3}{5}} + 2 = \tfrac{6}{25}\sqrt{\tfrac{3}{5}} + 2 \quad \text{(maximum)}$$

and similarly, $f\left(\sqrt{\tfrac{3}{5}}\right) = -\tfrac{6}{25}\sqrt{\tfrac{3}{5}} + 2$ (minimum).

59. (a)

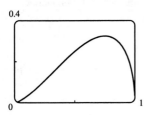

From the graph, it appears that the absolute maximum value is about $f(0.75) = 0.32$, and the absolute minimum value is $f(0) = f(1) = 0$; that is, at both endpoints.

(b) $f(x) = x\sqrt{x - x^2} \Rightarrow f'(x) = x \cdot \dfrac{1 - 2x}{2\sqrt{x - x^2}} + \sqrt{x - x^2} = \dfrac{(x - 2x^2) + (2x - 2x^2)}{2\sqrt{x - x^2}} = \dfrac{3x - 4x^2}{2\sqrt{x - x^2}}$.

So $f'(x) = 0 \Rightarrow 3x - 4x^2 = 0 \Rightarrow x(3 - 4x) = 0 \Rightarrow x = 0$ or $\frac{3}{4}$.

$f(0) = f(1) = 0$ (minimum), and $f(\frac{3}{4}) = \frac{3}{4}\sqrt{\frac{3}{4} - (\frac{3}{4})^2} = \frac{3}{4}\sqrt{\frac{3}{16}} = \frac{3\sqrt{3}}{16}$ (maximum).

61. The density is defined as $\rho = \dfrac{\text{mass}}{\text{volume}} = \dfrac{1000}{V(T)}$ (in g/cm^3). But a critical point of ρ will also be a critical point of V

[since $\dfrac{d\rho}{dT} = -1000V^{-2}\dfrac{dV}{dT}$ and V is never 0], and V is easier to differentiate than ρ.

$V(T) = 999.87 - 0.06426T + 0.0085043T^2 - 0.0000679T^3 \quad \Rightarrow \quad V'(T) = -0.06426 + 0.0170086T - 0.0002037T^2$.

Setting this equal to 0 and using the quadratic formula to find T, we get

$$T = \frac{-0.0170086 \pm \sqrt{0.0170086^2 - 4 \cdot 0.0002037 \cdot 0.06426}}{2(-0.0002037)} \approx 3.9665°C \text{ or } 79.5318°C. \text{ Since we are only interested}$$

in the region $0°C \le T \le 30°C$, we check the density ρ at the endpoints and at $3.9665°C$: $\rho(0) \approx \dfrac{1000}{999.87} \approx 1.00013$;

$\rho(30) \approx \dfrac{1000}{1003.7628} \approx 0.99625$; $\rho(3.9665) \approx \dfrac{1000}{999.7447} \approx 1.000255$. So water has its maximum density at

about $3.9665°C$.

63. Let $a = -0.000\,032\,37$, $b = 0.000\,903\,7$, $c = -0.008\,956$, $d = 0.03629$, $e = -0.04458$, and $f = 0.4074$.

Then $S(t) = at^5 + bt^4 + ct^3 + dt^2 + et + f$ and $S'(t) = 5at^4 + 4bt^3 + 3ct^2 + 2dt + e$.

We now apply the Closed Interval Method to the continuous function S on the interval $0 \le t \le 10$. Since S' exists for all t,

the only critical numbers of S occur when $S'(t) = 0$. We use a rootfinder on a CAS (or a graphing device) to find that

$S'(t) = 0$ when $t_1 \approx 0.855$, $t_2 \approx 4.618$, $t_3 \approx 7.292$, and $t_4 \approx 9.570$. The values of S at these critical numbers are

$S(t_1) \approx 0.39$, $S(t_2) \approx 0.43645$, $S(t_3) \approx 0.427$, and $S(t_4) \approx 0.43641$. The values of S at the endpoints of the interval are

$S(0) \approx 0.41$ and $S(10) \approx 0.435$. Comparing the six numbers, we see that sugar was most expensive at $t_2 \approx 4.618$

(corresponding roughly to March 1998) and cheapest at $t_1 \approx 0.855$ (June 1994).

65. (a) $v(r) = k(r_0 - r)r^2 = kr_0r^2 - kr^3 \quad \Rightarrow \quad v'(r) = 2kr_0r - 3kr^2$. $v'(r) = 0 \quad \Rightarrow \quad kr(2r_0 - 3r) = 0 \quad \Rightarrow$

$r = 0$ or $\frac{2}{3}r_0$ (but 0 is not in the interval). Evaluating v at $\frac{1}{2}r_0$, $\frac{2}{3}r_0$, and r_0, we get $v\left(\frac{1}{2}r_0\right) = \frac{1}{8}kr_0^3$, $v\left(\frac{2}{3}r_0\right) = \frac{4}{27}kr_0^3$,

and $v(r_0) = 0$. Since $\frac{4}{27} > \frac{1}{8}$, v attains its maximum value at $r = \frac{2}{3}r_0$. This supports the statement in the text.

(b) From part (a), the maximum value of v is $\frac{4}{27}kr_0^3$.

(c)

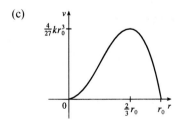

4.3 Derivatives and the Shapes of Curves

1. $\dfrac{f(8) - f(0)}{8 - 0} = \dfrac{6 - 4}{8} = \dfrac{1}{4}$. The values of c which satisfy

$f'(c) = \frac{1}{4}$ seem to be about $c = 0.8, 3.2, 4.4,$ and 6.1.

3. (a) Use the Increasing/Decreasing (I/D) Test.

(b) Use the Concavity Test.

(c) At any value of x where the concavity changes, we have an inflection point at $(x, f(x))$.

5. (a) There is an IP at $x = 3$ because the graph of f changes from CD to CU there. There is an IP at $x = 5$ because the graph

of f changes from CU to CD there.

(b) There is an IP at $x = 2$ and at $x = 6$ because $f'(x)$ has a maximum value there, and so $f''(x)$ changes from positive to

negative there. There is an IP at $x = 4$ because $f'(x)$ has a minimum value there and so $f''(x)$ changes from negative to

positive there.

(c) There is an inflection point at $x = 1$ because $f''(x)$ changes from negative to positive there, and so the graph of f changes

from concave downward to concave upward. There is an inflection point at $x = 7$ because $f''(x)$ changes from positive to

negative there, and so the graph of f changes from concave upward to concave downward.

7. (a) $f(x) = 2x^3 + 3x^2 - 36x \quad \Rightarrow \quad f'(x) = 6x^2 + 6x - 36 = 6(x^2 + x - 6) = 6(x + 3)(x - 2)$.

We don't need to include the "6" in the chart to determine the sign of $f'(x)$.

Interval	$x + 3$	$x - 2$	$f'(x)$	f
$x < -3$	$-$	$-$	$+$	increasing on $(-\infty, -3)$
$-3 < x < 2$	$+$	$-$	$-$	decreasing on $(-3, 2)$
$x > 2$	$+$	$+$	$+$	increasing on $(2, \infty)$

(b) f changes from increasing to decreasing at $x = -3$ and from decreasing to increasing at $x = 2$. Thus, $f(-3) = 81$ is a

local maximum value and $f(2) = -44$ is a local minimum value.

(c) $f'(x) = 6x^2 + 6x - 36 \quad \Rightarrow \quad f''(x) = 12x + 6$. $f''(x) = 0$ at $x = -\frac{1}{2}, f''(x) > 0 \quad \Leftrightarrow \quad x > -\frac{1}{2}$, and

$f''(x) < 0 \quad \Leftrightarrow \quad x < -\frac{1}{2}$. Thus, f is concave upward on $\left(-\frac{1}{2}, \infty\right)$ and concave downward on $\left(-\infty, -\frac{1}{2}\right)$. There is an

inflection point at $\left(-\frac{1}{2}, f\left(-\frac{1}{2}\right)\right) = \left(-\frac{1}{2}, \frac{37}{2}\right)$.

9. (a) $f(x) = x^4 - 2x^2 + 3 \Rightarrow f'(x) = 4x^3 - 4x = 4x(x^2 - 1) = 4x(x+1)(x-1)$.

Interval	$x+1$	x	$x-1$	$f'(x)$	f
$x < -1$	$-$	$-$	$-$	$-$	decreasing on $(-\infty, -1)$
$-1 < x < 0$	$+$	$-$	$-$	$+$	increasing on $(-1, 0)$
$0 < x < 1$	$+$	$+$	$-$	$-$	decreasing on $(0, 1)$
$x > 1$	$+$	$+$	$+$	$+$	increasing on $(1, \infty)$

(b) f changes from increasing to decreasing at $x = 0$ and from decreasing to increasing at $x = -1$ and $x = 1$. Thus,
$f(0) = 3$ is a local maximum value and $f(\pm 1) = 2$ are local minimum values.

(c) $f''(x) = 12x^2 - 4 = 12(x^2 - \frac{1}{3}) = 12(x + 1/\sqrt{3})(x - 1/\sqrt{3})$. $f''(x) > 0 \Leftrightarrow x < -1/\sqrt{3}$ or $x > 1/\sqrt{3}$ and
$f''(x) < 0 \Leftrightarrow -1/\sqrt{3} < x < 1/\sqrt{3}$. Thus, f is concave upward on $(-\infty, -\sqrt{3}/3)$ and $(\sqrt{3}/3, \infty)$ and concave
downward on $(-\sqrt{3}/3, \sqrt{3}/3)$. There are inflection points at $(\pm\sqrt{3}/3, \frac{22}{9})$.

11. (a) $f(x) = \sin x + \cos x$, $0 \le x \le 2\pi$. $f'(x) = \cos x - \sin x = 0 \Rightarrow \cos x = \sin x \Rightarrow 1 = \dfrac{\sin x}{\cos x} \Rightarrow$

$\tan x = 1 \Rightarrow x = \frac{\pi}{4}$ or $\frac{5\pi}{4}$. Thus, $f'(x) > 0 \Leftrightarrow \cos x - \sin x > 0 \Leftrightarrow \cos x > \sin x \Leftrightarrow 0 < x < \frac{\pi}{4}$ or

$\frac{5\pi}{4} < x < 2\pi$ and $f'(x) < 0 \Leftrightarrow \cos x < \sin x \Leftrightarrow \frac{\pi}{4} < x < \frac{5\pi}{4}$. So f is increasing on $(0, \frac{\pi}{4})$ and $(\frac{5\pi}{4}, 2\pi)$ and f
is decreasing on $(\frac{\pi}{4}, \frac{5\pi}{4})$.

(b) f changes from increasing to decreasing at $x = \frac{\pi}{4}$ and from decreasing to increasing at $x = \frac{5\pi}{4}$. Thus, $f(\frac{\pi}{4}) = \sqrt{2}$ is a
local maximum value and $f(\frac{5\pi}{4}) = -\sqrt{2}$ is a local minimum value.

(c) $f''(x) = -\sin x - \cos x = 0 \Rightarrow -\sin x = \cos x \Rightarrow \tan x = -1 \Rightarrow x = \frac{3\pi}{4}$ or $\frac{7\pi}{4}$. Divide the interval
$(0, 2\pi)$ into subintervals with these numbers as endpoints and complete a second derivative chart.

Interval	$f''(x) = -\sin x - \cos x$	Concavity
$(0, \frac{3\pi}{4})$	$f''(\frac{\pi}{2}) = -1 < 0$	downward
$(\frac{3\pi}{4}, \frac{7\pi}{4})$	$f''(\pi) = 1 > 0$	upward
$(\frac{7\pi}{4}, 2\pi)$	$f''(\frac{11\pi}{6}) = \frac{1}{2} - \frac{1}{2}\sqrt{3} < 0$	downward

There are inflection points at $(\frac{3\pi}{4}, 0)$ and $(\frac{7\pi}{4}, 0)$.

13. (a) $f(x) = e^{2x} + e^{-x} \Rightarrow f'(x) = 2e^{2x} - e^{-x}$. $f'(x) > 0 \Leftrightarrow 2e^{2x} > e^{-x} \Leftrightarrow e^{3x} > \frac{1}{2} \Leftrightarrow 3x > \ln\frac{1}{2} \Leftrightarrow$

$x > \frac{1}{3}(\ln 1 - \ln 2) \Leftrightarrow x > -\frac{1}{3}\ln 2 \ [\approx -0.23]$ and $f'(x) < 0$ if $x < -\frac{1}{3}\ln 2$. So f is increasing on $(-\frac{1}{3}\ln 2, \infty)$
and f is decreasing on $(-\infty, -\frac{1}{3}\ln 2)$.

(b) f changes from decreasing to increasing at $x = -\frac{1}{3}\ln 2$. Thus,

$f(-\frac{1}{3}\ln 2) = f\left(\ln\sqrt[3]{1/2}\right) = e^{2\ln\sqrt[3]{1/2}} + e^{-\ln\sqrt[3]{1/2}} = e^{\ln\sqrt[3]{1/4}} + e^{\ln\sqrt[3]{2}} = \sqrt[3]{1/4} + \sqrt[3]{2} = 2^{-2/3} + 2^{1/3} \ [\approx 1.89]$

is a local minimum value.

(c) $f''(x) = 4e^{2x} + e^{-x} > 0$ [the sum of two positive terms]. Thus, f is concave upward on $(-\infty, \infty)$ and there is no
point of inflection.

15. (a) $y = f(x) = \dfrac{\ln x}{\sqrt{x}}$. (Note that f is only defined for $x > 0$.)

$$f'(x) = \frac{\sqrt{x}\,(1/x) - \ln x\left(\frac{1}{2}x^{-1/2}\right)}{x} = \frac{\dfrac{1}{\sqrt{x}} - \dfrac{\ln x}{2\sqrt{x}}}{x} \cdot \frac{2\sqrt{x}}{2\sqrt{x}} = \frac{2 - \ln x}{2x^{3/2}} > 0 \quad \Leftrightarrow \quad 2 - \ln x > 0 \quad \Leftrightarrow$$

$\ln x < 2 \quad \Leftrightarrow \quad x < e^2$. Therefore f is increasing on $(0, e^2)$ and decreasing on (e^2, ∞).

(b) f changes from increasing to decreasing at $x = e^2$, so $f(e^2) = \dfrac{\ln e^2}{\sqrt{e^2}} = \dfrac{2}{e}$ is a local maximum value.

(c) $f''(x) = \dfrac{2x^{3/2}(-1/x) - (2 - \ln x)(3x^{1/2})}{(2x^{3/2})^2} = \dfrac{-2x^{1/2} + 3x^{1/2}(\ln x - 2)}{4x^3} = \dfrac{x^{1/2}(-2 + 3\ln x - 6)}{4x^3} = \dfrac{3\ln x - 8}{4x^{5/2}}$

$f''(x) = 0 \quad \Leftrightarrow \quad \ln x = \frac{8}{3} \quad \Leftrightarrow \quad x = e^{8/3}$. $f''(x) > 0 \quad \Leftrightarrow \quad x > e^{8/3}$, so f is concave upward on $(e^{8/3}, \infty)$ and

concave downward on $(0, e^{8/3})$. There is an inflection point at $\left(e^{8/3}, \frac{8}{3}e^{-4/3}\right) \approx (14.39, 0.70)$.

17. $f(x) = x + \sqrt{1-x} \quad \Rightarrow \quad f'(x) = 1 + \frac{1}{2}(1-x)^{-1/2}(-1) = 1 - \dfrac{1}{2\sqrt{1-x}}$. Note that f is defined for $1 - x \geq 0$; that is,

for $x \leq 1$. $f'(x) = 0 \quad \Rightarrow \quad 2\sqrt{1-x} = 1 \quad \Rightarrow \quad \sqrt{1-x} = \frac{1}{2} \quad \Rightarrow \quad 1 - x = \frac{1}{4} \quad \Rightarrow \quad x = \frac{3}{4}$. f' does not exist at $x = 1$,

but we can't have a local maximum or minimum at an endpoint.

First Derivative Test: $f'(x) > 0 \quad \Rightarrow \quad x < \frac{3}{4}$ and $f'(x) < 0 \quad \Rightarrow \quad \frac{3}{4} < x < 1$. Since f' changes from positive to

negative at $x = \frac{3}{4}$, $f\left(\frac{3}{4}\right) = \frac{5}{4}$ is a local maximum value.

Second Derivative Test: $f''(x) = -\frac{1}{2}\left(-\frac{1}{2}\right)(1-x)^{-3/2}(-1) = -\dfrac{1}{4\left(\sqrt{1-x}\right)^3}$.

$f''\left(\frac{3}{4}\right) = -2 < 0 \quad \Rightarrow \quad f\left(\frac{3}{4}\right) = \frac{5}{4}$ is a local maximum value.

Preference: The First Derivative Test may be slightly easier to apply in this case.

19. (a) By the Second Derivative Test, if $f'(2) = 0$ and $f''(2) = -5 < 0$, f has a local maximum at $x = 2$.

(b) If $f'(6) = 0$, we know that f has a horizontal tangent at $x = 6$. Knowing that $f''(6) = 0$ does not provide any additional

information since the Second Derivative Test fails. For example, the first and second derivatives of $y = (x - 6)^4$,

$y = -(x - 6)^4$, and $y = (x - 6)^3$ all equal zero for $x = 6$, but the first has a local minimum at $x = 6$, the second has a

local maximum at $x = 6$, and the third has an inflection point at $x = 6$.

21. (a) $f(x) = 2x^3 - 3x^2 - 12x \quad \Rightarrow \quad f'(x) = 6x^2 - 6x - 12 = 6(x^2 - x - 2) = 6(x - 2)(x + 1)$.

$f'(x) > 0 \quad \Leftrightarrow \quad x < -1$ or $x > 2$ and $f'(x) < 0 \quad \Leftrightarrow \quad -1 < x < 2$. So f is increasing on $(-\infty, -1)$ and $(2, \infty)$,

and f is decreasing on $(-1, 2)$.

(b) Since f changes from increasing to decreasing at $x = -1$, $f(-1) = 7$ is a local

maximum value. Since f changes from decreasing to increasing at $x = 2$,

$f(2) = -20$ is a local minimum value.

(d)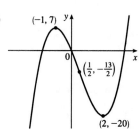

(c) $f''(x) = 6(2x - 1)$ $\Rightarrow$ $f''(x) > 0$ on $\left(\frac{1}{2}, \infty\right)$ and $f''(x) < 0$ on $\left(-\infty, \frac{1}{2}\right)$.

So f is concave upward on $\left(\frac{1}{2}, \infty\right)$ and concave downward on $\left(-\infty, \frac{1}{2}\right)$. There

is a change in concavity at $x = \frac{1}{2}$, and we have an inflection point at $\left(\frac{1}{2}, -\frac{13}{2}\right)$.

23. (a) $f(x) = 2 + 2x^2 - x^4$ $\Rightarrow$ $f'(x) = 4x - 4x^3 = 4x(1 - x^2) = 4x(1 + x)(1 - x)$. $f'(x) > 0$ $\Leftrightarrow$ $x < -1$ or

$0 < x < 1$ and $f'(x) < 0$ $\Leftrightarrow$ $-1 < x < 0$ or $x > 1$. So f is increasing on $(-\infty, -1)$ and $(0, 1)$ and f is decreasing

on $(-1, 0)$ and $(1, \infty)$.

(b) f changes from increasing to decreasing at $x = -1$ and $x = 1$, so $f(-1) = 3$ and $f(1) = 3$ are local maximum values.

f changes from decreasing to increasing at $x = 0$, so $f(0) = 2$ is a local minimum value.

(c) $f''(x) = 4 - 12x^2 = 4(1 - 3x^2)$. $f''(x) = 0$ $\Leftrightarrow$ $1 - 3x^2 = 0$ $\Leftrightarrow$ (d)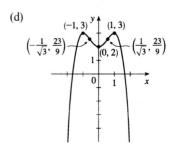

$x^2 = \frac{1}{3}$ $\Leftrightarrow$ $x = \pm 1/\sqrt{3}$. $f''(x) > 0$ on $\left(-1/\sqrt{3}, 1/\sqrt{3}\right)$ and $f''(x) < 0$

on $\left(-\infty, -1/\sqrt{3}\right)$ and $\left(1/\sqrt{3}, \infty\right)$. So f is concave upward on

$\left(-1/\sqrt{3}, 1/\sqrt{3}\right)$ and f is concave downward on $\left(-\infty, -1/\sqrt{3}\right)$ and

$\left(1/\sqrt{3}, \infty\right)$. $f\left(\pm 1/\sqrt{3}\right) = 2 + \frac{2}{3} - \frac{1}{9} = \frac{23}{9}$. There are points of inflection

at $\left(\pm 1/\sqrt{3}, \frac{23}{9}\right)$.

25. (a) $h(x) = (x + 1)^5 - 5x - 2$ $\Rightarrow$ $h'(x) = 5(x + 1)^4 - 5$. $h'(x) = 0$ $\Leftrightarrow$ $5(x + 1)^4 = 5$ $\Leftrightarrow$ $(x + 1)^4 = 1$ $\Rightarrow$

$(x + 1)^2 = 1$ $\Rightarrow$ $x + 1 = 1$ or $x + 1 = -1$ $\Rightarrow$ $x = 0$ or $x = -2$. $h'(x) > 0$ $\Leftrightarrow$ $x < -2$ or $x > 0$ and

$h'(x) < 0$ $\Leftrightarrow$ $-2 < x < 0$. So h is increasing on $(-\infty, -2)$ and $(0, \infty)$ and h is decreasing on $(-2, 0)$.

(b) $h(-2) = 7$ is a local maximum value and $h(0) = -1$ is a local minimum value. (d)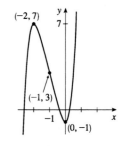

(c) $h''(x) = 20(x + 1)^3 = 0$ $\Leftrightarrow$ $x = -1$. $h''(x) > 0$ $\Leftrightarrow$ $x > -1$ and

$h''(x) < 0$ $\Leftrightarrow$ $x < -1$, so h is CU on $(-1, \infty)$ and h is CD on $(-\infty, -1)$.

There is a point of inflection at $(-1, h(-1)) = (-1, 3)$.

27. (a) $A(x) = x\sqrt{x + 3}$ $\Rightarrow$ $A'(x) = x \cdot \frac{1}{2}(x + 3)^{-1/2} + \sqrt{x + 3} \cdot 1 = \dfrac{x}{2\sqrt{x + 3}} + \sqrt{x + 3} = \dfrac{x + 2(x + 3)}{2\sqrt{x + 3}} = \dfrac{3x + 6}{2\sqrt{x + 3}}$.

The domain of A is $[-3, \infty)$. $A'(x) > 0$ for $x > -2$ and $A'(x) < 0$ for $-3 < x < -2$, so A is increasing on $(-2, \infty)$

and decreasing on $(-3, -2)$.

(b) $A(-2) = -2$ is a local minimum value. (d)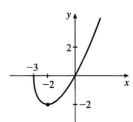

(c) $A''(x) = \dfrac{2\sqrt{x + 3} \cdot 3 - (3x + 6) \cdot \dfrac{1}{\sqrt{x + 3}}}{\left(2\sqrt{x + 3}\right)^2}$

$= \dfrac{6(x + 3) - (3x + 6)}{4(x + 3)^{3/2}} = \dfrac{3x + 12}{4(x + 3)^{3/2}} = \dfrac{3(x + 4)}{4(x + 3)^{3/2}}$

$A''(x) > 0$ for all $x > -3$, so A is concave upward on $(-3, \infty)$. There is no inflection point.

29. (a) $C(x) = x^{1/3}(x+4) = x^{4/3} + 4x^{1/3}$ $\Rightarrow$ $C'(x) = \frac{4}{3}x^{1/3} + \frac{4}{3}x^{-2/3} = \frac{4}{3}x^{-2/3}(x+1) = \frac{4(x+1)}{3\sqrt[3]{x^2}}$. $C'(x) > 0$ if

$-1 < x < 0$ or $x > 0$ and $C'(x) < 0$ for $x < -1$, so C is increasing on $(-1, \infty)$ and C is decreasing on $(-\infty, -1)$.

(b) $C(-1) = -3$ is a local minimum value.

(d)

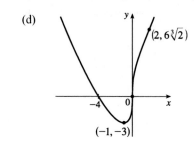

(c) $C''(x) = \frac{4}{9}x^{-2/3} - \frac{8}{9}x^{-5/3} = \frac{4}{9}x^{-5/3}(x-2) = \frac{4(x-2)}{9\sqrt[3]{x^5}}$.

$C''(x) < 0$ for $0 < x < 2$ and $C''(x) > 0$ for $x < 0$ and $x > 2$, so C is

concave downward on $(0, 2)$ and concave upward on $(-\infty, 0)$ and $(2, \infty)$.

There are inflection points at $(0, 0)$ and $(2, 6\sqrt[3]{2}) \approx (2, 7.56)$.

31. (a) $f(\theta) = 2\cos\theta + \cos^2\theta$, $0 \le \theta \le 2\pi$ $\Rightarrow$ $f'(\theta) = -2\sin\theta + 2\cos\theta(-\sin\theta) = -2\sin\theta(1+\cos\theta)$.

$f'(\theta) = 0$ $\Leftrightarrow$ $\theta = 0, \pi$, and 2π. $f'(\theta) > 0$ $\Leftrightarrow$ $\pi < \theta < 2\pi$ and $f'(\theta) < 0$ $\Leftrightarrow$ $0 < \theta < \pi$. So f is increasing

on $(\pi, 2\pi)$ and f is decreasing on $(0, \pi)$.

(b) $f(\pi) = -1$ is a local minimum value.

(c) $f'(\theta) = -2\sin\theta(1+\cos\theta)$ $\Rightarrow$

$$f''(\theta) = -2\sin\theta(-\sin\theta) + (1+\cos\theta)(-2\cos\theta) = 2\sin^2\theta - 2\cos\theta - 2\cos^2\theta$$

$$= 2(1 - \cos^2\theta) - 2\cos\theta - 2\cos^2\theta = -4\cos^2\theta - 2\cos\theta + 2$$

$$= -2(2\cos^2\theta + \cos\theta - 1) = -2(2\cos\theta - 1)(\cos\theta + 1)$$

Since $-2(\cos\theta + 1) < 0$ [for $\theta \ne \pi$], $f''(\theta) > 0$ $\Rightarrow$ $2\cos\theta - 1 < 0$ $\Rightarrow$ $\cos\theta < \frac{1}{2}$ $\Rightarrow$ $\frac{\pi}{3} < \theta < \frac{5\pi}{3}$ and

$f''(\theta) < 0$ $\Rightarrow$ $\cos\theta > \frac{1}{2}$ $\Rightarrow$ $0 < \theta < \frac{\pi}{3}$ or $\frac{5\pi}{3} < \theta < 2\pi$. So f is CU on $\left(\frac{\pi}{3}, \frac{5\pi}{3}\right)$ and f is CD on $\left(0, \frac{\pi}{3}\right)$ and

$\left(\frac{5\pi}{3}, 2\pi\right)$. There are points of inflection at $\left(\frac{\pi}{3}, f\left(\frac{\pi}{3}\right)\right) = \left(\frac{\pi}{3}, \frac{5}{4}\right)$ and $\left(\frac{5\pi}{3}, f\left(\frac{5\pi}{3}\right)\right) = \left(\frac{5\pi}{3}, \frac{5}{4}\right)$.

(d)

33. $f(x) = \dfrac{x^2}{x^2 - 1} = \dfrac{x^2}{(x+1)(x-1)}$ has domain $(-\infty, -1) \cup (-1, 1) \cup (1, \infty)$.

(a) $\displaystyle\lim_{x \to \pm\infty} f(x) = \lim_{x \to \pm\infty} \frac{x^2/x^2}{(x^2-1)/x^2} = \lim_{x \to \pm\infty} \frac{1}{1 - 1/x^2} = \frac{1}{1 - 0} = 1$, so $y = 1$ is a HA.

$\displaystyle\lim_{x \to -1^-} \frac{x^2}{x^2 - 1} = \infty$ since $x^2 \to 1$ and $(x^2 - 1) \to 0^+$ as $x \to -1^-$, so $x = -1$ is a VA.

$\displaystyle\lim_{x \to 1^+} \frac{x^2}{x^2 - 1} = \infty$ since $x^2 \to 1$ and $(x^2 - 1) \to 0^+$ as $x \to 1^+$, so $x = 1$ is a VA.

(b) $f(x) = \dfrac{x^2}{x^2 - 1}$ $\Rightarrow$ $f'(x) = \dfrac{(x^2 - 1)(2x) - x^2(2x)}{(x^2 - 1)^2} = \dfrac{2x[(x^2 - 1) - x^2]}{(x^2 - 1)^2} = \dfrac{-2x}{(x^2 - 1)^2}$. Since $(x^2 - 1)^2$ is

positive for all x in the domain of f, the sign of the derivative is determined by the sign of $-2x$. Thus, $f'(x) > 0$ if $x < 0$

$(x \ne -1)$ and $f'(x) < 0$ if $x > 0$ $(x \ne 1)$. So f is increasing on $(-\infty, -1)$ and $(-1, 0)$, and f is decreasing on $(0, 1)$

and $(1, \infty)$.

(c) $f'(x) = 0 \Rightarrow x = 0$ and $f(0) = 0$ is a local maximum value.

(d) $f''(x) = \dfrac{(x^2 - 1)^2(-2) - (-2x) \cdot 2(x^2 - 1)(2x)}{[(x^2 - 1)^2]^2}$

$= \dfrac{2(x^2 - 1)[-(x^2 - 1) + 4x^2]}{(x^2 - 1)^4} = \dfrac{2(3x^2 + 1)}{(x^2 - 1)^3}.$

The sign of $f''(x)$ is determined by the denominator; that is, $f''(x) > 0$ if

$|x| > 1$ and $f''(x) < 0$ if $|x| < 1$. Thus, f is CU on $(-\infty, -1)$ and $(1, \infty)$,

and f is CD on $(-1, 1)$. There are no inflection points.

(e)

35. (a) $\displaystyle \lim_{x \to -\infty} \left(\sqrt{x^2 + 1} - x\right) = \infty$ and

$\displaystyle \lim_{x \to \infty} \left(\sqrt{x^2 + 1} - x\right) = \lim_{x \to \infty} \left(\sqrt{x^2 + 1} - x\right) \dfrac{\sqrt{x^2 + 1} + x}{\sqrt{x^2 + 1} + x} = \lim_{x \to \infty} \dfrac{1}{\sqrt{x^2 + 1} + x} = 0$, so $y = 0$ is a HA.

(b) $f(x) = \sqrt{x^2 + 1} - x \Rightarrow f'(x) = \dfrac{x}{\sqrt{x^2 + 1}} - 1$. Since $\dfrac{x}{\sqrt{x^2 + 1}} < 1$ for all x, $f'(x) < 0$, so f is decreasing on $\mathbb{R}$.

(c) No minimum or maximum

(d) $f''(x) = \dfrac{(x^2 + 1)^{1/2}(1) - x \cdot \frac{1}{2}(x^2 + 1)^{-1/2}(2x)}{\left(\sqrt{x^2 + 1}\right)^2}$

$= \dfrac{(x^2 + 1)^{1/2} - \dfrac{x^2}{(x^2 + 1)^{1/2}}}{x^2 + 1} = \dfrac{(x^2 + 1) - x^2}{(x^2 + 1)^{3/2}} = \dfrac{1}{(x^2 + 1)^{3/2}} > 0,$

so f is CU on $\mathbb{R}$. No IP

(e)

37. $f(x) = \ln(1 - \ln x)$ is defined when $x > 0$ (so that $\ln x$ is defined) and $1 - \ln x > 0$ [so that $\ln(1 - \ln x)$ is defined].

The second condition is equivalent to $1 > \ln x \Leftrightarrow x < e$, so f has domain $(0, e)$.

(a) As $x \to 0^+$, $\ln x \to -\infty$, so $1 - \ln x \to \infty$ and $f(x) \to \infty$. As $x \to e^-$, $\ln x \to 1^-$, so $1 - \ln x \to 0^+$ and

$f(x) \to -\infty$. Thus, $x = 0$ and $x = e$ are vertical asymptotes. There is no horizontal asymptote.

(b) $f'(x) = \dfrac{1}{1 - \ln x}\left(-\dfrac{1}{x}\right) = -\dfrac{1}{x(1 - \ln x)} < 0$ on $(0, e)$. Thus, f is decreasing on its domain, $(0, e)$.

(c) $f'(x) \neq 0$ on $(0, e)$, so f has no local maximum or minimum value.

(e)

(d) $f''(x) = -\dfrac{-[x(1 - \ln x)]'}{[x(1 - \ln x)]^2} = \dfrac{x(-1/x) + (1 - \ln x)}{x^2(1 - \ln x)^2}$

$= -\dfrac{\ln x}{x^2(1 - \ln x)^2}$

so $f''(x) > 0 \Leftrightarrow \ln x < 0 \Leftrightarrow 0 < x < 1$. Thus, f is CU on $(0, 1)$

and CD on $(1, e)$. There is an inflection point at $(1, 0)$.

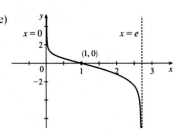

39. (a) $\displaystyle \lim_{x \to \pm\infty} e^{-1/(x+1)} = 1$ since $-1/(x + 1) \to 0$, so $y = 1$ is a HA. $\displaystyle \lim_{x \to -1^+} e^{-1/(x+1)} = 0$ since $-1/(x + 1) \to -\infty$,

$\displaystyle \lim_{x \to -1^-} e^{-1/(x+1)} = \infty$ since $-1/(x + 1) \to \infty$, so $x = -1$ is a VA.

(b) $f(x) = e^{-1/(x+1)}$ $\Rightarrow$ $f'(x) = e^{-1/(x+1)}\left[-(-1)\dfrac{1}{(x+1)^2}\right]$ [Reciprocal Rule] $= e^{-1/(x+1)}/(x+1)^2$ $\Rightarrow$

$f'(x) > 0$ for all x except -1, so f is increasing on $(-\infty, -1)$ and $(-1, \infty)$.

(c) There is no local maximum or minimum.

(d) $f''(x) = \dfrac{(x+1)^2 e^{-1/(x+1)}\left[1/(x+1)^2\right] - e^{-1/(x+1)}\left[2(x+1)\right]}{\left[(x+1)^2\right]^2}$

(e)

$\qquad = \dfrac{e^{-1/(x+1)}\left[1 - (2x+2)\right]}{(x+1)^4} = -\dfrac{e^{-1/(x+1)}(2x+1)}{(x+1)^4}$ $\Rightarrow$

$f''(x) > 0$ $\Leftrightarrow$ $2x + 1 < 0$ $\Leftrightarrow$ $x < -\tfrac{1}{2}$, so f is CU on $(-\infty, -1)$

and $\left(-1, -\tfrac{1}{2}\right)$, and CD on $\left(-\tfrac{1}{2}, \infty\right)$. f has an IP at $\left(-\tfrac{1}{2}, e^{-2}\right)$.

41. The nonnegative factors $(x+1)^2$ and $(x-6)^4$ do not affect the sign of $f'(x) = (x+1)^2(x-3)^5(x-6)^4$.

So $f'(x) > 0$ $\Rightarrow$ $(x-3)^5 > 0$ $\Rightarrow$ $x - 3 > 0$ $\Rightarrow$ $x > 3$. Thus, f is increasing on the interval $(3, \infty)$.

43. (a)

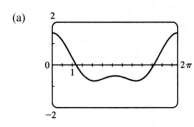

From the graph, we get an estimate of $f(1) \approx 1.41$ as a local maximum value, and no local minimum value.

$f(x) = \dfrac{x+1}{\sqrt{x^2+1}}$ $\Rightarrow$ $f'(x) = \dfrac{1-x}{(x^2+1)^{3/2}}$.

$f'(x) = 0$ $\Leftrightarrow$ $x = 1$. $f(1) = \tfrac{2}{\sqrt{2}} = \sqrt{2}$ is the exact value.

(b) From the graph in part (a), f increases most rapidly somewhere between $x = -\tfrac{1}{2}$ and $x = -\tfrac{1}{4}$. To find the exact value,

we need to find the maximum value of f', which we can do by finding the critical numbers of f'.

$f''(x) = \dfrac{2x^2 - 3x - 1}{(x^2+1)^{5/2}} = 0$ $\Leftrightarrow$ $x = \dfrac{3 \pm \sqrt{17}}{4}$. $x = \dfrac{3 + \sqrt{17}}{4}$ corresponds to the *minimum* value of f'.

The maximum value of f' occurs at $x = \tfrac{3 - \sqrt{17}}{4} \approx -0.28$.

45. $f(x) = \cos x + \tfrac{1}{2}\cos 2x$ $\Rightarrow$ $f'(x) = -\sin x - \sin 2x$ $\Rightarrow$ $f''(x) = -\cos x - 2\cos 2x$

(a)

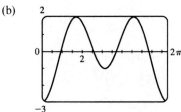

From the graph of f, it seems that f is CD on $(0, 1)$, CU on $(1, 2.5)$, CD on $(2.5, 3.7)$, CU on $(3.7, 5.3)$, and CD on $(5.3, 2\pi)$. The points of inflection appear to be at $(1, 0.4)$, $(2.5, -0.6)$, $(3.7, -0.6)$, and $(5.3, 0.4)$.

(b)

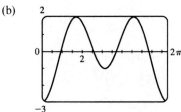

From the graph of f'' (and zooming in near the zeros), it seems that f is CD on $(0, 0.94)$, CU on $(0.94, 2.57)$, CD on $(2.57, 3.71)$, CU on $(3.71, 5.35)$, and CD on $(5.35, 2\pi)$. Refined estimates of the inflection points are $(0.94, 0.44)$, $(2.57, -0.63)$, $(3.71, -0.63)$, and $(5.35, 0.44)$.

47. In Maple, we define f and then use the command

`plot(diff(diff(f,x),x),x=-2..2);`. In Mathematica, we define f

and then use `Plot[Dt[Dt[f,x],x],{x,-2,2}]`. We see that $f'' > 0$ for

$x < -0.6$ and $x > 0.0$ [≈ 0.03] and $f'' < 0$ for $-0.6 < x < 0.0$. So f is CU

on $(-\infty, -0.6)$ and $(0.0, \infty)$ and CD on $(-0.6, 0.0)$.

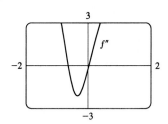

49. (a) I'm very unhappy. It's uncomfortably hot and $f'(3) = 2$ indicates that the

temperature is increasing, and $f''(3) = 4$ indicates that the rate of increase

is increasing. (The temperature is rapidly getting warmer.)

(b) I'm still unhappy, but not as unhappy as in part (a). It's uncomfortably hot

and $f'(3) = 2$ indicates that the temperature is increasing, but $f''(3) = -4$

indicates that the rate of increase is decreasing. (The temperature is slowly

getting warmer.)

(c) I'm somewhat happy. It's uncomfortably hot and $f'(3) = -2$ indicates that

the temperature is decreasing, but $f''(3) = 4$ indicates that the rate of

change is increasing. (The rate of change is negative but it's becoming less

negative. The temperature is slowly getting cooler.)

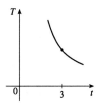

(d) I'm very happy. It's uncomfortably hot and $f'(3) = -2$ indicates that the

temperature is decreasing, and $f''(3) = -4$ indicates that the rate of change

is decreasing, that is, becoming more negative. (The temperature is rapidly

getting cooler.)

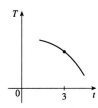

51. $x = t^3 - 12t, \quad y = t^2 - 1 \quad \Rightarrow \quad \dfrac{dy}{dx} = \dfrac{dy/dt}{dx/dt} = \dfrac{2t}{3t^2 - 12} \quad \Rightarrow$

$$\frac{d^2y}{dx^2} = \frac{\dfrac{d}{dt}\left(\dfrac{dy}{dx}\right)}{dx/dt} = \frac{\dfrac{(3t^2 - 12) \cdot 2 - 2t(6t)}{(3t^2 - 12)^2}}{3t^2 - 12} = \frac{-6t^2 - 24}{(3t^2 - 12)^3} = \frac{-6(t^2 + 4)}{3^3(t^2 - 4)^3} = \frac{-2(t^2 + 4)}{9(t^2 - 4)^3}.$$

Thus, the curve is CU when $t^2 - 4 < 0 \quad \Rightarrow \quad |t| < 2 \quad \Rightarrow \quad -2 < t < 2$.

53. $m = f(v) = \dfrac{m_0}{\sqrt{1 - v^2/c^2}}$. The m-intercept is $f(0) = m_0$. There are no v-intercepts. $\lim\limits_{v \to c^-} f(v) = \infty$, so $v = c$ is a VA.

$f'(v) = -\tfrac{1}{2}m_0(1 - v^2/c^2)^{-3/2}(-2v/c^2) = \dfrac{m_0 v}{c^2(1 - v^2/c^2)^{3/2}} = \dfrac{m_0 v}{\underbrace{c^2(c^2 - v^2)^{3/2}}_{c^3}} = \dfrac{m_0 c v}{(c^2 - v^2)^{3/2}} > 0$, so f is

increasing on $(0, c)$. There are no local extreme values.

$f''(v) = \dfrac{(c^2 - v^2)^{3/2}(m_0 c) - m_0 c v \cdot \frac{3}{2}(c^2 - v^2)^{1/2}(-2v)}{[(c^2 - v^2)^{3/2}]^2}$

$= \dfrac{m_0 c(c^2 - v^2)^{1/2}[(c^2 - v^2) + 3v^2]}{(c^2 - v^2)^3} = \dfrac{m_0 c(c^2 + 2v^2)}{(c^2 - v^2)^{5/2}} > 0,$

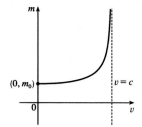

so f is CU on $(0, c)$. There are no inflection points.

55. $y = -\dfrac{W}{24EI}x^4 + \dfrac{WL}{12EI}x^3 - \dfrac{WL^2}{24EI}x^2 = -\dfrac{W}{24EI}x^2(x^2 - 2Lx + L^2)$

$= \dfrac{-W}{24EI}x^2(x - L)^2 = cx^2(x - L)^2$

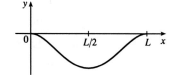

where $c = -\dfrac{W}{24EI}$ is a negative constant and $0 \le x \le L$. We sketch

$f(x) = cx^2(x - L)^2$ for $c = -1$. $f(0) = f(L) = 0$.

$f'(x) = cx^2[2(x - L)] + (x - L)^2(2cx) = 2cx(x - L)[x + (x - L)] = 2cx(x - L)(2x - L)$. So for $0 < x < L$,

$f'(x) > 0 \;\Leftrightarrow\; x(x - L)(2x - L) < 0$ [since $c < 0$] $\;\Leftrightarrow\; L/2 < x < L$ and $f'(x) < 0 \;\Leftrightarrow\; 0 < x < L/2$.

Thus, f is increasing on $(L/2, L)$ and decreasing on $(0, L/2)$, and there is a local and absolute

minimum at the point $(L/2, f(L/2)) = (L/2, cL^4/16)$. $f'(x) = 2c[x(x - L)(2x - L)] \;\Rightarrow$

$f''(x) = 2c[1(x - L)(2x - L) + x(1)(2x - L) + x(x - L)(2)] = 2c(6x^2 - 6Lx + L^2) = 0 \;\Leftrightarrow$

$x = \dfrac{6L \pm \sqrt{12L^2}}{12} = \tfrac{1}{2}L \pm \tfrac{\sqrt{3}}{6}L$, and these are the x-coordinates of the two inflection points.

57. $S(t) = At^p e^{-kt}$ with $A = 0.01$, $p = 4$, and $k = 0.07$. We will find the
zeros of f'' for $f(t) = t^p e^{-kt}$.

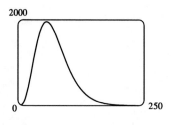

$f'(t) = t^p(-ke^{-kt}) + e^{-kt}(pt^{p-1}) = e^{-kt}(-kt^p + pt^{p-1})$

$f''(t) = e^{-kt}(-kpt^{p-1} + p(p - 1)t^{p-2}) + (-kt^p + pt^{p-1})(-ke^{-kt})$

$= t^{p-2}e^{-kt}[-kpt + p(p - 1) + k^2t^2 - kpt]$

$= t^{p-2}e^{-kt}(k^2t^2 - 2kpt + p^2 - p)$

Using the given values of p and k gives us $f''(t) = t^2 e^{-0.07t}(0.0049t^2 - 0.56t + 12)$. So $S''(t) = 0.01f''(t)$ and its zeros

are $t = 0$ and the solutions of $0.0049t^2 - 0.56t + 12 = 0$, which are $t_1 = \tfrac{200}{7} \approx 28.57$ and $t_2 = \tfrac{600}{7} \approx 85.71$.

At t_1 minutes, the rate of increase of the level of medication in the bloodstream is at its greatest and at t_2 minutes, the rate of

decrease is the greatest.

59. $f(x) = ax^3 + bx^2 + cx + d \Rightarrow f'(x) = 3ax^2 + 2bx + c$.

We are given that $f(1) = 0$ and $f(-2) = 3$, so $f(1) = a + b + c + d = 0$ and

$f(-2) = -8a + 4b - 2c + d = 3$. Also $f'(1) = 3a + 2b + c = 0$ and

$f'(-2) = 12a - 4b + c = 0$ by Fermat's Theorem. Solving these four equations, we get

$a = \frac{2}{9}, b = \frac{1}{3}, c = -\frac{4}{3}, d = \frac{7}{9}$, so the function is $f(x) = \frac{1}{9}(2x^3 + 3x^2 - 12x + 7)$.

61. $f(x) = \tan x - x \Rightarrow f'(x) = \sec^2 x - 1 > 0$ for $0 < x < \frac{\pi}{2}$ since $\sec^2 x > 1$ for $0 < x < \frac{\pi}{2}$. So f is increasing

on $\left(0, \frac{\pi}{2}\right)$. Thus, $f(x) > f(0) = 0$ for $0 < x < \frac{\pi}{2} \Rightarrow \tan x - x > 0 \Rightarrow \tan x > x$ for $0 < x < \frac{\pi}{2}$.

63. We are given that f is differentiable (and therefore continuous) everywhere. In particular, we can apply the Mean Value

Theorem on the interval $[0, 4]$. There exists a number c in $(0, 4)$ such that $f(4) - f(0) = f'(c)(4 - 0)$, so

$f(4) = f(0) + 4f'(c) = -3 + 4f'(c)$. We are given that $f'(x) \le 5$ for all x, so in particular we know that $f'(c) \le 5$.

Multiplying both sides of this inequality by 4, we have $4f'(c) \le 20$, so $f(4) = -3 + 4f'(c) \le -3 + 20 = 17$. The largest

possible value for $f(4)$ is 17.

65. Let $g(t)$ and $h(t)$ be the position functions of the two runners and let $f(t) = g(t) - h(t)$. By hypothesis,

$f(0) = g(0) - h(0) = 0$ and $f(b) = g(b) - h(b) = 0$, where b is the finishing time. Then by the Mean Value Theorem,

there is a time c, with $0 < c < b$, such that $f'(c) = \dfrac{f(b) - f(0)}{b - 0}$. But $f(b) = f(0) = 0$, so $f'(c) = 0$. Since

$f'(c) = g'(c) - h'(c) = 0$, we have $g'(c) = h'(c)$. So at time c, both runners have the same speed $g'(c) = h'(c)$.

67. $y = \dfrac{1 + x}{1 + x^2} \Rightarrow y' = \dfrac{(1 + x^2)(1) - (1 + x)(2x)}{(1 + x^2)^2} = \dfrac{1 - 2x - x^2}{(1 + x^2)^2} \Rightarrow$

$y'' = \dfrac{(1 + x^2)^2(-2 - 2x) - (1 - 2x - x^2) \cdot 2(1 + x^2)(2x)}{[(1 + x^2)^2]^2} = \dfrac{2(1 + x^2)[(1 + x^2)(-1 - x) - (1 - 2x - x^2)(2x)]}{(1 + x^2)^4}$

$= \dfrac{2(-1 - x - x^2 - x^3 - 2x + 4x^2 + 2x^3)}{(1 + x^2)^3} = \dfrac{2(x^3 + 3x^2 - 3x - 1)}{(1 + x^2)^3} = \dfrac{2(x - 1)(x^2 + 4x + 1)}{(1 + x^2)^3}$

So $y'' = 0 \Rightarrow x = 1, -2 \pm \sqrt{3}$. Let $a = -2 - \sqrt{3}, b = -2 + \sqrt{3}$, and $c = 1$. We can show that $f(a) = \frac{1}{4}(1 - \sqrt{3})$,

$f(b) = \frac{1}{4}(1 + \sqrt{3})$, and $f(c) = 1$. To show that these three points of inflection lie on one straight line, we'll show that the

slopes m_{ac} and m_{bc} are equal.

$$m_{ac} = \frac{f(c) - f(a)}{c - a} = \frac{1 - \frac{1}{4}(1 - \sqrt{3})}{1 - (-2 - \sqrt{3})} = \frac{\frac{3}{4} + \frac{1}{4}\sqrt{3}}{3 + \sqrt{3}} = \frac{1}{4}$$

$$m_{bc} = \frac{f(c) - f(b)}{c - b} = \frac{1 - \frac{1}{4}(1 + \sqrt{3})}{1 - (-2 + \sqrt{3})} = \frac{\frac{3}{4} - \frac{1}{4}\sqrt{3}}{3 - \sqrt{3}} = \frac{1}{4}$$

69. Let the cubic function be $f(x) = ax^3 + bx^2 + cx + d \Rightarrow f'(x) = 3ax^2 + 2bx + c \Rightarrow f''(x) = 6ax + 2b$.

So f is CU when $6ax + 2b > 0 \Leftrightarrow x > -b/(3a)$, CD when $x < -b/(3a)$, and so the only point of inflection occurs when

$x = -b/(3a)$. If the graph has three x-intercepts x_1, x_2 and x_3, then the expression for $f(x)$ must factor as

$f(x) = a(x - x_1)(x - x_2)(x - x_3)$. Multiplying these factors together gives us

$$f(x) = a[x^3 - (x_1 + x_2 + x_3)x^2 + (x_1 x_2 + x_1 x_3 + x_2 x_3)x - x_1 x_2 x_3]$$

Equating the coefficients of the x^2-terms for the two forms of f gives us $b = -a(x_1 + x_2 + x_3)$. Hence, the x-coordinate of

the point of inflection is $-\dfrac{b}{3a} = -\dfrac{-a(x_1 + x_2 + x_3)}{3a} = \dfrac{x_1 + x_2 + x_3}{3}$.

71. (a) $f(x) = x^3 + ax^2 + bx \Rightarrow f'(x) = 3x^2 + 2ax + b$. f has the local minimum value $-\frac{2}{9}\sqrt{3}$ at $x = 1/\sqrt{3}$, so

$f'\left(\frac{1}{\sqrt{3}}\right) = 0 \Rightarrow 1 + \frac{2}{\sqrt{3}}a + b = 0$ **(1)** and $f\left(\frac{1}{\sqrt{3}}\right) = -\frac{2}{9}\sqrt{3} \Rightarrow \frac{1}{9}\sqrt{3} + \frac{1}{3}a + \frac{1}{3}\sqrt{3}b = -\frac{2}{9}\sqrt{3}$ **(2)**.

Rewrite the system of equations as

$$\frac{2}{3}\sqrt{3}a \quad + \quad b \quad = \quad -1 \qquad \textbf{(3)}$$

$$\frac{1}{3}a \quad + \quad \frac{1}{3}\sqrt{3}b \quad = \quad -\frac{1}{3}\sqrt{3} \qquad \textbf{(4)}$$

and then multiplying **(4)** by $-2\sqrt{3}$ gives us the system

$$\frac{2}{3}\sqrt{3}a \quad + \quad b \quad = \quad -1$$

$$-\frac{2}{3}\sqrt{3}a \quad - \quad 2b \quad = \quad 2$$

Adding the equations gives us $-b = 1 \Rightarrow b = -1$. Substituting -1 for b into **(3)** gives us

$\frac{2}{3}\sqrt{3}a - 1 = -1 \Rightarrow \frac{2}{3}\sqrt{3}a = 0 \Rightarrow a = 0$. Thus, $f(x) = x^3 - x$.

(b) To find the smallest slope, we want to find the minimum of the slope function, f', so we'll find the critical

numbers of f'. $f(x) = x^3 - x \Rightarrow f'(x) = 3x^2 - 1 \Rightarrow f''(x) = 6x$. $f''(x) = 0 \Leftrightarrow x = 0$.

At $x = 0$, $y = 0$, $f'(x) = -1$, and f'' changes from negative to positive. Thus, we have a minimum for f' and

$y - 0 = -1(x - 0)$, or $y = -x$, is the tangent line that has the smallest slope.

4.4 Graphing with Calculus *and* Calculators

1. $f(x) = 4x^4 - 32x^3 + 89x^2 - 95x + 29 \Rightarrow f'(x) = 16x^3 - 96x^2 + 178x - 95 \Rightarrow f''(x) = 48x^2 - 192x + 178$.

$f(x) = 0 \Leftrightarrow x \approx 0.5, 1.60$; $f'(x) = 0 \Leftrightarrow x \approx 0.92, 2.5, 2.58$ and $f''(x) = 0 \Leftrightarrow x \approx 1.46, 2.54$.

From the graphs of f', we estimate that $f' < 0$ and that f is decreasing on $(-\infty, 0.92)$ and $(2.5, 2.58)$, and that $f' > 0$ and f

is increasing on $(0.92, 2.5)$ and $(2.58, \infty)$ with local minimum values $f(0.92) \approx -5.12$ and $f(2.58) \approx 3.998$ and local maximum value $f(2.5) = 4$. The graphs of f' make it clear that f has a maximum and a minimum near $x = 2.5$, shown more clearly in the fourth graph.

From the graph of f'', we estimate that $f'' > 0$ and that f is CU on $(-\infty, 1.46)$ and $(2.54, \infty)$, and that $f'' < 0$ and f is CD on $(1.46, 2.54)$. There are inflection points at about $(1.46, -1.40)$ and $(2.54, 3.999)$.

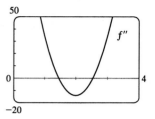

3. $f(x) = x^6 - 10x^5 - 400x^4 + 2500x^3 \quad \Rightarrow \quad f'(x) = 6x^5 - 50x^4 - 1600x^3 + 7500x^2 \quad \Rightarrow$

$f''(x) = 30x^4 - 200x^3 - 4800x^2 + 1500x$

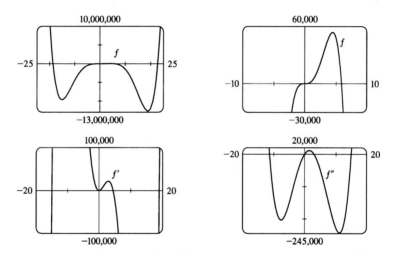

From the graph of f', we estimate that f is decreasing on $(-\infty, -15)$, increasing on $(-15, 4.40)$, decreasing on $(4.40, 18.93)$, and increasing on $(18.93, \infty)$, with local minimum values of $f(-15) \approx -9,700,000$ and $f(18.93) \approx -12,700,000$ and local maximum value $f(4.40) \approx 53,800$. From the graph of f'', we estimate that f is CU on $(-\infty, -11.34)$, CD on $(-11.34, 0)$, CU on $(0, 2.92)$, CD on $(2.92, 15.08)$, and CU on $(15.08, \infty)$. There is an inflection point at $(0, 0)$ and at about $(-11.34, -6,250,000)$, $(2.92, 31,800)$, and $(15.08, -8,150,000)$.

5. $f(x) = \dfrac{x}{x^3 - x^2 - 4x + 1} \quad \Rightarrow \quad f'(x) = \dfrac{-2x^3 + x^2 + 1}{(x^3 - x^2 - 4x + 1)^2} \quad \Rightarrow \quad f''(x) = \dfrac{2(3x^5 - 3x^4 + 5x^3 - 6x^2 + 3x + 4)}{(x^3 - x^2 - 4x + 1)^3}$

We estimate from the graph of f that $y = 0$ is a horizontal asymptote, and that there are vertical asymptotes at $x = -1.7$,

$x = 0.24$, and $x = 2.46$. From the graph of f', we estimate that f is increasing on $(-\infty, -1.7)$, $(-1.7, 0.24)$, and $(0.24, 1)$, and that f is decreasing on $(1, 2.46)$ and $(2.46, \infty)$. There is a local maximum value at $f(1) = -\frac{1}{3}$. From the graph of f'', we estimate that f is CU on $(-\infty, -1.7)$, $(-0.506, 0.24)$, and $(2.46, \infty)$, and that f is CD on $(-1.7, -0.506)$ and $(0.24, 2.46)$. There is an inflection point at $(-0.506, -0.192)$.

7. $f(x) = x^2 - 4x + 7\cos x$, $-4 \le x \le 4$. $f'(x) = 2x - 4 - 7\sin x$ $\Rightarrow$ $f''(x) = 2 - 7\cos x$.

$f(x) = 0$ $\Leftrightarrow$ $x \approx 1.10$; $f'(x) = 0$ $\Leftrightarrow$ $x \approx -1.49, -1.07$, or 2.89; $f''(x) = 0$ $\Leftrightarrow$ $x = \pm\cos^{-1}\left(\frac{2}{7}\right) \approx \pm 1.28$.

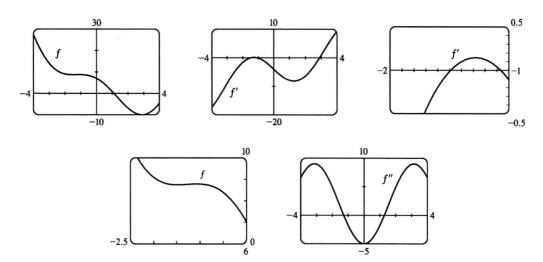

From the graphs of f', we estimate that f is decreasing ($f' < 0$) on $(-4, -1.49)$, increasing on $(-1.49, -1.07)$, decreasing on $(-1.07, 2.89)$, and increasing on $(2.89, 4)$, with local minimum values $f(-1.49) \approx 8.75$ and $f(2.89) \approx -9.99$ and local maximum value $f(-1.07) \approx 8.79$ (notice the second graph of f). From the graph of f'', we estimate that f is CU ($f'' > 0$) on $(-4, -1.28)$, CD on $(-1.28, 1.28)$, and CU on $(1.28, 4)$. There are inflection points at about $(-1.28, 8.77)$ and $(1.28, -1.48)$.

9. $f(x) = 1 + \dfrac{1}{x} + \dfrac{8}{x^2} + \dfrac{1}{x^3}$ $\Rightarrow$ $f'(x) = -\dfrac{1}{x^2} - \dfrac{16}{x^3} - \dfrac{3}{x^4} = -\dfrac{1}{x^4}(x^2 + 16x + 3)$ $\Rightarrow$

$f''(x) = \dfrac{2}{x^3} + \dfrac{48}{x^4} + \dfrac{12}{x^5} = \dfrac{2}{x^5}(x^2 + 24x + 6)$.

From the graphs, it appears that f increases on $(-15.8, -0.2)$ and decreases on $(-\infty, -15.8)$, $(-0.2, 0)$, and $(0, \infty)$; that f has a local minimum value of $f(-15.8) \approx 0.97$ and a local maximum value of $f(-0.2) \approx 72$; that f is CD on $(-\infty, -24)$ and $(-0.25, 0)$ and is CU on $(-24, -0.25)$ and $(0, \infty)$; and that f has IPs at $(-24, 0.97)$ and $(-0.25, 60)$.

To find the exact values, note that $f' = 0 \Rightarrow x = \dfrac{-16 \pm \sqrt{256 - 12}}{2} = -8 \pm \sqrt{61}$ $[\approx -0.19 \text{ and } -15.81]$.

f' is positive (f is increasing) on $\left(-8 - \sqrt{61}, -8 + \sqrt{61}\right)$ and f' is negative (f is decreasing) on $\left(-\infty, -8 - \sqrt{61}\right)$,

$\left(-8 + \sqrt{61}, 0\right)$, and $(0, \infty)$. $f'' = 0 \Rightarrow x = \dfrac{-24 \pm \sqrt{576 - 24}}{2} = -12 \pm \sqrt{138}$ $[\approx -0.25 \text{ and } -23.75]$. f'' is

positive (f is CU) on $\left(-12 - \sqrt{138}, -12 + \sqrt{138}\right)$ and $(0, \infty)$ and f'' is negative (f is CD) on $\left(-\infty, -12 - \sqrt{138}\right)$

and $\left(-12 + \sqrt{138}, 0\right)$.

11.

$f(x) = \dfrac{(x + 4)(x - 3)^2}{x^4(x - 1)}$ has VA at $x = 0$ and at $x = 1$ since $\lim\limits_{x \to 0} f(x) = -\infty$,

$\lim\limits_{x \to 1^-} f(x) = -\infty$ and $\lim\limits_{x \to 1^+} f(x) = \infty$.

$f(x) = \dfrac{\dfrac{x + 4}{x} \cdot \dfrac{(x - 3)^2}{x^2}}{\dfrac{x^4}{x^3} \cdot (x - 1)}$ $\begin{bmatrix}\text{dividing numerator} \\ \text{and denominator by } x^3\end{bmatrix} = \dfrac{(1 + 4/x)(1 - 3/x)^2}{x(x - 1)} \to 0$

as $x \to \pm\infty$, so f is asymptotic to the x-axis.

Since f is undefined at $x = 0$, it has no y-intercept. $f(x) = 0 \Rightarrow (x + 4)(x - 3)^2 = 0 \Rightarrow x = -4$ or $x = 3$, so f has

x-intercepts -4 and 3. Note, however, that the graph of f is only tangent to the x-axis and does not cross it at $x = 3$, since f is

positive as $x \to 3^-$ and as $x \to 3^+$.

From these graphs, it appears that f has three maximum values and one minimum value. The maximum values are

approximately $f(-5.6) = 0.0182$, $f(0.82) = -281.5$ and $f(5.2) = 0.0145$ and we know (since the graph is tangent to the

x-axis at $x = 3$) that the minimum value is $f(3) = 0$.

13. $f(x) = \dfrac{x^2(x + 1)^3}{(x - 2)^2(x - 4)^4} \Rightarrow f'(x) = -\dfrac{x(x + 1)^2(x^3 + 18x^2 - 44x - 16)}{(x - 2)^3(x - 4)^5}$ [from CAS].

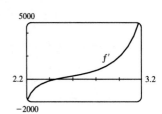

From the graphs of f', it seems that the critical points which indicate extrema occur at $x \approx -20$, -0.3, and 2.5, as estimated

in Example 3. (There is another critical point at $x = -1$, but the sign of f' does not change there.) We differentiate again,

obtaining $f''(x) = 2\dfrac{(x+1)(x^6 + 36x^5 + 6x^4 - 628x^3 + 684x^2 + 672x + 64)}{(x-2)^4(x-4)^6}$.

  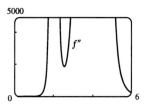

From the graphs of f'', it appears that f is CU on $(-35.3, -5.0)$, $(-1, -0.5)$, $(-0.1, 2)$, $(2, 4)$ and $(4, \infty)$ and CD on $(-\infty, -35.3)$, $(-5.0, -1)$ and $(-0.5, -0.1)$. We check back on the graphs of f to find the y-coordinates of the inflection points, and find that these points are approximately $(-35.3, -0.015)$, $(-5.0, -0.005)$, $(-1, 0)$, $(-0.5, 0.00001)$, and $(-0.1, 0.0000066)$.

15. $y = f(x) = \dfrac{\sqrt{x}}{x^2 + x + 1}$. From a CAS, $y' = -\dfrac{3x^2 + x - 1}{2\sqrt{x}\,(x^2 + x + 1)^2}$ and $y'' = \dfrac{15x^4 + 10x^3 - 15x^2 - 6x - 1}{4x^{3/2}(x^2 + x + 1)^3}$.

  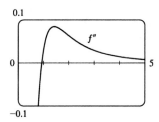

$f'(x) = 0 \iff x \approx 0.43$, so f is increasing on $(0, 0.43)$ and decreasing on $(0.43, \infty)$. There is a local maximum value of $f(0.43) \approx 0.41$. $f''(x) = 0 \iff x \approx 0.94$, so f is CD on $(0, 0.94)$ and CU on $(0.94, \infty)$. There is an inflection point at $(0.94, 0.34)$.

17. $y = f(x) = \sqrt{x + 5\sin x}$, $x \le 20$.

From a CAS, $y' = \dfrac{5\cos x + 1}{2\sqrt{x + 5\sin x}}$ and $y'' = -\dfrac{10\cos x + 25\sin^2 x + 10x\sin x + 26}{4(x + 5\sin x)^{3/2}}$.

We'll start with a graph of $g(x) = x + 5\sin x$. Note that $f(x) = \sqrt{g(x)}$ is only defined if $g(x) \ge 0$. $g(x) = 0 \iff x = 0$ or $x \approx -4.91, -4.10, 4.10$, and 4.91. Thus, the domain of f is $[-4.91, -4.10] \cup [0, 4.10] \cup [4.91, 20]$.

  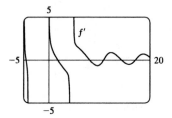

From the expression for y', we see that $y' = 0 \iff 5\cos x + 1 = 0 \implies x_1 = \cos^{-1}\left(-\tfrac{1}{5}\right) \approx 1.77$ and

$x_2 = 2\pi - x_1 \approx -4.51$ (not in the domain of f). The leftmost zero of f' is $x_1 - 2\pi \approx -4.51$. Moving to the right, the zeros of f' are x_1, $x_1 + 2\pi$, $x_2 + 2\pi$, $x_1 + 4\pi$, and $x_2 + 4\pi$. Thus, f is increasing on $(-4.91, -4.51)$, decreasing on $(-4.51, -4.10)$, increasing on $(0, 1.77)$, decreasing on $(1.77, 4.10)$, increasing on $(4.91, 8.06)$, decreasing on $(8.06, 10.79)$, increasing on $(10.79, 14.34)$, decreasing on $(14.34, 17.08)$, and increasing on $(17.08, 20)$. The local maximum values are $f(-4.51) \approx 0.62$, $f(1.77) \approx 2.58$, $f(8.06) \approx 3.60$, and $f(14.34) \approx 4.39$. The local minimum values are $f(10.79) \approx 2.43$ and $f(17.08) \approx 3.49$.

f is CD on $(-4.91, -4.10)$, $(0, 4.10)$, $(4.91, 9.60)$, CU on $(9.60, 12.25)$, CD on $(12.25, 15.81)$, CU on $(15.81, 18.65)$, and CD on $(18.65, 20)$. There are inflection points at $(9.60, 2.95)$, $(12.25, 3.27)$, $(15.81, 3.91)$, and $(18.65, 4.20)$.

19. $y = f(x) = \dfrac{1 - e^{1/x}}{1 + e^{1/x}}$.   From a CAS, $y' = \dfrac{2e^{1/x}}{x^2(1 + e^{1/x})^2}$ and $y'' = \dfrac{-2e^{1/x}(1 - e^{1/x} + 2x + 2xe^{1/x})}{x^4(1 + e^{1/x})^3}$.

f is an odd function defined on $(-\infty, 0) \cup (0, \infty)$. Its graph has no x- or y-intercepts. Since $\lim\limits_{x \to \pm\infty} f(x) = 0$, the x-axis is a HA. $f'(x) > 0$ for $x \neq 0$, so f is increasing on $(-\infty, 0)$ and $(0, \infty)$. It has no local extreme values.

$f''(x) = 0$ for $x \approx \pm 0.417$, so f is CU on $(-\infty, -0.417)$, CD on $(-0.417, 0)$, CU on $(0, 0.417)$, and CD on $(0.417, \infty)$. f has IPs at $(-0.417, 0.834)$ and $(0.417, -0.834)$.

21.

From the graph of $f(x) = \sin(x + \sin 3x)$ in the viewing rectangle $[0, \pi]$ by $[-1.2, 1.2]$, it looks like f has two maxima and two minima. If we calculate and graph $f'(x) = [\cos(x + \sin 3x)](1 + 3\cos 3x)$ on $[0, 2\pi]$, we see that the graph of f' appears to be almost tangent to the x-axis at about $x = 0.7$. The graph of

$$f'' = -[\sin(x + \sin 3x)](1 + 3\cos 3x)^2 + \cos(x + \sin 3x)(-9\sin 3x)$$

is even more interesting near this x-value: it seems to just touch the x-axis.

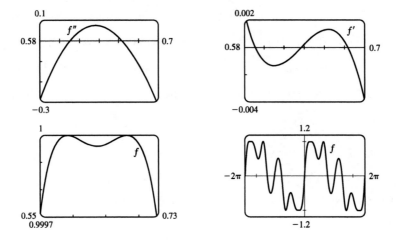

If we zoom in on this place on the graph of f'', we see that f'' actually does cross the axis twice near $x = 0.65$,

indicating a change in concavity for a very short interval. If we look at the graph of f' on the same interval, we see that it

changes sign three times near $x = 0.65$, indicating that what we had thought was a broad extremum at about $x = 0.7$ actually

consists of three extrema (two maxima and a minimum). These maximum values are roughly $f(0.59) = 1$ and $f(0.68) = 1$,

and the minimum value is roughly $f(0.64) = 0.99996$. There are also a maximum value of about $f(1.96) = 1$ and minimum

values of about $f(1.46) = 0.49$ and $f(2.73) = -0.51$. The points of inflection on $(0, \pi)$ are about $(0.61, 0.99998)$,

$(0.66, 0.99998)$, $(1.17, 0.72)$, $(1.75, 0.77)$, and $(2.28, 0.34)$. On $(\pi, 2\pi)$, they are about $(4.01, -0.34)$, $(4.54, -0.77)$,

$(5.11, -0.72)$, $(5.62, -0.99998)$, and $(5.67, -0.99998)$. There are also IP at $(0, 0)$ and $(\pi, 0)$. Note that the function is odd

and periodic with period 2π, and it is also rotationally symmetric about all points of the form $((2n + 1)\pi, 0)$, n an integer.

23.

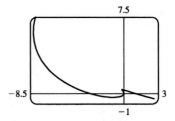

We graph the curve $x = t^4 - 2t^3 - 2t^2$, $y = t^3 - t$ in the viewing rectangle $[-2, 1.1]$ by $[-0.5, 0.5]$. This rectangle

corresponds approximately to $t \in [-1, 0.8]$. We estimate that the curve has horizontal tangents at about $(-1, -0.4)$ and

$(-0.17, 0.39)$ and vertical tangents at about $(0, 0)$ and $(-0.19, 0.37)$. We calculate $\dfrac{dy}{dx} = \dfrac{dy/dt}{dx/dt} = \dfrac{3t^2 - 1}{4t^3 - 6t^2 - 4t}$. The

horizontal tangents occur when $dy/dt = 3t^2 - 1 = 0 \iff t = \pm\frac{1}{\sqrt{3}}$, so both horizontal tangents are shown in our graph.

$t = \frac{1}{\sqrt{3}}$ corresponds to the point $\left(\frac{-2\sqrt{3}-5}{9}, \frac{-2\sqrt{3}}{9}\right) \approx (-0.94, -0.38)$ and $t = -\frac{1}{\sqrt{3}}$ corresponds to

$\left(\frac{2\sqrt{3}-5}{9}, \frac{2\sqrt{3}}{9}\right) \approx (-0.17, 0.38)$. The vertical tangents occur when $dx/dt = 2t(2t^2 - 3t - 2) = 0$ ⇔

$2t(2t + 1)(t - 2) = 0$ ⇔ $t = 0, -\frac{1}{2}$ or 2. It seems that we have missed one vertical tangent, and indeed if we plot the

curve on the t-interval $[-1.2, 2.2]$ we see that there is another vertical tangent at $(-8, 6)$. The t-values and points at which

there are vertical tangents are $t = 0, (0, 0); t = -\frac{1}{2}, \left(-\frac{3}{16}, \frac{3}{8}\right);$ and $t = 2, (-8, 6)$.

25. $x = t^3 - ct, y = t^2$. For $c = 0$, there is a cusp at $(0, 0)$. For $c < 0$, there is a local minimum at $(0, 0)$. For $c > 0$, there is a

loop whose size increases as c increases ($c = \frac{1}{2}$ and $c = 1$ are shown in the figure). The curve intersects itself on the y-axis;

that is, when $x = 0$ ⇔ $t^3 - ct = 0$ ⇔ $t(t^2 - c) = 0$ ⇔ $t = 0, \pm\sqrt{c}$. Substituting $\pm\sqrt{c}$ for t gives us $y = c$, so

the point of intersection is $(0, c)$.

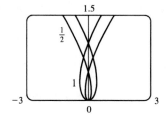

From the second figure, we see that the left- and rightmost points of the loop occur when there are vertical tangent lines.

$dx/dt = 0$ ⇒ $3t^2 - c = 0$ ⇒ $t = \pm\sqrt{c/3}$. The rightmost point occurs when $t = -\sqrt{c/3}$ and has coordinates

$\left(\frac{2c\sqrt{3c}}{9}, \frac{c}{3}\right)$. The leftmost point occurs when $t = \sqrt{c/3}$ and has coordinates $\left(-\frac{2c\sqrt{3c}}{9}, \frac{c}{3}\right)$.

27. $f(x) = x^4 + cx^2 = x^2(x^2 + c)$. Note that f is an even function. For $c \geq 0$, the only x-intercept is the point $(0, 0)$. We

calculate $f'(x) = 4x^3 + 2cx = 4x\left(x^2 + \frac{1}{2}c\right)$ ⇒ $f''(x) = 12x^2 + 2c$. If $c \geq 0$, $x = 0$ is the only critical point and there

is no inflection point. As we can see from the examples, there is no change in the basic shape of the graph for $c \geq 0$; it merely

becomes steeper as c increases. For $c = 0$, the graph is the simple curve $y = x^4$. For $c < 0$, there are x-intercepts at 0 and

at $\pm\sqrt{-c}$. Also, there is a maximum at $(0, 0)$, and there are minima

at $\left(\pm\sqrt{-\frac{1}{2}c}, -\frac{1}{4}c^2\right)$. As $c \to -\infty$, the x-coordinates of these

minima get larger in absolute value, and the minimum points move

downward. There are inflection points at $\left(\pm\sqrt{-\frac{1}{6}c}, -\frac{5}{36}c^2\right)$, which

also move away from the origin as $c \to -\infty$.

29. $f(x) = e^x + ce^{-x}$. $f = 0 \Rightarrow ce^{-x} = -e^x \Rightarrow c = -e^{2x} \Rightarrow 2x = \ln(-c) \Rightarrow x = \frac{1}{2}\ln(-c)$.

$f'(x) = e^x - ce^{-x}$. $f' = 0 \Rightarrow ce^{-x} = e^x \Rightarrow c = e^{2x} \Rightarrow 2x = \ln c \Rightarrow x = \frac{1}{2}\ln c$.

$f''(x) = e^x + ce^{-x} = f(x)$.

The only transitional value of c is 0. As c increases from $-\infty$ to 0, $\frac{1}{2}\ln(-c)$ is both the the x-intercept and inflection point,

and this decreases from ∞ to $-\infty$. Also $f' > 0$, so f is increasing. When $c = 0$, $f(x) = f'(x) = f''(x) = e^x$, f is positive,

increasing, and concave upward. As c increases from 0 to ∞, the absolute minimum occurs at $x = \frac{1}{2}\ln c$, which increases

from $-\infty$ to ∞. Also, $f = f'' > 0$, so f is positive and concave upward. The

value of the y-intercept is $f(0) = 1 + c$, and this increases as c increases from

$-\infty$ to ∞.

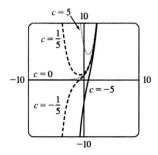

Note: The minimum point $\left(\frac{1}{2}\ln c, 2\sqrt{c}\right)$ can be parameterized by $x = \frac{1}{2}\ln c$,

$y = 2\sqrt{c}$, and after eliminating the parameter c, we see that the minimum point

lies on the graph of $y = 2e^x$.

31. Note that $c = 0$ is a transitional value at which the graph consists of the x-axis. Also, we can see that if we substitute $-c$ for c,

the function $f(x) = \dfrac{cx}{1 + c^2x^2}$ will be reflected in the x-axis, so we investigate only positive values of c (except $c = -1$, as a

demonstration of this reflective property). Also, f is an odd function. $\displaystyle\lim_{x \to \pm\infty} f(x) = 0$, so $y = 0$ is a horizontal asymptote

for all c. We calculate $f'(x) = \dfrac{(1 + c^2x^2)c - cx(2c^2x)}{(1 + c^2x^2)^2} = -\dfrac{c(c^2x^2 - 1)}{(1 + c^2x^2)^2}$. $f'(x) = 0 \Leftrightarrow c^2x^2 - 1 = 0 \Leftrightarrow$

$x = \pm 1/c$. So there is an absolute maximum value of $f(1/c) = \frac{1}{2}$ and an absolute minimum value of $f(-1/c) = -\frac{1}{2}$.

These extrema have the same value regardless of c, but the maximum points move closer to the y-axis as c increases.

$f''(x) = \dfrac{(-2c^3x)(1 + c^2x^2)^2 - (-c^3x^2 + c)[2(1 + c^2x^2)(2c^2x)]}{(1 + c^2x^2)^4}$

$= \dfrac{(-2c^3x)(1 + c^2x^2) + (c^3x^2 - c)(4c^2x)}{(1 + c^2x^2)^3} = \dfrac{2c^3x(c^2x^2 - 3)}{(1 + c^2x^2)^3}$

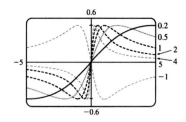

$f''(x) = 0 \Leftrightarrow x = 0$ or $\pm\sqrt{3}/c$, so there are inflection points at $(0, 0)$ and

at $\left(\pm\sqrt{3}/c, \pm\sqrt{3}/4\right)$. Again, the y-coordinate of the inflection points does not depend on c, but as c increases, both inflection

points approach the y-axis.

33. $f(x) = cx + \sin x \Rightarrow f'(x) = c + \cos x \Rightarrow f''(x) = -\sin x$

$f(-x) = -f(x)$, so f is an odd function and its graph is symmetric with respect to the origin.

$f(x) = 0 \Leftrightarrow \sin x = -cx$, so 0 is always an x-intercept.

$f'(x) = 0 \Leftrightarrow \cos x = -c$, so there is no critical number when $|c| > 1$. If $|c| \le 1$, then there are infinitely many critical numbers. If x_1 is the unique solution of $\cos x = -c$ in the interval $[0, \pi]$, then the critical numbers are $2n\pi \pm x_1$, where n ranges over the integers. (Special cases: When $c = -1$, $x_1 = 0$; when $c = 0$, $x = \frac{\pi}{2}$; and when $c = 1$, $x_1 = \pi$.)

$f''(x) < 0 \iff \sin x > 0$, so f is CD on intervals of the form $(2n\pi, (2n+1)\pi)$. f is CU on intervals of the form $((2n-1)\pi, 2n\pi)$. The inflection points of f are the points $(n\pi, n\pi c)$, where n is an integer.

If $c \geq 1$, then $f'(x) \geq 0$ for all x, so f is increasing and has no extremum. If $c \leq -1$, then $f'(x) \leq 0$ for all x, so f is decreasing and has no extremum. If $|c| < 1$, then $f'(x) > 0 \iff \cos x > -c \iff x$ is in an interval of the form $(2n\pi - x_1, 2n\pi + x_1)$ for some integer n. These are the intervals on which f is increasing. Similarly, we find that f is decreasing on the intervals of the form $(2n\pi + x_1, 2(n+1)\pi - x_1)$. Thus, f has local maxima at the points $2n\pi + x_1$, where f has the values $c(2n\pi + x_1) + \sin x_1 = c(2n\pi + x_1) + \sqrt{1 - c^2}$, and f has local minima at the points $2n\pi - x_1$, where we have $f(2n\pi - x_1) = c(2n\pi - x_1) - \sin x_1 = c(2n\pi - x_1) - \sqrt{1 - c^2}$.

The transitional values of c are -1 and 1. The inflection points move vertically, but not horizontally, when c changes.

When $|c| \geq 1$, there is no extremum. For $|c| < 1$, the maxima are spaced 2π apart horizontally, as are the minima. The horizontal spacing between maxima and adjacent minima is regular (and equals π) when $c = 0$, but the horizontal space between a local maximum and the nearest local minimum shrinks as $|c|$ approaches 1.

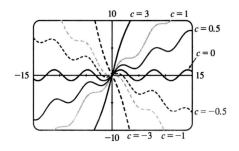

35. (a) $f(x) = cx^4 - 2x^2 + 1$. For $c = 0$, $f(x) = -2x^2 + 1$, a parabola whose vertex, $(0, 1)$, is the absolute maximum. For $c > 0$, $f(x) = cx^4 - 2x^2 + 1$ opens upward with two minimum points. As $c \to 0$, the minimum points spread apart and move downward; they are below the x-axis for $0 < c < 1$ and above for $c > 1$. For $c < 0$, the graph opens downward, and has an absolute maximum at $x = 0$ and no local minimum.

(b) $f'(x) = 4cx^3 - 4x = 4cx(x^2 - 1/c)$ $[c \neq 0]$. If $c \leq 0$, 0 is the only critical number.

$f''(x) = 12cx^2 - 4$, so $f''(0) = -4$ and there is a local maximum at $(0, f(0)) = (0, 1)$, which lies on $y = 1 - x^2$. If $c > 0$, the critical numbers are 0 and $\pm 1/\sqrt{c}$. As before, there is a local maximum at $(0, f(0)) = (0, 1)$, which lies on $y = 1 - x^2$.

$f''\left(\pm 1/\sqrt{c}\right) = 12 - 4 = 8 > 0$, so there is a local minimum at $x = \pm 1/\sqrt{c}$. Here $f\left(\pm 1/\sqrt{c}\right) = c(1/c^2) - 2/c + 1 = -1/c + 1$.

But $\left(\pm 1/\sqrt{c}, -1/c + 1\right)$ lies on $y = 1 - x^2$ since $1 - \left(\pm 1/\sqrt{c}\right)^2 = 1 - 1/c$.

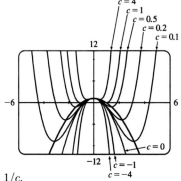

4.5 Indeterminate Forms and l'Hospital's Rule

Note: The use of l'Hospital's Rule is indicated by an H above the equal sign: $\overset{H}{=}$

1. (a) $\lim\limits_{x \to a} \dfrac{f(x)}{g(x)}$ is an indeterminate form of type $\dfrac{0}{0}$.

(b) $\lim\limits_{x \to a} \dfrac{f(x)}{p(x)} = 0$ because the numerator approaches 0 while the denominator becomes large.

(c) $\lim\limits_{x \to a} \dfrac{h(x)}{p(x)} = 0$ because the numerator approaches a finite number while the denominator becomes large.

(d) If $\lim\limits_{x \to a} p(x) = \infty$ and $f(x) \to 0$ through positive values, then $\lim\limits_{x \to a} \dfrac{p(x)}{f(x)} = \infty$. [For example, take $a = 0$, $p(x) = 1/x^2$,

and $f(x) = x^2$.] If $f(x) \to 0$ through negative values, then $\lim\limits_{x \to a} \dfrac{p(x)}{f(x)} = -\infty$. [For example, take $a = 0$, $p(x) = 1/x^2$,

and $f(x) = -x^2$.] If $f(x) \to 0$ through both positive and negative values, then the limit might not exist. [For example,

take $a = 0$, $p(x) = 1/x^2$, and $f(x) = x$.]

(e) $\lim\limits_{x \to a} \dfrac{p(x)}{q(x)}$ is an indeterminate form of type $\dfrac{\infty}{\infty}$.

3. (a) When x is near a, $f(x)$ is near 0 and $p(x)$ is large, so $f(x) - p(x)$ is large negative. Thus, $\lim\limits_{x \to a} [f(x) - p(x)] = -\infty$.

(b) $\lim\limits_{x \to a} [p(x) - q(x)]$ is an indeterminate form of type $\infty - \infty$.

(c) When x is near a, $p(x)$ and $q(x)$ are both large, so $p(x) + q(x)$ is large. Thus, $\lim\limits_{x \to a} [p(x) + q(x)] = \infty$.

5. This limit has the form $\frac{0}{0}$. We can simply factor and simplify to evaluate the limit.

$$\lim_{x \to 1} \frac{x^2 - 1}{x^2 - x} = \lim_{x \to 1} \frac{(x + 1)(x - 1)}{x(x - 1)} = \lim_{x \to 1} \frac{x + 1}{x} = \frac{1 + 1}{1} = 2$$

7. This limit has the form $\frac{0}{0}$. $\lim\limits_{x \to (\pi/2)^+} \dfrac{\cos x}{1 - \sin x} \overset{H}{=} \lim\limits_{x \to (\pi/2)^+} \dfrac{-\sin x}{-\cos x} = \lim\limits_{x \to (\pi/2)^+} \tan x = -\infty$.

9. This limit has the form $\frac{0}{0}$. $\lim\limits_{t \to 0} \dfrac{e^t - 1}{t^3} \overset{H}{=} \lim\limits_{t \to 0} \dfrac{e^t}{3t^2} = \infty$ since $e^t \to 1$ and $3t^2 \to 0^+$ as $t \to 0$.

11. This limit has the form $\frac{\infty}{\infty}$. $\lim\limits_{x \to \infty} \dfrac{\ln x}{\sqrt{x}} \overset{H}{=} \lim\limits_{x \to \infty} \dfrac{1/x}{\frac{1}{2} x^{-1/2}} = \lim\limits_{x \to \infty} \dfrac{2}{\sqrt{x}} = 0$

13. $\lim\limits_{x \to 0^+} [(\ln x)/x] = -\infty$ since $\ln x \to -\infty$ as $x \to 0^+$ and dividing by small values of x just increases the magnitude of the

quotient $(\ln x)/x$. L'Hospital's Rule does not apply.

15. This limit has the form $\frac{0}{0}$.

$$\lim_{x \to 0} \frac{\sqrt{1 + 2x} - \sqrt{1 - 4x}}{x} \overset{H}{=} \lim_{x \to 0} \frac{\frac{1}{2}(1 + 2x)^{-1/2} \cdot 2 - \frac{1}{2}(1 - 4x)^{-1/2}(-4)}{1}$$

$$= \lim_{x \to 0} \left(\frac{1}{\sqrt{1 + 2x}} + \frac{2}{\sqrt{1 - 4x}} \right) = \frac{1}{\sqrt{1}} + \frac{2}{\sqrt{1}} = 3$$

17. This limit has the form $\frac{0}{0}$. $\displaystyle\lim_{t\to0}\frac{5^t-3^t}{t}\overset{\text{H}}{=}\lim_{t\to0}\frac{5^t\ln5-3^t\ln3}{1}=\ln5-\ln3=\ln\tfrac{5}{3}$

19. This limit has the form $\frac{0}{0}$. $\displaystyle\lim_{x\to0}\frac{e^x-1-x}{x^2}\overset{\text{H}}{=}\lim_{x\to0}\frac{e^x-1}{2x}\overset{\text{H}}{=}\lim_{x\to0}\frac{e^x}{2}=\frac{1}{2}$

21. This limit has the form $\frac{0}{0}$. $\displaystyle\lim_{x\to1}\frac{1-x+\ln x}{1+\cos\pi x}\overset{\text{H}}{=}\lim_{x\to1}\frac{-1+1/x}{-\pi\sin\pi x}\overset{\text{H}}{=}\lim_{x\to1}\frac{-1/x^2}{-\pi^2\cos\pi x}=\frac{-1}{-\pi^2(-1)}=-\frac{1}{\pi^2}$

23. This limit has the form $\frac{0}{0}$. $\displaystyle\lim_{x\to1}\frac{x^a-ax+a-1}{(x-1)^2}\overset{\text{H}}{=}\lim_{x\to1}\frac{ax^{a-1}-a}{2(x-1)}\overset{\text{H}}{=}\lim_{x\to1}\frac{a(a-1)x^{a-2}}{2}=\frac{a(a-1)}{2}$

25. This limit has the form $\frac{0}{0}$. $\displaystyle\lim_{x\to0}\frac{\cos x-1+\frac{1}{2}x^2}{x^4}\overset{\text{H}}{=}\lim_{x\to0}\frac{-\sin x+x}{4x^3}\overset{\text{H}}{=}\lim_{x\to0}\frac{-\cos x+1}{12x^2}\overset{\text{H}}{=}\lim_{x\to0}\frac{\sin x}{24x}\overset{\text{H}}{=}\lim_{x\to0}\frac{\cos x}{24}=\frac{1}{24}$

27. This limit has the form $\infty\cdot0$.

$$\lim_{x\to\infty}x\sin(\pi/x)=\lim_{x\to\infty}\frac{\sin(\pi/x)}{1/x}\overset{\text{H}}{=}\lim_{x\to\infty}\frac{\cos(\pi/x)(-\pi/x^2)}{-1/x^2}=\pi\lim_{x\to\infty}\cos(\pi/x)=\pi(1)=\pi$$

29. This limit has the form $\infty\cdot0$. We'll change it to the form $\frac{0}{0}$.

$$\lim_{x\to0}\cot2x\sin6x=\lim_{x\to0}\frac{\sin6x}{\tan2x}\overset{\text{H}}{=}\lim_{x\to0}\frac{6\cos6x}{2\sec^22x}=\frac{6(1)}{2(1)^2}=3$$

31. This limit has the form $\infty\cdot0$. $\displaystyle\lim_{x\to\infty}x^3e^{-x^2}=\lim_{x\to\infty}\frac{x^3}{e^{x^2}}\overset{\text{H}}{=}\lim_{x\to\infty}\frac{3x^2}{2xe^{x^2}}=\lim_{x\to\infty}\frac{3x}{2e^{x^2}}\overset{\text{H}}{=}\lim_{x\to\infty}\frac{3}{4xe^{x^2}}=0$

33. This limit has the form $\infty-\infty$.

$$\lim_{x\to1}\left(\frac{x}{x-1}-\frac{1}{\ln x}\right)=\lim_{x\to1}\frac{x\ln x-(x-1)}{(x-1)\ln x}\overset{\text{H}}{=}\lim_{x\to1}\frac{x(1/x)+\ln x-1}{(x-1)(1/x)+\ln x}=\lim_{x\to1}\frac{\ln x}{1-(1/x)+\ln x}$$

$$\overset{\text{H}}{=}\lim_{x\to1}\frac{1/x}{1/x^2+1/x}\cdot\frac{x^2}{x^2}=\lim_{x\to1}\frac{x}{1+x}=\frac{1}{1+1}=\frac{1}{2}$$

35. We will multiply and divide by the conjugate of the expression to change the form of the expression.

$$\lim_{x\to\infty}\left(\sqrt{x^2+x}-x\right)=\lim_{x\to\infty}\left(\frac{\sqrt{x^2+x}-x}{1}\cdot\frac{\sqrt{x^2+x}+x}{\sqrt{x^2+x}+x}\right)=\lim_{x\to\infty}\frac{(x^2+x)-x^2}{\sqrt{x^2+x}+x}$$

$$=\lim_{x\to\infty}\frac{x}{\sqrt{x^2+x}+x}=\lim_{x\to\infty}\frac{1}{\sqrt{1+1/x}+1}=\frac{1}{\sqrt{1}+1}=\frac{1}{2}$$

As an alternate solution, write $\sqrt{x^2+x}-x$ as $\sqrt{x^2+x}-\sqrt{x^2}$, factor out $\sqrt{x^2}$, rewrite as $(\sqrt{1+1/x}-1)/(1/x)$, and apply l'Hospital's Rule.

37. The limit has the form $\infty-\infty$ and we will change the form to a product by factoring out x.

$$\lim_{x\to\infty}(x-\ln x)=\lim_{x\to\infty}x\left(1-\frac{\ln x}{x}\right)=\infty\text{ since }\lim_{x\to\infty}\frac{\ln x}{x}\overset{\text{H}}{=}\lim_{x\to\infty}\frac{1/x}{1}=0.$$

39. $y=x^{x^2}\ \Rightarrow\ \ln y=x^2\ln x$, so $\displaystyle\lim_{x\to0^+}\ln y=\lim_{x\to0^+}x^2\ln x=\lim_{x\to0^+}\frac{\ln x}{1/x^2}\overset{\text{H}}{=}\lim_{x\to0^+}\frac{1/x}{-2/x^3}=\lim_{x\to0^+}\left(-\frac{1}{2}x^2\right)=0\ \Rightarrow$

$$\lim_{x\to0^+}x^{x^2}=\lim_{x\to0^+}e^{\ln y}=e^0=1.$$

41. $y = (1 - 2x)^{1/x}$ $\Rightarrow$ $\ln y = \dfrac{1}{x}\ln(1 - 2x)$, so $\lim\limits_{x\to 0}\ln y = \lim\limits_{x\to 0}\dfrac{\ln(1 - 2x)}{x} \overset{\text{H}}{=} \lim\limits_{x\to 0}\dfrac{-2/(1 - 2x)}{1} = -2$ $\Rightarrow$

$\lim\limits_{x\to 0}(1 - 2x)^{1/x} = \lim\limits_{x\to 0}e^{\ln y} = e^{-2}$.

43. $y = x^{1/x}$ $\Rightarrow$ $\ln y = (1/x)\ln x$ $\Rightarrow$ $\lim\limits_{x\to\infty}\ln y = \lim\limits_{x\to\infty}\dfrac{\ln x}{x} \overset{\text{H}}{=} \lim\limits_{x\to\infty}\dfrac{1/x}{1} = 0$ $\Rightarrow$

$\lim\limits_{x\to\infty}x^{1/x} = \lim\limits_{x\to\infty}e^{\ln y} = e^0 = 1$

45. $y = (4x + 1)^{\cot x}$ $\Rightarrow$ $\ln y = \cot x\,\ln(4x + 1)$, so $\lim\limits_{x\to 0^+}\ln y = \lim\limits_{x\to 0^+}\dfrac{\ln(4x + 1)}{\tan x} \overset{\text{H}}{=} \lim\limits_{x\to 0^+}\dfrac{\dfrac{4}{4x + 1}}{\sec^2 x} = 4$ $\Rightarrow$

$\lim\limits_{x\to 0^+}(4x + 1)^{\cot x} = \lim\limits_{x\to 0^+}e^{\ln y} = e^4$.

47.

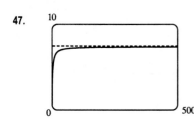

From the graph, if $x = 500$, $y \approx 7.36$. The limit has the form 1^∞.

Now $y = \left(1 + \dfrac{2}{x}\right)^x$ $\Rightarrow$ $\ln y = x\ln\left(1 + \dfrac{2}{x}\right)$ $\Rightarrow$

$\lim\limits_{x\to\infty}\ln y = \lim\limits_{x\to\infty}\dfrac{\ln(1 + 2/x)}{1/x} \overset{\text{H}}{=} \lim\limits_{x\to\infty}\dfrac{\dfrac{1}{1 + 2/x}\left(-\dfrac{2}{x^2}\right)}{-1/x^2}$

$= 2\lim\limits_{x\to\infty}\dfrac{1}{1 + 2/x} = 2(1) = 2$ $\Rightarrow$

$\lim\limits_{x\to\infty}\left(1 + \dfrac{2}{x}\right)^x = \lim\limits_{x\to\infty}e^{\ln y} = e^2$ $[\approx 7.39]$

49.

From the graph, it appears that $\lim\limits_{x\to 0}\dfrac{f(x)}{g(x)} = \lim\limits_{x\to 0}\dfrac{f'(x)}{g'(x)} = 0.25$.

We calculate $\lim\limits_{x\to 0}\dfrac{f(x)}{g(x)} = \lim\limits_{x\to 0}\dfrac{e^x - 1}{x^3 + 4x} \overset{\text{H}}{=} \lim\limits_{x\to 0}\dfrac{e^x}{3x^2 + 4} = \dfrac{1}{4}$.

51. $\lim\limits_{x\to\infty}xe^{-x} = \lim\limits_{x\to\infty}(x/e^x) \overset{\text{H}}{=} \lim\limits_{x\to\infty}(1/e^x) = 0$, so $y = 0$ is a HA. $\lim\limits_{x\to-\infty}xe^{-x} = -\infty$. $f(x) = xe^{-x}$ $\Rightarrow$

$f'(x) = x(-e^{-x}) + e^{-x}\cdot 1 = e^{-x}(1 - x) > 0$ $\Leftrightarrow$ $1 - x > 0$ $\Leftrightarrow$ $x < 1$,

so f is increasing on $(-\infty, 1)$ and decreasing on $(1, \infty)$. By the FDT, $f(1) = 1/e$ is a

local maximum.

$f''(x) = e^{-x}(-1) + (1 - x)(-e^{-x}) = e^{-x}(-1 - 1 + x) = e^{-x}(x - 2) > 0$ $\Leftrightarrow$

$x - 2 > 0$ $\Leftrightarrow$ $x > 2$, so f is CU on $(2, \infty)$ and CD on $(-\infty, 2)$. IP is $\left(2, 2/e^2\right)$.

53.

$\lim\limits_{x\to\infty}\dfrac{\ln x}{x} \overset{\text{H}}{=} \lim\limits_{x\to\infty}\dfrac{1/x}{1} = 0$, so $y = 0$ is a HA. Also $\lim\limits_{x\to 0^+}\dfrac{\ln x}{x} = -\infty$

since $\ln x \to -\infty$ and $x \to 0^+$, so $x = 0$ is a VA.

$f(x) = \dfrac{\ln x}{x}$ $\Rightarrow$ $f'(x) = \dfrac{x(1/x) - (\ln x)(1)}{x^2} = \dfrac{1 - \ln x}{x^2} = 0$

when $\ln x = 1$ $\Leftrightarrow$ $x = e$.

[continued]

$f'(x) > 0 \iff 1 - \ln x > 0 \iff \ln x < 1 \iff 0 < x < e. \quad f'(x) < 0 \iff x > e.$

So f is increasing on $(0, e)$ and decreasing on (e, ∞). By the FDT, $f(e) = 1/e$ is a local maximum.

$f''(x) = \dfrac{x^2(-1/x) - (1 - \ln x)(2x)}{(x^2)^2} = \dfrac{x(-1 - 2 + 2\ln x)}{x^4} = \dfrac{2\ln x - 3}{x^3}$, so $f''(x) > 0 \iff 2\ln x - 3 > 0 \iff$

$\ln x > \frac{3}{2} \iff x > e^{3/2}. \quad f''(x) < 0 \iff 0 < x < e^{3/2}$. So f is CU on $\left(e^{3/2}, \infty\right)$ and CD on $\left(0, e^{3/2}\right)$. There is an

inflection point at $\left(e^{3/2}, \frac{3}{2}e^{-3/2}\right)$.

55. (a) $f(x) = x^2 \ln x$. The domain of f is $(0, \infty)$.

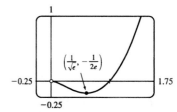

(b) $\lim\limits_{x \to 0^+} x^2 \ln x = \lim\limits_{x \to 0^+} \dfrac{\ln x}{1/x^2} \overset{\text{H}}{=} \lim\limits_{x \to 0^+} \dfrac{1/x}{-2/x^3} = \lim\limits_{x \to 0^+}\left(-\dfrac{x^2}{2}\right) = 0.$

There is a hole at $(0, 0)$.

(c) It appears that there is an IP at about $(0.2, -0.06)$ and a local minimum at $(0.6, -0.18)$. $f(x) = x^2 \ln x \Rightarrow$

$f'(x) = x^2(1/x) + (\ln x)(2x) = x(2\ln x + 1) > 0 \iff \ln x > -\frac{1}{2} \iff x > e^{-1/2}$, so f is increasing on

$\left(1/\sqrt{e}, \infty\right)$, decreasing on $\left(0, 1/\sqrt{e}\right)$. By the FDT, $f\left(1/\sqrt{e}\right) = -1/(2e)$ is a local minimum value. This point is

approximately $(0.6065, -0.1839)$, which agrees with our estimate.

$\qquad f''(x) = x(2/x) + (2\ln x + 1) = 2\ln x + 3 > 0 \iff \ln x > -\frac{3}{2} \iff x > e^{-3/2}$, so f is CU on $(e^{-3/2}, \infty)$

and CD on $(0, e^{-3/2})$. IP is $(e^{-3/2}, -3/(2e^3)) \approx (0.2231, -0.0747)$.

57. (a) $f(x) = x^{1/x}$

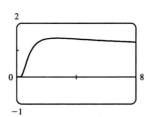

(b) Recall that $a^b = e^{b \ln a}$. $\lim\limits_{x \to 0^+} x^{1/x} = \lim\limits_{x \to 0^+} e^{(1/x)\ln x}$. As $x \to 0^+$, $\dfrac{\ln x}{x} \to -\infty$, so $x^{1/x} = e^{(1/x)\ln x} \to 0$. This

indicates that there is a hole at $(0, 0)$. As $x \to \infty$, we have the indeterminate form ∞^0. $\lim\limits_{x \to \infty} x^{1/x} = \lim\limits_{x \to \infty} e^{(1/x)\ln x}$,

but $\lim\limits_{x \to \infty} \dfrac{\ln x}{x} \overset{\text{H}}{=} \lim\limits_{x \to \infty} \dfrac{1/x}{1} = 0$, so $\lim\limits_{x \to \infty} x^{1/x} = e^0 = 1$. This indicates that $y = 1$ is a HA.

(c) Estimated maximum: $(2.72, 1.45)$. No estimated minimum. We use logarithmic differentiation to find any critical

numbers. $y = x^{1/x} \Rightarrow \ln y = \dfrac{1}{x}\ln x \Rightarrow \dfrac{y'}{y} = \dfrac{1}{x} \cdot \dfrac{1}{x} + (\ln x)\left(-\dfrac{1}{x^2}\right) \Rightarrow y' = x^{1/x}\left(\dfrac{1 - \ln x}{x^2}\right) = 0 \Rightarrow$

$\ln x = 1 \Rightarrow x = e$. For $0 < x < e$, $y' > 0$ and for $x > e$, $y' < 0$, so $f(e) = e^{1/e}$ is a local maximum value. This

point is approximately $(2.7183, 1.4447)$, which agrees with our estimate.

(d)

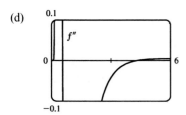

From the graph, we see that $f''(x) = 0$ at $x \approx 0.58$ and $x \approx 4.37$. Since f'' changes sign at these values, they are x-coordinates of inflection points.

59. If $c < 0$, then $\displaystyle\lim_{x \to -\infty} f(x) = \lim_{x \to -\infty} xe^{-cx} = \lim_{x \to -\infty} \frac{x}{e^{cx}} \overset{\text{H}}{=} \lim_{x \to -\infty} \frac{1}{ce^{cx}} = 0$, and $\displaystyle\lim_{x \to \infty} f(x) = \infty$.

If $c > 0$, then $\displaystyle\lim_{x \to -\infty} f(x) = -\infty$, and $\displaystyle\lim_{x \to \infty} f(x) \overset{\text{H}}{=} \lim_{x \to \infty} \frac{1}{ce^{cx}} = 0$.

If $c = 0$, then $f(x) = x$, so $\displaystyle\lim_{x \to \pm\infty} f(x) = \pm\infty$, respectively.

So we see that $c = 0$ is a transitional value. We now exclude the case $c = 0$, since we know how the function behaves in that case. To find the maxima and minima of f, we differentiate: $f(x) = xe^{-cx}$ $\Rightarrow$

$f'(x) = x(-ce^{-cx}) + e^{-cx} = (1 - cx)e^{-cx}$. This is 0 when $1 - cx = 0$ $\Leftrightarrow$ $x = 1/c$. If $c < 0$ then this represents a minimum value of $f(1/c) = 1/(ce)$, since $f'(x)$ changes from negative to positive at $x = 1/c$;

and if $c > 0$, it represents a maximum value. As $|c|$ increases, the maximum or minimum point gets closer to the origin. To find the inflection points, we differentiate again: $f'(x) = e^{-cx}(1 - cx)$ $\Rightarrow$

$f''(x) = e^{-cx}(-c) + (1 - cx)(-ce^{-cx}) = (cx - 2)ce^{-cx}$. This changes sign when $cx - 2 = 0$ $\Leftrightarrow$ $x = 2/c$. So as $|c|$ increases, the points of inflection get closer to the origin.

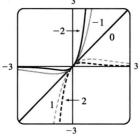

61. $\displaystyle\lim_{x \to \infty} \frac{x}{\sqrt{x^2 + 1}} \overset{\text{H}}{=} \lim_{x \to \infty} \frac{1}{\frac{1}{2}(x^2 + 1)^{-1/2}(2x)} = \lim_{x \to \infty} \frac{\sqrt{x^2 + 1}}{x}$. Repeated applications of l'Hospital's Rule result in the

original limit or the limit of the reciprocal of the function. Another method is to try dividing the numerator and denominator

by x: $\displaystyle\lim_{x \to \infty} \frac{x}{\sqrt{x^2 + 1}} = \lim_{x \to \infty} \frac{x/x}{\sqrt{x^2/x^2 + 1/x^2}} = \lim_{x \to \infty} \frac{1}{\sqrt{1 + 1/x^2}} = \frac{1}{1} = 1$

63. $\displaystyle\lim_{x \to \infty} \frac{e^x}{x^n} \overset{\text{H}}{=} \lim_{x \to \infty} \frac{e^x}{nx^{n-1}} \overset{\text{H}}{=} \lim_{x \to \infty} \frac{e^x}{n(n-1)x^{n-2}} \overset{\text{H}}{=} \cdots \overset{\text{H}}{=} \lim_{x \to \infty} \frac{e^x}{n!} = \infty$

65. First we will find $\displaystyle\lim_{n \to \infty} \left(1 + \frac{r}{n}\right)^{nt}$, which is of the form 1^{∞}. $y = \left(1 + \frac{r}{n}\right)^{nt}$ $\Rightarrow$ $\ln y = nt \ln\left(1 + \frac{r}{n}\right)$, so

$\displaystyle\lim_{n \to \infty} \ln y = \lim_{n \to \infty} nt \ln\left(1 + \frac{r}{n}\right) = t \lim_{n \to \infty} \frac{\ln(1 + r/n)}{1/n} \overset{\text{H}}{=} t \lim_{n \to \infty} \frac{(-r/n^2)}{(1 + r/n)(-1/n^2)} = t \lim_{n \to \infty} \frac{r}{1 + i/n} = tr$ $\Rightarrow$

$\displaystyle\lim_{n \to \infty} y = e^{rt}$. Thus, as $n \to \infty$, $A = A_0\left(1 + \frac{r}{n}\right)^{nt} \to A_0e^{rt}$.

67. $\lim\limits_{E \to 0^+} P(E) = \lim\limits_{E \to 0^+} \left(\dfrac{e^E + e^{-E}}{e^E - e^{-E}} - \dfrac{1}{E} \right)$

$= \lim\limits_{E \to 0^+} \dfrac{E(e^E + e^{-E}) - 1(e^E - e^{-E})}{(e^E - e^{-E})E} = \lim\limits_{E \to 0^+} \dfrac{Ee^E + Ee^{-E} - e^E + e^{-E}}{Ee^E - Ee^{-E}}$ $\quad$ $\left[\text{form is } \tfrac{0}{0}\right]$

$\overset{\text{H}}{=} \lim\limits_{E \to 0^+} \dfrac{Ee^E + e^E \cdot 1 + E(-e^{-E}) + e^{-E} \cdot 1 - e^E + (-e^{-E})}{Ee^E + e^E \cdot 1 - [E(-e^{-E}) + e^{-E} \cdot 1]}$

$= \lim\limits_{E \to 0^+} \dfrac{Ee^E - Ee^{-E}}{Ee^E + e^E + Ee^{-E} - e^{-E}} = \lim\limits_{E \to 0^+} \dfrac{e^E - e^{-E}}{e^E + \dfrac{e^E}{E} + e^{-E} - \dfrac{e^{-E}}{E}}$ $\quad$ $[\text{divide by } E]$

$= \dfrac{0}{2 + L}$, where $L = \lim\limits_{E \to 0^+} \dfrac{e^E - e^{-E}}{E}$ $\left[\text{form is } \tfrac{0}{0}\right]$ $\overset{\text{H}}{=} \lim\limits_{E \to 0^+} \dfrac{e^E + e^{-E}}{1} = \dfrac{1+1}{1} = 2$.

Thus, $\lim\limits_{E \to 0^+} P(E) = \dfrac{0}{2+2} = 0$.

69. We see that both numerator and denominator approach 0, so we can use l'Hospital's Rule:

$\lim\limits_{x \to a} \dfrac{\sqrt{2a^3 x - x^4} - a\sqrt[3]{aax}}{a - \sqrt[4]{ax^3}} \overset{\text{H}}{=} \lim\limits_{x \to a} \dfrac{\tfrac{1}{2}(2a^3 x - x^4)^{-1/2}(2a^3 - 4x^3) - a\left(\tfrac{1}{3}\right)(aax)^{-2/3}a^2}{-\tfrac{1}{4}(ax^3)^{-3/4}(3ax^2)}$

$= \dfrac{\tfrac{1}{2}(2a^3 a - a^4)^{-1/2}(2a^3 - 4a^3) - \tfrac{1}{3}a^3(a^2 a)^{-2/3}}{-\tfrac{1}{4}(aa^3)^{-3/4}(3aa^2)}$

$= \dfrac{(a^4)^{-1/2}(-a^3) - \tfrac{1}{3}a^3(a^3)^{-2/3}}{-\tfrac{3}{4}a^3(a^4)^{-3/4}} = \dfrac{-a - \tfrac{1}{3}a}{-\tfrac{3}{4}} = \tfrac{4}{3}\left(\tfrac{4}{3}a\right) = \tfrac{16}{9}a$

71. The limit, $L = \lim\limits_{x \to \infty}\left[x - x^2 \ln\left(\dfrac{1+x}{x}\right)\right] = \lim\limits_{x \to \infty}\left[x - x^2 \ln\left(\dfrac{1}{x} + 1\right)\right]$. Let $t = 1/x$, so as $x \to \infty$, $t \to 0^+$.

$L = \lim\limits_{t \to 0^+}\left[\dfrac{1}{t} - \dfrac{1}{t^2}\ln(t+1)\right] = \lim\limits_{t \to 0^+} \dfrac{t - \ln(t+1)}{t^2} \overset{\text{H}}{=} \lim\limits_{t \to 0^+} \dfrac{1 - \dfrac{1}{t+1}}{2t} = \lim\limits_{t \to 0^+} \dfrac{t/(t+1)}{2t} = \lim\limits_{t \to 0^+} \dfrac{1}{2(t+1)} = \dfrac{1}{2}$

Note: Starting the solution by factoring x or x^2 leads to a more complicated solution.

73. Since $f(2) = 0$, the given limit has the form $\tfrac{0}{0}$.

$\lim\limits_{x \to 0} \dfrac{f(2+3x) + f(2+5x)}{x} \overset{\text{H}}{=} \lim\limits_{x \to 0} \dfrac{f'(2+3x) \cdot 3 + f'(2+5x) \cdot 5}{1} = f'(2) \cdot 3 + f'(2) \cdot 5 = 8f'(2) = 8 \cdot 7 = 56$

75. Since $\lim\limits_{h \to 0}[f(x+h) - f(x-h)] = f(x) - f(x) = 0$ (f is differentiable and hence continuous) and $\lim\limits_{h \to 0} 2h = 0$, we use

l'Hospital's Rule:

$\lim\limits_{h \to 0} \dfrac{f(x+h) - f(x-h)}{2h} \overset{\text{H}}{=} \lim\limits_{h \to 0} \dfrac{f'(x+h)(1) - f'(x-h)(-1)}{2} = \dfrac{f'(x) + f'(x)}{2} = \dfrac{2f'(x)}{2} = f'(x)$

$\dfrac{f(x+h) - f(x-h)}{2h}$ is the slope of the secant line between

$(x-h, f(x-h))$ and $(x+h, f(x+h))$. As $h \to 0$, this line gets closer

to the tangent line and its slope approaches $f'(x)$.

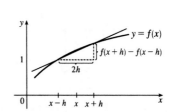

4.6 Optimization Problems

1. (a)

First Number	Second Number	Product
1	22	22
2	21	42
3	20	60
4	19	76
5	18	90
6	17	102
7	16	112
8	15	120
9	14	126
10	13	130
11	12	132

We needn't consider pairs where the first number is larger than the second, since we can just interchange the numbers in such cases. The answer appears to be 11 and 12, but we have considered only integers in the table.

(b) Call the two numbers x and y. Then $x + y = 23$, so $y = 23 - x$. Call the product P. Then

$P = xy = x(23 - x) = 23x - x^2$, so we wish to maximize the function $P(x) = 23x - x^2$. Since $P'(x) = 23 - 2x$,

we see that $P'(x) = 0$ $\Leftrightarrow$ $x = \frac{23}{2} = 11.5$. Thus, the maximum value of P is $P(11.5) = (11.5)^2 = 132.25$ and it

occurs when $x = y = 11.5$.

Or: Note that $P''(x) = -2 < 0$ for all x, so P is everywhere concave downward and the local maximum at $x = 11.5$

must be an absolute maximum.

3. The two numbers are x and $\dfrac{100}{x}$, where $x > 0$. Minimize $f(x) = x + \dfrac{100}{x}$. $f'(x) = 1 - \dfrac{100}{x^2} = \dfrac{x^2 - 100}{x^2}$. The critical

number is $x = 10$. Since $f'(x) < 0$ for $0 < x < 10$ and $f'(x) > 0$ for $x > 10$, there is an absolute minimum at $x = 10$.

The numbers are 10 and 10.

5. If the rectangle has dimensions x and y, then its perimeter is $2x + 2y = 100$ m, so $y = 50 - x$. Thus, the area is

$A = xy = x(50 - x)$. We wish to maximize the function $A(x) = x(50 - x) = 50x - x^2$, where $0 < x < 50$. Since

$A'(x) = 50 - 2x = -2(x - 25)$, $A'(x) > 0$ for $0 < x < 25$ and $A'(x) < 0$ for $25 < x < 50$. Thus, A has an absolute

maximum at $x = 25$, and $A(25) = 25^2 = 625$ m². The dimensions of the rectangle that maximize its area are $x = y = 25$ m.

(The rectangle is a square.)

7. We need to maximize Y for $N \geq 0$. $Y(N) = \dfrac{kN}{1 + N^2}$ $\Rightarrow$

$Y'(N) = \dfrac{(1 + N^2)k - kN(2N)}{(1 + N^2)^2} = \dfrac{k(1 - N^2)}{(1 + N^2)^2} = \dfrac{k(1 + N)(1 - N)}{(1 + N^2)^2}$. $Y'(N) > 0$ for $0 < N < 1$ and $Y'(N) < 0$

for $N > 1$. Thus, Y has an absolute maximum of $Y(1) = \frac{1}{2}k$ at $N = 1$.

9. (a)

The areas of the three figures are 12,500, 12,500, and 9000 ft². There appears to be a maximum area of at least 12,500 ft².

(b) Let x denote the length of each of two sides and three dividers.

Let y denote the length of the other two sides.

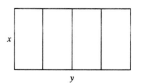

(c) Area $A =$ length $\times$ width $= y \cdot x$

(d) Length of fencing $= 750 \quad \Rightarrow \quad 5x + 2y = 750$

(e) $5x + 2y = 750 \quad \Rightarrow \quad y = 375 - \frac{5}{2}x \quad \Rightarrow \quad A(x) = \left(375 - \frac{5}{2}x\right)x = 375x - \frac{5}{2}x^2$

(f) $A'(x) = 375 - 5x = 0 \quad \Rightarrow \quad x = 75$. Since $A''(x) = -5 < 0$ there is an absolute maximum when $x = 75$. Then

$y = \frac{375}{2} = 187.5$. The largest area is $75\left(\frac{375}{2}\right) = 14{,}062.5$ ft^2. These values of x and y are between the values in the first

and second figures in part (a). Our original estimate was low.

11. Let b be the length of the base of the box and h the height. The surface area is $1200 = b^2 + 4hb \quad \Rightarrow \quad h = (1200 - b^2)/(4b)$.

The volume is $V = b^2h = b^2(1200 - b^2)/4b = 300b - b^3/4 \quad \Rightarrow \quad V'(b) = 300 - \frac{3}{4}b^2$.

$V'(b) = 0 \quad \Rightarrow \quad 300 = \frac{3}{4}b^2 \quad \Rightarrow \quad b^2 = 400 \quad \Rightarrow \quad b = \sqrt{400} = 20$. Since $V'(b) > 0$ for $0 < b < 20$ and $V'(b) < 0$ for

$b > 20$, there is an absolute maximum when $b = 20$ by the First Derivative Test for Absolute Extreme Values (see page 302).

If $b = 20$, then $h = (1200 - 20^2)/(4 \cdot 20) = 10$, so the largest possible volume is $b^2h = (20)^2(10) = 4000$ cm^3.

13. (a) Let the rectangle have sides x and y and area A, so $A = xy$ or $y = A/x$. The problem is to minimize the

perimeter $= 2x + 2y = 2x + 2A/x = P(x)$. Now $P'(x) = 2 - 2A/x^2 = 2(x^2 - A)/x^2$. So the critical number is

$x = \sqrt{A}$. Since $P'(x) < 0$ for $0 < x < \sqrt{A}$ and $P'(x) > 0$ for $x > \sqrt{A}$, there is an absolute minimum at $x = \sqrt{A}$.

The sides of the rectangle are $\sqrt{A}$ and $A/\sqrt{A} = \sqrt{A}$, so the rectangle is a square.

(b) Let p be the perimeter and x and y the lengths of the sides, so $p = 2x + 2y \quad \Rightarrow \quad 2y = p - 2x \quad \Rightarrow \quad y = \frac{1}{2}p - x$.

The area is $A(x) = x\left(\frac{1}{2}p - x\right) = \frac{1}{2}px - x^2$. Now $A'(x) = 0 \quad \Rightarrow \quad \frac{1}{2}p - 2x = 0 \quad \Rightarrow \quad 2x = \frac{1}{2}p \quad \Rightarrow \quad x = \frac{1}{4}p$. Since

$A''(x) = -2 < 0$, there is an absolute maximum for A when $x = \frac{1}{4}p$ by the Second Derivative Test. The sides of the

rectangle are $\frac{1}{4}p$ and $\frac{1}{2}p - \frac{1}{4}p = \frac{1}{4}p$, so the rectangle is a square.

15.

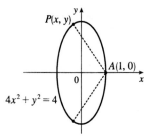

From the figure, we see that there are two points that are farthest away from

$A(1, 0)$. The distance d from A to an arbitrary point $P(x, y)$ on the ellipse is

$d = \sqrt{(x - 1)^2 + (y - 0)^2}$ and the square of the distance is

$S = d^2 = x^2 - 2x + 1 + y^2 = x^2 - 2x + 1 + (4 - 4x^2) = -3x^2 - 2x + 5$.

$S' = -6x - 2$ and $S' = 0 \quad \Rightarrow \quad x = -\frac{1}{3}$. Now $S'' = -6 < 0$, so we know

that S has a maximum at $x = -\frac{1}{3}$. Since $-1 \le x \le 1$, $S(-1) = 4$,

$S\left(-\frac{1}{3}\right) = \frac{16}{3}$, and $S(1) = 0$, we see that the maximum distance is $\sqrt{\frac{16}{3}}$. The corresponding y-values are

$y = \pm\sqrt{4 - 4\left(-\frac{1}{3}\right)^2} = \pm\sqrt{\frac{32}{9}} = \pm\frac{4}{3}\sqrt{2} \approx \pm 1.89$. The points are $\left(-\frac{1}{3}, \pm\frac{4}{3}\sqrt{2}\right)$.

17.

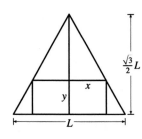

The height h of the equilateral triangle with sides of length L is $\frac{\sqrt{3}}{2}L$,

since $h^2 + (L/2)^2 = L^2 \Rightarrow h^2 = L^2 - \frac{1}{4}L^2 = \frac{3}{4}L^2 \Rightarrow$

$h = \frac{\sqrt{3}}{2}L$. Using similar triangles, $\dfrac{\frac{\sqrt{3}}{2}L - y}{x} = \dfrac{\frac{\sqrt{3}}{2}L}{L/2} = \sqrt{3} \Rightarrow$

$\sqrt{3}\,x = \dfrac{\sqrt{3}}{2}L - y \Rightarrow y = \dfrac{\sqrt{3}}{2}L - \sqrt{3}\,x \Rightarrow y = \dfrac{\sqrt{3}}{2}(L - 2x).$

The area of the inscribed rectangle is $A(x) = (2x)y = \sqrt{3}\,x(L - 2x) = \sqrt{3}\,Lx - 2\sqrt{3}\,x^2$, where $0 \le x \le L/2$. Now

$0 = A'(x) = \sqrt{3}\,L - 4\sqrt{3}\,x \Rightarrow x = \sqrt{3}\,L/(4\sqrt{3}) = L/4$. Since $A(0) = A(L/2) = 0$, the maximum occurs when

$x = L/4$, and $y = \frac{\sqrt{3}}{2}L - \frac{\sqrt{3}}{4}L = \frac{\sqrt{3}}{4}L$, so the dimensions are $L/2$ and $\frac{\sqrt{3}}{4}L$.

19.

The cylinder has volume $V = \pi y^2(2x)$. Also $x^2 + y^2 = r^2 \Rightarrow y^2 = r^2 - x^2$, so

$V(x) = \pi(r^2 - x^2)(2x) = 2\pi(r^2x - x^3)$, where $0 \le x \le r$.

$V'(x) = 2\pi(r^2 - 3x^2) = 0 \Rightarrow x = r/\sqrt{3}$. Now $V(0) = V(r) = 0$, so there is a

maximum when $x = r/\sqrt{3}$ and $V(r/\sqrt{3}) = \pi(r^2 - r^2/3)(2r/\sqrt{3}) = 4\pi r^3/(3\sqrt{3})$.

21.

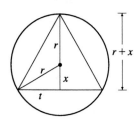

The area of the triangle is

$$A(x) = \tfrac{1}{2}(2t)(r + x) = t(r + x) = \sqrt{r^2 - x^2}\,(r + x). \text{ Then}$$

$$0 = A'(x) = r\frac{-2x}{2\sqrt{r^2 - x^2}} + \sqrt{r^2 - x^2} + x\frac{-2x}{2\sqrt{r^2 - x^2}}$$

$$= -\frac{x^2 + rx}{\sqrt{r^2 - x^2}} + \sqrt{r^2 - x^2} \Rightarrow$$

$\dfrac{x^2 + rx}{\sqrt{r^2 - x^2}} = \sqrt{r^2 - x^2} \Rightarrow x^2 + rx = r^2 - x^2 \Rightarrow 0 = 2x^2 + rx - r^2 = (2x - r)(x + r) \Rightarrow$

$x = \frac{1}{2}r$ or $x = -r$. Now $A(r) = 0 = A(-r) \Rightarrow$ the maximum occurs where $x = \frac{1}{2}r$, so the triangle has

height $r + \frac{1}{2}r = \frac{3}{2}r$ and base $2\sqrt{r^2 - \left(\frac{1}{2}r\right)^2} = 2\sqrt{\frac{3}{4}r^2} = \sqrt{3}\,r.$

23.

Perimeter $= 30 \Rightarrow 2y + x + \pi\left(\dfrac{x}{2}\right) = 30 \Rightarrow$

$y = \dfrac{1}{2}\left(30 - x - \dfrac{\pi x}{2}\right) = 15 - \dfrac{x}{2} - \dfrac{\pi x}{4}$. The area is the area of the rectangle plus the area of

the semicircle, or $xy + \frac{1}{2}\pi\left(\dfrac{x}{2}\right)^2$, so $A(x) = x\left(15 - \dfrac{x}{2} - \dfrac{\pi x}{4}\right) + \frac{1}{8}\pi x^2 = 15x - \frac{1}{2}x^2 - \frac{\pi}{8}x^2.$

$A'(x) = 15 - \left(1 + \frac{\pi}{4}\right)x = 0 \Rightarrow x = \dfrac{15}{1 + \pi/4} = \dfrac{60}{4 + \pi}.\quad A''(x) = -\left(1 + \dfrac{\pi}{4}\right) < 0$, so this gives a maximum.

The dimensions are $x = \dfrac{60}{4 + \pi}$ ft and $y = 15 - \dfrac{30}{4 + \pi} - \dfrac{15\pi}{4 + \pi} = \dfrac{60 + 15\pi - 30 - 15\pi}{4 + \pi} = \dfrac{30}{4 + \pi}$ ft, so the height of the

rectangle is half the base.

25.

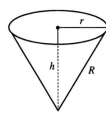

Let x be the length of the wire used for the square. The total area is

$$A(x) = \left(\frac{x}{4}\right)^2 + \frac{1}{2}\left(\frac{10-x}{3}\right)\frac{\sqrt{3}}{2}\left(\frac{10-x}{3}\right)$$

$$= \tfrac{1}{16}x^2 + \tfrac{\sqrt{3}}{36}(10-x)^2, \ 0 \le x \le 10$$

$A'(x) = \tfrac{1}{8}x - \tfrac{\sqrt{3}}{18}(10-x) = 0 \ \Leftrightarrow \ \tfrac{9}{72}x + \tfrac{4\sqrt{3}}{72}x - \tfrac{40\sqrt{3}}{72} = 0 \ \Leftrightarrow \ x = \tfrac{40\sqrt{3}}{9+4\sqrt{3}}.$

Now $A(0) = \left(\tfrac{\sqrt{3}}{36}\right)100 \approx 4.81$, $A(10) = \tfrac{100}{16} = 6.25$ and $A\left(\tfrac{40\sqrt{3}}{9+4\sqrt{3}}\right) \approx 2.72$, so

(a) The maximum area occurs when $x = 10$ m, and all the wire is used for the square.

(b) The minimum area occurs when $x = \tfrac{40\sqrt{3}}{9+4\sqrt{3}} \approx 4.35$ m.

27.

$h^2 + r^2 = R^2 \ \Rightarrow \ V = \tfrac{\pi}{3}r^2h = \tfrac{\pi}{3}(R^2 - h^2)h = \tfrac{\pi}{3}(R^2h - h^3).$

$V'(h) = \tfrac{\pi}{3}(R^2 - 3h^2) = 0$ when $h = \tfrac{1}{\sqrt{3}}R.$ This gives an absolute maximum, since

$V'(h) > 0$ for $0 < h < \tfrac{1}{\sqrt{3}}R$ and $V'(h) < 0$ for $h > \tfrac{1}{\sqrt{3}}R.$ The maximum volume is

$$V\left(\tfrac{1}{\sqrt{3}}R\right) = \tfrac{\pi}{3}\left(\tfrac{1}{\sqrt{3}}R^3 - \tfrac{1}{3\sqrt{3}}R^3\right) = \tfrac{2}{9\sqrt{3}}\pi R^3.$$

29.

By similar triangles, $\dfrac{H}{R} = \dfrac{H-h}{r}$ (1). The volume of the inner cone is $V = \tfrac{1}{3}\pi r^2 h$,

so we'll solve (1) for h. $\dfrac{Hr}{R} = H - h \ \Rightarrow$

$h = H - \dfrac{Hr}{R} = \dfrac{HR - Hr}{R} = \dfrac{H}{R}(R - r)$ (2).

Thus, $V(r) = \dfrac{\pi}{3}r^2 \cdot \dfrac{H}{R}(R - r) = \dfrac{\pi H}{3R}(Rr^2 - r^3) \ \Rightarrow$

$V'(r) = \dfrac{\pi H}{3R}(2Rr - 3r^2) = \dfrac{\pi H}{3R}r(2R - 3r).$

$V'(r) = 0 \ \Rightarrow \ r = 0$ or $2R = 3r \ \Rightarrow \ r = \tfrac{2}{3}R$ and from (2), $h = \dfrac{H}{R}\left(R - \tfrac{2}{3}R\right) = \dfrac{H}{R}\left(\tfrac{1}{3}R\right) = \tfrac{1}{3}H.$

$V'(r)$ changes from positive to negative at $r = \tfrac{2}{3}R$, so the inner cone has a maximum volume of

$V = \tfrac{1}{3}\pi r^2 h = \tfrac{1}{3}\pi\left(\tfrac{2}{3}R\right)^2\left(\tfrac{1}{3}H\right) = \tfrac{4}{27} \cdot \tfrac{1}{3}\pi R^2 H$, which is approximately 15% of the volume of the larger cone.

31. $P(R) = \dfrac{E^2 R}{(R+r)^2} \ \Rightarrow$

$$P'(R) = \dfrac{(R+r)^2 \cdot E^2 - E^2 R \cdot 2(R+r)}{[(R+r)^2]^2} = \dfrac{(R^2 + 2Rr + r^2)E^2 - 2E^2R^2 - 2E^2Rr}{(R+r)^4}$$

$$= \dfrac{E^2 r^2 - E^2 R^2}{(R+r)^4} = \dfrac{E^2(r^2 - R^2)}{(R+r)^4} = \dfrac{E^2(r+R)(r-R)}{(R+r)^4} = \dfrac{E^2(r-R)}{(R+r)^3}$$

$P'(R) = 0 \ \Rightarrow \ R = r \ \Rightarrow \ P(r) = \dfrac{E^2 r}{(r+r)^2} = \dfrac{E^2 r}{4r^2} = \dfrac{E^2}{4r}.$

The expression for $P'(R)$ shows that $P'(R) > 0$ for $R < r$ and $P'(R) < 0$ for $R > r$. Thus, the maximum value of the power is $E^2/(4r)$, and this occurs when $R = r$.

33. $S = 6sh - \frac{3}{2}s^2 \cot \theta + 3s^2 \frac{\sqrt{3}}{2} \csc \theta$

(a) $\frac{dS}{d\theta} = \frac{3}{2}s^2 \csc^2 \theta - 3s^2 \frac{\sqrt{3}}{2} \csc \theta \cot \theta$ or $\frac{3}{2}s^2 \csc \theta \left(\csc \theta - \sqrt{3} \cot \theta \right)$.

(b) $\frac{dS}{d\theta} = 0$ when $\csc \theta - \sqrt{3} \cot \theta = 0 \Rightarrow \frac{1}{\sin \theta} - \sqrt{3} \frac{\cos \theta}{\sin \theta} = 0 \Rightarrow \cos \theta = \frac{1}{\sqrt{3}}$. The First Derivative Test shows

that the minimum surface area occurs when $\theta = \cos^{-1}\left(\frac{1}{\sqrt{3}}\right) \approx 55°$.

(c)   If $\cos \theta = \frac{1}{\sqrt{3}}$, then $\cot \theta = \frac{1}{\sqrt{2}}$ and $\csc \theta = \frac{\sqrt{3}}{\sqrt{2}}$, so the surface area is

$$S = 6sh - \frac{3}{2}s^2 \frac{1}{\sqrt{2}} + 3s^2 \frac{\sqrt{3}}{2} \frac{\sqrt{3}}{\sqrt{2}} = 6sh - \frac{3}{2\sqrt{2}}s^2 + \frac{9}{2\sqrt{2}}s^2$$

$$= 6sh + \frac{6}{2\sqrt{2}}s^2 = 6s\left(h + \frac{1}{2\sqrt{2}}s\right)$$

35. There are $(6 - x)$ km over land and $\sqrt{x^2 + 4}$ km under the river.

We need to minimize the cost C (measured in $100,000) of the pipeline.

$C(x) = (6 - x)(4) + \left(\sqrt{x^2 + 4}\right)(8) \Rightarrow$

$C'(x) = -4 + 8 \cdot \frac{1}{2}(x^2 + 4)^{-1/2}(2x) = -4 + \frac{8x}{\sqrt{x^2 + 4}}$.

$C'(x) = 0 \Rightarrow 4 = \frac{8x}{\sqrt{x^2 + 4}} \Rightarrow \sqrt{x^2 + 4} = 2x \Rightarrow x^2 + 4 = 4x^2 \Rightarrow 4 = 3x^2 \Rightarrow x^2 = \frac{4}{3} \Rightarrow$

$x = 2/\sqrt{3}$ $[0 \leq x \leq 6]$. Compare the costs for $x = 0, 2/\sqrt{3},$ and $6.$ $C(0) = 24 + 16 = 40,$

$C(2/\sqrt{3}) = 24 - 8/\sqrt{3} + 32/\sqrt{3} = 24 + 24/\sqrt{3} \approx 37.9,$ and $C(6) = 0 + 8\sqrt{40} \approx 50.6.$ So the minimum cost is about

$3.79 million when P is $6 - 2/\sqrt{3} \approx 4.85$ km east of the refinery.

37. 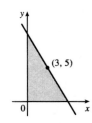   The total illumination is $I(x) = \frac{3k}{x^2} + \frac{k}{(10 - x)^2},$ $0 < x < 10.$ Then

$$I'(x) = \frac{-6k}{x^3} + \frac{2k}{(10 - x)^3} = 0 \Rightarrow 6k(10 - x)^3 = 2kx^3 \Rightarrow$$

$3(10 - x)^3 = x^3 \Rightarrow \sqrt[3]{3}(10 - x) = x \Rightarrow 10\sqrt[3]{3} - \sqrt[3]{3}x = x \Rightarrow 10\sqrt[3]{3} = x + \sqrt[3]{3}x \Rightarrow$

$10\sqrt[3]{3} = (1 + \sqrt[3]{3})x \Rightarrow x = \frac{10\sqrt[3]{3}}{1 + \sqrt[3]{3}} \approx 5.9$ ft. This gives a minimum since $I''(x) > 0$ for $0 < x < 10.$

39.   The line with slope m (where $m < 0$) through $(3, 5)$ has equation $y - 5 = m(x - 3)$ or

$y = mx + (5 - 3m).$ The y-intercept is $5 - 3m$ and the x-intercept is $-5/m + 3.$ So the

triangle has area $A(m) = \frac{1}{2}(5 - 3m)(-5/m + 3) = 15 - 25/(2m) - \frac{9}{2}m.$ Now

$$A'(m) = \frac{25}{2m^2} - \frac{9}{2} = 0 \Leftrightarrow m^2 = \frac{25}{9} \Rightarrow m = -\frac{5}{3} \text{ (since } m < 0).$$

$A''(m) = -\frac{25}{m^3} > 0,$ so there is an absolute minimum when $m = -\frac{5}{3}.$ Thus, an equation of the line is $y - 5 = -\frac{5}{3}(x - 3)$

or $y = -\frac{5}{3}x + 10.$

41. $y = \dfrac{3}{x}$ $\Rightarrow$ $y' = -\dfrac{3}{x^2}$, so an equation of the tangent line at the point $\left(a, \frac{3}{a}\right)$ is

$y - \dfrac{3}{a} = -\dfrac{3}{a^2}(x - a)$, or $y = -\dfrac{3}{a^2}x + \dfrac{6}{a}$. The y-intercept $[x = 0]$ is $6/a$. The

x-intercept $[y = 0]$ is $2a$. The distance d of the line segment that has endpoints at the

intercepts is $d = \sqrt{(2a - 0)^2 + (0 - 6/a)^2}$. Let $S = d^2$, so $S = 4a^2 + \dfrac{36}{a^2}$ $\Rightarrow$

$S' = 8a - \dfrac{72}{a^3}$. $S' = 0$ $\Leftrightarrow$ $\dfrac{72}{a^3} = 8a$ $\Leftrightarrow$ $a^4 = 9$ $\Leftrightarrow$ $a^2 = 3$ $\Rightarrow$ $a = \sqrt{3}$.

$S'' = 8 + \dfrac{216}{a^4} > 0$, so there is an absolute minimum at $a = \sqrt{3}$. Thus, $S = 4(3) + \frac{36}{3} = 12 + 12 = 24$ and

hence, $d = \sqrt{24} = 2\sqrt{6}$.

43. (a) If $c(x) = \dfrac{C(x)}{x}$, then, by Quotient Rule, we have $c'(x) = \dfrac{xC'(x) - C(x)}{x^2}$. Now $c'(x) = 0$ when $xC'(x) - C(x) = 0$

and this gives $C'(x) = \dfrac{C(x)}{x} = c(x)$. Therefore, the marginal cost equals the average cost.

(b) (i) $C(x) = 16{,}000 + 200x + 4x^{3/2}$, $C(1000) = 16{,}000 + 200{,}000 + 40{,}000\sqrt{10} \approx 216{,}000 + 126{,}491$, so

$C(1000) \approx \$342{,}491$. $c(x) = C(x)/x = \dfrac{16{,}000}{x} + 200 + 4x^{1/2}$, $c(1000) \approx \$342.49/\text{unit}$. $C'(x) = 200 + 6x^{1/2}$,

$C'(1000) = 200 + 60\sqrt{10} \approx \$389.74/\text{unit}$.

(ii) We must have $C'(x) = c(x)$ $\Leftrightarrow$ $200 + 6x^{1/2} = \dfrac{16{,}000}{x} + 200 + 4x^{1/2}$ $\Leftrightarrow$ $2x^{3/2} = 16{,}000$ $\Leftrightarrow$

$x = (8{,}000)^{2/3} = 400$ units. To check that this is a minimum, we calculate

$c'(x) = \dfrac{-16{,}000}{x^2} + \dfrac{2}{\sqrt{x}} = \dfrac{2}{x^2}\left(x^{3/2} - 8000\right)$. This is negative for $x < (8000)^{2/3} = 400$, zero at $x = 400$,

and positive for $x > 400$, so c is decreasing on $(0, 400)$ and increasing on $(400, \infty)$. Thus, c has an absolute minimum

at $x = 400$. [*Note:* $c''(x)$ is *not* positive for all $x > 0$.]

(iii) The minimum average cost is $c(400) = 40 + 200 + 80 = \$320/\text{unit}$.

45. (a) We are given that the demand function p is linear and $p(27{,}000) = 10$, $p(33{,}000) = 8$, so the slope is

$\frac{10 - 8}{27{,}000 - 33{,}000} = -\frac{1}{3000}$ and an equation of the line is $y - 10 = \left(-\frac{1}{3000}\right)(x - 27{,}000)$ $\Rightarrow$

$y = p(x) = -\frac{1}{3000}x + 19 = 19 - (x/3000)$.

(b) The revenue is $R(x) = xp(x) = 19x - (x^2/3000)$ $\Rightarrow$ $R'(x) = 19 - (x/1500) = 0$ when $x = 28{,}500$. Since

$R''(x) = -1/1500 < 0$, the maximum revenue occurs when $x = 28{,}500$ $\Rightarrow$ the price is $p(28{,}500) = \$9.50$.

47. (a) As in Example 6, we see that the demand function p is linear. We are given that $p(1000) = 450$ and deduce that

$p(1100) = 440$, since a \$10 reduction in price increases sales by 100 per week. The slope for p is $\frac{440 - 450}{1100 - 1000} = -\frac{1}{10}$,

so an equation is $p - 450 = -\frac{1}{10}(x - 1000)$ or $p(x) = -\frac{1}{10}x + 550$.

(b) $R(x) = xp(x) = -\frac{1}{10}x^2 + 550x$. $R'(x) = -\frac{1}{5}x + 550 = 0$ when $x = 5(550) = 2750$.

$p(2750) = 275$, so the rebate should be $450 - 275 = \$175$.

(c) $C(x) = 68{,}000 + 150x \Rightarrow P(x) = R(x) - C(x) = -\frac{1}{10}x^2 + 550x - 68{,}000 - 150x = -\frac{1}{10}x^2 + 400x - 68{,}000$,

$P'(x) = -\frac{1}{5}x + 400 = 0$ when $x = 2000$. $p(2000) = 350$. Therefore, the rebate to maximize profits should be

$450 - 350 = \$100$.

49.

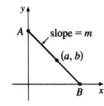

Every line segment in the first quadrant passing through (a, b) with endpoints on the x-
and y-axes satisfies an equation of the form $y - b = m(x - a)$, where $m < 0$. By setting
$x = 0$ and then $y = 0$, we find its endpoints, $A(0, b - am)$ and $B\left(a - \frac{b}{m}, 0\right)$. The
distance d from A to B is given by $d = \sqrt{[(a - \frac{b}{m}) - 0]^2 + [0 - (b - am)]^2}$.

It follows that the square of the length of the line segment, as a function of m, is given by

$$S(m) = \left(a - \frac{b}{m}\right)^2 + (am - b)^2 = a^2 - \frac{2ab}{m} + \frac{b^2}{m^2} + a^2m^2 - 2abm + b^2. \text{ Thus,}$$

$$S'(m) = \frac{2ab}{m^2} - \frac{2b^2}{m^3} + 2a^2m - 2ab = \frac{2}{m^3}(abm - b^2 + a^2m^4 - abm^3)$$

$$= \frac{2}{m^3}[b(am - b) + am^3(am - b)] = \frac{2}{m^3}(am - b)(b + am^3)$$

Thus, $S'(m) = 0 \Leftrightarrow m = b/a$ or $m = -\sqrt[3]{\frac{b}{a}}$. Since $b/a > 0$ and $m < 0$, m must equal $-\sqrt[3]{\frac{b}{a}}$. Since $\frac{2}{m^3} < 0$, we see

that $S'(m) < 0$ for $m < -\sqrt[3]{\frac{b}{a}}$ and $S'(m) > 0$ for $m > -\sqrt[3]{\frac{b}{a}}$. Thus, S has its absolute minimum value when $m = -\sqrt[3]{\frac{b}{a}}$.

That value is

$$S\left(-\sqrt[3]{\frac{b}{a}}\right) = \left(a + b\sqrt[3]{\frac{a}{b}}\right)^2 + \left(-a\sqrt[3]{\frac{b}{a}} - b\right)^2 = \left(a + \sqrt[3]{ab^2}\right)^2 + \left(\sqrt[3]{a^2b} + b\right)^2$$

$$= a^2 + 2a^{4/3}b^{2/3} + a^{2/3}b^{4/3} + a^{4/3}b^{2/3} + 2a^{2/3}b^{4/3} + b^2 = a^2 + 3a^{4/3}b^{2/3} + 3a^{2/3}b^{4/3} + b^2$$

The last expression is of the form $x^3 + 3x^2y + 3xy^2 + y^3 \quad [= (x + y)^3]$ with $x = a^{2/3}$ and $y = b^{2/3}$,

so we can write it as $(a^{2/3} + b^{2/3})^3$ and the shortest such line segment has length $\sqrt{S} = (a^{2/3} + b^{2/3})^{3/2}$.

51.

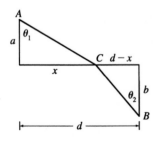

The total time is

$T(x) = (\text{time from } A \text{ to } C) + (\text{time from } C \text{ to } B)$

$$= \frac{\sqrt{a^2 + x^2}}{v_1} + \frac{\sqrt{b^2 + (d - x)^2}}{v_2}, \quad 0 < x < d$$

$$T'(x) = \frac{x}{v_1\sqrt{a^2 + x^2}} - \frac{d - x}{v_2\sqrt{b^2 + (d - x)^2}} = \frac{\sin\theta_1}{v_1} - \frac{\sin\theta_2}{v_2}$$

The minimum occurs when $T'(x) = 0 \Rightarrow \dfrac{\sin\theta_1}{v_1} = \dfrac{\sin\theta_2}{v_2}$.

[*Note:* $T''(x) > 0$]

53.

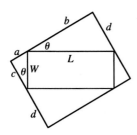

$y^2 = x^2 + z^2$, but triangles CDE and BCA are similar, so

$z/8 = x/\left(4\sqrt{x-4}\right)$ $\Rightarrow$ $z = 2x/\sqrt{x-4}$. Thus, we minimize

$f(x) = y^2 = x^2 + 4x^2/(x-4) = x^3/(x-4)$, $4 < x \le 8$.

$f'(x) = \dfrac{(x-4)(3x^2) - x^3}{(x-4)^2} = \dfrac{x^2[3(x-4) - x]}{(x-4)^2} = \dfrac{2x^2(x-6)}{(x-4)^2} = 0$

when $x = 6$. $f'(x) < 0$ when $x < 6$, $f'(x) > 0$ when $x > 6$, so the minimum

occurs when $x = 6$ in.

55.

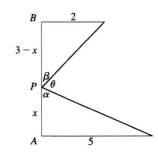

In the small triangle with sides a and c and hypotenuse W, $\sin\theta = \dfrac{a}{W}$ and

$\cos\theta = \dfrac{c}{W}$. In the triangle with sides b and d and hypotenuse L, $\sin\theta = \dfrac{d}{L}$ and

$\cos\theta = \dfrac{b}{L}$. Thus, $a = W\sin\theta$, $c = W\cos\theta$, $d = L\sin\theta$, and $b = L\cos\theta$, so the

area of the circumscribed rectangle is

$$A(\theta) = (a+b)(c+d) = (W\sin\theta + L\cos\theta)(W\cos\theta + L\sin\theta)$$

$$= W^2\sin\theta\,\cos\theta + WL\sin^2\theta + LW\cos^2\theta + L^2\sin\theta\,\cos\theta$$

$$= LW\sin^2\theta + LW\cos^2\theta + (L^2 + W^2)\sin\theta\,\cos\theta$$

$$= LW(\sin^2\theta + \cos^2\theta) + (L^2 + W^2)\cdot\tfrac{1}{2}\cdot 2\sin\theta\,\cos\theta = LW + \tfrac{1}{2}(L^2 + W^2)\sin 2\theta,\ 0 \le \theta \le \tfrac{\pi}{2}$$

This expression shows, without calculus, that the maximum value of $A(\theta)$ occurs when $\sin 2\theta = 1$ $\Leftrightarrow$ $2\theta = \tfrac{\pi}{2}$ $\Rightarrow$

$\theta = \tfrac{\pi}{4}$. So the maximum area is $A\left(\tfrac{\pi}{4}\right) = LW + \tfrac{1}{2}(L^2 + W^2) = \tfrac{1}{2}(L^2 + 2LW + W^2) = \tfrac{1}{2}(L+W)^2$.

57.

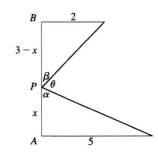

From the figure, $\tan\alpha = \dfrac{5}{x}$ and $\tan\beta = \dfrac{2}{3-x}$. Since

$\alpha + \beta + \theta = 180° = \pi$, $\theta = \pi - \tan^{-1}\left(\dfrac{5}{x}\right) - \tan^{-1}\left(\dfrac{2}{3-x}\right)$ $\Rightarrow$

$\dfrac{d\theta}{dx} = -\dfrac{1}{1 + \left(\dfrac{5}{x}\right)^2}\left(-\dfrac{5}{x^2}\right) - \dfrac{1}{1 + \left(\dfrac{2}{3-x}\right)^2}\left[\dfrac{2}{(3-x)^2}\right]$

$= \dfrac{x^2}{x^2 + 25}\cdot\dfrac{5}{x^2} - \dfrac{(3-x)^2}{(3-x)^2 + 4}\cdot\dfrac{2}{(3-x)^2}$.

Now $\dfrac{d\theta}{dx} = 0$ $\Rightarrow$ $\dfrac{5}{x^2 + 25} = \dfrac{2}{x^2 - 6x + 13}$ $\Rightarrow$ $2x^2 + 50 = 5x^2 - 30x + 65$ $\Rightarrow$

$3x^2 - 30x + 15 = 0$ $\Rightarrow$ $x^2 - 10x + 5 = 0$ $\Rightarrow$ $x = 5 \pm 2\sqrt{5}$. We reject the root with the $+$ sign, since it is

larger than 3. $d\theta/dx > 0$ for $x < 5 - 2\sqrt{5}$ and $d\theta/dx < 0$ for $x > 5 - 2\sqrt{5}$, so θ is maximized when

$|AP| = x = 5 - 2\sqrt{5} \approx 0.53$.

59. (a)

If k = energy/km over land, then energy/km over water = $1.4k$.

So the total energy is $E = 1.4k\sqrt{25 + x^2} + k(13 - x)$, $0 \le x \le 13$,

and so $\dfrac{dE}{dx} = \dfrac{1.4kx}{(25 + x^2)^{1/2}} - k$.

Set $\dfrac{dE}{dx} = 0$: $1.4kx = k(25 + x^2)^{1/2}$ $\Rightarrow$ $1.96x^2 = x^2 + 25$ $\Rightarrow$ $0.96x^2 = 25$ $\Rightarrow$ $x = \dfrac{5}{\sqrt{0.96}} \approx 5.1$.

Testing against the value of E at the endpoints: $E(0) = 1.4k(5) + 13k = 20k$, $E(5.1) \approx 17.9k$, $E(13) \approx 19.5k$.

Thus, to minimize energy, the bird should fly to a point about 5.1 km from B.

(b) If W/L is large, the bird would fly to a point C that is closer to B than to D to minimize the energy used flying over water.

If W/L is small, the bird would fly to a point C that is closer to D than to B to minimize the distance of the flight.

$E = W\sqrt{25 + x^2} + L(13 - x)$ $\Rightarrow$ $\dfrac{dE}{dx} = \dfrac{Wx}{\sqrt{25 + x^2}} - L = 0$ when $\dfrac{W}{L} = \dfrac{\sqrt{25 + x^2}}{x}$. By the same sort of

argument as in part (a), this ratio will give the minimal expenditure of energy if the bird heads for the point x km from B.

(c) For flight direct to D, $x = 13$, so from part (b), $W/L = \dfrac{\sqrt{25 + 13^2}}{13} \approx 1.07$. There is no value of W/L for which the bird

should fly directly to B. But note that $\lim\limits_{x \to 0^+} (W/L) = \infty$, so if the point at which E is a minimum is close to B, then

W/L is large.

(d) Assuming that the birds instinctively choose the path that minimizes the energy expenditure, we can use the equation for

$dE/dx = 0$ from part (a) with $1.4k = c$, $x = 4$, and $k = 1$: $c(4) = 1 \cdot (25 + 4^2)^{1/2}$ $\Rightarrow$ $c = \sqrt{41}/4 \approx 1.6$.

61. (a) Distance = rate × time, so time = distance/rate. $T_1 = \dfrac{D}{c_1}$, $T_2 = \dfrac{2|PR|}{c_1} + \dfrac{|RS|}{c_2} = \dfrac{2h\sec\theta}{c_1} + \dfrac{D - 2h\tan\theta}{c_2}$,

$T_3 = \dfrac{2\sqrt{h^2 + D^2/4}}{c_1} = \dfrac{\sqrt{4h^2 + D^2}}{c_1}$.

(b) $\dfrac{dT_2}{d\theta} = \dfrac{2h}{c_1} \cdot \sec\theta\tan\theta - \dfrac{2h}{c_2}\sec^2\theta = 0$ when $2h\sec\theta\left(\dfrac{1}{c_1}\tan\theta - \dfrac{1}{c_2}\sec\theta\right) = 0$ $\Rightarrow$

$\dfrac{1}{c_1}\dfrac{\sin\theta}{\cos\theta} - \dfrac{1}{c_2}\dfrac{1}{\cos\theta} = 0$ $\Rightarrow$ $\dfrac{\sin\theta}{c_1\cos\theta} = \dfrac{1}{c_2\cos\theta}$ $\Rightarrow$ $\sin\theta = \dfrac{c_1}{c_2}$. The First Derivative Test shows that this gives

a minimum.

(c) Using part (a) with $D = 1$ and $T_1 = 0.26$, we have $T_1 = \dfrac{D}{c_1}$ $\Rightarrow$ $c_1 = \dfrac{1}{0.26} \approx 3.85$ km/s. $T_3 = \dfrac{\sqrt{4h^2 + D^2}}{c_1}$ $\Rightarrow$

$4h^2 + D^2 = T_3^2c_1^2$ $\Rightarrow$ $h = \frac{1}{2}\sqrt{T_3^2c_1^2 - D^2} = \frac{1}{2}\sqrt{(0.34)^2(1/0.26)^2 - 1^2} \approx 0.42$ km. To find c_2, we use $\sin\theta = \dfrac{c_1}{c_2}$

from part (b) and $T_2 = \dfrac{2h\sec\theta}{c_1} + \dfrac{D - 2h\tan\theta}{c_2}$ from part (a). From the figure,

$\sin\theta = \dfrac{c_1}{c_2}$ $\Rightarrow$ $\sec\theta = \dfrac{c_2}{\sqrt{c_2^2 - c_1^2}}$ and $\tan\theta = \dfrac{c_1}{\sqrt{c_2^2 - c_1^2}}$, so

$T_2 = \dfrac{2hc_2}{c_1\sqrt{c_2^2 - c_1^2}} + \dfrac{D\sqrt{c_2^2 - c_1^2} - 2hc_1}{c_2\sqrt{c_2^2 - c_1^2}}$. Using the values for T_2 [given as 0.32],

17.

$f(x) = x^6 - x^5 - 6x^4 - x^2 + x + 10 \Rightarrow$

$f'(x) = 6x^5 - 5x^4 - 24x^3 - 2x + 1 \Rightarrow$

$$x_{n+1} = x_n - \frac{x_n^6 - x_n^5 - 6x_n^4 - x_n^2 + x_n + 10}{6x_n^5 - 5x_n^4 - 24x_n^3 - 2x_n + 1}.$$

From the graph of f, there appear to be roots near -1.9, -1.2, 1.1, and 3.

$x_1 = -1.9$	$x_1 = -1.2$	$x_1 = 1.1$	$x_1 = 3$
$x_2 \approx -1.94278290$	$x_2 \approx -1.22006245$	$x_2 \approx 1.14111662$	$x_2 \approx 2.99$
$x_3 \approx -1.93828380$	$x_3 \approx -1.21997997 \approx x_4$	$x_3 \approx 1.13929741$	$x_3 \approx 2.98984106$
$x_4 \approx -1.93822884$		$x_4 \approx 1.13929375 \approx x_5$	$x_4 \approx 2.98984102 \approx x_5$
$x_5 \approx -1.93822883 \approx x_6$			

To eight decimal places, the roots of the equation are -1.93822883, -1.21997997, 1.13929375, and 2.98984102.

19.

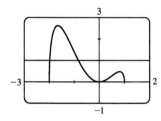

From the graph, $y = x^2\sqrt{2 - x - x^2}$ and $y = 1$ intersect twice, at $x \approx -2$ and at $x \approx -1$. $f(x) = x^2\sqrt{2 - x - x^2} - 1 \Rightarrow$

$f'(x) = x^2 \cdot \frac{1}{2}(2 - x - x^2)^{-1/2}(-1 - 2x) + (2 - x - x^2)^{1/2} \cdot 2x$

$\quad\quad = \frac{1}{2}x(2 - x - x^2)^{-1/2}[x(-1 - 2x) + 4(2 - x - x^2)]$

$\quad\quad = \frac{x(8 - 5x - 6x^2)}{2\sqrt{(2 + x)(1 - x)}}.$

so $x_{n+1} = x_n - \dfrac{x_n^2\sqrt{2 - x_n - x_n^2} - 1}{\dfrac{x_n(8 - 5x_n - 6x_n^2)}{2\sqrt{(2 + x_n)(1 - x_n)}}}$. Trying $x_1 = -2$ won't work because $f'(-2)$ is undefined, so we'll

try $x_1 = -1.95$.

$x_1 = -1.95$	$x_1 = -0.8$
$x_2 \approx -1.98580357$	$x_2 \approx -0.82674444$
$x_3 \approx -1.97899778$	$x_3 \approx -0.82646236$
$x_4 \approx -1.97807848$	$x_4 \approx -0.82646233 \approx x_5$
$x_5 \approx -1.97806682$	
$x_6 \approx -1.97806681 \approx x_7$	

To eight decimal places, the roots of the equation are -1.97806681 and -0.82646233.

21.

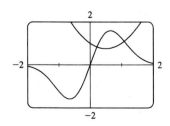

Solving $4e^{-x^2}\sin x = x^2 - x + 1$ is the same as solving

$f(x) = 4e^{-x^2}\sin x - x^2 + x - 1 = 0.$

$f'(x) = 4e^{-x^2}(\cos x - 2x\sin x) - 2x + 1 \Rightarrow$

$$x_{n+1} = x_n - \frac{4e^{-x_n^2}\sin x_n - x_n^2 + x_n - 1}{4e^{-x_n^2}(\cos x_n - 2x_n\sin x_n) - 2x_n + 1}.$$

From the figure, we see that the graphs intersect at approximately $x = 0.2$ and $x = 1.1$.

$$x_1 = 0.2 \qquad\qquad x_1 = 1.1$$

$$x_2 \approx 0.21883273 \qquad\qquad x_2 \approx 1.08432830$$

$$x_3 \approx 0.21916357 \qquad\qquad x_3 \approx 1.08422462 \approx x_4$$

$$x_4 \approx 0.21916368 \approx x_5$$

To eight decimal places, the roots of the equation are 0.21916368 and 1.08422462.

23. (a) $f(x) = x^2 - a \;\Rightarrow\; f'(x) = 2x$, so Newton's method gives

$$x_{n+1} = x_n - \frac{x_n^2 - a}{2x_n} = x_n - \frac{1}{2}x_n + \frac{a}{2x_n} = \frac{1}{2}x_n + \frac{a}{2x_n} = \frac{1}{2}\left(x_n + \frac{a}{x_n}\right).$$

(b) Using (a) with $a = 1000$ and $x_1 = \sqrt{900} = 30$, we get $x_2 \approx 31.666667$, $x_3 \approx 31.622807$, and $x_4 \approx 31.622777 \approx x_5$.

So $\sqrt{1000} \approx 31.622777$.

25. $f(x) = x^3 - 3x + 6 \;\Rightarrow\; f'(x) = 3x^2 - 3$. If $x_1 = 1$, then $f'(x_1) = 0$ and the tangent line used for approximating x_2 is horizontal. Attempting to find x_2 results in trying to divide by zero.

27. For $f(x) = x^{1/3}$, $f'(x) = \frac{1}{3}x^{-2/3}$ and

$$x_{n+1} = x_n - \frac{f(x_n)}{f'(x_n)} = x_n - \frac{x_n^{1/3}}{\frac{1}{3}x_n^{-2/3}} = x_n - 3x_n = -2x_n.$$

Therefore, each successive approximation becomes twice as large as the previous one in absolute value, so the sequence of approximations fails to converge to the root, which is 0. In the figure, we have $x_1 = 0.5$, $x_2 = -2(0.5) = -1$, and $x_3 = -2(-1) = 2$.

29.

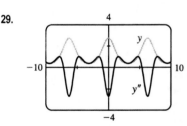

From the figure, we see that $y = f(x) = e^{\cos x}$ is periodic with period 2π. To find the x-coordinates of the IP, we only need to approximate the zeros of y'' on $[0, \pi]$. $f'(x) = -e^{\cos x}\sin x \;\Rightarrow\; f''(x) = e^{\cos x}(\sin^2 x - \cos x)$. Since $e^{\cos x} \neq 0$, we will use Newton's method with $g(x) = \sin^2 x - \cos x$, $g'(x) = 2\sin x \cos x + \sin x$, and $x_1 = 1$. $x_2 \approx 0.904173$, $x_3 \approx 0.904557 \approx x_4$. Thus, $(0.904557, 1.855277)$ is the IP.

31. We need to minimize the distance from $(0, 0)$ to an arbitrary point (x, y) on the curve $y = (x - 1)^2$. $d = \sqrt{x^2 + y^2} \;\Rightarrow\;$

$d(x) = \sqrt{x^2 + [(x-1)^2]^2} = \sqrt{x^2 + (x-1)^4}$. When $d' = 0$, d will be minimized and equivalently, $s = d^2$ will be minimized, so we will use Newton's method with $f = s'$ and $f' = s''$.

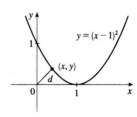

$f(x) = 2x + 4(x-1)^3 \;\Rightarrow\; f'(x) = 2 + 12(x-1)^2$, so $x_{n+1} = x_n - \dfrac{2x_n + 4(x_n - 1)^3}{2 + 12(x_n - 1)^2}$. Try $x_1 = 0.5 \;\Rightarrow\;$

$x_2 = 0.4$, $x_3 \approx 0.410127$, $x_4 \approx 0.410245 \approx x_5$. Now $d(0.410245) \approx 0.537841$ is the minimum distance and the point on the parabola is $(0.410245, 0.347810)$, correct to six decimal places.

33. In this case, $A = 18,000$, $R = 375$, and $n = 5(12) = 60$. So the formula $A = \dfrac{R}{i}[1 - (1+i)^{-n}]$ becomes

$$18,000 = \frac{375}{x}[1 - (1+x)^{-60}] \quad \Leftrightarrow \quad 48x = 1 - (1+x)^{-60} \quad \text{[multiply each term by } (1+x)^{60}] \quad \Leftrightarrow$$

$48x(1+x)^{60} - (1+x)^{60} + 1 = 0$. Let the LHS be called $f(x)$, so that

$$f'(x) = 48x(60)(1+x)^{59} + 48(1+x)^{60} - 60(1+x)^{59}$$

$$= 12(1+x)^{59}[4x(60) + 4(1+x) - 5] = 12(1+x)^{59}(244x - 1)$$

$x_{n+1} = x_n - \dfrac{48x_n(1+x_n)^{60} - (1+x_n)^{60} + 1}{12(1+x_n)^{59}(244x_n - 1)}$. An interest rate of 1% per month seems like a reasonable estimate for

$x = i$. So let $x_1 = 1\% = 0.01$, and we get $x_2 \approx 0.0082202$, $x_3 \approx 0.0076802$, $x_4 \approx 0.0076291$, $x_5 \approx 0.0076286 \approx x_6$.

Thus, the dealer is charging a monthly interest rate of 0.76286% (or 9.55% per year, compounded monthly).

4.8 Antiderivatives

1. $f(x) = \frac{1}{2} + \frac{3}{4}x^2 - \frac{4}{5}x^3 \;\Rightarrow\; F(x) = \frac{1}{2}x + \frac{3}{4}\dfrac{x^{2+1}}{2+1} - \frac{4}{5}\dfrac{x^{3+1}}{3+1} + C = \frac{1}{2}x + \frac{1}{4}x^3 - \frac{1}{5}x^4 + C$

   Check: $F'(x) = \frac{1}{2} + \frac{1}{4}(3x^2) - \frac{1}{5}(4x^3) + 0 = \frac{1}{2} + \frac{3}{4}x^2 - \frac{4}{5}x^3 = f(x)$

3. $f(x) = (x+1)(2x-1) = 2x^2 + x - 1 \;\Rightarrow\; F(x) = 2\left(\frac{1}{3}x^3\right) + \frac{1}{2}x^2 - x + C = \frac{2}{3}x^3 + \frac{1}{2}x^2 - x + C$

5. $f(x) = 5x^{1/4} - 7x^{3/4} \;\Rightarrow\; F(x) = 5\dfrac{x^{1/4+1}}{\frac{1}{4}+1} - 7\dfrac{x^{3/4+1}}{\frac{3}{4}+1} + C = 5\dfrac{x^{5/4}}{5/4} - 7\dfrac{x^{7/4}}{7/4} + C = 4x^{5/4} - 4x^{7/4} + C$

7. $f(x) = 6\sqrt{x} - \sqrt[6]{x} = 6x^{1/2} - x^{1/6} \;\Rightarrow$

$$F(x) = 6\frac{x^{1/2+1}}{\frac{1}{2}+1} - \frac{x^{1/6+1}}{\frac{1}{6}+1} + C = 6\frac{x^{3/2}}{3/2} - \frac{x^{7/6}}{7/6} + C = 4x^{3/2} - \frac{6}{7}x^{7/6} + C$$

9. $f(x) = \dfrac{10}{x^9} = 10x^{-9}$ has domain $(-\infty, 0) \cup (0, \infty)$, so $F(x) = \begin{cases} \dfrac{10x^{-8}}{-8} + C_1 = -\dfrac{5}{4x^8} + C_1 & \text{if } x < 0 \\[2mm] -\dfrac{5}{4x^8} + C_2 & \text{if } x > 0 \end{cases}$

   See Example 1(b) for a similar problem.

11. $f(u) = \dfrac{u^4 + 3\sqrt{u}}{u^2} = \dfrac{u^4}{u^2} + \dfrac{3u^{1/2}}{u^2} = u^2 + 3u^{-3/2} \;\Rightarrow$

$$F(u) = \frac{u^3}{3} + 3\frac{u^{-3/2+1}}{-3/2+1} + C = \frac{1}{3}u^3 + 3\frac{u^{-1/2}}{-1/2} + C = \frac{1}{3}u^3 - \frac{6}{\sqrt{u}} + C$$

13. $g(\theta) = \cos\theta - 5\sin\theta \;\Rightarrow\; G(\theta) = \sin\theta - 5(-\cos\theta) + C = \sin\theta + 5\cos\theta + C$

15. $f(x) = \dfrac{x^5 - x^3 + 2x}{x^4} = x - \dfrac{1}{x} + \dfrac{2}{x^3} = x - \dfrac{1}{x} + 2x^{-3} \;\Rightarrow$

$$F(x) = \frac{x^2}{2} - \ln|x| + 2\left(\frac{x^{-3+1}}{-3+1}\right) + C = \frac{1}{2}x^2 - \ln|x| - \frac{1}{x^2} + C$$

17. $f(x) = 5x^4 - 2x^5 \ \Rightarrow \ F(x) = 5 \cdot \dfrac{x^5}{5} - 2 \cdot \dfrac{x^6}{6} + C = x^5 - \frac{1}{3}x^6 + C.$

$F(0) = 4 \ \Rightarrow \ 0^5 - \frac{1}{3} \cdot 0^6 + C = 4 \ \Rightarrow \ C = 4,$ so $F(x) = x^5 - \frac{1}{3}x^6 + 4.$

The graph confirms our answer since $f(x) = 0$ when F has a local maximum, f is

positive when F is increasing, and f is negative when F is decreasing.

19. $f''(x) = 6x + 12x^2 \ \Rightarrow \ f'(x) = 6 \cdot \dfrac{x^2}{2} + 12 \cdot \dfrac{x^3}{3} + C = 3x^2 + 4x^3 + C \ \Rightarrow$

$f(x) = 3 \cdot \dfrac{x^3}{3} + 4 \cdot \dfrac{x^4}{4} + Cx + D = x^3 + x^4 + Cx + D$ [C and D are just arbitrary constants]

21. $f''(x) = \frac{2}{3}x^{2/3} \ \Rightarrow \ f'(x) = \dfrac{2}{3}\left(\dfrac{x^{5/3}}{5/3}\right) + C = \frac{2}{5}x^{5/3} + C \ \Rightarrow \ f(x) = \dfrac{2}{5}\left(\dfrac{x^{8/3}}{8/3}\right) + Cx + D = \frac{3}{20}x^{8/3} + Cx + D$

23. $f'(x) = 1 - 6x \ \Rightarrow \ f(x) = x - 3x^2 + C. \ f(0) = C$ and $f(0) = 8 \ \Rightarrow \ C = 8,$ so $f(x) = x - 3x^2 + 8.$

25. $f'(x) = \sqrt{x}(6 + 5x) = 6x^{1/2} + 5x^{3/2} \ \Rightarrow \ f(x) = 4x^{3/2} + 2x^{5/2} + C.$

$f(1) = 6 + C$ and $f(1) = 10 \ \Rightarrow \ C = 4,$ so $f(x) = 4x^{3/2} + 2x^{5/2} + 4.$

27. $f'(t) = 2\cos t + \sec^2 t \ \Rightarrow \ f(t) = 2\sin t + \tan t + C$ because $-\pi/2 < t < \pi/2.$

$f\left(\frac{\pi}{3}\right) = 2(\sqrt{3}/2) + \sqrt{3} + C = 2\sqrt{3} + C$ and $f\left(\frac{\pi}{3}\right) = 4 \ \Rightarrow \ C = 4 - 2\sqrt{3},$ so $f(t) = 2\sin t + \tan t + 4 - 2\sqrt{3}.$

29. $f''(x) = -2 + 12x - 12x^2 \ \Rightarrow \ f'(x) = -2x + 6x^2 - 4x^3 + C. \ f'(0) = C$ and $f'(0) = 12 \ \Rightarrow \ C = 12,$ so

$f'(x) = -2x + 6x^2 - 4x^3 + 12$ and hence, $f(x) = -x^2 + 2x^3 - x^4 + 12x + D. \ f(0) = D$ and $f(0) = 4 \ \Rightarrow \ D = 4,$

so $f(x) = -x^2 + 2x^3 - x^4 + 12x + 4.$

31. $f''(\theta) = \sin\theta + \cos\theta \ \Rightarrow \ f'(\theta) = -\cos\theta + \sin\theta + C. \ f'(0) = -1 + C$ and $f'(0) = 4 \ \Rightarrow \ C = 5,$ so

$f'(\theta) = -\cos\theta + \sin\theta + 5$ and hence, $f(\theta) = -\sin\theta - \cos\theta + 5\theta + D. \ f(0) = -1 + D$ and $f(0) = 3 \ \Rightarrow \ D = 4,$

so $f(\theta) = -\sin\theta - \cos\theta + 5\theta + 4.$

33. $f''(x) = 2 - 12x \ \Rightarrow \ f'(x) = 2x - 6x^2 + C \ \Rightarrow \ f(x) = x^2 - 2x^3 + Cx + D.$

$f(0) = D$ and $f(0) = 9 \ \Rightarrow \ D = 9. \ f(2) = 4 - 16 + 2C + 9 = 2C - 3$ and $f(2) = 15 \ \Rightarrow \ 2C = 18 \ \Rightarrow$

$C = 9,$ so $f(x) = x^2 - 2x^3 + 9x + 9.$

35. $f''(x) = 2 + \cos x \ \Rightarrow \ f'(x) = 2x + \sin x + C \ \Rightarrow \ f(x) = x^2 - \cos x + Cx + D.$

$f(0) = -1 + D$ and $f(0) = -1 \ \Rightarrow \ D = 0. \ f\left(\frac{\pi}{2}\right) = \pi^2/4 + \left(\frac{\pi}{2}\right)C$ and $f\left(\frac{\pi}{2}\right) = 0 \ \Rightarrow \ \left(\frac{\pi}{2}\right)C = -\pi^2/4 \ \Rightarrow$

$C = -\frac{\pi}{2},$ so $f(x) = x^2 - \cos x - \left(\frac{\pi}{2}\right)x.$

37. Given $f'(x) = 2x + 1,$ we have $f(x) = x^2 + x + C.$ Since f passes through $(1, 6),$ $f(1) = 6 \ \Rightarrow \ 1^2 + 1 + C = 6 \ \Rightarrow$

$C = 4.$ Therefore, $f(x) = x^2 + x + 4$ and $f(2) = 2^2 + 2 + 4 = 10.$

39.

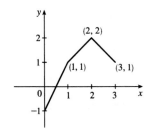

$$f'(x) = \begin{cases} 2 & \text{if } 0 \le x < 1 \\ 1 & \text{if } 1 < x < 2 \\ -1 & \text{if } 2 < x \le 3 \end{cases} \quad \Rightarrow \quad f(x) = \begin{cases} 2x + C & \text{if } 0 \le x < 1 \\ x + D & \text{if } 1 < x < 2 \\ -x + E & \text{if } 2 < x \le 3 \end{cases}$$

$f(0) = -1 \quad \Rightarrow \quad 2(0) + C = -1 \quad \Rightarrow \quad C = -1$. Starting at the point

$(0, -1)$ and moving to the right on a line with slope 2 gets us to the point $(1, 1)$.

The slope for $1 < x < 2$ is 1, so we get to the point $(2, 2)$. Here we have used the fact that f is continuous. We can include the

point $x = 1$ on either the first or the second part of f. The line connecting $(1, 1)$ to $(2, 2)$ is $y = x$, so $D = 0$. The slope for

$2 < x \le 3$ is -1, so we get to $(3, 1)$. $f(3) = 1 \quad \Rightarrow \quad -3 + E = 1 \quad \Rightarrow \quad E = 4$. Thus

$$f(x) = \begin{cases} 2x - 1 & \text{if } 0 \le x \le 1 \\ x & \text{if } 1 < x < 2 \\ -x + 4 & \text{if } 2 \le x \le 3 \end{cases}$$

Note that $f'(x)$ does not exist at $x = 1$ or at $x = 2$.

41. $v(t) = s'(t) = \sin t - \cos t \quad \Rightarrow \quad s(t) = -\cos t - \sin t + C$. $s(0) = -1 + C$ and $s(0) = 0 \quad \Rightarrow \quad C = 1$, so

$s(t) = -\cos t - \sin t + 1$.

43. (a) We first observe that since the stone is dropped 450 m above the ground, $v(0) = 0$ and $s(0) = 450$.

$v'(t) = a(t) = -9.8 \quad \Rightarrow \quad v(t) = -9.8t + C$. Now $v(0) = 0 \quad \Rightarrow \quad C = 0$, so $v(t) = -9.8t \quad \Rightarrow$

$s(t) = -4.9t^2 + D$. Last, $s(0) = 450 \quad \Rightarrow \quad D = 450 \quad \Rightarrow \quad s(t) = 450 - 4.9t^2$.

(b) The stone reaches the ground when $s(t) = 0$. $450 - 4.9t^2 = 0 \quad \Rightarrow \quad t^2 = 450/4.9 \quad \Rightarrow \quad t_1 = \sqrt{450/4.9} \approx 9.58$ s.

(c) The velocity with which the stone strikes the ground is $v(t_1) = -9.8\sqrt{450/4.9} \approx -93.9$ m/s.

(d) This is just reworking parts (a) and (b) with $v(0) = -5$. Using $v(t) = -9.8t + C$. $v(0) = -5 \quad \Rightarrow \quad 0 + C = -5 \quad \Rightarrow$

$v(t) = -9.8t - 5$. So $s(t) = -4.9t^2 - 5t + D$ and $s(0) = 450 \quad \Rightarrow \quad D = 450 \quad \Rightarrow \quad s(t) = -4.9t^2 - 5t + 450$.

Solving $s(t) = 0$ by using the quadratic formula gives us $t = \left(5 \pm \sqrt{8845}\right)/(-9.8) \quad \Rightarrow \quad t_1 \approx 9.09$ s.

45. By Exercise 44 with $a = -9.8$, $s(t) = -4.9t^2 + v_0 t + s_0$ and $v(t) = s'(t) = -9.8t + v_0$. So

$[v(t)]^2 = (-9.8t + v_0)^2 = (9.8)^2 t^2 - 19.6v_0 t + v_0^2 = v_0^2 + 96.04t^2 - 19.6v_0 t = v_0^2 - 19.6(-4.9t^2 + v_0 t)$.

But $-4.9t^2 + v_0 t$ is just $s(t)$ without the s_0 term; that is, $s(t) - s_0$. Thus, $[v(t)]^2 = v_0^2 - 19.6[s(t) - s_0]$.

47. Marginal cost $= 1.92 - 0.002x = C'(x) \quad \Rightarrow \quad C(x) = 1.92x - 0.001x^2 + K$. But $C(1) = 1.92 - 0.001 + K = 562 \quad \Rightarrow$

$K = 560.081$. Therefore, $C(x) = 1.92x - 0.001x^2 + 560.081 \quad \Rightarrow \quad C(100) = 742.081$, so the cost of producing

100 items is $742.08.

49. Using Exercise 44 with $a = -32$, $v_0 = 0$, and $s_0 = h$ (the height of the cliff), we know that the height at time t is

$s(t) = -16t^2 + h$. $v(t) = s'(t) = -32t$ and $v(t) = -120 \quad \Rightarrow \quad -32t = -120 \quad \Rightarrow \quad t = 3.75$, so

$0 = s(3.75) = -16(3.75)^2 + h \quad \Rightarrow \quad h = 16(3.75)^2 = 225$ ft.

51. $a(t) = k$, the initial velocity is $30 \text{ mi/h} = 30 \cdot \frac{5280}{3600} = 44 \text{ ft/s}$, and the final velocity (after 5 seconds) is

$50 \text{ mi/h} = 50 \cdot \frac{5280}{3600} = \frac{220}{3} \text{ ft/s}$. So $v(t) = kt + C$ and $v(0) = 44 \quad \Rightarrow \quad C = 44$. Thus, $v(t) = kt + 44 \quad \Rightarrow$

$v(5) = 5k + 44$. But $v(5) = \frac{220}{3}$, so $5k + 44 = \frac{220}{3} \quad \Rightarrow \quad 5k = \frac{88}{3} \quad \Rightarrow \quad k = \frac{88}{15} \approx 5.87 \text{ ft/s}^2$.

53. Let the acceleration be $a(t) = k \text{ km/h}^2$. We have $v(0) = 100 \text{ km/h}$ and we can take the initial position $s(0)$ to be 0.

We want the time t_f for which $v(t) = 0$ to satisfy $s(t) < 0.08 \text{ km}$. In general, $v'(t) = a(t) = k$, so $v(t) = kt + C$,

where $C = v(0) = 100$. Now $s'(t) = v(t) = kt + 100$, so $s(t) = \frac{1}{2}kt^2 + 100t + D$, where $D = s(0) = 0$.

Thus, $s(t) = \frac{1}{2}kt^2 + 100t$. Since $v(t_f) = 0$, we have $kt_f + 100 = 0$ or $t_f = -100/k$, so

$$s(t_f) = \frac{1}{2}k\left(-\frac{100}{k}\right)^2 + 100\left(-\frac{100}{k}\right) = 10{,}000\left(\frac{1}{2k} - \frac{1}{k}\right) = -\frac{5{,}000}{k}. \text{ The condition } s(t_f) \text{ must satisfy is}$$

$-\frac{5{,}000}{k} < 0.08 \quad \Rightarrow \quad -\frac{5{,}000}{0.08} > k \quad [k \text{ is negative}] \quad \Rightarrow \quad k < -62{,}500 \text{ km/h}^2, \text{ or equivalently,}$

$k < -\frac{3125}{648} \approx -4.82 \text{ m/s}^2$.

55. (a) The Mean Value Theorem says that there exists a number c in the interval (x_1, x_2) such that $H'(c) = \dfrac{H(x_2) - H(x_1)}{x_2 - x_1}$.

Since $H = G - F$ and G and F are antiderivatives of f, $H'(c) = G'(c) - F'(c) = f(c) - f(c) = 0$. So now

$\dfrac{H(x_2) - H(x_1)}{x_2 - x_1} = 0 \quad \Rightarrow \quad H(x_2) - H(x_1) = 0 \, (x_2 \neq x_1) \quad \Rightarrow \quad H(x_2) = H(x_1)$. Since this is true for any

$x_1 < x_2$ in I, H must be a constant function.

(b) We have $H = G - F$ and $H(x) = C$, so $C = G - F \quad \Rightarrow \quad G(x) = F(x) + C$. Thus, any antiderivative G can be

expressed as $F(x) + C$.

57. (a) First note that $90 \text{ mi/h} = 90 \times \frac{5280}{3600} \text{ ft/s} = 132 \text{ ft/s}$. Then $a(t) = 4 \text{ ft/s}^2 \quad \Rightarrow \quad v(t) = 4t + C$, but $v(0) = 0 \quad \Rightarrow$

$C = 0$. Now $4t = 132$ when $t = \frac{132}{4} = 33 \text{ s}$, so it takes 33 s to reach 132 ft/s. Therefore, taking $s(0) = 0$, we have

$s(t) = 2t^2, 0 \le t \le 33$. So $s(33) = 2178 \text{ ft}$. 15 minutes $= 15(60) = 900 \text{ s}$, so for $33 < t \le 933$ we have

$v(t) = 132 \text{ ft/s} \quad \Rightarrow \quad s(933) = 132(900) + 2178 = 120{,}978 \text{ ft} = 22.9125 \text{ mi}$.

(b) As in part (a), the train accelerates for 33 s and travels 2178 ft while doing so. Similarly, it decelerates for 33 s and travels

2178 ft at the end of its trip. During the remaining $900 - 66 = 834 \text{ s}$ it travels at 132 ft/s, so the distance traveled is

$132 \cdot 834 = 110{,}088 \text{ ft}$. Thus, the total distance is $2178 + 110{,}088 + 2178 = 114{,}444 \text{ ft} = 21.675 \text{ mi}$.

(c) $45 \text{ mi} = 45(5280) = 237{,}600 \text{ ft}$. Subtract $2(2178)$ to take care of the speeding up and slowing down, and we have

$233{,}244 \text{ ft at } 132 \text{ ft/s for a trip of } 233{,}244/132 = 1767 \text{ s at } 90 \text{ mi/h}$. The total time is

$1767 + 2(33) = 1833 \text{ s} = 30 \text{ min } 33 \text{ s} = 30.55 \text{ min}$.

(d) $37.5(60) = 2250 \text{ s}$. $2250 - 2(33) = 2184 \text{ s at maximum speed}$. $2184(132) + 2(2178) = 292{,}644 \text{ total feet or}$

$292{,}644/5280 = 55.425 \text{ mi}$.

4 Review

CONCEPT CHECK

1. A function f has an **absolute maximum** at $x = c$ if $f(c)$ is the largest function value on the entire domain of f, whereas f has a **local maximum** at c if $f(c)$ is the largest function value when x is near c. See Figure 6 in Section 4.2.

2. (a) See Theorem 4.2.3.

(b) See the Closed Interval Method before Example 6 in Section 4.2.

3. (a) See Theorem 4.2.4.

(b) See Definition 4.2.5.

4. See the Mean Value Theorem in Section 4.3. Geometric interpretation—there is some point P on the graph of a function f [on the interval (a, b)] where the tangent line is parallel to the secant line that connects $(a, f(a))$ and $(b, f(b))$.

5. (a) See the I/D Test before Example 2 in Section 4.3.

(b) A function f is concave upward on an interval I if f' is an increasing function on I (or, equivalently, the graph of f lies above all of its tangent lines on I).

(c) See the Concavity Test before Example 4 in Section 4.3.

(d) An inflection point is a point where a curve changes its direction of concavity. They can be found by determining the points at which the second derivative changes sign.

6. (a) See the First Derivative Test after Example 2 in Section 4.3.

(b) See the Second Derivative Test before Example 4 in Section 4.3.

(c) See the note before Example 5 in Section 4.3.

7. (a) See l'Hospital's Rule and the three notes that follow it in Section 4.5.

(b) Write fg as $\dfrac{f}{1/g}$ or $\dfrac{g}{1/f}$.

(c) Convert the difference into a quotient using a common denominator, rationalizing, factoring, or some other method.

(d) Convert the power to a product by taking the natural logarithm of both sides of $y = f^g$ or by writing f^g as $e^{g \ln f}$.

8. Without calculus you could get misleading graphs that fail to show the most interesting features of a function. See Example 1 in Section 4.4.

9. (a) See Figure 3 in Section 4.7.

(b) $x_2 = x_1 - \dfrac{f(x_1)}{f'(x_1)}$

(c) $x_{n+1} = x_n - \dfrac{f(x_n)}{f'(x_n)}$

(d) Newton's method is likely to fail or to work very slowly when $f'(x_1)$ is close to 0. It also fails when $f'(x_i)$ is undefined, such as with $f(x) = 1/x - 2$ and $x_1 = 1$.

10. (a) See the definition at the beginning of Section 4.8.

(b) If F_1 and F_2 are both antiderivatives of f on an interval I, then they differ by a constant.

TRUE-FALSE QUIZ

1. False. For example, take $f(x) = x^3$, then $f'(x) = 3x^2$ and $f'(0) = 0$, but $f(0) = 0$ is not a maximum or minimum; $(0, 0)$ is an inflection point.

3. False. For example, $f(x) = x$ is continuous on $(0, 1)$ but attains neither a maximum nor a minimum value on $(0, 1)$. Don't confuse this with f being continuous on the *closed* interval $[a, b]$, which would make the statement true.

5. True. This is an example of part (b) of the I/D Test.

7. False. $f'(x) = g'(x) \Rightarrow f(x) = g(x) + C$. For example, if $f(x) = x + 2$ and $g(x) = x + 1$, then $f'(x) = g'(x) = 1$, but $f(x) \neq g(x)$.

9. True. The graph of one such function is sketched.

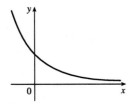

11. True. Let $x_1 < x_2$ where $x_1, x_2 \in I$. Then $f(x_1) < f(x_2)$ and $g(x_1) < g(x_2)$ [since f and g are increasing on I], so $(f + g)(x_1) = f(x_1) + g(x_1) < f(x_2) + g(x_2) = (f + g)(x_2)$.

13. False. Take $f(x) = x$ and $g(x) = x - 1$. Then both f and g are increasing on $(0, 1)$. But $f(x) g(x) = x(x - 1)$ is not increasing on $(0, 1)$.

15. True. Let $x_1, x_2 \in I$ and $x_1 < x_2$. Then $f(x_1) < f(x_2)$ [f is increasing] $\Rightarrow \dfrac{1}{f(x_1)} > \dfrac{1}{f(x_2)}$ [f is positive] $\Rightarrow$ $g(x_1) > g(x_2) \Rightarrow g(x) = 1/f(x)$ is decreasing on I.

17. True. If f is periodic, then there is a number p such that $f(x + p) = f(p)$ for all x. Differentiating gives $f'(x) = f'(x + p) \cdot (x + p)' = f'(x + p) \cdot 1 = f'(x + p)$, so f' is periodic.

19. True. By the Mean Value Theorem, there exists a number c in $(0, 1)$ such that $f(1) - f(0) = f'(c)(1 - 0) = f'(c)$. Since $f'(c)$ is nonzero, $f(1) - f(0) \neq 0$, so $f(1) \neq f(0)$.

EXERCISES

1. $f(x) = x^3 - 6x^2 + 9x + 1$, $[2, 4]$. $f'(x) = 3x^2 - 12x + 9 = 3(x^2 - 4x + 3) = 3(x - 1)(x - 3)$. $f'(x) = 0 \Rightarrow$ $x = 1$ or $x = 3$, but 1 is not in the interval. $f'(x) > 0$ for $3 < x < 4$ and $f'(x) < 0$ for $2 < x < 3$, so $f(3) = 1$ is a local minimum value. Checking the endpoints, we find $f(2) = 3$ and $f(4) = 5$. Thus, $f(3) = 1$ is the absolute minimum value and $f(4) = 5$ is the absolute maximum value.

3. $f(x) = \dfrac{3x-4}{x^2+1}$, $[-2, 2]$. $f'(x) = \dfrac{(x^2+1)(3) - (3x-4)(2x)}{(x^2+1)^2} = \dfrac{-(3x^2 - 8x - 3)}{(x^2+1)^2} = \dfrac{-(3x+1)(x-3)}{(x^2+1)^2}$.

$f'(x) = 0 \Rightarrow x = -\frac{1}{3}$ or $x = 3$, but 3 is not in the interval. $f'(x) > 0$ for $-\frac{1}{3} < x < 2$ and $f'(x) < 0$ for

$-2 < x < -\frac{1}{3}$, so $f\left(-\frac{1}{3}\right) = \frac{-5}{10/9} = -\frac{9}{2}$ is a local minimum value. Checking the endpoints, we find $f(-2) = -2$ and

$f(2) = \frac{2}{5}$. Thus, $f\left(-\frac{1}{3}\right) = -\frac{9}{2}$ is the absolute minimum value and $f(2) = \frac{2}{5}$ is the absolute maximum value.

5. $f(x) = x + \sin 2x$, $[0, \pi]$. $f'(x) = 1 + 2\cos 2x = 0 \Leftrightarrow \cos 2x = -\frac{1}{2} \Leftrightarrow 2x = \frac{2\pi}{3}$ or $\frac{4\pi}{3} \Leftrightarrow x = \frac{\pi}{3}$ or $\frac{2\pi}{3}$.

$f''(x) = -4\sin 2x$, so $f''\left(\frac{\pi}{3}\right) = -4\sin\frac{2\pi}{3} = -2\sqrt{3} < 0$ and $f''\left(\frac{2\pi}{3}\right) = -4\sin\frac{4\pi}{3} = 2\sqrt{3} > 0$, so

$f\left(\frac{\pi}{3}\right) = \frac{\pi}{3} + \frac{\sqrt{3}}{2} \approx 1.91$ is a local maximum value and $f\left(\frac{2\pi}{3}\right) = \frac{2\pi}{3} - \frac{\sqrt{3}}{2} \approx 1.23$ is a local minimum value. Also $f(0) = 0$

and $f(\pi) = \pi$, so $f(0) = 0$ is the absolute minimum value and $f(\pi) = \pi$ is the absolute maximum value.

7. (a) $f(x) = 2 - 2x - x^3$ is a polynomial, so there is no asymptote.

(b) $f'(x) = -2 - 3x^2 = -1(3x^2 + 2) < 0$, so f is decreasing on $\mathbb{R}$.

(c) No local extrema

(d) $f''(x) = -6x < 0$ on $(0, \infty)$ and $f''(x) > 0$ on $(-\infty, 0)$, so f is

CD on $(0, \infty)$ and CU on $(-\infty, 0)$. IP at $(0, 2)$

(e)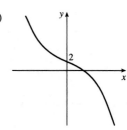

9. (a) $f(x) = x + \sqrt{1-x}$ has no asymptote.

(b) $f'(x) = 1 - 1/\left(2\sqrt{1-x}\right) = 0 \Leftrightarrow 2\sqrt{1-x} = 1 \Leftrightarrow$

$1 - x = \frac{1}{4} \Leftrightarrow x = \frac{3}{4}$ and $f'(x) > 0 \Leftrightarrow x < \frac{3}{4}$, so f is increasing

on $\left(-\infty, \frac{3}{4}\right)$ and decreasing on $\left(\frac{3}{4}, 1\right)$.

(c) $f\left(\frac{3}{4}\right) = \frac{3}{4} + \sqrt{1 - \frac{3}{4}} = \frac{3}{4} + \sqrt{\frac{1}{4}} = \frac{3}{4} + \frac{1}{2} = \frac{5}{4}$ is a local maximum.

(d) $f''(x) = -\dfrac{1}{4(1-x)^{3/2}} < 0$ on the domain of f, so f is CD on $(-\infty, 1)$. No IP

(e)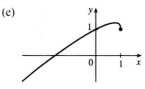

11. (a) $y = f(x) = \sin^2 x - 2\cos x$ has no asymptote.

(b) $y' = 2\sin x \cos x + 2\sin x = 2\sin x(\cos x + 1)$. $y' = 0 \Leftrightarrow \sin x = 0$ or $\cos x = -1 \Leftrightarrow x = n\pi$ or

$x = (2n+1)\pi$. $y' > 0$ when $\sin x > 0$, since $\cos x + 1 \geq 0$ for all x. Therefore, $y' > 0$ (and so f is increasing) on

$(2n\pi, (2n+1)\pi)$; $y' < 0$ (and so f is decreasing) on $((2n-1)\pi, 2n\pi)$ or equivalently, $((2n+1)\pi, (2n+2)\pi)$.

(c) Local maxima are $f((2n+1)\pi) = 2$; local minima are $f(2n\pi) = -2$.

(d) $y' = \sin 2x + 2\sin x \Rightarrow$

$y'' = 2\cos 2x + 2\cos x = 2(2\cos^2 x - 1) + 2\cos x$

$\quad = 4\cos^2 x + 2\cos x - 2 = 2(2\cos^2 x + \cos x - 1)$

$\quad = 2(2\cos x - 1)(\cos x + 1)$

$y'' = 0 \Leftrightarrow \cos x = \frac{1}{2}$ or $-1 \Leftrightarrow x = 2n\pi \pm \frac{\pi}{3}$ or $x = (2n+1)\pi$.

$y'' > 0$ (and so f is CU) on $\left(2n\pi - \frac{\pi}{3}, 2n\pi + \frac{\pi}{3}\right)$; $y'' \leq 0$ (and so f is

CD) on $\left(2n\pi + \frac{\pi}{3}, 2n\pi + \frac{5\pi}{3}\right)$. IPs at $\left(2n\pi \pm \frac{\pi}{3}, -\frac{1}{4}\right)$

(e)

13. (a) $\lim_{x \to \pm\infty} \left(e^x + e^{-3x}\right) = \infty$, no asymptote.

(e)

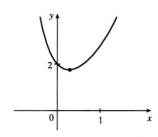

(b) $y = f(x) = e^x + e^{-3x} \Rightarrow f'(x) = e^x - 3e^{-3x} = e^{-3x}(e^{4x} - 3) > 0 \Leftrightarrow$

$e^{4x} > 3 \Leftrightarrow 4x > \ln 3 \Leftrightarrow x > \frac{1}{4}\ln 3$, so f is increasing on $\left(\frac{1}{4}\ln 3, \infty\right)$

and decreasing on $\left(-\infty, \frac{1}{4}\ln 3\right)$.

(c) $f\left(\frac{1}{4}\ln 3\right) = 3^{1/4} + 3^{-3/4} \approx 1.75$ is a local and absolute minimum.

(d) $f''(x) = e^x + 9e^{-3x} > 0$, so f is CU on $(-\infty, \infty)$.   No IP

15. $f(x) = \dfrac{x^2 - 1}{x^3} \Rightarrow f'(x) = \dfrac{x^3(2x) - (x^2 - 1)3x^2}{x^6} = \dfrac{3 - x^2}{x^4} \Rightarrow$

$f''(x) = \dfrac{x^4(-2x) - (3 - x^2)4x^3}{x^8} = \dfrac{2x^2 - 12}{x^5}$

Estimates: From the graphs of f' and f'', it appears that f is increasing on

$(-1.73, 0)$ and $(0, 1.73)$ and decreasing on $(-\infty, -1.73)$ and $(1.73, \infty)$;

f has a local maximum of about $f(1.73) = 0.38$ and a local minimum of about

$f(-1.7) = -0.38$; f is CU on $(-2.45, 0)$ and $(2.45, \infty)$, and CD on

$(-\infty, -2.45)$ and $(0, 2.45)$; and f has inflection points at about

$(-2.45, -0.34)$ and $(2.45, 0.34)$.

Exact: Now $f'(x) = \dfrac{3 - x^2}{x^4}$ is positive for $0 < x^2 < 3$, that is, f is increasing

on $(-\sqrt{3}, 0)$ and $(0, \sqrt{3})$; and $f'(x)$ is negative (and so f is decreasing) on

$(-\infty, -\sqrt{3})$ and $(\sqrt{3}, \infty)$. $f'(x) = 0$ when $x = \pm\sqrt{3}$.

f' goes from positive to negative at $x = \sqrt{3}$, so f has a local maximum of

$f(\sqrt{3}) = \dfrac{(\sqrt{3})^2 - 1}{(\sqrt{3})^3} = \dfrac{2\sqrt{3}}{9}$; and since f is odd, we know that maxima on the

interval $(0, \infty)$ correspond to minima on $(-\infty, 0)$, so f has a local minimum of

$f(-\sqrt{3}) = -\dfrac{2\sqrt{3}}{9}$. Also, $f''(x) = \dfrac{2x^2 - 12}{x^5}$ is positive (so f is CU) on

$(-\sqrt{6}, 0)$ and $(\sqrt{6}, \infty)$, and negative (so f is CD) on $(-\infty, -\sqrt{6})$ and

$(0, \sqrt{6})$. There are IP at $\left(\sqrt{6}, \dfrac{5\sqrt{6}}{36}\right)$ and $\left(-\sqrt{6}, -\dfrac{5\sqrt{6}}{36}\right)$.

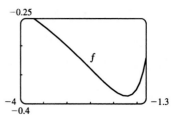

17. $f(x) = 3x^6 - 5x^5 + x^4 - 5x^3 - 2x^2 + 2 \Rightarrow f'(x) = 18x^5 - 25x^4 + 4x^3 - 15x^2 - 4x \Rightarrow$

$f''(x) = 90x^4 - 100x^3 + 12x^2 - 30x - 4$

  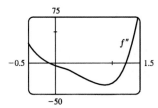

From the graphs of f' and f'', it appears that f is increasing on $(-0.23, 0)$ and $(1.62, \infty)$ and decreasing on $(-\infty, -0.23)$

and $(0, 1.62)$; f has a local maximum of $f(0) = 2$ and local minima of about $f(-0.23) = 1.96$ and $f(1.62) = -19.2$;

f is CU on $(-\infty, -0.12)$ and $(1.24, \infty)$ and CD on $(-0.12, 1.24)$; and f has inflection points at about $(-0.12, 1.98)$ and

$(1.24, -12.1)$.

19.

From the graph, we estimate the points of inflection to be about $(\pm 0.82, 0.22)$.

$f(x) = e^{-1/x^2} \Rightarrow f'(x) = 2x^{-3}e^{-1/x^2} \Rightarrow$

$f''(x) = 2[x^{-3}(2x^{-3})e^{-1/x^2} + e^{-1/x^2}(-3x^{-4})] = 2x^{-6}e^{-1/x^2}(2 - 3x^2)$.

This is 0 when $2 - 3x^2 = 0 \Leftrightarrow x = \pm\sqrt{\frac{2}{3}}$, so the inflection points

are $\left(\pm\sqrt{\frac{2}{3}}, e^{-3/2}\right)$.

21. $f(x) = \dfrac{\cos^2 x}{\sqrt{x^2 + x + 1}}, \quad -\pi \le x \le \pi \Rightarrow f'(x) = -\dfrac{\cos x\,[(2x+1)\cos x + 4(x^2 + x + 1)\sin x]}{2(x^2 + x + 1)^{3/2}} \Rightarrow$

$f''(x) = -\dfrac{(8x^4 + 16x^3 + 16x^2 + 8x + 9)\cos^2 x - 8(x^2 + x + 1)(2x + 1)\sin x \cos x - 8(x^2 + x + 1)^2 \sin^2 x}{4(x^2 + x + 1)^{5/2}}$

$f(x) = 0 \Leftrightarrow x = \pm\frac{\pi}{2}; \quad f'(x) = 0 \Leftrightarrow x \approx -2.96, -1.57, -0.18, 1.57, 3.01;$

$f''(x) = 0 \Leftrightarrow x \approx -2.16, -0.75, 0.46,$ and 2.21.

  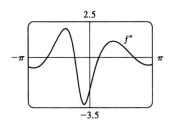

The x-coordinates of the maximum points are the values at which f' changes from positive to negative, that is, -2.96,

-0.18, and 3.01. The x-coordinates of the minimum points are the values at which f' changes from negative to positive, that is, -1.57 and 1.57. The x-coordinates of the inflection points are the values at which f'' changes sign, that is, -2.16, -0.75, 0.46, and 2.21.

23. The family of functions $f(x) = \ln(\sin x + C)$ all have the same period and all have maximum values at $x = \frac{\pi}{2} + 2\pi n$. Since the domain of $\ln$ is $(0, \infty)$, f has a graph only if $\sin x + C > 0$ somewhere. Since $-1 \le \sin x \le 1$, this happens if $C > -1$, that is, f has no graph if $C \le -1$. Similarly, if $C > 1$, then $\sin x + C > 0$ and f is continuous on $(-\infty, \infty)$. As C increases, the graph of

f is shifted vertically upward and flattens out. If $-1 < C \le 1$, f is defined where $\sin x + C > 0$ $\Leftrightarrow$ $\sin x > -C$ $\Leftrightarrow$ $\sin^{-1}(-C) < x < \pi - \sin^{-1}(-C)$. Since the period is 2π, the domain of f is $\left(2n\pi + \sin^{-1}(-C), (2n+1)\pi - \sin^{-1}(-C)\right)$, n an integer.

25. For $(1, 6)$ to be on the curve $y = x^3 + ax^2 + bx + 1$, we have that $6 = 1 + a + b + 1$ $\Rightarrow$ $b = 4 - a$. Now $y' = 3x^2 + 2ax + b$ and $y'' = 6x + 2a$. Also, for $(1, 6)$ to be an inflection point it must be true that $y''(1) = 6(1) + 2a = 0$ $\Rightarrow$ $a = -3$ $\Rightarrow$ $b = 4 - (-3) = 7$. Note that with $a = -3$, we have $y'' = 6x - 6 = 6(x-1)$, so y'' changes sign at $x = 1$, proving that $(1, 6)$ is a point of inflection. [This does not follow from the fact that $y''(1) = 0$.]

27. This limit has the form $\frac{0}{0}$. $\displaystyle\lim_{x \to 0} \frac{\tan \pi x}{\ln(1+x)} \overset{\text{H}}{=} \lim_{x \to 0} \frac{\pi \sec^2 \pi x}{1/(1+x)} = \frac{\pi \cdot 1^2}{1/1} = \pi$

29. This limit has the form $\frac{0}{0}$. $\displaystyle\lim_{x \to 0} \frac{e^{4x} - 1 - 4x}{x^2} \overset{\text{H}}{=} \lim_{x \to 0} \frac{4e^{4x} - 4}{2x} \overset{\text{H}}{=} \lim_{x \to 0} \frac{16e^{4x}}{2} = \lim_{x \to 0} 8e^{4x} = 8 \cdot 1 = 8$

31. This limit has the form $\infty \cdot 0$. $\displaystyle\lim_{x \to \infty} x^3 e^{-x} = \lim_{x \to \infty} \frac{x^3}{e^x} \overset{\text{H}}{=} \lim_{x \to \infty} \frac{3x^2}{e^x} \overset{\text{H}}{=} \lim_{x \to \infty} \frac{6x}{e^x} \overset{\text{H}}{=} \lim_{x \to \infty} \frac{6}{e^x} = 0$

33. This limit has the form $\infty - \infty$.

$$\lim_{x \to 1^+} \left(\frac{x}{x-1} - \frac{1}{\ln x} \right) = \lim_{x \to 1^+} \left(\frac{x \ln x - x + 1}{(x-1)\ln x} \right) \overset{\text{H}}{=} \lim_{x \to 1^+} \frac{x \cdot (1/x) + \ln x - 1}{(x-1) \cdot (1/x) + \ln x} = \lim_{x \to 1^+} \frac{\ln x}{1 - 1/x + \ln x}$$

$$\overset{\text{H}}{=} \lim_{x \to 1^+} \frac{1/x}{1/x^2 + 1/x} = \frac{1}{1+1} = \frac{1}{2}$$

35. We are given $d\theta/dt = -0.25$ rad/h. $\tan \theta = 400/x$ $\Rightarrow$

$x = 400 \cot \theta$ $\Rightarrow$ $\dfrac{dx}{dt} = -400 \csc^2 \theta \, \dfrac{d\theta}{dt}$. When $\theta = \frac{\pi}{6}$,

$\dfrac{dx}{dt} = -400(2)^2(-0.25) = 400$ ft/h.

37. Given $dh/dt = 5$ and $dx/dt = 15$, find dz/dt. $z^2 = x^2 + h^2$ $\Rightarrow$

$$2z\frac{dz}{dt} = 2x\frac{dx}{dt} + 2h\frac{dh}{dt} \quad \Rightarrow \quad \frac{dz}{dt} = \frac{1}{z}(15x + 5h). \text{ When } t = 3,$$

$h = 45 + 3(5) = 60$ and $x = 15(3) = 45$ $\Rightarrow$ $z = \sqrt{45^2 + 60^2} = 75$,

so $\dfrac{dz}{dt} = \dfrac{1}{75}[15(45) + 5(60)] = 13$ ft/s.

39. Call the two integers x and y. Then $x + 4y = 1000$, so $x = 1000 - 4y$. Their product is $P = xy = (1000 - 4y)y$, so our

problem is to maximize the function $P(y) = 1000y - 4y^2$, where $0 < y < 250$ and y is an integer. $P'(y) = 1000 - 8y$, so

$P'(y) = 0$ $\Leftrightarrow$ $y = 125$. $P''(y) = -8 < 0$, so $P(125) = 62,500$ is an absolute maximum. Since the optimal y turned

out to be an integer, we have found the desired pair of numbers, namely $x = 1000 - 4(125) = 500$ and $y = 125$.

41.

By similar triangles, $\dfrac{y}{x} = \dfrac{r}{\sqrt{x^2 - 2rx}}$, so the area of the triangle is

$$A(x) = \tfrac{1}{2}(2y)x = xy = \frac{rx^2}{\sqrt{x^2 - 2rx}} \quad \Rightarrow$$

$$A'(x) = \frac{2rx\sqrt{x^2 - 2rx} - rx^2(x-r)/\sqrt{x^2 - 2rx}}{x^2 - 2rx} = \frac{rx^2(x - 3r)}{(x^2 - 2rx)^{3/2}} = 0$$

when $x = 3r$.

$A'(x) < 0$ when $2r < x < 3r$, $A'(x) > 0$ when $x > 3r$. So $x = 3r$ gives a minimum and $A(3r) = \dfrac{r(9r^2)}{\sqrt{3}\,r} = 3\sqrt{3}\,r^2$.

43.

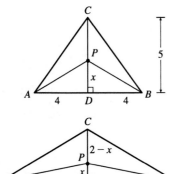

We minimize $L(x) = |PA| + |PB| + |PC| = 2\sqrt{x^2 + 16} + (5 - x)$,

$0 \le x \le 5$. $L'(x) = 2x/\sqrt{x^2 + 16} - 1 = 0$ $\Leftrightarrow$ $2x = \sqrt{x^2 + 16}$ $\Leftrightarrow$

$4x^2 = x^2 + 16$ $\Leftrightarrow$ $x = \frac{4}{\sqrt{3}}$. $L(0) = 13$, $L\left(\frac{4}{\sqrt{3}}\right) \approx 11.9$, $L(5) \approx 12.8$, so the

minimum occurs when $x = \frac{4}{\sqrt{3}} \approx 2.3$.

If $|CD| = 2$, $L(x)$ changes from $(5 - x)$ to $(2 - x)$ with $0 \le x \le 2$. But we still

get $L'(x) = 0$ $\Leftrightarrow$ $x = \frac{4}{\sqrt{3}}$, which isn't in the interval $[0, 2]$. Now $L(0) = 10$

and $L(2) = 2\sqrt{20} = 4\sqrt{5} \approx 8.9$. The minimum occurs when $P = C$.

45. $v = K\sqrt{\dfrac{L}{C} + \dfrac{C}{L}}$ $\Rightarrow$ $\dfrac{dv}{dL} = \dfrac{K}{2\sqrt{(L/C) + (C/L)}}\left(\dfrac{1}{C} - \dfrac{C}{L^2}\right) = 0$ $\Leftrightarrow$ $\dfrac{1}{C} = \dfrac{C}{L^2}$ $\Leftrightarrow$ $L^2 = C^2$ $\Leftrightarrow$ $L = C$.

This gives the minimum velocity since $v' < 0$ for $0 < L < C$ and $v' > 0$ for $L > C$.

47. Let x denote the number of $1 decreases in ticket price. Then the ticket price is $12 - 1(x)$, and the average attendance is

$11,000 + 1000(x)$. Now the revenue per game is

$$R(x) = (\text{price per person}) \times (\text{number of people per game})$$

$$= (12 - x)(11,000 + 1000x) = -1000x^2 + 1000x + 132,000$$

for $0 \le x \le 4$ [since the seating capacity is 15,000] $\Rightarrow$ $R'(x) = -2000x + 1000 = 0$ $\Leftrightarrow$ $x = 0.5$. This is a maximum since $R''(x) = -2000 < 0$ for all x. Now we must check the value of $R(x) = (12 - x)(11{,}000 + 1000x)$ at $x = 0.5$ and at the endpoints of the domain to see which value of x gives the maximum value of R.

$R(0) = (12)(11{,}000) = 132{,}000$, $R(0.5) = (11.5)(11{,}500) = 132{,}250$, and $R(4) = (8)(15{,}000) = 120{,}000$. Thus, the maximum revenue of \$132,250 per game occurs when the average attendance is 11,500 and the ticket price is \$11.50.

49. $f(t) = \cos t + t - t^2$ $\Rightarrow$ $f'(t) = -\sin t + 1 - 2t$. $f'(t)$ exists for all

t, so to find the maximum of f, we can examine the zeros of f'.

From the graph of f', we see that a good choice for t_1 is $t_1 = 0.3$.

Use $g(t) = -\sin t + 1 - 2t$ and $g'(t) = -\cos t - 2$ to obtain

$t_2 \approx 0.33535293$, $t_3 \approx 0.33541803 \approx t_4$. Since $f''(t) = -\cos t - 2 < 0$

for all t, $f(0.33541803) \approx 1.16718557$ is the absolute maximum.

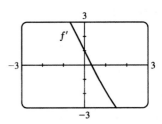

51. $f(x) = e^x - (2/\sqrt{x}) = e^x - 2x^{-1/2}$ $\Rightarrow$ $F(x) = e^x - 2\dfrac{x^{-1/2+1}}{-1/2+1} + C = e^x - 2\dfrac{x^{1/2}}{1/2} + C = e^x - 4\sqrt{x} + C$

53. $f'(t) = 2t - 3\sin t$ $\Rightarrow$ $f(t) = t^2 + 3\cos t + C$.

$f(0) = 3 + C$ and $f(0) = 5$ $\Rightarrow$ $C = 2$, so $f(t) = t^2 + 3\cos t + 2$.

55. $f''(x) = 1 - 6x + 48x^2$ $\Rightarrow$ $f'(x) = x - 3x^2 + 16x^3 + C$. $f'(0) = C$ and $f'(0) = 2$ $\Rightarrow$ $C = 2$, so

$f'(x) = x - 3x^2 + 16x^3 + 2$ and hence, $f(x) = \frac{1}{2}x^2 - x^3 + 4x^4 + 2x + D$.

$f(0) = D$ and $f(0) = 1$ $\Rightarrow$ $D = 1$, so $f(x) = \frac{1}{2}x^2 - x^3 + 4x^4 + 2x + 1$.

57. $v(t) = s'(t) = 2t - \dfrac{1}{1+t^2}$ $\Rightarrow$ $s(t) = t^2 - \tan^{-1} t + C$.

$s(0) = 0 - 0 + C = C$ and $s(0) = 1$ $\Rightarrow$ $C = 1$, so $s(t) = t^2 - \tan^{-1} t + 1$.

59. (a) Since f is 0 just to the left of the y-axis, we must have a minimum of F at the same place since we are increasing through

$(0, 0)$ on F. There must be a local maximum to the left of $x = -3$, since f changes from positive to negative there.

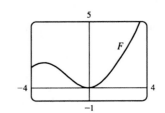

(b) $f(x) = 0.1e^x + \sin x$ $\Rightarrow$

$\quad F(x) = 0.1e^x - \cos x + C$. $F(0) = 0$ $\Rightarrow$

$\quad 0.1 - 1 + C = 0$ $\Rightarrow$ $C = 0.9$, so

$\quad F(x) = 0.1e^x - \cos x + 0.9$.

(c)

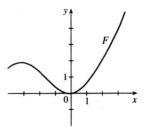

61. Choosing the positive direction to be upward, we have $a(t) = -9.8 \Rightarrow v(t) = -9.8t + v_0$, but $v(0) = 0 = v_0 \Rightarrow$

$v(t) = -9.8t = s'(t) \Rightarrow s(t) = -4.9t^2 + s_0$, but $s(0) = s_0 = 500 \Rightarrow s(t) = -4.9t^2 + 500$. When $s = 0$,

$-4.9t^2 + 500 = 0 \Rightarrow t_1 = \sqrt{\frac{500}{4.9}} \approx 10.1 \Rightarrow v(t_1) = -9.8\sqrt{\frac{500}{4.9}} \approx -98.995$ m/s. Since the canister has been

designed to withstand an impact velocity of 100 m/s, the canister will *not burst*.

63. (a)

The cross-sectional area of the rectangular beam is

$A = 2x \cdot 2y = 4xy = 4x\sqrt{100 - x^2}, 0 \le x \le 10$, so

$$\frac{dA}{dx} = 4x\left(\tfrac{1}{2}\right)(100 - x^2)^{-1/2}(-2x) + (100 - x^2)^{1/2} \cdot 4$$

$$= \frac{-4x^2}{(100 - x^2)^{1/2}} + 4(100 - x^2)^{1/2} = \frac{4[-x^2 + (100 - x^2)]}{(100 - x^2)^{1/2}}.$$

$\frac{dA}{dx} = 0$ when $-x^2 + (100 - x^2) = 0 \Rightarrow x^2 = 50 \Rightarrow x = \sqrt{50} \approx 7.07 \Rightarrow y = \sqrt{100 - (\sqrt{50})^2} = \sqrt{50}$.

Since $A(0) = A(10) = 0$, the rectangle of maximum area is a square.

(b)

The cross-sectional area of each rectangular plank (shaded in the figure) is

$A = 2x(y - \sqrt{50}) = 2x[\sqrt{100 - x^2} - \sqrt{50}], 0 \le x \le \sqrt{50}$, so

$$\frac{dA}{dx} = 2(\sqrt{100 - x^2} - \sqrt{50}) + 2x\left(\tfrac{1}{2}\right)(100 - x^2)^{-1/2}(-2x)$$

$$= 2(100 - x^2)^{1/2} - 2\sqrt{50} - \frac{2x^2}{(100 - x^2)^{1/2}}$$

Set $\frac{dA}{dx} = 0$: $(100 - x^2) - \sqrt{50}(100 - x^2)^{1/2} - x^2 = 0 \Rightarrow 100 - 2x^2 = \sqrt{50}(100 - x^2)^{1/2} \Rightarrow$

$10{,}000 - 400x^2 + 4x^4 = 50(100 - x^2) \Rightarrow 4x^4 - 350x^2 + 5000 = 0 \Rightarrow 2x^4 - 175x^2 + 2500 = 0 \Rightarrow$

$x^2 = \dfrac{175 \pm \sqrt{10{,}625}}{4} \approx 69.52$ or $17.98 \Rightarrow x \approx 8.34$ or 4.24. But $8.34 > \sqrt{50}$, so $x_1 \approx 4.24 \Rightarrow$

$y - \sqrt{50} = \sqrt{100 - x_1^2} - \sqrt{50} \approx 1.99$. Each plank should have dimensions about $8\frac{1}{2}$ inches by 2 inches.

(c) From the figure in part (a), the width is $2x$ and the depth is $2y$, so the strength is

$S = k(2x)(2y)^2 = 8kxy^2 = 8kx(100 - x^2) = 800kx - 8kx^3, 0 \le x \le 10$. $dS/dx = 800k - 24kx^2 = 0$ when

$24kx^2 = 800k \Rightarrow x^2 = \frac{100}{3} \Rightarrow x = \frac{10}{\sqrt{3}} \Rightarrow y = \sqrt{\frac{200}{3}} = \frac{10\sqrt{2}}{\sqrt{3}} = \sqrt{2}\,x$. Since $S(0) = S(10) = 0$, the

maximum strength occurs when $x = \frac{10}{\sqrt{3}}$. The dimensions should be $\frac{20}{\sqrt{3}} \approx 11.55$ inches by $\frac{20\sqrt{2}}{\sqrt{3}} \approx 16.33$ inches.

65. (a) $I = \dfrac{k\cos\theta}{d^2} = \dfrac{k(h/d)}{d^2} = k\dfrac{h}{d^3} = k\dfrac{h}{\left(\sqrt{40^2 + h^2}\,\right)^3} = k\dfrac{h}{(1600 + h^2)^{3/2}} \quad\Rightarrow$

$$\frac{dI}{dh} = k\,\frac{(1600 + h^2)^{3/2} - h\frac{3}{2}(1600 + h^2)^{1/2}\cdot 2h}{[(1600 + h^2)^{3/2}]^2} = \frac{k(1600 + h^2)^{1/2}(1600 + h^2 - 3h^2)}{(1600 + h^2)^{3/2}}$$

$$= \frac{k(1600 - 2h^2)}{(1600 + h^2)^{5/2}} \qquad [k \text{ is the constant of proportionality}]$$

Set $dI/dh = 0$: $1600 - 2h^2 = 0 \;\Rightarrow\; h^2 = 800 \;\Rightarrow\; h = \sqrt{800} = 20\sqrt{2}$. By the First Derivative Test, I has a local

maximum at $h = 20\sqrt{2} \approx 28$ ft.

(b)

$\dfrac{dx}{dt} = 4 \text{ ft/s}$

$I = \dfrac{k\cos\theta}{d^2} = \dfrac{k[(h-4)/d]}{d^2} = \dfrac{k(h-4)}{d^3}$

$= \dfrac{k(h-4)}{[(h-4)^2 + x^2]^{3/2}} = k(h-4)[(h-4)^2 + x^2]^{-3/2}$

$$\frac{dI}{dt} = \frac{dI}{dx}\cdot\frac{dx}{dt} = k(h-4)\left(-\tfrac{3}{2}\right)[(h-4)^2 + x^2]^{-5/2}\cdot 2x\cdot\frac{dx}{dt}$$

$$= k(h-4)(-3x)[(h-4)^2 + x^2]^{-5/2}\cdot 4 = \frac{-12xk(h-4)}{[(h-4)^2 + x^2]^{5/2}}$$

$$\left.\frac{dI}{dt}\right|_{x=40} = -\frac{480k(h-4)}{[(h-4)^2 + 1600]^{5/2}}$$

☐ FOCUS ON PROBLEM SOLVING

1. Let $y = f(x) = e^{-x^2}$. The area of the rectangle under the curve from $-x$ to x is $A(x) = 2xe^{-x^2}$ where $x \geq 0$. We maximize

$A(x)$: $A'(x) = 2e^{-x^2} - 4x^2e^{-x^2} = 2e^{-x^2}\left(1 - 2x^2\right) = 0 \quad \Rightarrow \quad x = \frac{1}{\sqrt{2}}$. This gives a maximum since $A'(x) > 0$

for $0 \leq x < \frac{1}{\sqrt{2}}$ and $A'(x) < 0$ for $x > \frac{1}{\sqrt{2}}$. We next determine the points of inflection of $f(x)$. Notice that

$f'(x) = -2xe^{-x^2} = -A(x)$. So $f''(x) = -A'(x)$ and hence, $f''(x) < 0$ for $-\frac{1}{\sqrt{2}} < x < \frac{1}{\sqrt{2}}$ and $f''(x) > 0$ for $x < -\frac{1}{\sqrt{2}}$

and $x > \frac{1}{\sqrt{2}}$. So $f(x)$ changes concavity at $x = \pm\frac{1}{\sqrt{2}}$, and the two vertices of the rectangle of largest area are at the inflection

points.

3. First, we recognize some symmetry in the inequality: $\dfrac{e^{x+y}}{xy} \geq e^2 \quad \Leftrightarrow \quad \dfrac{e^x}{x} \cdot \dfrac{e^y}{y} \geq e \cdot e$. This suggests that we need to show

that $\dfrac{e^x}{x} \geq e$ for $x > 0$. If we can do this, then the inequality $\dfrac{e^y}{y} \geq e$ is true, and the given inequality follows. $f(x) = \dfrac{e^x}{x} \quad \Rightarrow$

$f'(x) = \dfrac{xe^x - e^x}{x^2} = \dfrac{e^x(x-1)}{x^2} = 0 \quad \Rightarrow \quad x = 1$. By the First Derivative Test, we have a minimum of $f(1) = e$, so

$e^x/x \geq e$ for all x.

5. $f(x)$ has the form $e^{g(x)}$, so it will have an absolute maximum (minimum) where g has an absolute maximum (minimum).

$$g(x) = 10|x-2| - x^2 = \begin{cases} 10(x-2) - x^2 & \text{if } x-2 > 0 \\ 10[-(x-2)] - x^2 & \text{if } x-2 < 0 \end{cases} = \begin{cases} -x^2 + 10x - 20 & \text{if } x > 2 \\ -x^2 - 10x + 20 & \text{if } x < 2 \end{cases} \Rightarrow$$

$$g'(x) = \begin{cases} -2x + 10 & \text{if } x > 2 \\ -2x - 10 & \text{if } x < 2 \end{cases}$$

$g'(x) = 0$ if $x = -5$ or $x = 5$, and $g'(2)$ does not exist, so the critical numbers of g are -5, 2, and 5. Since $g''(x) = -2$ for

all $x \neq 2$, g is concave downward on $(-\infty, 2)$ and $(2, \infty)$, and g will attain its absolute maximum at one of the critical

numbers. Since $g(-5) = 45$, $g(2) = -4$, and $g(5) = 5$, we see that $f(-5) = e^{45}$ is the absolute maximum value of f. Also,

$\lim\limits_{x \to \infty} g(x) = -\infty$, so $\lim\limits_{x \to \infty} f(x) = \lim\limits_{x \to \infty} e^{g(x)} = 0$. But $f(x) > 0$ for all x, so there is no absolute minimum value of f.

7. Differentiating $x^2 + xy + y^2 = 12$ implicitly with respect to x gives $2x + y + x\dfrac{dy}{dx} + 2y\dfrac{dy}{dx} = 0$, so $\dfrac{dy}{dx} = -\dfrac{2x+y}{x+2y}$.

At a highest or lowest point, $\dfrac{dy}{dx} = 0 \quad \Leftrightarrow \quad y = -2x$. Substituting $-2x$ for y in the original equation gives

$x^2 + x(-2x) + (-2x)^2 = 12$, so $3x^2 = 12$ and $x = \pm 2$. If $x = 2$, then $y = -2x = -4$, and if $x = -2$ then $y = 4$. Thus,

the highest and lowest points are $(-2, 4)$ and $(2, -4)$.

9. Let $L = \lim\limits_{x \to 0} \dfrac{ax^2 + \sin bx + \sin cx + \sin dx}{3x^2 + 5x^4 + 7x^6}$. Now L has the indeterminate form of type $\frac{0}{0}$, so we can apply l'Hospital's

Rule. $L = \lim\limits_{x \to 0} \dfrac{2ax + b \cos bx + c \cos cx + d \cos dx}{6x + 20x^3 + 42x^5}$. The denominator approaches 0 as $x \to 0$, so the numerator must also

approach 0 (because the limit exists). But the numerator approaches $0 + b + c + d$, so $b + c + d = 0$. Apply l'Hospital's Rule

again. $L = \lim\limits_{x \to 0} \dfrac{2a - b^2 \sin bx - c^2 \sin cx - d^2 \sin dx}{6 + 60x^2 + 210x^4} = \dfrac{2a - 0}{6 + 0} = \dfrac{2a}{6}$, which must equal 8.

$\dfrac{2a}{6} = 8 \;\Rightarrow\; a = 24$. Thus, $a + b + c + d = a + (b + c + d) = 24 + 0 = 24$.

11. $f(x) = \left(a^2 + a - 6\right) \cos 2x + (a - 2)x + \cos 1 \;\Rightarrow\; f'(x) = -\left(a^2 + a - 6\right) \sin 2x \,(2) + (a - 2)$. The derivative exists

for all x, so the only possible critical points will occur where $f'(x) = 0 \;\Leftrightarrow\; 2(a - 2)(a + 3) \sin 2x = a - 2 \;\Leftrightarrow\;$

either $a = 2$ or $2(a + 3) \sin 2x = 1$, with the latter implying that $\sin 2x = \dfrac{1}{2(a + 3)}$. Since the range of $\sin 2x$ is $[-1, 1]$,

this equation has no solution whenever either $\dfrac{1}{2(a + 3)} < -1$ or $\dfrac{1}{2(a + 3)} > 1$. Solving these inequalities, we get

$-\frac{7}{2} < a < -\frac{5}{2}$.

13.

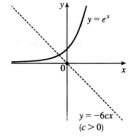

$y = cx^3 + e^x \;\Rightarrow\; y' = 3cx^2 + e^x \;\Rightarrow\; y'' = 6cx + e^x$. The curve will have

inflection points when y'' changes sign. $y'' = 0 \;\Rightarrow\; -6cx = e^x$, so y'' will change

sign when the line $y = -6cx$ intersects the curve $y = e^x$ (but is not tangent to it).

Note that if $c = 0$, the curve is just $y = e^x$, which has no inflection point.

The first figure shows that for $c > 0$, $y = -6cx$ will intersect $y = e^x$ once, so

$y = cx^3 + e^x$ will have one inflection point.

The second figure shows that for $c < 0$, the line $y = -6cx$ can intersect the curve

$y = e^x$ in two points (two inflection points), be tangent to it (no inflection point), or not

intersect it (no inflection point). The tangent line at (a, e^a) has slope e^a, but from the

diagram we see that the slope is $\dfrac{e^a}{a}$. So $\dfrac{e^a}{a} = e^a \;\Rightarrow\; a = 1$. Thus, the slope is e.

The line $y = -6cx$ must have slope greater than e, so $-6c > e \;\Rightarrow\; c < -e/6$.

Therefore, the curve $y = cx^3 + e^x$ will have one inflection point if $c > 0$ and two inflection points if $c < -e/6$.

15. $y = x^2 \;\Rightarrow\; y' = 2x$, so the slope of the tangent line at $P(a, a^2)$ is $2a$ and the slope of the normal line is $-\dfrac{1}{2a}$ for $a \neq 0$.

An equation of the normal line is $y - a^2 = -\dfrac{1}{2a}(x - a)$. Substitute x^2 for y to find the x-coordinates of the two points of intersection of the parabola and the normal line. $x^2 - a^2 = -\dfrac{x}{2a} + \dfrac{1}{2}$ $\Rightarrow$ $2ax^2 + x - 2a^3 - a = 0$ $\Rightarrow$

$$x = \frac{-1 \pm \sqrt{1 - 4(2a)(-2a^3 - a)}}{2(2a)} = \frac{-1 \pm \sqrt{1 + 16a^4 + 8a^2}}{4a} = \frac{-1 \pm \sqrt{(4a^2 + 1)^2}}{4a} = \frac{-1 \pm (4a^2 + 1)}{4a}$$

$$= \frac{4a^2}{4a} \quad \text{or} \quad \frac{-4a^2 - 2}{4a}, \text{ or equivalently, } a \text{ or } -a - \frac{1}{2a}.$$

So the point Q has coordinates $\left(-a - \dfrac{1}{2a}, \left(-a - \dfrac{1}{2a}\right)^2\right)$. The square S of the distance from P to Q is given by

$$S = \left(-a - \frac{1}{2a} - a\right)^2 + \left[\left(-a - \frac{1}{2a}\right)^2 - a^2\right]^2 = \left(-2a - \frac{1}{2a}\right)^2 + \left[\left(a^2 + 1 + \frac{1}{4a^2}\right) - a^2\right]^2$$

$$= \left(4a^2 + 2 + \frac{1}{4a^2}\right) + \left(1 + \frac{1}{4a^2}\right)^2 = \left(4a^2 + 2 + \frac{1}{4a^2}\right) + 1 + \frac{2}{4a^2} + \frac{1}{16a^4} = 4a^2 + 3 + \frac{3}{4a^2} + \frac{1}{16a^4}$$

$S' = 8a - \dfrac{6}{4a^3} - \dfrac{4}{16a^5} = 8a - \dfrac{3}{2a^3} - \dfrac{1}{4a^5} = \dfrac{32a^6 - 6a^2 - 1}{4a^5}$. The only real positive zero of the equation $S' = 0$ is

$a = \dfrac{1}{\sqrt{2}}$. Since $S'' = 8 + \dfrac{9}{2a^4} + \dfrac{5}{4a^6} > 0$, $a = \dfrac{1}{\sqrt{2}}$ corresponds to the shortest possible length of the line segment PQ.

17. $A = (x_1, x_1^2)$ and $B = (x_2, x_2^2)$, where x_1 and x_2 are the solutions of the quadratic equation $x^2 = mx + b$. Let $P = (x, x^2)$ and set $A_1 = (x_1, 0)$, $B_1 = (x_2, 0)$, and $P_1 = (x, 0)$. Let $f(x)$ denote the area of triangle PAB. Then $f(x)$ can be expressed in terms of the areas of three trapezoids as follows:

$$f(x) = \text{area}\,(A_1 A B B_1) - \text{area}\,(A_1 A P P_1) - \text{area}\,(B_1 B P P_1)$$
$$= \tfrac{1}{2}\left(x_1^2 + x_2^2\right)(x_2 - x_1) - \tfrac{1}{2}\left(x_1^2 + x^2\right)(x - x_1) - \tfrac{1}{2}\left(x^2 + x_2^2\right)(x_2 - x)$$

After expanding and canceling terms, we get

$$f(x) = \tfrac{1}{2}\left(x_2 x_1^2 - x_1 x_2^2 - x x_1^2 + x_1 x^2 - x_2 x^2 + x x_2^2\right) = \tfrac{1}{2}\left[x_1^2(x_2 - x) + x_2^2(x - x_1) + x^2(x_1 - x_2)\right]$$

$$f'(x) = \tfrac{1}{2}\left[-x_1^2 + x_2^2 + 2x(x_1 - x_2)\right]. \quad f''(x) = \tfrac{1}{2}[2(x_1 - x_2)] = x_1 - x_2 < 0 \text{ since } x_2 > x_1.$$

$$f'(x) = 0 \quad \Rightarrow \quad 2x(x_1 - x_2) = x_1^2 - x_2^2 \quad \Rightarrow \quad x_P = \tfrac{1}{2}(x_1 + x_2).$$

$$f(x_P) = \tfrac{1}{2}\left(x_1^2\left[\tfrac{1}{2}(x_2 - x_1)\right] + x_2^2\left[\tfrac{1}{2}(x_2 - x_1)\right] + \tfrac{1}{4}(x_1 + x_2)^2(x_1 - x_2)\right)$$

$$= \tfrac{1}{2}\left[\tfrac{1}{2}(x_2 - x_1)(x_1^2 + x_2^2) - \tfrac{1}{4}(x_2 - x_1)(x_1 + x_2)^2\right] = \tfrac{1}{8}(x_2 - x_1)\left[2(x_1^2 + x_2^2) - (x_1^2 + 2x_1 x_2 + x_2^2)\right]$$

$$= \tfrac{1}{8}(x_2 - x_1)(x_1^2 - 2x_1 x_2 + x_2^2) = \tfrac{1}{8}(x_2 - x_1)(x_1 - x_2)^2 = \tfrac{1}{8}(x_2 - x_1)(x_2 - x_1)^2 = \tfrac{1}{8}(x_2 - x_1)^3$$

To put this in terms of m and b, we solve the system $y = x_1^2$ and $y = mx_1 + b$, giving us $x_1^2 - mx_1 - b = 0$ $\Rightarrow$

$x_1 = \tfrac{1}{2}\left(m - \sqrt{m^2 + 4b}\right)$. Similarly, $x_2 = \tfrac{1}{2}\left(m + \sqrt{m^2 + 4b}\right)$. The area is then $\tfrac{1}{8}(x_2 - x_1)^3 = \tfrac{1}{8}\left(\sqrt{m^2 + 4b}\right)^3$,

and is attained at the point $P\left(x_P, x_P^2\right) = P\left(\tfrac{1}{2}m, \tfrac{1}{4}m^2\right)$.

Note: Another way to get an expression for $f(x)$ is to use the formula for an area of a triangle in terms of the coordinates of the vertices: $f(x) = \tfrac{1}{2}\left[\left(x_2 x_1^2 - x_1 x_2^2\right) + \left(x_1 x^2 - x x_1^2\right) + \left(x x_2^2 - x_2 x^2\right)\right]$.

19. Let $s_A(t)$ and $s_B(t)$ be the position functions for cars A and B and let $f(t) = s_A(t) - s_B(t)$. Since A passed B twice (B passed A once), there must be three values of t such that $f(t) = 0$. Let these times be denoted t_1, t_2, and t_3. By the Mean Value Theorem, we know that for some number c_1 in (t_1, t_2), $f'(c_1) = \dfrac{f(t_2) - f(t_1)}{t_2 - t_1}$, but $f(t_2) - f(t_1) = 0$, so $f'(c_1) = 0 \iff s'_A(c_1) - s'_B(c_1) = 0 \iff v_A(c_1) - v_B(c_1) = 0 \implies v_A(c_1) = v_B(c_1)$.

By a similar argument, there exists some number c_2 in (t_2, t_3) such that $v_A(c_2) = v_B(c_2)$. Now let $g(t) = v_A(t) - v_B(t)$ and apply the Mean Value Theorem on $[c_1, c_2]$ with $c_1 < c < c_2$. $g'(c) = \dfrac{g(c_2) - g(c_1)}{c_2 - c_1}$, but $g(c_2) = g(c_1) = 0$, so $g'(c) = 0 \implies v'_A(c) - v'_B(c) = 0 \implies a_A(c) - a_B(c) = 0 \implies a_A(c) = a_B(c)$; that is, A and B had equal accelerations at $t = c$.

21.

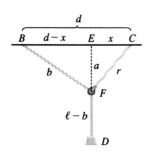

Let $a = |EF|$ and $b = |BF|$ as shown in the figure. Since $\ell = |BF| + |FD|$, $|FD| = \ell - b$. Now

$$|ED| = |EF| + |FD| = a + \ell - b$$

$$\sqrt{r^2 - x^2} + \ell - \sqrt{(d-x)^2 + a^2}$$

$$= \sqrt{r^2 - x^2} + \ell - \sqrt{(d-x)^2 + \left(\sqrt{r^2 - x^2}\right)^2}$$

$$= \sqrt{r^2 - x^2} + \ell - \sqrt{d^2 - 2dx + x^2 + r^2 - x^2}$$

Let $f(x) = \sqrt{r^2 - x^2} + \ell - \sqrt{d^2 + r^2 - 2dx}$.

$$f'(x) = \tfrac{1}{2}\left(r^2 - x^2\right)^{-1/2}(-2x) - \tfrac{1}{2}\left(d^2 + r^2 - 2dx\right)^{-1/2}(-2d) = \frac{-x}{\sqrt{r^2 - x^2}} + \frac{d}{\sqrt{d^2 + r^2 - 2dx}}.$$

$$f'(x) = 0 \implies \frac{x}{\sqrt{r^2 - x^2}} = \frac{d}{\sqrt{d^2 + r^2 - 2dx}} \implies \frac{x^2}{r^2 - x^2} = \frac{d^2}{d^2 + r^2 - 2dx} \implies$$

$$d^2 x^2 + r^2 x^2 - 2dx^3 = d^2 r^2 - d^2 x^2 \implies 0 = 2dx^3 - 2d^2 x^2 - r^2 x^2 + d^2 r^2 \implies$$

$$0 = 2dx^2(x - d) - r^2(x^2 - d^2) \implies 0 = 2dx^2(x - d) - r^2(x + d)(x - d) \implies 0 = (x - d)\left[2dx^2 - r^2(x + d)\right]$$

But $d > r > x$, so $x \ne d$. Thus, we solve $2dx^2 - r^2 x - dr^2 = 0$ for x:

$$x = \frac{-(-r^2) \pm \sqrt{(-r^2)^2 - 4(2d)(-dr^2)}}{2(2d)} = \frac{r^2 \pm \sqrt{r^4 + 8d^2 r^2}}{4d}. \text{ Because } \sqrt{r^4 + 8d^2 r^2} > r^2, \text{ the "negative" can be}$$

discarded. Thus, $x = \dfrac{r^2 + \sqrt{r^2}\sqrt{r^2 + 8d^2}}{4d} = \dfrac{r^2 + r\sqrt{r^2 + 8d^2}}{4d} \quad [r > 0] \quad = \dfrac{r}{4d}\left(r + \sqrt{r^2 + 8d^2}\right)$. The maximum value of $|ED|$ occurs at this value of x.

23.

By similar triangles, $\dfrac{r}{5} = \dfrac{h}{16} \;\Rightarrow\; r = \dfrac{5h}{16}$. The volume of the cone is

$$V = \tfrac{1}{3}\pi r^2 h = \tfrac{1}{3}\pi\left(\frac{5h}{16}\right)^2 h = \frac{25\pi}{768}h^3, \text{ so } \frac{dV}{dt} = \frac{25\pi}{256}h^2\frac{dh}{dt}. \text{ Now the rate of}$$

change of the volume is also equal to the difference of what is being added

($2\ \mathrm{cm^3/min}$) and what is oozing out ($k\pi rl$, where πrl is the area of the cone and k

is a proportionality constant). Thus, $\dfrac{dV}{dt} = 2 - k\pi rl$.

Equating the two expressions for $\dfrac{dV}{dt}$ and substituting $h = 10$, $\dfrac{dh}{dt} = -0.3$, $r = \dfrac{5(10)}{16} = \dfrac{25}{8}$, and $\dfrac{l}{\sqrt{281}} = \dfrac{10}{16} \;\Leftrightarrow\;$

$l = \dfrac{5}{8}\sqrt{281}$, we get $\dfrac{25\pi}{256}(10)^2(-0.3) = 2 - k\pi\dfrac{25}{8}\cdot\dfrac{5}{8}\sqrt{281} \;\Leftrightarrow\; \dfrac{125k\pi\sqrt{281}}{64} = 2 + \dfrac{750\pi}{256}$. Solving for k gives us

$k = \dfrac{256 + 375\pi}{250\pi\sqrt{281}}$. To maintain a certain height, the rate of oozing, $k\pi rl$, must equal the rate of the liquid being poured in;

that is, $\dfrac{dV}{dt} = 0$. Thus, the rate at which we should pour the liquid into the container is

$$k\pi rl = \frac{256 + 375\pi}{250\pi\sqrt{281}}\cdot\pi\cdot\frac{25}{8}\cdot\frac{5\sqrt{281}}{8} = \frac{256 + 375\pi}{128} \approx 11.204\ \mathrm{cm^3/min}$$

5 ☐ INTEGRALS

5.1 Areas and Distances

1. (a) Since f is *increasing*, we can obtain a *lower* estimate by using *left* endpoints. We are instructed to use four rectangles, so $n = 4$.

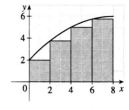

$$L_4 = \sum_{i=1}^{4} f(x_{i-1})\,\Delta x \quad \left[\Delta x = \frac{b-a}{n} = \frac{8-0}{4} = 2\right]$$

$$= f(x_0)\cdot 2 + f(x_1)\cdot 2 + f(x_2)\cdot 2 + f(x_3)\cdot 2$$

$$= 2[f(0) + f(2) + f(4) + f(6)]$$

$$= 2(2 + 3.75 + 5 + 5.75) = 2(16.5) = 33$$

Since f is *increasing*, we can obtain a *upper* estimate by using *right* endpoints.

$$R_4 = \sum_{i=1}^{4} f(x_i)\,\Delta x$$

$$= f(x_1)\cdot 2 + f(x_2)\cdot 2 + f(x_3)\cdot 2 + f(x_4)\cdot 2$$

$$= 2[f(2) + f(4) + f(6) + f(8)]$$

$$= 2(3.75 + 5 + 5.75 + 6) = 2(20.5) = 41$$

Comparing R_4 to L_4, we see that we have added the area of the rightmost upper rectangle, $f(8)\cdot 2$, to the sum and subtracted the area of the leftmost lower rectangle, $f(0)\cdot 2$, from the sum.

(b) $L_8 = \sum_{i=1}^{8} f(x_{i-1})\Delta x \quad \left[\Delta x = \frac{8-0}{8} = 1\right]$

$$= 1[f(x_0) + f(x_1) + \cdots + f(x_7)]$$

$$= f(0) + f(1) + \cdots + f(7)$$

$$\approx 2 + 2.9 + 3.75 + 4.4 + 5 + 5.4 + 5.75 + 5.9$$

$$= 35.1$$

$$R_8 = \sum_{i=1}^{8} f(x_i)\Delta x = f(1) + f(2) + \cdots + f(8)$$

$$= L_8 + 1\cdot f(8) - 1\cdot f(0) \quad \begin{bmatrix} \text{add rightmost upper rectangle,} \\ \text{subtract leftmost lower rectangle} \end{bmatrix}$$

$$= 35.1 + 6 - 2 = 39.1$$

3. (a) $R_4 = \sum_{i=1}^{4} f(x_i)\,\Delta x \quad \left[\Delta x = \frac{\pi/2 - 0}{4} = \frac{\pi}{8}\right] \quad = \left[\sum_{i=1}^{4} f(x_i)\right]\Delta x$

$$= [f(x_1) + f(x_2) + f(x_3) + f(x_4)]\,\Delta x$$

$$= \left[\cos\frac{\pi}{8} + \cos\frac{2\pi}{8} + \cos\frac{3\pi}{8} + \cos\frac{4\pi}{8}\right]\frac{\pi}{8}$$

$$\approx (0.9239 + 0.7071 + 0.3827 + 0)\frac{\pi}{8} \approx 0.7908$$

Since f is *decreasing* on $[0, \pi/2]$, an *underestimate* is obtained by using the *right* endpoint approximation, R_4.

(b) $L_4 = \sum_{i=1}^{4} f(x_{i-1}) \Delta x = \left[\sum_{i=1}^{4} f(x_{i-1}) \right] \Delta x$

$= [f(x_0) + f(x_1) + f(x_2) + f(x_3)] \, \Delta x$

$= \left[\cos 0 + \cos \frac{\pi}{8} + \cos \frac{2\pi}{8} + \cos \frac{3\pi}{8} \right] \frac{\pi}{8}$

$\approx (1 + 0.9239 + 0.7071 + 0.3827) \frac{\pi}{8} \approx 1.1835$

L_4 is an overestimate. Alternatively, we could just add the area of the leftmost upper rectangle and subtract the area of the rightmost lower rectangle; that is, $L_4 = R_4 + f(0) \cdot \frac{\pi}{8} - f\left(\frac{\pi}{2}\right) \cdot \frac{\pi}{8}$.

5. (a) $f(x) = 1 + x^2$ and $\Delta x = \dfrac{2 - (-1)}{3} = 1 \;\Rightarrow$

$R_3 = 1 \cdot f(0) + 1 \cdot f(1) + 1 \cdot f(2) = 1 \cdot 1 + 1 \cdot 2 + 1 \cdot 5 = 8.$

$\Delta x = \dfrac{2 - (-1)}{6} = 0.5 \;\Rightarrow$

$R_6 = 0.5[f(-0.5) + f(0) + f(0.5) + f(1) + f(1.5) + f(2)]$

$= 0.5(1.25 + 1 + 1.25 + 2 + 3.25 + 5)$

$= 0.5(13.75) = 6.875$

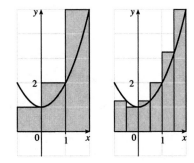

(b) $L_3 = 1 \cdot f(-1) + 1 \cdot f(0) + 1 \cdot f(1) = 1 \cdot 2 + 1 \cdot 1 + 1 \cdot 2 = 5$

$L_6 = 0.5[f(-1) + f(-0.5) + f(0) + f(0.5) + f(1) + f(1.5)]$

$= 0.5(2 + 1.25 + 1 + 1.25 + 2 + 3.25)$

$= 0.5(10.75) = 5.375$

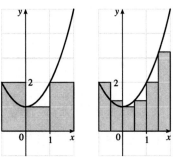

(c) $M_3 = 1 \cdot f(-0.5) + 1 \cdot f(0.5) + 1 \cdot f(1.5)$

$= 1 \cdot 1.25 + 1 \cdot 1.25 + 1 \cdot 3.25 = 5.75$

$M_6 = 0.5[f(-0.75) + f(-0.25) + f(0.25)$
$\qquad + f(0.75) + f(1.25) + f(1.75)]$

$= 0.5(1.5625 + 1.0625 + 1.0625 + 1.5625 + 2.5625 + 4.0625)$

$= 0.5(11.875) = 5.9375$

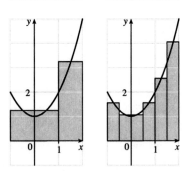

(d) M_6 appears to be the best estimate.

7. Here is one possible algorithm (ordered sequence of operations) for calculating the sums:

1 Let SUM = 0, X_MIN = 0, X_MAX = 1, N = 10 (depending on which sum we are calculating),

DELTA_X = (X_MAX - X_MIN)/N, and RIGHT_ENDPOINT = X_MIN + DELTA_X.

2 Repeat steps 2a, 2b in sequence until RIGHT_ENDPOINT > X_MAX.

2a Add (RIGHT_ENDPOINT)^4 to SUM.

Add DELTA_X to RIGHT_ENDPOINT.

At the end of this procedure, (DELTA_X)·(SUM) is equal to the answer we are looking for. We find that

$$R_{10} = \frac{1}{10} \sum_{i=1}^{10} \left(\frac{i}{10} \right)^4 \approx 0.2533, \; R_{30} = \frac{1}{30} \sum_{i=1}^{30} \left(\frac{i}{30} \right)^4 \approx 0.2170, \; R_{50} = \frac{1}{50} \sum_{i=1}^{50} \left(\frac{i}{50} \right)^4 \approx 0.2101, \text{ and}$$

$$R_{100} = \frac{1}{100} \sum_{i=1}^{100} \left(\frac{i}{100} \right)^4 \approx 0.2050. \text{ It appears that the exact area is } 0.2. \quad \text{The following display shows the program}$$

SUMRIGHT and its output from a TI-83 Plus calculator. To generalize the program, we have input (rather than assign)

values for Xmin, Xmax, and N. Also, the function, x^4, is assigned to Y_1, enabling us to evaluate any right sum merely

by changing Y_1 and running the program.

9. In Maple, we have to perform a number of steps before getting a numerical answer. After loading the student package

   [command: `with(student);`] we use the command

   `left_sum:=leftsum(1/(x^2+1),x=0..1,10` [or 30, or 50]); which gives us the expression in summation

   notation. To get a numerical approximation to the sum, we use `evalf(left_sum);`. Mathematica does not have a special

   command for these sums, so we must type them in manually. For example, the first left sum is given by

   `(1/10)*Sum[1/(((i-1)/10)^2+1)],{i,1,10}]`, and we use the N command on the resulting output to get a

   numerical approximation.

   In Derive, we use the `LEFT_RIEMANN` command to get the left sums, but must define the right sums ourselves.

   (We can define a new function using `LEFT_RIEMANN` with k ranging from 1 to n instead of from 0 to $n-1$.)

   (a) With $f(x) = \dfrac{1}{x^2 + 1}, 0 \le x \le 1$, the left sums are of the form $L_n = \dfrac{1}{n} \sum\limits_{i=1}^{n} \dfrac{1}{\left(\frac{i-1}{n} \right)^2 + 1}$. Specifically, $L_{10} \approx 0.8100$,

   $L_{30} \approx 0.7937$, and $L_{50} \approx 0.7904$. The right sums are of the form $R_n = \dfrac{1}{n} \sum\limits_{i=1}^{n} \dfrac{1}{\left(\frac{i}{n} \right)^2 + 1}$. Specifically, $R_{10} \approx 0.7600$,

   $R_{30} \approx 0.7770$, and $R_{50} \approx 0.7804$.

(b) In Maple, we use the `leftbox` (with the same arguments as `left_sum`) and `rightbox` commands to generate the graphs.

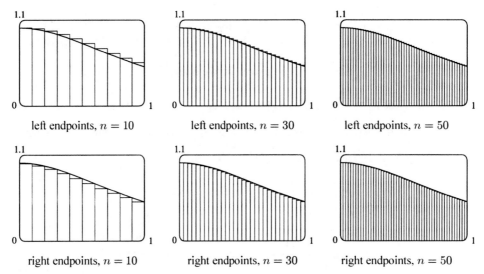

left endpoints, $n = 10$ left endpoints, $n = 30$ left endpoints, $n = 50$

right endpoints, $n = 10$ right endpoints, $n = 30$ right endpoints, $n = 50$

(c) We know that since $y = 1/(x^2 + 1)$ is a decreasing function on $(0, 1)$, all of the left sums are larger than the actual area, and all of the right sums are smaller than the actual area. Since the left sum with $n = 50$ is about $0.7904 < 0.791$ and the right sum with $n = 50$ is about $0.7804 > 0.780$, we conclude that $0.780 < R_{50} <$ exact area $< L_{50} < 0.791$, so the exact area is between 0.780 and 0.791.

11. Since v is an increasing function, L_6 will give us a lower estimate and R_6 will give us an upper estimate.

$$L_6 = (0 \text{ ft/s})(0.5 \text{ s}) + (6.2)(0.5) + (10.8)(0.5) + (14.9)(0.5) + (18.1)(0.5) + (19.4)(0.5) = 0.5(69.4) = 34.7 \text{ ft}$$

$$R_6 = 0.5(6.2 + 10.8 + 14.9 + 18.1 + 19.4 + 20.2) = 0.5(89.6) = 44.8 \text{ ft}$$

13. Lower estimate for oil leakage: $R_5 = (7.6 + 6.8 + 6.2 + 5.7 + 5.3)(2) = (31.6)(2) = 63.2$ L.

Upper estimate for oil leakage: $L_5 = (8.7 + 7.6 + 6.8 + 6.2 + 5.7)(2) = (35)(2) = 70$ L.

15. For a decreasing function, using left endpoints gives us an overestimate and using right endpoints results in an underestimate. We will use M_6 to get an estimate. $\Delta t = 1$, so

$$M_6 = 1[v(0.5) + v(1.5) + v(2.5) + v(3.5) + v(4.5) + v(5.5)] \approx 55 + 40 + 28 + 18 + 10 + 4 = 155 \text{ ft}$$

For a very rough check on the above calculation, we can draw a line from $(0, 70)$ to $(6, 0)$ and calculate the area of the triangle: $\frac{1}{2}(70)(6) = 210$. This is clearly an overestimate, so our midpoint estimate of 155 is reasonable.

17. $f(x) = \dfrac{2x}{x^2 + 1}$, $1 \le x \le 3$. $\Delta x = (3 - 1)/n = 2/n$ and $x_i = 1 + i\Delta x = 1 + 2i/n$.

$$A = \lim_{n \to \infty} R_n = \lim_{n \to \infty} \sum_{i=1}^{n} f(x_i)\Delta x = \lim_{n \to \infty} \sum_{i=1}^{n} \frac{2(1 + 2i/n)}{(1 + 2i/n)^2 + 1} \cdot \frac{2}{n}.$$

19. $f(x) = x \cos x$, $0 \le x \le \frac{\pi}{2}$. $\Delta x = (\frac{\pi}{2} - 0)/n = \frac{\pi}{2}/n$ and $x_i = 0 + i\,\Delta x = \frac{\pi}{2} i/n$.

$$A = \lim_{n \to \infty} R_n = \lim_{n \to \infty} \sum_{i=1}^{n} f(x_i)\,\Delta x = \lim_{n \to \infty} \sum_{i=1}^{n} \frac{i\pi}{2n} \cos\left(\frac{i\pi}{2n}\right) \cdot \frac{\pi}{2n}.$$

21. $\displaystyle\lim_{n \to \infty} \sum_{i=1}^{n} \frac{\pi}{4n} \tan \frac{i\pi}{4n}$ can be interpreted as the area of the region lying under the graph of $y = \tan x$ on the interval $\left[0, \frac{\pi}{4}\right]$,

since for $y = \tan x$ on $\left[0, \frac{\pi}{4}\right]$ with $\Delta x = \dfrac{\pi/4 - 0}{n} = \dfrac{\pi}{4n}$, $x_i = 0 + i\,\Delta x = \dfrac{i\pi}{4n}$, and $x_i^* = x_i$, the expression for the area is

$$A = \lim_{n \to \infty} \sum_{i=1}^{n} f(x_i^*)\,\Delta x = \lim_{n \to \infty} \sum_{i=1}^{n} \tan\left(\frac{i\pi}{4n}\right) \frac{\pi}{4n}.$$ Note that this answer is not unique, since the expression for the area is

the same for the function $y = \tan(x - k\pi)$ on the interval $\left[k\pi, k\pi + \frac{\pi}{4}\right]$, where k is any integer.

23. (a) Since f is an increasing function, L_n is an underestimate of A and R_n is an overestimate of A. Thus, A, L_n, and R_n are

related by the inequality $L_n < A < R_n$.

(b) $R_n = f(x_1)\Delta x + f(x_2)\Delta x + \cdots + f(x_n)\Delta x$

$\quad\quad L_n = f(x_0)\Delta x + f(x_1)\Delta x + \cdots + f(x_{n-1})\Delta x$

$\quad R_n - L_n = f(x_n)\Delta x - f(x_0)\Delta x = \Delta x[f(x_n) - f(x_0)] = \dfrac{b-a}{n}[f(b) - f(a)]$

(c) $A > L_n$, so $R_n - A < R_n - L_n$; that is,

$$R_n - A < \frac{b-a}{n}[f(b) - f(a)]$$

25. (a) $y = f(x) = x^5$. $\Delta x = \dfrac{2-0}{n} = \dfrac{2}{n}$ and $x_i = 0 + i\,\Delta x = \dfrac{2i}{n}$.

$$A = \lim_{n \to \infty} R_n = \lim_{n \to \infty} \sum_{i=1}^{n} f(x_i)\,\Delta x = \lim_{n \to \infty} \sum_{i=1}^{n} \left(\frac{2i}{n}\right)^5 \cdot \frac{2}{n} = \lim_{n \to \infty} \sum_{i=1}^{n} \frac{32i^5}{n^5} \cdot \frac{2}{n} = \lim_{n \to \infty} \frac{64}{n^6} \sum_{i=1}^{n} i^5.$$

(b) $\displaystyle\sum_{i=1}^{n} i^5 \overset{\text{CAS}}{=} \frac{n^2(n+1)^2(2n^2 + 2n - 1)}{12}$

(c) $\displaystyle\lim_{n \to \infty} \frac{64}{n^6} \cdot \frac{n^2(n+1)^2(2n^2+2n-1)}{12} = \frac{64}{12} \lim_{n \to \infty} \frac{(n^2+2n+1)(2n^2+2n-1)}{n^2 \cdot n^2}$

$$= \frac{16}{3} \lim_{n \to \infty} \left(1 + \frac{2}{n} + \frac{1}{n^2}\right)\left(2 + \frac{2}{n} - \frac{1}{n^2}\right) = \tfrac{16}{3} \cdot 1 \cdot 2 = \tfrac{32}{3}$$

27. $y = f(x) = \cos x$. $\Delta x = \dfrac{b-0}{n} = \dfrac{b}{n}$ and $x_i = 0 + i\,\Delta x = \dfrac{bi}{n}$.

$$A = \lim_{n \to \infty} R_n = \lim_{n \to \infty} \sum_{i=1}^{n} f(x_i)\,\Delta x = \lim_{n \to \infty} \sum_{i=1}^{n} \cos\left(\frac{bi}{n}\right) \cdot \frac{b}{n} \overset{\text{CAS}}{=} \lim_{n \to \infty} \left[\frac{b \sin\left(b\left(\frac{1}{2n} + 1\right)\right)}{2n \sin\left(\dfrac{b}{2n}\right)} - \frac{b}{2n}\right] \overset{\text{CAS}}{=} \sin b$$

If $b = \frac{\pi}{2}$, then $A = \sin \frac{\pi}{2} = 1$.

5.2   The Definite Integral

1. $f(x) = 3 - \frac{1}{2}x,\ 2 \le x \le 14.$ $\Delta x = \dfrac{b-a}{n} = \dfrac{14-2}{6} = 2.$

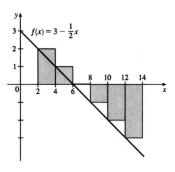

Since we are using left endpoints, $x_i^* = x_{i-1}$.

$$L_6 = \sum_{i=1}^{6} f(x_{i-1})\,\Delta x$$

$$= (\Delta x)\,[f(x_0) + f(x_1) + f(x_2) + f(x_3) + f(x_4) + f(x_5)]$$

$$= 2[f(2) + f(4) + f(6) + f(8) + f(10) + f(12)]$$

$$= 2[2 + 1 + 0 + (-1) + (-2) + (-3)] = 2(-3) = -6$$

The Riemann sum represents the sum of the areas of the two rectangles above the x-axis minus the sum of the areas of the

three rectangles below the x-axis; that is, the *net area* of the rectangles with respect to the x-axis.

3. $f(x) = e^x - 2,\ 0 \le x \le 2.$ $\Delta x = \dfrac{b-a}{n} = \dfrac{2-0}{4} = \dfrac{1}{2}.$

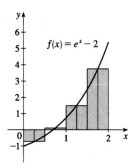

Since we are using midpoints, $x_i^* = \overline{x}_i = \frac{1}{2}(x_{i-1} + x_i).$

$$M_4 = \sum_{i=1}^{4} f(\overline{x}_i)\,\Delta x = (\Delta x)\,[f(\overline{x}_1) + f(\overline{x}_2) + f(\overline{x}_3) + f(\overline{x}_4)]$$

$$= \tfrac{1}{2}\big[f\big(\tfrac{1}{4}\big) + f\big(\tfrac{3}{4}\big) + f\big(\tfrac{5}{4}\big) + f\big(\tfrac{7}{4}\big)\big]$$

$$= \tfrac{1}{2}\big[(e^{1/4} - 2) + (e^{3/4} - 2) + (e^{5/4} - 2) + (e^{7/4} - 2)\big]$$

$$\approx 2.322986$$

The Riemann sum represents the sum of the areas of the three rectangles above the x-axis minus the area of the rectangle

below the x-axis; that is, the *net area* of the rectangles with respect to the x-axis.

5. $\Delta x = (b-a)/n = (8-0)/4 = 8/4 = 2.$

(a) Using the right endpoints to approximate $\int_0^8 f(x)\,dx$, we have

$$\sum_{i=1}^{4} f(x_i)\,\Delta x = 2[f(2) + f(4) + f(6) + f(8)] \approx 2[1 + 2 + (-2) + 1] = 4.$$

(b) Using the left endpoints to approximate $\int_0^8 f(x)\,dx$, we have

$$\sum_{i=1}^{4} f(x_{i-1})\,\Delta x = 2[f(0) + f(2) + f(4) + f(6)] \approx 2[2 + 1 + 2 + (-2)] = 6.$$

(c) Using the midpoint of each subinterval to approximate $\int_0^8 f(x)\,dx$, we have

$$\sum_{i=1}^{4} f(\overline{x}_i)\,\Delta x = 2[f(1) + f(3) + f(5) + f(7)] \approx 2[3 + 2 + 1 + (-1)] = 10.$$

7. Since f is increasing, $L_5 \leq \int_{10}^{30} f(x)\,dx \leq R_5$.

$$\text{Lower estimate} = L_5 = \sum_{i=1}^{5} f(x_{i-1})\Delta x = 4[f(10) + f(14) + f(18) + f(22) + f(26)]$$

$$= 4[-12 + (-6) + (-2) + 1 + 3] = 4(-16) = -64$$

$$\text{Upper estimate} = R_5 = \sum_{i=1}^{5} f(x)\Delta x = 4[f(14) + f(18) + f(22) + f(26) + f(30)]$$

$$= 4[-6 + (-2) + 1 + 3 + 8] = 4(4) = 16$$

9. $\Delta x = (10 - 2)/4 = 2$, so the endpoints are 2, 4, 6, 8, and 10, and the midpoints are 3, 5, 7, and 9. The Midpoint Rule

gives $\int_{2}^{10} \sqrt{x^3 + 1}\,dx \approx \sum_{i=1}^{4} f(\overline{x}_i)\,\Delta x = 2(\sqrt{3^3 + 1} + \sqrt{5^3 + 1} + \sqrt{7^3 + 1} + \sqrt{9^3 + 1}) \approx 124.1644$.

11. $\Delta x = (1 - 0)/5 = 0.2$, so the endpoints are 0, 0.2, 0.4, 0.6, 0.8, and 1, and the midpoints are 0.1, 0.3, 0.5, 0.7, and 0.9.

The Midpoint Rule gives

$$\int_{0}^{1} \sin(x^2)\,dx \approx \sum_{i=1}^{5} f(\overline{x}_i)\,\Delta x = 0.2[\sin(0.1)^2 + \sin(0.3)^2 + \sin(0.5)^2 + \sin(0.7)^2 + \sin(0.9)^2] \approx 0.3084.$$

13. In Maple, we use the command `with(student);` to load the sum and box commands, then

`m:=middlesum(sin(x^2),x=0..1,5);` which gives us the sum in summation notation, then `M:=evalf(m);` which

gives $M_5 \approx 0.30843908$, confirming the result of Exercise 11. The command `middlebox(sin(x^2),x=0..1,5)`

generates the graph. Repeating for $n = 10$ and $n = 20$ gives $M_{10} \approx 0.30981629$ and $M_{20} \approx 0.31015563$.

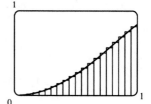

15. We'll create the table of values to approximate $\int_{0}^{\pi} \sin x\,dx$ by using the

program in the solution to Exercise 5.1.7 with $Y_1 = \sin x$, $X\min = 0$,

$X\max = \pi$, and $n = 5$, 10, 50, and 100.

The values of R_n appear to be approaching 2.

n	R_n
5	1.933766
10	1.983524
50	1.999342
100	1.999836

17. On $[2, 6]$, $\lim\limits_{n \to \infty} \sum\limits_{i=1}^{n} x_i \ln(1 + x_i^2)\,\Delta x = \int_{2}^{6} x \ln(1 + x^2)\,dx$.

19. On $[1, 8]$, $\lim\limits_{n \to \infty} \sum\limits_{i=1}^{n} \sqrt{2x_i^* + (x_i^*)^2}\,\Delta x = \int_{1}^{8} \sqrt{2x + x^2}\,dx$.

21. Note that $\Delta x = \dfrac{5 - (-1)}{n} = \dfrac{6}{n}$ and $x_i = -1 + i\,\Delta x = -1 + \dfrac{6i}{n}$.

$$\int_{-1}^{5} (1 + 3x)\,dx = \lim_{n\to\infty} \sum_{i=1}^{n} f(x_i)\,\Delta x = \lim_{n\to\infty} \sum_{i=1}^{n} \left[1 + 3\left(-1 + \frac{6i}{n}\right)\right]\frac{6}{n} = \lim_{n\to\infty} \frac{6}{n} \sum_{i=1}^{n} \left[-2 + \frac{18i}{n}\right]$$

$$= \lim_{n\to\infty} \frac{6}{n} \left[\sum_{i=1}^{n}(-2) + \sum_{i=1}^{n}\frac{18i}{n}\right] = \lim_{n\to\infty} \frac{6}{n}\left[-2n + \frac{18}{n}\sum_{i=1}^{n} i\right]$$

$$= \lim_{n\to\infty} \frac{6}{n}\left[-2n + \frac{18}{n}\cdot\frac{n(n+1)}{2}\right] = \lim_{n\to\infty}\left[-12 + \frac{108}{n^2}\cdot\frac{n(n+1)}{2}\right]$$

$$= \lim_{n\to\infty}\left[-12 + 54\frac{n+1}{n}\right] = \lim_{n\to\infty}\left[-12 + 54\left(1 + \frac{1}{n}\right)\right] = -12 + 54\cdot 1 = 42$$

23. Note that $\Delta x = \dfrac{2 - 0}{n} = \dfrac{2}{n}$ and $x_i = 0 + i\,\Delta x = \dfrac{2i}{n}$.

$$\int_{0}^{2} (2 - x^2)\,dx = \lim_{n\to\infty} \sum_{i=1}^{n} f(x_i)\,\Delta x = \lim_{n\to\infty} \sum_{i=1}^{n}\left(2 - \frac{4i^2}{n^2}\right)\left(\frac{2}{n}\right) = \lim_{n\to\infty}\frac{2}{n}\left[\sum_{i=1}^{n} 2 - \frac{4}{n^2}\sum_{i=1}^{n} i^2\right]$$

$$= \lim_{n\to\infty}\frac{2}{n}\left(2n - \frac{4}{n^2}\sum_{i=1}^{n} i^2\right) = \lim_{n\to\infty}\left[4 - \frac{8}{n^3}\cdot\frac{n(n+1)(2n+1)}{6}\right]$$

$$= \lim_{n\to\infty}\left(4 - \frac{4}{3}\cdot\frac{n+1}{n}\cdot\frac{2n+1}{n}\right) = \lim_{n\to\infty}\left[4 - \frac{4}{3}\left(1 + \frac{1}{n}\right)\left(2 + \frac{1}{n}\right)\right] = 4 - \frac{4}{3}\cdot 1\cdot 2 = \frac{4}{3}$$

25. Note that $\Delta x = \dfrac{2 - 1}{n} = \dfrac{1}{n}$ and $x_i = 1 + i\,\Delta x = 1 + i(1/n) = 1 + i/n$.

$$\int_{1}^{2} x^3\,dx = \lim_{n\to\infty} \sum_{i=1}^{n} f(x_i)\,\Delta x = \lim_{n\to\infty} \sum_{i=1}^{n}\left(1 + \frac{i}{n}\right)^3\left(\frac{1}{n}\right) = \lim_{n\to\infty}\frac{1}{n}\sum_{i=1}^{n}\left(\frac{n+i}{n}\right)^3$$

$$= \lim_{n\to\infty}\frac{1}{n^4}\sum_{i=1}^{n}\left(n^3 + 3n^2 i + 3ni^2 + i^3\right) = \lim_{n\to\infty}\frac{1}{n^4}\left[\sum_{i=1}^{n} n^3 + \sum_{i=1}^{n} 3n^2 i + \sum_{i=1}^{n} 3ni^2 + \sum_{i=1}^{n} i^3\right]$$

$$= \lim_{n\to\infty}\frac{1}{n^4}\left[n\cdot n^3 + 3n^2\sum_{i=1}^{n} i + 3n\sum_{i=1}^{n} i^2 + \sum_{i=1}^{n} i^3\right]$$

$$= \lim_{n\to\infty}\left[1 + \frac{3}{n^2}\cdot\frac{n(n+1)}{2} + \frac{3}{n^3}\cdot\frac{n(n+1)(2n+1)}{6} + \frac{1}{n^4}\cdot\frac{n^2(n+1)^2}{4}\right]$$

$$= \lim_{n\to\infty}\left[1 + \frac{3}{2}\cdot\frac{n+1}{n} + \frac{1}{2}\cdot\frac{n+1}{n}\cdot\frac{2n+1}{n} + \frac{1}{4}\cdot\frac{(n+1)^2}{n^2}\right]$$

$$= \lim_{n\to\infty}\left[1 + \frac{3}{2}\left(1 + \frac{1}{n}\right) + \frac{1}{2}\left(1 + \frac{1}{n}\right)\left(2 + \frac{1}{n}\right) + \frac{1}{4}\left(1 + \frac{1}{n}\right)^2\right] = 1 + \frac{3}{2} + \frac{1}{2}\cdot 2 + \frac{1}{4} = 3.75$$

27. $f(x) = \dfrac{x}{1 + x^5}$, $a = 2$, $b = 6$, and $\Delta x = \dfrac{6 - 2}{n} = \dfrac{4}{n}$. Using Theorem 4, we get $x_i^* = x_i = 2 + i\,\Delta x = 2 + \dfrac{4i}{n}$,

so $\displaystyle\int_{2}^{6} \frac{x}{1 + x^5}\,dx = \lim_{n\to\infty} R_n = \lim_{n\to\infty} \sum_{i=1}^{n} \frac{2 + \dfrac{4i}{n}}{1 + \left(2 + \dfrac{4i}{n}\right)^5}\cdot\frac{4}{n}$.

29. $\Delta x = (\pi - 0)/n = \pi/n$ and $x_i^* = x_i = \pi i/n$.

$$\int_0^\pi \sin 5x\, dx = \lim_{n\to\infty} \sum_{i=1}^n (\sin 5x_i)\left(\frac{\pi}{n}\right) = \lim_{n\to\infty} \sum_{i=1}^n \left(\sin \frac{5\pi i}{n}\right)\frac{\pi}{n} \overset{\text{CAS}}{=} \pi \lim_{n\to\infty} \frac{1}{n}\cot\left(\frac{5\pi}{2n}\right) \overset{\text{CAS}}{=} \pi\left(\frac{2}{5\pi}\right) = \frac{2}{5}$$

31. (a) Think of $\int_0^2 f(x)\, dx$ as the area of a trapezoid with bases 1 and 3 and height 2. The area of a trapezoid is $A = \frac{1}{2}(b + B)h$,

so $\int_0^2 f(x)\, dx = \frac{1}{2}(1+3)2 = 4$.

(b) $\int_0^5 f(x)\, dx = \int_0^2 f(x)\, dx + \int_2^3 f(x)\, dx + \int_3^5 f(x)\, dx$

           trapezoid   rectangle   triangle

    $= \frac{1}{2}(1+3)2 + \quad 3\cdot 1 \quad + \quad \frac{1}{2}\cdot 2\cdot 3 \quad = 4+3+3 = 10$

(c) $\int_5^7 f(x)\, dx$ is the negative of the area of the triangle with base 2 and height 3. $\int_5^7 f(x)\, dx = -\frac{1}{2}\cdot 2\cdot 3 = -3$.

(d) $\int_7^9 f(x)\, dx$ is the negative of the area of a trapezoid with bases 3 and 2 and height 2, so it equals

$-\frac{1}{2}(B + b)h = -\frac{1}{2}(3 + 2)2 = -5$. Thus,

$\int_0^9 f(x)\, dx = \int_0^5 f(x)\, dx + \int_5^7 f(x)\, dx + \int_7^9 f(x)\, dx = 10 + (-3) + (-5) = 2$.

33. $\int_0^3 \left(\frac{1}{2}x - 1\right) dx$ can be interpreted as the area of the triangle above the x-axis

minus the area of the triangle below the x-axis; that is,

$\frac{1}{2}(1)\left(\frac{1}{2}\right) - \frac{1}{2}(2)(1) = \frac{1}{4} - 1 = -\frac{3}{4}$.

35. $\int_{-3}^0 \left(1 + \sqrt{9 - x^2}\right) dx$ can be interpreted as the area under the graph of

$f(x) = 1 + \sqrt{9 - x^2}$ between $x = -3$ and $x = 0$. This is equal to one-quarter

the area of the circle with radius 3, plus the area of the rectangle, so

$\int_{-3}^0 \left(1 + \sqrt{9 - x^2}\right) dx = \frac{1}{4}\pi\cdot 3^2 + 1\cdot 3 = 3 + \frac{9}{4}\pi$.

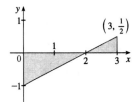

37. $\int_{-1}^2 |x|\, dx$ can be interpreted as the sum of the areas of the two shaded

triangles; that is, $\frac{1}{2}(1)(1) + \frac{1}{2}(2)(2) = \frac{1}{2} + \frac{4}{2} = \frac{5}{2}$.

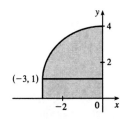

39. $\int_\pi^\pi \sin^2 x \cos^4 x\, dx = 0$ since the limits of intergration are equal.

41. $\int_{-2}^2 f(x)\, dx + \int_2^5 f(x)\, dx - \int_{-2}^{-1} f(x)\, dx = \int_{-2}^5 f(x)\, dx + \int_{-1}^{-2} f(x)\, dx$ [by Property 5 and reversing limits]

                      $= \int_{-1}^5 f(x)\, dx$ [Property 5]

43. $\int_0^9 [2f(x) + 3g(x)]\, dx = 2\int_0^9 f(x)\, dx + 3\int_0^9 g(x)\, dx = 2(37) + 3(16) = 122$

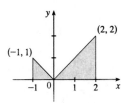

45. $\int_1^3 e^{x+2}\,dx = \int_1^3 e^x \cdot e^2\,dx = e^2\int_1^3 e^x\,dx = e^2(e^3 - e) = e^5 - e^3$

47. $\int_0^3 f(x)\,dx$ is clearly less than -1 and has the smallest value. The slope of the tangent line of f at $x = 1$, $f'(1)$, has a value

between -1 and 0, so it has the next smallest value. The largest value is $\int_3^8 f(x)\,dx$, followed by $\int_4^8 f(x)\,dx$, which has a

value about 1 unit less than $\int_3^8 f(x)\,dx$. Still positive, but with a smaller value than $\int_4^8 f(x)\,dx$, is $\int_0^8 f(x)\,dx$. Ordering these

quantities from smallest to largest gives us

$$\int_0^3 f(x)\,dx < f'(1) < \int_0^8 f(x)\,dx < \int_4^8 f(x)\,dx < \int_3^8 f(x)\,dx \quad \text{or} \quad B < E < A < D < C$$

49. $I = \int_{-4}^2 [f(x) + 2x + 5]\,dx = \int_{-4}^2 f(x)\,dx + 2\int_{-4}^2 x\,dx + \int_{-4}^2 5\,dx$

$\quad = I_1 + 2I_2 + I_3$

$I_1 = -3$ [area below x-axis] $+3 - 3 = -3$

$I_2 = -\frac{1}{2}(4)(4)$ [area of triangle, see figure] $+\frac{1}{2}(2)(2)$

$\quad = -8 + 2 = -6$

$I_3 = 5[2 - (-4)] = 5(6) = 30$

Thus, $I = -3 + 2(-6) + 30 = 15$.

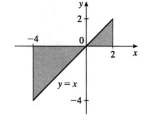

51. If $-1 \le x \le 1$, then $0 \le x^2 \le 1$ and $1 \le 1 + x^2 \le 2$, so $1 \le \sqrt{1 + x^2} \le \sqrt{2}$ and

$1[1 - (-1)] \le \int_{-1}^1 \sqrt{1 + x^2}\,dx \le \sqrt{2}[1 - (-1)]$ [Property 8]; that is, $2 \le \int_{-1}^1 \sqrt{1 + x^2}\,dx \le 2\sqrt{2}$.

53. $\displaystyle\lim_{n\to\infty} \sum_{i=1}^n \frac{i^4}{n^5} = \lim_{n\to\infty} \sum_{i=1}^n \frac{i^4}{n^4} \cdot \frac{1}{n} = \lim_{n\to\infty} \sum_{i=1}^n \left(\frac{i}{n}\right)^4 \frac{1}{n}$. At this point, we need to recognize the limit as being of the form

$\displaystyle\lim_{n\to\infty} \sum_{i=1}^n f(x_i)\,\Delta x$, where $\Delta x = (1 - 0)/n = 1/n$, $x_i = 0 + i\,\Delta x = i/n$, and $f(x) = x^4$. Thus, the definite integral

is $\int_0^1 x^4\,dx$.

55. Suppose that f is integrable on $[0, 1]$, that is, $\displaystyle\lim_{n\to\infty} \sum_{i=1}^n f(x_i^*)\,\Delta x$ exists for any choice of x_i^* in $[x_{i-1}, x_i]$. Let n denote a

positive integer and divide the interval $[0, 1]$ into n equal subintervals $\left[0, \dfrac{1}{n}\right]$, $\left[\dfrac{1}{n}, \dfrac{2}{n}\right]$, ..., $\left[\dfrac{n-1}{n}, 1\right]$. If we choose x_i^* to be

a rational number in the ith subinterval, then we obtain the Riemann sum $\displaystyle\sum_{i=1}^n f(x_i^*) \cdot \frac{1}{n} = 0$, so

$\displaystyle\lim_{n\to\infty} \sum_{i=1}^n f(x_i^*) \cdot \frac{1}{n} = \lim_{n\to\infty} 0 = 0$. Now suppose we choose x_i^* to be an irrational number. Then we get

$\displaystyle\sum_{i=1}^n f(x_i^*) \cdot \frac{1}{n} = \sum_{i=1}^n 1 \cdot \frac{1}{n} = n \cdot \frac{1}{n} = 1$ for each n, so $\displaystyle\lim_{n\to\infty} \sum_{i=1}^n f(x_i^*) \cdot \frac{1}{n} = \lim_{n\to\infty} 1 = 1$. Since the value of

$\displaystyle\lim_{n\to\infty} \sum_{i=1}^n f(x_i^*)\,\Delta x$ depends on the choice of the sample points x_i^*, the limit does not exist, and f is not integrable on $[0, 1]$.

5.3   Evaluating Definite Integrals

1. $\int_{-2}^{3}(x^2-3)\,dx = \left[\frac{1}{3}x^3 - 3x\right]_{-2}^{3} = (9-9) - \left(-\frac{8}{3}+6\right) = \frac{8}{3} - \frac{18}{3} = -\frac{10}{3}$

3. $\int_{0}^{2}\left(x^4 - \frac{3}{4}x^2 + \frac{2}{3}x - 1\right)dx = \left[\frac{1}{5}x^5 - \frac{1}{4}x^3 + \frac{1}{3}x^2 - x\right]_{0}^{2} = \left(\frac{32}{5} - 2 + \frac{4}{3} - 2\right) - 0 = \frac{96-30+20-30}{15} = \frac{56}{15}$

5. $\int_{0}^{1} x^{4/5}\,dx = \left[\frac{5}{9}x^{9/5}\right]_{0}^{1} = \frac{5}{9} - 0 = \frac{5}{9}$

7. $\int_{-1}^{0}(2x - e^x)\,dx = \left[x^2 - e^x\right]_{-1}^{0} = (0-1) - \left(1 - e^{-1}\right) = -2 + 1/e$

9. $\int_{1}^{2}(1+2y)^2\,dy = \int_{1}^{2}(1 + 4y + 4y^2)\,dy = \left[y + 2y^2 + \frac{4}{3}y^3\right]_{1}^{2} = \left(2 + 8 + \frac{32}{3}\right) - \left(1 + 2 + \frac{4}{3}\right) = \frac{62}{3} - \frac{13}{3} = \frac{49}{3}$

11. $\displaystyle\int_{1}^{9}\frac{x-1}{\sqrt{x}}\,dx = \int_{1}^{9}\left(\frac{x}{\sqrt{x}} - \frac{1}{\sqrt{x}}\right)dx = \int_{1}^{9}(x^{1/2} - x^{-1/2})\,dx = \left[\frac{2}{3}x^{3/2} - 2x^{1/2}\right]_{1}^{9}$

$= \left(\frac{2}{3}\cdot 27 - 2\cdot 3\right) - \left(\frac{2}{3} - 2\right) = 12 - \left(-\frac{4}{3}\right) = \frac{40}{3}$

13. $\int_{0}^{1} x\left(\sqrt[3]{x} + \sqrt[4]{x}\right)dx = \int_{0}^{1}(x^{4/3} + x^{5/4})\,dx = \left[\frac{3}{7}x^{7/3} + \frac{4}{9}x^{9/4}\right]_{0}^{1} = \left(\frac{3}{7} + \frac{4}{9}\right) - 0 = \frac{55}{63}$

15. $\int_{0}^{\pi/4}\sec^2 t\,dt = \left[\tan t\right]_{0}^{\pi/4} = \tan\frac{\pi}{4} - \tan 0 = 1 - 0 = 1$

17. $\displaystyle\int_{1}^{9}\frac{1}{2x}\,dx = \frac{1}{2}\int_{1}^{9}\frac{1}{x}\,dx = \frac{1}{2}\left[\ln|x|\right]_{1}^{9} = \frac{1}{2}(\ln 9 - \ln 1) = \frac{1}{2}\ln 9 - 0 = \ln 9^{1/2} = \ln 3$

19. $\displaystyle\int_{1/2}^{\sqrt{3}/2}\frac{6}{\sqrt{1-t^2}}\,dt = 6\int_{1/2}^{\sqrt{3}/2}\frac{1}{\sqrt{1-t^2}}\,dt = 6\left[\sin^{-1}t\right]_{1/2}^{\sqrt{3}/2} = 6\left[\sin^{-1}\left(\frac{\sqrt{3}}{2}\right) - \sin^{-1}\left(\frac{1}{2}\right)\right] = 6\left(\frac{\pi}{3} - \frac{\pi}{6}\right) = 6\left(\frac{\pi}{6}\right) = \pi$

21. $\int_{-1}^{1} e^{u+1}\,du = \left[e^{u+1}\right]_{-1}^{1} = e^2 - e^0 = e^2 - 1$ [or start with $e^{u+1} = e^u e^1$]

23. $\displaystyle\int_{1}^{2}\frac{v^3 + 3v^6}{v^4} = \int_{1}^{2}\left(\frac{1}{v} + 3v^2\right)dv = \left[\ln|v| + v^3\right]_{1}^{2} = (\ln 2 + 8) - (\ln 1 + 1) = \ln 2 + 7$

25. $\displaystyle\int_{0}^{\pi/4}\frac{1+\cos^2\theta}{\cos^2\theta}\,d\theta = \int_{0}^{\pi/4}\left(\frac{1}{\cos^2\theta} + \frac{\cos^2\theta}{\cos^2\theta}\right)d\theta = \int_{0}^{\pi/4}(\sec^2\theta + 1)\,d\theta$

$= \left[\tan\theta + \theta\right]_{0}^{\pi/4} = \left(\tan\frac{\pi}{4} + \frac{\pi}{4}\right) - (0+0) = 1 + \frac{\pi}{4}$

27. $\displaystyle\int_{0}^{1/\sqrt{3}}\frac{t^2-1}{t^4-1}\,dt = \int_{0}^{1/\sqrt{3}}\frac{t^2-1}{(t^2+1)(t^2-1)}\,dt = \int_{0}^{1/\sqrt{3}}\frac{1}{t^2+1}\,dt = \left[\arctan t\right]_{0}^{1/\sqrt{3}} = \arctan\left(1/\sqrt{3}\right) - \arctan 0$

$= \frac{\pi}{6} - 0 = \frac{\pi}{6}$

29. $\int_{-1}^{2}(x - 2|x|)\,dx = \int_{-1}^{0}[x - 2(-x)]\,dx + \int_{0}^{2}[x - 2(x)]\,dx = \int_{-1}^{0} 3x\,dx + \int_{0}^{2}(-x)\,dx = 3\left[\frac{1}{2}x^2\right]_{-1}^{0} - \left[\frac{1}{2}x^2\right]_{0}^{2}$

$= 3\left(0 - \frac{1}{2}\right) - (2-0) = -\frac{7}{2} = -3.5$

31. $f(x) = 1/x^2$ is not continuous on the interval $[-1,3]$, so the Evaluation Theorem does not apply. In fact, f has an infinite

discontinuity at $x = 0$, so $\int_{-1}^{3}(1/x^2)\,dx$ does not exist.

33. It appears that the area under the graph is about $\frac{2}{3}$ of the area of the viewing

rectangle, or about $\frac{2}{3}\pi \approx 2.1$. The actual area is

$\int_0^\pi \sin x \, dx = [-\cos x]_0^\pi = (-\cos \pi) - (-\cos 0) = -(-1) + 1 = 2$.

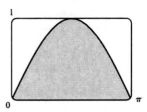

35. The graph shows that $y = 1 - 2x - 5x^4$ has x-intercepts at

$x = a \approx -0.86$ and at $x = b \approx 0.42$. So the area of the region that

lies under the curve and above the x-axis is

$\int_a^b (1 - 2x - 5x^4) \, dx = \left[x - x^2 - x^5\right]_a^b$

$\qquad = (b - b^2 - b^5) - (a - a^2 - a^5)$

$\qquad \approx 1.36$

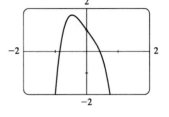

37. $\int_{-1}^2 x^3 \, dx = \left[\frac{1}{4}x^4\right]_{-1}^2 = 4 - \frac{1}{4} = \frac{15}{4} = 3.75$

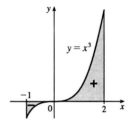

39. $\dfrac{d}{dx}\left[\sin x - \frac{1}{3}\sin^3 x + C\right] = \dfrac{d}{dx}\left[\sin x - \frac{1}{3}(\sin x)^3 + C\right] = \cos x - \frac{1}{3}\cdot 3(\sin x)^2(\cos x) + 0$

$\qquad\qquad\qquad\qquad = \cos x(1 - \sin^2 x) = \cos x(\cos^2 x) = \cos^3 x$

41. $\int \left(\cos x + \frac{1}{2}x\right) dx = \sin x + \frac{1}{4}x^2 + C$. The members of the family

in the figure correspond to $C = -5, 0, 5$, and 10.

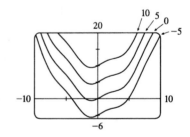

43. $\displaystyle\int (1 - t)(2 + t^2) \, dt = \int (2 - 2t + t^2 - t^3) \, dt = 2t - 2\frac{t^2}{2} + \frac{t^3}{3} - \frac{t^4}{4} + C = 2t - t^2 + \frac{1}{3}t^3 - \frac{1}{4}t^4 + C$

45. $\int (1 + \tan^2 \alpha) \, d\alpha = \int \sec^2 \alpha \, d\alpha = \tan \alpha + C$

47. $\displaystyle\int \frac{\sin x}{1 - \sin^2 x} \, dx = \int \frac{\sin x}{\cos^2 x} \, dx = \int \frac{1}{\cos x} \cdot \frac{\sin x}{\cos x} \, dx = \int \sec x \tan x \, dx = \sec x + C$

49. $A = \int_0^2 (2y - y^2) \, dy = \left[y^2 - \frac{1}{3}y^3\right]_0^2 = \left(4 - \frac{8}{3}\right) - 0 = \frac{4}{3}$

51. If $w'(t)$ is the rate of change of weight in pounds per year, then $w(t)$ represents the weight in pounds of the child at age t. We

know from the Net Change Theorem that $\int_5^{10} w'(t) \, dt = w(10) - w(5)$, so the integral represents the increase in the child's

weight (in pounds) between the ages of 5 and 10.

53. Since $r(t)$ is the rate at which oil leaks, we can write $r(t) = -V'(t)$, where $V(t)$ is the volume of oil at time t. [Note that the minus sign is needed because V is decreasing, so $V'(t)$ is negative, but $r(t)$ is positive.] Thus, by the Net Change Theorem, $\int_0^{120} r(t)\, dt = -\int_0^{120} V'(t)\, dt = -[V(120) - V(0)] = V(0) - V(120)$, which is the number of gallons of oil that leaked from the tank in the first two hours (120 minutes).

55. By the Net Change Theorem, $\int_{1000}^{5000} R'(x)\, dx = R(5000) - R(1000)$, so it represents the increase in revenue when production is increased from 1000 units to 5000 units.

57. In general, the unit of measurement for $\int_a^b f(x)\, dx$ is the product of the unit for $f(x)$ and the unit for x. Since $f(x)$ is measured in newtons and x is measured in meters, the units for $\int_0^{100} f(x)\, dx$ are newton-meters. (A newton-meter is abbreviated N·m.)

59. (a) Displacement $= \int_0^3 (3t - 5)\, dt = \left[\frac{3}{2}t^2 - 5t\right]_0^3 = \frac{27}{2} - 15 = -\frac{3}{2}$ m

(b) Distance traveled $= \int_0^3 |3t - 5|\, dt = \int_0^{5/3}(5 - 3t)\, dt + \int_{5/3}^3 (3t - 5)\, dt$

$\qquad = \left[5t - \frac{3}{2}t^2\right]_0^{5/3} + \left[\frac{3}{2}t^2 - 5t\right]_{5/3}^3 = \frac{25}{3} - \frac{3}{2}\cdot\frac{25}{9} + \frac{27}{2} - 15 - \left(\frac{3}{2}\cdot\frac{25}{9} - \frac{25}{3}\right) = \frac{41}{6}$ m

61. (a) $v'(t) = a(t) = t + 4 \ \Rightarrow\ v(t) = \frac{1}{2}t^2 + 4t + C \ \Rightarrow\ v(0) = C = 5 \ \Rightarrow\ v(t) = \frac{1}{2}t^2 + 4t + 5$ m/s

(b) Distance traveled $= \int_0^{10} |v(t)|\, dt = \int_0^{10} \left|\frac{1}{2}t^2 + 4t + 5\right| dt = \int_0^{10}\left(\frac{1}{2}t^2 + 4t + 5\right) dt = \left[\frac{1}{6}t^3 + 2t^2 + 5t\right]_0^{10}$

$\qquad = \frac{500}{3} + 200 + 50 = 416\frac{2}{3}$ m

63. Since $m'(x) = \rho(x)$, $m = \int_0^4 \rho(x)\, dx = \int_0^4\left(9 + 2\sqrt{x}\right) dx = \left[9x + \frac{4}{3}x^{3/2}\right]_0^4 = 36 + \frac{32}{3} - 0 = \frac{140}{3} = 46\frac{2}{3}$ kg.

65. Let s be the position of the car. We know from Equation 2 that $s(100) - s(0) = \int_0^{100} v(t)\, dt$. We use the Midpoint Rule for $0 \le t \le 100$ with $n = 5$. Note that the length of each of the five time intervals is 20 seconds $= \frac{20}{3600}$ hour $= \frac{1}{180}$ hour.

So the distance traveled is

$\int_0^{100} v(t)\, dt \approx \frac{1}{180}[v(10) + v(30) + v(50) + v(70) + v(90)] = \frac{1}{180}(38 + 58 + 51 + 53 + 47) = \frac{247}{180} \approx 1.4$ miles.

67. From the Net Change Theorem, the increase in cost if the production level is raised

from 2000 yards to 4000 yards is $C(4000) - C(2000) = \int_{2000}^{4000} C'(x)\, dx$.

$\int_{2000}^{4000} C'(x)\, dx = \int_{2000}^{4000}\left(3 - 0.01x + 0.000006x^2\right) dx = \left[3x - 0.005x^2 + 0.000002x^3\right]_{2000}^{4000} = 60{,}000 - 2{,}000 = \$58{,}000$

69. (a) We can find the area between the Lorenz curve and the line $y = x$ by subtracting the area under $y = L(x)$ from the area under $y = x$. Thus,

$$\text{coefficient of inequality} = \frac{\text{area between Lorenz curve and line } y = x}{\text{area under line } y = x} = \frac{\int_0^1 [x - L(x)]\, dx}{\int_0^1 x\, dx}$$

$$= \frac{\int_0^1 [x - L(x)]\, dx}{\left[x^2/2\right]_0^1} = \frac{\int_0^1 [x - L(x)]\, dx}{1/2} = 2\int_0^1 [x - L(x)]\, dx$$

(b) $L(x) = \frac{5}{12}x^2 + \frac{7}{12}x \ \Rightarrow\ L(50\%) = L\left(\frac{1}{2}\right) = \frac{5}{48} + \frac{7}{24} = \frac{19}{48} = 0.3958\overline{3}$, so the bottom 50% of the households receive at most about 40% of the income. Using the result in part (a),

$$\text{coefficient of inequality} = 2\int_0^1 [x - L(x)]\, dx = 2\int_0^1 \left(x - \tfrac{5}{12}x^2 - \tfrac{7}{12}x\right) dx = 2\int_0^1 \left(\tfrac{5}{12}x - \tfrac{5}{12}x^2\right) dx$$

$$= 2\int_0^1 \tfrac{5}{12}(x - x^2)\, dx = \tfrac{5}{6}\left[\tfrac{1}{2}x^2 - \tfrac{1}{3}x^3\right]_0^1 = \tfrac{5}{6}\left(\tfrac{1}{2} - \tfrac{1}{3}\right) = \tfrac{5}{6}\left(\tfrac{1}{6}\right) = \tfrac{5}{36}$$

71. (a) Let $f(x) = \sqrt{x}$ $\Rightarrow$ $f'(x) = 1/(2\sqrt{x}) > 0$ for $x > 0$ $\Rightarrow$ f is increasing on $(0, \infty)$. If $x \geq 0$, then $x^3 \geq 0$, so

$1 + x^3 \geq 1$ and since f is increasing, this means that $f(1 + x^3) \geq f(1)$ $\Rightarrow$ $\sqrt{1 + x^3} \geq 1$ for $x \geq 0$. Next let

$g(t) = t^2 - t$ $\Rightarrow$ $g'(t) = 2t - 1$ $\Rightarrow$ $g'(t) > 0$ when $t \geq 1$. Thus, g is increasing on $(1, \infty)$. And since $g(1) = 0$,

$g(t) \geq 0$ when $t \geq 1$. Now let $t = \sqrt{1 + x^3}$, where $x \geq 0$. $\sqrt{1 + x^3} \geq 1$ (from above) $\Rightarrow$ $t \geq 1$ $\Rightarrow$ $g(t) \geq 0$ $\Rightarrow$

$(1 + x^3) - \sqrt{1 + x^3} \geq 0$ for $x \geq 0$. Therefore, $1 \leq \sqrt{1 + x^3} \leq 1 + x^3$ for $x \geq 0$.

(b) From part (a) and Property 7: $\int_0^1 1\, dx \leq \int_0^1 \sqrt{1 + x^3}\, dx \leq \int_0^1 (1 + x^3)\, dx$ $\Leftrightarrow$

$\left[x\right]_0^1 \leq \int_0^1 \sqrt{1 + x^3}\, dx \leq \left[x + \tfrac{1}{4}x^4\right]_0^1$ $\Leftrightarrow$ $1 \leq \int_0^1 \sqrt{1 + x^3}\, dx \leq 1 + \tfrac{1}{4} = 1.25$.

73. The second derivative is the derivative of the first derivative, so we'll apply the Net Change Theorem with $F = h'$.

$\int_1^2 h''(u)\, du = \int_1^2 (h')'(u)\, du = h'(2) - h'(1) = 5 - 2 = 3$. The other information is unnecessary.

75. $\displaystyle\lim_{n \to \infty} \sum_{i=1}^n \frac{i^3}{n^4} = \lim_{n \to \infty} \frac{1 - 0}{n} \sum_{i=1}^n \left(\frac{i}{n}\right)^3 = \int_0^1 x^3\, dx = \left[\frac{x^4}{4}\right]_0^1 = \frac{1}{4}$

5.4 The Fundamental Theorem of Calculus

1. One process undoes what the other one does. The precise version of this statement is given by the Fundamental Theorem of Calculus. See the statement of this theorem and the paragraph that follows it on page 371.

3. (a) $g(x) = \int_0^x f(t)\, dt$.

$g(0) = \int_0^0 f(t)\, dt = 0$

$g(1) = \int_0^1 f(t)\, dt = 1 \cdot 2 = 2$ [rectangle],

$g(2) = \int_0^2 f(t)\, dt = \int_0^1 f(t)\, dt + \int_1^2 f(t)\, dt = g(1) + \int_1^2 f(t)\, dt$

$\quad = 2 + 1 \cdot 2 + \tfrac{1}{2} \cdot 1 \cdot 2 = 5$ [rectangle plus triangle],

$g(3) = \int_0^3 f(t)\, dt = g(2) + \int_2^3 f(t)\, dt = 5 + \tfrac{1}{2} \cdot 1 \cdot 4 = 7$,

$g(6) = g(3) + \int_3^6 f(t)\, dt$ [the integral is negative since f lies under the x-axis]

$\quad = 7 + \left[-\left(\tfrac{1}{2} \cdot 2 \cdot 2 + 1 \cdot 2\right)\right] = 7 - 4 = 3$

(d)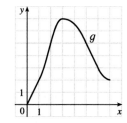

(b) g is increasing on $(0, 3)$ because as x increases from 0 to 3, we keep adding more area.

(c) g has a maximum value when we start subtracting area; that is, at $x = 3$.

5.

(a) By FTC1, $g(x) = \int_0^x (1 + t^2)\, dt \;\Rightarrow\; g'(x) = f(x) = 1 + x^2$.

(b) By FTC2, $g(x) = \int_0^x (1 + t^2)\, dt = \left[t + \frac{1}{3} t^3 \right]_0^x = \left(x + \frac{1}{3} x^3 \right) - 0 \;\Rightarrow\; g'(x) = 1 + x^2$.

7. $f(t) = \dfrac{1}{t^3 + 1}$ and $g(x) = \displaystyle\int_1^x \dfrac{1}{t^3 + 1}\, dt$, so by FTC1, $g'(x) = f(x) = \dfrac{1}{x^3 + 1}$. Note that the lower limit, 1, could be any

real number greater than -1 and not affect this answer.

9. $f(t) = t^2 \sin t$ and $g(y) = \int_2^y t^2 \sin t\, dt$, so by FTC1, $g'(y) = f(y) = y^2 \sin y$.

11. $F(x) = \displaystyle\int_x^\pi \sqrt{1 + \sec t}\, dt = -\int_\pi^x \sqrt{1 + \sec t}\, dt \;\Rightarrow\; F'(x) = -\dfrac{d}{dx}\int_\pi^x \sqrt{1 + \sec t}\, dt = -\sqrt{1 + \sec x}$

13. Let $u = \dfrac{1}{x}$. Then $\dfrac{du}{dx} = -\dfrac{1}{x^2}$. Also, $\dfrac{dh}{dx} = \dfrac{dh}{du}\dfrac{du}{dx}$, so

$$h'(x) = \dfrac{d}{dx}\int_2^{1/x} \arctan t\, dt = \dfrac{d}{du}\int_2^u \arctan t\, dt \cdot \dfrac{du}{dx} = \arctan u\, \dfrac{du}{dx} = -\dfrac{\arctan(1/x)}{x^2}.$$

15. Let $u = \tan x$. Then $\dfrac{du}{dx} = \sec^2 x$. Also, $\dfrac{dy}{dx} = \dfrac{dy}{du}\dfrac{du}{dx}$, so

$$y' = \dfrac{d}{dx}\int_0^{\tan x} \sqrt{t + \sqrt{t}}\, dt = \dfrac{d}{du}\int_0^u \sqrt{t + \sqrt{t}}\, dt \cdot \dfrac{du}{dx} = \sqrt{u + \sqrt{u}}\, \dfrac{du}{dx} = \sqrt{\tan x + \sqrt{\tan x}}\, \sec^2 x.$$

17. $g(x) = \displaystyle\int_{2x}^{3x} \dfrac{u^2 - 1}{u^2 + 1}\, du = \int_{2x}^0 \dfrac{u^2 - 1}{u^2 + 1}\, du + \int_0^{3x} \dfrac{u^2 - 1}{u^2 + 1}\, du = -\int_0^{2x} \dfrac{u^2 - 1}{u^2 + 1}\, du + \int_0^{3x} \dfrac{u^2 - 1}{u^2 + 1}\, du \;\Rightarrow$

$$g'(x) = -\dfrac{(2x)^2 - 1}{(2x)^2 + 1} \cdot \dfrac{d}{dx}(2x) + \dfrac{(3x)^2 - 1}{(3x)^2 + 1} \cdot \dfrac{d}{dx}(3x) = -2 \cdot \dfrac{4x^2 - 1}{4x^2 + 1} + 3 \cdot \dfrac{9x^2 - 1}{9x^2 + 1}$$

19. (a) By FTC1, $g'(x) = f(x)$. So $g'(x) = f(x) = 0$ at $x = 1, 3, 5, 7,$ and 9. g has local maxima at $x = 1$ and 5 (since $f = g'$

changes from positive to negative there) and local minima at $x = 3$ and 7. There is no local maximum or minimum at

$x = 9$, since f is not defined for $x > 9$.

(b) We can see from the graph that $\left| \int_0^1 f\, dt \right| < \left| \int_1^3 f\, dt \right| < \left| \int_3^5 f\, dt \right| < \left| \int_5^7 f\, dt \right| < \left| \int_7^9 f\, dt \right|$. So $g(1) = \left| \int_0^1 f\, dt \right|$,

$g(5) = \int_0^5 f\, dt = g(1) - \left| \int_1^3 f\, dt \right| + \left| \int_3^5 f\, dt \right|$, and $g(9) = \int_0^9 f\, dt = g(5) - \left| \int_5^7 f\, dt \right| + \left| \int_7^9 f\, dt \right|$. Thus,

$g(1) < g(5) < g(9)$, and so the absolute maximum of $g(x)$ occurs at $x = 9$.

(c) g is concave downward on those intervals where $g'' < 0$. But $g'(x) = f(x)$,

so $g''(x) = f'(x)$, which is negative on (approximately) $\left(\frac{1}{2}, 2 \right)$, $(4, 6)$ and

$(8, 9)$. So g is concave downward on these intervals.

(d)

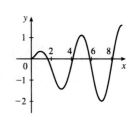

21. $f(x) = \int_0^x (1 - t^2)e^{t^2}\, dt$ is increasing when $f'(x) = (1 - x^2)e^{x^2}$ is positive.

Since $e^{x^2} > 0$, $f'(x) > 0 \iff 1 - x^2 > 0 \iff |x| < 1$, so f is increasing on $(-1, 1)$.

23. $y = \int_0^x \dfrac{t^2}{t^2 + t + 2}\, dt \implies y' = \dfrac{x^2}{x^2 + x + 2} \implies$

$$y'' = \frac{(x^2 + x + 2)(2x) - x^2(2x + 1)}{(x^2 + x + 2)^2} = \frac{2x^3 + 2x^2 + 4x - 2x^3 - x^2}{(x^2 + x + 2)^2} = \frac{x^2 + 4x}{(x^2 + x + 2)^2} = \frac{x(x + 4)}{(x^2 + x + 2)^2}.$$

The curve y is concave downward when $y'' < 0$; that is, on the interval $(-4, 0)$.

25. By FTC2, $\int_1^4 f'(x)\, dx = f(4) - f(1)$, so $17 = f(4) - 12 \implies f(4) = 17 + 12 = 29$.

27. (a) The Fresnel function $S(x) = \int_0^x \sin\left(\frac{\pi}{2}t^2\right) dt$ has local maximum values where $0 = S'(x) = \sin\left(\frac{\pi}{2}t^2\right)$ and

S' changes from positive to negative. For $x > 0$, this happens when $\frac{\pi}{2}x^2 = (2n - 1)\pi$ [odd multiples of π] $\iff$

$x^2 = 2(2n - 1) \iff x = \sqrt{4n - 2}$, n any positive integer. For $x < 0$, S' changes from positive to negative where

$\frac{\pi}{2}x^2 = 2n\pi$ [even multiples of π] $\iff x^2 = 4n \iff x = -2\sqrt{n}$. S' does not change sign at $x = 0$.

(b) S is concave upward on those intervals where $S''(x) > 0$. Differentiating our expression for $S'(x)$, we get

$S''(x) = \cos\left(\frac{\pi}{2}x^2\right)\left(2\frac{\pi}{2}x\right) = \pi x \cos\left(\frac{\pi}{2}x^2\right)$. For $x > 0$, $S''(x) > 0$ where $\cos\left(\frac{\pi}{2}x^2\right) > 0 \iff 0 < \frac{\pi}{2}x^2 < \frac{\pi}{2}$ or

$\left(2n - \frac{1}{2}\right)\pi < \frac{\pi}{2}x^2 < \left(2n + \frac{1}{2}\right)\pi$, n any integer $\iff 0 < x < 1$ or $\sqrt{4n - 1} < x < \sqrt{4n + 1}$, n any positive integer.

For $x < 0$, $S''(x) > 0$ where $\cos\left(\frac{\pi}{2}x^2\right) < 0 \iff \left(2n - \frac{3}{2}\right)\pi < \frac{\pi}{2}x^2 < \left(2n - \frac{1}{2}\right)\pi$, n any integer $\iff$

$4n - 3 < x^2 < 4n - 1 \iff \sqrt{4n - 3} < |x| < \sqrt{4n - 1} \implies \sqrt{4n - 3} < -x < \sqrt{4n - 1} \implies$

$-\sqrt{4n - 3} > x > -\sqrt{4n - 1}$, so the intervals of upward concavity for $x < 0$ are $\left(-\sqrt{4n - 1}, -\sqrt{4n - 3}\right)$, n any

positive integer. To summarize: S is concave upward on the intervals $(0, 1)$, $\left(-\sqrt{3}, -1\right)$, $\left(\sqrt{3}, \sqrt{5}\right)$, $\left(-\sqrt{7}, -\sqrt{5}\right)$,

$\left(\sqrt{7}, 3\right), \ldots$.

(c) In Maple, we use `plot({int(sin(Pi*t^2/2),t=0..x),0.2},x=0..2);`. Note that

Maple recognizes the Fresnel function, calling it `FresnelS(x)`. In Mathematica, we use

`Plot[{Integrate[Sin[Pi*t^2/2],{t,0,x}],0.2},{x,0,2}]`. In Derive, we load the utility file

FRESNEL and plot FRESNEL_SIN(x). From the graphs, we see that $\int_0^x \sin\left(\frac{\pi}{2}t^2\right) dt = 0.2$ at $x \approx 0.74$.

29. By FTC2, $\int_1^x f'(t)\, dt = f(x) - f(1) \implies f(x) = f(1) + \int_1^x f'(t)\, dt = f(1) + \int_1^x \left(2^t/t\right) dt$. This integral cannot be

expressed in a simpler form. Since we want $f(1) = 0$, we have $f(x) = \int_1^x \left(2^t/t\right) dt$.

31. Using FTC1, we differentiate both sides of $6 + \int_a^x \frac{f(t)}{t^2}\, dt = 2\sqrt{x}$ to get $\frac{f(x)}{x^2} = 2\frac{1}{2\sqrt{x}}$ $\Rightarrow$ $f(x) = x^{3/2}$.

To find a, we substitute $x = a$ in the original equation to obtain $6 + \int_a^a \frac{f(t)}{t^2}\, dt = 2\sqrt{a}$ $\Rightarrow$ $6 + 0 = 2\sqrt{a}$ $\Rightarrow$

$3 = \sqrt{a}$ $\Rightarrow$ $a = 9$.

33. (a) Let $F(t) = \int_0^t f(s)\, ds$. Then, by FTC1, $F'(t) = f(t) = $ rate of depreciation, so $F(t)$ represents the loss in value over the interval $[0, t]$.

(b) $C(t) = \frac{1}{t}\left[A + \int_0^t f(s)\, ds\right] = \frac{A + F(t)}{t}$ represents the average expenditure per unit of t during the interval $[0, t]$,

assuming that there has been only one overhaul during that time period. The company wants to minimize average expenditure.

(c) $C(t) = \frac{1}{t}\left[A + \int_0^t f(s)\, ds\right]$. Using FTC1, we have $C'(t) = -\frac{1}{t^2}\left[A + \int_0^t f(s)\, ds\right] + \frac{1}{t}f(t)$.

$C'(t) = 0$ $\Rightarrow$ $t\, f(t) = A + \int_0^t f(s)\, ds$ $\Rightarrow$ $f(t) = \frac{1}{t}\left[A + \int_0^t f(s)\, ds\right] = C(t)$.

5.5   The Substitution Rule

1. Let $u = -x$. Then $du = -dx$, so $dx = -du$. Thus, $\int e^{-x}\, dx = \int e^u(-du) = -e^u + C = -e^{-x} + C$. Don't forget that it is often very easy to check an indefinite integration by differentiating your answer. In this case,

$\frac{d}{dx}(-e^{-x} + C) = -[e^{-x}(-1)] = e^{-x}$, the desired result.

3. Let $u = x^3 + 1$. Then $du = 3x^2\, dx$ and $x^2\, dx = \frac{1}{3}\, du$, so

$\int x^2\sqrt{x^3 + 1}\, dx = \int \sqrt{u}\left(\frac{1}{3}\, du\right) = \frac{1}{3}\frac{u^{3/2}}{3/2} + C = \frac{1}{3}\cdot\frac{2}{3}u^{3/2} + C = \frac{2}{9}(x^3 + 1)^{3/2} + C$.

5. Let $u = \cos\theta$. Then $du = -\sin\theta\, d\theta$ and $\sin\theta\, d\theta = -du$, so

$\int \cos^3\theta \sin\theta\, d\theta = \int u^3(-du) = -\frac{u^4}{4} + C = -\frac{1}{4}\cos^4\theta + C$.

7. Let $u = x^2$. Then $du = 2x\, dx$ and $x\, dx = \frac{1}{2}\, du$, so $\int x\sin(x^2)\, dx = \int \sin u\left(\frac{1}{2}\, du\right) = -\frac{1}{2}\cos u + C = -\frac{1}{2}\cos(x^2) + C$.

9. Let $u = 3x - 2$. Then $du = 3\, dx$ and $dx = \frac{1}{3}\, du$, so $\int (3x - 2)^{20}\, dx = \int u^{20}\left(\frac{1}{3}\, du\right) = \frac{1}{3}\cdot\frac{1}{21}u^{21} + C = \frac{1}{63}(3x - 2)^{21} + C$.

11. Let $u = \pi t$. Then $du = \pi\, dt$ and $dt = \frac{1}{\pi}\, du$, so $\int \sin\pi t\, dt = \int \sin u\left(\frac{1}{\pi}\, du\right) = \frac{1}{\pi}(-\cos u) + C = -\frac{1}{\pi}\cos\pi t + C$.

13. Let $u = \ln x$. Then $du = \frac{dx}{x}$, so $\int \frac{(\ln x)^2}{x}\, dx = \int u^2\, du = \frac{1}{3}u^3 + C = \frac{1}{3}(\ln x)^3 + C$.

15. Let $u = 5 - 3x$. Then $du = -3\, dx$ and $dx = -\frac{1}{3}\, du$, so

$\int \frac{dx}{5 - 3x} = \int \frac{1}{u}\left(-\frac{1}{3}\, du\right) = -\frac{1}{3}\ln|u| + C = -\frac{1}{3}\ln|5 - 3x| + C$.

17. Let $u = 3ax + bx^3$. Then $du = (3a + 3bx^2)\,dx = 3(a + bx^2)\,dx$, so

$$\int \frac{a + bx^2}{\sqrt{3ax + bx^3}}\,dx = \int \frac{\frac{1}{3}\,du}{u^{1/2}} = \frac{1}{3}\int u^{-1/2}\,du = \frac{1}{3}\cdot 2u^2 + C = \frac{2}{3}\sqrt{3ax + bx^3} + C.$$

19. Let $u = 1 + e^x$. Then $du = e^x\,dx$, so $\int e^x\sqrt{1 + e^x}\,dx = \int \sqrt{u}\,du = \frac{2}{3}u^{3/2} + C = \frac{2}{3}(1 + e^x)^{3/2} + C$.

Or: Let $u = \sqrt{1 + e^x}$. Then $u^2 = 1 + e^x$ and $2u\,du = e^x\,dx$, so

$\int e^x\sqrt{1 + e^x}\,dx = \int u \cdot 2u\,du = \frac{2}{3}u^3 + C = \frac{2}{3}(1 + e^x)^{3/2} + C$.

21. Let $u = \sin x$. Then $du = \cos x\,dx$, so $\displaystyle\int \frac{\cos x}{\sin^2 x}\,dx = \int \frac{1}{u^2}\,du = \int u^{-2}\,du = \frac{u^{-1}}{-1} + C = -\frac{1}{u} + C = -\frac{1}{\sin x} + C$

[or $-\csc x + C$].

23. Let $u = x^3 + 3x$. Then $du = (3x^2 + 3)\,dx$ and $\frac{1}{3}\,du = (x^2 + 1)\,dx$, so

$\int (x^2 + 1)(x^3 + 3x)^4\,dx = \int u^4\left(\frac{1}{3}\,du\right) = \frac{1}{3}\cdot\frac{1}{5}u^5 + C = \frac{1}{15}(x^3 + 3x)^5 + C$.

25. Let $u = \cot x$. Then $du = -\csc^2 x\,dx$ and $\csc^2 x\,dx = -du$, so

$$\int \sqrt{\cot x}\,\csc^2 x\,dx = \int \sqrt{u}\,(-du) = -\frac{u^{3/2}}{3/2} + C = -\frac{2}{3}(\cot x)^{3/2} + C.$$

27. Let $u = \sin^{-1} x$. Then $du = \dfrac{1}{\sqrt{1 - x^2}}\,dx$, so $\displaystyle\int \frac{dx}{\sqrt{1 - x^2}\,\sin^{-1} x} = \int \frac{1}{u}\,du = \ln|u| + C = \ln\left|\sin^{-1} x\right| + C$.

29. Let $u = \sec x$. Then $du = \sec x\,\tan x\,dx$, so

$\int \sec^3 x\,\tan x\,dx = \int \sec^2 x\,(\sec x\,\tan x)\,dx = \int u^2\,du = \frac{1}{3}u^3 + C = \frac{1}{3}\sec^3 x + C$.

31. Let $u = 2x + 5$. Then $du = 2\,dx$ and $x = \frac{1}{2}(u - 5)$, so

$\int x(2x + 5)^8\,dx = \int \frac{1}{2}(u - 5)u^8\left(\frac{1}{2}\,du\right) = \frac{1}{4}\int(u^9 - 5u^8)\,du$

$$= \frac{1}{4}\left(\frac{1}{10}u^{10} - \frac{5}{9}u^9\right) + C = \frac{1}{40}(2x + 5)^{10} - \frac{5}{36}(2x + 5)^9 + C$$

33. $\displaystyle\int \frac{\sin 2x}{1 + \cos^2 x}\,dx = 2\int \frac{\sin x\,\cos x}{1 + \cos^2 x}\,dx = 2I$. Let $u = \cos x$. Then $du = -\sin x\,dx$, so

$2I = -2\displaystyle\int \frac{u\,du}{1 + u^2} = -2\cdot\frac{1}{2}\ln(1 + u^2) + C = -\ln(1 + u^2) + C = -\ln(1 + \cos^2 x) + C$.

Or: Let $u = 1 + \cos^2 x$.

35. Let $u = 1 + x^2$. Then $du = 2x\,dx$, so

$$\int \frac{1 + x}{1 + x^2}\,dx = \int \frac{1}{1 + x^2}\,dx + \int \frac{x}{1 + x^2}\,dx = \tan^{-1} x + \int \frac{\frac{1}{2}\,du}{u} = \tan^{-1} x + \frac{1}{2}\ln|u| + C$$

$$= \tan^{-1} x + \frac{1}{2}\ln\left|1 + x^2\right| + C = \tan^{-1} x + \frac{1}{2}\ln\left(1 + x^2\right) + C \quad \text{[since } 1 + x^2 > 0\text{]}.$$

37. $f(x) = x(x^2 - 1)^3$. $u = x^2 - 1$ $\Rightarrow$ $du = 2x\,dx$, so

$$\int x(x^2 - 1)^3\,dx = \int u^3\left(\tfrac{1}{2}\,du\right) = \tfrac{1}{8}u^4 + C = \tfrac{1}{8}(x^2 - 1)^4 + C$$

Where f is positive (negative), F is increasing (decreasing). Where f changes from negative to positive (positive to negative), F has a local minimum (maximum).

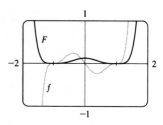

39. $f(x) = e^{\cos x}\sin x$. $u = \cos x$ $\Rightarrow$ $du = -\sin x\,dx$, so

$$\int e^u\,(-du) = -e^u + C = -e^{\cos x} + C$$

Note that at $x = \pi$, f changes from positive to negative and F has a local maximum. Also, both f and F are periodic with period 2π, so at $x = 0$ and at $x = 2\pi$, f changes from negative to positive and F has a local minimum.

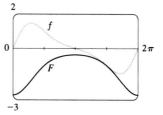

41. Let $u = \tfrac{\pi}{2}t$, so $du = \tfrac{\pi}{2}\,dt$. When $t = 0$, $u = 0$; when $t = 1$, $u = \tfrac{\pi}{2}$. Thus,

$$\int_0^1 \cos(\pi t/2)\,dt = \int_0^{\pi/2} \cos u\left(\tfrac{2}{\pi}\,du\right) = \tfrac{2}{\pi}\left[\sin u\right]_0^{\pi/2} = \tfrac{2}{\pi}\left(\sin\tfrac{\pi}{2} - \sin 0\right) = \tfrac{2}{\pi}(1 - 0) = \tfrac{2}{\pi}$$

43. Let $u = 1 + 7x$, so $du = 7\,dx$. When $x = 0$, $u = 1$; when $x = 1$, $u = 8$. Thus,

$$\int_0^1 \sqrt[3]{1 + 7x}\,dx = \int_1^8 u^{1/3}\left(\tfrac{1}{7}\,du\right) = \tfrac{1}{7}\left[\tfrac{3}{4}u^{4/3}\right]_1^8 = \tfrac{3}{28}(8^{4/3} - 1^{4/3}) = \tfrac{3}{28}(16 - 1) = \tfrac{45}{28}$$

45. Let $u = 1 + 2x^3$, so $du = 6x^2\,dx$. When $x = 0$, $u = 1$; when $x = 1$, $u = 3$. Thus,

$$\int_0^1 x^2\left(1 + 2x^3\right)^5\,dx = \int_1^3 u^5\left(\tfrac{1}{6}\,du\right) = \tfrac{1}{6}\left[\tfrac{1}{6}u^6\right]_1^3 = \tfrac{1}{36}(3^6 - 1^6) = \tfrac{1}{36}(729 - 1) = \tfrac{728}{36} = \tfrac{182}{9}.$$

47. Let $u = \sqrt{x}$, so $du = \dfrac{1}{2\sqrt{x}}\,dx$. When $x = 1$, $u = 1$; when $x = 4$, $u = 2$.

Thus, $\displaystyle\int_1^4 \frac{e^{\sqrt{x}}}{\sqrt{x}}\,dx = \int_1^2 e^u(2\,du) = 2\left[e^u\right]_1^2 = 2(e^2 - e).$

49. $\int_{-\pi/4}^{\pi/4} (x^3 + x^4\tan x)\,dx = 0$ by Theorem 6(b), since $f(x) = x^3 + x^4\tan x$ is an odd function.

51. Let $u = x - 1$, so $u + 1 = x$ and $du = dx$. When $x = 1$, $u = 0$; when $x = 2$, $u = 1$. Thus,

$$\int_1^2 x\sqrt{x - 1}\,dx = \int_0^1 (u + 1)\sqrt{u}\,du = \int_0^1 (u^{3/2} + u^{1/2})\,du = \left[\tfrac{2}{5}u^{5/2} + \tfrac{2}{3}u^{3/2}\right]_0^1 = \tfrac{2}{5} + \tfrac{2}{3} = \tfrac{16}{15}.$$

53. Let $u = e^z + z$, so $du = (e^z + 1)\,dz$. When $z = 0$, $u = 1$; when $z = 1$, $u = e + 1$. Thus,

$$\int_0^1 \frac{e^z + 1}{e^z + z}\,dz = \int_1^{e+1} \frac{1}{u}\,du = \left[\ln|u|\right]_1^{e+1} = \ln|e + 1| - \ln|1| = \ln(e + 1).$$

55. Let $u = \ln x$, so $du = \dfrac{dx}{x}$. When $x = e$, $u = 1$; when $x = e^4$; $u = 4$. Thus,

$$\int_e^{e^4} \frac{dx}{x\sqrt{\ln x}} = \int_1^4 u^{-1/2}\,du = 2\left[u^{1/2}\right]_1^4 = 2(2 - 1) = 2.$$

57. Let $u = 1 + \sqrt{x}$, so $du = \dfrac{1}{2\sqrt{x}}\,dx$ ⇒ $2\sqrt{x}\,du = dx$ ⇒ $2(u-1)\,du = dx$. When $x = 0$, $u = 1$; when $x = 1$,

$u = 2$. Thus,

$$\int_0^1 \frac{dx}{(1+\sqrt{x})^4} = \int_1^2 \frac{1}{u^4}\cdot[2(u-1)\,du] = 2\int_1^2\left(\frac{1}{u^3}-\frac{1}{u^4}\right)du = 2\left[-\frac{1}{2u^2}+\frac{1}{3u^3}\right]_1^2$$

$$= 2\left[\left(-\tfrac{1}{8}+\tfrac{1}{24}\right)-\left(-\tfrac{1}{2}+\tfrac{1}{3}\right)\right] = 2\left(\tfrac{1}{12}\right) = \tfrac{1}{6}$$

59. First write the integral as a sum of two integrals:

$I = \int_{-2}^2 (x+3)\sqrt{4-x^2}\,dx = I_1 + I_2 = \int_{-2}^2 x\sqrt{4-x^2}\,dx + \int_{-2}^2 3\sqrt{4-x^2}\,dx$. $I_1 = 0$ by Theorem 6(b), since

$f(x) = x\sqrt{4-x^2}$ is an odd function and we are integrating from $x = -2$ to $x = 2$. We interpret I_2 as three times the area of

a semicircle with radius 2, so $I = 0 + 3\cdot\tfrac{1}{2}(\pi\cdot 2^2) = 6\pi$.

61. First Figure   Let $u = \sqrt{x}$, so $x = u^2$ and $dx = 2u\,du$. When $x = 0$, $u = 0$; when $x = 1$, $u = 1$. Thus,

$A_1 = \int_0^1 e^{\sqrt{x}}\,dx = \int_0^1 e^u(2u\,du) = 2\int_0^1 ue^u\,du$.

Second Figure $A_2 = \int_0^1 2xe^x\,dx = 2\int_0^1 ue^u\,du$.

Third Figure   Let $u = \sin x$, so $du = \cos x\,dx$. When $x = 0$, $u = 0$; when $x = \tfrac{\pi}{2}$, $u = 1$. Thus,

$A_3 = \int_0^{\pi/2} e^{\sin x}\sin 2x\,dx = \int_0^{\pi/2} e^{\sin x}(2\sin x\,\cos x)\,dx = \int_0^1 e^u(2u\,du) = 2\int_0^1 ue^u\,du$.

Since $A_1 = A_2 = A_3$, all three areas are equal.

63. The rate is measured in liters per minute. Integrating from $t = 0$ minutes to $t = 60$ minutes will give us the total amount of oil

that leaks out (in liters) during the first hour.

$\int_0^{60} r(t)\,dt = \int_0^{60} 100e^{-0.01t}\,dt$ $[u = -0.01t, du = -0.01dt]$

$= 100\int_0^{-0.6} e^u(-100\,du) = -10{,}000\left[e^u\right]_0^{-0.6} = -10{,}000(e^{-0.6}-1) \approx 4511.9 \approx 4512$ liters

65. The volume of inhaled air in the lungs at time t is

$$V(t) = \int_0^t f(u)\,du = \int_0^t \frac{1}{2}\sin\left(\frac{2\pi}{5}u\right)du = \int_0^{2\pi t/5}\frac{1}{2}\sin v\left(\frac{5}{2\pi}\,dv\right) \qquad \left[\text{substitute } v = \tfrac{2\pi}{5}u,\ dv = \tfrac{2\pi}{5}\,du\right]$$

$$= \frac{5}{4\pi}\left[-\cos v\right]_0^{2\pi t/5} = \frac{5}{4\pi}\left[-\cos\left(\frac{2\pi}{5}t\right)+1\right] = \frac{5}{4\pi}\left[1-\cos\left(\frac{2\pi}{5}t\right)\right]\ \text{liters}$$

67. Let $u = 2x$. Then $du = 2\,dx$, so $\int_0^2 f(2x)\,dx = \int_0^4 f(u)\left(\tfrac{1}{2}\,du\right) = \tfrac{1}{2}\int_0^4 f(u)\,du = \tfrac{1}{2}(10) = 5$.

69. Let $u = -x$. Then $du = -dx$, so

$\int_a^b f(-x)\,dx = \int_{-a}^{-b} f(u)(-du) = \int_{-b}^{-a} f(u)\,du = \int_{-b}^{-a} f(x)\,dx$

From the diagram, we see that the equality follows from the fact that we are

reflecting the graph of f, and the limits of integration, about the y-axis.

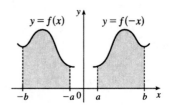

71. Let $u = 1 - x$. Then $x = 1 - u$ and $dx = -du$, so

$\int_0^1 x^a(1-x)^b\,dx = \int_1^0 (1-u)^a\,u^b(-du) = \int_0^1 u^b(1-u)^a\,du = \int_0^1 x^b(1-x)^a\,dx$.

5.6   Integration by Parts

1. Let $u = \ln x$, $dv = x^2\,dx$ $\Rightarrow$ $du = \frac{1}{x}\,dx$, $v = \frac{1}{3}x^3$. Then by Equation 2,

$$\int x^2 \ln x\,dx = (\ln x)\left(\tfrac{1}{3}x^3\right) - \int \left(\tfrac{1}{3}x^3\right)\left(\tfrac{1}{x}\right)dx = \tfrac{1}{3}x^3 \ln x - \tfrac{1}{3}\int x^2\,dx = \tfrac{1}{3}x^3 \ln x - \tfrac{1}{3}\left(\tfrac{1}{3}x^3\right) + C$$

$$= \tfrac{1}{3}x^3 \ln x - \tfrac{1}{9}x^3 + C \quad \left[\text{or } \tfrac{1}{3}x^3\left(\ln x - \tfrac{1}{3}\right) + C\right]$$

Note: A mnemonic device which is helpful for selecting u when using integration by parts is the LIATE principle of precedence for u:

Logarithmic

Inverse trigonometric

Algebraic

Trigonometric

Exponential

If the integrand has several factors, then we try to choose among them a u which appears as high as possible on the list. For example, in $\int xe^{2x}\,dx$ the integrand is xe^{2x}, which is the product of an algebraic function (x) and an exponential function (e^{2x}). Since Algebraic appears before Exponential, we choose $u = x$. Sometimes the integration turns out to be similar regardless of the selection of u and dv, but it is advisable to refer to LIATE when in doubt.

3. Let $u = x$, $dv = \cos 5x\,dx$ $\Rightarrow$ $du = dx$, $v = \frac{1}{5}\sin 5x$. Then by Equation 2,

$$\int x \cos 5x\,dx = \tfrac{1}{5}x \sin 5x - \int \tfrac{1}{5}\sin 5x\,dx = \tfrac{1}{5}x \sin 5x + \tfrac{1}{25}\cos 5x + C.$$

5. Let $u = r$, $dv = e^{r/2}\,dr$ $\Rightarrow$ $du = dr$, $v = 2e^{r/2}$.

Then $\int re^{r/2}\,dr = 2re^{r/2} - \int 2e^{r/2}\,dr = 2re^{r/2} - 4e^{r/2} + C = 2(r-2)e^{r/2} + C.$

7. Let $u = x^2$, $dv = \sin \pi x\,dx$ $\Rightarrow$ $du = 2x\,dx$ and $v = -\frac{1}{\pi}\cos \pi x$. Then

$$I = \int x^2 \sin \pi x\,dx = -\tfrac{1}{\pi}x^2 \cos \pi x + \tfrac{2}{\pi}\int x \cos \pi x\,dx \;(\star). \text{ Next let } U = x, dV = \cos \pi x\,dx \;\Rightarrow\; dU = dx,$$

$V = \frac{1}{\pi}\sin \pi x$, so $\int x \cos \pi x\,dx = \frac{1}{\pi}x \sin \pi x - \frac{1}{\pi}\int \sin \pi x\,dx = \frac{1}{\pi}x \sin \pi x + \frac{1}{\pi^2}\cos \pi x + C_1.$

Substituting for $\int x \cos \pi x\,dx$ in $(\star)$, we get

$$I = -\tfrac{1}{\pi}x^2 \cos \pi x + \tfrac{2}{\pi}\left(\tfrac{1}{\pi}x \sin \pi x + \tfrac{1}{\pi^2}\cos \pi x + C_1\right) = -\tfrac{1}{\pi}x^2 \cos \pi x + \tfrac{2}{\pi^2}x \sin \pi x + \tfrac{2}{\pi^3}\cos \pi x + C, \text{ where } C = \tfrac{2}{\pi}C_1.$$

9. Let $u = \ln \sqrt[3]{x}$, $dv = dx$ $\Rightarrow$ $du = \dfrac{1}{\sqrt[3]{x}}\left(\dfrac{1}{3}x^{-2/3}\right)dx = \dfrac{1}{3x}\,dx$, $v = x$. Then

$$\int \ln \sqrt[3]{x}\,dx = x \ln \sqrt[3]{x} - \int x \cdot \frac{1}{3x}\,dx = x \ln \sqrt[3]{x} - \frac{1}{3}x + C.$$

Second solution: Rewrite $\int \ln \sqrt[3]{x}\,dx = \frac{1}{3}\int \ln x\,dx$, and apply Example 2.

Third solution: Substitute $y = \sqrt[3]{x}$, to obtain $\int \ln \sqrt[3]{x}\,dx = 3\int y^2 \ln y\,dy$, and apply Exercise 1.

11. Let $u = \arctan 4t$, $dv = dt$ $\Rightarrow$ $du = \dfrac{4}{1+(4t)^2}\,dt = \dfrac{4}{1+16t^2}\,dt$, $v = t$. Then

$$\int \arctan 4t\,dt = t \arctan 4t - \int \frac{4t}{1+16t^2}\,dt = t \arctan 4t - \frac{1}{8}\int \frac{32t}{1+16t^2}\,dt = t \arctan 4t - \tfrac{1}{8}\ln(1+16t^2) + C.$$

13. First let $u = \sin 3\theta$, $dv = e^{2\theta}\,d\theta$ $\Rightarrow$ $du = 3\cos 3\theta\,d\theta$, $v = \frac{1}{2}e^{2\theta}$. Then

$$I = \int e^{2\theta} \sin 3\theta\,d\theta = \tfrac{1}{2}e^{2\theta} \sin 3\theta - \tfrac{3}{2}\int e^{2\theta} \cos 3\theta\,d\theta. \text{ Next let } U = \cos 3\theta, dV = e^{2\theta}\,d\theta \;\Rightarrow\; dU = -3\sin 3\theta\,d\theta,$$

$V = \frac{1}{2}e^{2\theta}$ to get $\int e^{2\theta} \cos 3\theta\,d\theta = \frac{1}{2}e^{2\theta} \cos 3\theta + \frac{3}{2}\int e^{2\theta} \sin 3\theta\,d\theta.$ Substituting in the previous formula gives

$I = \frac{1}{2}e^{2\theta}\sin 3\theta - \frac{3}{4}e^{2\theta}\cos 3\theta - \frac{9}{4}\int e^{2\theta}\sin 3\theta\, d\theta = \frac{1}{2}e^{2\theta}\sin 3\theta - \frac{3}{4}e^{2\theta}\cos 3\theta - \frac{9}{4}I \quad \Rightarrow$

$\frac{13}{4}I = \frac{1}{2}e^{2\theta}\sin 3\theta - \frac{3}{4}e^{2\theta}\cos 3\theta + C_1$. Hence, $I = \frac{1}{13}e^{2\theta}(2\sin 3\theta - 3\cos 3\theta) + C$, where $C = \frac{4}{13}C_1$.

15. Let $u = t$, $dv = \sin 3t\, dt \quad \Rightarrow \quad du = dt$, $v = -\frac{1}{3}\cos 3t$. Then

$\int_0^\pi t\sin 3t\, dt = \left[-\frac{1}{3}t\cos 3t\right]_0^\pi + \frac{1}{3}\int_0^\pi \cos 3t\, dt = \left(\frac{1}{3}\pi - 0\right) + \frac{1}{9}\left[\sin 3t\right]_0^\pi = \frac{\pi}{3}$.

17. Let $u = \ln x$, $dv = x^{-2}\, dx \quad \Rightarrow \quad du = \dfrac{1}{x}\, dx$, $v = -x^{-1}$. By (6),

$\displaystyle\int_1^2 \frac{\ln x}{x^2}\, dx = \left[-\frac{\ln x}{x}\right]_1^2 + \int_1^2 x^{-2}\, dx = -\frac{1}{2}\ln 2 + \ln 1 + \left[-\frac{1}{x}\right]_1^2 = -\frac{1}{2}\ln 2 + 0 - \frac{1}{2} + 1 = \frac{1}{2} - \frac{1}{2}\ln 2$.

19. Let $u = y$, $dv = \dfrac{dy}{e^{2y}} = e^{-2y}dy \quad \Rightarrow \quad du = dy$, $v = -\frac{1}{2}e^{-2y}$. Then

$\displaystyle\int_0^1 \frac{y}{e^{2y}}\, dy = \left[-\frac{1}{2}ye^{-2y}\right]_0^1 + \frac{1}{2}\int_0^1 e^{-2y}dy = \left(-\frac{1}{2}e^{-2} + 0\right) - \frac{1}{4}\left[e^{-2y}\right]_0^1 = -\frac{1}{2}e^{-2} - \frac{1}{4}e^{-2} + \frac{1}{4} = \frac{1}{4} - \frac{3}{4}e^{-2}$.

21. Let $u = \cos^{-1} x$, $dv = dx \quad \Rightarrow \quad du = -\dfrac{dx}{\sqrt{1-x^2}}$, $v = x$. Then

$I = \displaystyle\int_0^{1/2} \cos^{-1} x\, dx = \left[x\cos^{-1} x\right]_0^{1/2} + \int_0^{1/2} \frac{x\, dx}{\sqrt{1-x^2}} = \frac{1}{2}\cdot\frac{\pi}{3} + \int_1^{3/4} t^{-1/2}\left[-\frac{1}{2}dt\right]$, where $t = 1 - x^2 \quad \Rightarrow$

$dt = -2x\, dx$. Thus, $I = \frac{\pi}{6} + \frac{1}{2}\int_{3/4}^1 t^{-1/2}\, dt = \frac{\pi}{6} + \left[\sqrt{t}\right]_{3/4}^1 = \frac{\pi}{6} + 1 - \frac{\sqrt{3}}{2} = \frac{1}{6}\left(\pi + 6 - 3\sqrt{3}\right)$.

23. Let $u = (\ln x)^2$, $dv = dx \quad \Rightarrow \quad du = \dfrac{2}{x}\ln x\, dx$, $v = x$. By (6), $I = \int_1^2 (\ln x)^2\, dx = \left[x(\ln x)^2\right]_1^2 - 2\int_1^2 \ln x\, dx$.

To evaluate the last integral, let $U = \ln x$, $dV = dx \quad \Rightarrow \quad dU = \dfrac{1}{x}\, dx$, $V = x$. Thus,

$$I = \left[x(\ln x)^2\right]_1^2 - 2\left(\left[x\ln x\right]_1^2 - \int_1^2 dx\right) = \left[x(\ln x)^2 - 2x\ln x + 2x\right]_1^2$$

$$= \left(2(\ln 2)^2 - 4\ln 2 + 4\right) - (0 - 0 + 2) = 2(\ln 2)^2 - 4\ln 2 + 2$$

25. Let $y = \sqrt{x}$, so that $dy = \frac{1}{2}x^{-1/2}\, dx = \dfrac{1}{2\sqrt{x}}\, dx = \dfrac{1}{2y}\, dx$. Thus, $\int \cos\sqrt{x}\, dx = \int \cos y\, (2y\, dy) = 2\int y\cos y\, dy$. Now

use parts with $u = y$, $dv = \cos y\, dy$, $du = dy$, $v = \sin y$ to get $\int y\cos y\, dy = y\sin y - \int \sin y\, dy = y\sin y + \cos y + C_1$,

so $\int \cos\sqrt{x}\, dx = 2y\sin y + 2\cos y + C = 2\sqrt{x}\sin\sqrt{x} + 2\cos\sqrt{x} + C$.

27. Let $x = \theta^2$, so that $dx = 2\theta\, d\theta$. Thus, $\displaystyle\int_{\sqrt{\pi/2}}^{\sqrt{\pi}} \theta^3\cos(\theta^2)\, d\theta = \int_{\sqrt{\pi/2}}^{\sqrt{\pi}} \theta^2\cos(\theta^2)\cdot\frac{1}{2}(2\theta\, d\theta) = \frac{1}{2}\int_{\pi/2}^{\pi} x\cos x\, dx$. Now use

parts with $u = x$, $dv = \cos x\, dx$, $du = dx$, $v = \sin x$ to get

$$\frac{1}{2}\int_{\pi/2}^{\pi} x\cos x\, dx = \frac{1}{2}\left(\left[x\sin x\right]_{\pi/2}^{\pi} - \int_{\pi/2}^{\pi} \sin x\, dx\right) = \frac{1}{2}\left[x\sin x + \cos x\right]_{\pi/2}^{\pi}$$

$$= \frac{1}{2}(\pi\sin\pi + \cos\pi) - \frac{1}{2}\left(\frac{\pi}{2}\sin\frac{\pi}{2} + \cos\frac{\pi}{2}\right) = \frac{1}{2}(\pi\cdot 0 - 1) - \frac{1}{2}\left(\frac{\pi}{2}\cdot 1 + 0\right) = -\frac{1}{2} - \frac{\pi}{4}$$

29. Let $y = 1 + x$, so that $dy = dx$. Thus, $\int x\ln(1+x)\, dx = \int (y-1)\ln y\, dy$. Now use parts with $u = \ln y$, $dv = (y-1)\, dy$,

$du = \frac{1}{y}\, dy$, $v = \frac{1}{2}y^2 - y$ to get

$$\int (y-1)\ln y\, dy = \left(\frac{1}{2}y^2 - y\right)\ln y - \int \left(\frac{1}{2}y - 1\right) dy = \frac{1}{2}y(y-2)\ln y - \frac{1}{4}y^2 + y + C$$

$$= \frac{1}{2}(1+x)(x-1)\ln(1+x) - \frac{1}{4}(1+x)^2 + 1 + x + C,$$

which can be written as $\frac{1}{2}(x^2 - 1)\ln(1+x) - \frac{1}{4}x^2 + \frac{1}{2}x + \frac{3}{4} + C$.

31. Let $u = x$, $dv = e^{-2x}\,dx$ $\Rightarrow$ $du = dx$, $v = -\frac{1}{2}e^{-2x}$. Then

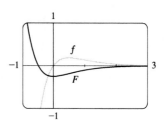

$\int xe^{-2x}\,dx = -\frac{1}{2}xe^{-2x} + \int \frac{1}{2}e^{-2x}\,dx = -\frac{1}{2}xe^{-2x} - \frac{1}{4}e^{-2x} + C$. We

see from the graph that this is reasonable, since F has a minimum where f

changes from negative to positive. Also, F increases where f is positive and

F decreases where f is negative.

33. Let $u = \frac{1}{2}x^2$, $dv = 2x\sqrt{1+x^2}\,dx$ $\Rightarrow$ $du = x\,dx$, $v = \frac{2}{3}(1+x^2)^{3/2}$.

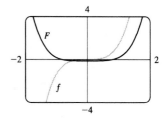

Then

$$\int x^3\sqrt{1+x^2}\,dx = \frac{1}{2}x^2\left[\frac{2}{3}(1+x^2)^{3/2}\right] - \frac{2}{3}\int x(1+x^2)^{3/2}\,dx$$

$$= \frac{1}{3}x^2(1+x^2)^{3/2} - \frac{2}{3}\cdot\frac{2}{5}\cdot\frac{1}{2}(1+x^2)^{5/2} + C$$

$$= \frac{1}{3}x^2(1+x^2)^{3/2} - \frac{2}{15}(1+x^2)^{5/2} + C$$

We see from the graph that this is reasonable, since F increases where f is positive and F decreases where f is negative.

Note also that f is an odd function and F is an even function.

Another method: Use substitution with $u = 1 + x^2$ to get $\frac{1}{5}(1+x^2)^{5/2} - \frac{1}{3}(1+x^2)^{3/2} + C$.

35. (a) Take $n = 2$ in Example 6 to get $\displaystyle\int \sin^2 x\,dx = -\frac{1}{2}\cos x\sin x + \frac{1}{2}\int 1\,dx = \frac{x}{2} - \frac{\sin 2x}{4} + C$.

(b) $\int \sin^4 x\,dx = -\frac{1}{4}\cos x\sin^3 x + \frac{3}{4}\int \sin^2 x\,dx = -\frac{1}{4}\cos x\sin^3 x + \frac{3}{8}x - \frac{3}{16}\sin 2x + C$.

37. (a) From Example 6, $\displaystyle\int \sin^n x\,dx = -\frac{1}{n}\cos x\,\sin^{n-1}x + \frac{n-1}{n}\int \sin^{n-2}x\,dx$. Using (6),

$$\int_0^{\pi/2} \sin^n x\,dx = \left[-\frac{\cos x\,\sin^{n-1}x}{n}\right]_0^{\pi/2} + \frac{n-1}{n}\int_0^{\pi/2} \sin^{n-2}x\,dx$$

$$= (0 - 0) + \frac{n-1}{n}\int_0^{\pi/2} \sin^{n-2}x\,dx = \frac{n-1}{n}\int_0^{\pi/2} \sin^{n-2}x\,dx$$

(b) Using $n = 3$ in part (a), we have $\int_0^{\pi/2} \sin^3 x\,dx = \frac{2}{3}\int_0^{\pi/2}\sin x\,dx = \left[-\frac{2}{3}\cos x\right]_0^{\pi/2} = \frac{2}{3}$.

Using $n = 5$ in part (a), we have $\int_0^{\pi/2} \sin^5 x\,dx = \frac{4}{5}\int_0^{\pi/2}\sin^3 x\,dx = \frac{4}{5}\cdot\frac{2}{3} = \frac{8}{15}$.

(c) The formula holds for $n = 1$ (that is, $2n + 1 = 3$) by (b). Assume it holds for some $k \geq 1$. Then

$$\int_0^{\pi/2} \sin^{2k+1}x\,dx = \frac{2\cdot 4\cdot 6\cdot\cdots\cdot(2k)}{3\cdot 5\cdot 7\cdot\cdots\cdot(2k+1)}.$$ By Example 6,

$$\int_0^{\pi/2} \sin^{2k+3}x\,dx = \frac{2k+2}{2k+3}\int_0^{\pi/2} \sin^{2k+1}x\,dx = \frac{2k+2}{2k+3}\cdot\frac{2\cdot 4\cdot 6\cdot\cdots\cdot(2k)}{3\cdot 5\cdot 7\cdot\cdots\cdot(2k+1)}$$

$$= \frac{2\cdot 4\cdot 6\cdot\cdots\cdot(2k)[2\,(k+1)]}{3\cdot 5\cdot 7\cdot\cdots\cdot(2k+1)[2\,(k+1)+1]}.$$

so the formula holds for $n = k + 1$. By induction, the formula holds for all $n \geq 1$.

39. Let $u = (\ln x)^n$, $dv = dx$ $\Rightarrow$ $du = n(\ln x)^{n-1}(dx/x)$, $v = x$. By Equation 2,

$\int (\ln x)^n\,dx = x(\ln x)^n - \int nx(\ln x)^{n-1}(dx/x) = x(\ln x)^n - n\int (\ln x)^{n-1}\,dx$.

41. By repeated applications of the reduction formula in Exercise 39,

$$\int (\ln x)^3 \, dx = x \, (\ln x)^3 - 3 \int (\ln x)^2 \, dx = x (\ln x)^3 - 3 \left[x (\ln x)^2 - 2 \int (\ln x)^1 \, dx \right]$$

$$= x \, (\ln x)^3 - 3x (\ln x)^2 + 6 \left[x (\ln x)^1 - 1 \int (\ln x)^0 \, dx \right]$$

$$= x \, (\ln x)^3 - 3x (\ln x)^2 + 6x \ln x - 6 \int 1 \, dx = x \, (\ln x)^3 - 3x (\ln x)^2 + 6x \ln x - 6x + C$$

43. Since $v(t) > 0$ for all t, the desired distance is $s(t) = \int_0^t v(w) \, dw = \int_0^t w^2 e^{-w} \, dw$.

First let $u = w^2$, $dv = e^{-w} \, dw$ ⇒ $du = 2w \, dw$, $v = -e^{-w}$. Then $s(t) = \left[-w^2 e^{-w} \right]_0^t + 2 \int_0^t w e^{-w} \, dw$.

Next let $U = w$, $dV = e^{-w} \, dw$ ⇒ $dU = dw$, $V = -e^{-w}$. Then

$$s(t) = -t^2 e^{-t} + 2 \left(\left[-w e^{-w} \right]_0^t + \int_0^t e^{-w} \, dw \right) = -t^2 e^{-t} + 2 \left(-t e^{-t} + 0 + \left[-e^{-w} \right]_0^t \right)$$

$$= -t^2 e^{-t} + 2(-t e^{-t} - e^{-t} + 1) = -t^2 e^{-t} - 2t e^{-t} - 2e^{-t} + 2 = 2 - e^{-t}(t^2 + 2t + 2) \text{ meters}$$

45. For $I = \int_1^4 x f''(x) \, dx$, let $u = x$, $dv = f''(x) \, dx$ ⇒ $du = dx$, $v = f'(x)$. Then

$$I = \left[x f'(x) \right]_1^4 - \int_1^4 f'(x) \, dx = 4 f'(4) - 1 \cdot f'(1) - [f(4) - f(1)] = 4 \cdot 3 - 1 \cdot 5 - (7 - 2) = 12 - 5 - 5 = 2.$$

We used the fact that f'' is continuous to guarantee that I exists.

47. Suppose $f(0) = g(0) = 0$ and let $u = f(x)$, $dv = g''(x) \, dx$ ⇒ $du = f'(x) \, dx$, $v = g'(x)$.

Then $\int_0^a f(x) g''(x) \, dx = \left[f(x) g'(x) \right]_0^a - \int_0^a f'(x) g'(x) \, dx = f(a) g'(a) - \int_0^a f'(x) g'(x) \, dx$.

Now let $U = f'(x)$, $dV = g'(x) \, dx$ ⇒ $dU = f''(x) \, dx$ and $V = g(x)$, so

$$\int_0^a f'(x) g'(x) \, dx = \left[f'(x) g(x) \right]_0^a - \int_0^a f''(x) g(x) \, dx = f'(a) g(a) - \int_0^a f''(x) g(x) \, dx.$$

Combining the two results, we get $\int_0^a f(x) g''(x) \, dx = f(a) g'(a) - f'(a) g(a) + \int_0^a f''(x) g(x) \, dx$.

5.7 Additional Techniques of Integration

1. $\int \sin^3 x \cos^2 x \, dx = \int \sin^2 x \cos^2 x \sin x \, dx = \int (1 - \cos^2 x) \cos^2 x \sin x \, dx$

$$= \int (1 - u^2) u^2 (-du) \qquad [u = \cos x, \, du = -\sin x \, dx]$$

$$= \int (u^2 - 1) u^2 \, du = \int (u^4 - u^2) \, du = \tfrac{1}{5} u^5 - \tfrac{1}{3} u^3 + C = \tfrac{1}{5} \cos^5 x - \tfrac{1}{3} \cos^3 x + C$$

3. $\int_{\pi/2}^{3\pi/4} \sin^5 x \cos^3 x \, dx = \int_{\pi/2}^{3\pi/4} \sin^5 x \cos^2 x \cos x \, dx = \int_{\pi/2}^{3\pi/4} \sin^5 x \, (1 - \sin^2 x) \cos x \, dx$

$$= \int_1^{\sqrt{2}/2} u^5 (1 - u^2) \, du \qquad [u = \sin x, \, du = \cos x \, dx]$$

$$= \int_1^{\sqrt{2}/2} (u^5 - u^7) \, du = \left[\tfrac{1}{6} u^6 - \tfrac{1}{8} u^8 \right]_1^{\sqrt{2}/2} = \left(\tfrac{1/8}{6} - \tfrac{1/16}{8} \right) - \left(\tfrac{1}{6} - \tfrac{1}{8} \right) = -\tfrac{11}{384}$$

5. $\int_0^{2\pi} \cos^2 (6\theta) \, d\theta = \tfrac{1}{2} \int_0^{2\pi} [1 + \cos(12\theta)] \, d\theta = \tfrac{1}{2} \left[\theta + \tfrac{1}{12} \sin(12\theta) \right]_0^{2\pi} = \tfrac{1}{2} [(2\pi + 0) - (0 + 0)] = \pi$

7. Let $u = \sec x$. Then $du = \sec x \tan x \, dx$, so

$$\int \tan^3 x \sec x \, dx = \int (\tan^2 x)(\tan x \sec x) \, dx = \int (\sec^2 x - 1)(\sec x \tan x \, dx)$$

$$= \int (u^2 - 1) \, du = \tfrac{1}{3} u^3 - u + C = \tfrac{1}{3} \sec^3 x - \sec x + C$$

9. Let $u = \tan x$. Then $du = \sec^2 x\, dx$, so

$$\int_0^{\pi/4} \tan^2 x \sec^4 x\, dx = \int_0^{\pi/4} \tan^2 x \sec^2 x\, (\sec^2 x\, dx) = \int_0^{\pi/4} \tan^2 x\, (1 + \tan^2 x)(\sec^2 x\, dx)$$

$$= \int_0^1 u^2 (1 + u^2)\, du = \int_0^1 (u^2 + u^4)\, du = \left[\tfrac{1}{3} u^3 + \tfrac{1}{5} u^5\right]_0^1 = \tfrac{1}{3} + \tfrac{1}{5} = \tfrac{8}{15}$$

11. $x = 3 \sin\theta$, where $-\pi/2 \le \theta \le \pi/2$. Then $dx = 3 \cos\theta\, d\theta$ and

$$\sqrt{9 - x^2} = \sqrt{9 - 9\sin^2\theta} = \sqrt{9\cos^2\theta} = 3\,|\cos\theta| = 3\cos\theta. \text{ (Note that } \cos\theta \ge 0 \text{ because } -\pi/2 \le \theta \le \pi/2.\text{)}$$

Thus, substitution gives

$$\int \frac{\sqrt{9 - x^2}}{x^2}\, dx = \int \frac{3\cos\theta}{9\sin^2\theta}\, 3\cos\theta\, d\theta = \int \frac{\cos^2\theta}{\sin^2\theta}\, d\theta = \int \cot^2\theta\, d\theta$$

$$= \int (\csc^2\theta - 1)\, d\theta = -\cot\theta - \theta + C$$

Since this is an indefinite integral, we must return to the original
variable x. This can be done either by using trigonometric identities to
express $\cot\theta$ in terms of $\sin\theta = x/3$ or by drawing a diagram, as
shown, where θ is interpreted as an angle of a right triangle.

Since $\sin\theta = x/3$, we label the opposite side and the hypotenuse as having lengths x and 3. Then the Pythagorean Theorem

gives the length of the adjacent side as $\sqrt{9 - x^2}$, so we can simply read the value of $\cot\theta$ from the figure: $\cot\theta = \dfrac{\sqrt{9 - x^2}}{x}$.

(Although $\theta > 0$ in the diagram, this expression for $\cot\theta$ is valid even when $\theta < 0$.) Since $\sin\theta = x/3$, we have

$$\theta = \sin^{-1}(x/3) \text{ and so } \int \frac{\sqrt{9 - x^2}}{x^2}\, dx = -\frac{\sqrt{9 - x^2}}{x} - \sin^{-1}\left(\frac{x}{3}\right) + C.$$

13. $x = 2\tan\theta$, where $-\pi/2 < \theta < \pi/2$. Then $dx = 2\sec^2\theta\, d\theta$ and

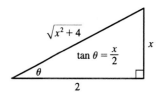

$$\sqrt{x^2 + 4} = \sqrt{(2\tan\theta)^2 + 4} = \sqrt{4\tan^2\theta + 4}$$

$$= \sqrt{4(\tan^2\theta + 1)} = 2\sqrt{\sec^2\theta} = 2\,|\sec\theta|$$

$$= 2\sec\theta \quad [\text{since } \sec\theta \ge 0 \text{ for } -\pi/2 < \theta < \pi/2].$$

Thus, substitution gives

$$\int \frac{1}{x^2\sqrt{x^2 + 4}}\, dx = \int \frac{1}{4\tan^2\theta\,(2\sec\theta)}\, (2\sec^2\theta\, d\theta) = \frac{1}{4}\int \frac{\sec\theta}{\tan^2\theta}\, d\theta = \frac{1}{4}\int \frac{1}{\cos\theta}\cdot\frac{\cos^2\theta}{\sin^2\theta}\, d\theta$$

$$= \frac{1}{4}\int \frac{\cos\theta}{\sin^2\theta}\, d\theta = \frac{1}{4}\int \frac{1}{u^2}\, du \quad [u = \sin\theta,\ du = \cos\theta\, d\theta]$$

$$= \frac{1}{4}\left(-\frac{1}{u}\right) + C = -\frac{1}{4}\frac{1}{\sin\theta} + C = -\frac{1}{4}\cdot\frac{\sqrt{x^2 + 4}}{x} + C = -\frac{\sqrt{x^2 + 4}}{4x} + C$$

15. Let $t = \sec\theta$, so $dt = \sec\theta \tan\theta\, d\theta$, $t = \sqrt{2} \Rightarrow \theta = \frac{\pi}{4}$, and $t = 2 \Rightarrow \theta = \frac{\pi}{3}$. Then

$$\int_{\sqrt{2}}^2 \frac{1}{t^3\sqrt{t^2 - 1}}\, dt = \int_{\pi/4}^{\pi/3} \frac{1}{\sec^3\theta \tan\theta}\, \sec\theta \tan\theta\, d\theta = \int_{\pi/4}^{\pi/3} \frac{1}{\sec^2\theta}\, d\theta = \int_{\pi/4}^{\pi/3} \cos^2\theta\, d\theta$$

$$= \int_{\pi/4}^{\pi/3} \tfrac{1}{2}(1 + \cos 2\theta)\, d\theta = \tfrac{1}{2}\left[\theta + \tfrac{1}{2}\sin 2\theta\right]_{\pi/4}^{\pi/3}$$

$$= \tfrac{1}{2}\left[\left(\tfrac{\pi}{3} + \tfrac{1}{2}\tfrac{\sqrt{3}}{2}\right) - \left(\tfrac{\pi}{4} + \tfrac{1}{2}\cdot 1\right)\right] = \tfrac{1}{2}\left(\tfrac{\pi}{12} + \tfrac{\sqrt{3}}{4} - \tfrac{1}{2}\right) = \tfrac{\pi}{24} + \tfrac{\sqrt{3}}{8} - \tfrac{1}{4}$$

17. Let $x = 2\sin\theta$, where $-\pi/2 \le \theta \le \pi/2$. Then $dx = 2\cos\theta\,d\theta$ and

$$\sqrt{4 - x^2} = \sqrt{4 - 4\sin^2\theta} = \sqrt{4\cos^2\theta} = 2\,|\cos\theta| = 2\cos\theta.$$

Thus, $\displaystyle \int \frac{dx}{x^2\sqrt{4 - x^2}} = \int \frac{2\cos\theta}{4\sin^2\theta(2\cos\theta)}\,d\theta = \frac{1}{4}\int \csc^2\theta\,d\theta$

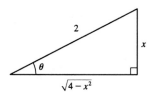

$$= -\frac{1}{4}\cot\theta + C = -\frac{\sqrt{4 - x^2}}{4x} + C \quad \text{[see figure]}$$

19. (a) $\displaystyle \frac{2x}{(x + 3)(3x + 1)} = \frac{A}{x + 3} + \frac{B}{3x + 1}$

(b) $\displaystyle \frac{1}{x^3 + 2x^2 + x} = \frac{1}{x(x^2 + 2x + 1)} = \frac{1}{x(x + 1)^2} = \frac{A}{x} + \frac{B}{x + 1} + \frac{C}{(x + 1)^2}$

21. $\displaystyle \frac{5x + 1}{(2x + 1)(x - 1)} = \frac{A}{2x + 1} + \frac{B}{x - 1}$. Multiply both sides by $(2x + 1)(x - 1)$ to get $5x + 1 = A(x - 1) + B(2x + 1)$ $\Rightarrow$

$5x + 1 = Ax - A + 2Bx + B$ $\Rightarrow$ $5x + 1 = (A + 2B)x + (-A + B)$.

The coefficients of x must be equal and the constant terms are also equal, so $A + 2B = 5$ and

$-A + B = 1$. Adding these equations gives us $3B = 6$ $\Leftrightarrow$ $B = 2$, and hence, $A = 1$. Thus,

$$\int \frac{5x + 1}{(2x + 1)(x - 1)}\,dx = \int \left(\frac{1}{2x + 1} + \frac{2}{x - 1}\right)dx = \tfrac{1}{2}\ln|2x + 1| + 2\ln|x - 1| + C.$$

Another method: Substituting 1 for x in the equation $5x + 1 = A(x - 1) + B(2x + 1)$ gives $6 = 3B$ $\Leftrightarrow$ $B = 2$.

Substituting $-\frac{1}{2}$ for x gives $-\frac{3}{2} = -\frac{3}{2}A$ $\Leftrightarrow$ $A = 1$.

23. $\displaystyle \frac{1}{x^2 - 1} = \frac{1}{(x + 1)(x - 1)} = \frac{A}{x + 1} + \frac{B}{x - 1}$. Multiply both sides by $(x + 1)(x - 1)$ to get $1 = A(x - 1) + B(x + 1)$ $\Rightarrow$

$1 = Ax - A + Bx + B$ $\Rightarrow$ $1 = (A + B)x + (-A + B)$. The coefficients of x must be equal and the constant terms are

also equal, so $A + B = 0$ and $-A + B = 1$. Adding these equations gives us $2B = 1$ $\Leftrightarrow$ $B = \frac{1}{2}$, and hence, $A = -\frac{1}{2}$.

Thus

$$\int_2^3 \frac{1}{x^2 - 1}\,dx = \int_2^3 \left(\frac{-1/2}{x + 1} + \frac{1/2}{x - 1}\right)dx = \left[-\tfrac{1}{2}\ln|x + 1| + \tfrac{1}{2}\ln|x - 1|\right]_2^3$$

$$= \left(-\tfrac{1}{2}\ln 4 + \tfrac{1}{2}\ln 2\right) - \left(-\tfrac{1}{2}\ln 3 + \tfrac{1}{2}\ln 1\right) = \tfrac{1}{2}(\ln 2 + \ln 3 - \ln 4) \quad \left[\text{or } \tfrac{1}{2}\ln\tfrac{3}{2}\right]$$

Another method: Substituting 1 for x in the equation $1 = A(x - 1) + B(x + 1)$ gives $1 = 2B$ $\Leftrightarrow$ $B = \frac{1}{2}$.

Substituting -1 for x gives $1 = -2A$ $\Leftrightarrow$ $A = -\frac{1}{2}$.

25. $\displaystyle \frac{10}{(x - 1)(x^2 + 9)} = \frac{A}{x - 1} + \frac{Bx + C}{x^2 + 9}$. Multiply both sides by $(x - 1)(x^2 + 9)$ to get

$10 = A(x^2 + 9) + (Bx + C)(x - 1)$ $(\star)$. Substituting 1 for x gives $10 = 10A$ $\Leftrightarrow$ $A = 1$. Substituting 0 for x gives

$10 = 9A - C$ $\Rightarrow$ $C = 9(1) - 10 = -1$. The coefficients of the x^2-terms in $(\star)$ must be equal, so $0 = A + B$ $\Rightarrow$

$B = -1$. Thus,

$$\int \frac{10}{(x - 1)(x^2 + 9)}\,dx = \int \left(\frac{1}{x - 1} + \frac{-x - 1}{x^2 + 9}\right)dx = \int \left(\frac{1}{x - 1} - \frac{x}{x^2 + 9} - \frac{1}{x^2 + 9}\right)dx$$

$$= \ln|x - 1| - \tfrac{1}{2}\ln(x^2 + 9) - \tfrac{1}{3}\tan^{-1}\left(\tfrac{x}{3}\right) + C$$

In the second term we used the substitution $u = x^2 + 9$ and in the last term we used Formula 10.

27. $\dfrac{x^3 + x^2 + 2x + 1}{(x^2 + 1)(x^2 + 2)} = \dfrac{Ax + B}{x^2 + 1} + \dfrac{Cx + D}{x^2 + 2}$. Multiply both sides by $(x^2 + 1)(x^2 + 2)$ to get

$x^3 + x^2 + 2x + 1 = (Ax + B)(x^2 + 2) + (Cx + D)(x^2 + 1) \quad \Leftrightarrow$

$x^3 + x^2 + 2x + 1 = (Ax^3 + Bx^2 + 2Ax + 2B) + (Cx^3 + Dx^2 + Cx + D) \quad \Leftrightarrow$

$x^3 + x^2 + 2x + 1 = (A + C)x^3 + (B + D)x^2 + (2A + C)x + (2B + D)$. Comparing coefficients gives us the following system of equations:

$$A + C = 1 \quad \textbf{(1)} \qquad\qquad B + D = 1 \quad \textbf{(2)}$$
$$2A + C = 2 \quad \textbf{(3)} \qquad\qquad 2B + D = 1 \quad \textbf{(4)}$$

Subtracting equation **(1)** from equation **(3)** gives us $A = 1$, so $C = 0$. Subtracting equation **(2)** from equation **(4)** gives us

$B = 0$, so $D = 1$. Thus, $I = \displaystyle\int \dfrac{x^3 + x^2 + 2x + 1}{(x^2 + 1)(x^2 + 2)}\, dx = \int \left(\dfrac{x}{x^2 + 1} + \dfrac{1}{x^2 + 2} \right) dx$. For $\displaystyle\int \dfrac{x}{x^2 + 1}\, dx$, let $u = x^2 + 1$

so $du = 2x\, dx$ and then $\displaystyle\int \dfrac{x}{x^2 + 1}\, dx = \dfrac{1}{2}\int \dfrac{1}{u}\, du = \dfrac{1}{2}\ln|u| + C = \dfrac{1}{2}\ln(x^2 + 1) + C$. For $\displaystyle\int \dfrac{1}{x^2 + 2}\, dx$, use

Formula 10 with $a = \sqrt{2}$. So $\displaystyle\int \dfrac{1}{x^2 + 2}\, dx = \int \dfrac{1}{x^2 + (\sqrt{2})^2}\, dx = \dfrac{1}{\sqrt{2}}\tan^{-1}\dfrac{x}{\sqrt{2}} + C$.

Thus, $I = \dfrac{1}{2}\ln(x^2 + 1) + \dfrac{1}{\sqrt{2}}\tan^{-1}\dfrac{x}{\sqrt{2}} + C$.

29. $\displaystyle\int \dfrac{x}{x - 6}\, dx = \int \dfrac{(x - 6) + 6}{x - 6}\, dx = \int \left(1 + \dfrac{6}{x - 6} \right) dx = x + 6\ln|x - 6| + C$

31.

$$\begin{array}{r} x \\ x^2 + 4 \overline{\smash{\big)}\, x^3\ +\ 0x^2\ +\ 0x\ +\ 4} \\ \underline{x^3 + 4x } \\ -4x\ +\ 4 \end{array}$$

By long division, $\dfrac{x^3 + 4}{x^2 + 4} = x + \dfrac{-4x + 4}{x^2 + 4}$. Thus,

$\displaystyle\int \dfrac{x^3 + 4}{x^2 + 4}\, dx = \int \left(x + \dfrac{-4x + 4}{x^2 + 4} \right) dx = \int \left(x - \dfrac{4x}{x^2 + 4} + \dfrac{4}{x^2 + 2^2} \right) dx$

$\qquad = \dfrac{1}{2}x^2 - 4 \cdot \dfrac{1}{2}\ln|x^2 + 4| + 4 \cdot \dfrac{1}{2}\tan^{-1}\left(\dfrac{x}{2} \right) + C = \dfrac{1}{2}x^2 - 2\ln(x^2 + 4) + 2\tan^{-1}\left(\dfrac{x}{2} \right) + C$

33. Let $u = \sqrt{x}$, so $u^2 = x$ and $dx = 2u\, du$. Thus,

$$\int_9^{16} \dfrac{\sqrt{x}}{x - 4}\, dx = \int_3^4 \dfrac{u}{u^2 - 4}\, 2u\, du = 2\int_3^4 \dfrac{u^2}{u^2 - 4}\, du = 2\int_3^4 \left(1 + \dfrac{4}{u^2 - 4} \right) du \qquad \text{[by long division]}$$

$$= 2 + 8\int_3^4 \dfrac{du}{(u + 2)(u - 2)} \quad (\star)$$

Multiply $\dfrac{1}{(u + 2)(u - 2)} = \dfrac{A}{u + 2} + \dfrac{B}{u - 2}$ by $(u + 2)(u - 2)$ to get $1 = A(u - 2) + B(u + 2)$. Equating coefficients we

get $A + B = 0$ and $-2A + 2B = 1$. Solving gives us $B = \frac{1}{4}$ and $A = -\frac{1}{4}$, so $\dfrac{1}{(u + 2)(u - 2)} = \dfrac{-1/4}{u + 2} + \dfrac{1/4}{u - 2}$ and $(\star)$ is

$$2 + 8\int_3^4 \left(\dfrac{-1/4}{u + 2} + \dfrac{1/4}{u - 2} \right) du = 2 + 8\left[-\tfrac{1}{4}\ln|u + 2| + \tfrac{1}{4}\ln|u - 2| \right]_3^4 = 2 + \left[2\ln|u - 2| - 2\ln|u + 2| \right]_3^4$$

$$= 2 + 2\left[\ln\left| \dfrac{u - 2}{u + 2} \right| \right]_3^4 = 2 + 2\left(\ln \tfrac{2}{6} - \ln \tfrac{1}{5} \right) = 2 + 2\ln \tfrac{2/6}{1/5}$$

$$= 2 + 2\ln \tfrac{5}{3} \quad \text{or} \quad 2 + \ln\left(\tfrac{5}{3} \right)^2 = 2 + \ln \tfrac{25}{9}$$

35. $x^2 + x + 1 = x^2 + x + \frac{1}{4} + 1 - \frac{1}{4}$ [add and subtract the square of one-half the
                                              coefficient of x to complete the square]

$$= x^2 + x + \frac{1}{4} + \frac{3}{4} = \left(x + \frac{1}{2}\right)^2 + \left(\frac{\sqrt{3}}{2}\right)^2$$

So $I = \displaystyle\int \frac{dx}{x^2 + x + 1} = \int \frac{1}{\left(x + \frac{1}{2}\right)^2 + \left(\frac{\sqrt{3}}{2}\right)^2}\, dx$. Now let $u = x + \frac{1}{2}$ $\Rightarrow$ $du = dx$ and

$$I = \int \frac{1}{u^2 + \left(\frac{\sqrt{3}}{2}\right)^2}\, du = \frac{1}{\frac{\sqrt{3}}{2}} \tan^{-1} \frac{u}{\frac{\sqrt{3}}{2}} + C = \frac{2}{\sqrt{3}} \tan^{-1} \frac{2\left(x + \frac{1}{2}\right)}{\sqrt{3}} + C = \frac{2}{\sqrt{3}} \tan^{-1} \frac{2x + 1}{\sqrt{3}} + C.$$

5.8   Integration Using Tables and Computer Algebra Systems

Keep in mind that there are several ways to approach many of these exercises, and different methods can lead to different forms of the answer.

1. Let $u = \pi x$, so that $du = \pi\, dx$. Then

$$\int \tan^3(\pi x)\, dx = \int \tan^3 u\left(\tfrac{1}{\pi}\, du\right) = \tfrac{1}{\pi} \int \tan^3 u\, du \overset{69}{=} \tfrac{1}{\pi}\left[\tfrac{1}{2}\tan^2 u + \ln|\cos u|\right] + C$$

$$= \tfrac{1}{2\pi}\tan^2(\pi x) + \tfrac{1}{\pi}\ln|\cos(\pi x)| + C$$

3. Let $u = 2x$ and $a = 3$. Then $du = 2\, dx$ and

$$\int \frac{dx}{x^2\sqrt{4x^2 + 9}} = \int \frac{\frac{1}{2}\, du}{\frac{u^2}{4}\sqrt{u^2 + a^2}} = 2\int \frac{du}{u^2\sqrt{a^2 + u^2}} \overset{28}{=} -2\frac{\sqrt{a^2 + u^2}}{a^2 u} + C$$

$$= -2\frac{\sqrt{4x^2 + 9}}{9 \cdot 2x} + C = -\frac{\sqrt{4x^2 + 9}}{9x} + C$$

5. Let $u = e^x$, so that $du = e^x\, dx$ and $e^{2x} = u^2$. Then

$$\int e^{2x} \arctan(e^x)\, dx = \int u^2 \arctan u \left(\frac{du}{u}\right) = \int u \arctan u\, du$$

$$\overset{92}{=} \frac{u^2 + 1}{2}\arctan u - \frac{u}{2} + C = \frac{1}{2}(e^{2x} + 1)\arctan(e^x) - \frac{1}{2}e^x + C$$

7. $\int x^3 \sin x\, dx \overset{84}{=} -x^3 \cos x + 3\int x^2 \cos x\, dx$, $\int x^2 \cos x\, dx \overset{85}{=} x^2 \sin x - 2\int x \sin x\, dx$, and

$\int x \sin x\, dx \overset{82}{=} \sin x - x \cos x + C$. Substituting, we get

$\int x^3 \sin x\, dx = -x^3 \cos x + 3\left[x^2 \sin x - 2(\sin x - x \cos x)\right] + C = -x^3 \cos x + 3x^2 \sin x - 6\sin x + 6x\cos x + C$.

So $\int_0^\pi x^3 \sin x\, dx = \left[-x^3 \cos x + 3x^2 \sin x - 6\sin x + 6x\cos x\right]_0^\pi = (-\pi^3 \cdot -1 + 6\pi \cdot -1) - (0) = \pi^3 - 6\pi.$

9. $\displaystyle\int \frac{\tan^3(1/z)}{z^2}\, dz$ $\begin{bmatrix} u = 1/z, \\ du = -dz/z^2 \end{bmatrix}$ $= -\displaystyle\int \tan^3 u\, du \overset{69}{=} -\tfrac{1}{2}\tan^2 u - \ln|\cos u| + C = -\tfrac{1}{2}\tan^2\left(\tfrac{1}{z}\right) - \ln\left|\cos\left(\tfrac{1}{z}\right)\right| + C$

11. Let $z = 6 + 4y - 4y^2 = 6 - (4y^2 - 4y + 1) + 1 = 7 - (2y - 1)^2$, $u = 2y - 1$, and $a = \sqrt{7}$. Then $z = a^2 - u^2$, $du = 2\, dy$, and

$$\int y\sqrt{6 + 4y - 4y^2}\, dy = \int y\sqrt{z}\, dy = \int \tfrac{1}{2}(u + 1)\sqrt{a^2 - u^2}\, \tfrac{1}{2}\, du = \tfrac{1}{4}\int u\sqrt{a^2 - u^2}\, du + \tfrac{1}{4}\int \sqrt{a^2 - u^2}\, du$$

$$= \tfrac{1}{4}\int \sqrt{a^2 - u^2}\, du - \tfrac{1}{8}\int(-2u)\sqrt{a^2 - u^2}\, du$$

$$\overset{30}{=} \frac{u}{8}\sqrt{a^2 - u^2} + \frac{a^2}{8}\sin^{-1}\left(\frac{u}{a}\right) - \frac{1}{8}\int \sqrt{w}\, dw \qquad \begin{bmatrix} w = a^2 - u^2, \\ dw = -2u\, du \end{bmatrix}$$

$$= \frac{2y - 1}{8}\sqrt{6 + 4y - 4y^2} + \frac{7}{8}\sin^{-1}\frac{2y - 1}{\sqrt{7}} - \frac{1}{8}\cdot\frac{2}{3}w^{3/2} + C$$

$$= \frac{2y - 1}{8}\sqrt{6 + 4y - 4y^2} + \frac{7}{8}\sin^{-1}\frac{2y - 1}{\sqrt{7}} - \frac{1}{12}(6 + 4y - 4y^2)^{3/2} + C$$

This can be rewritten as

$$\sqrt{6 + 4y - 4y^2}\left[\frac{1}{8}(2y - 1) - \frac{1}{12}(6 + 4y - 4y^2)\right] + \frac{7}{8}\sin^{-1}\frac{2y - 1}{\sqrt{7}} + C$$

$$= \left(\frac{1}{3}y^2 - \frac{1}{12}y - \frac{5}{8}\right)\sqrt{6 + 4y - 4y^2} + \frac{7}{8}\sin^{-1}\left(\frac{2y - 1}{\sqrt{7}}\right) + C$$

$$= \frac{1}{24}(8y^2 - 2y - 15)\sqrt{6 + 4y - 4y^2} + \frac{7}{8}\sin^{-1}\left(\frac{2y - 1}{\sqrt{7}}\right) + C$$

13. Let $u = \sin x$. Then $du = \cos x\, dx$, so

$$\int \sin^2 x \cos x \ln(\sin x)\, dx = \int u^2 \ln u\, du \overset{101}{=} \frac{u^{2+1}}{(2+1)^2}[(2+1)\ln u - 1] + C = \frac{1}{9}u^3(3\ln u - 1) + C$$

$$= \frac{1}{9}\sin^3 x\,[3\ln(\sin x) - 1] + C$$

15. Let $u = e^x$ and $a = \sqrt{3}$. Then $du = e^x\, dx$ and

$$\int \frac{e^x}{3 - e^{2x}}\, dx = \int \frac{du}{a^2 - u^2} \overset{19}{=} \frac{1}{2a}\ln\left|\frac{u + a}{u - a}\right| + C = \frac{1}{2\sqrt{3}}\ln\left|\frac{e^x + \sqrt{3}}{e^x - \sqrt{3}}\right| + C.$$

17. $\displaystyle \int \frac{x^4\, dx}{\sqrt{x^{10} - 2}} = \int \frac{x^4\, dx}{\sqrt{(x^5)^2 - 2}} = \frac{1}{5}\int \frac{du}{\sqrt{u^2 - 2}} \qquad \left[\begin{array}{l} u = x^5, \\ du = 5x^4\, dx \end{array}\right]$

$$\overset{43}{=} \frac{1}{5}\ln\left|u + \sqrt{u^2 - 2}\right| + C = \frac{1}{5}\ln\left|x^5 + \sqrt{x^{10} - 2}\right| + C$$

19. Let $u = \ln x$ and $a = 2$. Then $du = dx/x$ and

$$\int \frac{\sqrt{4 + (\ln x)^2}}{x}\, dx = \int \sqrt{a^2 + u^2}\, du \overset{21}{=} \frac{u}{2}\sqrt{a^2 + u^2} + \frac{a^2}{2}\ln\left(u + \sqrt{a^2 + u^2}\right) + C$$

$$= \frac{1}{2}(\ln x)\sqrt{4 + (\ln x)^2} + 2\ln\left[\ln x + \sqrt{4 + (\ln x)^2}\right] + C$$

21. Let $u = e^x$. Then $x = \ln u$, $dx = du/u$, so

$$\int \sqrt{e^{2x} - 1}\, dx = \int \frac{\sqrt{u^2 - 1}}{u}\, du \overset{41}{=} \sqrt{u^2 - 1} - \cos^{-1}(1/u) + C = \sqrt{e^{2x} - 1} - \cos^{-1}(e^{-x}) + C.$$

23. (a) $\displaystyle \frac{d}{du}\left[\frac{1}{b^3}\left(a + bu - \frac{a^2}{a + bu} - 2a\ln|a + bu|\right) + C\right] = \frac{1}{b^3}\left[b + \frac{ba^2}{(a + bu)^2} - \frac{2ab}{(a + bu)}\right]$

$$= \frac{1}{b^3}\left[\frac{b(a + bu)^2 + ba^2 - (a + bu)2ab}{(a + bu)^2}\right] = \frac{1}{b^3}\left[\frac{b^3 u^2}{(a + bu)^2}\right] = \frac{u^2}{(a + bu)^2}$$

(b) Let $t = a + bu \Rightarrow dt = b\, du$. Note that $u = \dfrac{t - a}{b}$ and $du = \dfrac{1}{b}\, dt$.

$$\int \frac{u^2\, du}{(a + bu)^2} = \frac{1}{b^3}\int \frac{(t - a)^2}{t^2}\, dt = \frac{1}{b^3}\int \frac{t^2 - 2at + a^2}{t^2}\, dt = \frac{1}{b^3}\int\left(1 - \frac{2a}{t} + \frac{a^2}{t^2}\right)dt$$

$$= \frac{1}{b^3}\left(t - 2a\ln|t| - \frac{a^2}{t}\right) + C = \frac{1}{b^3}\left(a + bu - \frac{a^2}{a + bu} - 2a\ln|a + bu|\right) + C$$

25. Maple and Mathematica both give $\int \sec^4 x\, dx = \frac{2}{3}\tan x + \frac{1}{3}\tan x \sec^2 x$, while Derive gives the second

term as $\dfrac{\sin x}{3\cos^3 x} = \dfrac{1}{3}\dfrac{\sin x}{\cos x}\dfrac{1}{\cos^2 x} = \dfrac{1}{3}\tan x \sec^2 x$. Using Formula 77, we get

$\int \sec^4 x\, dx = \frac{1}{3}\tan x \sec^2 x + \frac{2}{3}\int \sec^2 x\, dx = \frac{1}{3}\tan x \sec^2 x + \frac{2}{3}\tan x + C.$

27. Derive gives $\int x^2 \sqrt{x^2+4}\,dx = \frac{1}{4}x(x^2+2)\sqrt{x^2+4} - 2\ln(\sqrt{x^2+4}+x)$. Maple gives

$\frac{1}{4}x(x^2+4)^{3/2} - \frac{1}{2}x\sqrt{x^2+4} - 2\operatorname{arcsinh}(\frac{1}{2}x)$. Applying the command `convert(%,ln);` yields

$$\frac{1}{4}x(x^2+4)^{3/2} - \frac{1}{2}x\sqrt{x^2+4} - 2\ln(\frac{1}{2}x + \frac{1}{2}\sqrt{x^2+4}) = \frac{1}{4}x(x^2+4)^{1/2}[(x^2+4)-2] - 2\ln[(x+\sqrt{x^2+4})/2]$$
$$= \frac{1}{4}x(x^2+2)\sqrt{x^2+4} - 2\ln(\sqrt{x^2+4}+x) + 2\ln 2$$

Mathematica gives $\frac{1}{4}x(2+x^2)\sqrt{3+x^2} - 2\operatorname{arcsinh}(x/2)$. Applying the `TrigToExp` and `Simplify` commands gives

$\frac{1}{4}[x(2+x^2)\sqrt{4+x^2} - 8\log(\frac{1}{2}(x+\sqrt{4+x^2}))] = \frac{1}{4}x(x^2+2)\sqrt{x^2+4} - 2\ln(x+\sqrt{4+x^2}) + 2\ln 2$, so all are

equivalent (without constant).

Now use Formula 22 to get

$$\int x^2\sqrt{2^2+x^2}\,dx = \frac{x}{8}(2^2+2x^2)\sqrt{2^2+x^2} - \frac{2^4}{8}\ln(x+\sqrt{2^2+x^2}) + C$$
$$= \frac{x}{8}(2)(2+x^2)\sqrt{4+x^2} - 2\ln(x+\sqrt{4+x^2}) + C$$
$$= \frac{1}{4}x(x^2+2)\sqrt{x^2+4} - 2\ln(\sqrt{x^2+4}+x) + C$$

29. Maple gives $\int x\sqrt{1+2x}\,dx = \frac{1}{10}(1+2x)^{5/2} - \frac{1}{6}(1+2x)^{3/2}$, Mathematica gives $\sqrt{1+2x}(\frac{2}{5}x^2 + \frac{1}{15}x - \frac{1}{15})$, and Derive

gives $\frac{1}{15}(1+2x)^{3/2}(3x-1)$. The first two expressions can be simplified to Derive's result. If we use Formula 54, we get

$$\int x\sqrt{1+2x}\,dx = \frac{2}{15(2)^2}(3\cdot 2x - 2\cdot 1)(1+2x)^{3/2} + C = \frac{1}{30}(6x-2)(1+2x)^{3/2} + C = \frac{1}{15}(3x-1)(1+2x)^{3/2}.$$

31. Maple gives $\int \tan^5 x\,dx = \frac{1}{4}\tan^4 x - \frac{1}{2}\tan^2 x + \frac{1}{2}\ln(1+\tan^2 x)$, Mathematica gives

$\int \tan^5 x\,dx = \frac{1}{4}[-1-2\cos(2x)]\sec^4 x - \ln(\cos x)$, and Derive gives $\int \tan^5 x\,dx = \frac{1}{4}\tan^4 x - \frac{1}{2}\tan^2 x - \ln(\cos x)$.

These expressions are equivalent, and none includes absolute value bars or a constant of integration. Note that Mathematica's

and Derive's expressions suggest that the integral is undefined where $\cos x < 0$, which is not the case. Using Formula 75,

$\int \tan^5 x\,dx = \frac{1}{5-1}\tan^{5-1} x - \int \tan^{5-2} x\,dx = \frac{1}{4}\tan^4 x - \int \tan^3 x\,dx$. Using Formula 69,

$\int \tan^3 x\,dx = \frac{1}{2}\tan^2 x + \ln|\cos x| + C$, so $\int \tan^5 x\,dx = \frac{1}{4}\tan^4 x - \frac{1}{2}\tan^2 x - \ln|\cos x| + C$.

33. (a) $F(x) = \int f(x)\,dx = \int \dfrac{1}{x\sqrt{1-x^2}}\,dx \overset{35}{=} -\frac{1}{1}\ln\left|\dfrac{1+\sqrt{1-x^2}}{x}\right| + C = -\ln\left|\dfrac{1+\sqrt{1-x^2}}{x}\right| + C$.

f has domain $\{x \mid x \neq 0, 1-x^2 > 0\} = \{x \mid x \neq 0, |x| < 1\} = (-1,0) \cup (0,1)$. F has the same domain.

(b) Derive gives $F(x) = \ln(\sqrt{1-x^2} - 1) - \ln x$ and Mathematica gives $F(x) = \ln x - \ln(1+\sqrt{1-x^2})$.

Both are correct if you take absolute values of the logarithm arguments, and both would then have the

same domain. Maple gives $F(x) = -\operatorname{arctanh}(1/\sqrt{1-x^2})$. This function has domain

$\{x \mid |x| < 1, -1 < 1/\sqrt{1-x^2} < 1\} = \{x \mid |x| < 1, 1/\sqrt{1-x^2} < 1\} = \{x \mid |x| < 1, \sqrt{1-x^2} > 1\} = \emptyset$,

the empty set! If we apply the command `convert(%,ln);` to Maple's answer, we get

$-\dfrac{1}{2}\ln\left(\dfrac{1}{\sqrt{1-x^2}} + 1\right) + \dfrac{1}{2}\ln\left(1 - \dfrac{1}{\sqrt{1-x^2}}\right)$, which has the same domain, $\emptyset$.

5.9 Approximate Integration

1. (a) $\Delta x = (b-a)/n = (4-0)/2 = 2$

$$L_2 = \sum_{i=1}^{2} f(x_{i-1})\,\Delta x = f(x_0) \cdot 2 + f(x_1) \cdot 2 = 2\left[f(0) + f(2)\right] = 2(0.5 + 2.5) = 6$$

$$R_2 = \sum_{i=1}^{2} f(x_i)\,\Delta x = f(x_1) \cdot 2 + f(x_2) \cdot 2 = 2\left[f(2) + f(4)\right] = 2(2.5 + 3.5) = 12$$

$$M_2 = \sum_{i=1}^{2} f(\overline{x}_i)\Delta x = f(\overline{x}_1) \cdot 2 + f(\overline{x}_2) \cdot 2 = 2\left[f(1) + f(3)\right] \approx 2(1.6 + 3.2) = 9.6$$

(b)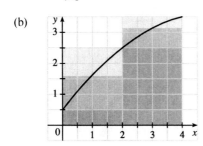

L_2 is an underestimate, since the area under the small rectangles is less than the area under the curve, and R_2 is an overestimate, since the area under the large rectangles is greater than the area under the curve. It appears that M_2 is an overestimate, though it is fairly close to I. See the solution to Exercise 39 for a proof of the fact that if f is concave down on $[a, b]$, then the Midpoint Rule is an overestimate of $\int_a^b f(x)\,dx$.

(c) $T_2 = \left(\frac{1}{2}\Delta x\right)[f(x_0) + 2f(x_1) + f(x_2)] = \frac{2}{2}[f(0) + 2f(2) + f(4)] = 0.5 + 2(2.5) + 3.5 = 9.$

This approximation is an underestimate, since the graph is concave down. Thus, $T_2 = 9 < I$. See the solution to Exercise 39 for a general proof of this conclusion.

(d) For any n, we will have $L_n < T_n < I < M_n < R_n$.

3. $f(x) = \cos(x^2)$, $\Delta x = \frac{1-0}{4} = \frac{1}{4}$

(a) $T_4 = \frac{1}{4 \cdot 2}\left[f(0) + 2f\left(\frac{1}{4}\right) + 2f\left(\frac{2}{4}\right) + 2f\left(\frac{3}{4}\right) + f(1)\right] \approx 0.895759$

(b) $M_4 = \frac{1}{4}\left[f\left(\frac{1}{8}\right) + f\left(\frac{3}{8}\right) + f\left(\frac{5}{8}\right) + f\left(\frac{7}{8}\right)\right] \approx 0.908907$

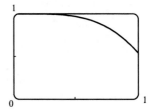

The graph shows that f is concave down on $[0, 1]$. So T_4 is an underestimate and M_4 is an overestimate. We can conclude that $0.895759 < \int_0^1 \cos(x^2)\,dx < 0.908907$.

5. (a) $f(x) = \dfrac{x}{1+x^2}$, $\quad \Delta x = \dfrac{b-a}{n} = \dfrac{2-0}{10} = \dfrac{1}{5}$

$$M_{10} = \frac{1}{5}\left[f\left(\frac{1}{10}\right) + f\left(\frac{3}{10}\right) + f\left(\frac{5}{10}\right) + \cdots + f\left(\frac{19}{10}\right)\right] \approx 0.806598$$

(b) $S_{10} = \dfrac{1}{5 \cdot 3}\left[f(0) + 4f\left(\frac{1}{5}\right) + 2f\left(\frac{2}{5}\right) + 4f\left(\frac{3}{5}\right) + 2f\left(\frac{4}{5}\right) + \cdots + 4f\left(\frac{9}{5}\right) + f(2)\right] \approx 0.804779$

$$\textit{Actual: } I = \int_0^2 \frac{x}{1+x^2}\,dx = \left[\tfrac{1}{2}\ln\left|1+x^2\right|\right]_0^2 \qquad [u = 1 + x^2,\, du = 2x\,dx]$$

$$= \tfrac{1}{2}\ln 5 - \tfrac{1}{2}\ln 1 = \tfrac{1}{2}\ln 5 \approx 0.804719$$

$$\textit{Errors: } E_M = \text{ actual} - M_{10} = I - M_{10} \approx -0.001879$$

$$E_S = \text{ actual} - S_{10} = I - S_{10} \approx -0.000060$$

7. $f(x) = \sqrt[4]{1+x^2}$, $\quad \Delta x = \dfrac{2-0}{8} = \dfrac{1}{4}$

(a) $T_8 = \frac{1}{4 \cdot 2}\left[f(0) + 2f\left(\frac{1}{4}\right) + 2f\left(\frac{1}{2}\right) + \cdots + 2f\left(\frac{3}{2}\right) + 2f\left(\frac{7}{4}\right) + f(2)\right] \approx 2.413790$

(b) $M_8 = \frac{1}{4}\left[f\left(\frac{1}{8}\right) + f\left(\frac{3}{8}\right) + \cdots + f\left(\frac{13}{8}\right) + f\left(\frac{15}{8}\right)\right] \approx 2.411453$

(c) $S_8 = \frac{1}{4 \cdot 3}\left[f(0) + 4f\left(\frac{1}{4}\right) + 2f\left(\frac{1}{2}\right) + 4f\left(\frac{3}{4}\right) + 2f(1) + 4f\left(\frac{5}{4}\right) + 2f\left(\frac{3}{2}\right) + 4f\left(\frac{7}{4}\right) + f(2)\right] \approx 2.412232$

9. $f(x) = \dfrac{\ln x}{1+x}$, $\Delta x = \dfrac{2-1}{10} = \dfrac{1}{10}$

(a) $T_{10} = \frac{1}{10 \cdot 2}[f(1) + 2f(1.1) + 2f(1.2) + \cdots + 2f(1.8) + 2f(1.9) + f(2)] \approx 0.146879$

(b) $M_{10} = \frac{1}{10}[f(1.05) + f(1.15) + \cdots + f(1.85) + f(1.95)] \approx 0.147391$

(c) $S_{10} = \frac{1}{10 \cdot 3}[f(1) + 4f(1.1) + 2f(1.2) + 4f(1.3) + 2f(1.4) + 4f(1.5) + 2f(1.6) + 4f(1.7)$
$\qquad\qquad + 2f(1.8) + 4f(1.9) + f(2)]$

$\qquad \approx 0.147219$

11. $f(t) = \sin(e^{t/2})$, $\Delta t = \dfrac{\frac{1}{2}-0}{8} = \dfrac{1}{16}$

(a) $T_8 = \frac{1}{16 \cdot 2}\left[f(0) + 2f\left(\frac{1}{16}\right) + 2f\left(\frac{2}{16}\right) + \cdots + 2f\left(\frac{7}{16}\right) + f\left(\frac{1}{2}\right)\right] \approx 0.451948$

(b) $M_8 = \frac{1}{16}\left[f\left(\frac{1}{32}\right) + f\left(\frac{3}{32}\right) + f\left(\frac{5}{32}\right) + \cdots + f\left(\frac{13}{32}\right) + f\left(\frac{15}{32}\right)\right] \approx 0.451991$

(c) $S_8 = \frac{1}{16 \cdot 3}\left[f(0) + 4f\left(\frac{1}{16}\right) + 2f\left(\frac{2}{16}\right) + \cdots + 4f\left(\frac{7}{16}\right) + f\left(\frac{1}{2}\right)\right] \approx 0.451976$

13. $f(t) = e^{\sqrt{t}}\sin t$, $\Delta t = \dfrac{4-0}{8} = \dfrac{1}{2}$

(a) $T_8 = \frac{1}{2 \cdot 2}\left[f(0) + 2f\left(\frac{1}{2}\right) + 2f(1) + 2f\left(\frac{3}{2}\right) + 2f(2) + 2f\left(\frac{5}{2}\right) + 2f(3) + 2f\left(\frac{7}{2}\right) + f(4)\right] \approx 4.513618$

(b) $M_8 = \frac{1}{2}\left[f\left(\frac{1}{4}\right) + f\left(\frac{3}{4}\right) + f\left(\frac{5}{4}\right) + f\left(\frac{7}{4}\right) + f\left(\frac{9}{4}\right) + f\left(\frac{11}{4}\right) + f\left(\frac{13}{4}\right) + f\left(\frac{15}{4}\right)\right] \approx 4.748256$

(c) $S_8 = \frac{1}{2 \cdot 3}\left[f(0) + 4f\left(\frac{1}{2}\right) + 2f(1) + 4f\left(\frac{3}{2}\right) + 2f(2) + 4f\left(\frac{5}{2}\right) + 2f(3) + 4f\left(\frac{7}{2}\right) + f(4)\right] \approx 4.675111$

15. $f(x) = \dfrac{\cos x}{x}$, $\Delta x = \dfrac{5-1}{8} = \dfrac{1}{2}$

(a) $T_8 = \frac{1}{2 \cdot 2}\left[f(1) + 2f\left(\frac{3}{2}\right) + 2f(2) + \cdots + 2f(4) + 2f\left(\frac{9}{2}\right) + f(5)\right] \approx -0.495333$

(b) $M_8 = \frac{1}{2}\left[f\left(\frac{5}{4}\right) + f\left(\frac{7}{4}\right) + f\left(\frac{9}{4}\right) + f\left(\frac{11}{4}\right) + f\left(\frac{13}{4}\right) + f\left(\frac{15}{4}\right) + f\left(\frac{17}{4}\right) + f\left(\frac{19}{4}\right)\right] \approx -0.543321$

(c) $S_8 = \frac{1}{2 \cdot 3}\left[f(1) + 4f\left(\frac{3}{2}\right) + 2f(2) + 4f\left(\frac{5}{2}\right) + 2f(3) + 4f\left(\frac{7}{2}\right) + 2f(4) + 4f\left(\frac{9}{2}\right) + f(5)\right] \approx -0.526123$

17. $f(x) = \cos(x^2)$, $\Delta x = \dfrac{1-0}{8} = \dfrac{1}{8}$

(a) $T_8 = \frac{1}{8 \cdot 2}\left\{f(0) + 2\left[f\left(\frac{1}{8}\right) + f\left(\frac{2}{8}\right) + \cdots + f\left(\frac{7}{8}\right)\right] + f(1)\right\} \approx 0.902333$

$\quad M_8 = \frac{1}{8}\left[f\left(\frac{1}{16}\right) + f\left(\frac{3}{16}\right) + f\left(\frac{5}{16}\right) + \cdots + f\left(\frac{15}{16}\right)\right] = 0.905620$

(b) $f(x) = \cos(x^2)$, $f'(x) = -2x\sin(x^2)$, $f''(x) = -2\sin(x^2) - 4x^2\cos(x^2)$. For $0 \le x \le 1$, sin and cos are positive,

so $|f''(x)| = 2\sin(x^2) + 4x^2\cos(x^2) \le 2\cdot 1 + 4\cdot 1\cdot 1 = 6$ since $\sin(x^2) \le 1$ and $\cos(x^2) \le 1$ for all x,

and $x^2 \le 1$ for $0 \le x \le 1$. So for $n = 8$, we take $K = 6$, $a = 0$, and $b = 1$ in Theorem 3, to get

$|E_T| \le 6 \cdot 1^3/(12 \cdot 8^2) = \frac{1}{128} = 0.0078125$ and $|E_M| \le \frac{1}{256} = 0.00390625$. [A better estimate is obtained by noting

from a graph of f'' that $|f''(x)| \le 4$ for $0 \le x \le 1$.]

(c) Take $K = 6$ [as in part (b)] in Theorem 3. $|E_T| \le \dfrac{K(b-a)^3}{12n^2} \le 0.0001 \;\Leftrightarrow\; \dfrac{6(1-0)^3}{12n^2} \le 10^{-4} \;\Leftrightarrow\;$

$\dfrac{1}{2n^2} \le \dfrac{1}{10^4} \;\Leftrightarrow\; 2n^2 \ge 10^4 \;\Leftrightarrow\; n^2 \ge 5000 \;\Leftrightarrow\; n \ge 71$. Take $n = 71$ for T_n. For E_M, again take $K = 6$ in

Theorem 3 to get $|E_M| \le 10^{-4} \;\Leftrightarrow\; 4n^2 \ge 10^4 \;\Leftrightarrow\; n^2 \ge 2500 \;\Leftrightarrow\; n \ge 50$. Take $n = 50$ for M_n.

19. $f(x) = \sin x$, $\Delta x = \frac{\pi - 0}{10} = \frac{\pi}{10}$

(a) $T_{10} = \frac{\pi}{10\cdot 2}\left[f(0) + 2f\left(\frac{\pi}{10}\right) + 2f\left(\frac{2\pi}{10}\right) + \cdots + 2f\left(\frac{9\pi}{10}\right) + f(\pi)\right] \approx 1.983524$

$M_{10} = \frac{\pi}{10}\left[f\left(\frac{\pi}{20}\right) + f\left(\frac{3\pi}{20}\right) + f\left(\frac{5\pi}{20}\right) + \cdots + f\left(\frac{19\pi}{20}\right)\right] \approx 2.008248$

$S_{10} = \frac{\pi}{10\cdot 3}\left[f(0) + 4f\left(\frac{\pi}{10}\right) + 2f\left(\frac{2\pi}{10}\right) + 4f\left(\frac{3\pi}{10}\right) + \cdots + 4f\left(\frac{9\pi}{10}\right) + f(\pi)\right] \approx 2.000110$

Since $I = \int_0^\pi \sin x\, dx = \left[-\cos x\right]_0^\pi = 1 - (-1) = 2$, $E_T = I - T_{10} \approx 0.016476$, $E_M = I - M_{10} \approx -0.008248$,

and $E_S = I - S_{10} \approx -0.000110$.

(b) $f(x) = \sin x \;\Rightarrow\; \left|f^{(n)}(x)\right| \le 1$, so take $K = 1$ for all error estimates.

$|E_T| \le \dfrac{K(b-a)^3}{12n^2} = \dfrac{1(\pi - 0)^3}{12(10)^2} = \dfrac{\pi^3}{1200} \approx 0.025839. \quad |E_M| \le \dfrac{|E_T|}{2} = \dfrac{\pi^3}{2400} \approx 0.012919.$

$|E_S| \le \dfrac{K(b-a)^5}{180n^4} = \dfrac{1(\pi - 0)^5}{180(10)^4} = \dfrac{\pi^5}{1{,}800{,}000} \approx 0.000170.$

The actual error is about 64% of the error estimate in all three cases.

(c) $|E_T| \le 0.00001 \;\Leftrightarrow\; \dfrac{\pi^3}{12n^2} \le \dfrac{1}{10^5} \;\Leftrightarrow\; n^2 \ge \dfrac{10^5\pi^3}{12} \;\Rightarrow\; n \ge 508.3$. Take $n = 509$ for T_n.

$|E_M| \le 0.00001 \;\Leftrightarrow\; \dfrac{\pi^3}{24n^2} \le \dfrac{1}{10^5} \;\Leftrightarrow\; n^2 \ge \dfrac{10^5\pi^3}{24} \;\Rightarrow\; n \ge 359.4$. Take $n = 360$ for M_n.

$|E_S| \le 0.00001 \;\Leftrightarrow\; \dfrac{\pi^5}{180n^4} \le \dfrac{1}{10^5} \;\Leftrightarrow\; n^4 \ge \dfrac{10^5\pi^5}{180} \;\Rightarrow\; n \ge 20.3.$

Take $n = 22$ for S_n (since n must be even).

21. (a) Using a CAS, we differentiate $f(x) = e^{\cos x}$ twice, and find that

$f''(x) = e^{\cos x}(\sin^2 x - \cos x)$. From the graph, we see that the maximum

value of $|f''(x)|$ occurs at the endpoints of the interval $[0, 2\pi]$.

Since $f''(0) = -e$, we can use $K = e$ or $K = 2.8$.

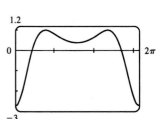

(b) A CAS gives $M_{10} \approx 7.954926518$. (In Maple, use `student[middlesum]`.)

(c) Using Theorem 3 for the Midpoint Rule, with $K = e$, we get $|E_M| \le \dfrac{e(2\pi - 0)^3}{24 \cdot 10^2} \approx 0.280945995.$

With $K = 2.8$, we get $|E_M| \le \dfrac{2.8(2\pi - 0)^3}{24 \cdot 10^2} = 0.289391916.$

(d) A CAS gives $I \approx 7.954926521$.

(e) The actual error is only about 3×10^{-9}, much less than the estimate in part (c).

(f) We use the CAS to differentiate twice more, and then graph

$$f^{(4)}(x) = e^{\cos x}(\sin^4 x - 6\sin^2 x \, \cos x + 3 - 7\sin^2 x + \cos x).$$

From the graph, we see that the maximum value of $\left| f^{(4)}(x) \right|$ occurs at the

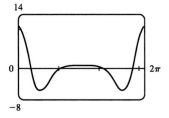

endpoints of the interval $[0, 2\pi]$. Since $f^{(4)}(0) = 4e$, we can use $K = 4e$

or $K = 10.9$.

(g) A CAS gives $S_{10} \approx 7.953789422$. (In Maple, use `student[simpson]`.)

(h) Using Theorem 4 with $K = 4e$, we get $|E_S| \le \dfrac{4e(2\pi - 0)^5}{180 \cdot 10^4} \approx 0.059153618$.

With $K = 10.9$, we get $|E_S| \le \dfrac{10.9(2\pi - 0)^5}{180 \cdot 10^4} \approx 0.059299814$.

(i) The actual error is about $7.954926521 - 7.953789422 \approx 0.00114$. This is quite a bit smaller than the estimate in part (h), though the difference is not nearly as great as it was in the case of the Midpoint Rule.

(j) To ensure that $|E_S| \le 0.0001$, we use Theorem 4: $|E_S| \le \dfrac{4e(2\pi)^5}{180 \cdot n^4} \le 0.0001 \quad \Rightarrow \quad \dfrac{4e(2\pi)^5}{180 \cdot 0.0001} \le n^4 \quad \Rightarrow$

$n^4 \ge 5{,}915{,}362 \quad \Leftrightarrow \quad n \ge 49.3$. So we must take $n \ge 50$ to ensure that $|I - S_n| \le 0.0001$.

$(K = 10.9$ leads to the same value of n.)

23. $I = \int_0^1 x e^x \, dx = [(x-1)e^x]_0^1$ [parts or Formula 96] $= 0 - (-1) = 1$, $f(x) = xe^x$, $\Delta x = 1/n$

$n = 5$: $L_5 = \frac{1}{5}[f(0) + f(0.2) + f(0.4) + f(0.6) + f(0.8)] \approx 0.742943$

 $R_5 = \frac{1}{5}[f(0.2) + f(0.4) + f(0.6) + f(0.8) + f(1)] \approx 1.286599$

 $T_5 = \frac{1}{5 \cdot 2}[f(0) + 2f(0.2) + 2f(0.4) + 2f(0.6) + 2f(0.8) + f(1)] \approx 1.014771$

 $M_5 = \frac{1}{5}[f(0.1) + f(0.3) + f(0.5) + f(0.7) + f(0.9)] \approx 0.992621$

 $E_L = I - L_5 \approx 1 - 0.742943 = 0.257057$

 $E_R \approx 1 - 1.286599 = -0.286599$

 $E_T \approx 1 - 1.014771 = -0.014771$

 $E_M \approx 1 - 0.992621 = 0.007379$

$n = 10$: $L_{10} = \frac{1}{10}[f(0) + f(0.1) + f(0.2) + \cdots + f(0.9)] \approx 0.867782$

 $R_{10} = \frac{1}{10}[f(0.1) + f(0.2) + \cdots + f(0.9) + f(1)] \approx 1.139610$

 $T_{10} = \frac{1}{10 \cdot 2}\{f(0) + 2[f(0.1) + f(0.2) + \cdots + f(0.9)] + f(1)\} \approx 1.003696$

 $M_{10} = \frac{1}{10}[f(0.05) + f(0.15) + \cdots + f(0.85) + f(0.95)] \approx 0.998152$

 $E_L = I - L_{10} \approx 1 - 0.867782 = 0.132218$

 $E_R \approx 1 - 1.139610 = -0.139610$

 $E_T \approx 1 - 1.003696 = -0.003696$

 $E_M \approx 1 - 0.998152 = 0.001848$

$n = 20$: $L_{20} = \frac{1}{20}[f(0) + f(0.05) + f(0.10) + \cdots + f(0.95)] \approx 0.932967$

$R_{20} = \frac{1}{20}[f(0.05) + f(0.10) + \cdots + f(0.95) + f(1)] \approx 1.068881$

$T_{20} = \frac{1}{20 \cdot 2}\{f(0) + 2[f(0.05) + f(0.10) + \cdots + f(0.95)] + f(1)\} \approx 1.000924$

$M_{20} = \frac{1}{20}[f(0.025) + f(0.075) + f(0.125) + \cdots + f(0.975)] \approx 0.999538$

$E_L = I - L_{20} \approx 1 - 0.932967 = 0.067033$

$E_R \approx 1 - 1.068881 = -0.068881$

$E_T \approx 1 - 1.000924 = -0.000924$

$E_M \approx 1 - 0.999538 = 0.000462$

n	L_n	R_n	T_n	M_n
5	0.742943	1.286599	1.014771	0.992621
10	0.867782	1.139610	1.003696	0.998152
20	0.932967	1.068881	1.000924	0.999538

n	E_L	E_R	E_T	E_M
5	0.257057	−0.286599	−0.014771	0.007379
10	0.132218	−0.139610	−0.003696	0.001848
20	0.067033	−0.068881	−0.000924	0.000462

Observations:

1. E_L and E_R are always opposite in sign, as are E_T and E_M.

2. As n is doubled, E_L and E_R are decreased by about a factor of 2, and E_T and E_M are decreased by a factor of about 4.

3. The Midpoint approximation is about twice as accurate as the Trapezoidal approximation.

4. All the approximations become more accurate as the value of n increases.

5. The Midpoint and Trapezoidal approximations are much more accurate than the endpoint approximations.

25. $I = \int_0^2 x^4 \, dx = \left[\frac{1}{5}x^5\right]_0^2 = \frac{32}{5} - 0 = 6.4$, $f(x) = x^4$, $\Delta x = \frac{2-0}{n} = \frac{2}{n}$

$n = 6$: $T_6 = \frac{2}{6 \cdot 2}\{f(0) + 2[f(\frac{1}{3}) + f(\frac{2}{3}) + f(\frac{3}{3}) + f(\frac{4}{3}) + f(\frac{5}{3})] + f(2)\} \approx 6.695473$

$M_6 = \frac{2}{6}[f(\frac{1}{6}) + f(\frac{3}{6}) + f(\frac{5}{6}) + f(\frac{7}{6}) + f(\frac{9}{6}) + f(\frac{11}{6})] \approx 6.252572$

$S_6 = \frac{2}{6 \cdot 3}[f(0) + 4f(\frac{1}{3}) + 2f(\frac{2}{3}) + 4f(\frac{3}{3}) + 2f(\frac{4}{3}) + 4f(\frac{5}{3}) + f(2)] \approx 6.403292$

$E_T = I - T_6 \approx 6.4 - 6.695473 = -0.295473$

$E_M \approx 6.4 - 6.252572 = 0.147428$

$E_S \approx 6.4 - 6.403292 = -0.003292$

$n = 12$: $T_{12} = \frac{2}{12 \cdot 2}\{f(0) + 2[f(\frac{1}{6}) + f(\frac{2}{6}) + f(\frac{3}{6}) + \cdots + f(\frac{11}{6})] + f(2)\} \approx 6.474023$

$M_6 = \frac{2}{12}[f(\frac{1}{12}) + f(\frac{3}{12}) + f(\frac{5}{12}) + \cdots + f(\frac{23}{12})] \approx 6.363008$

$S_6 = \frac{2}{12 \cdot 3}[f(0) + 4f(\frac{1}{6}) + 2f(\frac{2}{6}) + 4f(\frac{3}{6}) + 2f(\frac{4}{6}) + \cdots + 4f(\frac{11}{6}) + f(2)] \approx 6.400206$

$E_T = I - T_{12} \approx 6.4 - 6.474023 = -0.074023$

$E_M \approx 6.4 - 6.363008 = 0.036992$

$E_S \approx 6.4 - 6.400206 = -0.000206$

n	T_n	M_n	S_n
6	6.695473	6.252572	6.403292
12	6.474023	6.363008	6.400206

n	E_T	E_M	E_S
6	−0.295473	0.147428	−0.003292
12	−0.074023	0.036992	−0.000206

[continued]

Observations:

1. E_T and E_M are opposite in sign and decrease by a factor of about 4 as n is doubled.

2. The Simpson's approximation is much more accurate than the Midpoint and Trapezoidal approximations, and E_S seems to decrease by a factor of about 16 as n is doubled.

27. $\Delta x = (b-a)/n = (6-0)/6 = 1$

(a) $T_6 = \frac{\Delta x}{2}[f(0) + 2f(1) + 2f(2) + 2f(3) + 2f(4) + 2f(5) + f(6)]$

$\approx \frac{1}{2}[3 + 2(5) + 2(4) + 2(2) + 2(2.8) + 2(4) + 1]$

$= \frac{1}{2}(39.6) = 19.8$

(b) $M_6 = \Delta x[f(0.5) + f(1.5) + f(2.5) + f(3.5) + f(4.5) + f(5.5)]$

$\approx 1[4.5 + 4.7 + 2.6 + 2.2 + 3.4 + 3.2]$

$= 20.6$

(c) $S_6 = \frac{\Delta x}{3}[f(0) + 4f(1) + 2f(2) + 4f(3) + 2f(4) + 4f(5) + f(6)]$

$\approx \frac{1}{3}[3 + 4(5) + 2(4) + 4(2) + 2(2.8) + 4(4) + 1]$

$= \frac{1}{3}(61.6) = 20.5\overline{3}$

29. By the Net Change Theorem, the increase in velocity is equal to $\int_0^6 a(t)\,dt$. We use Simpson's Rule with $n = 6$ and $\Delta t = (6-0)/6 = 1$ to estimate this integral:

$\int_0^6 a(t)\,dt \approx S_6 = \frac{1}{3}[a(0) + 4a(1) + 2a(2) + 4a(3) + 2a(4) + 4a(5) + a(6)]$

$\approx \frac{1}{3}[0 + 4(0.5) + 2(4.1) + 4(9.8) + 2(12.9) + 4(9.5) + 0] = \frac{1}{3}(113.2) = 37.7\overline{3}$ ft/s

31. By the Net Change Theorem, the energy used is equal to $\int_0^6 P(t)\,dt$. We use Simpson's Rule with $n = 12$ and $\Delta t = \frac{6-0}{12} = \frac{1}{2}$ to estimate this integral:

$\int_0^6 P(t)\,dt \approx S_{12} = \frac{1/2}{3}[P(0) + 4P(0.5) + 2P(1) + 4P(1.5) + 2P(2) + 4P(2.5) + 2P(3)$

$+ 4P(3.5) + 2P(4) + 4P(4.5) + 2P(5) + 4P(5.5) + P(6)]$

$= \frac{1}{6}[1814 + 4(1735) + 2(1686) + 4(1646) + 2(1637) + 4(1609) + 2(1604)$

$+ 4(1611) + 2(1621) + 4(1666) + 2(1745) + 4(1886) + 2052]$

$= \frac{1}{6}(61{,}064) = 10{,}177.\overline{3}$ megawatt-hours

33. (a) We are given the function values at the endpoints of 8 intervals of length 0.4, so we'll use the Midpoint Rule with $n = 8/2 = 4$ and $\Delta x = (3.2 - 0)/4 = 0.8$.

$\int_0^{3.2} f(x)\,dx \approx M_4 = 0.8[f(0.4) + f(1.2) + f(2.0) + f(2.8)] = 0.8[6.5 + 6.4 + 7.6 + 8.8]$

$= 0.8(29.3) = 23.44$

(b) $-4 \le f''(x) \le 1 \;\Rightarrow\; |f''(x)| \le 4$, so use $K = 4$, $a = 0$, $b = 3.2$, and $n = 4$ in Theorem 3.

So $|E_M| \le \dfrac{4(3.2 - 0)^3}{24(4)^2} = \dfrac{128}{375} = 0.341\overline{3}$.

35. $I(\theta) = \dfrac{N^2 \sin^2 k}{k^2}$, where $k = \dfrac{\pi N d \sin \theta}{\lambda}$, $N = 10{,}000$, $d = 10^{-4}$, and $\lambda = 632.8 \times 10^{-9}$. So $I(\theta) = \dfrac{(10^4)^2 \sin^2 k}{k^2}$,

where $k = \dfrac{\pi (10^4)(10^{-4}) \sin \theta}{632.8 \times 10^{-9}}$. Now $n = 10$ and $\Delta\theta = \dfrac{10^{-6} - (-10^{-6})}{10} = 2 \times 10^{-7}$, so

$$M_{10} = 2 \times 10^{-7}[I(-0.0000009) + I(-0.0000007) + \cdots + I(0.0000009)] \approx 59.4.$$

37. Consider the function f whose graph is shown. The area $\int_0^2 f(x)\, dx$

is close to 2. The Trapezoidal Rule gives

$$T_2 = \tfrac{2-0}{2 \cdot 2} \left[f(0) + 2f(1) + f(2) \right] = \tfrac{1}{2}\left[1 + 2 \cdot 1 + 1 \right] = 2.$$

The Midpoint Rule gives $M_2 = \tfrac{2-0}{2}\left[f(0.5) + f(1.5) \right] = 1[0 + 0] = 0$,

so the Trapezoidal Rule is more accurate.

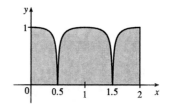

39. Since the Trapezoidal and Midpoint approximations on the interval $[a, b]$ are the sums of the Trapezoidal and Midpoint

approximations on the subintervals $[x_{i-1}, x_i]$, $i = 1, 2, \ldots, n$, we can focus our attention on one such interval. The condition

$f''(x) < 0$ for $a \le x \le b$ means that the graph of f is concave down as in Figure 5. In that figure, T_n is the area of the

trapezoid $AQRD$, $\int_a^b f(x)\, dx$ is the area of the region $AQPRD$, and M_n is the area of the trapezoid $ABCD$, so

$T_n < \int_a^b f(x)\, dx < M_n$. In general, the condition $f'' < 0$ implies that the graph of f on $[a, b]$ lies above the chord joining the

points $(a, f(a))$ and $(b, f(b))$. Thus, $\int_a^b f(x)\, dx > T_n$. Since M_n is the area under a tangent to the graph, and since $f'' < 0$

implies that the tangent lies above the graph, we also have $M_n > \int_a^b f(x)\, dx$. Thus, $T_n < \int_a^b f(x)\, dx < M_n$.

41. $T_n = \tfrac{1}{2} \Delta x \left[f(x_0) + 2f(x_1) + \cdots + 2f(x_{n-1}) + f(x_n) \right]$ and

$M_n = \Delta x \left[f(\overline{x}_1) + f(\overline{x}_2) + \cdots + f(\overline{x}_{n-1}) + f(\overline{x}_n) \right]$, where $\overline{x}_i = \tfrac{1}{2}(x_{i-1} + x_i)$. Now

$T_{2n} = \tfrac{1}{2}\left(\tfrac{1}{2}\Delta x \right)[f(x_0) + 2f(\overline{x}_1) + 2f(x_1) + 2f(\overline{x}_2) + 2f(x_2) + \cdots + 2f(\overline{x}_{n-1}) + 2f(x_{n-1}) + 2f(\overline{x}_n) + f(x_n)]$ so

$\tfrac{1}{2}(T_n + M_n) = \tfrac{1}{2}T_n + \tfrac{1}{2}M_n$

$\qquad = \tfrac{1}{4}\Delta x [f(x_0) + 2f(x_1) + \cdots + 2f(x_{n-1}) + f(x_n)] + \tfrac{1}{4}\Delta x [2f(\overline{x}_1) + 2f(\overline{x}_2) + \cdots + 2f(\overline{x}_{n-1}) + 2f(\overline{x}_n)]$

$\qquad = T_{2n}$

5.10   Improper Integrals

1. (a) Since $\int_1^\infty x^4 e^{-x^4}\, dx$ has an infinite interval of integration, it is an improper integral of Type I.

(b) Since $y = \sec x$ has an infinite discontinuity at $x = \tfrac{\pi}{2}$, $\int_0^{\pi/2} \sec x\, dx$ is a Type II improper integral.

(c) Since $y = \dfrac{x}{(x-2)(x-3)}$ has an infinite discontinuity at $x = 2$, $\displaystyle\int_0^2 \dfrac{x}{x^2 - 5x + 6}\, dx$ is a Type II improper integral.

(d) Since $\displaystyle\int_{-\infty}^0 \dfrac{1}{x^2 + 5}\, dx$ has an infinite interval of integration, it is an improper integral of Type I.

3. The area under the graph of $y = 1/x^3 = x^{-3}$ between $x = 1$ and $x = t$ is

$$A(t) = \int_1^t x^{-3}\, dx = \left[-\tfrac{1}{2}x^{-2} \right]_1^t = -\tfrac{1}{2}t^{-2} - \left(-\tfrac{1}{2} \right) = \tfrac{1}{2} - 1/(2t^2).$$ So the area for $1 \le x \le 10$ is

$A(10) = 0.5 - 0.005 = 0.495$, the area for $1 \le x \le 100$ is $A(100) = 0.5 - 0.00005 = 0.49995$, and the area for

$1 \le x \le 1000$ is $A(1000) = 0.5 - 0.0000005 = 0.4999995$. The total area under the curve for $x \ge 1$ is

$\lim_{t\to\infty} A(t) = \lim_{t\to\infty} \left[\frac{1}{2} - 1/(2t^2) \right] = \frac{1}{2}$.

5. $\int_3^\infty \frac{1}{(x-2)^{3/2}} \, dx = \lim_{t\to\infty} \int_3^t (x-2)^{-3/2} \, dx = \lim_{t\to\infty} \left[-2(x-2)^{-1/2} \right]_3^t$ $\quad [u = x - 2, \, du = dx]$

$= \lim_{t\to\infty} \left(\frac{-2}{\sqrt{t-2}} + \frac{2}{\sqrt{1}} \right) = 0 + 2 = 2.$ Convergent

7. $\int_{-\infty}^{-1} \frac{1}{\sqrt{2-w}} \, dw = \lim_{t\to-\infty} \int_t^{-1} \frac{1}{\sqrt{2-w}} \, dw = \lim_{t\to-\infty} \left[-2\sqrt{2-w} \right]_t^{-1}$ $\quad [u = 2 - w, \, du = -dw]$

$= \lim_{t\to-\infty} \left[-2\sqrt{3} + 2\sqrt{2-t} \right] = \infty.$ Divergent

9. $\int_4^\infty e^{-y/2} \, dy = \lim_{t\to\infty} \int_4^t e^{-y/2} \, dy = \lim_{t\to\infty} \left[-2e^{-y/2} \right]_4^t = \lim_{t\to\infty} (-2e^{-t/2} + 2e^{-2}) = 0 + 2e^{-2} = 2e^{-2}.$

Convergent

11. $\int_{2\pi}^\infty \sin\theta \, d\theta = \lim_{t\to\infty} \int_{2\pi}^t \sin\theta \, d\theta = \lim_{t\to\infty} \left[-\cos\theta \right]_{2\pi}^t = \lim_{t\to\infty} (-\cos t + 1).$ This limit does not exist, so the integral is

divergent.   Divergent

13. $\int_{-\infty}^\infty xe^{-x^2} \, dx = \int_{-\infty}^0 xe^{-x^2} \, dx + \int_0^\infty xe^{-x^2} \, dx.$

$\int_{-\infty}^0 xe^{-x^2} \, dx = \lim_{t\to-\infty} \left(-\frac{1}{2} \right) \left[e^{-x^2} \right]_t^0 = \lim_{t\to-\infty} \left(-\frac{1}{2} \right) \left(1 - e^{-t^2} \right) = -\frac{1}{2} \cdot 1 = -\frac{1}{2},$ and

$\int_0^\infty xe^{-x^2} \, dx = \lim_{t\to\infty} \left(-\frac{1}{2} \right) \left[e^{-x^2} \right]_0^t = \lim_{t\to\infty} \left(-\frac{1}{2} \right) \left(e^{-t^2} - 1 \right) = -\frac{1}{2} \cdot (-1) = \frac{1}{2}.$

Therefore, $\int_{-\infty}^\infty xe^{-x^2} \, dx = -\frac{1}{2} + \frac{1}{2} = 0.$ Convergent

15. $\int_1^\infty \frac{x+1}{x^2+2x} \, dx = \lim_{t\to\infty} \int_1^t \frac{\frac{1}{2}(2x+2)}{x^2+2x} \, dx = \frac{1}{2} \lim_{t\to\infty} \left[\ln(x^2 + 2x) \right]_1^t = \frac{1}{2} \lim_{t\to\infty} \left[\ln(t^2 + 2t) - \ln 3 \right] = \infty.$

Divergent

17. $\int_0^\infty se^{-5s} \, ds = \lim_{t\to\infty} \int_0^t se^{-5s} \, ds = \lim_{t\to\infty} \left[-\frac{1}{5}se^{-5s} - \frac{1}{25}e^{-5s} \right]$ $\quad \begin{bmatrix} \text{by integration by} \\ \text{parts with } u = s \end{bmatrix}$

$= \lim_{t\to\infty} \left(-\frac{1}{5}te^{-5t} - \frac{1}{25}e^{-5t} + \frac{1}{25} \right) = 0 - 0 + \frac{1}{25}$ $\quad$ [by l'Hospital's Rule]

$= \frac{1}{25}.$ Convergent

19. $\int_1^\infty \frac{\ln x}{x} \, dx = \lim_{t\to\infty} \left[\frac{(\ln x)^2}{2} \right]_1^t$ $\quad \begin{bmatrix} \text{by substitution with} \\ u = \ln x, \, du = dx/x \end{bmatrix} = \lim_{t\to\infty} \frac{(\ln t)^2}{2} = \infty.$ Divergent

21. $\int_{-\infty}^\infty \frac{x^2}{9+x^6} \, dx = \int_{-\infty}^0 \frac{x^2}{9+x^6} \, dx + \int_0^\infty \frac{x^2}{9+x^6} \, dx = 2 \int_0^\infty \frac{x^2}{9+x^6} \, dx$ $\quad$ [since the integrand is even].

Now $\int \frac{x^2 \, dx}{9+x^6} \begin{bmatrix} u = x^3 \\ du = 3x^2 \, dx \end{bmatrix} = \int \frac{\frac{1}{3} du}{9+u^2} \begin{bmatrix} u = 3v \\ du = 3 \, dv \end{bmatrix} = \int \frac{\frac{1}{3}(3 \, dv)}{9+9v^2} = \frac{1}{9} \int \frac{dv}{1+v^2}$

$= \frac{1}{9} \tan^{-1} v + C = \frac{1}{9} \tan^{-1} \left(\frac{u}{3} \right) + C = \frac{1}{9} \tan^{-1} \left(\frac{x^3}{3} \right) + C,$

so $2 \int_0^\infty \dfrac{x^2}{9+x^6}\, dx = 2 \lim_{t\to\infty} \int_0^t \dfrac{x^2}{9+x^6}\, dx = 2 \lim_{t\to\infty} \left[\dfrac{1}{9}\tan^{-1}\left(\dfrac{x^3}{3}\right)\right]_0^t = 2 \lim_{t\to\infty} \dfrac{1}{9}\tan^{-1}\left(\dfrac{t^3}{3}\right) = \dfrac{2}{9}\cdot\dfrac{\pi}{2} = \dfrac{\pi}{9}.$

Convergent

23. $\displaystyle\int_e^\infty \dfrac{1}{x(\ln x)^3}\, dx = \lim_{t\to\infty} \int_e^t \dfrac{1}{x(\ln x)^3}\, dx = \lim_{t\to\infty} \int_1^{\ln t} u^{-3}\, du \quad \begin{bmatrix} u = \ln x, \\ du = dx/x \end{bmatrix} = \lim_{t\to\infty}\left[-\dfrac{1}{2u^2}\right]_1^{\ln t}$

$= \lim_{t\to\infty}\left[-\dfrac{1}{2\,(\ln t)^2} + \dfrac{1}{2}\right] = 0 + \dfrac{1}{2} = \dfrac{1}{2}.$ Convergent

25. $\displaystyle\int_0^1 \dfrac{3}{x^5}\, dx = \lim_{t\to 0^+} \int_t^1 3x^{-5}\, dx = \lim_{t\to 0^+}\left[-\dfrac{3}{4x^4}\right]_t^1 = -\dfrac{3}{4}\lim_{t\to 0^+}\left(1 - \dfrac{1}{t^4}\right) = \infty.$ Divergent

27. $\displaystyle\int_{-2}^{14} \dfrac{dx}{\sqrt[4]{x+2}} = \lim_{t\to -2^+} \int_t^{14} (x+2)^{-1/4}\, dx = \lim_{t\to -2^+}\left[\dfrac{4}{3}(x+2)^{3/4}\right]_t^{14} = \dfrac{4}{3}\lim_{t\to -2^+}\left[16^{3/4} - (t+2)^{3/4}\right]$

$= \dfrac{4}{3}(8 - 0) = \dfrac{32}{3}.$ Convergent

29. There is an infinite discontinuity at $x = 1$. $\displaystyle\int_0^{33}(x-1)^{-1/5}\, dx = \int_0^1 (x-1)^{-1/5}\, dx + \int_1^{33}(x-1)^{-1/5}\, dx.$ Here

$\displaystyle\int_0^1 (x-1)^{-1/5}\, dx = \lim_{t\to 1^-} \int_0^t (x-1)^{-1/5}\, dx = \lim_{t\to 1^-}\left[\dfrac{5}{4}(x-1)^{4/5}\right]_0^t = \lim_{t\to 1^-}\left[\dfrac{5}{4}(t-1)^{4/5} - \dfrac{5}{4}\right] = -\dfrac{5}{4}$ and

$\displaystyle\int_1^{33}(x-1)^{-1/5}\, dx = \lim_{t\to 1^+} \int_t^{33}(x-1)^{-1/5}\, dx = \lim_{t\to 1^+}\left[\dfrac{5}{4}(x-1)^{4/5}\right]_t^{33} = \lim_{t\to 1^+}\left[\dfrac{5}{4}\cdot 16 - \dfrac{5}{4}(t-1)^{4/5}\right] = 20.$

Thus, $\displaystyle\int_0^{33}(x-1)^{-1/5}\, dx = -\dfrac{5}{4} + 20 = \dfrac{75}{4}.$ Convergent

31. There is an infinite discontinuity at $x = 0$. $\displaystyle\int_{-1}^1 \dfrac{e^x}{e^x - 1}\, dx = \int_{-1}^0 \dfrac{e^x}{e^x - 1}\, dx + \int_0^1 \dfrac{e^x}{e^x - 1}\, dx.$

$\displaystyle\int_{-1}^0 \dfrac{e^x}{e^x - 1}\, dx = \lim_{t\to 0^-} \int_{-1}^t \dfrac{e^x}{e^x - 1}\, dx = \lim_{t\to 0^-}\left[\ln\left|e^x - 1\right|\right]_{-1}^t = \lim_{t\to 0^-}\left[\ln\left|e^t - 1\right| - \ln\left|e^{-1} - 1\right|\right] = -\infty,$

so $\displaystyle\int_{-1}^1 \dfrac{e^x}{e^x - 1}\, dx$ is divergent. The integral $\displaystyle\int_0^1 \dfrac{e^x}{e^x - 1}\, dx$ also diverges since

$\displaystyle\int_0^1 \dfrac{e^x}{e^x - 1}\, dx = \lim_{t\to 0^+} \int_t^1 \dfrac{e^x}{e^x - 1}\, dx = \lim_{t\to 0^+}\left[\ln\left|e^x - 1\right|\right]_t^1 = \lim_{t\to 0^+}\left[\ln\left|e - 1\right| - \ln\left|e^t - 1\right|\right] = \infty.$ Divergent

33. $I = \displaystyle\int_0^2 z^2 \ln z\, dz = \lim_{t\to 0^+} \int_t^2 z^2 \ln z\, dz = \lim_{t\to 0^+}\left[\dfrac{z^3}{3^2}(3\ln z - 1)\right]_t^2$ $\begin{bmatrix} \text{integrate by parts} \\ \text{or use Formula 101} \end{bmatrix}$

$= \lim_{t\to 0^+}\left[\dfrac{8}{9}(3\ln 2 - 1) - \dfrac{1}{9}t^3(3\ln t - 1)\right] = \dfrac{8}{3}\ln 2 - \dfrac{8}{9} - \dfrac{1}{9}\lim_{t\to 0^+}\left[t^3(3\ln t - 1)\right] = \dfrac{8}{3}\ln 2 - \dfrac{8}{9} - \dfrac{1}{9}L.$

Now $L = \displaystyle\lim_{t\to 0^+}\left[t^3(3\ln t - 1)\right] = \lim_{t\to 0^+}\dfrac{3\ln t - 1}{t^{-3}} \overset{\text{H}}{=} \lim_{t\to 0^+}\dfrac{3/t}{-3/t^4} = \lim_{t\to 0^+}\left(-t^3\right) = 0.$

Thus, $L = 0$ and $I = \dfrac{8}{3}\ln 2 - \dfrac{8}{9}.$ Convergent

35.

Area $= \displaystyle\int_{-\infty}^1 e^x\, dx = \lim_{t\to -\infty}\left[e^x\right]_t^1 = e - \lim_{t\to -\infty} e^t = e$

37.

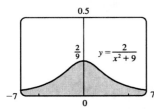

$$\text{Area} = \int_{-\infty}^{\infty} \frac{2}{x^2+9}\,dx = 2\cdot 2\int_0^{\infty} \frac{1}{x^2+9}\,dx = 4\lim_{t\to\infty}\int_0^t \frac{1}{x^2+9}\,dx$$

$$= 4\lim_{t\to\infty}\left[\frac{1}{3}\tan^{-1}\frac{x}{3}\right]_0^t = \frac{4}{3}\lim_{t\to\infty}\left[\tan^{-1}\frac{t}{3}-0\right] = \frac{4}{3}\cdot\frac{\pi}{2} = \frac{2\pi}{3}$$

39.

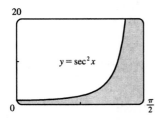

$$\text{Area} = \int_0^{\pi/2}\sec^2 x\,dx = \lim_{t\to(\pi/2)^-}\int_0^t \sec^2 x\,dx = \lim_{t\to(\pi/2)^-}[\tan x]_0^t$$

$$= \lim_{t\to(\pi/2)^-}(\tan t - 0) = \infty$$

Infinite area

41. (a)

t	$\int_1^t g(x)\,dx$
2	0.447453
5	0.577101
10	0.621306
100	0.668479
1000	0.672957
10,000	0.673407

$$g(x) = \frac{\sin^2 x}{x^2}.$$

It appears that the integral is convergent.

(b) $-1 \le \sin x \le 1 \;\Rightarrow\; 0 \le \sin^2 x \le 1 \;\Rightarrow\; 0 \le \dfrac{\sin^2 x}{x^2} \le \dfrac{1}{x^2}$. Since $\displaystyle\int_1^{\infty}\frac{1}{x^2}\,dx$ is convergent

[Equation 2 with $p = 2 > 1$], $\displaystyle\int_1^{\infty}\frac{\sin^2 x}{x^2}\,dx$ is convergent by the Comparison Theorem.

(c)

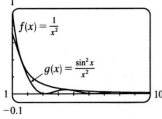

Since $\int_1^{\infty} f(x)\,dx$ is finite and the area under $g(x)$ is less than the area under $f(x)$ on any interval $[1, t]$, $\int_1^{\infty} g(x)\,dx$ must be finite; that is, the integral is convergent.

43. For $x > 0$, $\dfrac{x}{x^3+1} < \dfrac{x}{x^3} = \dfrac{1}{x^2}$. $\displaystyle\int_1^{\infty}\frac{1}{x^2}\,dx$ is convergent by Equation 2 with $p = 2 > 1$, so $\displaystyle\int_1^{\infty}\frac{x}{x^3+1}\,dx$ is convergent

by the Comparison Theorem. $\displaystyle\int_0^1 \frac{x}{x^3+1}\,dx$ is a constant, so $\displaystyle\int_0^{\infty}\frac{x}{x^3+1}\,dx = \int_0^1\frac{x}{x^3+1}\,dx + \int_1^{\infty}\frac{x}{x^3+1}\,dx$ is also

convergent.

45. For $x > 1$, $f(x) = \dfrac{x+1}{\sqrt{x^4-x}} > \dfrac{x+1}{\sqrt{x^4}} > \dfrac{x}{x^2} = \dfrac{1}{x}$, so $\displaystyle\int_2^{\infty} f(x)\,dx$ diverges by comparison with $\displaystyle\int_2^{\infty}\frac{1}{x}\,dx$, which diverges

by Equation 2 with $p = 1 \le 1$. Thus, $\int_1^{\infty} f(x)\,dx = \int_1^2 f(x)\,dx + \int_2^{\infty} f(x)\,dx$ also diverges.

47. For $0 < x \leq 1$, $\dfrac{\sec^2 x}{x\sqrt{x}} > \dfrac{1}{x^{3/2}}$. Now

$$I = \int_0^1 x^{-3/2}\, dx = \lim_{t \to 0^+} \int_t^1 x^{-3/2}\, dx = \lim_{t \to 0^+} \left[-2x^{-1/2} \right]_t^1 = \lim_{t \to 0^+} \left(-2 + \frac{2}{\sqrt{t}} \right) = \infty,\ \text{so } I \text{ is divergent, and by}$$

comparison, $\displaystyle\int_0^1 \frac{\sec^2 x}{x\sqrt{x}}$ is divergent.

49. $\displaystyle\int_0^\infty \frac{dx}{\sqrt{x}\,(1+x)} = \int_0^1 \frac{dx}{\sqrt{x}\,(1+x)} + \int_1^\infty \frac{dx}{\sqrt{x}\,(1+x)} = \lim_{t \to 0^+} \int_t^1 \frac{dx}{\sqrt{x}\,(1+x)} + \lim_{t \to \infty} \int_1^t \frac{dx}{\sqrt{x}\,(1+x)}.$ Now

$$\int \frac{dx}{\sqrt{x}\,(1+x)} = \int \frac{2u\, du}{u(1+u^2)} \quad \begin{bmatrix} u = \sqrt{x},\, x = u^2, \\ dx = 2u\, du \end{bmatrix} = 2\int \frac{du}{1+u^2} = 2\tan^{-1} u + C = 2\tan^{-1}\sqrt{x} + C,\ \text{so}$$

$$\int_0^\infty \frac{dx}{\sqrt{x}\,(1+x)} = \lim_{t \to 0^+} \left[2\tan^{-1}\sqrt{x}\right]_t^1 + \lim_{t \to \infty} \left[2\tan^{-1}\sqrt{x}\right]_1^t$$

$$= \lim_{t \to 0^+} \left[2\left(\tfrac{\pi}{4}\right) - 2\tan^{-1}\sqrt{t}\right] + \lim_{t \to \infty} \left[2\tan^{-1}\sqrt{t} - 2\left(\tfrac{\pi}{4}\right)\right] = \tfrac{\pi}{2} - 0 + 2\left(\tfrac{\pi}{2}\right) - \tfrac{\pi}{2} = \pi.$$

51. If $p = 1$, then $\displaystyle\int_0^1 \frac{dx}{x^p} = \lim_{t \to 0^+} \int_t^1 \frac{dx}{x} = \lim_{t \to 0^+} \left[\ln x\right]_t^1 = \infty.$ Divergent.

If $p \neq 1$, then $\displaystyle\int_0^1 \frac{dx}{x^p} = \lim_{t \to 0^+} \int_t^1 \frac{dx}{x^p}$ [note that the integral is not improper if $p < 0$]

$$= \lim_{t \to 0^+} \left[\frac{x^{-p+1}}{-p+1}\right]_t^1 = \lim_{t \to 0^+} \frac{1}{1-p}\left[1 - \frac{1}{t^{p-1}}\right]$$

If $p > 1$, then $p - 1 > 0$, so $\dfrac{1}{t^{p-1}} \to \infty$ as $t \to 0^+$, and the integral diverges.

If $p < 1$, then $p - 1 < 0$, so $\dfrac{1}{t^{p-1}} \to 0$ as $t \to 0^+$ and $\displaystyle\int_0^1 \frac{dx}{x^p} = \frac{1}{1-p}\left[\lim_{t \to 0^+}\left(1 - t^{1-p}\right)\right] = \frac{1}{1-p}.$

Thus, the integral converges if and only if $p < 1$, and in that case its value is $\dfrac{1}{1-p}$.

53. (a) $I = \int_{-\infty}^\infty x\, dx = \int_{-\infty}^0 x\, dx + \int_0^\infty x\, dx$, and $\int_0^\infty x\, dx = \lim_{t \to \infty} \int_0^t x\, dx = \lim_{t \to \infty}\left[\tfrac{1}{2}x^2\right]_0^t = \lim_{t \to \infty}\left[\tfrac{1}{2}t^2 - 0\right] = \infty$,

so I is divergent.

(b) $\int_{-t}^t x\, dx = \left[\tfrac{1}{2}x^2\right]_{-t}^t = \tfrac{1}{2}t^2 - \tfrac{1}{2}t^2 = 0$, so $\lim_{t \to \infty}\int_{-t}^t x\, dx = 0.$ Therefore, $\int_{-\infty}^\infty x\, dx \neq \lim_{t \to \infty}\int_{-t}^t x\, dx.$

55. We would expect a small percentage of bulbs to burn out in the first few hundred hours, most of the bulbs to burn out after close to 700 hours, and a few overachievers to burn on and on.

(a)

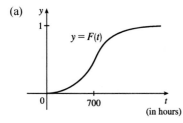

(b) $r(t) = F'(t)$ is the rate at which the fraction $F(t)$ of burnt-out bulbs increases as t increases. This could be interpreted as a fractional burnout rate.

(c) $\int_0^\infty r(t)\,dt = \lim\limits_{x\to\infty} F(x) = 1$, since all of the bulbs will eventually burn out.

57. $I = \displaystyle\int_0^\infty te^{kt}\,dt = \lim_{s\to\infty}\left[\frac{1}{k^2}(kt-1)e^{kt}\right]_0^s$ [Formula 96, or parts] $= \lim\limits_{s\to\infty}\left[\left(\frac{1}{k}se^{ks} - \frac{1}{k^2}e^{ks}\right) - \left(-\frac{1}{k^2}\right)\right]$.

Since $k < 0$ the first two terms approach 0 (you can verify that the first term does so with l'Hospital's Rule), so the limit is equal to $1/k^2$. Thus, $M = -kI = -k\left(1/k^2\right) = -1/k = -1/(-0.000121) \approx 8264.5$ years.

59. $I = \displaystyle\int_a^\infty \frac{1}{x^2+1}\,dx = \lim_{t\to\infty}\int_a^t \frac{1}{x^2+1}\,dx = \lim_{t\to\infty}\left[\tan^{-1}x\right]_a^t = \lim_{t\to\infty}\left(\tan^{-1}t - \tan^{-1}a\right) = \frac{\pi}{2} - \tan^{-1}a$.

$I < 0.001 \quad\Rightarrow\quad \frac{\pi}{2} - \tan^{-1}a < 0.001 \quad\Rightarrow\quad \tan^{-1}a > \frac{\pi}{2} - 0.001 \quad\Rightarrow\quad a > \tan\left(\frac{\pi}{2} - 0.001\right) \approx 1000$.

61. We use integration by parts: let $u = x$, $dv = xe^{-x^2}\,dx \quad\Rightarrow\quad du = dx$, $v = -\frac{1}{2}e^{-x^2}$. So

$$\int_0^\infty x^2 e^{-x^2}\,dx = \lim_{t\to\infty}\left[-\frac{1}{2}xe^{-x^2}\right]_0^t + \frac{1}{2}\int_0^\infty e^{-x^2}\,dx = \lim_{t\to\infty}\left[-\frac{t}{2e^{t^2}}\right] + \frac{1}{2}\int_0^\infty e^{-x^2}\,dx = \frac{1}{2}\int_0^\infty e^{-x^2}\,dx$$

(The limit is 0 by l'Hospital's Rule.)

63. For the first part of the integral, let $x = 2\tan\theta \quad\Rightarrow\quad dx = 2\sec^2\theta\,d\theta$.

$$\int \frac{1}{\sqrt{x^2+4}}\,dx = \int \frac{2\sec^2\theta}{2\sec\theta}\,d\theta = \int \sec\theta\,d\theta = \ln|\sec\theta + \tan\theta|.$$

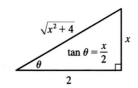

From the figure, $\tan\theta = \dfrac{x}{2}$, and $\sec\theta = \dfrac{\sqrt{x^2+4}}{2}$. So

$$I = \int_0^\infty \left(\frac{1}{\sqrt{x^2+4}} - \frac{C}{x+2}\right)dx = \lim_{t\to\infty}\left[\ln\left|\frac{\sqrt{x^2+4}}{2} + \frac{x}{2}\right| - C\ln|x+2|\right]_0^t$$

$$= \lim_{t\to\infty}\left[\ln\frac{\sqrt{t^2+4}+t}{2} - C\ln(t+2) - (\ln 1 - C\ln 2)\right]$$

$$= \lim_{t\to\infty}\left[\ln\left(\frac{\sqrt{t^2+4}+t}{2\,(t+2)^C}\right) + \ln 2^C\right] = \ln\left(\lim_{t\to\infty}\frac{t+\sqrt{t^2+4}}{(t+2)^C}\right) + \ln 2^{C-1}$$

Now $L = \displaystyle\lim_{t\to\infty}\frac{t+\sqrt{t^2+4}}{(t+2)^C} \overset{H}{=} \lim_{t\to\infty}\frac{1+t/\sqrt{t^2+4}}{C(t+2)^{C-1}} = \frac{2}{C\lim\limits_{t\to\infty}(t+2)^{C-1}}$.

If $C < 1$, $L = \infty$ and I diverges.

If $C = 1$, $L = 2$ and I converges to $\ln 2 + \ln 2^0 = \ln 2$.

If $C > 1$, $L = 0$ and I diverges to $-\infty$.

65. No, $I = \int_0^\infty f(x)\,dx$ must be *divergent*. Since $\lim\limits_{x\to\infty} f(x) = 1$, there must exist an N such that if $x \geq N$, then $f(x) \geq \frac{1}{2}$.

Thus, $I = I_1 + I_2 = \int_0^N f(x)\,dx + \int_N^\infty f(x)\,dx$, where I_1 is an ordinary definite integral that has a finite value, and I_2 is improper and diverges by comparison with the divergent integral $\int_N^\infty \frac{1}{2}\,dx$.

5 Review

CONCEPT CHECK

1. (a) $\sum_{i=1}^{n} f(x_i^*) \, \Delta x$ is an expression for a Riemann sum of a function f.

   x_i^* is a point in the ith subinterval $[x_{i-1}, x_i]$ and Δx is the length of the subintervals.

   (b) See Figure 1 in Section 5.2.

   (c) In Section 5.2, see Figure 3 and the paragraph beside it.

2. (a) See Definition 5.2.2.

   (b) See Figure 2 in Section 5.2.

   (c) In Section 5.2, see Figure 4 and the paragraph by it (contains "**net area**").

3. (a) See the Evaluation Theorem at the beginning of Section 5.3.

   (b) See the Net Change Theorem after Example 6 in Section 5.3.

4. $\int_{t_1}^{t_2} r(t) \, dt$ represents the change in the amount of water in the reservoir between time t_1 and time t_2.

5. (a) $\int_{60}^{120} v(t) \, dt$ represents the change in position of the particle from $t = 60$ to $t = 120$ seconds.

   (b) $\int_{60}^{120} |v(t)| \, dt$ represents the total distance traveled by the particle from $t = 60$ to 120 seconds.

   (c) $\int_{60}^{120} a(t) \, dt$ represents the change in the velocity of the particle from $t = 60$ to $t = 120$ seconds.

6. (a) $\int f(x) \, dx$ is the family of functions $\{F \mid F' = f\}$. Any two such functions differ by a constant.

   (b) The connection is given by the Evaluation Theorem: $\int_a^b f(x) \, dx = \left[\int f(x) \, dx \right]_a^b$ if f is continuous.

7. See the Fundamental Theorem of Calculus after Example 5 in Section 5.4.

8. (a) See the Substitution Rule (5.5.4). This says that it is permissible to operate with the dx after an integral sign as if it were a differential.

   (b) See Formula 5.6.1 or 5.6.2. We try to choose $u = f(x)$ to be a function that becomes simpler when differentiated (or at least not more complicated) as long as $dv = g'(x) \, dx$ can be readily integrated to give v.

9. See the Midpoint Rule, the Trapezoidal Rule, and Simpson's Rule, as well as their associated error bounds, all in Section 5.9. We would expect the best estimate to be given by Simpson's Rule.

10. See Definitions 1(a), (b), and (c) in Section 5.10.

11. See Definitions 3(b), (a), and (c) in Section 5.10.

12. See the Comparison Theorem after Example 8 in Section 5.10.

13. The precise version of this statement is given by the Fundamental Theorem of Calculus. See the statement of this theorem and the paragraph that follows it in Section 5.4.

TRUE-FALSE QUIZ

1. True by Property 2 of the Integral in Section 5.2.

3. True by Property 3 of the Integral in Section 5.2.

5. False.   For example, let $f(x) = x^2$. Then $\int_0^1 \sqrt{x^2}\, dx = \int_0^1 x\, dx = \frac{1}{2}$, but $\sqrt{\int_0^1 x^2\, dx} = \sqrt{\frac{1}{3}} = \frac{1}{\sqrt{3}}$.

7. True by Comparison Property 7 of the Integral in Section 5.2.

9. True.   The integrand is an odd function that is continuous on $[-1, 1]$, so the result follows from Theorem 5.5.6(b).

11. False.   This is an improper integral, since the denominator vanishes at $x = 1$.

$$\int_0^4 \frac{x}{x^2 - 1}\, dx = \int_0^1 \frac{x}{x^2 - 1}\, dx + \int_1^4 \frac{x}{x^2 - 1}\, dx \text{ and}$$

$$\int_0^1 \frac{x}{x^2 - 1}\, dx = \lim_{t \to 1^-} \int_0^t \frac{x}{x^2 - 1}\, dx = \lim_{t \to 1^-} \left[\frac{1}{2}\ln\left|x^2 - 1\right|\right]_0^t = \lim_{t \to 1^-} \frac{1}{2}\ln\left|t^2 - 1\right| = \infty$$

So the integral diverges.

13. False.   See the paragraph before Note 4 and Figure 4 in Section 5.2, and notice that $y = x - x^3 < 0$ for $1 < x \le 2$.

15. False.   For example, the function $y = |x|$ is continuous on $\mathbb{R}$, but has no derivative at $x = 0$.

17. False.   See Exercise 53 in Section 5.10.

19. False.   If $f(x) = 1/x$, then f is continuous and decreasing on $[1, \infty)$ with $\lim_{x \to \infty} f(x) = 0$, but $\int_1^\infty f(x)\, dx$ is divergent.

21. False.   Take $f(x) = 1$ for all x and $g(x) = -1$ for all x. Then $\int_a^\infty f(x)\, dx = \infty$ [divergent]

and $\int_a^\infty g(x)\, dx = -\infty$ [divergent],   but $\int_a^\infty [f(x) + g(x)]\, dx = 0$ [convergent].

23. False. $\int_a^b f(x)\, dx$ is a constant, so $\dfrac{d}{dx}\left(\int_a^b f(x)\, dx\right) = 0$, not $f(x)$ [unless $f(x) = 0$]. Compare the given statement carefully with FTC1, in which the upper limit in the integral is x.

EXERCISES

1. (a)

$L_6 = \sum_{i=1}^6 f(x_{i-1})\, \Delta x$ $[\Delta x = \frac{6 - 0}{6} = 1]$

$= f(x_0) \cdot 1 + f(x_1) \cdot 1 + f(x_2) \cdot 1 + f(x_3) \cdot 1 + f(x_4) \cdot 1 + f(x_5) \cdot 1$

$\approx 2 + 3.5 + 4 + 2 + (-1) + (-2.5) = 8$

The Riemann sum represents the sum of the areas of the four rectangles above the x-axis minus the sum of the areas of the two rectangles below the x-axis.

(b)

$$M_6 = \sum_{i=1}^{6} f(\overline{x}_i)\,\Delta x \quad [\Delta x = \tfrac{6-0}{6} = 1]$$

$$= f(\overline{x}_1)\cdot 1 + f(\overline{x}_2)\cdot 1 + f(\overline{x}_3)\cdot 1 + f(\overline{x}_4)\cdot 1 + f(\overline{x}_5)\cdot 1 + f(\overline{x}_6)\cdot 1$$

$$= f(0.5) + f(1.5) + f(2.5) + f(3.5) + f(4.5) + f(5.5)$$

$$\approx 3 + 3.9 + 3.4 + 0.3 + (-2) + (-2.9) = 5.7$$

3. $\int_0^1 \left(x + \sqrt{1-x^2}\right)dx = \int_0^1 x\,dx + \int_0^1 \sqrt{1-x^2}\,dx = I_1 + I_2.$

I_1 can be interpreted as the area of the triangle shown in the figure

and I_2 can be interpreted as the area of the quarter-circle.

Area $= \tfrac{1}{2}(1)(1) + \tfrac{1}{4}(\pi)(1)^2 = \tfrac{1}{2} + \tfrac{\pi}{4}.$

5. $\int_0^6 f(x)\,dx = \int_0^4 f(x)\,dx + \int_4^6 f(x)\,dx \quad \Rightarrow \quad 10 = 7 + \int_4^6 f(x)\,dx \quad \Rightarrow \quad \int_4^6 f(x)\,dx = 10 - 7 = 3$

7. First note that either a or b must be the graph of $\int_0^x f(t)\,dt$, since $\int_0^0 f(t)\,dt = 0$, and $c(0) \neq 0$. Now notice that $b > 0$ when c is increasing, and that $c > 0$ when a is increasing. It follows that c is the graph of $f(x)$, b is the graph of $f'(x)$, and a is the graph of $\int_0^x f(t)\,dt$.

9. $\int_1^2 (8x^3 + 3x^2)\,dx = \left[8\cdot\tfrac{1}{4}x^4 + 3\cdot\tfrac{1}{3}x^3\right]_1^2 = \left[2x^4 + x^3\right]_1^2 = (2\cdot 2^4 + 2^3) - (2+1) = 40 - 3 = 37$

11. $\int_0^1 (1 - x^9)\,dx = \left[x - \tfrac{1}{10}x^{10}\right]_0^1 = \left(1 - \tfrac{1}{10}\right) - 0 = \tfrac{9}{10}$

13. $\int \left(\dfrac{1-x}{x}\right)^2 dx = \int \left(\dfrac{1}{x} - 1\right)^2 dx = \int \left(\dfrac{1}{x^2} - \dfrac{2}{x} + 1\right)dx = -\dfrac{1}{x} - 2\ln|x| + x + C$

15. $u = x^2 + 1$, $du = 2x\,dx$, so $\displaystyle\int_0^1 \dfrac{x}{x^2+1}\,dx = \int_1^2 \dfrac{1}{u}\left(\tfrac{1}{2}\,du\right) = \tfrac{1}{2}\left[\ln u\right]_1^2 = \tfrac{1}{2}\ln 2.$

17. Let $u = v^3$, so $du = 3v^2\,dv$. When $v = 0$, $u = 0$; when $v = 1$, $u = 1$. Thus,

$\int_0^1 v^2 \cos(v^3)\,dv = \int_0^1 \cos u\,\left(\tfrac{1}{3}\,du\right) = \tfrac{1}{3}\left[\sin u\right]_0^1 = \tfrac{1}{3}(\sin 1 - 0) = \tfrac{1}{3}\sin 1.$

19. $\int_0^1 e^{\pi t}\,dt = \left[\tfrac{1}{\pi}e^{\pi t}\right]_0^1 = \tfrac{1}{\pi}(e^\pi - 1)$

21. Let $u = x^2 + 4x$. Then $du = (2x + 4)\,dx = 2(x + 2)\,dx$, so

$\int \dfrac{x+2}{\sqrt{x^2+4x}}\,dx = \int u^{-1/2}\left(\tfrac{1}{2}\,du\right) = \tfrac{1}{2}\cdot 2u^{1/2} + C = \sqrt{u} + C = \sqrt{x^2 + 4x} + C.$

23. $\displaystyle\int_0^5 \dfrac{x}{x+10}\,dx = \int_0^5 \left(1 - \dfrac{10}{x+10}\right)dx = \left[x - 10\ln(x+10)\right]_0^5 = 5 - 10\ln 15 + 10\ln 10$

$\qquad = 5 + 10\ln\tfrac{10}{15} = 5 + 10\ln\tfrac{2}{3}$

25. $\displaystyle\int_{-\pi/4}^{\pi/4} \frac{t^4 \tan t}{2 + \cos t}\, dt = 0$ by Theorem 5.5.6(b), since $f(t) = \dfrac{t^4 \tan t}{2 + \cos t}$ is an odd function.

27. $\displaystyle\int_1^4 x^{3/2} \ln x\, dx \quad \begin{bmatrix} u = \ln x, & dv = x^{3/2}\, dx, \\ du = dx/x & v = \frac{2}{5}x^{5/2} \end{bmatrix} = \frac{2}{5}\left[x^{5/2} \ln x\right]_1^4 - \frac{2}{5}\int_1^4 x^{3/2}\, dx = \frac{2}{5}(32 \ln 4 - \ln 1) - \frac{2}{5}\left[\frac{2}{5}x^{5/2}\right]_1^4$

$\qquad = \frac{2}{5}(64 \ln 2) - \frac{4}{25}(32 - 1) = \frac{128}{5}\ln 2 - \frac{124}{25} \quad \left[\text{or } \frac{64}{5}\ln 4 - \frac{124}{25}\right]$

29. $\dfrac{1}{t^2 + 6t + 8} = \dfrac{1}{(t+2)(t+4)} = \dfrac{A}{t+2} + \dfrac{B}{t+4}$. Multiply both sides by $(t+2)(t+4)$ to get $1 = A(t+4) + B(t+2)$.

Substituting -4 for t gives $1 = -2B \quad \Leftrightarrow \quad B = -\frac{1}{2}$. Substituting -2 for t gives $1 = 2A \quad \Leftrightarrow \quad A = \frac{1}{2}$. Thus,

$\displaystyle\int \frac{dt}{t^2 + 6t + 8} = \int \left(\frac{1/2}{t+2} - \frac{1/2}{t+4}\right) dt = \frac{1}{2}\ln|t+2| - \frac{1}{2}\ln|t+4| + C = \frac{1}{2}\ln\left|\frac{t+2}{t+4}\right| + C.$

31. Let $w = \sqrt[3]{x}$. Then $w^3 = x$ and $3w^2\, dw = dx$, so $\int e^{\sqrt[3]{x}}\, dx = \int e^w \cdot 3w^2\, dw = 3I$. To evaluate I, let $u = w^2$,

$dv = e^w\, dw \quad \Rightarrow \quad du = 2w\, dw, v = e^w$, so $I = \int w^2 e^w\, dw = w^2 e^w - \int 2w e^w\, dw$. Now let $U = w, dV = e^w\, dw \quad \Rightarrow$

$dU = dw, V = e^w$. Thus, $I = w^2 e^w - 2\left[w e^w - \int e^w\, dw\right] = w^2 e^w - 2w e^w + 2e^w + C_1$, and hence

$3I = 3e^w (w^2 - 2w + 2) + C = 3e^{\sqrt[3]{x}}(x^{2/3} - 2x^{1/3} + 2) + C.$

33. Let $u = 1 + \sec\theta$. Then $du = \sec\theta \tan\theta\, d\theta$, so

$\displaystyle\int \frac{\sec\theta \tan\theta}{1 + \sec\theta}\, d\theta = \int \frac{1}{1 + \sec\theta}(\sec\theta \tan\theta\, d\theta) = \int \frac{1}{u}\, du = \ln|u| + C = \ln|1 + \sec\theta| + C.$

35. Let $u = 1 + \sin x$. Then $du = \cos x\, dx$, so

$\displaystyle\int \frac{\cos x\, dx}{\sqrt{1 + \sin x}} = \int u^{-1/2}\, du = 2u^{1/2} + C = 2\sqrt{1 + \sin x} + C.$

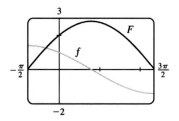

37. From the graph, it appears that the area under the curve $y = x\sqrt{x}$ between $x = 0$ and $x = 4$ is somewhat less than half the area of an 8×4 rectangle, so perhaps about 13 or 14. To find the exact value, we evaluate

$\int_0^4 x\sqrt{x}\, dx = \int_0^4 x^{3/2}\, dx = \left[\frac{2}{5}x^{5/2}\right]_0^4 = \frac{2}{5}(4)^{5/2} = \frac{64}{5} = 12.8.$

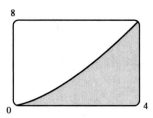

39. $F(x) = \displaystyle\int_0^x \frac{t^2}{1 + t^3}\, dt \quad \Rightarrow \quad F'(x) = \frac{d}{dx}\int_0^x \frac{t^2}{1 + t^3}\, dt = \frac{x^2}{1 + x^3}$

41. $y = \int_{\sqrt{x}}^{x} \frac{e^t}{t}\,dt = \int_{\sqrt{x}}^{1} \frac{e^t}{t}\,dt + \int_{1}^{x} \frac{e^t}{t}\,dt = -\int_{1}^{\sqrt{x}} \frac{e^t}{t}\,dt + \int_{1}^{x} \frac{e^t}{t}\,dt \quad \Rightarrow$

$\dfrac{dy}{dx} = -\dfrac{d}{dx}\left(\int_{1}^{\sqrt{x}} \frac{e^t}{t}\,dt\right) + \dfrac{d}{dx}\left(\int_{1}^{x} \frac{e^t}{t}\,dt\right)$. Let $u = \sqrt{x}$. Then

$$\frac{d}{dx}\int_{1}^{\sqrt{x}} \frac{e^t}{t}\,dt = \frac{d}{dx}\int_{1}^{u} \frac{e^t}{t}\,dt = \frac{d}{du}\left(\int_{1}^{u} \frac{e^t}{t}\,dt\right)\frac{du}{dx} = \frac{e^u}{u}\cdot\frac{1}{2\sqrt{x}} = \frac{e^{\sqrt{x}}}{\sqrt{x}}\cdot\frac{1}{2\sqrt{x}} = \frac{e^{\sqrt{x}}}{2x}.$$

so $\dfrac{dy}{dx} = -\dfrac{e^{\sqrt{x}}}{2x} + \dfrac{e^x}{x}$.

43. $u = e^x \quad \Rightarrow \quad du = e^x\,dx$, so

$\int e^x \sqrt{1 - e^{2x}}\,dx = \int \sqrt{1 - u^2}\,du \overset{30}{=} \frac{1}{2}u\sqrt{1-u^2} + \frac{1}{2}\sin^{-1}u + C = \frac{1}{2}\left[e^x\sqrt{1-e^{2x}} + \sin^{-1}(e^x)\right] + C$.

45. $\displaystyle\int \sqrt{x^2 + x + 1}\,dx = \int \sqrt{x^2 + x + \frac{1}{4} + \frac{3}{4}}\,dx = \int \sqrt{\left(x + \frac{1}{2}\right)^2 + \frac{3}{4}}\,dx$

$= \displaystyle\int \sqrt{u^2 + \left(\frac{\sqrt{3}}{2}\right)^2}\,du \qquad [u = x + \frac{1}{2},\ du = dx]$

$\overset{21}{=} \frac{1}{2}u\sqrt{u^2 + \frac{3}{4}} + \frac{3}{8}\ln\left(u + \sqrt{u^2 + \frac{3}{4}}\right) + C$

$= \dfrac{2x+1}{4}\sqrt{x^2 + x + 1} + \frac{3}{8}\ln\left(x + \frac{1}{2} + \sqrt{x^2 + x + 1}\right) + C$

47. $f(x) = \sqrt{1 + x^4}, \quad \Delta x = \dfrac{b-a}{n} = \dfrac{1-0}{10} = \dfrac{1}{10}$.

(a) $T_{10} = \frac{1}{10\cdot 2}\{f(0) + 2[f(0.1) + f(0.2) + \cdots + f(0.9)] + f(1)\} \approx 1.090608$

(b) $M_{10} = \frac{1}{10}\left[f\left(\frac{1}{20}\right) + f\left(\frac{3}{20}\right) + f\left(\frac{5}{20}\right) + \cdots + f\left(\frac{19}{20}\right)\right] \approx 1.088840$

(c) $S_{10} = \frac{1}{10\cdot 3}[f(0) + 4f(0.1) + 2f(0.2) + \cdots + 4f(0.9) + f(1)] \approx 1.089429$

f is concave upward, so the Trapezoidal Rule gives us an overestimate, the Midpoint Rule gives an underestimate, and we cannot tell whether Simpson's Rule gives us an overestimate or an underestimate.

49. $f(x) = (1 + x^4)^{1/2},\ f'(x) = \frac{1}{2}(1 + x^4)^{-1/2}(4x^3) = 2x^3(1 + x^4)^{-1/2},\ f''(x) = (2x^6 + 6x^2)(1 + x^4)^{-3/2}$.

A graph of f'' on $[0, 1]$ shows that it has its maximum at $x = 1$, so $|f''(x)| \le f''(1) = \sqrt{8}$ on $[0, 1]$. By taking $K = \sqrt{8}$, we find that the error in Exercise 47(a) is bounded by $\dfrac{K(b-a)^3}{12n^2} = \dfrac{\sqrt{8}}{1200} \approx 0.0024$, and in (b) by about $\frac{1}{2}(0.0024) = 0.0012$.

Note: Another way to estimate K is to let $x = 1$ in the factor $2x^6 + 6x^2$ (maximizing the numerator) and let $x = 0$ in the factor $(1 + x^4)^{-3/2}$ (minimizing the denominator). Doing so gives us $K = 8$ and errors of $0.00\overline{6}$ and $0.00\overline{3}$. Using $K = 8$ for the Trapezoidal Rule, we have $|E_T| \le \dfrac{K(b-a)^3}{12n^2} \le 0.00001 \quad \Leftrightarrow \quad \dfrac{8(1-0)^3}{12n^2} \le \dfrac{1}{100{,}000} \quad \Leftrightarrow \quad n^2 \ge \dfrac{800{,}000}{12} \quad \Leftrightarrow$

$n \gtrsim 258.2$, so we should take $n = 259$. For the Midpoint Rule, $|E_M| \le \dfrac{K(b-a)^3}{24n^2} \le 0.00001 \quad \Leftrightarrow \quad n^2 \ge \dfrac{800{,}000}{24} \quad \Leftrightarrow$

$n \gtrsim 182.6$, so we should take $n = 183$.

51. (a) $f(x) = \sin(\sin x)$. A CAS gives

$$f^{(4)}(x) = \sin(\sin x)[\cos^4 x + 7\cos^2 x - 3]$$
$$+ \cos(\sin x)\left[6\cos^2 x \sin x + \sin x\right]$$

From the graph, we see that $\left|f^{(4)}(x)\right| < 3.8$ for $x \in [0, \pi]$.

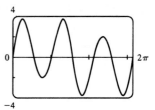

(b) We use Simpson's Rule with $f(x) = \sin(\sin x)$ and $\Delta x = \frac{\pi}{10}$:

$$\int_0^\pi f(x)\, dx \approx \frac{\pi}{10 \cdot 3}\left[f(0) + 4f\left(\tfrac{\pi}{10}\right) + 2f\left(\tfrac{2\pi}{10}\right) + \cdots + 4f\left(\tfrac{9\pi}{10}\right) + f(\pi)\right] \approx 1.786721$$

From part (a), we know that $\left|f^{(4)}(x)\right| < 3.8$ on $[0, \pi]$, so we use Theorem 8.7.4 with $K = 3.8$, and estimate the error

as $|E_S| \le \dfrac{3.8(\pi - 0)^5}{180(10)^4} \approx 0.000646$.

(c) If we want the error to be less than 0.00001, we must have $|E_S| \le \dfrac{3.8\pi^5}{180n^4} \le 0.00001$,

so $n^4 \ge \dfrac{3.8\pi^5}{180(0.00001)} \approx 646{,}041.6 \;\Rightarrow\; n \ge 28.35$. Since n must be even for Simpson's Rule, we must have $n \ge 30$

to ensure the desired accuracy.

53. If $1 \le x \le 3$, then $\sqrt{1^2 + 3} \le \sqrt{x^2 + 3} \le \sqrt{3^2 + 3} \;\Rightarrow\; 2 \le \sqrt{x^2 + 3} \le 2\sqrt{3}$, so

$2(3 - 1) \le \int_1^3 \sqrt{x^2 + 3}\, dx \le 2\sqrt{3}(3 - 1)$; that is, $4 \le \int_1^3 \sqrt{x^2 + 3}\, dx \le 4\sqrt{3}$.

55. $\displaystyle\int_1^\infty \frac{1}{(2x+1)^3}\, dx = \lim_{t\to\infty}\int_1^t \frac{1}{(2x+1)^3}\, dx = \lim_{t\to\infty}\int_1^t \tfrac{1}{2}(2x+1)^{-3}\, 2\, dx = \lim_{t\to\infty}\left[-\frac{1}{4(2x+1)^2}\right]_1^t$

$$= -\frac{1}{4}\lim_{t\to\infty}\left[\frac{1}{(2t+1)^2} - \frac{1}{9}\right] = -\frac{1}{4}\left(0 - \frac{1}{9}\right) = \frac{1}{36}$$

57. $\displaystyle\int_{-\infty}^0 e^{-2x}\, dx = \lim_{t\to-\infty}\int_t^0 e^{-2x}\, dx = \lim_{t\to-\infty}\left[-\tfrac{1}{2}e^{-2x}\right]_t^0 = \lim_{t\to-\infty}\left(-\tfrac{1}{2} + \tfrac{1}{2}e^{-2t}\right) = \infty.$ Divergent

59. Let $u = \ln x$. Then $du = dx/x$, so $\displaystyle\int \frac{dx}{x\sqrt{\ln x}} = \int \frac{du}{\sqrt{u}} = 2\sqrt{u} + C = 2\sqrt{\ln x} + C.$

Thus, $\displaystyle\int_1^e \frac{dx}{x\sqrt{\ln x}} = \lim_{t\to1^+}\int_t^e \frac{dx}{x\sqrt{\ln x}} = \lim_{t\to1^+}\left[2\sqrt{\ln x}\right]_t^e = \lim_{t\to1^+}\left(2\sqrt{\ln e} - 2\sqrt{\ln t}\right) = 2\cdot 1 - 2\cdot 0 = 2.$

61. $\dfrac{x^3}{x^5 + 2} \le \dfrac{x^3}{x^5} = \dfrac{1}{x^2}$ for x in $[1, \infty)$. $\displaystyle\int_1^\infty \frac{1}{x^2}\, dx$ is convergent by (5.10.2) with $p = 2 > 1$. Therefore, $\displaystyle\int_1^\infty \frac{x^3}{x^5 + 2}\, dx$ is

convergent by the Comparison Theorem.

63. (a) Displacement $= \displaystyle\int_0^5 (t^2 - t)\, dt = \left[\tfrac{1}{3}t^3 - \tfrac{1}{2}t^2\right]_0^5 = \dfrac{125}{3} - \dfrac{25}{2} = \dfrac{175}{6} = 29.1\overline{6}$ meters

(b) Distance traveled $= \displaystyle\int_0^5 \left|t^2 - t\right|\, dt = \int_0^5 |t(t - 1)|\, dt = \int_0^1 (t - t^2)\, dt + \int_1^5 (t^2 - t)\, dt$

$$= \left[\tfrac{1}{2}t^2 - \tfrac{1}{3}t^3\right]_0^1 + \left[\tfrac{1}{3}t^3 - \tfrac{1}{2}t^2\right]_1^5 = \tfrac{1}{2} - \tfrac{1}{3} - 0 + \left(\dfrac{125}{3} - \dfrac{25}{2}\right) - \left(\tfrac{1}{3} - \tfrac{1}{2}\right) = \dfrac{177}{6} = 29.5 \text{ meters}$$

65. Note that $r(t) = b'(t)$, where $b(t) =$ the number of barrels of oil consumed up to time t. So, by the Net Change Theorem,
$\int_0^8 r(t)\,dt = b(8) - b(0)$ represents the number of barrels of oil consumed from Jan. 1, 2000, through Jan. 1, 2008.

67. Both numerator and denominator approach 0 as $a \to 0$, so we use l'Hospital's Rule. (Note that we are differentiating *with respect to* a, since that is the quantity which is changing.) We also use FTC1:

$$\lim_{a \to 0} T(x, t) = \lim_{a \to 0} \frac{C \int_0^a e^{-(x-u)^2/(4kt)}\,du}{a\sqrt{4\pi kt}} \overset{\text{H}}{=} \lim_{a \to 0} \frac{Ce^{-(x-a)^2/(4kt)}}{\sqrt{4\pi kt}} = \frac{Ce^{-x^2/(4kt)}}{\sqrt{4\pi kt}}$$

69. Using FTC1, we differentiate both sides of the given equation, $\int_0^x f(t)\,dt = xe^{2x} + \int_0^x e^{-t}f(t)\,dt$, and get

$$f(x) = e^{2x} + 2xe^{2x} + e^{-x}f(x) \quad \Rightarrow \quad f(x)(1 - e^{-x}) = e^{2x} + 2xe^{2x} \quad \Rightarrow \quad f(x) = \frac{e^{2x}(1 + 2x)}{1 - e^{-x}}.$$

71. Let $u = f(x)$ and $du = f'(x)\,dx$. So $2\int_a^b f(x)f'(x)\,dx = 2\int_{f(a)}^{f(b)} u\,du = [u^2]_{f(a)}^{f(b)} = [f(b)]^2 - [f(a)]^2$.

73. By the Fundamental Theorem of Calculus,

$$\int_0^\infty f'(x)\,dx = \lim_{t \to \infty} \int_0^t f'(x)\,dx = \lim_{t \to \infty} [f(t) - f(0)] = \lim_{t \to \infty} f(t) - f(0) = 0 - f(0) = -f(0).$$

FOCUS ON PROBLEM SOLVING

1.

 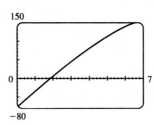

By symmetry, the problem can be reduced to finding the line $x = c$ such that the shaded area is one-third of the area of the quarter-circle. An equation of the semicircle is $y = \sqrt{49 - x^2}$, so we require that $\int_0^c \sqrt{49 - x^2}\,dx = \frac{1}{3} \cdot \frac{1}{4}\pi(7)^2$ $\Leftrightarrow$
$\left[\frac{1}{2}x\sqrt{49 - x^2} + \frac{49}{2}\sin^{-1}(x/7)\right]_0^c = \frac{49}{12}\pi$ [by Formula 30] $\Leftrightarrow$ $\frac{1}{2}c\sqrt{49 - c^2} + \frac{49}{2}\sin^{-1}(c/7) = \frac{49}{12}\pi$.

This equation would be difficult to solve exactly, so we plot the left-hand side as a function of c, and find that the equation holds for $c \approx 1.85$. So the cuts should be made at distances of about 1.85 inches from the center of the pizza.

3. Differentiating both sides of the equation $x\sin\pi x = \int_0^{x^2} f(t)\,dt$ (using FTC1 and the Chain Rule for the right side) gives

$\sin\pi x + \pi x\cos\pi x = 2xf(x^2)$. Letting $x = 2$ so that $f(x^2) = f(4)$, we obtain $\sin 2\pi + 2\pi\cos 2\pi = 4f(4)$, so

$f(4) = \frac{1}{4}(0 + 2\pi \cdot 1) = \frac{\pi}{2}$.

5. Differentiating the given equation, $\int_0^x f(t)\,dt = [f(x)]^2$, using FTC1 gives $f(x) = 2f(x)f'(x) \Rightarrow$

$f(x)[2f'(x) - 1] = 0$, so $f(x) = 0$ or $f'(x) = \frac{1}{2}$. Since $f(x)$ is never 0, we must have $f'(x) = \frac{1}{2}$ and $f'(x) = \frac{1}{2} \Rightarrow$

$f(x) = \frac{1}{2}x + C$. To find C, we substitute into the given equation to get $\int_0^x \left(\frac{1}{2}t + C\right)dt = \left(\frac{1}{2}x + C\right)^2$ $\Leftrightarrow$

$\frac{1}{4}x^2 + Cx = \frac{1}{4}x^2 + Cx + C^2$. It follows that $C^2 = 0$, so $C = 0$, and $f(x) = \frac{1}{2}x$.

7. By l'Hospital's Rule and the Fundamental Theorem, using the notation $\exp(y) = e^y$,

$$\lim_{x\to 0} \frac{\int_0^x (1 - \tan 2t)^{1/t}\,dt}{x} \overset{H}{=} \lim_{x\to 0} \frac{(1 - \tan 2x)^{1/x}}{1} = \exp\left(\lim_{x\to 0} \frac{\ln(1 - \tan 2x)}{x}\right)$$

$$\overset{H}{=} \exp\left(\lim_{x\to 0} \frac{-2\sec^2 2x}{1 - \tan 2x}\right) = \exp\left(\frac{-2 \cdot 1^2}{1 - 0}\right) = e^{-2}$$

9. For $I = \int_0^4 xe^{(x-2)^4}\,dx$, let $u = x - 2$ so that $x = u + 2$ and $dx = du$. Then

$I = \int_{-2}^2 (u + 2)e^{u^4}\,du = \int_{-2}^2 ue^{u^4}\,du + \int_{-2}^2 2e^{u^4}\,du$

$= 0$ [by 5.5.6(b)] $+ 2\int_0^4 e^{(x-2)^4}\,dx = 2k$.

11. Such a function cannot exist. $f'(x) > 3$ for all x means that f is differentiable (and hence continuous) for all x. So by FTC2,

$\int_1^4 f'(x)\,dx = f(4) - f(1) = 7 - (-1) = 8$. However, if $f'(x) > 3$ for all x, then $\int_1^4 f'(x)\,dx \geq 3 \cdot (4 - 1) = 9$ by

Comparison Property 8 in Section 5.2.

[continued]

Another solution: By the Mean Value Theorem, there exists a number $c \in (1, 4)$ such that

$$f'(c) = \frac{f(4) - f(1)}{4 - 1} = \frac{7 - (-1)}{3} = \frac{8}{3} \quad \Rightarrow \quad 8 = 3f'(c). \text{ But } f'(x) > 3 \quad \Rightarrow \quad 3f'(c) > 9, \text{ so such a function cannot}$$

exist.

13. $f(x) = 2 + x - x^2 = (-x + 2)(x + 1) = 0 \quad \Leftrightarrow \quad x = 2 \text{ or } x = -1.$ $f(x) \geq 0$ for $x \in [-1, 2]$ and $f(x) < 0$ everywhere

else. The integral $\int_a^b (2 + x - x^2)\, dx$ has a maximum on the interval where the integrand is positive, which is $[-1, 2]$. So

$a = -1, b = 2$. (Any larger interval gives a smaller integral since $f(x) < 0$ outside $[-1, 2]$. Any smaller interval also gives a

smaller integral since $f(x) \geq 0$ in $[-1, 2]$.)

15. By FTC1, $\dfrac{d}{dx} \int_0^x \left(\int_1^{\sin t} \sqrt{1 + u^4}\, du \right) dt = \int_1^{\sin x} \sqrt{1 + u^4}\, du.$ Again using FTC1,

$$\frac{d^2}{dx^2} \int_0^x \left(\int_1^{\sin t} \sqrt{1 + u^4}\, du \right) dt = \frac{d}{dx} \int_1^{\sin x} \sqrt{1 + u^4}\, du = \sqrt{1 + \sin^4 x}\, \cos x.$$

17. Write $I = \displaystyle\int \frac{x^8}{(1 + x^6)^2}\, dx = \int x^3 \cdot \frac{x^5}{(1 + x^6)^2}\, dx.$ Integrate by parts with $u = x^3$, $dv = \dfrac{x^5}{(1 + x^6)^2}\, dx.$ Then

$du = 3x^2\, dx, v = -\dfrac{1}{6(1 + x^6)} \quad \Rightarrow \quad I = -\dfrac{x^3}{6(1 + x^6)} + \dfrac{1}{2} \displaystyle\int \dfrac{x^2}{1 + x^6}\, dx.$ Substitute $t = x^3$ in this latter integral.

$\displaystyle\int \frac{x^2}{1 + x^6}\, dx = \frac{1}{3} \int \frac{dt}{1 + t^2} = \frac{1}{3} \tan^{-1} t + C = \frac{1}{3} \tan^{-1}(x^3) + C.$ Therefore $I = -\dfrac{x^3}{6(1 + x^6)} + \dfrac{1}{6} \tan^{-1}(x^3) + C.$

Returning to the improper integral,

$$\int_{-1}^{\infty} \left(\frac{x^4}{1 + x^6} \right)^2 dx = \lim_{t \to \infty} \int_{-1}^{t} \frac{x^8}{(1 + x^6)^2}\, dx = \lim_{t \to \infty} \left[-\frac{x^3}{6(1 + x^6)} + \frac{1}{6} \tan^{-1}(x^3) \right]_{-1}^{t}$$

$$= \lim_{t \to \infty} \left(-\frac{t^3}{6(1 + t^6)} + \frac{1}{6} \tan^{-1}(t^3) + \frac{-1}{6(1 + 1)} - \frac{1}{6} \tan^{-1}(-1) \right)$$

$$= 0 + \frac{1}{6} \left(\frac{\pi}{2} \right) - \frac{1}{12} - \frac{1}{6} \left(-\frac{\pi}{4} \right) = \frac{\pi}{12} - \frac{1}{12} + \frac{\pi}{24} = \frac{\pi}{8} - \frac{1}{12}$$

19. The given integral represents the difference of the shaded areas, which appears to

be 0. It can be calculated by integrating with respect to either x or y, so we find x

in terms of y for each curve: $y = \sqrt[3]{1 - x^7} \quad \Rightarrow \quad x = \sqrt[7]{1 - y^3}$ and

$y = \sqrt[7]{1 - x^3} \quad \Rightarrow \quad x = \sqrt[3]{1 - y^7}$, so

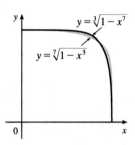

$\int_0^1 \left(\sqrt[3]{1 - y^7} - \sqrt[7]{1 - y^3} \right) dy = \int_0^1 \left(\sqrt[7]{1 - x^3} - \sqrt[3]{1 - x^7} \right) dx.$ But this

equation is of the form $z = -z$. So $\int_0^1 \left(\sqrt[3]{1 - x^7} - \sqrt[7]{1 - x^3} \right) dx = 0.$

21. In accordance with the hint, we let $I_k = \int_0^1 (1 - x^2)^k\, dx$, and we find an expression for I_{k+1} in terms of I_k. We integrate

I_{k+1} by parts with $u = (1 - x^2)^{k+1}, dv = dx \quad \Rightarrow \quad du = (k + 1)(1 - x^2)^k (-2x), v = x$, and then split the remaining integral into identifiable quantities:

$$I_{k+1} = \left[x(1-x^2)^{k+1}\right]_0^1 + 2(k+1)\int_0^1 x^2(1-x^2)^k \, dx = (2k+2)\int_0^1 (1-x^2)^k \left[1-(1-x^2)\right] dx$$

$$= (2k+2)(I_k - I_{k+1})$$

So $I_{k+1}\left[1+(2k+2)\right] = (2k+2)I_k \Rightarrow I_{k+1} = \dfrac{2k+2}{2k+3}I_k.$

Now to complete the proof, we use induction: $I_0 = 1 = \dfrac{2^0(0!)^2}{1!}$, so the formula holds for $n = 0$. Now suppose it holds

for $n = k$. Then

$$I_{k+1} = \frac{2k+2}{2k+3}I_k = \frac{2k+2}{2k+3}\left[\frac{2^{2k}(k!)^2}{(2k+1)!}\right] = \frac{2(k+1)2^{2k}(k!)^2}{(2k+3)(2k+1)!} = \frac{2(k+1)2^{2k}(k!)^2}{(2k+3)(2k+1)!} \cdot \frac{2(k+1)}{2k+2}$$

$$= \frac{[2(k+1)]^2 \, 2^{2k}(k!)^2}{(2k+3)(2k+2)(2k+1)!} = \frac{2^{2(k+1)}\left[(k+1)!\right]^2}{[2(k+1)+1]!}$$

So by induction, the formula holds for all integers $n \geq 0$.

23. (a) The tangent to the curve $y = f(x)$ at $x = x_0$ has the equation $y - f(x_0) = f'(x_0)(x - x_0)$. The y-intercept

of this tangent line is $f(x_0) - f'(x_0)x_0$. Thus, L is the distance from the point $(0, f(x_0) - f'(x_0)x_0)$ to

the point $(x_0, f(x_0))$; that is, $L^2 = x_0^2 + [f'(x_0)]^2 x_0^2$, so $[f'(x_0)]^2 = \dfrac{L^2 - x_0^2}{x_0^2}$ and $f'(x_0) = -\dfrac{\sqrt{L^2 - x_0^2}}{x_0}$

for $0 < x_0 < L$.

(b) $\dfrac{dy}{dx} = -\dfrac{\sqrt{L^2 - x^2}}{x} \Rightarrow y = \displaystyle\int\left(-\dfrac{\sqrt{L^2 - x^2}}{x}\right) dx.$

Let $x = L\sin\theta$. Then $dx = L\cos\theta \, d\theta$ and

$$y = \int \frac{-L\cos\theta \, L\cos\theta \, d\theta}{L\sin\theta} = L\int \frac{\sin^2\theta - 1}{\sin\theta}\, d\theta = L\int (\sin\theta - \csc\theta)\, d\theta$$

$$= -L\cos\theta - L\ln|\csc\theta - \cot\theta| + C = -\sqrt{L^2 - x^2} - L\ln\left(\frac{L}{x} - \frac{\sqrt{L^2 - x^2}}{x}\right) + C$$

When $x = L$, $y = 0$, and $0 = -0 - L\ln(1-0) + C$, so $C = 0$. Therefore, $y = -\sqrt{L^2 - x^2} - L\ln\left(\dfrac{L - \sqrt{L^2 - x^2}}{x}\right).$

6 □ APPLICATIONS OF INTEGRATION

6.1 More about Areas

1. $A = \int_{x=0}^{x=4} (y_T - y_B)\, dx = \int_0^4 \left[(5x - x^2) - x \right] dx = \int_0^4 (4x - x^2)\, dx = \left[2x^2 - \tfrac{1}{3}x^3 \right]_0^4 = \left(32 - \tfrac{64}{3} \right) - (0) = \tfrac{32}{3}$

3. $A = \int_{y=-1}^{y=1} (x_R - x_L)\, dy = \int_{-1}^1 \left[e^y - (y^2 - 2) \right] dy = \int_{-1}^1 \left(e^y - y^2 + 2 \right) dy$

$= \left[e^y - \tfrac{1}{3}y^3 + 2y \right]_{-1}^1 = \left(e^1 - \tfrac{1}{3} + 2 \right) - \left(e^{-1} + \tfrac{1}{3} - 2 \right) = e - \dfrac{1}{e} + \dfrac{10}{3}$

5. $A = \int_{-1}^1 \left[e^x - (x^2 - 1) \right] dx = \left[e^x - \tfrac{1}{3}x^3 + x \right]_{-1}^1$

$= (e - \tfrac{1}{3} + 1) - (e^{-1} + \tfrac{1}{3} - 1) = e - \tfrac{1}{e} + \tfrac{4}{3}$

7. $A = \int_0^1 \left(\sqrt{x} - x^2 \right) dx$

$= \left[\tfrac{2}{3}x^{3/2} - \tfrac{1}{3}x^3 \right]_0^1$

$= \tfrac{2}{3} - \tfrac{1}{3} = \tfrac{1}{3}$

9. The curves intersect when $1 - y^2 = y^2 - 1 \iff 2 = 2y^2 \iff y^2 = 1 \iff y = \pm 1$.

$A = \int_{-1}^1 \left[(1 - y^2) - (y^2 - 1) \right] dy$

$= \int_{-1}^1 2(1 - y^2)\, dy$

$= 2 \cdot 2 \int_0^1 (1 - y^2)\, dy$

$= 4 \left[y - \tfrac{1}{3}y^3 \right]_0^1 = 4 \left(1 - \tfrac{1}{3} \right) = \tfrac{8}{3}$

11. $2y^2 = 4 + y^2 \iff y^2 = 4 \iff y = \pm 2$, so

$$A = \int_{-2}^{2} \left[(4 + y^2) - 2y^2 \right] dy$$

$$= 2 \int_{0}^{2} (4 - y^2) \, dy \qquad \text{[by symmetry]}$$

$$= 2 \left[4y - \tfrac{1}{3} y^3 \right]_0^2 = 2 \left(8 - \tfrac{8}{3} \right) = \tfrac{32}{3}$$

13. $12 - x^2 = x^2 - 6 \iff 2x^2 = 18 \iff$

$x^2 = 9 \iff x = \pm 3$, so

$$A = \int_{-3}^{3} \left[(12 - x^2) - (x^2 - 6) \right] dx$$

$$= 2 \int_{0}^{3} (18 - 2x^2) \, dx \qquad \text{[by symmetry]}$$

$$= 2 \left[18x - \tfrac{2}{3} x^3 \right]_0^3 = 2 \left[(54 - 18) - 0 \right]$$

$$= 2(36) = 72$$

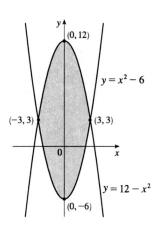

15. $e^x = xe^x \iff e^x - xe^x = 0 \iff e^x(1 - x) = 0 \iff x = 1$.

$$A = \int_{0}^{1} (e^x - xe^x) \, dx$$

$$= \left[e^x - (xe^x - e^x) \right]_0^1 \qquad \text{[use parts with } u = x \text{ and } dv = e^x \, dx]$$

$$= \left[2e^x - xe^x \right]_0^1 = (2e - e) - (2 - 0) = e - 2$$

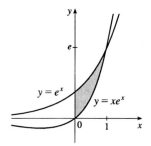

17. $1/x = x \iff 1 = x^2 \iff x = \pm 1$ and $1/x = \tfrac{1}{4} x \iff$

$4 = x^2 \iff x = \pm 2$, so for $x > 0$,

$$A = \int_{0}^{1} \left(x - \tfrac{1}{4} x \right) dx + \int_{1}^{2} \left(\tfrac{1}{x} - \tfrac{1}{4} x \right) dx$$

$$= \int_{0}^{1} \left(\tfrac{3}{4} x \right) dx + \int_{1}^{2} \left(\tfrac{1}{x} - \tfrac{1}{4} x \right) dx$$

$$= \left[\tfrac{3}{8} x^2 \right]_0^1 + \left[\ln |x| - \tfrac{1}{8} x^2 \right]_1^2$$

$$= \tfrac{3}{8} + \left(\ln 2 - \tfrac{1}{2} \right) - \left(0 - \tfrac{1}{8} \right) = \ln 2$$

19.

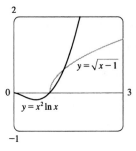

From the graph, we see that the curves intersect at $x = 0$ and $x = a \approx 0.896$, with $x \sin(x^2) > x^4$ on $(0, a)$. So the area A of the region bounded by the curves is

$$A = \int_0^a \left[x \sin(x^2) - x^4 \right] dx = \left[-\tfrac{1}{2} \cos(x^2) - \tfrac{1}{5} x^5 \right]_0^a$$

$$= -\tfrac{1}{2} \cos(a^2) - \tfrac{1}{5} a^5 + \tfrac{1}{2} \approx 0.037$$

21. From the graph, we see that the curves intersect at $x = 1$ and $x = a \approx 1.382$, with $\sqrt{x-1} > x^2 \ln x$ on $(1, a)$. The area A of the region bounded by the curves is

$$A = \int_1^a \left[\sqrt{x-1} - x^2 \ln x \right] dx$$

$$= \left[\tfrac{2}{3}(x-1)^{3/2} - \left(\tfrac{1}{3}x^3 \ln x - \tfrac{1}{9}x^3 \right) \right]_1^a \quad \text{[by Exercise 5.6.1]}$$

$$\approx 0.05$$

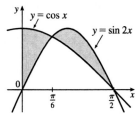

23. $\cos x = \sin 2x = 2 \sin x \cos x \quad \Leftrightarrow \quad 2 \sin x \cos x - \cos x = 0 \quad \Leftrightarrow \quad \cos x \left(2 \sin x - 1 \right) = 0 \quad \Leftrightarrow$

$2 \sin x = 1$ or $\cos x = 0 \quad \Leftrightarrow \quad x = \tfrac{\pi}{6}$ or $\tfrac{\pi}{2}$.

$$A = \int_0^{\pi/6} (\cos x - \sin 2x) \, dx + \int_{\pi/6}^{\pi/2} (\sin 2x - \cos x) \, dx$$

$$= \left[\sin x + \tfrac{1}{2} \cos 2x \right]_0^{\pi/6} + \left[-\tfrac{1}{2} \cos 2x - \sin x \right]_{\pi/6}^{\pi/2}$$

$$= \left(\tfrac{1}{2} + \tfrac{1}{2} \cdot \tfrac{1}{2} \right) - \left(0 + \tfrac{1}{2} \cdot 1 \right) + \left[-\tfrac{1}{2} \cdot (-1) - 1 \right] - \left(-\tfrac{1}{2} \cdot \tfrac{1}{2} - \tfrac{1}{2} \right)$$

$$= \tfrac{3}{4} - \tfrac{1}{2} - \tfrac{1}{2} + \tfrac{3}{4} = \tfrac{1}{2}$$

25. As in Example 4, we approximate the distance between the two cars after ten seconds using Simpson's Rule with $\Delta t = 1 \, \text{s} = \tfrac{1}{3600} \, \text{h}$.

$$\text{distance}_{\text{Kelly}} - \text{distance}_{\text{Chris}} = \int_0^{10} v_K \, dt - \int_0^{10} v_C \, dt = \int_0^{10} (v_K - v_C) \, dt \approx S_{10}$$

$$= \tfrac{1}{3 \cdot 3600} [(0 - 0) + 4(22 - 20) + 2(37 - 32) + 4(52 - 46) + 2(61 - 54) + 4(71 - 62)$$

$$+ \, 2(80 - 69) + 4(86 - 75) + 2(93 - 81) + 4(98 - 86) + (102 - 90)]$$

$$= \tfrac{1}{10,800} (242) = \tfrac{121}{5400} \, \text{mi}$$

So after 10 seconds, Kelly's car is about $\dfrac{121}{5400} \, \text{mi} \left(5280 \dfrac{\text{ft}}{\text{mi}} \right) \approx 118 \, \text{ft}$ ahead of Chris's.

27. If x = distance from left end of pool and $w = w(x)$ = width at x, then Simpson's Rule with $n = 8$ and $\Delta x = 2$ gives

$$\text{Area} = \int_0^{16} w \, dx \approx \tfrac{2}{3} [0 + 4(6.2) + 2(7.2) + 4(6.8) + 2(5.6) + 4(5.0) + 2(4.8) + 4(4.8) + 0] = \tfrac{2}{3}(126.4) \approx 84 \, \text{m}^2.$$

29. For $0 \le t \le 10$, $b(t) > d(t)$, so the area between the curves is given by

$$\int_0^{10} [b(t) - d(t)]\, dt = \int_0^{10} (2200e^{0.024t} - 1460e^{0.018t})\, dt = \left[\frac{2200}{0.024}e^{0.024t} - \frac{1460}{0.018}e^{0.018t} \right]_0^{10}$$

$$= \left(\frac{275{,}000}{3}e^{0.24} - \frac{730{,}000}{9}e^{0.18} \right) - \left(\frac{275{,}000}{3} - \frac{730{,}000}{9} \right) \approx 8868 \text{ people}$$

This area A represents the increase in population over a 10-year period.

31. Let the equation of the large circle be $x^2 + y^2 = R^2$. Then the equation of
the small circle is $x^2 + (y - b)^2 = r^2$, where $b = \sqrt{R^2 - r^2}$ is the distance
between the centers of the circles. The desired area is

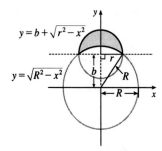

$$A = \int_{-r}^{r} \left[(b + \sqrt{r^2 - x^2}) - \sqrt{R^2 - x^2} \right] dx$$

$$= 2\int_0^r (b + \sqrt{r^2 - x^2} - \sqrt{R^2 - x^2})\, dx$$

$$= 2\int_0^r b\, dx + 2\int_0^r \sqrt{r^2 - x^2}\, dx - 2\int_0^r \sqrt{R^2 - x^2}\, dx$$

The first integral is just $2br = 2r\sqrt{R^2 - r^2}$. The second integral represents the area of a quarter-circle of radius r, so its value
is $\frac{1}{4}\pi r^2$. To evaluate the other integral, note that

$$\int \sqrt{a^2 - x^2}\, dx = \int a^2 \cos^2\theta\, d\theta \quad [x = a\sin\theta,\, dx = a\cos\theta\, d\theta] \quad = \left(\tfrac{1}{2}a^2\right)\int (1 + \cos 2\theta)\, d\theta$$

$$= \tfrac{1}{2}a^2\left(\theta + \tfrac{1}{2}\sin 2\theta\right) + C = \tfrac{1}{2}a^2(\theta + \sin\theta\,\cos\theta) + C$$

$$= \frac{a^2}{2}\arcsin\left(\frac{x}{a}\right) + \frac{a^2}{2}\left(\frac{x}{a}\right)\frac{\sqrt{a^2 - x^2}}{a} + C = \frac{a^2}{2}\arcsin\left(\frac{x}{a}\right) + \frac{x}{2}\sqrt{a^2 - x^2} + C$$

Thus, the desired area is

$$A = 2r\sqrt{R^2 - r^2} + 2\left(\tfrac{1}{4}\pi r^2\right) - \left[R^2\arcsin(x/R) + x\sqrt{R^2 - x^2} \right]_0^r$$

$$= 2r\sqrt{R^2 - r^2} + \tfrac{1}{2}\pi r^2 - \left[R^2\arcsin(r/R) + r\sqrt{R^2 - r^2} \right] = r\sqrt{R^2 - r^2} + \tfrac{\pi}{2}r^2 - R^2\arcsin(r/R)$$

33. By symmetry of the ellipse about the x- and y-axes,

$$A = 4\int_0^a y\, dx = 4\int_{\pi/2}^0 b\sin\theta\,(-a\sin\theta)\, d\theta \qquad \begin{bmatrix} x = a\cos\theta = 0 \;\Rightarrow\; \theta = \frac{\pi}{2}, \\ x = a\cos\theta = a \;\Rightarrow\; \theta = 0, \end{bmatrix}$$

$$= 4ab\int_0^{\pi/2} \sin^2\theta\, d\theta = 4ab\int_0^{\pi/2} \tfrac{1}{2}(1 - \cos 2\theta)\, d\theta$$

$$= 2ab\left[\theta - \tfrac{1}{2}\sin 2\theta\right]_0^{\pi/2} = 2ab\left(\tfrac{\pi}{2}\right) = \pi ab$$

Note that the formula for the area of a circle, $A = \pi r^2$, is just a special case of this formula with $a = b = r$.

35. The curve $x = 1 + e^t$, $y = t - t^2 = t(1 - t)$ intersects the x-axis when $y = 0$,

that is, when $t = 0$ and $t = 1$. The corresponding values of x are 2 and $1 + e$.

The shaded area is given by

$$\int_{x=2}^{x=1+e} (y_T - y_B)\,dx = \int_{t=0}^{t=1} [y(t) - 0]\,x'(t)\,dt = \int_0^1 (t - t^2)e^t\,dt$$

$$= \int_0^1 te^t\,dt - \int_0^1 t^2 e^t\,dt = \int_0^1 te^t\,dt - \left[t^2 e^t\right]_0^1 + 2\int_0^1 te^t\,dt \qquad \text{[Formula 97 or parts]}$$

$$= 3\int_0^1 te^t\,dt - (e - 0) = 3\left[(t-1)e^t\right]_0^1 - e \qquad \text{[Formula 96 or parts]}$$

$$= 3[0 - (-1)] - e = 3 - e$$

37. By symmetry, the area of the region enclosed by the loop is twice the area above

the x-axis inside the loop. $y = 0 \iff t^3 - 3t = 0 \iff t(t^2 - 3) = 0 \iff$

$t = 0, \pm\sqrt{3}$. The top half of the loop is described by $x = t^2$, $y = t^3 - 3t$,

$-\sqrt{3} \le t \le 0$, so, using the Substitution Rule with $y = t^3 - 3t$ and $dx = 2t\,dt$,

we find that

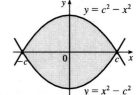

$$\text{Area} = 2\int_0^3 y\,dx = 2\int_0^{-\sqrt{3}} (t^3 - 3t)2t\,dt = 4\int_0^{-\sqrt{3}} (t^4 - 3t^2)\,dt$$

$$= 4\left[\tfrac{1}{5}t^5 - t^3\right]_0^{-\sqrt{3}} = 4\left[\tfrac{1}{5}\left(-3^{1/2}\right)^5 - \left(-3^{1/2}\right)^3\right]$$

$$= 4\left[\tfrac{1}{5}\left(-9\sqrt{3}\right) - \left(-3\sqrt{3}\right)\right] = \tfrac{24}{5}\sqrt{3} \approx 8.31.$$

39. We first assume that $c > 0$, since c can be replaced by $-c$ in both equations without changing the graphs, and if $c = 0$ the

curves do not enclose a region. We see from the graph that the enclosed area A lies between $x = -c$ and $x = c$, and by

symmetry, it is equal to four times the area in the first quadrant. The enclosed area is

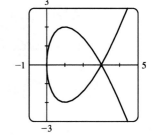

$$A = 4\int_0^c (c^2 - x^2)\,dx = 4\left[c^2 x - \tfrac{1}{3}x^3\right]_0^c = 4\left(c^3 - \tfrac{1}{3}c^3\right) = 4\left(\tfrac{2}{3}c^3\right) = \tfrac{8}{3}c^3$$

So $A = 576 \iff \tfrac{8}{3}c^3 = 576 \iff c^3 = 216 \iff c = \sqrt[3]{216} = 6.$

Note that $c = -6$ is another solution, since the graphs are the same.

41.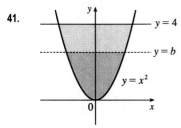

By the symmetry of the problem, we consider only the first quadrant, where

$y = x^2 \implies x = \sqrt{y}$. We are looking for a number b such that

$$\int_0^b \sqrt{y}\,dy = \int_b^4 \sqrt{y}\,dy \implies \tfrac{2}{3}\left[y^{3/2}\right]_0^b = \tfrac{2}{3}\left[y^{3/2}\right]_b^4 \implies$$

$$b^{3/2} = 4^{3/2} - b^{3/2} \implies 2b^{3/2} = 8 \implies b^{3/2} = 4 \implies b = 4^{2/3} \approx 2.52.$$

43. The area under the graph of f from 0 to t is equal to $\int_0^t f(x)\,dx$, so the requirement is that $\int_0^t f(x)\,dx = t^3$ for all t. We

differentiate both sides of this equation with respect to t (with the help of FTC1) to get $f(t) = 3t^2$. This function is positive

and continuous, as required.

45. The curve and the line will determine a region when they intersect at two or

more points. So we solve the equation $x/(x^2 + 1) = mx$ ⇒

$x = x(mx^2 + m)$ ⇒ $x(mx^2 + m) - x = 0$ ⇒

$x(mx^2 + m - 1) = 0$ ⇒ $x = 0$ or $mx^2 + m - 1 = 0$ ⇒

$x = 0$ or $x^2 = \dfrac{1-m}{m}$ ⇒ $x = 0$ or $x = \pm\sqrt{\dfrac{1}{m} - 1}$. Note that if $m = 1$, this has only the solution $x = 0$, and no region

is determined. But if $1/m - 1 > 0$ ⇔ $1/m > 1$ ⇔ $0 < m < 1$, then there are two solutions. [Another way of seeing

this is to observe that the slope of the tangent to $y = x/(x^2 + 1)$ at the origin is $y'(0) = 1$ and therefore we must have

$0 < m < 1$.] Note that we cannot just integrate between the positive and negative roots, since the curve and the line cross at

the origin. Since mx and $x/(x^2 + 1)$ are both odd functions, the total area is twice the area between the curves on the interval

$\left[0, \sqrt{1/m - 1}\right]$. So the total area enclosed is

$$2\int_0^{\sqrt{1/m-1}} \left[\frac{x}{x^2+1} - mx\right] dx = 2\left[\tfrac{1}{2}\ln(x^2+1) - \tfrac{1}{2}mx^2\right]_0^{\sqrt{1/m-1}} = \left[\ln(1/m - 1 + 1) - m(1/m - 1)\right] - (\ln 1 - 0)$$

$$= \ln(1/m) - 1 + m = m - \ln m - 1$$

6.2 Volumes

1. A cross-section is a disk with radius $2 - \tfrac{1}{2}x$, so its area is $A(x) = \pi\left(2 - \tfrac{1}{2}x\right)^2$.

$$V = \int_1^2 A(x)\,dx = \int_1^2 \pi\left(2 - \tfrac{1}{2}x\right)^2 dx$$

$$= \pi\int_1^2 \left(4 - 2x + \tfrac{1}{4}x^2\right) dx$$

$$= \pi\left[4x - x^2 + \tfrac{1}{12}x^3\right]_1^2$$

$$= \pi\left[\left(8 - 4 + \tfrac{8}{12}\right) - \left(4 - 1 + \tfrac{1}{12}\right)\right]$$

$$= \pi\left(1 + \tfrac{7}{12}\right) = \tfrac{19}{12}\pi$$

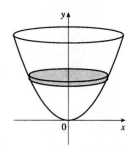

3. A cross-section is a disk with radius $2\sqrt{y}$, so its area is

$$A(y) = \pi\left(2\sqrt{y}\right)^2.$$

$$V = \int_0^9 A(y)\,dy = \int_0^9 \pi\left(2\sqrt{y}\right)^2 dy = 4\pi\int_0^9 y\,dy$$

$$= 4\pi\left[\tfrac{1}{2}y^2\right]_0^9 = 2\pi(81) = 162\pi$$

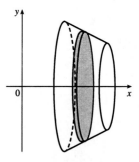

5. A cross-section is a washer (annulus) with inner radius x^3 and outer radius x, so its area is

$$A(x) = \pi(x)^2 - \pi(x^3)^2 = \pi(x^2 - x^6).$$

$$V = \int_0^1 A(x)\,dx = \int_0^1 \pi(x^2 - x^6)\,dx$$

$$= \pi\left[\tfrac{1}{3}x^3 - \tfrac{1}{7}x^7\right]_0^1 = \pi\left(\tfrac{1}{3} - \tfrac{1}{7}\right) = \tfrac{4}{21}\pi$$

 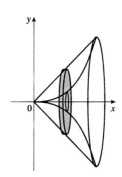

7. A cross-section is a washer with inner radius y^2 and outer radius $2y$, so its area is

$$A(y) = \pi(2y)^2 - \pi(y^2)^2 = \pi(4y^2 - y^4).$$

$$V = \int_0^2 A(y)\,dy = \pi\int_0^2 (4y^2 - y^4)\,dy$$

$$= \pi\left[\tfrac{4}{3}y^3 - \tfrac{1}{5}y^5\right]_0^2 = \pi\left(\tfrac{32}{3} - \tfrac{32}{5}\right) = \tfrac{64}{15}\pi$$

 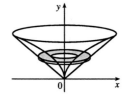

9. A cross-section is a washer with inner radius $1 - \sqrt{x}$ and outer radius $1 - x$, so its area is

$$A(x) = \pi(1-x)^2 - \pi\left(1 - \sqrt{x}\right)^2$$

$$= \pi\left[(1 - 2x + x^2) - \left(1 - 2\sqrt{x} + x\right)\right]$$

$$= \pi\left(-3x + x^2 + 2\sqrt{x}\right).$$

$$V = \int_0^1 A(x)\,dx = \pi\int_0^1 \left(-3x + x^2 + 2\sqrt{x}\right)dx$$

$$= \pi\left[-\tfrac{3}{2}x^2 + \tfrac{1}{3}x^3 + \tfrac{4}{3}x^{3/2}\right]_0^1 = \pi\left(-\tfrac{3}{2} + \tfrac{5}{3}\right) = \tfrac{\pi}{6}$$

 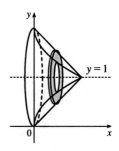

11. A cross-section is a washer with inner radius $(1 + \sec x) - 1 = \sec x$ and outer radius $3 - 1 = 2$, so its area is

$$A(x) = \pi\left[2^2 - (\sec x)^2\right] = \pi(4 - \sec^2 x).$$

$$V = \int_{-\pi/3}^{\pi/3} A(x)\,dx = \int_{-\pi/3}^{\pi/3} \pi(4 - \sec^2 x)\,dx$$

$$= 2\pi\int_0^{\pi/3} (4 - \sec^2 x)\,dx \qquad \text{[by symmetry]}$$

$$= 2\pi\left[4x - \tan x\right]_0^{\pi/3} = 2\pi\left[\left(\tfrac{4\pi}{3} - \sqrt{3}\right) - 0\right]$$

$$= 2\pi\left(\tfrac{4\pi}{3} - \sqrt{3}\right)$$

  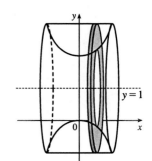

13. A cross-section is a disk with radius $1/x$, so its area is $A(x) = \pi(1/x)^2$.

$$V = \int_1^2 A(x)\, dx = \int_1^2 \pi\left(\frac{1}{x}\right)^2 dx$$

$$= \pi \int_1^2 \frac{1}{x^2}\, dx = \pi\left[-\frac{1}{x}\right]_1^2$$

$$= \pi\left[-\frac{1}{2} - (-1)\right] = \frac{\pi}{2}$$

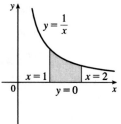

15. The curves $x - y = 1$ and $y = x^2 - 4x + 3$ intersect when

$x - 1 = x^2 - 4x + 3 \quad\Leftrightarrow\quad 0 = x^2 - 5x + 4 \quad\Leftrightarrow$

$0 = (x - 1)(x - 4) \quad\Leftrightarrow\quad x = 1$ or 4. A cross-section is a washer with

inner radius $3 - (x - 1)$ and outer radius $3 - (x^2 - 4x + 3)$, so its area is

$A(x) = \pi[3 - (x^2 - 4x + 3)]^2 - \pi[3 - (x - 1)]^2$.

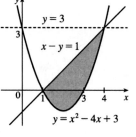

$$V = \int_1^4 A(x)\, dx = \pi \int_1^4 \left\{[3 - (x^2 - 4x + 3)]^2 - [3 - (x - 1)]^2\right\} dx$$

$$= \pi \int_1^4 \left[(4x - x^2)^2 - (4 - x)^2\right] dx = \pi \int_1^4 (16x^2 - 8x^3 + x^4 - 16 + 8x - x^2)\, dx$$

$$= \pi \int_1^4 (x^4 - 8x^3 + 15x^2 + 8x - 16)\, dx = \pi\left[\tfrac{1}{5}x^5 - 2x^4 + 5x^3 + 4x^2 - 16x\right]_1^4$$

$$= \pi\left[\left(\tfrac{1024}{5} - 512 + 320 + 64 - 64\right) - \left(\tfrac{1}{5} - 2 + 5 + 4 - 16\right)\right] = \pi\left(\tfrac{1023}{5} - 183\right) = \tfrac{108}{5}\pi$$

17. $y = \sqrt{x} \;\Rightarrow\; x = y^2$ and $y = x^3 \;\Rightarrow\; x = \sqrt[3]{y}$. A cross-section is a

washer with inner radius $1 - \sqrt[3]{y}$ and outer radius $1 - y^2$, so its area is

$A(y) = \pi(1 - y^2)^2 - \pi\left(1 - \sqrt[3]{y}\right)^2$.

$$V = \int_0^1 A(y)\, dy = \int_0^1 \left[\pi(1 - y^2)^2 - \pi\left(1 - \sqrt[3]{y}\right)^2\right] dy$$

$$= \pi \int_0^1 \left[(1 - 2y^2 + y^4) - (1 - 2y^{1/3} + y^{2/3})\right] dy$$

$$= \pi \int_0^1 (-2y^2 + y^4 + 2y^{1/3} - y^{2/3})\, dy = \pi\left[-\tfrac{2}{3}y^3 + \tfrac{1}{5}y^5 + \tfrac{3}{2}y^{4/3} - \tfrac{3}{5}y^{5/3}\right]_0^1 = \pi\left(-\tfrac{2}{3} + \tfrac{1}{5} + \tfrac{3}{2} - \tfrac{3}{5}\right) = \tfrac{13\pi}{30}$$

19. $V = \pi \displaystyle\int_{-\sqrt{8}}^{\sqrt{8}} \left\{[3 - (-2)]^2 - \left[\sqrt{y^2 + 1} - (-2)\right]^2\right\} dy$

$$= \pi \int_{-2\sqrt{2}}^{2\sqrt{2}} \left[5^2 - \left(\sqrt{1 + y^2} + 2\right)^2\right] dy$$

21.

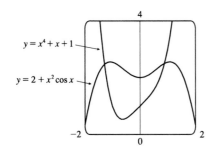

$y = 2 + x^2 \cos x$ and $y = x^4 + x + 1$ intersect at

$x = a \approx -1.288$ and $x = b \approx 0.884$.

$$V = \pi \int_a^b \left[(2 + x^2 \cos x)^2 - (x^4 + x + 1)^2 \right] dx \approx 23.780$$

23. $V = \pi \int_0^\pi \left\{ \left[\sin^2 x - (-1) \right]^2 - \left[0 - (-1) \right]^2 \right\} dx$

$\overset{\text{CAS}}{=} \frac{11}{8} \pi^2$

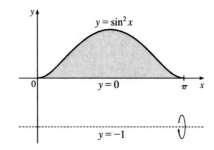

25. (a) $\pi \int_0^{\pi/2} \cos^2 x \, dx$ describes the volume of the solid obtained by rotating the region

$\mathcal{R} = \left\{ (x, y) \mid 0 \le x \le \frac{\pi}{2}, 0 \le y \le \cos x \right\}$ of the xy-plane about the x-axis.

(b) $\pi \int_0^1 (y^4 - y^8) \, dy = \pi \int_0^1 \left[(y^2)^2 - (y^4)^2 \right] dy$ describes the volume of the solid obtained by rotating the region

$\mathcal{R} = \left\{ (x, y) \mid 0 \le y \le 1, y^4 \le x \le y^2 \right\}$ of the xy-plane about the y-axis.

27. There are 10 subintervals over the 15-cm length, so we'll use $n = 10/2 = 5$ for the Midpoint Rule.

$$V = \int_0^{15} A(x) \, dx \approx M_5 = \frac{15-0}{5} [A(1.5) + A(4.5) + A(7.5) + A(10.5) + A(13.5)]$$

$$= 3(18 + 79 + 106 + 128 + 39) = 3 \cdot 370 = 1110 \text{ cm}^3$$

29. (a) Using disks, $V = \int_2^{10} \pi [f(x)]^2 \, dx = \pi \int_2^{10} [f(x)]^2 \, dx = \pi I_2$. Now use Simpson's Rule to approximate I_2:

$$I_2 \approx S_8 = \frac{10-2}{3(8)} \left\{ [f(2)]^2 + 4 [f(3)]^2 + 2 [f(4)]^2 + 4 [f(5)]^2 + 2 [f(6)]^2 \right.$$
$$\left. + 4 [f(7)]^2 + 2 [f(8)]^2 + 4 [f(9)]^2 + [f(10)]^2 \right\}$$
$$\approx \frac{1}{3} \left[(0)^2 + 4(1.5)^2 + 2(1.9)^2 + 4(2.2)^2 + 2(3.0)^2 + 4(3.8)^2 + 2(4.0)^2 + 4(3.1)^2 + (0)^2 \right]$$
$$= \frac{1}{3}(181.78)$$

Thus, $V \approx \pi \cdot \frac{1}{3}(181.78) \approx 190.4$ or 190 cubic units.

(b) $V = \int_0^4 \pi \left[(\text{outer radius})^2 - (\text{inner radius})^2 \right] dy$

$$\approx \pi \frac{4-0}{4 \cdot 3} \left\{ [10^2 - 2^2] + 4 [(9.8)^2 - (2.5)^2] + 2 [(9.5)^2 - (4.3)^2] + 4 [(9.1)^2 - (6.0)^2] + [8^2 - 8^2] \right\}$$

$$= \frac{\pi}{3}(785.92) \approx 823 \text{ units}^3$$

31. We'll form a right circular cone with height h and base radius r by revolving the line $y = \frac{r}{h}x$ about the x-axis.

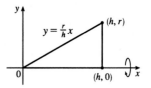

$$V = \pi \int_0^h \left(\frac{r}{h}x\right)^2 dx = \pi \int_0^h \frac{r^2}{h^2}x^2\,dx = \pi\frac{r^2}{h^2}\left[\frac{1}{3}x^3\right]_0^h$$

$$= \pi\frac{r^2}{h^2}\left(\frac{1}{3}h^3\right) = \frac{1}{3}\pi r^2 h$$

Another solution: Revolve $x = -\dfrac{r}{h}y + r$ about the y-axis.

$$V = \pi \int_0^h \left(-\frac{r}{h}y + r\right)^2 dy \overset{*}{=} \pi \int_0^h \left[\frac{r^2}{h^2}y^2 - \frac{2r^2}{h}y + r^2\right] dy$$

$$= \pi \left[\frac{r^2}{3h^2}y^3 - \frac{r^2}{h}y^2 + r^2 y\right]_0^h = \pi\left(\tfrac{1}{3}r^2 h - r^2 h + r^2 h\right) = \tfrac{1}{3}\pi r^2 h$$

* Or use substitution with $u = r - \dfrac{r}{h}y$ and $du = -\dfrac{r}{h}\,dy$ to get

$$\pi \int_r^0 u^2\left(-\frac{h}{r}\,du\right) = -\pi\frac{h}{r}\left[\frac{1}{3}u^3\right]_r^0 = -\pi\frac{h}{r}\left(-\frac{1}{3}r^3\right) = \frac{1}{3}\pi r^2 h.$$

33. $x^2 + y^2 = r^2 \;\Leftrightarrow\; x^2 = r^2 - y^2$

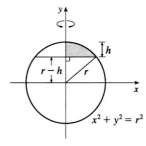

$$V = \pi \int_{r-h}^r (r^2 - y^2)\,dy = \pi\left[r^2 y - \frac{y^3}{3}\right]_{r-h}^r$$

$$= \pi\left\{\left[r^3 - \frac{r^3}{3}\right] - \left[r^2(r-h) - \frac{(r-h)^3}{3}\right]\right\}$$

$$= \pi\left\{\tfrac{2}{3}r^3 - \tfrac{1}{3}(r-h)\left[3r^2 - (r-h)^2\right]\right\}$$

$$= \tfrac{1}{3}\pi\left\{2r^3 - (r-h)\left[3r^2 - (r^2 - 2rh + h^2)\right]\right\}$$

$$= \tfrac{1}{3}\pi\left\{2r^3 - (r-h)\left[2r^2 + 2rh - h^2\right]\right\}$$

$$= \tfrac{1}{3}\pi\left(2r^3 - 2r^3 - 2r^2 h + rh^2 + 2r^2 h + 2rh^2 - h^3\right)$$

$$= \tfrac{1}{3}\pi\left(3rh^2 - h^3\right) = \tfrac{1}{3}\pi h^2(3r - h), \text{ or, equivalently, } \pi h^2\left(r - \frac{h}{3}\right)$$

35. For a cross-section at height y, we see from similar triangles that $\dfrac{\alpha/2}{b/2} = \dfrac{h-y}{h}$, so $\alpha = b\left(1 - \dfrac{y}{h}\right)$.

Similarly, for cross-sections having $2b$ as their base and β replacing α, $\beta = 2b\left(1 - \dfrac{y}{h}\right)$. So

$$V = \int_0^h A(y)\,dy = \int_0^h \left[b\left(1 - \frac{y}{h}\right)\right]\left[2b\left(1 - \frac{y}{h}\right)\right] dy$$

$$= \int_0^h 2b^2\left(1 - \frac{y}{h}\right)^2 dy = 2b^2 \int_0^h \left(1 - \frac{2y}{h} + \frac{y^2}{h^2}\right) dy$$

$$= 2b^2\left[y - \frac{y^2}{h} + \frac{y^3}{3h^2}\right]_0^h = 2b^2\left[h - h + \tfrac{1}{3}h\right]$$

$$= \tfrac{2}{3}b^2 h \quad [\,= \tfrac{1}{3}Bh \text{ where } B \text{ is the area of the base, as with any pyramid.}]$$

37. A cross-section at height z is a triangle similar to the base, so we'll multiply the legs of the base triangle, 3 and 4, by a proportionality factor of $(5 - z)/5$. Thus, the triangle at height z has area

$$A(z) = \frac{1}{2} \cdot 3\left(\frac{5-z}{5}\right) \cdot 4\left(\frac{5-z}{5}\right) = 6\left(1 - \frac{z}{5}\right)^2, \text{ so}$$

$$V = \int_0^5 A(z)\,dz = 6 \int_0^5 \left(1 - \frac{z}{5}\right)^2 dz = 6 \int_1^0 u^2(-5\,du) \qquad \begin{bmatrix} u = 1 - z/5, \\ du = -\frac{1}{5}\,dz \end{bmatrix}$$

$$= -30\left[\tfrac{1}{3}u^3\right]_1^0 = -30\left(-\tfrac{1}{3}\right) = 10 \text{ cm}^3$$

39. If l is a leg of the isosceles right triangle and $2y$ is the hypotenuse,

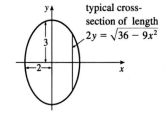

typical cross-section of length
$2y = \sqrt{36 - 9x^2}$

then $l^2 + l^2 = (2y)^2 \;\Rightarrow\; 2l^2 = 4y^2 \;\Rightarrow\; l^2 = 2y^2$.

$$V = \int_{-2}^2 A(x)\,dx = 2\int_0^2 A(x)\,dx = 2\int_0^2 \tfrac{1}{2}(l)(l)\,dx = 2\int_0^2 y^2\,dx$$

$$= 2\int_0^2 \tfrac{1}{4}(36 - 9x^2)\,dx = \tfrac{9}{2}\int_0^2(4 - x^2)\,dx$$

$$= \tfrac{9}{2}\left[4x - \tfrac{1}{3}x^3\right]_0^2 = \tfrac{9}{2}\left(8 - \tfrac{8}{3}\right) = 24$$

41. The cross-section of the base corresponding to the coordinate x has length $y = 1 - x$. The corresponding square with side s has area

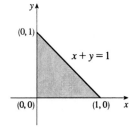

$$A(x) = s^2 = (1 - x)^2 = 1 - 2x + x^2. \text{ Therefore,}$$

$$V = \int_0^1 A(x)\,dx = \int_0^1 (1 - 2x + x^2)\,dx$$

$$= \left[x - x^2 + \tfrac{1}{3}x^3\right]_0^1 = \left(1 - 1 + \tfrac{1}{3}\right) - 0 = \tfrac{1}{3}$$

Or: $\displaystyle \int_0^1 (1 - x)^2\,dx = \int_1^0 u^2(-du) \quad [u = 1 - x] = \left[\tfrac{1}{3}u^3\right]_0^1 = \tfrac{1}{3}$

43. The cross-section of the base b corresponding to the coordinate x has length $1 - x^2$. The height h also has length $1 - x^2$, so the corresponding isosceles triangle has area $A(x) = \frac{1}{2}bh = \frac{1}{2}(1 - x^2)^2$. Therefore,

$$V = \int_{-1}^1 \tfrac{1}{2}(1 - x^2)^2\,dx$$

$$= 2 \cdot \tfrac{1}{2}\int_0^1 (1 - 2x^2 + x^4)\,dx \qquad \text{[by symmetry]}$$

$$= \left[x - \tfrac{2}{3}x^3 + \tfrac{1}{5}x^5\right]_0^1 = \left(1 - \tfrac{2}{3} + \tfrac{1}{5}\right) - 0 = \tfrac{8}{15}$$

45. (a) The torus is obtained by rotating the circle $(x - R)^2 + y^2 = r^2$ about the y-axis. Solving for x, we see that the right half of the circle is given by

$$x = R + \sqrt{r^2 - y^2} = f(y) \text{ and the left half by } x = R - \sqrt{r^2 - y^2} = g(y). \text{ So}$$

$$V = \pi \int_{-r}^{r} \left\{ [f(y)]^2 - [g(y)]^2 \right\} dy$$

$$= 2\pi \int_{0}^{r} \left[\left(R^2 + 2R\sqrt{r^2 - y^2} + r^2 - y^2 \right) - \left(R^2 - 2R\sqrt{r^2 - y^2} + r^2 - y^2 \right) \right] dy$$

$$= 2\pi \int_{0}^{r} 4R\sqrt{r^2 - y^2}\, dy = 8\pi R \int_{0}^{r} \sqrt{r^2 - y^2}\, dy$$

(b) Observe that the integral represents a quarter of the area of a circle with radius r, so

$$8\pi R \int_{0}^{r} \sqrt{r^2 - y^2}\, dy = 8\pi R \cdot \tfrac{1}{4}\pi r^2 = 2\pi^2 r^2 R.$$

47. (a) Volume$(S_1) = \int_{0}^{h} A(z)\, dz =$ Volume(S_2) since the cross-sectional area $A(z)$ at height z is the same for both solids.

(b) By Cavalieri's Principle, the volume of the cylinder in the figure is the same as that of a right circular cylinder with radius r

and height h, that is, $\pi r^2 h$.

49. The volume is obtained by rotating the area common to two circles of radius r, as

shown. The volume of the right half is

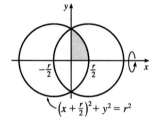

$$V_{\text{right}} = \pi \int_{0}^{r/2} y^2\, dx = \pi \int_{0}^{r/2} \left[r^2 - \left(\tfrac{1}{2}r + x \right)^2 \right] dx$$

$$= \pi \left[r^2 x - \tfrac{1}{3}\left(\tfrac{1}{2}r + x \right)^3 \right]_{0}^{r/2} = \pi \left[\left(\tfrac{1}{2}r^3 - \tfrac{1}{3}r^3 \right) - \left(0 - \tfrac{1}{24}r^3 \right) \right] = \tfrac{5}{24}\pi r^3$$

$(x + \tfrac{r}{2})^2 + y^2 = r^2$

So by symmetry, the total volume is twice this, or $\tfrac{5}{12}\pi r^3$.

Another solution: We observe that the volume is the twice the volume of a cap of a sphere, so we can use the formula from

Exercise 33 with $h = \tfrac{1}{2}r$: $V = 2 \cdot \tfrac{1}{3}\pi h^2 (3r - h) = \tfrac{2}{3}\pi \left(\tfrac{1}{2}r \right)^2 (3r - \tfrac{1}{2}r) = \tfrac{5}{12}\pi r^3$.

51. Take the x-axis to be the axis of the cylindrical hole of radius r.

A quarter of the cross-section through y, perpendicular to the

y-axis, is the rectangle shown. Using the Pythagorean Theorem

twice, we see that the dimensions of this rectangle are

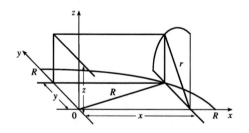

$x = \sqrt{R^2 - y^2}$ and $z = \sqrt{r^2 - y^2}$, so

$\tfrac{1}{4}A(y) = xz = \sqrt{r^2 - y^2}\sqrt{R^2 - y^2}$, and

$$V = \int_{-r}^{r} A(y)\, dy = \int_{-r}^{r} 4\sqrt{r^2 - y^2}\sqrt{R^2 - y^2}\, dy$$

$$= 8\int_{0}^{r} \sqrt{r^2 - y^2}\sqrt{R^2 - y^2}\, dy$$

53. (a) The radius of the barrel is the same at each end by symmetry, since the

function $y = R - cx^2$ is even. Since the barrel is obtained by rotating

the graph of the function y about the x-axis, this radius is equal to the

value of y at $x = \tfrac{1}{2}h$, which is $R - c\left(\tfrac{1}{2}h \right)^2 = R - d = r$.

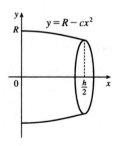

(b) The barrel is symmetric about the y-axis, so its volume is twice the volume of that part of the barrel for $x > 0$. Also, the barrel is a volume of rotation, so

$$V = 2\int_0^{h/2} \pi y^2 \, dx = 2\pi \int_0^{h/2} \left(R - cx^2\right)^2 dx = 2\pi\left[R^2 x - \tfrac{2}{3}Rcx^3 + \tfrac{1}{5}c^2 x^5\right]_0^{h/2}$$
$$= 2\pi\left(\tfrac{1}{2}R^2 h - \tfrac{1}{12}Rch^3 + \tfrac{1}{160}c^2 h^5\right)$$

Trying to make this look more like the expression we want, we rewrite it as $V = \tfrac{1}{3}\pi h\left[2R^2 + \left(R^2 - \tfrac{1}{2}Rch^2 + \tfrac{3}{80}c^2 h^4\right)\right]$.

But $R^2 - \tfrac{1}{2}Rch^2 + \tfrac{3}{80}c^2 h^4 = \left(R - \tfrac{1}{4}ch^2\right)^2 - \tfrac{1}{40}c^2 h^4 = (R - d)^2 - \tfrac{2}{5}\left(\tfrac{1}{4}ch^2\right)^2 = r^2 - \tfrac{2}{5}d^2$.

Substituting this back into V, we see that $V = \tfrac{1}{3}\pi h\left(2R^2 + r^2 - \tfrac{2}{5}d^2\right)$, as required.

6.3   Volumes by Cylindrical Shells

1.

If we were to use the "washer" method, we would first have to locate the local maximum point (a, b) of $y = x(x - 1)^2$ using the methods of Chapter 4. Then we would have to solve the equation $y = x(x - 1)^2$ for x in terms of y to obtain the functions $x = g_1(y)$ and $x = g_2(y)$ shown in the first figure. This step would be difficult because it involves the cubic formula. Finally we would find the volume using

$$V = \pi \int_0^b \left\{\left[g_1(y)\right]^2 - \left[g_2(y)\right]^2\right\} dy.$$

Using shells, we find that a typical approximating shell has radius x, so its circumference is $2\pi x$. Its height is y, that is, $x(x - 1)^2$. So the total volume is

$$V = \int_0^1 2\pi x\left[x(x-1)^2\right] dx = 2\pi \int_0^1 \left(x^4 - 2x^3 + x^2\right) dx = 2\pi\left[\frac{x^5}{5} - 2\frac{x^4}{4} + \frac{x^3}{3}\right]_0^1 = \frac{\pi}{15}$$

3. $V = \displaystyle\int_1^2 2\pi x \cdot \frac{1}{x} \, dx = 2\pi \int_1^2 1 \, dx$

$\qquad = 2\pi\left[x\right]_1^2 = 2\pi(2 - 1) = 2\pi$

5. $V = \int_0^1 2\pi x e^{-x^2} \, dx$. Let $u = x^2$.

Thus, $du = 2x \, dx$, so

$V = \pi \int_0^1 e^{-u} \, du = \pi\left[-e^{-u}\right]_0^1 = \pi(1 - 1/e)$.

7. The curves intersect when $4(x-2)^2 = x^2 - 4x + 7 \Leftrightarrow 4x^2 - 16x + 16 = x^2 - 4x + 7 \Leftrightarrow$

$3x^2 - 12x + 9 = 0 \Leftrightarrow 3(x^2 - 4x + 3) = 0 \Leftrightarrow 3(x-1)(x-3) = 0$, so $x = 1$ or 3.

$V = 2\pi \int_1^3 \left\{ x \left[(x^2 - 4x + 7) - 4(x-2)^2 \right] \right\} dx = 2\pi \int_1^3 \left[x(x^2 - 4x + 7 - 4x^2 + 16x - 16) \right] dx$

$\quad = 2\pi \int_1^3 \left[x(-3x^2 + 12x - 9) \right] dx = 2\pi(-3) \int_1^3 (x^3 - 4x^2 + 3x) \, dx = -6\pi \left[\frac{1}{4}x^4 - \frac{4}{3}x^3 + \frac{3}{2}x^2 \right]_1^3$

$\quad = -6\pi \left[\left(\frac{81}{4} - 36 + \frac{27}{2} \right) - \left(\frac{1}{4} - \frac{4}{3} + \frac{3}{2} \right) \right] = -6\pi \left(20 - 36 + 12 + \frac{4}{3} \right) = -6\pi \left(-\frac{8}{3} \right) = 16\pi$

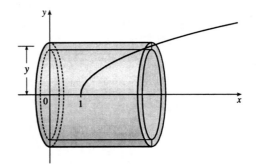

9. $V = \int_1^2 2\pi y (1 + y^2) \, dy = 2\pi \int_1^2 (y + y^3) \, dy = 2\pi \left[\frac{1}{2}y^2 + \frac{1}{4}y^4 \right]_1^2$

$\quad = 2\pi \left[(2 + 4) - \left(\frac{1}{2} + \frac{1}{4} \right) \right] = 2\pi \left(\frac{21}{4} \right) = \frac{21}{2}\pi$

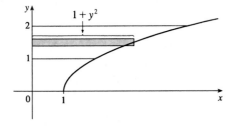

11. The height of the shell is $2 - \left[1 + (y-2)^2 \right] = 1 - (y-2)^2 = 1 - (y^2 - 4y + 4) = -y^2 + 4y - 3$.

$V = 2\pi \int_1^3 y(-y^2 + 4y - 3) \, dy$

$\quad = 2\pi \int_1^3 (-y^3 + 4y^2 - 3y) \, dy$

$\quad = 2\pi \left[-\frac{1}{4}y^4 + \frac{4}{3}y^3 - \frac{3}{2}y^2 \right]_1^3$

$\quad = 2\pi \left[\left(-\frac{81}{4} + 36 - \frac{27}{2} \right) - \left(-\frac{1}{4} + \frac{4}{3} - \frac{3}{2} \right) \right]$

$\quad = 2\pi \left(\frac{8}{3} \right) = \frac{16}{3}\pi$

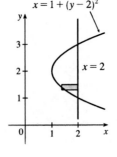

13. The shell has radius $2 - x$, circumference $2\pi(2 - x)$, and height x^4.

$$V = \int_0^1 2\pi(2 - x)x^4 \, dx$$

$$= 2\pi \int_0^1 (2x^4 - x^5) \, dx$$

$$= 2\pi \left[\frac{2}{5}x^5 - \frac{1}{6}x^6 \right]_0^1$$

$$= 2\pi \left[\left(\frac{2}{5} - \frac{1}{6} \right) - 0 \right] = 2\pi \left(\frac{7}{30} \right) = \frac{7}{15}\pi$$

15. The shell has radius $x - 1$, circumference $2\pi(x - 1)$, and height $(4x - x^2) - 3 = -x^2 + 4x - 3$.

$$V = \int_1^3 2\pi(x - 1)(-x^2 + 4x - 3) \, dx$$

$$= 2\pi \int_1^3 (-x^3 + 5x^2 - 7x + 3) \, dx$$

$$= 2\pi \left[-\frac{1}{4}x^4 + \frac{5}{3}x^3 - \frac{7}{2}x^2 + 3x \right]_1^3$$

$$= 2\pi \left[\left(-\frac{81}{4} + 45 - \frac{63}{2} + 9 \right) - \left(-\frac{1}{4} + \frac{5}{3} - \frac{7}{2} + 3 \right) \right]$$

$$= 2\pi \left(\frac{4}{3} \right) = \frac{8}{3}\pi$$

17. The shell has radius $1 - y$, circumference $2\pi(1 - y)$, and height $1 - \sqrt[3]{y}$ $\left[y = x^3 \quad \Leftrightarrow \quad x = \sqrt[3]{y} \right]$.

$$V = \int_0^1 2\pi(1 - y)(1 - y^{1/3}) \, dy$$

$$= 2\pi \int_0^1 (1 - y - y^{1/3} + y^{4/3}) \, dy$$

$$= 2\pi \left[y - \frac{1}{2}y^2 - \frac{3}{4}y^{4/3} + \frac{3}{7}y^{7/3} \right]_0^1$$

$$= 2\pi \left[\left(1 - \frac{1}{2} - \frac{3}{4} + \frac{3}{7} \right) - 0 \right]$$

$$= 2\pi \left(\frac{5}{28} \right) = \frac{5}{14}\pi$$

19. $V = \int_0^\pi 2\pi(4 - y) \sqrt{\sin y} \, dy$

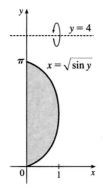

21. $V = \int_0^1 2\pi x \sqrt{1 + x^3}\, dx$. Let $f(x) = x\sqrt{1+x^3}$. Then Simpson's Rule

with $n = 10$ gives

$\int_0^1 f(x)\, dx \approx \frac{1-0}{10 \cdot 3}[f(0) + 4f(0.1) + 2f(0.2) + \cdots + 4f(0.9) + f(1)]$

$\approx \frac{1}{30}(17.65)$

Multiplying by 2π gives $V \approx 3.70$.

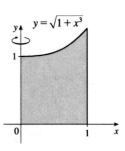

23. (a) $\int_0^3 2\pi x^5\, dx = 2\pi \int_0^3 x(x^4)\, dx$. The solid is obtained by rotating the region $0 \le y \le x^4$, $0 \le x \le 3$ about the y-axis using

cylindrical shells.

(b) $\int_0^1 2\pi(3-y)(1-y^2)\, dy$. The solid is obtained by rotating the region bounded by (i) $x = 1 - y^2$, $x = 0$, and $y = 0$ or

(ii) $x = y^2$, $x = 1$, and $y = 0$ about the line $y = 3$ using cylindrical shells.

25.

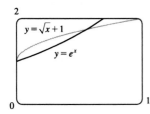

From the graph, the curves intersect at $x = 0$ and $x = a \approx 0.56$,

with $\sqrt{x} + 1 > e^x$ on the interval $(0, a)$. So the volume of the solid

obtained by rotating the region about the y-axis is

$$V = 2\pi \int_0^a x\left[\left(\sqrt{x} + 1\right) - e^x\right]\, dx \approx 0.13.$$

27. $V = 2\pi \int_0^{\pi/2} \left[\left(\frac{\pi}{2} - x\right)\left(\sin^2 x - \sin^4 x\right)\right]\, dx$

$\overset{\text{CAS}}{=} \frac{1}{32}\pi^3$

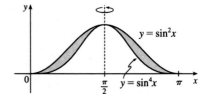

29. Use shells:

$V = \int_2^4 2\pi x(-x^2 + 6x - 8)\, dx = 2\pi \int_2^4 (-x^3 + 6x^2 - 8x)\, dx$

$= 2\pi \left[-\frac{1}{4}x^4 + 2x^3 - 4x^2\right]_2^4$

$= 2\pi[(-64 + 128 - 64) - (-4 + 16 - 16)]$

$= 2\pi(4) = 8\pi$

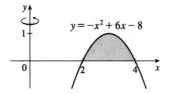

31. Use disks: $x^2 + (y-1)^2 = 1 \quad \Leftrightarrow \quad x = \pm\sqrt{1 - (y-1)^2}$

$V = \pi \int_0^2 \left[\sqrt{1 - (y-1)^2}\right]^2\, dy = \pi \int_0^2 (2y - y^2)\, dy$

$= \pi\left[y^2 - \frac{1}{3}y^3\right]_0^2 = \pi\left(4 - \frac{8}{3}\right) = \frac{4}{3}\pi$

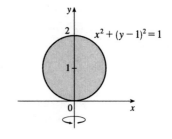

33. Use shells:

$$V = \int_1^4 2\pi[x - (-1)][5 - (x + 4/x)]\, dx$$

$$= 2\pi \int_1^4 (x + 1)(5 - x - 4/x)\, dx$$

$$= 2\pi \int_1^4 (5x - x^2 - 4 + 5 - x - 4/x)\, dx$$

$$= 2\pi \int_1^4 (-x^2 + 4x + 1 - 4/x)\, dx = 2\pi \left[-\tfrac{1}{3}x^3 + 2x^2 + x - 4\ln x \right]_1^4$$

$$= 2\pi \left[\left(-\tfrac{64}{3} + 32 + 4 - 4\ln 4 \right) - \left(-\tfrac{1}{3} + 2 + 1 - 0 \right) \right]$$

$$= 2\pi(12 - 4\ln 4) = 8\pi(3 - \ln 4)$$

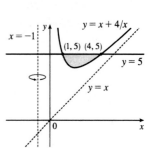

35. Use shells:

$$V = 2 \int_0^r 2\pi x \sqrt{r^2 - x^2}\, dx = -2\pi \int_0^r (r^2 - x^2)^{1/2}(-2x)\, dx$$

$$= \left[-2\pi \cdot \tfrac{2}{3}(r^2 - x^2)^{3/2} \right]_0^r = -\tfrac{4}{3}\pi(0 - r^3) = \tfrac{4}{3}\pi r^3$$

37. $V = 2\pi \displaystyle\int_0^r x\left(-\frac{h}{r}x + h \right) dx = 2\pi h \int_0^r \left(-\frac{x^2}{r} + x \right) dx$

$$= 2\pi h \left[-\frac{x^3}{3r} + \frac{x^2}{2} \right]_0^r = 2\pi h\, \frac{r^2}{6} = \frac{\pi r^2 h}{3}$$

6.4 Arc Length

1. $y = 2x - 5 \;\Rightarrow\; L = \int_{-1}^3 \sqrt{1 + (dy/dx)^2}\, dx = \int_{-1}^3 \sqrt{1 + (2)^2}\, dx = \sqrt{5}\,[3 - (-1)] = 4\sqrt{5}.$

The arc length can be calculated using the distance formula, since the curve is a line segment, so

$$L = [\text{distance from } (-1, -7) \text{ to } (3, 1)] = \sqrt{[3 - (-1)]^2 + [1 - (-7)]^2} = \sqrt{80} = 4\sqrt{5}$$

3. $y = \sin x \;\Rightarrow\; dy/dx = \cos x \;\Rightarrow\; 1 + (dy/dx)^2 = 1 + \cos^2 x.$ So $L = \int_0^\pi \sqrt{1 + \cos^2 x}\, dx \approx 3.8202.$

5. $x = t + \cos t,\; y = t - \sin t,\; 0 \le t \le 2\pi.$ $dx/dt = 1 - \sin t$ and $dy/dt = 1 - \cos t,$ so

$$\left(\tfrac{dx}{dt}\right)^2 + \left(\tfrac{dy}{dt}\right)^2 = (1 - \sin t)^2 + (1 - \cos t)^2 = (1 - 2\sin t + \sin^2 t) + (1 - 2\cos t + \cos^2 t) = 3 - 2\sin t - 2\cos t.$$

Thus, $L = \int_a^b \sqrt{(dx/dt)^2 + (dy/dt)^2}\, dt = \int_0^{2\pi} \sqrt{3 - 2\sin t - 2\cos t}\, dt \approx 10.0367.$

7. $x = 1 + 3t^2$, $y = 4 + 2t^3$, $0 \le t \le 1$. $dx/dt = 6t$ and $dy/dt = 6t^2$, so $(dx/dt)^2 + (dy/dt)^2 = 36t^2 + 36t^4$

Thus, $L = \int_0^1 \sqrt{36t^2 + 36t^4}\,dt = \int_0^1 6t\sqrt{1 + t^2}\,dt = 6\int_1^2 \sqrt{u}\,\left(\tfrac{1}{2}du\right)$ $\quad [u = 1 + t^2, du = 2t\,dt]$

$\qquad = 3\left[\tfrac{2}{3}u^{3/2}\right]_1^2 = 2(2^{3/2} - 1) = 2(2\sqrt{2} - 1)$

9. $x = y^{3/2}$ $\Rightarrow$ $1 + (dx/dy)^2 = 1 + \left(\tfrac{3}{2}y^{1/2}\right)^2 = 1 + \tfrac{9}{4}y$.

$L = \int_0^1 \sqrt{1 + \tfrac{9}{4}y}\,dy = \int_1^{13/4} \sqrt{u}\,\left(\tfrac{4}{9}du\right)$ $\quad \left[u = 1 + \tfrac{9}{4}y, du = \tfrac{9}{4}dy\right]$

$\qquad = \tfrac{4}{9} \cdot \tfrac{2}{3}\left[u^{3/2}\right]_1^{13/4} = \tfrac{8}{27}\left(\tfrac{13\sqrt{13}}{8} - 1\right) = \tfrac{13\sqrt{13} - 8}{27}$.

11. $y = \tfrac{1}{4}x^2 - \tfrac{1}{2}\ln x$ $\Rightarrow$ $y' = \tfrac{1}{2}x - \tfrac{1}{2x}$ $\Rightarrow$ $1 + (y')^2 = 1 + \left(\tfrac{1}{4}x^2 - \tfrac{1}{2} + \tfrac{1}{4x^2}\right) = \tfrac{1}{4}x^2 + \tfrac{1}{2} + \tfrac{1}{4x^2} = \left(\tfrac{1}{2}x + \tfrac{1}{2x}\right)^2$.

So

$$L = \int_1^2 \sqrt{1 + (y')^2}\,dx = \int_1^2 \left|\tfrac{1}{2}x + \tfrac{1}{2x}\right|\,dx = \int_1^2 \left(\tfrac{1}{2}x + \tfrac{1}{2x}\right)\,dx$$

$$= \left[\tfrac{1}{4}x^2 + \tfrac{1}{2}\ln|x|\right]_1^2 = \left(1 + \tfrac{1}{2}\ln 2\right) - \left(\tfrac{1}{4} + 0\right) = \tfrac{3}{4} + \tfrac{1}{2}\ln 2$$

13.

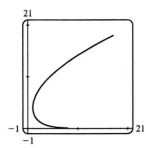

$x = e^t - t$, $y = 4e^{t/2}$, $-8 \le t \le 3$

$\left(\tfrac{dx}{dt}\right)^2 + \left(\tfrac{dy}{dt}\right)^2 = (e^t - 1)^2 + (2e^{t/2})^2 = e^{2t} - 2e^t + 1 + 4e^t$

$\qquad\qquad\qquad\quad = e^{2t} + 2e^t + 1 = (e^t + 1)^2$

$L = \int_{-8}^3 \sqrt{(e^t + 1)^2}\,dt = \int_{-8}^3 (e^t + 1)\,dt = \left[e^t + t\right]_{-8}^{3t}$

$\quad = (e^3 + 3) - (e^{-8} - 8) = e^3 - e^{-8} + 11$

15.

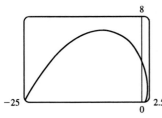

$x = e^t\cos t$, $y = e^t\sin t$, $0 \le t \le \pi$.

$\left(\tfrac{dx}{dt}\right)^2 + \left(\tfrac{dy}{dt}\right)^2 = [e^t(\cos t - \sin t)]^2 + [e^t(\sin t + \cos t)]^2$

$\qquad\qquad = (e^t)^2(\cos^2 t - 2\cos t\sin t + \sin^2 t)$

$\qquad\qquad\qquad + (e^t)^2(\sin^2 t + 2\sin t\cos t + \cos^2 t$

$\qquad\qquad = e^{2t}(2\cos^2 t + 2\sin^2 t) = 2e^{2t}$

Thus, $L = \int_0^\pi \sqrt{2e^{2t}}\,dt = \int_0^\pi \sqrt{2}\,e^t\,dt = \sqrt{2}\left[e^t\right]_0^\pi = \sqrt{2}(e^\pi - 1)$.

17. $y = xe^{-x}$ $\Rightarrow$ $dy/dx = e^{-x} - xe^{-x} = e^{-x}(1 - x)$ $\Rightarrow$ $1 + (dy/dx)^2 = 1 + e^{-2x}(1 - x)^2$. Let

$f(x) = \sqrt{1 + (dy/dx)^2} = \sqrt{1 + e^{-2x}(1 - x)^2}$. Then $L = \int_0^5 f(x)\,dx$. Since $n = 10$, $\Delta x = \tfrac{5 - 0}{10} = \tfrac{1}{2}$. Now

$L \approx S_{10} = \tfrac{1/2}{3}[f(0) + 4f\left(\tfrac{1}{2}\right) + 2f(1) + 4f\left(\tfrac{3}{2}\right) + 2f(2) + 4f\left(\tfrac{5}{2}\right) + 2f(3) + 4f\left(\tfrac{7}{2}\right) + 2f(4) + 4f\left(\tfrac{9}{2}\right) + f(5)]$

$\qquad \approx 5.115840$

The value of the integral produced by a calculator is 5.113568 (to six decimal places).

19. $x = \sin t$, $y = t^2$ $\Rightarrow$ $(dx/dt)^2 + (dy/dt)^2 = (\cos t)^2 + (2t)^2 = \cos^2 t + 4t^2$ $\Rightarrow$ $L = \int_0^{2\pi} \sqrt{\cos^2 t + 4t^2}\, dt$.

Using Simpson's Rule with $n = 10$, $\Delta t = \dfrac{2\pi - 0}{10} = \dfrac{\pi}{5}$, and $f(t) = \sqrt{\cos^2 t + 4t^2}\, dt$, we get

$$L \approx S_{10} = \tfrac{2\pi-0}{3(10)} \left[f(0) + 4f\left(\tfrac{\pi}{5}\right) + 2f\left(\tfrac{2\pi}{5}\right) + 4f\left(\tfrac{3\pi}{5}\right) + 2f\left(\tfrac{4\pi}{5}\right) + 4f(\pi) + 2f\left(\tfrac{6\pi}{5}\right) \right.$$
$$\left. + 4f\left(\tfrac{7\pi}{5}\right) + 2f\left(\tfrac{8\pi}{5}\right) + 4f\left(\tfrac{9\pi}{5}\right) + f(2\pi) \right]$$
$$\approx 40.056222$$

The value of the integral produced by a calculator is 40.051156 (to six decimal places).

21. (a)

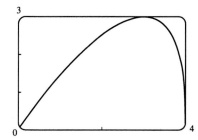

(b)

Let $f(x) = y = x\sqrt[3]{4-x}$. The polygon with one side is just the line segment joining the points $(0, f(0)) = (0, 0)$ and $(4, f(4)) = (4, 0)$, and its length $L_1 = 4$.

The polygon with two sides joins the points $(0, 0)$, $(2, f(2)) = (2, 2\sqrt[3]{2})$ and $(4, 0)$. Its length

$$L_2 = \sqrt{(2-0)^2 + \left(2\sqrt[3]{2} - 0\right)^2} + \sqrt{(4-2)^2 + \left(0 - 2\sqrt[3]{2}\right)^2} = 2\sqrt{4 + 2^{8/3}} \approx 6.43$$

Similarly, the inscribed polygon with four sides joins the points $(0, 0)$, $\left(1, \sqrt[3]{3}\right)$, $\left(2, 2\sqrt[3]{2}\right)$, $(3, 3)$, and $(4, 0)$, so its length

$$L_4 = \sqrt{1 + \left(\sqrt[3]{3}\right)^2} + \sqrt{1 + \left(2\sqrt[3]{2} - \sqrt[3]{3}\right)^2} + \sqrt{1 + \left(3 - 2\sqrt[3]{2}\right)^2} + \sqrt{1 + 9} \approx 7.50$$

(c) Using the arc length formula with $\dfrac{dy}{dx} = x\left[\tfrac{1}{3}(4-x)^{-2/3}(-1)\right] + \sqrt[3]{4-x} = \dfrac{12 - 4x}{3(4-x)^{2/3}}$, the length of the curve is

$$L = \int_0^4 \sqrt{1 + \left(\dfrac{dy}{dx}\right)^2}\, dx = \int_0^4 \sqrt{1 + \left[\dfrac{12 - 4x}{3(4-x)^{2/3}}\right]^2}\, dx.$$

(d) According to a calculator, the length of the curve is $L \approx 7.7988$. The actual value is larger than any of the approximations in part (b). This is always true, since any approximating straight line between two points on the curve is shorter than the length of the curve between the two points.

23. $x = t^3 \Rightarrow dx/dt = 3t^2$ and $y = t^4 \Rightarrow dy/dt = 4t^3$. So

$L = \int_0^1 \sqrt{9t^4 + 16t^6}\, dt = \int_0^1 \sqrt{t^4(9 + 16t^2)}\, dt = \int_0^1 t^2\sqrt{9 + 16t^2}\, dt$.

Now use Formula 22 from the table of integrals to evaluate L.

$$L = \int_0^4 \left(\tfrac{1}{4}u\right)^2 \sqrt{a^2 + u^2}\left(\tfrac{1}{4}du\right) \qquad [a = 3,\, u = 4t,\, du = 4\,dt]$$

$$= \tfrac{1}{64}\int_0^4 u^2\sqrt{a^2 + u^2}\, du = \tfrac{1}{64}\left[\tfrac{u}{8}(9 + 2u^2)\sqrt{9 + u^2} - \tfrac{81}{8}\ln\left(u + \sqrt{9 + u^2}\right)\right]_0^4$$

$$= \tfrac{1}{64}\left\{\left[\tfrac{1}{2}\cdot 41 \cdot 5 - \tfrac{81}{8}\ln(4 + 5)\right] - \left[0 - \tfrac{81}{8}\ln 3\right]\right\}$$

$$= \tfrac{1}{64}\left[\tfrac{205}{2} - \tfrac{81}{8}(2\ln 3) + \tfrac{81}{8}\ln 3\right] \qquad \left[\ln 9 = \ln 3^2 = 2\ln 3\right]$$

$$= \tfrac{1}{64}\left(\tfrac{205}{2} - \tfrac{81}{8}\ln 3\right) = \tfrac{205}{128} - \tfrac{81}{512}\ln 3 \approx 1.428.$$

25. $y = \ln(\cos x) \Rightarrow y' = \dfrac{1}{\cos x}(-\sin x) = -\tan x \Rightarrow 1 + (y')^2 = 1 + \tan^2 x = \sec^2 x$.

So $L = \int_0^{\pi/4}\sec x\, dx \overset{14}{=} \Big[\ln|\sec x + \tan x|\Big]_0^{\pi/4} = \ln(\sqrt{2} + 1) - \ln(1 + 0) = \ln(\sqrt{2} + 1) \approx 0.881$.

27. The prey hits the ground when $y = 0 \Leftrightarrow 180 - \tfrac{1}{45}x^2 = 0 \Leftrightarrow x^2 = 45 \cdot 180 \Rightarrow x = \sqrt{8100} = 90$,

since x must be positive. $y' = -\tfrac{2}{45}x \Rightarrow 1 + (y')^2 = 1 + \tfrac{4}{45^2}x^2$, so the distance traveled by the prey is

$$L = \int_0^{90}\sqrt{1 + \frac{4}{45^2}x^2}\, dx = \int_0^4 \sqrt{1 + u^2}\left(\tfrac{45}{2}\, du\right) \qquad \begin{bmatrix} u = \tfrac{2}{45}x, \\ du = \tfrac{2}{45}\, dx \end{bmatrix}$$

$$\overset{21}{=} \tfrac{45}{2}\left[\tfrac{1}{2}u\sqrt{1 + u^2} + \tfrac{1}{2}\ln\left(u + \sqrt{1 + u^2}\right)\right]_0^4 = \tfrac{45}{2}\left[2\sqrt{17} + \tfrac{1}{2}\ln\left(4 + \sqrt{17}\right)\right] = 45\sqrt{17} + \tfrac{45}{4}\ln\left(4 + \sqrt{17}\right) \approx 209.1 \text{ m}$$

29. The sine wave has amplitude 1 and period 14, since it goes through two periods in a distance of 28 in., so its equation is

$y = 1\sin\left(\tfrac{2\pi}{14}x\right) = \sin\left(\tfrac{\pi}{7}x\right)$. The width w of the flat metal sheet needed to make the panel is the arc length of the sine curve

from $x = 0$ to $x = 28$. We set up the integral to evaluate w using the arc length formula with $\tfrac{dy}{dx} = \tfrac{\pi}{7}\cos\left(\tfrac{\pi}{7}x\right)$:

$L = \int_0^{28}\sqrt{1 + \left[\tfrac{\pi}{7}\cos\left(\tfrac{\pi}{7}x\right)\right]^2}\, dx = 2\int_0^{14}\sqrt{1 + \left[\tfrac{\pi}{7}\cos\left(\tfrac{\pi}{7}x\right)\right]^2}\, dx$. This integral would be very difficult to evaluate exactly,

so we use a CAS, and find that $L \approx 29.36$ inches.

31. $x = a\sin\theta$, $y = b\cos\theta$, $0 \le \theta \le 2\pi$.

$$\left(\tfrac{dx}{dt}\right)^2 + \left(\tfrac{dy}{dt}\right)^2 = (a\cos\theta)^2 + (-b\sin\theta)^2 = a^2\cos^2\theta + b^2\sin^2\theta = a^2(1 - \sin^2\theta) + b^2\sin^2\theta$$

$$= a^2 - (a^2 - b^2)\sin^2\theta = a^2 - c^2\sin^2\theta = a^2\left(1 - \frac{c^2}{a^2}\sin^2\theta\right) = a^2(1 - e^2\sin^2\theta)$$

So $L = 4\int_0^{\pi/2}\sqrt{a^2\left(1 - e^2\sin^2\theta\right)}\, d\theta$ [by symmetry] $= 4a\int_0^{\pi/2}\sqrt{1 - e^2\sin^2\theta}\, d\theta$.

33. (a) $x = 11\cos t - 4\cos(11t/2)$, $y = 11\sin t - 4\sin(11t/2)$.

Notice that $0 \le t \le 2\pi$ does not give the complete curve because

$x(0) \ne x(2\pi)$. In fact, we must take $t \in [0, 4\pi]$ in order to obtain the

complete curve, since the first term in each of the parametric equations has

period 2π and the second has period $\frac{2\pi}{11/2} = \frac{4\pi}{11}$, and the least common

integer multiple of these two numbers is 4π.

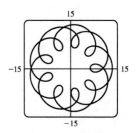

(b) We use the CAS to find the derivatives dx/dt and dy/dt, and then use Formula 1 to find the arc length. Recent versions

of Maple express the integral $\int_0^{4\pi} \sqrt{(dx/dt)^2 + (dy/dt)^2}\, dt$ as $88E\left(2\sqrt{2}\, i\right)$, where $E(x)$ is the elliptic integral

$\int_0^1 \dfrac{\sqrt{1 - x^2 t^2}}{\sqrt{1 - t^2}}\, dt$ and i is the imaginary number $\sqrt{-1}$.

Some earlier versions of Maple (as well as Mathematica) cannot do the integral exactly, so we use the command

`evalf(Int(sqrt(diff(x,t)^2+diff(y,t)^2),t=0..4*Pi));` to estimate the length, and find that the arc

length is approximately 294.03. Derive's `Para_arc_length` function in the utility file `Int_apps` simplifies the

integral to $11 \int_0^{4\pi} \sqrt{-4\cos t\, \cos\left(\frac{11t}{2}\right) - 4\sin t\, \sin\left(\frac{11t}{2}\right) + 5}\, dt$.

6.5 Average Value of a Function

1. $f_{\text{ave}} = \frac{1}{b-a}\int_a^b f(x)\, dx = \frac{1}{4-0}\int_0^4 (4x - x^2)\, dx = \frac{1}{4}\left[2x^2 - \frac{1}{3}x^3\right]_0^4 = \frac{1}{4}\left[(32 - \frac{64}{3}) - 0\right] = \frac{1}{4}\left(\frac{32}{3}\right) = \frac{8}{3}$

3. $g_{\text{ave}} = \frac{1}{b-a}\int_a^b g(x)\, dx = \frac{1}{8-1}\int_1^8 \sqrt[3]{x}\, dx = \frac{1}{7}\left[\frac{3}{4}x^{4/3}\right]_1^8 = \frac{3}{28}(16 - 1) = \frac{45}{28}$

5. $h_{\text{ave}} = \frac{1}{\pi - 0}\int_0^\pi \cos^4 x \sin x\, dx = \frac{1}{\pi}\int_1^{-1} u^4(-du)$ $[u = \cos x, du = -\sin x\, dx]$

$= \frac{1}{\pi}\int_{-1}^1 u^4\, du = \frac{1}{\pi} \cdot 2\int_0^1 u^4\, du$ [by Theorem 5.5.6(a)] $= \frac{2}{\pi}\left[\frac{1}{5}u^5\right]_0^1 = \frac{2}{5\pi}$

7. (a) $f_{\text{ave}} = \dfrac{1}{5-2}\int_2^5 (x-3)^2\, dx = \frac{1}{3}\left[\frac{1}{3}(x-3)^3\right]_2^5$

$= \frac{1}{9}\left[2^3 - (-1)^3\right] = \frac{1}{9}(8 + 1) = 1$

(b) $f(c) = f_{\text{ave}} \Leftrightarrow (c - 3)^2 = 1 \Leftrightarrow$

$c - 3 = \pm 1 \Leftrightarrow c = 2$ or 4

(c)

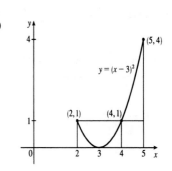

9. (a) $f_{\text{ave}} = \dfrac{1}{\pi - 0} \displaystyle\int_0^\pi (2\sin x - \sin 2x)\,dx$

$\qquad = \frac{1}{\pi}\left[-2\cos x + \frac{1}{2}\cos 2x\right]_0^\pi$

$\qquad = \frac{1}{\pi}\left[\left(2 + \frac{1}{2}\right) - \left(-2 + \frac{1}{2}\right)\right] = \frac{4}{\pi}$

(c)

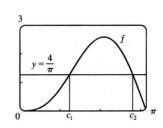

(b) $f(c) = f_{\text{ave}} \;\Leftrightarrow\; 2\sin c - \sin 2c = \frac{4}{\pi} \;\Leftrightarrow$

$\qquad c_1 \approx 1.238 \text{ or } c_2 \approx 2.808$

11. f is continuous on $[1,3]$, so by the Mean Value Theorem for Integrals there exists a number c in $[1,3]$ such that

$\displaystyle\int_1^3 f(x)\,dx = f(c)(3 - 1) \;\Rightarrow\; 8 = 2f(c)$; that is, there is a number c such that $f(c) = \frac{8}{2} = 4$.

13. $f_{\text{ave}} = \dfrac{1}{b - a}\displaystyle\int_a^b f(x)\,dx \approx \dfrac{1}{50 - 20} S_6$

$\qquad = \frac{1}{30} \cdot \frac{50 - 20}{6 \cdot 3}[f(20) + 4f(25) + 2f(30) + 4f(35) + 2f(40) + 4f(45) + f(50)]$

$\qquad = \frac{1}{18}[42 + 4(38) + 2(31) + 4(29) + 2(35) + 4(48) + 60] = \frac{1}{18}(694) = \frac{347}{9} \approx 38.6$

15. Let $t = 0$ and $t = 12$ correspond to 9 AM and 9 PM, respectively.

$$T_{\text{ave}} = \frac{1}{12 - 0}\int_0^{12}\left[50 + 14\sin\frac{1}{12}\pi t\right]dt = \frac{1}{12}\left[50t - 14 \cdot \frac{12}{\pi}\cos\frac{1}{12}\pi t\right]_0^{12}$$

$$= \frac{1}{12}\left[50 \cdot 12 + 14 \cdot \frac{12}{\pi} + 14 \cdot \frac{12}{\pi}\right] = \left(50 + \frac{28}{\pi}\right){}^\circ F \approx 59\,^\circ F$$

17. $\rho_{\text{ave}} = \dfrac{1}{8}\displaystyle\int_0^8 \dfrac{12}{\sqrt{x + 1}}\,dx = \dfrac{3}{2}\int_0^8 (x + 1)^{-1/2}\,dx = \left[3\sqrt{x + 1}\right]_0^8 = 9 - 3 = 6 \text{ kg/m}$

19. $V_{\text{ave}} = \frac{1}{5}\displaystyle\int_0^5 V(t)\,dt = \frac{1}{5}\int_0^5 \frac{5}{4\pi}\left[1 - \cos\left(\frac{2}{5}\pi t\right)\right]dt = \frac{1}{4\pi}\int_0^5 \left[1 - \cos\left(\frac{2}{5}\pi t\right)\right]dt$

$\qquad = \frac{1}{4\pi}\left[t - \frac{5}{2\pi}\sin\left(\frac{2}{5}\pi t\right)\right]_0^5 = \frac{1}{4\pi}\left[(5 - 0) - 0\right] = \frac{5}{4\pi} \approx 0.4 \text{ L}$

21. Let $F(x) = \displaystyle\int_a^x f(t)\,dt$ for x in $[a, b]$. Then F is continuous on $[a, b]$ and differentiable on (a, b), so by the Mean Value

Theorem there is a number c in (a, b) such that $F(b) - F(a) = F'(c)(b - a)$. But $F'(x) = f(x)$ by the Fundamental

Theorem of Calculus. Therefore, $\displaystyle\int_a^b f(t)\,dt - 0 = f(c)(b - a)$.

6.6 Applications to Physics and Engineering

1. $W = \displaystyle\int_a^b f(x)\,dx = \int_0^9 \frac{10}{(1 + x)^2}\,dx = 10\int_1^{10}\frac{1}{u^2}\,du \quad [u = 1 + x,\ du = dx] = 10\left[-\frac{1}{u}\right]_1^{10} = 10\left(-\frac{1}{10} + 1\right) = 9 \text{ ft-lb}$

3. The force function is given by $F(x)$ (in newtons) and the work (in joules) is the area under the curve, given by

$\displaystyle\int_0^8 F(x)\,dx = \int_0^4 F(x)\,dx + \int_4^8 F(x)\,dx = \frac{1}{2}(4)(30) + (4)(30) = 180 \text{ J}$.

5. According to Hooke's Law, the force required to maintain a spring stretched x units beyond its natural length is proportional to x, that is, $f(x) = kx$. Here, the amount stretched is 4 in. $= \frac{1}{3}$ ft and the force is 10 lb. Thus, $10 = k\left(\frac{1}{3}\right)$ $\Rightarrow$ $k = 30$ lb/ft, and $f(x) = 30x$. The work done in stretching the spring from its natural length to 6 in. $= \frac{1}{2}$ ft beyond its natural length is $W = \int_0^{1/2} 30x \, dx = \left[15x^2\right]_0^{1/2} = \frac{15}{4}$ ft-lb.

7. (a) If $\int_0^{0.12} kx \, dx = 2$ J, then $2 = \left[\frac{1}{2}kx^2\right]_0^{0.12} = \frac{1}{2}k(0.0144) = 0.0072k$ and $k = \frac{2}{0.0072} = \frac{2500}{9} \approx 277.78$ N/m.

Thus, the work needed to stretch the spring from 35 cm to 40 cm is

$\int_{0.05}^{0.10} \frac{2500}{9}x \, dx = \left[\frac{1250}{9}x^2\right]_{1/20}^{1/10} = \frac{1250}{9}\left(\frac{1}{100} - \frac{1}{400}\right) = \frac{25}{24} \approx 1.04$ J.

(b) $f(x) = kx$, so $30 = \frac{2500}{9}x$ and $x = \frac{270}{2500}$ m $= 10.8$ cm

9. The distance from 20 cm to 30 cm is 0.1 m, so with $f(x) = kx$, we get $W_1 = \int_0^{0.1} kx \, dx = k\left[\frac{1}{2}x^2\right]_0^{0.1} = \frac{1}{200}k$.

Now $W_2 = \int_{0.1}^{0.2} kx \, dx = k\left[\frac{1}{2}x^2\right]_{0.1}^{0.2} = k\left(\frac{4}{200} - \frac{1}{200}\right) = \frac{3}{200}k$. Thus, $W_2 = 3W_1$.

In Exercises 11 – 18, n is the number of subintervals of length Δx, and x_i^* is a sample point in the ith subinterval $[x_{i-1}, x_i]$.

11. (a) The portion of the rope from x ft to $(x + \Delta x)$ ft below the top of the building weighs $\frac{1}{2}\Delta x$ lb and must be lifted x_i^* ft, so its contribution to the total work is $\frac{1}{2}x_i^* \Delta x$ ft-lb. The total work is

$$W = \lim_{n \to \infty} \sum_{i=1}^{n} \frac{1}{2}x_i^* \Delta x = \int_0^{50} \frac{1}{2}x \, dx = \left[\frac{1}{4}x^2\right]_0^{50} = \frac{2500}{4} = 625 \text{ ft-lb}$$

Notice that the exact height of the building does not matter (as long as it is more than 50 ft).

(b) When half the rope is pulled to the top of the building, the work to lift the top half of the rope is

$W_1 = \int_0^{25} \frac{1}{2}x \, dx = \left[\frac{1}{4}x^2\right]_0^{25} = \frac{625}{4}$ ft-lb. The bottom half of the rope is lifted 25 ft and the work needed to accomplish

that is $W_2 = \int_{25}^{50} \frac{1}{2} \cdot 25 \, dx = \frac{25}{2}\left[x\right]_{25}^{50} = \frac{625}{2}$ ft-lb. The total work done in pulling half the rope to the top of the building

is $W = W_1 + W_2 = \frac{625}{2} + \frac{625}{4} = \frac{3}{4} \cdot 625 = \frac{1875}{4}$ ft-lb.

13. The work needed to lift the cable is $\lim_{n \to \infty} \sum_{i=1}^{n} 2x_i^* \Delta x = \int_0^{500} 2x \, dx = \left[x^2\right]_0^{500} = 250{,}000$ ft-lb. The work needed to lift the coal is 800 lb $\cdot 500$ ft $= 400{,}000$ ft-lb. Thus, the total work required is $250{,}000 + 400{,}000 = 650{,}000$ ft-lb.

15. At a height of x meters ($0 \le x \le 12$), the mass of the rope is $(0.8 \text{ kg/m})(12 - x \text{ m}) = (9.6 - 0.8x)$ kg and the mass of the water is $\left(\frac{36}{12} \text{ kg/m}\right)(12 - x \text{ m}) = (36 - 3x)$ kg. The mass of the bucket is 10 kg, so the total mass is $(9.6 - 0.8x) + (36 - 3x) + 10 = (55.6 - 3.8x)$ kg, and hence, the total force is $9.8(55.6 - 3.8x)$ N. The work needed to lift the bucket Δx m through the ith subinterval of $[0, 12]$ is $9.8(55.6 - 3.8x_i^*)\Delta x$, so the total work is

$$W = \lim_{n \to \infty} \sum_{i=1}^{n} 9.8(55.6 - 3.8x_i^*)\Delta x = \int_0^{12}(9.8)(55.6 - 3.8x)\, dx = 9.8\left[55.6x - 1.9x^2\right]_0^{12} = 9.8(393.6) \approx 3857 \text{ J}$$

17. A "slice" of water Δx m thick and lying at a depth of x_i^* m (where $0 \le x_i^* \le \frac{1}{2}$) has volume $(2 \times 1 \times \Delta x)$ m³, a mass of

$2000\,\Delta x$ kg, weighs about $(9.8)(2000\,\Delta x) = 19{,}600\,\Delta x$ N, and thus requires about $19{,}600 x_i^*\,\Delta x$ J of work for its removal.

So $W = \lim\limits_{n\to\infty} \sum\limits_{i=1}^{n} 19{,}600 x_i^*\,\Delta x = \int_0^{1/2} 19{,}600 x\,dx = \left[9800 x^2\right]_0^{1/2} = 2450$ J.

19. A rectangular "slice" of water Δx m thick and lying x m above the bottom has width x m and volume $8x\,\Delta x$ m³. It weighs

about $(9.8 \times 1000)(8x\,\Delta x)$ N, and must be lifted $(5 - x)$ m by the pump, so the work needed is about

$(9.8 \times 10^3)(5 - x)(8x\,\Delta x)$ J. The total work required is

$$W \approx \int_0^3 (9.8 \times 10^3)(5 - x)8x\,dx = (9.8 \times 10^3)\int_0^3 (40x - 8x^2)\,dx = (9.8 \times 10^3)\left[20x^2 - \tfrac{8}{3}x^3\right]_0^3$$
$$= (9.8 \times 10^3)(180 - 72) = (9.8 \times 10^3)(108) = 1058.4 \times 10^3 \approx 1.06 \times 10^6 \text{ J}$$

21. Let x measure depth (in feet) below the spout at the top of the tank. A horizontal

disk-shaped "slice" of water Δx ft thick and lying at coordinate x has radius

$\frac{3}{8}(16 - x)$ ft $(\star)$ and volume $\pi r^2 \Delta x = \pi \cdot \frac{9}{64}(16 - x)^2\,\Delta x$ ft³. It weighs

about $(62.5)\frac{9\pi}{64}(16 - x)^2\,\Delta x$ lb and must be lifted x ft by the pump, so the

work needed to pump it out is about $(62.5)x\,\frac{9\pi}{64}(16 - x)^2\,\Delta x$ ft-lb. The total

work required is

$$W \approx \int_0^8 (62.5)x\,\frac{9\pi}{64}(16 - x)^2\,dx = (62.5)\frac{9\pi}{64}\int_0^8 x(256 - 32x + x^2)\,dx$$

$$= (62.5)\frac{9\pi}{64}\int_0^8 (256x - 32x^2 + x^3)\,dx = (62.5)\frac{9\pi}{64}\left[128x^2 - \tfrac{32}{3}x^3 + \tfrac{1}{4}x^4\right]_0^8$$

$$= (62.5)\frac{9\pi}{64}\left(\frac{11{,}264}{3}\right) = 33{,}000\pi \approx 1.04 \times 10^5 \text{ ft-lb}$$

$(\star)$ From similar triangles, $\dfrac{d}{8 - x} = \dfrac{3}{8}$.

So $r = 3 + d = 3 + \tfrac{3}{8}(8 - x)$

$= \dfrac{3(8)}{8} + \dfrac{3}{8}(8 - x)$

$= \tfrac{3}{8}(16 - x)$

23. If only 4.7×10^5 J of work is done, then only the water above a certain level (call

it h) will be pumped out. So we use the same formula as in Exercise 19, except that

the work is fixed, and we are trying to find the lower limit of integration:

$4.7 \times 10^5 \approx \int_h^3 (9.8 \times 10^3)(5 - x)8x\,dx = (9.8 \times 10^3)\left[20x^2 - \tfrac{8}{3}x^3\right]_h^3 \quad \Leftrightarrow$

$\frac{4.7}{9.8} \times 10^2 \approx 48 = \left(20 \cdot 3^2 - \tfrac{8}{3}\cdot 3^3\right) - \left(20h^2 - \tfrac{8}{3}h^3\right) \quad \Leftrightarrow$

$2h^3 - 15h^2 + 45 = 0$. To find the solution of this equation, we plot $2h^3 - 15h^2 + 45$ between $h = 0$ and $h = 3$.

We see that the equation is satisfied for $h \approx 2.0$. So the depth of water remaining in the tank is about 2.0 m.

25. $V = \pi r^2 x$, so V is a function of x and P can also be regarded as a function of x. If $V_1 = \pi r^2 x_1$ and $V_2 = \pi r^2 x_2$, then

$$W = \int_{x_1}^{x_2} F(x)\,dx = \int_{x_1}^{x_2} \pi r^2 P(V(x))\,dx = \int_{x_1}^{x_2} P(V(x))\,dV(x) \qquad [\text{Let } V(x) = \pi r^2 x, \text{ so } dV(x) = \pi r^2\,dx.]$$

$$= \int_{V_1}^{V_2} P(V)\,dV \quad \text{by the Substitution Rule.}$$

27. (a) $W = \int_a^b F(r)\,dr = \int_a^b G\frac{m_1 m_2}{r^2}\,dr = Gm_1 m_2 \left[\frac{-1}{r}\right]_a^b = Gm_1 m_2 \left(\frac{1}{a} - \frac{1}{b}\right)$

(b) By part (a), $W = GMm\left(\dfrac{1}{R} - \dfrac{1}{R + 1{,}000{,}000}\right)$ where $M =$ mass of the earth in kg, $R =$ radius of the earth in m,

and $m =$ mass of satellite in kg. (Note that 1000 km $= 1{,}000{,}000$ m.) Thus,

$$W = (6.67 \times 10^{-11})(5.98 \times 10^{24})(1000) \times \left(\frac{1}{6.37 \times 10^6} - \frac{1}{7.37 \times 10^6}\right) \approx 8.50 \times 10^9 \text{ J}$$

29. The weight density of water is $\delta = 62.5$ lb/ft^3.

(a) $P = \delta d \approx (62.5 \text{ lb/ft}^3)(3 \text{ ft}) = 187.5 \text{ lb/ft}^2$

(b) $F = PA \approx (187.5 \text{ lb/ft}^2)(5 \text{ ft})(2 \text{ ft}) = 1875 \text{ lb}$. ($A$ is the area of the bottom of the tank.)

(c) As in Example 1, the area of the ith strip is $2\,(\Delta x)$ and the pressure is $\delta d = \delta x_i$. Thus,

$$F = \int_0^3 \delta x \cdot 2\,dx \approx (62.5)(2)\int_0^3 x\,dx = 125\left[\tfrac{1}{2}x^2\right]_0^3 = 125\left(\tfrac{9}{2}\right) = 562.5 \text{ lb}.$$

31. Set up a vertical x-axis as shown. The base of the triangle shown in the figure

has length $\sqrt{3^2 - (x_i^*)^2}$, so $w_i = 2\sqrt{9 - (x_i^*)^2}$, and the area of the ith

rectangular strip is $2\sqrt{9 - (x_i^*)^2}\,\Delta x$. The ith rectangular strip is $(x_i^* - 1)$ m

below the surface level of the water, so the pressure on the strip is $\rho g(x_i^* - 1)$.

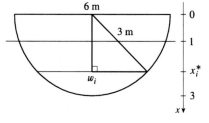

The hydrostatic force on the strip is $\rho g(x_i^* - 1) \cdot 2\sqrt{9 - (x_i^*)^2}\,\Delta x$ and the total

force on the plate $\approx \sum_{i=1}^n \rho g(x_i^* - 1) \cdot 2\sqrt{9 - (x_i^*)^2}\,\Delta x$. The total force

$$F = \lim \sum_{i=1}^n \rho g(x_i^* - 1) \cdot 2\sqrt{9 - (x_i^*)^2}\,\Delta x = 2\rho g\int_1^3 (x - 1)\sqrt{9 - x^2}\,dx$$

$$= 2\rho g\int_1^3 x\sqrt{9 - x^2}\,dx - 2\rho g\int_1^3 \sqrt{9 - x^2}\,dx \stackrel{30}{=} 2\rho g\left[-\tfrac{1}{3}(9 - x^2)^{3/2}\right]_1^3 - 2\rho g\left[\frac{x}{2}\sqrt{9 - x^2} + \tfrac{9}{2}\sin^{-1}\left(\frac{x}{3}\right)\right]_1^3$$

$$= 2\rho g\left[0 + \tfrac{1}{3}(8\sqrt{8})\right] - 2\rho g\left[(0 + \tfrac{9}{2}\cdot\tfrac{\pi}{2}) - (\tfrac{1}{2}\sqrt{8} + \tfrac{9}{2}\sin^{-1}(\tfrac{1}{3}))\right]$$

$$= \tfrac{32}{3}\sqrt{2}\,\rho g - \tfrac{9\pi}{2}\rho g + 2\sqrt{2}\,\rho g + 9\left[\sin^{-1}(\tfrac{1}{3})\right]\rho g = \left(\tfrac{38}{3}\sqrt{2} - \tfrac{9\pi}{2} + 9\sin^{-1}(\tfrac{1}{3})\right)\rho g$$

$$\approx 6.835 \cdot 1000 \cdot 9.8 \approx 6.7 \times 10^4 \text{ N}$$

Note: If you set up a typical coordinate system with the water level at $y = -1$, then $F = \int_{-3}^{-1} \rho g(-1 - y)2\sqrt{9 - y^2}\,dy$.

33. Set up a vertical x-axis as shown. Then the area of the ith rectangular strip is

$$\left(2 - \frac{2}{\sqrt{3}}\, x_i^*\right)\Delta x. \quad \left[\text{By similar triangles,}\ \frac{w_i}{2} = \frac{\sqrt{3} - x_i^*}{\sqrt{3}},\ \text{so}\ w_i = 2 - \frac{2}{\sqrt{3}}\, x_i^*.\right]$$

The pressure on the strip is $\rho g x_i^*$, so the hydrostatic force on the strip is

$\rho g x_i^*\left(2 - \dfrac{2}{\sqrt{3}}\, x_i^*\right)\Delta x$ and the hydrostatic force on the plate $\approx \displaystyle\sum_{i=1}^{n}\rho g x_i^*\left(2 - \dfrac{2}{\sqrt{3}}\, x_i^*\right)\Delta x.$

The total force

$$F = \lim_{n\to\infty}\sum_{i=1}^{n}\rho g x_i^*\left(2 - \frac{2}{\sqrt{3}}\, x_i^*\right)\Delta x = \int_0^{\sqrt{3}}\rho g x\left(2 - \frac{2}{\sqrt{3}}\, x\right)dx = \rho g\int_0^{\sqrt{3}}\left(2x - \frac{2}{\sqrt{3}}\, x^2\right)dx$$

$$= \rho g\left[x^2 - \frac{2}{3\sqrt{3}}\, x^3\right]_0^{\sqrt{3}} = \rho g\left[(3 - 2) - 0\right] = \rho g \approx 1000\cdot 9.8 = 9.8\times 10^3\ \text{N}$$

35. Set up coordinate axes as shown in the figure. The length of the ith strip is

$2\sqrt{25 - (y_i^*)^2}$ and its area is $2\sqrt{25 - (y_i^*)^2}\,\Delta y$. The pressure on this strip is

approximately $\delta d_i = 62.5(7 - y_i^*)$ and so the force on the strip is approximately

$62.5(7 - y_i^*)2\sqrt{25 - (y_i^*)^2}\,\Delta y$. The total force

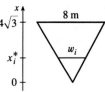

$$F = \lim_{n\to\infty}\sum_{i=1}^{n}62.5(7 - y_i^*)2\sqrt{25 - (y_i^*)^2}\,\Delta y = 125\int_0^5(7 - y)\sqrt{25 - y^2}\,dy$$

$$= 125\left\{\int_0^5 7\sqrt{25 - y^2}\,dy - \int_0^5 y\sqrt{25 - y^2}\,dy\right\} = 125\left\{7\int_0^5\sqrt{25 - y^2}\,dy - \left[-\tfrac{1}{3}(25 - y^2)^{3/2}\right]_0^5\right\}$$

$$= 125\left\{7\left(\tfrac{1}{4}\pi\cdot 5^2\right) + \tfrac{1}{3}(0 - 125)\right\} = 125\left(\tfrac{175\pi}{4} - \tfrac{125}{3}\right) \approx 11{,}972 \approx 1.2\times 10^4\ \text{lb}$$

37. By similar triangles, $\dfrac{8}{4\sqrt{3}} = \dfrac{w_i}{x_i^*} \;\Rightarrow\; w_i = \dfrac{2x_i^*}{\sqrt{3}}$. The area of the ith

rectangular strip is $\dfrac{2x_i^*}{\sqrt{3}}\,\Delta x$ and the pressure on it is $\rho g(4\sqrt{3} - x_i^*)$.

$$F = \int_0^{4\sqrt{3}}\rho g\left(4\sqrt{3} - x\right)\frac{2x}{\sqrt{3}}\,dx = 8\rho g\int_0^{4\sqrt{3}}x\,dx - \frac{2\rho g}{\sqrt{3}}\int_0^{4\sqrt{3}}x^2\,dx$$

$$= 4\rho g\left[x^2\right]_0^{4\sqrt{3}} - \frac{2\rho g}{3\sqrt{3}}\left[x^3\right]_0^{4\sqrt{3}} = 192\rho g - \frac{2\rho g}{3\sqrt{3}}\,64\cdot 3\sqrt{3} = 192\rho g - 128\rho g = 64\rho g$$

$$\approx 64(840)(9.8) \approx 5.27\times 10^5\ \text{N}$$

39. (a) The area of a strip is $20\,\Delta x$ and the pressure on it is δx_i.

$$F = \int_0^3 \delta x 20\,dx = 20\delta\left[\tfrac{1}{2}x^2\right]_0^3 = 20\delta\cdot\tfrac{9}{2} = 90\delta$$

$$= 90(62.5) = 5625\ \text{lb} \approx 5.63\times 10^3\ \text{lb}$$

(b) $F = \int_0^9 \delta x 20\,dx = 20\delta\left[\tfrac{1}{2}x^2\right]_0^9 = 20\delta\cdot\tfrac{81}{2} = 810\delta = 810(62.5) = 50{,}625\ \text{lb} \approx 5.06\times 10^4\ \text{lb}.$

(c) For the first 3 ft, the length of the side is constant at 40 ft. For $3 < x \le 9$, we can use similar triangles to find the length a:

$$\frac{a}{40} = \frac{9-x}{6} \quad \Rightarrow \quad a = 40 \cdot \frac{9-x}{6}.$$

$$F = \int_0^3 \delta x 40 \, dx + \int_3^9 \delta x (40) \frac{9-x}{6} \, dx = 40\delta \left[\tfrac{1}{2}x^2\right]_0^3 + \tfrac{20}{3}\delta \int_3^9 (9x - x^2) \, dx = 180\delta + \tfrac{20}{3}\delta \left[\tfrac{9}{2}x^2 - \tfrac{1}{3}x^3\right]_3^9$$

$$= 180\delta + \tfrac{20}{3}\delta \left[\left(\tfrac{729}{2} - 243\right) - \left(\tfrac{81}{2} - 9\right)\right] = 180\delta + 600\delta = 780\delta = 780(62.5) = 48{,}750 \text{ lb} \approx 4.88 \times 10^4 \text{ lb}$$

(d) For any right triangle with hypotenuse on the bottom,

$$\sin\theta = \frac{\Delta x}{\text{hypotenuse}} \quad \Rightarrow$$

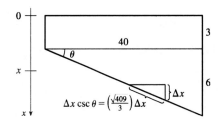

$$\text{hypotenuse} = \Delta x \csc\theta = \Delta x \frac{\sqrt{40^2 + 6^2}}{6} = \frac{\sqrt{409}}{3}\Delta x.$$

$$F = \int_3^9 \delta x 20 \frac{\sqrt{409}}{3} \, dx = \tfrac{1}{3}\left(20\sqrt{409}\right)\delta\left[\tfrac{1}{2}x^2\right]_3^9$$

$$= \tfrac{1}{3} \cdot 10\sqrt{409}\,\delta(81 - 9) \approx 303{,}356 \text{ lb} \approx 3.03 \times 10^5 \text{ lb}$$

41. $F = \int_2^5 \rho g x \cdot w(x) \, dx$, where $w(x)$ is the width of the plate at depth x. Since $n = 6$, $\Delta x = \frac{5-2}{6} = \frac{1}{2}$, and

$$F \approx S_6$$

$$= \rho g \cdot \frac{1/2}{3}\left[2 \cdot w(2) + 4 \cdot 2.5 \cdot w(2.5) + 2 \cdot 3 \cdot w(3) + 4 \cdot 3.5 \cdot w(3.5) + 2 \cdot 4 \cdot w(4) + 4 \cdot 4.5 \cdot w(4.5) + 5 \cdot w(5)\right]$$

$$= \tfrac{1}{6}\rho g(2 \cdot 0 + 10 \cdot 0.8 + 6 \cdot 1.7 + 14 \cdot 2.4 + 8 \cdot 2.9 + 18 \cdot 3.3 + 5 \cdot 3.6)$$

$$= \tfrac{1}{6}(1000)(9.8)(152.4) \approx 2.5 \times 10^5 \text{ N}$$

43. $m = \sum\limits_{i=1}^{3} m_i = 6 + 5 + 10 = 21.$

$$M_x = \sum\limits_{i=1}^{3} m_i\, y_i = 6(5) + 5(-2) + 10(-1) = 10; \quad M_y = \sum\limits_{i=1}^{3} m_i\, x_i = 6(1) + 5(3) + 10(-2) = 1.$$

$$\overline{x} = \frac{M_y}{m} = \frac{1}{21} \text{ and } \overline{y} = \frac{M_x}{m} = \frac{10}{21}, \text{ so the center of mass of the system is } \left(\tfrac{1}{21}, \tfrac{10}{21}\right).$$

45. Since the region in the figure is symmetric about the y-axis, we know

that $\overline{x} = 0$. The region is "bottom-heavy," so we know that $\overline{y} < 2$,

and we might guess that $\overline{y} = 1.5$.

$$A = \int_{-2}^2 (4 - x^2) \, dx = 2\int_0^2 (4 - x^2) \, dx = 2\left[4x - \tfrac{1}{3}x^3\right]_0^2$$

$$= 2\left(8 - \tfrac{8}{3}\right) = \tfrac{32}{3}.$$

$\overline{x} = \tfrac{1}{A}\int_{-2}^2 x(4 - x^2) \, dx = 0$ since $f(x) = x(4 - x^2)$ is an odd

function (or since the region is symmetric about the y-axis).

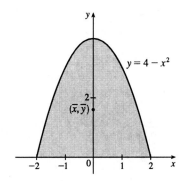

$$\overline{y} = \tfrac{1}{A}\int_{-2}^2 \tfrac{1}{2}(4 - x^2)^2 \, dx = \tfrac{3}{32} \cdot \tfrac{1}{2} \cdot 2\int_0^2 (16 - 8x^2 + x^4) \, dx = \tfrac{3}{32}\left[16x - \tfrac{8}{3}x^3 + \tfrac{1}{5}x^5\right]_0^2$$

$$= \tfrac{3}{32}\left(32 - \tfrac{64}{3} + \tfrac{32}{5}\right) = 3\left(1 - \tfrac{2}{3} + \tfrac{1}{5}\right) = 3\left(\tfrac{8}{15}\right) = \tfrac{8}{5}$$

Thus, the centroid is $(\overline{x}, \overline{y}) = \left(0, \tfrac{8}{5}\right)$.

47. The region in the figure is "right-heavy" and "bottom-heavy," so we know

$\overline{x} > 0.5$ and $\overline{y} < 1$, and we might guess that $\overline{x} = 0.6$ and $\overline{y} = 0.9$.

$A = \int_0^1 e^x \, dx = [e^x]_0^1 = e - 1.$

$\overline{x} = \frac{1}{A} \int_0^1 x e^x \, dx = \frac{1}{e-1}[xe^x - e^x]_0^1$ [by parts]

$\quad = \frac{1}{e-1}[0 - (-1)] = \frac{1}{e-1}.$

$\overline{y} = \frac{1}{A} \int_0^1 \frac{1}{2}(e^x)^2 \, dx = \frac{1}{e-1} \cdot \frac{1}{4}[e^{2x}]_0^1 = \frac{1}{4(e-1)}(e^2 - 1) = \frac{e+1}{4}.$

Thus, the centroid is $(\overline{x}, \overline{y}) = \left(\frac{1}{e-1}, \frac{e+1}{4}\right) \approx (0.58, 0.93).$

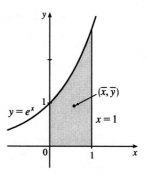

49. The line has equation $y = \frac{3}{4}x$. $A = \frac{1}{2}(4)(3) = 6$, so $m = \rho A = 10(6) = 60.$

$$M_x = \rho \int_0^4 \frac{1}{2}\left(\frac{3}{4}x\right)^2 dx = 10 \int_0^4 \frac{9}{32}x^2 \, dx = \frac{45}{16}\left[\frac{1}{3}x^3\right]_0^4 = \frac{45}{16}\left(\frac{64}{3}\right) = 60$$

$$M_y = \rho \int_0^4 x\left(\frac{3}{4}x\right) dx = \frac{15}{2}\int_0^4 x^2 \, dx = \frac{15}{2}\left[\frac{1}{3}x^3\right]_0^4 = \frac{15}{2}\left(\frac{64}{3}\right) = 160$$

$\overline{x} = \dfrac{M_y}{m} = \dfrac{160}{60} = \dfrac{8}{3}$ and $\overline{y} = \dfrac{M_x}{m} = \dfrac{60}{60} = 1$. Thus, the centroid is $(\overline{x}, \overline{y}) = \left(\frac{8}{3}, 1\right).$

51. (a)

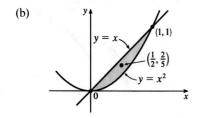

Suppose the region lies between two curves $y = f(x)$ and $y = g(x)$

where $f(x) \geq g(x)$, as illustrated in the figure. Use n subintervals

determined by points x_i with $a = x_0 < x_1 < \cdots < x_n = b$ and

choose $x_i^* = \overline{x}_i$ to be the midpoint of the ith subinterval; that is,

$\overline{x}_i = \frac{1}{2}(x_{i-1} + x_i)$. Then the centroid of the ith approximating

rectangle R_i is its center $C_i = \left(\overline{x}_i, \frac{1}{2}[f(\overline{x}_i) + g(\overline{x}_i)]\right).$

Its area is $[f(\overline{x}_i) - g(\overline{x}_i)] \, \Delta x$, so its mass is $\rho[f(\overline{x}_i) - g(\overline{x}_i)] \, \Delta x.$

Thus, $M_y(R_i) = \rho[f(\overline{x}_i) - g(\overline{x}_i)] \, \Delta x \cdot \overline{x}_i = \rho\overline{x}_i[f(\overline{x}_i) - g(\overline{x}_i)] \, \Delta x$ and

$M_x(R_i) = \rho[f(\overline{x}_i) - g(\overline{x}_i)] \, \Delta x \cdot \frac{1}{2}[f(\overline{x}_i) + g(\overline{x}_i)] = \rho \cdot \frac{1}{2}\left\{[f(\overline{x}_i)]^2 - [g(\overline{x}_i)]^2\right\} \Delta x.$ Summing over i and taking

the limit as $n \to \infty$, we get $M_y = \lim\limits_{n\to\infty} \sum\limits_{i=1}^n \rho\overline{x}_i[f(\overline{x}_i) - g(\overline{x}_i)] \, \Delta x = \rho \int_a^b x[f(x) - g(x)] \, dx$ and

$M_x = \lim\limits_{n\to\infty} \sum\limits_{i=1}^n \rho \cdot \frac{1}{2}\left[f(\overline{x}_i)^2 - g(\overline{x}_i)^2\right] \Delta x = \rho \int_a^b \frac{1}{2}\left\{[f(x)]^2 - [g(x)]^2\right\} dx.$ Thus,

$\overline{x} = \dfrac{M_y}{m} = \dfrac{M_y}{\rho A} = \dfrac{1}{A}\displaystyle\int_a^b x[f(x) - g(x)] \, dx$ and $\overline{y} = \dfrac{M_x}{m} = \dfrac{M_x}{\rho A} = \dfrac{1}{A}\displaystyle\int_a^b \frac{1}{2}\left\{[f(x)]^2 - [g(x)]^2\right\} dx.$

(b)

The region is sketched in the figure. We take $f(x) = x$, $g(x) = x^2$, $a = 0$,

and $b = 1$ in the formulas in part (a). First we note that the area of the

region is $A = \int_0^1 (x - x^2) \, dx = \left[\frac{1}{2}x^2 - \frac{1}{3}x^3\right]_0^1 = \frac{1}{6}.$

Therefore, $\bar{x} = \frac{1}{A} \int_0^1 x[f(x) - g(x)]\,dx = \frac{1}{1/6} \int_0^1 x(x - x^2)\,dx = 6 \int_0^1 (x^2 - x^3)\,dx = 6\left[\frac{1}{3}x^3 - \frac{1}{4}x^4\right]_0^1 = \frac{1}{2}$

and $\bar{y} = \frac{1}{A} \int_0^1 \frac{1}{2}\left\{[f(x)]^2 - [g(x)]^2\right\}dx = \frac{1}{1/6} \int_0^1 \frac{1}{2}(x^2 - x^4)\,dx = 3\left[\frac{1}{3}x^3 - \frac{1}{5}x^5\right]_0^1 = \frac{2}{5}.$

The centroid is $\left(\frac{1}{2}, \frac{2}{5}\right)$.

6.7   Applications to Economics and Biology

1. By the Net Change Theorem, $C(2000) - C(0) = \int_0^{2000} C'(x)\,dx$ ⇒

$$C(2000) = 20{,}000 + \int_0^{2000}(5 - 0.008x + 0.000009x^2)\,dx = 20{,}000 + \left[5x - 0.004x^2 + 0.000003x^3\right]_0^{2000}$$

$$= 20{,}000 + 10{,}000 - 0.004(4{,}000{,}000) + 0.000003(8{,}000{,}000{,}000) = 30{,}000 - 16{,}000 + 24{,}000$$

$$= \$38{,}000$$

3. If the production level is raised from 1200 units to 1600 units, then the increase in cost is

$$C(1600) - C(1200) = \int_{1200}^{1600} C'(x)\,dx = \int_{1200}^{1600}(74 + 1.1x - 0.002x^2 + 0.00004x^3)\,dx$$

$$= \left[74x + 0.55x^2 - \frac{0.002}{3}x^3 + 0.00001x^4\right]_{1200}^{1600} = 64{,}331{,}733.33 - 20{,}464{,}800 = \$43{,}866{,}933.33$$

5. $p(x) = 10$ ⇒ $\dfrac{450}{x + 8} = 10$ ⇒ $x + 8 = 45$ ⇒ $x = 37.$

Consumer surplus $= \displaystyle\int_0^{37}[p(x) - 10]\,dx = \int_0^{37}\left(\dfrac{450}{x + 8} - 10\right)dx$

$$= \left[450\ln(x + 8) - 10x\right]_0^{37} = (450\ln 45 - 370) - 450\ln 8$$

$$= 450\ln\left(\frac{45}{8}\right) - 370 \approx \$407.25$$

7. $P = p_S(x)$ ⇒ $400 = 200 + 0.2x^{3/2}$ ⇒ $200 = 0.2x^{3/2}$ ⇒ $1000 = x^{3/2}$ ⇒ $x = 1000^{2/3} = 100.$

Producer surplus $= \int_0^{100}[P - p_S(x)]\,dx = \int_0^{100}[400 - (200 + 0.2x^{3/2})]\,dx = \int_0^{100}\left(200 - \frac{1}{5}x^{3/2}\right)dx$

$$= \left[200x - \frac{2}{25}x^{5/2}\right]_0^{100} = 20{,}000 - 8{,}000 = \$12{,}000$$

9. $p(x) = \dfrac{800{,}000e^{-x/5000}}{x + 20{,}000} = 16$ ⇒ $x = x_1 \approx 3727.04.$

Consumer surplus $= \int_0^{x_1}[p(x) - 16]\,dx \approx \$37{,}753$

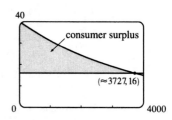

11. $f(8) - f(4) = \int_4^8 f'(t)\,dt = \int_4^8 \sqrt{t}\,dt = \left[\frac{2}{3}t^{3/2}\right]_4^8 = \frac{2}{3}\left(16\sqrt{2} - 8\right) \approx \$9.75$ million

13. $N = \int_a^b Ax^{-k}\, dx = A\left[\dfrac{x^{-k+1}}{-k+1}\right]_a^b = \dfrac{A}{1-k}\left(b^{1-k} - a^{1-k}\right).$

Similarly, $\displaystyle\int_a^b Ax^{1-k}\, dx = A\left[\dfrac{x^{2-k}}{2-k}\right]_a^b = \dfrac{A}{2-k}\left(b^{2-k} - a^{2-k}\right).$

Thus, $\bar{x} = \dfrac{1}{N}\int_a^b Ax^{1-k}\, dx = \dfrac{[A/(2-k)](b^{2-k} - a^{2-k})}{[A/(1-k)](b^{1-k} - a^{1-k})} = \dfrac{(1-k)(b^{2-k} - a^{2-k})}{(2-k)(b^{1-k} - a^{1-k})}.$

15. $F = \dfrac{\pi PR^4}{8\eta l} = \dfrac{\pi(4000)(0.008)^4}{8(0.027)(2)} \approx 1.19 \times 10^{-4}\ \text{cm}^3/\text{s}$

17. From (3), $F = \dfrac{A}{\int_0^T c(t)\, dt} = \dfrac{6}{20I}$, where

$$I = \int_0^{10} te^{-0.6t}\, dt = \left[\dfrac{1}{(-0.6)^2}(-0.6t - 1)\,e^{-0.6t}\right]_0^{10} \begin{bmatrix}\text{integrating}\\ \text{by parts}\end{bmatrix} = \tfrac{1}{0.36}(-7e^{-6} + 1)$$

Thus, $F = \dfrac{6(0.36)}{20(1 - 7e^{-6})} = \dfrac{0.108}{1 - 7e^{-6}} \approx 0.1099\ \text{L/s}$ or $6.594\ \text{L/min}.$

19. As in Example 2, we will estimate the cardiac output using Simpson's Rule with $\Delta t = (16 - 0)/8 = 2.$

$$\int_0^{16} c(t)\, dt \approx \tfrac{2}{3}[c(0) + 4c(2) + 2c(4) + 4c(6) + 2c(8) + 4c(10) + 2c(12) + 4c(14) + c(16)]$$

$$\approx \tfrac{2}{3}[0 + 4(6.1) + 2(7.4) + 4(6.7) + 2(5.4) + 4(4.1) + 2(3.0) + 4(2.1) + 1.5]$$

$$= \tfrac{2}{3}(109.1) = 72.7\overline{3}\ \text{mg}\cdot\text{s/L}$$

Therefore, $F \approx \dfrac{A}{72.7\overline{3}} = \dfrac{7}{72.7\overline{3}} \approx 0.0962\ \text{L/s}$ or $5.77\ \text{L/min}.$

6.8 Probability

1. (a) $\int_{30{,}000}^{40{,}000} f(x)\, dx$ is the probability that a randomly chosen tire will have a lifetime between 30,000 and 40,000 miles.

(b) $\int_{25{,}000}^{\infty} f(x)\, dx$ is the probability that a randomly chosen tire will have a lifetime of at least 25,000 miles.

3. (a) In general, we must satisfy the two conditions that are mentioned before Example 1—namely, (1) $f(x) \geq 0$ for all x, and (2) $\int_{-\infty}^{\infty} f(x)\, dx = 1.$ For $0 \leq x \leq 4$, we have $f(x) = \tfrac{3}{64}x\sqrt{16 - x^2} \geq 0$, so $f(x) \geq 0$ for all x. Also,

$$\int_{-\infty}^{\infty} f(x)\, dx = \int_0^4 \tfrac{3}{64}x\sqrt{16 - x^2}\, dx = -\tfrac{3}{128}\int_0^4 (16 - x^2)^{1/2}(-2x)\, dx = -\tfrac{3}{128}\left[\tfrac{2}{3}(16 - x^2)^{3/2}\right]_0^4$$

$$= -\tfrac{1}{64}\left[(16 - x^2)^{3/2}\right]_0^4 = -\tfrac{1}{64}(0 - 64) = 1.$$

Therefore, f is a probability density function.

(b) $P(X < 2) = \int_{-\infty}^{2} f(x)\, dx = \int_0^2 \tfrac{3}{64}x\sqrt{16 - x^2}\, dx = -\tfrac{3}{128}\int_0^2 (16 - x^2)^{1/2}(-2x)\, dx$

$$= -\tfrac{3}{128}\left[\tfrac{2}{3}(16 - x^2)^{3/2}\right]_0^2 = -\tfrac{1}{64}\left[(16 - x^2)^{3/2}\right]_0^2 = -\tfrac{1}{64}(12^{3/2} - 16^{3/2})$$

$$= \tfrac{1}{64}(64 - 12\sqrt{12}) = \tfrac{1}{64}(64 - 24\sqrt{3}) = 1 - \tfrac{3}{8}\sqrt{3} \approx 0.350481$$

5. (a) In general, we must satisfy the two conditions that are mentioned before Example 1—namely, (1) $f(x) \geq 0$ for all x, and (2) $\int_{-\infty}^{\infty} f(x)\,dx = 1$. If $c \geq 0$, then $f(x) \geq 0$, so condition (1) is satisfied. For condition (2), we see that

$$\int_{-\infty}^{\infty} f(x)\,dx = \int_{-\infty}^{\infty} \frac{c}{1+x^2}\,dx \text{ and}$$

$$\int_0^{\infty} \frac{c}{1+x^2}\,dx = \lim_{t \to \infty} \int_0^t \frac{c}{1+x^2}\,dx = c \lim_{t \to \infty} \left[\tan^{-1} x\right]_0^t = c \lim_{t \to \infty} \tan^{-1} t = c\left(\frac{\pi}{2}\right)$$

Similarly, $\displaystyle \int_{-\infty}^0 \frac{c}{1+x^2}\,dx = c\left(\frac{\pi}{2}\right)$, so $\displaystyle \int_{-\infty}^{\infty} \frac{c}{1+x^2}\,dx = 2c\left(\frac{\pi}{2}\right) = c\pi.$

Since $c\pi$ must equal 1, we must have $c = 1/\pi$ so that f is a probability density function.

(b) $P(-1 < X < 1) = \int_{-1}^1 \frac{1/\pi}{1+x^2}\,dx = \frac{2}{\pi} \int_0^1 \frac{1}{1+x^2}\,dx = \frac{2}{\pi}\left[\tan^{-1} x\right]_0^1 = \frac{2}{\pi}\left(\frac{\pi}{4} - 0\right) = \frac{1}{2}$

7. (a) In general, we must satisfy the two conditions that are mentioned before Example 1—namely, (1) $f(x) \geq 0$ for all x, and (2) $\int_{-\infty}^{\infty} f(x)\,dx = 1$. Since $f(x) = 0$ or $f(x) = 0.1$, condition (1) is satisfied. For condition (2), we see that

$\int_{-\infty}^{\infty} f(x)\,dx = \int_0^{10} 0.1\,dx = \left[\frac{1}{10}x\right]_0^{10} = 1$. Thus, $f(x)$ is a probability density function for the spinner's values.

(b) Since all the numbers between 0 and 10 are equally likely to be selected, we expect the mean to be halfway between the endpoints of the interval; that is, $x = 5$.

$$\mu = \int_{-\infty}^{\infty} xf(x)\,dx = \int_0^{10} x(0.1)\,dx = \left[\frac{1}{20}x^2\right]_0^{10} = \frac{100}{20} = 5, \quad \text{as expected.}$$

9. We need to find m so that $\int_m^{\infty} f(t)\,dt = \frac{1}{2}$ $\Rightarrow$ $\lim_{x \to \infty} \int_m^x \frac{1}{5}e^{-t/5}\,dt = \frac{1}{2}$ $\Rightarrow$ $\lim_{x \to \infty} \left[\frac{1}{5}(-5)e^{-t/5}\right]_m^x = \frac{1}{2}$ $\Rightarrow$
$(-1)(0 - e^{-m/5}) = \frac{1}{2}$ $\Rightarrow$ $e^{-m/5} = \frac{1}{2}$ $\Rightarrow$ $-m/5 = \ln\frac{1}{2}$ $\Rightarrow$ $m = -5\ln\frac{1}{2} = 5\ln 2 \approx 3.47$ min.

11. We use an exponential density function with $\mu = 2.5$ min.

(a) $P(X > 4) = \int_4^{\infty} f(t)\,dt = \lim_{x \to \infty} \int_4^x \frac{1}{2.5}e^{-t/2.5}\,dt = \lim_{x \to \infty}\left[-e^{-t/2.5}\right]_4^x = 0 + e^{-4/2.5} \approx 0.202$

(b) $P(0 \leq X \leq 2) = \int_0^2 f(t)\,dt = \left[-e^{-t/2.5}\right]_0^2 = -e^{-2/2.5} + 1 \approx 0.551$

(c) We need to find a value a so that $P(X \geq a) = 0.02$, or, equivalently, $P(0 \leq X \leq a) = 0.98$ $\Leftrightarrow$

$\int_0^a f(t)\,dt = 0.98$ $\Leftrightarrow$ $\left[-e^{-t/2.5}\right]_0^a = 0.98$ $\Leftrightarrow$ $-e^{-a/2.5} + 1 = 0.98$ $\Leftrightarrow$ $e^{-a/2.5} = 0.02$ $\Leftrightarrow$

$-a/2.5 = \ln 0.02$ $\Leftrightarrow$ $a = -2.5\ln\frac{1}{50} = 2.5\ln 50 \approx 9.78$ min ≈ 10 min. The ad should say that if you aren't served within 10 minutes, you get a free hamburger.

13. $P(X \geq 10) = \int_{10}^{\infty} \frac{1}{4.2\sqrt{2\pi}} \exp\left(-\frac{(x-9.4)^2}{2 \cdot 4.2^2}\right) dx$. To avoid the improper integral we approximate it by the integral from

10 to 100. Thus, $P(X \geq 10) \approx \int_{10}^{100} \frac{1}{4.2\sqrt{2\pi}} \exp\left(-\frac{(x-9.4)^2}{2 \cdot 4.2^2}\right) dx \approx 0.443$ (using a calculator or computer to estimate

the integral), so about 44 percent of the households throw out at least 10 lb of paper a week.

Note: We can't evaluate $1 - P(0 \leq X \leq 10)$ for this problem since a significant amount of area lies to the left of $X = 0$.

15. (a) $P(0 \le X \le 100) = \int_0^{100} \frac{1}{8\sqrt{2\pi}} \exp\left(-\frac{(x-112)^2}{2 \cdot 8^2}\right) dx \approx 0.0668$ (using a calculator or computer to estimate the

integral), so there is about a 6.68% chance that a randomly chosen vehicle is traveling at a legal speed.

(b) $P(X \ge 125) = \int_{125}^{\infty} \frac{1}{8\sqrt{2\pi}} \exp\left(-\frac{(x-112)^2}{2 \cdot 8^2}\right) dx = \int_{125}^{\infty} f(x)\, dx$. In this case, we could use a calculator or computer

to estimate either $\int_{125}^{300} f(x)\, dx$ or $1 - \int_0^{125} f(x)\, dx$. Both are approximately 0.0521, so about 5.21% of the motorists are

targeted.

17. $P(\mu - 2\sigma \le X \le \mu + 2\sigma) = \int_{\mu-2\sigma}^{\mu+2\sigma} \frac{1}{\sigma\sqrt{2\pi}} \exp\left(-\frac{(x-\mu)^2}{2\sigma^2}\right) dx$. Substituting $t = \frac{x-\mu}{\sigma}$ and $dt = \frac{1}{\sigma} dx$ gives us

$\int_{-2}^{2} \frac{1}{\sigma\sqrt{2\pi}} e^{-t^2/2}(\sigma\, dt) = \frac{1}{\sqrt{2\pi}} \int_{-2}^{2} e^{-t^2/2}\, dt \approx 0.9545$.

6 Review

CONCEPT CHECK

1. (a) See Section 6.1, Figure 2 and Equations 6.1.1 and 6.1.2.

(b) Instead of using "top minus bottom" and integrating from left to right, we use "right minus left" and integrate from bottom to top. See Figures 9 and 10 in Section 6.1.

2. The numerical value of the area represents the number of meters by which Sue is ahead of Kathy after 1 minute.

3. (a) See the discussion in Section 6.2, near Figures 2 and 3, ending in the Definition of Volume.

(b) See the discussion between Examples 5 and 6 in Section 6.2. If the cross-section is a disk, find the radius in terms of x or y and use $A = \pi(\text{radius})^2$. If the cross-section is a washer, find the inner radius r_{in} and outer radius r_{out} and use $A = \pi\left(r_{out}^2\right) - \pi\left(r_{in}^2\right)$.

4. (a) $V = 2\pi r h\, \Delta r = (\text{circumference})(\text{height})(\text{thickness})$

(b) For a typical shell, find the circumference and height in terms of x or y and calculate $V = \int_a^b (\text{circumference})(\text{height})(dx \text{ or } dy)$, where a and b are the limits on x or y.

(c) Sometimes slicing produces washers or disks whose radii are difficult (or impossible) to find explicitly. On other occasions, the cylindrical shell method leads to an easier integral than slicing does.

5. (a) The length of a curve is defined to be the limit of the lengths of the inscribed polygons, as described near Figure 3 in Section 6.4.

(b) See Equation 6.4.1.

(c) See Equations 6.4.2 and 6.4.3.

6. (a) The average value of a function f on an interval $[a, b]$ is $f_{\text{ave}} = \dfrac{1}{b-a} \displaystyle\int_a^b f(x)\,dx$.

(b) The Mean Value Theorem for Integrals says that there is a number c at which the value of f is exactly equal to the average value of the function, that is, $f(c) = f_{\text{ave}}$. For a geometric interpretation of the Mean Value Theorem for Integrals, see Figure 2 in Section 6.4 and the discussion that accompanies it.

7. $\int_0^6 f(x)\,dx$ represents the amount of work done. Its units are newton-meters, or joules.

8. Let $c(x)$ be the cross-sectional length of the wall (measured parallel to the surface of the fluid) at depth x. Then the hydrostatic force against the wall is given by $F = \int_a^b \delta x c(x)\,dx$, where a and b are the lower and upper limits for x at points of the wall and δ is the weight density of the fluid.

9. (a) The center of mass is the point at which the plate balances horizontally.

(b) See Equations 6.6.12.

10. See Figure 3 in Section 6.6, and the discussion which precedes it.

11. (a) See the definition in the first paragraph of the subsection *Cardiac Output* in Section 6.7.

(b) See the discussion in the second paragraph of the subsection *Cardiac Output* in Section 6.7.

12. A probability density function f is a function on the domain of a continuous random variable X such that $\int_a^b f(x)\,dx$ measures the probability that X lies between a and b. Such a function f has nonnegative values and satisfies the relation $\int_D f(x)\,dx = 1$, where D is the domain of the corresponding random variable X. If $D = \mathbb{R}$, or if we define $f(x) = 0$ for real numbers $x \notin D$, then $\int_{-\infty}^{\infty} f(x)\,dx = 1$. (Of course, to work with f in this way, we must assume that the integrals of f exist.)

13. (a) $\int_0^{130} f(x)\,dx$ represents the probability that the weight of a randomly chosen female college student is less than 130 pounds.

(b) $\mu = \int_{-\infty}^{\infty} x f(x)\,dx = \int_0^{\infty} x f(x)\,dx$

(c) The median of f is the number m such that $\int_m^{\infty} f(x)\,dx = \tfrac{1}{2}$.

14. See the discussion near Equation 3 in Section 6.8.

EXERCISES

1. The curves intersect when $x^2 = 4x - x^2 \iff 2x^2 - 4x = 0 \iff$
$2x(x - 2) = 0 \iff x = 0$ or 2.
$A = \int_0^2 \left[(4x - x^2) - x^2\right]dx = \int_0^2 (4x - 2x^2)\,dx$
$\quad = \left[2x^2 - \tfrac{2}{3}x^3\right]_0^2 = \left[(8 - \tfrac{16}{3}) - 0\right] = \tfrac{8}{3}$

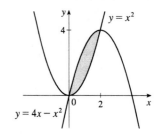

3. If $x \geq 0$, then $|x| = x$, and the graphs intersect when $x = 1 - 2x^2$ ⟺ $2x^2 + x - 1 = 0$ ⟺ $(2x - 1)(x + 1) = 0$ ⟺ $x = \frac{1}{2}$ or -1, but $-1 < 0$. By symmetry, we can double the area from $x = 0$ to $x = \frac{1}{2}$.

$$A = 2 \int_0^{1/2} [(1 - 2x^2) - x] \, dx = 2 \int_0^{1/2} (-2x^2 - x + 1) \, dx$$

$$= 2\left[-\tfrac{2}{3}x^3 - \tfrac{1}{2}x^2 + x\right]_0^{1/2} = 2\left[\left(-\tfrac{1}{12} - \tfrac{1}{8} + \tfrac{1}{2}\right) - 0\right]$$

$$= 2\left(\tfrac{7}{24}\right) = \tfrac{7}{12}$$

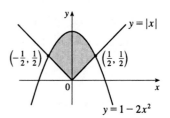

5. $x = 2\theta - \sin\theta$ ⟹ $dx = (2 - \cos\theta) \, d\theta$

$$A = \int_0^{2\pi} y \, dx = \int_0^{2\pi} [(2 - \cos\theta)(2 - \cos\theta)] \, d\theta$$

$$= \int_0^{2\pi} \left(4 - 4\cos\theta + \cos^2\theta\right) d\theta = \int_0^{2\pi} \left(4 - 4\cos\theta + \tfrac{1}{2} + \tfrac{1}{2}\cos 2\theta\right) d\theta$$

$$= \left[4\theta - 4\sin\theta + \tfrac{1}{2}\theta + \tfrac{1}{4}\sin 2\theta\right]_0^{2\pi} = (8\pi - 0 + \pi + 0) - (0) = 9\pi$$

7. (a) Using the Midpoint Rule on $[0, 1]$ with $f(x) = \tan(x^2)$ and $n = 4$, we estimate

$$A = \int_0^1 \tan(x^2) \, dx \approx \tfrac{1}{4}\left[\tan\left(\left(\tfrac{1}{8}\right)^2\right) + \tan\left(\left(\tfrac{3}{8}\right)^2\right) + \tan\left(\left(\tfrac{5}{8}\right)^2\right) + \tan\left(\left(\tfrac{7}{8}\right)^2\right)\right] \approx \tfrac{1}{4}(1.53) \approx 0.38$$

(b) Using the Midpoint Rule on $[0, 1]$ with $f(x) = \pi\tan^2(x^2)$ (for disks) and $n = 4$, we estimate

$$V = \int_0^1 f(x) \, dx \approx \tfrac{1}{4}\pi\left[\tan^2\left(\left(\tfrac{1}{8}\right)^2\right) + \tan^2\left(\left(\tfrac{3}{8}\right)^2\right) + \tan^2\left(\left(\tfrac{5}{8}\right)^2\right) + \tan^2\left(\left(\tfrac{7}{8}\right)^2\right)\right] \approx \tfrac{\pi}{4}(1.114) \approx 0.87$$

9. (a) A cross-section is a washer with inner radius x^2 and outer radius x.

$$V = \int_0^1 \pi\left[(x)^2 - (x^2)^2\right] dx = \int_0^1 \pi(x^2 - x^4) \, dx = \pi\left[\tfrac{1}{3}x^3 - \tfrac{1}{5}x^5\right]_0^1 = \pi\left[\tfrac{1}{3} - \tfrac{1}{5}\right] = \tfrac{2}{15}\pi$$

(b) A cross-section is a washer with inner radius y and outer radius $\sqrt{y}$.

$$V = \int_0^1 \pi\left[\left(\sqrt{y}\right)^2 - y^2\right] dy = \int_0^1 \pi(y - y^2) \, dy = \pi\left[\tfrac{1}{2}y^2 - \tfrac{1}{3}y^3\right]_0^1 = \pi\left[\tfrac{1}{2} - \tfrac{1}{3}\right] = \tfrac{\pi}{6}$$

(c) A cross-section is a washer with inner radius $2 - x$ and outer radius $2 - x^2$.

$$V = \int_0^1 \pi\left[(2 - x^2)^2 - (2 - x)^2\right] dx = \int_0^1 \pi(x^4 - 5x^2 + 4x) \, dx = \pi\left[\tfrac{1}{5}x^5 - \tfrac{5}{3}x^3 + 2x^2\right]_0^1 = \pi\left[\tfrac{1}{5} - \tfrac{5}{3} + 2\right] = \tfrac{8}{15}\pi$$

11. $V = \pi \int_{-3}^{3} \left\{\left[(9 - y^2) - (-1)\right]^2 - [0 - (-1)]^2\right\} dy$

$$= 2\pi \int_0^3 \left[(10 - y^2)^2 - 1\right] dy = 2\pi \int_0^3 (100 - 20y^2 + y^4 - 1) \, dy$$

$$= 2\pi \int_0^3 (99 - 20y^2 + y^4) \, dy = 2\pi\left[99y - \tfrac{20}{3}y^3 + \tfrac{1}{5}y^5\right]_0^3$$

$$= 2\pi\left(297 - 180 + \tfrac{243}{5}\right) = \tfrac{1656}{5}\pi$$

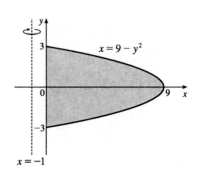

13. The graph of $x^2 - y^2 = a^2$ is a hyperbola with right and left branches.

Solving for y gives us $y^2 = x^2 - a^2 \Rightarrow y = \pm\sqrt{x^2 - a^2}$.

We'll use shells and the height of each shell is

$\sqrt{x^2 - a^2} - \left(-\sqrt{x^2 - a^2}\right) = 2\sqrt{x^2 - a^2}$.

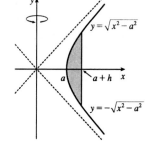

The volume is $V = \int_a^{a+h} 2\pi x \cdot 2\sqrt{x^2 - a^2}\, dx$. To evaluate, let $u = x^2 - a^2$,

so $du = 2x\, dx$ and $x\, dx = \frac{1}{2}\, du$. When $x = a$, $u = 0$, and when $x = a + h$,

$u = (a + h)^2 - a^2 = a^2 + 2ah + h^2 - a^2 = 2ah + h^2$.

Thus, $V = 4\pi \int_0^{2ah+h^2} \sqrt{u}\left(\frac{1}{2}\, du\right) = 2\pi\left[\frac{2}{3}u^{3/2}\right]_0^{2ah+h^2} = \frac{4}{3}\pi\left(2ah + h^2\right)^{3/2}$.

15. A shell has radius $\frac{\pi}{2} - x$, circumference $2\pi\left(\frac{\pi}{2} - x\right)$, and height $\cos^2 x - \frac{1}{4}$.

$y = \cos^2 x$ intersects $y = \frac{1}{4}$ when $\cos^2 x = \frac{1}{4} \Leftrightarrow$

$\cos x = \pm\frac{1}{2} \quad [\,|x| \leq \pi/2\,] \quad \Leftrightarrow \quad x = \pm\frac{\pi}{3}$.

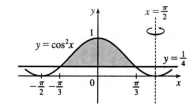

$V = \int_{-\pi/3}^{\pi/3} 2\pi\left(\frac{\pi}{2} - x\right)\left(\cos^2 x - \frac{1}{4}\right) dx$

17. (a) $\int_0^{\pi/2} 2\pi \cos^2 x\, dx = \int_0^{\pi/2} \pi\left(\sqrt{2}\cos x\right)^2 dx$

The solid is obtained by rotating the region $\mathcal{R} = \left\{(x,y) \mid 0 \leq x \leq \frac{\pi}{2}, 0 \leq y \leq \sqrt{2}\cos x\right\}$ about the x-axis.

(b) The solid is obtained by rotating the region $\mathcal{R} = \left\{(x,y) \mid 0 \leq x \leq 1, 2 - \sqrt{x} \leq y \leq 2 - x^2\right\}$ about the x-axis.

Or: The solid is obtained by rotating the region $\mathcal{R} = \left\{(x,y) \mid 0 \leq x \leq 1, x^2 \leq y \leq \sqrt{x}\right\}$ about the line $y = 2$.

19. Take the base to be the disk $x^2 + y^2 \leq 9$. Then $V = \int_{-3}^3 A(x)\, dx$, where $A(x_0)$ is the area of the isosceles right triangle

whose hypotenuse lies along the line $x = x_0$ in the xy-plane. The length of the hypotenuse is $2\sqrt{9 - x^2}$ and the length of

each leg is $\sqrt{2}\sqrt{9 - x^2}$. $A(x) = \frac{1}{2}\left(\sqrt{2}\sqrt{9 - x^2}\right)^2 = 9 - x^2$, so

$$V = 2\int_0^3 A(x)\, dx = 2\int_0^3 (9 - x^2)\, dx = 2\left[9x - \frac{1}{3}x^3\right]_0^3 = 2(27 - 9) = 36$$

21. Equilateral triangles with sides measuring $\frac{1}{4}x$ meters have height $\frac{1}{4}x \sin 60° = \frac{\sqrt{3}}{8}x$. Therefore,

$A(x) = \frac{1}{2} \cdot \frac{1}{4}x \cdot \frac{\sqrt{3}}{8}x = \frac{\sqrt{3}}{64}x^2$. $V = \int_0^{20} A(x)\, dx = \frac{\sqrt{3}}{64}\int_0^{20} x^2\, dx = \frac{\sqrt{3}}{64}\left[\frac{1}{3}x^3\right]_0^{20} = \frac{8000\sqrt{3}}{64 \cdot 3} = \frac{125\sqrt{3}}{3}$ m³.

23. $x = 3t^2, y = 2t^3, 0 \leq t \leq 2$.

$L = \int_0^2 \sqrt{(dx/dt)^2 + (dy/dt)^2}\, dt = \int_0^2 \sqrt{(6t)^2 + (6t^2)^2}\, dt = \int_0^2 \sqrt{36t^2 + 36t^4}\, dt = 6\int_0^2 t\sqrt{1 + t^2}\, dt$

$= 6\int_1^5 \sqrt{u}\left(\frac{1}{2}du\right) \quad [u = 1 + t^2, du = 2t\, dt] = 3\left[\frac{2}{3}u^{3/2}\right]_1^5 = 2(5\sqrt{5} - 1)$

25. $y = \frac{1}{6}(x^2 + 4)^{3/2}$ $\Rightarrow$ $dy/dx = \frac{1}{4}(x^2 + 4)^{1/2}(2x)$ $\Rightarrow$

$1 + (dy/dx)^2 = 1 + \left[\frac{1}{2}x(x^2 + 4)^{1/2}\right]^2 = 1 + \frac{1}{4}x^2(x^2 + 4) = \frac{1}{4}x^4 + x^2 + 1 = \left(\frac{1}{2}x^2 + 1\right)^2$.

Thus, $L = \int_0^3 \sqrt{\left(\frac{1}{2}x^2 + 1\right)^2}\,dx = \int_0^3 \left(\frac{1}{2}x^2 + 1\right)dx = \left[\frac{1}{6}x^3 + x\right]_0^3 = \frac{15}{2}$.

27. $f(x) = kx$ $\Rightarrow$ $30\,\text{N} = k(15 - 12)\,\text{cm}$ $\Rightarrow$ $k = 10\,\text{N/cm} = 1000\,\text{N/m}$. $20\,\text{cm} - 12\,\text{cm} = 0.08\,\text{m}$ $\Rightarrow$

$W = \int_0^{0.08} kx\,dx = 1000\int_0^{0.08} x\,dx = 500\left[x^2\right]_0^{0.08} = 500(0.08)^2 = 3.2\,\text{N·m} = 3.2\,\text{J}$.

29. (a) The parabola has equation $y = ax^2$ with vertex at the origin and passing through

$(4, 4)$. $4 = a \cdot 4^2$ $\Rightarrow$ $a = \frac{1}{4}$ $\Rightarrow$ $y = \frac{1}{4}x^2$ $\Rightarrow$ $x^2 = 4y$ $\Rightarrow$

 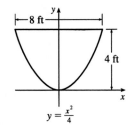

$x = 2\sqrt{y}$. Each circular disk has radius $2\sqrt{y}$ and is moved $4 - y$ ft.

$$W = \int_0^4 \pi\left(2\sqrt{y}\right)^2 62.5(4 - y)\,dy = 250\pi\int_0^4 y(4 - y)\,dy$$

$$= 250\pi\left[2y^2 - \frac{1}{3}y^3\right]_0^4 = 250\pi\left(32 - \frac{64}{3}\right) = \frac{8000\pi}{3} \approx 8378\,\text{ft-lb}$$

(b) In part (a) we knew the final water level (0) but not the amount of work done. Here

we use the same equation, except with the work fixed, and the lower limit of

integration (that is, the final water level — call it h) unknown: $W = 4000$ $\Leftrightarrow$

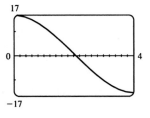

$250\pi\left[2y^2 - \frac{1}{3}y^3\right]_h^4 = 4000$ $\Leftrightarrow$ $\frac{16}{\pi} = \left[\left(32 - \frac{64}{3}\right) - \left(2h^2 - \frac{1}{3}h^3\right)\right]$ $\Leftrightarrow$

$h^3 - 6h^2 + 32 - \frac{48}{\pi} = 0$. We graph the function $f(h) = h^3 - 6h^2 + 32 - \frac{48}{\pi}$

on the interval $[0, 4]$ to see where it is 0. From the graph, $f(h) = 0$ for $h \approx 2.1$.

So the depth of water remaining is about 2.1 ft.

31. As in Example 4 of Section 6.6, $\dfrac{a}{2 - x} = \dfrac{1}{2}$ $\Rightarrow$ $2a = 2 - x$ and $w = 2(1.5 + a) = 3 + 2a = 3 + 2 - x = 5 - x$.

Thus, $F = \int_0^2 \rho gx(5 - x)\,dx = \rho g\left[\frac{5}{2}x^2 - \frac{1}{3}x^3\right]_0^2 = \rho g\left(10 - \frac{8}{3}\right) = \frac{22}{3}\delta$ $[\rho g = \delta]$ $\approx \frac{22}{3} \cdot 62.5 \approx 458\,\text{lb}$.

33. $x = 100$ $\Rightarrow$ $P = 2000 - 0.1(100) - 0.01(100)^2 = 1890$

$$\text{Consumer surplus} = \int_0^{100}[p(x) - P]\,dx = \int_0^{100}\left(2000 - 0.1x - 0.01x^2 - 1890\right)dx$$

$$= \left[110x - 0.05x^2 - \frac{0.01}{3}x^3\right]_0^{100} = 11{,}000 - 500 - \frac{10{,}000}{3} \approx \$7166.67$$

35. $\displaystyle\lim_{h \to 0} f_{\text{ave}} = \lim_{h \to 0}\frac{1}{(x + h) - x}\int_x^{x+h} f(t)\,dt = \lim_{h \to 0}\frac{F(x + h) - F(x)}{h}$, where $F(x) = \int_a^x f(t)\,dt$. But we recognize this

limit as being $F'(x)$ by the definition of a derivative. Therefore, $\displaystyle\lim_{h \to 0} f_{\text{ave}} = F'(x) = f(x)$ by FTC1.

37. $f(x) = \begin{cases} \frac{\pi}{20}\sin\left(\frac{\pi}{10}x\right) & \text{if } 0 \le x \le 10 \\ 0 & \text{if } x < 0 \text{ or } x > 10 \end{cases}$

(a) $f(x) \ge 0$ for all real numbers x and

$$\int_{-\infty}^{\infty} f(x)\, dx = \int_{0}^{10} \frac{\pi}{20}\sin\left(\frac{\pi}{10}x\right) dx = \frac{\pi}{20} \cdot \frac{10}{\pi}\left[-\cos\left(\frac{\pi}{10}x\right)\right]_{0}^{10} = \frac{1}{2}(-\cos\pi + \cos 0) = \frac{1}{2}(1+1) = 1$$

Therefore, f is a probability density function.

(b) $P(X < 4) = \int_{-\infty}^{4} f(x)\, dx = \int_{0}^{4} \frac{\pi}{20}\sin\left(\frac{\pi}{10}x\right) dx = \frac{1}{2}\left[-\cos\left(\frac{\pi}{10}x\right)\right]_{0}^{4} = \frac{1}{2}\left(-\cos\frac{2\pi}{5} + \cos 0\right)$

$\approx \frac{1}{2}(-0.309017 + 1) \approx 0.3455$

(c) $\mu = \int_{-\infty}^{\infty} x f(x)\, dx = \int_{0}^{10} \frac{\pi}{20} x \sin\left(\frac{\pi}{10}x\right) dx$

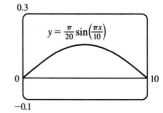

$= \int_{0}^{\pi} \frac{\pi}{20} \cdot \frac{10}{\pi} u(\sin u)\left(\frac{10}{\pi}\right) du \qquad \left[u = \frac{\pi}{10}x,\, du = \frac{\pi}{10}\,dx\right]$

$= \frac{5}{\pi}\int_{0}^{\pi} u \sin u\, du \overset{82}{=} \frac{5}{\pi}\left[\sin u - u\cos u\right]_{0}^{\pi} = \frac{5}{\pi}[0 - \pi(-1)] = 5$

This answer is expected because the graph of f is symmetric about the

line $x = 5$.

39. (a) The probability density function is $f(t) = \begin{cases} 0 & \text{if } t < 0 \\ \frac{1}{8}e^{-t/8} & \text{if } t \ge 0 \end{cases}$

$P(0 \le X \le 3) = \int_{0}^{3} \frac{1}{8}e^{-t/8}\, dt = \left[-e^{-t/8}\right]_{0}^{3} = -e^{-3/8} + 1 \approx 0.3127$

(b) $P(X > 10) = \int_{10}^{\infty} \frac{1}{8}e^{-t/8}\, dt = \lim_{x \to \infty}\left[-e^{-t/8}\right]_{10}^{x} = \lim_{x \to \infty}\left(-e^{-x/8} + e^{-10/8}\right) = 0 + e^{-5/4} \approx 0.2865$

(c) We need to find m such that $P(X \ge m) = \frac{1}{2} \Rightarrow \int_{m}^{\infty} \frac{1}{8}e^{-t/8}\, dt = \frac{1}{2} \Rightarrow \lim_{x \to \infty}\left[-e^{-t/8}\right]_{m}^{x} = \frac{1}{2} \Rightarrow$

$\lim_{x \to \infty}\left(-e^{-x/8} + e^{-m/8}\right) = \frac{1}{2} \Rightarrow e^{-m/8} = \frac{1}{2} \Rightarrow -m/8 = \ln\frac{1}{2} \Rightarrow m = -8\ln\frac{1}{2} = 8\ln 2 \approx 5.55$ minutes.

☐ FOCUS ON PROBLEM SOLVING

1. The volume generated from $x = 0$ to $x = b$ is $\int_0^b \pi[f(x)]^2 \, dx$. Hence, we are given that $b^2 = \int_0^b \pi[f(x)]^2 \, dx$ for all $b > 0$. Differentiating both sides of this equation with respect to b using the Fundamental Theorem of Calculus gives

   $2b = \pi[f(b)]^2 \quad \Rightarrow \quad f(b) = \sqrt{2b/\pi}$, since f is positive. Therefore, $f(x) = \sqrt{2x/\pi}$.

3. (a) $V = \pi h^2(r - h/3) = \frac{1}{3}\pi h^2(3r - h)$. See the solution to Exercise 6.2.33.

   (b) The smaller segment has height $h = 1 - x$ and so by part (a) its volume is

   $V = \frac{1}{3}\pi(1 - x)^2 \, [3(1) - (1 - x)] = \frac{1}{3}\pi(x - 1)^2(x + 2)$. This volume must be $\frac{1}{3}$ of the total volume of the sphere,

   which is $\frac{4}{3}\pi(1)^3$. So $\frac{1}{3}\pi(x - 1)^2(x + 2) = \frac{1}{3}\left(\frac{4}{3}\pi\right) \quad \Rightarrow \quad (x^2 - 2x + 1)(x + 2) = \frac{4}{3} \quad \Rightarrow \quad x^3 - 3x + 2 = \frac{4}{3} \quad \Rightarrow$

   $3x^3 - 9x + 2 = 0$. Using Newton's method with $f(x) = 3x^3 - 9x + 2$, $f'(x) = 9x^2 - 9$, we get

   $x_{n+1} = x_n - \dfrac{3x_n^3 - 9x_n + 2}{9x_n^2 - 9}$. Taking $x_1 = 0$, we get $x_2 \approx 0.2222$, and $x_3 \approx 0.2261 \approx x_4$, so, correct to four decimal

   places, $x \approx 0.2261$.

   (c) With $r = 0.5$ and $s = 0.75$, the equation $x^3 - 3rx^2 + 4r^3s = 0$ becomes $x^3 - 3(0.5)x^2 + 4(0.5)^3(0.75) = 0 \quad \Rightarrow$

   $x^3 - \frac{3}{2}x^2 + 4\left(\frac{1}{8}\right)\frac{3}{4} = 0 \quad \Rightarrow \quad 8x^3 - 12x^2 + 3 = 0$. We use Newton's method with $f(x) = 8x^3 - 12x^2 + 3$,

   $f'(x) = 24x^2 - 24x$, so $x_{n+1} = x_n - \dfrac{8x_n^3 - 12x_n^2 + 3}{24x_n^2 - 24x_n}$. Take $x_1 = 0.5$. Then $x_2 \approx 0.6667$, and $x_3 \approx 0.6736 \approx x_4$.

   So to four decimal places the depth is 0.6736 m.

   (d) (i) From part (a) with $r = 5$ in., the volume of water in the bowl is

   $V = \frac{1}{3}\pi h^2(3r - h) = \frac{1}{3}\pi h^2(15 - h) = 5\pi h^2 - \frac{1}{3}\pi h^3$. We are given that $\dfrac{dV}{dt} = 0.2$ in^3/s and we want to find $\dfrac{dh}{dt}$

   when $h = 3$. Now $\dfrac{dV}{dt} = 10\pi h \dfrac{dh}{dt} - \pi h^2 \dfrac{dh}{dt}$, so $\dfrac{dh}{dt} = \dfrac{0.2}{\pi(10h - h^2)}$. When $h = 3$, we have

   $\dfrac{dh}{dt} = \dfrac{0.2}{\pi(10 \cdot 3 - 3^2)} = \dfrac{1}{105\pi} \approx 0.003$ in/s.

   (ii) From part (a), the volume of water required to fill the bowl from the instant that the water is 4 in. deep is

   $V = \frac{1}{2} \cdot \frac{4}{3}\pi(5)^3 - \frac{1}{3}\pi(4)^2(15 - 4) = \frac{2}{3} \cdot 125\pi - \frac{16}{3} \cdot 11\pi = \frac{74}{3}\pi$. To find the time required to fill the bowl we divide

   this volume by the rate: Time $= \frac{74\pi/3}{0.2} = \frac{370\pi}{3} \approx 387$ s ≈ 6.5 min.

5. We are given that the rate of change of the volume of water is $\dfrac{dV}{dt} = -kA(x)$, where k is some positive constant and $A(x)$ is

   the area of the surface when the water has depth x. Now we are concerned with the rate of change of the depth of the water

   with respect to time, that is, $\dfrac{dx}{dt}$. But by the Chain Rule, $\dfrac{dV}{dt} = \dfrac{dV}{dx}\dfrac{dx}{dt}$, so the first equation can be written

$\dfrac{dV}{dx}\dfrac{dx}{dt} = -kA(x)$ $(\star)$. Also, we know that the total volume of water up to a depth x is $V(x) = \int_0^x A(s)\,ds$, where $A(s)$ is

the area of a cross-section of the water at a depth s. Differentiating this equation with respect to x, we get $dV/dx = A(x)$.

Substituting this into equation $\star$, we get $A(x)(dx/dt) = -kA(x)$ $\Rightarrow$ $dx/dt = -k$, a constant.

7. To find the height of the pyramid, we use similar triangles. The first figure shows a cross-section of the pyramid passing

through the top and through two opposite corners of the square base. Now $|BD| = b$, since it is a radius of the sphere, which

has diameter $2b$ since it is tangent to the opposite sides of the square base. Also, $|AD| = b$ since $\triangle ADB$ is isosceles. So the

height is $|AB| = \sqrt{b^2 + b^2} = \sqrt{2}\,b$.

  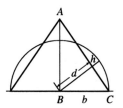

We first observe that the shared volume is equal to half the volume of the sphere, minus the sum of the four equal volumes

(caps of the sphere) cut off by the triangular faces of the pyramid. See Exercise 6.2.33 for a derivation of the formula for the

volume of a cap of a sphere. To use the formula, we need to find the perpendicular distance h of each triangular face from the

surface of the sphere. We first find the distance d from the center of the sphere to one of the triangular faces. The third figure

shows a cross-section of the pyramid through the top and through the midpoints of opposite sides of the square base. From

similar triangles we find that

$$\frac{d}{b} = \frac{|AB|}{|AC|} = \frac{\sqrt{2}\,b}{\sqrt{b^2 + \left(\sqrt{2}\,b\right)^2}} \quad \Rightarrow \quad d = \frac{\sqrt{2}\,b^2}{\sqrt{3b^2}} = \frac{\sqrt{6}}{3}\,b$$

So $h = b - d = b - \frac{\sqrt{6}}{3}b = \frac{3 - \sqrt{6}}{3}b$. So, using the formula $V = \pi h^2(r - h/3)$ from Exercise 6.2.33 with $r = b$, we find that

the volume of each of the caps is $\pi\left(\frac{3-\sqrt{6}}{3}b\right)^2\left(b - \frac{3-\sqrt{6}}{3\cdot3}b\right) = \frac{15-6\sqrt{6}}{9}\cdot\frac{6+\sqrt{6}}{9}\pi b^3 = \left(\frac{2}{3} - \frac{7}{27}\sqrt{6}\right)\pi b^3$. So, using our first

observation, the shared volume is $V = \frac{1}{2}\left(\frac{4}{3}\pi b^3\right) - 4\left(\frac{2}{3} - \frac{7}{27}\sqrt{6}\right)\pi b^3 = \left(\frac{28}{27}\sqrt{6} - 2\right)\pi b^3$.

9. $x = \displaystyle\int_1^t \frac{\cos u}{u}\,du,\ y = \int_1^t \frac{\sin u}{u}\,du$, so by FTC1, we have $\dfrac{dx}{dt} = \dfrac{\cos t}{t}$ and $\dfrac{dy}{dt} = \dfrac{\sin t}{t}$. Vertical tangent lines occur when

$dx/dt = 0$ $\Leftrightarrow$ $\cos t = 0$ $\Leftrightarrow$ $t = \frac{\pi}{2} + n\pi$. The parameter value corresponding to the origin, $(x, y) = (0, 0)$, is $t = 1$,

so the nearest vertical tangent occurs when $t = \frac{\pi}{2}$. Therefore, the arc length between these points is

$$L = \int_1^{\pi/2} \sqrt{\left(\frac{dx}{dt}\right)^2 + \left(\frac{dy}{dt}\right)^2}\,dt = \int_1^{\pi/2} \sqrt{\frac{\cos^2 t}{t^2} + \frac{\sin^2 t}{t^2}}\,dt = \int_1^{\pi/2} \frac{dt}{t} = \left[\ln t\right]_1^{\pi/2} = \ln \frac{\pi}{2}$$

11.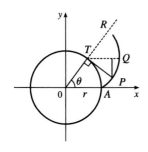

The coordinates of T are $(x_1, y_1) = (r \cos \theta, r \sin \theta)$. Since TP was unwound from arc TA, TP has length $r\theta$. Also $\angle PTQ = \angle PTR - \angle QTR = \frac{1}{2}\pi - \theta$, so P has coordinates

$$x = x_1 + |TP| \cos \angle PTQ = r \cos \theta + r\theta \cos \left(\tfrac{1}{2}\pi - \theta\right) = r(\cos \theta + \theta \sin \theta),$$

$$y = y_1 + |TP| \sin \angle PTQ = r \sin \theta - r\theta \sin \left(\tfrac{1}{2}\pi - \theta\right) = r(\sin \theta - \theta \cos \theta).$$

13. We can assume that the cut is made along a vertical line $x = b > 0$, that the disk's boundary is the circle $x^2 + y^2 = 1$, and that the center of mass of the smaller piece (to the right of $x = b$) is $\left(\frac{1}{2}, 0\right)$. We wish to find b to two decimal places. We have $\dfrac{1}{2} = \overline{x} = \dfrac{\int_b^1 x \cdot 2\sqrt{1 - x^2}\,dx}{\int_b^1 2\sqrt{1 - x^2}\,dx}$. Evaluating the

numerator gives us $-\int_b^1 (1 - x^2)^{1/2}(-2x)\,dx = -\frac{2}{3}\left[\left(1 - x^2\right)^{3/2}\right]_b^1 = -\frac{2}{3}\left[0 - \left(1 - b^2\right)^{3/2}\right] = \frac{2}{3}(1 - b^2)^{3/2}$.

Using Formula 30 in the table of integrals, we find that the denominator is

$$\left[x\sqrt{1 - x^2} + \sin^{-1} x\right]_b^1 = \left(0 + \tfrac{\pi}{2}\right) - \left(b\sqrt{1 - b^2} + \sin^{-1} b\right). \text{ Thus, we have } \dfrac{1}{2} = \overline{x} = \dfrac{\frac{2}{3}(1 - b^2)^{3/2}}{\frac{\pi}{2} - b\sqrt{1 - b^2} - \sin^{-1} b}, \text{ or,}$$

equivalently, $\frac{2}{3}(1 - b^2)^{3/2} = \frac{\pi}{4} - \frac{1}{2}b\sqrt{1 - b^2} - \frac{1}{2}\sin^{-1} b$. Solving this equation numerically with a calculator or CAS, we obtain $b \approx 0.138173$, or $b = 0.14$ m to two decimal places.

15. The cubic polynomial passes through the origin, so let its equation be $y = px^3 + qx^2 + rx$. The curves intersect when $px^3 + qx^2 + rx = x^2 \Leftrightarrow px^3 + (q - 1)x^2 + rx = 0$. Call the left side $f(x)$. Since $f(a) = f(b) = 0$, another form of f is

$$f(x) = px(x - a)(x - b) = px[x^2 - (a + b)x + ab]$$
$$= p[x^3 - (a + b)x^2 + abx]$$

Since the two areas are equal, we must have $\int_0^a f(x)\,dx = -\int_a^b f(x)\,dx \Rightarrow$

$[F(x)]_0^a = [F(x)]_b^a \Rightarrow F(a) - F(0) = F(a) - F(b) \Rightarrow F(0) = F(b)$, where F is an antiderivative of f.

Now $F(x) = \int f(x)\,dx = \int p[x^3 - (a + b)x^2 + abx]\,dx = p\left[\frac{1}{4}x^4 - \frac{1}{3}(a + b)x^3 + \frac{1}{2}abx^2\right] + C$, so

$F(0) = F(b) \Rightarrow C = p\left[\frac{1}{4}b^4 - \frac{1}{3}(a + b)b^3 + \frac{1}{2}ab^3\right] + C \Rightarrow 0 = p\left[\frac{1}{4}b^4 - \frac{1}{3}(a + b)b^3 + \frac{1}{2}ab^3\right] \Rightarrow$

$0 = 3b - 4(a + b) + 6a$ [multiply by $12/(pb^3), b \neq 0$] $\Rightarrow 0 = 3b - 4a - 4b + 6a \Rightarrow b = 2a$.

Hence, b is twice the value of a.

7 □ DIFFERENTIAL EQUATIONS

7.1 Modeling with Differential Equations

1. $y = \frac{2}{3}e^x + e^{-2x}$ $\Rightarrow$ $y' = \frac{2}{3}e^x - 2e^{-2x}$. To show that y is a solution of the differential equation, we will substitute the expressions for y and y' in the left-hand side of the equation and show that the left-hand side is equal to the right hand side.

$$\text{LHS} = y' + 2y = \frac{2}{3}e^x - 2e^{-2x} + 2\left(\frac{2}{3}e^x + e^{-2x}\right) = \frac{2}{3}e^x - 2e^{-2x} + \frac{4}{3}e^x + 2e^{-2x}$$
$$= \frac{6}{3}e^x = 2e^x = \text{RHS}$$

3. (a) $y = e^{rx}$ $\Rightarrow$ $y' = re^{rx}$ $\Rightarrow$ $y'' = r^2 e^{rx}$. Substituting these expressions into the differential equation

$2y'' + y' - y = 0$, we get $2r^2 e^{rx} + re^{rx} - e^{rx} = 0$ $\Rightarrow$ $(2r^2 + r - 1)e^{rx} = 0$ $\Rightarrow$

$(2r - 1)(r + 1) = 0$ [since e^{rx} is never zero] $\Rightarrow$ $r = \frac{1}{2}$ or -1.

(b) Let $r_1 = \frac{1}{2}$ and $r_2 = -1$, so we need to show that every member of the family of functions $y = ae^{x/2} + be^{-x}$ is a

solution of the differential equation $2y'' + y' - y = 0$.

$y = ae^{x/2} + be^{-x}$ $\Rightarrow$ $y' = \frac{1}{2}ae^{x/2} - be^{-x}$ $\Rightarrow$ $y'' = \frac{1}{4}ae^{x/2} + be^{-x}$.

$$\text{LHS} = 2y'' + y' - y = 2\left(\frac{1}{4}ae^{x/2} + be^{-x}\right) + \left(\frac{1}{2}ae^{x/2} - be^{-x}\right) - \left(ae^{x/2} + be^{-x}\right)$$
$$= \frac{1}{2}ae^{x/2} + 2be^{-x} + \frac{1}{2}ae^{x/2} - be^{-x} - ae^{x/2} - be^{-x}$$
$$= \left(\frac{1}{2}a + \frac{1}{2}a - a\right)e^{x/2} + (2b - b - b)e^{-x}$$
$$= 0 = \text{RHS}$$

5. (a) $y = \sin x$ $\Rightarrow$ $y' = \cos x$ $\Rightarrow$ $y'' = -\sin x$.

$\text{LHS} = y'' + y = -\sin x + \sin x = 0 \neq \sin x$, so $y = \sin x$ **is not** a solution of the differential equation.

(b) $y = \cos x$ $\Rightarrow$ $y' = -\sin x$ $\Rightarrow$ $y'' = -\cos x$.

$\text{LHS} = y'' + y = -\cos x + \cos x = 0 \neq \sin x$, so $y = \cos x$ **is not** a solution of the differential equation.

(c) $y = \frac{1}{2}x \sin x$ $\Rightarrow$ $y' = \frac{1}{2}(x \cos x + \sin x)$ $\Rightarrow$ $y'' = \frac{1}{2}(-x \sin x + \cos x + \cos x)$.

$\text{LHS} = y'' + y = \frac{1}{2}(-x \sin x + 2 \cos x) + \frac{1}{2}x \sin x = \cos x \neq \sin x$, so $y = \frac{1}{2}x \sin x$ **is not** a solution of the

differential equation.

(d) $y = -\frac{1}{2}x \cos x$ $\Rightarrow$ $y' = -\frac{1}{2}(-x \sin x + \cos x)$ $\Rightarrow$ $y'' = -\frac{1}{2}(-x \cos x - \sin x - \sin x)$.

$\text{LHS} = y'' + y = -\frac{1}{2}(-x \cos x - 2 \sin x) + \left(-\frac{1}{2}x \cos x\right) = \sin x = \text{RHS}$, so $y = -\frac{1}{2}x \cos x$ **is** a solution of the

differential equation.

7. (a) Since the derivative $y' = -y^2$ is always negative (or 0 if $y = 0$), the function y must be decreasing (or equal to 0) on any interval on which it is defined.

(b) $y = \dfrac{1}{x+C}$ $\Rightarrow$ $y' = -\dfrac{1}{(x+C)^2}$. LHS $= y' = -\dfrac{1}{(x+C)^2} = -\left(\dfrac{1}{x+C}\right)^2 = -y^2 = $ RHS

(c) $y = 0$ is a solution of $y' = -y^2$ that is not a member of the family in part (b).

(d) If $y(x) = \dfrac{1}{x+C}$, then $y(0) = \dfrac{1}{0+C} = \dfrac{1}{C}$. Since $y(0) = 0.5$, $\dfrac{1}{C} = \dfrac{1}{2}$ $\Rightarrow$ $C = 2$, so $y = \dfrac{1}{x+2}$.

9. (a) $\dfrac{dP}{dt} = 1.2P\left(1 - \dfrac{P}{4200}\right)$. Now $\dfrac{dP}{dt} > 0$ $\Rightarrow$ $1 - \dfrac{P}{4200} > 0$ [assuming that $P > 0$] $\Rightarrow$ $\dfrac{P}{4200} < 1$ $\Rightarrow$

$P < 4200$ $\Rightarrow$ the population is increasing for $0 < P < 4200$.

(b) $\dfrac{dP}{dt} < 0$ $\Rightarrow$ $P > 4200$

(c) $\dfrac{dP}{dt} = 0$ $\Rightarrow$ $P = 4200$ or $P = 0$

11. (a) This function is increasing *and* also decreasing. But $dy/dt = e^t(y-1)^2 \geq 0$ for all t, implying that the graph of the solution of the differential equation cannot be decreasing on any interval.

(b) When $y = 1$, $dy/dt = 0$, but the graph does not have a horizontal tangent line.

13. (a) $y' = 1 + x^2 + y^2 \geq 1$ and $y' \to \infty$ as $x \to \infty$. The only curve satisfying these conditions is labeled III.

(b) $y' = xe^{-x^2-y^2} > 0$ if $x > 0$ and $y' < 0$ if $x < 0$. The only curve with negative tangent slopes when $x < 0$ and positive tangent slopes when $x > 0$ is labeled I.

(c) $y' = \dfrac{1}{1 + e^{x^2+y^2}} > 0$ and $y' \to 0$ as $x \to \infty$. The only curve satisfying these conditions is labeled IV.

(d) $y' = \sin(xy)\cos(xy) = 0$ if $y = 0$, which is the solution graph labeled II.

15. (a) P increases most rapidly at the beginning, since there are usually many simple, easily-learned sub-skills associated with learning a skill. As t increases, we would expect dP/dt to remain positive, but decrease. This is because as time progresses, the only points left to learn are the more difficult ones.

(b) $\dfrac{dP}{dt} = k(M - P)$ is always positive, so the level of performance P is increasing. As P gets close to M, dP/dt gets close to 0; that is, the performance levels off, as explained in part (a).

(c)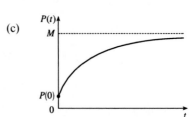

7.2 Direction Fields and Euler's Method

1. (a)

(b) It appears that the constant functions $y = 0.5$ and $y = 1.5$ are equilibrium solutions. Note that these two values of y satisfy the given differential equation $y' = x \cos \pi y$.

3. $y' = 2 - y$. The slopes at each point are independent of x, so the slopes are the same along each line parallel to the x-axis. Thus, III is the direction field for this equation. Note that for $y = 2$, $y' = 0$.

5. $y' = x + y - 1 = 0$ on the line $y = -x + 1$. Direction field IV satisfies this condition. Notice also that on the line $y = -x$ we have $y' = -1$, which is true in IV.

7. (a) $y(0) = 1$

(b) $y(0) = 2$

(c) $y(0) = -1$

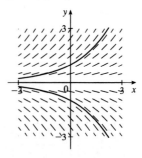

9.

x	y	$y' = \frac{1}{2}y$
0	0	0
0	1	0.5
0	2	1
0	-3	-1.5
0	-2	-1

Note that for $y = 0$, $y' = 0$. The three solution curves sketched go through $(0, 0)$, $(0, 1)$, and $(0, -1)$.

11.

x	y	$y' = y - 2x$
-2	-2	2
-2	2	6
2	2	-2
2	-2	-6

Note that $y' = 0$ for any point on the line $y = 2x$. The slopes are positive to the left of the line and negative to the right of the line. The solution curve in the graph passes through $(1, 0)$.

13.

x	y	$y' = y + xy$
0	± 2	± 2
1	± 2	± 4
-3	± 2	∓ 4

Note that $y' = y(x + 1) = 0$ for any point on $y = 0$ or on $x = -1$. The slopes are positive when the factors y and $x + 1$ have the same sign and negative when they have opposite signs. The solution curve in the graph passes through $(0, 1)$.

15. In Maple, we can use either `directionfield` (in Maple's share library) or `DEtools[DEplot]` to plot the direction field. To plot the solution, we can either use the initial-value option in `directionfield`, or actually solve the equation.

In Mathematica, we use `PlotVectorField` for the direction field, and the `Plot[Evaluate[...]]` construction to plot the solution, which is

$$y = 2 \arctan\left(e^{x^3/3} \cdot \tan \tfrac{1}{2}\right).$$

In Derive, use `Direction_Field` (in utility file `ODE_APPR`) to plot the direction field. Then use `DSOLVE1(-x^2*SIN(y),1,x,y,0,1)` (in utility file `ODE1`) to solve the equation. Simplify each result.

17.

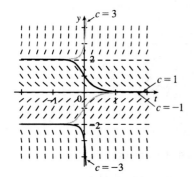

The direction field is for the differential equation $y' = y^3 - 4y$.

$L = \lim\limits_{t \to \infty} y(t)$ exists for $-2 \le c \le 2$;

$L = \pm 2$ for $c = \pm 2$ and $L = 0$ for $-2 < c < 2$.

For other values of c, L does not exist.

19. (a) $y' = F(x, y) = y$ and $y(0) = 1 \quad \Rightarrow \quad x_0 = 0, y_0 = 1$.

(i) $h = 0.4$ and $y_1 = y_0 + hF(x_0, y_0) \quad \Rightarrow \quad y_1 = 1 + 0.4 \cdot 1 = 1.4$. $x_1 = x_0 + h = 0 + 0.4 = 0.4$,

so $y_1 = y(0.4) = 1.4$.

(ii) $h = 0.2 \quad \Rightarrow \quad x_1 = 0.2$ and $x_2 = 0.4$, so we need to find y_2.

$y_1 = y_0 + hF(x_0, y_0) = 1 + 0.2y_0 = 1 + 0.2 \cdot 1 = 1.2$,

$y_2 = y_1 + hF(x_1, y_1) = 1.2 + 0.2y_1 = 1.2 + 0.2 \cdot 1.2 = 1.44$.

(iii) $h = 0.1$ $\Rightarrow$ $x_4 = 0.4$, so we need to find y_4. $y_1 = y_0 + hF(x_0, y_0) = 1 + 0.1y_0 = 1 + 0.1 \cdot 1 = 1.1$,

$$y_2 = y_1 + hF(x_1, y_1) = 1.1 + 0.1y_1 = 1.1 + 0.1 \cdot 1.1 = 1.21,$$

$$y_3 = y_2 + hF(x_2, y_2) = 1.21 + 0.1y_2 = 1.21 + 0.1 \cdot 1.21 = 1.331,$$

$$y_4 = y_3 + hF(x_3, y_3) = 1.331 + 0.1y_3 = 1.331 + 0.1 \cdot 1.331 = 1.4641.$$

(b) 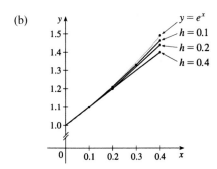   We see that the estimates are underestimates since they are all below the graph of $y = e^x$.

(c) (i) For $h = 0.4$:   (exact value) $-$ (approximate value) $= e^{0.4} - 1.4 \approx 0.0918$

   (ii) For $h = 0.2$:   (exact value) $-$ (approximate value) $= e^{0.4} - 1.44 \approx 0.0518$

   (iii) For $h = 0.1$:   (exact value) $-$ (approximate value) $= e^{0.4} - 1.4641 \approx 0.0277$

Each time the step size is halved, the error estimate also appears to be halved (approximately).

21. $h = 0.5$, $x_0 = 1$, $y_0 = 0$, and $F(x, y) = y - 2x$.

Note that $x_1 = x_0 + h = 1 + 0.5 = 1.5$, $x_2 = 2$, and $x_3 = 2.5$.

$y_1 = y_0 + hF(x_0, y_0) = 0 + 0.5F(1, 0) = 0.5[0 - 2(1)] = -1$.

$y_2 = y_1 + hF(x_1, y_1) = -1 + 0.5F(1.5, -1) = -1 + 0.5[-1 - 2(1.5)] = -3$.

$y_3 = y_2 + hF(x_2, y_2) = -3 + 0.5F(2, -3) = -3 + 0.5[-3 - 2(2)] = -6.5$.

$y_4 = y_3 + hF(x_3, y_3) = -6.5 + 0.5F(2.5, -6.5) = -6.5 + 0.5[-6.5 - 2(2.5)] = -12.25$.

23. $h = 0.1$, $x_0 = 0$, $y_0 = 1$, and $F(x, y) = y + xy$.

Note that $x_1 = x_0 + h = 0 + 0.1 = 0.1$, $x_2 = 0.2$, $x_3 = 0.3$, and $x_4 = 0.4$.

$y_1 = y_0 + hF(x_0, y_0) = 1 + 0.1F(0, 1) = 1 + 0.1[1 + (0)(1)] = 1.1$.

$y_2 = y_1 + hF(x_1, y_1) = 1.1 + 0.1F(0.1, 1.1) = 1.1 + 0.1[1.1 + (0.1)(1.1)] = 1.221$.

$y_3 = y_2 + hF(x_2, y_2) = 1.221 + 0.1F(0.2, 1.221) = 1.221 + 0.1[1.221 + (0.2)(1.221)] = 1.36752$.

$y_4 = y_3 + hF(x_3, y_3) = 1.36752 + 0.1F(0.3, 1.36752) = 1.36752 + 0.1[1.36752 + (0.3)(1.36752)]$
$$= 1.5452976.$$

$y_5 = y_4 + hF(x_4, y_4) = 1.5452976 + 0.1F(0.4, 1.5452976)$
$$= 1.5452976 + 0.1[1.5452976 + (0.4)(1.5452976)] = 1.761639264.$$

Thus, $y(0.5) \approx 1.7616$.

25. (a) $dy/dx + 3x^2 y = 6x^2 \Rightarrow y' = 6x^2 - 3x^2 y$. Store this expression in Y_1 and use the following simple program to

evaluate $y(1)$ for each part, using $H = h = 1$ and $N = 1$ for part (i), $H = 0.1$ and $N = 10$ for part (ii), and so forth.

$$h \to H: 0 \to X: 3 \to Y:$$

$$\text{For}(I, 1, N): Y + H \times Y_1 \to Y: X + H \to X:$$

$$\text{End(loop)}:$$

$$\text{Display Y.} \quad [\text{To see all iterations, include this statement in the loop.}]$$

(i) $H = 1, N = 1 \Rightarrow y(1) = 3$

(ii) $H = 0.1, N = 10 \Rightarrow y(1) \approx 2.3928$

(iii) $H = 0.01, N = 100 \Rightarrow y(1) \approx 2.3701$

(iv) $H = 0.001, N = 1000 \Rightarrow y(1) \approx 2.3681$

(b) $y = 2 + e^{-x^3} \Rightarrow y' = -3x^2 e^{-x^3}$

$$\text{LHS} = y' + 3x^2 y = -3x^2 e^{-x^3} + 3x^2\left(2 + e^{-x^3}\right) = -3x^2 e^{-x^3} + 6x^2 + 3x^2 e^{-x^3} = 6x^2 = \text{RHS}$$

$$y(0) = 2 + e^{-0} = 2 + 1 = 3$$

(c) The exact value of $y(1)$ is $2 + e^{-1^3} = 2 + e^{-1}$.

(i) For $h = 1$: (exact value) − (approximate value) $= 2 + e^{-1} - 3 \approx -0.6321$

(ii) For $h = 0.1$: (exact value) − (approximate value) $= 2 + e^{-1} - 2.3928 \approx -0.0249$

(iii) For $h = 0.01$: (exact value) − (approximate value) $= 2 + e^{-1} - 2.3701 \approx -0.0022$

(iv) For $h = 0.001$: (exact value) − (approximate value) $= 2 + e^{-1} - 2.3681 \approx -0.0002$

In (ii)–(iv), it seems that when the step size is divided by 10, the error estimate is also divided by 10 (approximately).

27. (a) $R\dfrac{dQ}{dt} + \dfrac{1}{C}Q = E(t)$ becomes $5Q' + \dfrac{1}{0.05}Q = 60$

or $Q' + 4Q = 12$.

(b) From the graph, it appears that the limiting value of the charge Q is about 3.

(c) If $Q' = 0$, then $4Q = 12 \Rightarrow Q = 3$ is an equilibrium solution.

(d)

(e) $Q' + 4Q = 12 \Rightarrow Q' = 12 - 4Q$. Now $Q(0) = 0$, so $t_0 = 0$ and $Q_0 = 0$.

$$Q_1 = Q_0 + hF(t_0, Q_0) = 0 + 0.1(12 - 4 \cdot 0) = 1.2$$

$$Q_2 = Q_1 + hF(t_1, Q_1) = 1.2 + 0.1(12 - 4 \cdot 1.2) = 1.92$$

$$Q_3 = Q_2 + hF(t_2, Q_2) = 1.92 + 0.1(12 - 4 \cdot 1.92) = 2.352$$

$$Q_4 = Q_3 + hF(t_3, Q_3) = 2.352 + 0.1(12 - 4 \cdot 2.352) = 2.6112$$

$$Q_5 = Q_4 + hF(t_4, Q_4) = 2.6112 + 0.1(12 - 4 \cdot 2.6112) = 2.76672$$

Thus, $Q_5 = Q(0.5) \approx 2.77$ C.

7.3 Separable Equations

1. $\dfrac{dy}{dx} = xy^2$ $\Rightarrow$ $\dfrac{dy}{y^2} = x\,dx$ $[y \neq 0]$ $\Rightarrow$ $\displaystyle\int y^{-2}\,dy = \int x\,dx$ $\Rightarrow$ $-y^{-1} = \tfrac{1}{2}x^2 + C$ $\Rightarrow$

$\dfrac{1}{y} = -\tfrac{1}{2}x^2 - C$ $\Rightarrow$ $y = \dfrac{1}{-\tfrac{1}{2}x^2 - C} = \dfrac{2}{K - x^2}$, where $K = -2C$. $y = 0$ is also a solution.

3. $(x^2 + 1)y' = xy$ $\Rightarrow$ $\dfrac{dy}{dx} = \dfrac{xy}{x^2 + 1}$ $\Rightarrow$ $\dfrac{dy}{y} = \dfrac{x\,dx}{x^2 + 1}$ $[y \neq 0]$ $\Rightarrow$ $\displaystyle\int \dfrac{dy}{y} = \int \dfrac{x\,dx}{x^2 + 1}$ $\Rightarrow$

$\ln|y| = \tfrac{1}{2}\ln(x^2 + 1) + C$ $[u = x^2 + 1, du = 2x\,dx]$ $= \ln(x^2 + 1)^{1/2} + \ln e^C = \ln\left(e^C \sqrt{x^2 + 1}\right)$ $\Rightarrow$

$|y| = e^C \sqrt{x^2 + 1}$ $\Rightarrow$ $y = K\sqrt{x^2 + 1}$, where $K = \pm e^C$ is a constant. (In our derivation, K was nonzero, but we can

restore the excluded case $y = 0$ by allowing K to be zero.)

5. $(y + \sin y)\,y' = x + x^3$ $\Rightarrow$ $(y + \sin y)\dfrac{dy}{dx} = x + x^3$ $\Rightarrow$ $\displaystyle\int (y + \sin y)\,dy = \int (x + x^3)\,dx$ $\Rightarrow$

$\tfrac{1}{2}y^2 - \cos y = \tfrac{1}{2}x^2 + \tfrac{1}{4}x^4 + C$. We cannot solve explicitly for y.

7. $\dfrac{dy}{dt} = \dfrac{te^t}{y\sqrt{1 + y^2}}$ $\Rightarrow$ $y\sqrt{1 + y^2}\,dy = te^t\,dt$ $\Rightarrow$ $\int y\sqrt{1 + y^2}\,dy = \int te^t\,dt$ $\Rightarrow$ $\tfrac{1}{3}(1 + y^2)^{3/2} = te^t - e^t + C$

[where the first integral is evaluated by substitution and the second by parts] $\Rightarrow$ $1 + y^2 = [3(te^t - e^t + C)]^{2/3}$ $\Rightarrow$

$y = \pm\sqrt{[3(te^t - e^t + C)]^{2/3} - 1}$

9. $\dfrac{du}{dt} = 2 + 2u + t + tu$ $\Rightarrow$ $\dfrac{du}{dt} = (1 + u)(2 + t)$ $\Rightarrow$ $\displaystyle\int \dfrac{du}{1 + u} = \int (2 + t)dt$ $[u \neq -1]$ $\Rightarrow$

$\ln|1 + u| = \tfrac{1}{2}t^2 + 2t + C$ $\Rightarrow$ $|1 + u| = e^{t^2/2 + 2t + C} = Ke^{t^2/2 + 2t}$, where $K = e^C$ $\Rightarrow$ $1 + u = \pm Ke^{t^2/2 + 2t}$ $\Rightarrow$

$u = -1 \pm Ke^{t^2/2 + 2t}$ where $K > 0$. $u = -1$ is also a solution, so $u = -1 + Ae^{t^2/2 + 2t}$, where A is an arbitrary constant.

11. $\dfrac{dy}{dx} = \dfrac{x}{y}$ $\Rightarrow$ $y\,dy = x\,dx$ $\Rightarrow$ $\int y\,dy = \int x\,dx$ $\Rightarrow$ $\tfrac{1}{2}y^2 = \tfrac{1}{2}x^2 + C$. $y(0) = -3$ $\Rightarrow$

$\tfrac{1}{2}(-3)^2 = \tfrac{1}{2}(0)^2 + C$ $\Rightarrow$ $C = \tfrac{9}{2}$, so $\tfrac{1}{2}y^2 = \tfrac{1}{2}x^2 + \tfrac{9}{2}$ $\Rightarrow$ $y^2 = x^2 + 9$ $\Rightarrow$ $y = -\sqrt{x^2 + 9}$ since $y(0) = -3 < 0$.

13. $\dfrac{du}{dt} = \dfrac{2t + \sec^2 t}{2u}$, $u(0) = -5$. $\int 2u\,du = \int (2t + \sec^2 t)\,dt$ $\Rightarrow$ $u^2 = t^2 + \tan t + C$,

where $[u(0)]^2 = 0^2 + \tan 0 + C$ $\Rightarrow$ $C = (-5)^2 = 25$. Therefore, $u^2 = t^2 + \tan t + 25$, so $u = \pm\sqrt{t^2 + \tan t + 25}$.

Since $u(0) = -5$, we must have $u = -\sqrt{t^2 + \tan t + 25}$.

15. $x \ln x = y\left(1 + \sqrt{3 + y^2}\right)y'$, $y(1) = 1$. $\int x \ln x\,dx = \int \left(y + y\sqrt{3 + y^2}\right)dy$ $\Rightarrow$ $\tfrac{1}{2}x^2 \ln x - \int \tfrac{1}{2}x\,dx$

[use parts with $u = \ln x$, $dv = xdx$] $= \tfrac{1}{2}y^2 + \tfrac{1}{3}(3 + y^2)^{3/2}$ $\Rightarrow$ $\tfrac{1}{2}x^2 \ln x - \tfrac{1}{4}x^2 + C = \tfrac{1}{2}y^2 + \tfrac{1}{3}(3 + y^2)^{3/2}$.

Now $y(1) = 1$ $\Rightarrow$ $0 - \tfrac{1}{4} + C = \tfrac{1}{2} + \tfrac{1}{3}(4)^{3/2}$ $\Rightarrow$ $C = \tfrac{1}{2} + \tfrac{8}{3} + \tfrac{1}{4} = \tfrac{41}{12}$, so

$\tfrac{1}{2}x^2 \ln x - \tfrac{1}{4}x^2 + \tfrac{41}{12} = \tfrac{1}{2}y^2 + \tfrac{1}{3}(3 + y^2)^{3/2}$. We do not solve explicitly for y.

17. $y' \tan x = a + y$, $0 < x < \pi/2$ $\Rightarrow$ $\dfrac{dy}{dx} = \dfrac{a+y}{\tan x}$ $\Rightarrow$ $\dfrac{dy}{a+y} = \cot x \, dx$ $[a + y \neq 0]$ $\Rightarrow$

$\displaystyle\int \dfrac{dy}{a+y} = \int \dfrac{\cos x}{\sin x} \, dx$ $\Rightarrow$ $\ln|a + y| = \ln|\sin x| + C$ $\Rightarrow$ $|a + y| = e^{\ln|\sin x|+C} = e^{\ln|\sin x|} \cdot e^C = e^C |\sin x|$ $\Rightarrow$

$a + y = K \sin x$, where $K = \pm e^C$. (In our derivation, K was nonzero, but we can restore the excluded case

$y = -a$ by allowing K to be zero.) $y(\pi/3) = a$ $\Rightarrow$ $a + a = K \sin\left(\dfrac{\pi}{3}\right)$ $\Rightarrow$ $2a = K \dfrac{\sqrt{3}}{2}$ $\Rightarrow$ $K = \dfrac{4a}{\sqrt{3}}$.

Thus, $a + y = \dfrac{4a}{\sqrt{3}} \sin x$ and so $y = \dfrac{4a}{\sqrt{3}} \sin x - a$.

19. If the slope at the point (x, y) is xy, then we have $\dfrac{dy}{dx} = xy$ $\Rightarrow$ $\dfrac{dy}{y} = x \, dx$ $[y \neq 0]$ $\Rightarrow$ $\displaystyle\int \dfrac{dy}{y} = \int x \, dx$ $\Rightarrow$

$\ln|y| = \frac{1}{2}x^2 + C$. $y(0) = 1$ $\Rightarrow$ $\ln 1 = 0 + C$ $\Rightarrow$ $C = 0$. Thus, $|y| = e^{x^2/2}$ $\Rightarrow$ $y = \pm e^{x^2/2}$, so $y = e^{x^2/2}$

since $y(0) = 1 > 0$. Note that $y = 0$ is not a solution because it doesn't satisfy the initial condition $y(0) = 1$.

21. $u = x + y$ $\Rightarrow$ $\dfrac{d}{dx}(u) = \dfrac{d}{dx}(x + y)$ $\Rightarrow$ $\dfrac{du}{dx} = 1 + \dfrac{dy}{dx}$, but $\dfrac{dy}{dx} = x + y = u$, so $\dfrac{du}{dx} = 1 + u$ $\Rightarrow$

$\dfrac{du}{1+u} = dx$ $[u \neq -1]$ $\Rightarrow$ $\displaystyle\int \dfrac{du}{1+u} = \int dx$ $\Rightarrow$ $\ln|1 + u| = x + C$ $\Rightarrow$ $|1 + u| = e^{x+C}$ $\Rightarrow$

$1 + u = \pm e^C e^x$ $\Rightarrow$ $u = \pm e^C e^x - 1$ $\Rightarrow$ $x + y = \pm e^C e^x - 1$ $\Rightarrow$ $y = K e^x - x - 1$, where $K = \pm e^C \neq 0$.

If $u = -1$, then $-1 = x + y$ $\Rightarrow$ $y = -x - 1$, which is just $y = K e^x - x - 1$ with $K = 0$. Thus, the general solution

is $y = K e^x - x - 1$, where $K \in \mathbb{R}$.

23. (a) $y' = 2x \sqrt{1 - y^2}$ $\Rightarrow$ $\dfrac{dy}{dx} = 2x \sqrt{1 - y^2}$ $\Rightarrow$ $\dfrac{dy}{\sqrt{1 - y^2}} = 2x \, dx$ $\Rightarrow$ $\displaystyle\int \dfrac{dy}{\sqrt{1 - y^2}} = \int 2x \, dx$ $\Rightarrow$

$\sin^{-1} y = x^2 + C$ for $-\frac{\pi}{2} \leq x^2 + C \leq \frac{\pi}{2}$.

(b) $y(0) = 0$ $\Rightarrow$ $\sin^{-1} 0 = 0^2 + C$ $\Rightarrow$ $C = 0$,

so $\sin^{-1} y = x^2$ and $y = \sin(x^2)$ for $-\sqrt{\pi/2} \leq x \leq \sqrt{\pi/2}$.

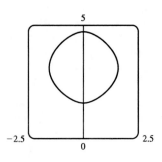

(c) For $\sqrt{1 - y^2}$ to be a real number, we must have $-1 \leq y \leq 1$; that is, $-1 \leq y(0) \leq 1$. Thus, the initial-value problem

$y' = 2x \sqrt{1 - y^2}$, $y(0) = 2$ does *not* have a solution.

25. $\dfrac{dy}{dx} = \dfrac{\sin x}{\sin y}$, $y(0) = \dfrac{\pi}{2}$. So $\int \sin y \, dy = \int \sin x \, dx$ $\Leftrightarrow$

$-\cos y = -\cos x + C$ $\Leftrightarrow$ $\cos y = \cos x - C$. From the initial condition,

we need $\cos \frac{\pi}{2} = \cos 0 - C$ $\Rightarrow$ $0 = 1 - C$ $\Rightarrow$ $C = 1$, so the solution is

$\cos y = \cos x - 1$. Note that we cannot take $\cos^{-1}$ of both sides, since that would

unnecessarily restrict the solution to the case where $-1 \leq \cos x - 1$ $\Leftrightarrow$ $0 \leq \cos x$,

as $\cos^{-1}$ is defined only on $[-1, 1]$. Instead we plot the graph using Maple's

plots[implicitplot] or Mathematica's Plot[Evaluate[···]].

27. (a) , (c)

(b) $y' = y^2 \Rightarrow \dfrac{dy}{dx} = y^2 \Rightarrow \displaystyle\int y^{-2}\,dy = \int dx \Rightarrow$

$-y^{-1} = x + C \Rightarrow \dfrac{1}{y} = -x - C \Rightarrow$

$y = \dfrac{1}{K - x}$, where $K = -C$. $y = 0$ is also a solution.

29. The curves $x^2 + 2y^2 = k^2$ form a family of ellipses with major axis on the x-axis. Differentiating gives

$\dfrac{d}{dx}\left(x^2 + 2y^2\right) = \dfrac{d}{dx}\left(k^2\right) \Rightarrow 2x + 4yy' = 0 \Rightarrow 4yy' = -2x \Rightarrow y' = \dfrac{-x}{2y}$. Thus, the slope of the tangent line

at any point (x, y) on one of the ellipses is $y' = \dfrac{-x}{2y}$, so the orthogonal trajectories

must satisfy $y' = \dfrac{2y}{x} \Leftrightarrow \dfrac{dy}{dx} = \dfrac{2y}{x} \Leftrightarrow \dfrac{dy}{y} = 2 = \dfrac{dx}{x} \Leftrightarrow$

$\displaystyle\int \dfrac{dy}{y} = 2\int \dfrac{dx}{x} \Leftrightarrow \ln|y| = 2\ln|x| + C_1 \Leftrightarrow \ln|y| = \ln|x|^2 + C_1 \Leftrightarrow$

$|y| = e^{\ln x^2 + C_1} \Leftrightarrow y = \pm x^2 \cdot e^{C_1} = Cx^2$. This is a family of parabolas.

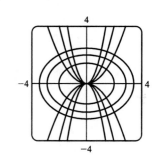

31. The curves $y = k/x$ form a family of hyperbolas with asymptotes $x = 0$ and $y = 0$. Differentiating gives

$\dfrac{d}{dx}(y) = \dfrac{d}{dx}\left(\dfrac{k}{x}\right) \Rightarrow y' = -\dfrac{k}{x^2} \Rightarrow y' = -\dfrac{xy}{x^2}$ [since $y = k/x \Rightarrow xy = k$] $\Rightarrow y' = -\dfrac{y}{x}$. Thus, the slope

of the tangent line at any point (x, y) on one of the hyperbolas is $y' = -y/x$,

so the orthogonal trajectories must satisfy $y' = x/y \Leftrightarrow \dfrac{dy}{dx} = \dfrac{x}{y} \Leftrightarrow$

$y\,dy = x\,dx \Leftrightarrow \displaystyle\int y\,dy = \int x\,dx \Leftrightarrow \tfrac{1}{2}y^2 = \tfrac{1}{2}x^2 + C_1 \Leftrightarrow$

$y^2 = x^2 + C_2 \Leftrightarrow x^2 - y^2 = C$. This is a family of hyperbolas with

asymptotes $y = \pm x$.

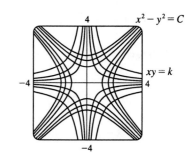

33. $y(x) = 2 + \displaystyle\int_2^x [t - ty(t)]\,dt \Rightarrow y'(x) = x - xy(x)$ [by FTC 1] $\Rightarrow \dfrac{dy}{dx} = x(1 - y) \Rightarrow$

$\displaystyle\int \dfrac{dy}{1 - y} = \int x\,dx \Rightarrow -\ln|1 - y| = \tfrac{1}{2}x^2 + C$. Letting $x = 2$ in the original integral equation

gives us $y(2) = 2 + 0 = 2$. Thus, $-\ln|1 - 2| = \tfrac{1}{2}(2)^2 + C \Rightarrow 0 = 2 + C \Rightarrow C = -2$.

Thus, $-\ln|1 - y| = \tfrac{1}{2}x^2 - 2 \Rightarrow \ln|1 - y| = 2 - \tfrac{1}{2}x^2 \Rightarrow |1 - y| = e^{2 - x^2/2} \Rightarrow$

$1 - y = \pm e^{2 - x^2/2} \Rightarrow y = 1 + e^{2 - x^2/2}$ [$y(2) = 2$].

35. $y(x) = 4 + \int_0^x 2t\sqrt{y(t)}\,dt \;\Rightarrow\; y'(x) = 2x\sqrt{y(x)} \;\Rightarrow\; \dfrac{dy}{dx} = 2x\sqrt{y} \;\Rightarrow\; \int \dfrac{dy}{\sqrt{y}} = \int 2x\,dx \;\Rightarrow$

$2\sqrt{y} = x^2 + C$. Letting $x = 0$ in the original integral equation gives us $y(0) = 4 + 0 = 4$.

Thus, $2\sqrt{4} = 0^2 + C \;\Rightarrow\; C = 4.\; 2\sqrt{y} = x^2 + 4 \;\Rightarrow\; \sqrt{y} = \frac{1}{2}x^2 + 2 \;\Rightarrow\; y = \left(\frac{1}{2}x^2 + 2\right)^2.$

37. From Exercise 7.2.27, $\dfrac{dQ}{dt} = 12 - 4Q \;\Leftrightarrow\; \int \dfrac{dQ}{12 - 4Q} = \int dt \;\Leftrightarrow\; -\frac{1}{4}\ln|12 - 4Q| = t + C \;\Leftrightarrow$

$\ln|12 - 4Q| = -4t - 4C \;\Leftrightarrow\; |12 - 4Q| = e^{-4t - 4C} \;\Leftrightarrow\; 12 - 4Q = Ke^{-4t}\; [K = \pm e^{-4C}] \;\Leftrightarrow$

$4Q = 12 - Ke^{-4t} \;\Leftrightarrow\; Q = 3 - Ae^{-4t}\; [A = K/4].\; Q(0) = 0 \;\Leftrightarrow\; 0 = 3 - A \;\Leftrightarrow\; A = 3 \;\Leftrightarrow$

$Q(t) = 3 - 3e^{-4t}$. As $t \to \infty$, $Q(t) \to 3 - 0 = 3$ (the limiting value).

39. $\dfrac{dP}{dt} = k(M - P) \;\Leftrightarrow\; \int \dfrac{dP}{P - M} = \int (-k)\,dt \;\Leftrightarrow\; \ln|P - M| = -kt + C \;\Leftrightarrow\; |P - M| = e^{-kt + C} \;\Leftrightarrow$

$P - M = Ae^{-kt}\; [A = \pm e^C] \;\Leftrightarrow\; P = M + Ae^{-kt}$. If we assume that performance is at level 0 when $t = 0$, then

$P(0) = 0 \;\Leftrightarrow\; 0 = M + A \;\Leftrightarrow\; A = -M \;\Leftrightarrow\; P(t) = M - Me^{-kt}.\;\; \lim\limits_{t \to \infty} P(t) = M - M \cdot 0 = M.$

41. (a) If $a = b$, then $\dfrac{dx}{dt} = k(a - x)(b - x)^{1/2}$ becomes $\dfrac{dx}{dt} = k(a - x)^{3/2} \;\Rightarrow\; (a - x)^{-3/2}\,dx = k\,dt \;\Rightarrow$

$\int (a - x)^{-3/2}\,dx = \int k\,dt \;\Rightarrow\; 2(a - x)^{-1/2} = kt + C$ [by substitution] $\;\Rightarrow\; \dfrac{2}{kt + C} = \sqrt{a - x} \;\Rightarrow$

$\left(\dfrac{2}{kt + C}\right)^2 = a - x \;\Rightarrow\; x(t) = a - \dfrac{4}{(kt + C)^2}$. The initial concentration of HBr is 0, so $x(0) = 0 \;\Rightarrow$

$0 = a - \dfrac{4}{C^2} \;\Rightarrow\; \dfrac{4}{C^2} = a \;\Rightarrow\; C^2 = \dfrac{4}{a} \;\Rightarrow\; C = 2/\sqrt{a}$ $[C$ is positive since $kt + C = 2(a - x)^{-1/2} > 0]$.

Thus, $x(t) = a - \dfrac{4}{(kt + 2/\sqrt{a})^2}$.

(b) $\dfrac{dx}{dt} = k(a - x)(b - x)^{1/2} \;\Rightarrow\; \dfrac{dx}{(a - x)\sqrt{b - x}} = k\,dt \;\Rightarrow\; \int \dfrac{dx}{(a - x)\sqrt{b - x}} = \int k\,dt$ $(\star)$.

From the hint, $u = \sqrt{b - x} \;\Rightarrow\; u^2 = b - x \;\Rightarrow\; 2u\,du = -dx$, so

$$\int \dfrac{dx}{(a - x)\sqrt{b - x}} = \int \dfrac{-2u\,du}{[a - (b - u^2)]u} = -2\int \dfrac{du}{a - b + u^2} = -2\int \dfrac{du}{\left(\sqrt{a - b}\right)^2 + u^2}$$

$$\overset{17}{=} -2\left(\dfrac{1}{\sqrt{a - b}}\tan^{-1}\dfrac{u}{\sqrt{a - b}}\right)$$

So $(\star)$ becomes $\dfrac{-2}{\sqrt{a - b}}\tan^{-1}\dfrac{\sqrt{b - x}}{\sqrt{a - b}} = kt + C$. Now $x(0) = 0 \;\Rightarrow\; C = \dfrac{-2}{\sqrt{a - b}}\tan^{-1}\dfrac{\sqrt{b}}{\sqrt{a - b}}$ and we have

$\dfrac{-2}{\sqrt{a - b}}\tan^{-1}\dfrac{\sqrt{b - x}}{\sqrt{a - b}} = kt - \dfrac{2}{\sqrt{a - b}}\tan^{-1}\dfrac{\sqrt{b}}{\sqrt{a - b}} \;\Rightarrow\; \dfrac{2}{\sqrt{a - b}}\left(\tan^{-1}\sqrt{\dfrac{b}{a - b}} - \tan^{-1}\sqrt{\dfrac{b - x}{a - b}}\right) = kt \;\Rightarrow$

$t(x) = \dfrac{2}{k\sqrt{a - b}}\left(\tan^{-1}\sqrt{\dfrac{b}{a - b}} - \tan^{-1}\sqrt{\dfrac{b - x}{a - b}}\right).$

43. (a) $\dfrac{dC}{dt} = r - kC \;\Rightarrow\; \dfrac{dC}{dt} = -(kC - r) \;\Rightarrow\; \displaystyle\int \dfrac{dC}{kC - r} = \int -dt \;\Rightarrow\; (1/k)\ln|kC - r| = -t + M_1 \;\Rightarrow$

$\ln|kC - r| = -kt + M_2 \;\Rightarrow\; |kC - r| = e^{-kt + M_2} \;\Rightarrow\; kC - r = M_3 e^{-kt} \;\Rightarrow\; kC = M_3 e^{-kt} + r \;\Rightarrow$

$C(t) = M_4 e^{-kt} + r/k.\; C(0) = C_0 \;\Rightarrow\; C_0 = M_4 + r/k \;\Rightarrow\; M_4 = C_0 - r/k \;\Rightarrow$

$C(t) = (C_0 - r/k)e^{-kt} + r/k.$

(b) If $C_0 < r/k$, then $C_0 - r/k < 0$ and the formula for $C(t)$ shows that $C(t)$ increases and $\lim\limits_{t \to \infty} C(t) = r/k$.

As t increases, the formula for $C(t)$ shows how the role of C_0 steadily diminishes as that of r/k increases.

45. (a) Let $y(t)$ be the amount of salt (in kg) after t minutes. Then $y(0) = 15$. The amount of liquid in the tank is 1000 L at all

times, so the concentration at time t (in minutes) is $y(t)/1000$ kg/L and $\dfrac{dy}{dt} = -\left[\dfrac{y(t)}{1000}\,\dfrac{\text{kg}}{\text{L}}\right]\left(10\,\dfrac{\text{L}}{\text{min}}\right) = -\dfrac{y(t)}{100}\,\dfrac{\text{kg}}{\text{min}}.$

$\displaystyle\int \dfrac{dy}{y} = -\dfrac{1}{100}\int dt \;\Rightarrow\; \ln y = -\dfrac{t}{100} + C,$ and $y(0) = 15 \;\Rightarrow\; \ln 15 = C,$ so $\ln y = \ln 15 - \dfrac{t}{100}.$

It follows that $\ln\!\left(\dfrac{y}{15}\right) = -\dfrac{t}{100}$ and $\dfrac{y}{15} = e^{-t/100},$ so $y = 15e^{-t/100}$ kg.

(b) After 20 minutes, $y = 15e^{-20/100} = 15e^{-0.2} \approx 12.3$ kg.

47. Let $y(t)$ be the amount of alcohol in the vat after t minutes. Then $y(0) = 0.04(500) = 20$ gal. The amount of beer in the vat

is 500 gallons at all times, so the percentage at time t (in minutes) is $y(t)/500 \times 100$, and the change in the amount of alcohol

with respect to time t is $\dfrac{dy}{dt} = \text{rate in} - \text{rate out} = 0.06\left(5\,\dfrac{\text{gal}}{\text{min}}\right) - \dfrac{y(t)}{500}\left(5\,\dfrac{\text{gal}}{\text{min}}\right) = 0.3 - \dfrac{y}{100} = \dfrac{30 - y}{100}\,\dfrac{\text{gal}}{\text{min}}.$

Hence, $\displaystyle\int \dfrac{dy}{30 - y} = \int \dfrac{dt}{100}$ and $-\ln|30 - y| = \tfrac{1}{100}t + C.$ Because $y(0) = 20$, we have $-\ln 10 = C$, so

$-\ln|30 - y| = \tfrac{1}{100}t - \ln 10 \;\Rightarrow\; \ln|30 - y| = -t/100 + \ln 10 \;\Rightarrow\; \ln|30 - y| = \ln e^{-t/100} + \ln 10 \;\Rightarrow$

$\ln|30 - y| = \ln(10e^{-t/100}) \;\Rightarrow\; |30 - y| = 10e^{-t/100}.$ Since y is continuous, $y(0) = 20$, and the right-hand side is

never zero, we deduce that $30 - y$ is always positive. Thus, $30 - y = 10e^{-t/100} \;\Rightarrow\; y = 30 - 10e^{-t/100}.$ The

percentage of alcohol is $p(t) = y(t)/500 \times 100 = y(t)/5 = 6 - 2e^{-t/100}.$ The percentage of alcohol after one hour is

$p(60) = 6 - 2e^{-60/100} \approx 4.9.$

49. Assume that the raindrop begins at rest, so that $v(0) = 0$. $dm/dt = km$ and $(mv)' = gm \;\Rightarrow\; mv' + vm' = gm \;\Rightarrow$

$mv' + v(km) = gm \;\Rightarrow\; v' + vk = g \;\Rightarrow\; \dfrac{dv}{dt} = g - kv \;\Rightarrow\; \displaystyle\int \dfrac{dv}{g - kv} = \int dt \;\Rightarrow$

$-(1/k)\ln|g - kv| = t + C \;\Rightarrow\; \ln|g - kv| = -kt - kC \;\Rightarrow\; g - kv = Ae^{-kt}.\; v(0) = 0 \;\Rightarrow\; A = g.$

So $kv = g - ge^{-kt} \;\Rightarrow\; v = (g/k)(1 - e^{-kt}).$ Since $k > 0$, as $t \to \infty$, $e^{-kt} \to 0$ and therefore, $\lim\limits_{t \to \infty} v(t) = g/k.$

51. (a) $\dfrac{1}{L_1}\dfrac{dL_1}{dt} = k\dfrac{1}{L_2}\dfrac{dL_2}{dt} \;\Rightarrow\; \dfrac{d}{dt}(\ln L_1) = \dfrac{d}{dt}(k\ln L_2) \;\Rightarrow\; \displaystyle\int \dfrac{d}{dt}(\ln L_1)\,dt = \int \dfrac{d}{dt}(\ln L_2^k)\,dt \;\Rightarrow$

$\ln L_1 = \ln L_2^k + C \;\Rightarrow\; L_1 = e^{\ln L_2^k + C} = e^{\ln L_2^k}e^C \;\Rightarrow\; L_1 = KL_2^k,$ where $K = e^C.$

(b) From part (a) with $L_1 = B$, $L_2 = V$, and $k = 0.0794$, we have $B = KV^{0.0794}.$

53. (a) The rate of growth of the area is jointly proportional to $\sqrt{A(t)}$ and $M - A(t)$; that is, the rate is proportional to the product of those two quantities. So for some constant k, $dA/dt = k\sqrt{A}(M - A)$. We are interested in the maximum of the function dA/dt (when the tissue grows the fastest), so we differentiate, using the Chain Rule and then substituting for dA/dt from the differential equation:

$$\frac{d}{dt}\left(\frac{dA}{dt}\right) = k\left[\sqrt{A}(-1)\frac{dA}{dt} + (M - A)\cdot\tfrac{1}{2}A^{-1/2}\frac{dA}{dt}\right] = \tfrac{1}{2}kA^{-1/2}\frac{dA}{dt}[-2A + (M - A)]$$

$$= \tfrac{1}{2}kA^{-1/2}\left[k\sqrt{A}(M - A)\right][M - 3A] = \tfrac{1}{2}k^2(M - A)(M - 3A)$$

This is 0 when $M - A = 0$ [this situation never actually occurs, since the graph of $A(t)$ is asymptotic to the line $y = M$, as in the logistic model] and when $M - 3A = 0 \;\Leftrightarrow\; A(t) = M/3$. This represents a maximum by the First Derivative Test, since $\dfrac{d}{dt}\left(\dfrac{dA}{dt}\right)$ goes from positive to negative when $A(t) = M/3$.

(b) From the CAS, we get $A(t) = M\left(\dfrac{Ce^{\sqrt{M}kt} - 1}{Ce^{\sqrt{M}kt} + 1}\right)^2$. To get C in terms of the initial area A_0 and the maximum area M, we substitute $t = 0$ and $A = A_0 = A(0)$: $A_0 = M\left(\dfrac{C - 1}{C + 1}\right)^2 \;\Leftrightarrow\; (C + 1)\sqrt{A_0} = (C - 1)\sqrt{M} \;\Leftrightarrow$

$C\sqrt{A_0} + \sqrt{A_0} = C\sqrt{M} - \sqrt{M} \;\Leftrightarrow\; \sqrt{M} + \sqrt{A_0} = C\sqrt{M} - C\sqrt{A_0} \;\Leftrightarrow$

$\sqrt{M} + \sqrt{A_0} = C\left(\sqrt{M} - \sqrt{A_0}\right) \;\Leftrightarrow\; C = \dfrac{\sqrt{M} + \sqrt{A_0}}{\sqrt{M} - \sqrt{A_0}}$. [Notice that if $A_0 = 0$, then $C = 1$.]

7.4 Exponential Growth and Decay

1. The relative growth rate is $\dfrac{1}{P}\dfrac{dP}{dt} = 0.7944$, so $\dfrac{dP}{dt} = 0.7944P$ and, by Theorem 2, $P(t) = P(0)e^{0.7944t} = 2e^{0.7944t}$.

Thus, $P(6) = 2e^{0.7944(6)} \approx 234.99$ or about 235 members.

3. (a) By Theorem 2, $P(t) = P(0)e^{kt} = 100e^{kt}$. Now $P(1) = 100e^{k(1)} = 420 \;\Rightarrow\; e^k = \frac{420}{100} \;\Rightarrow\; k = \ln 4.2$.

So $P(t) = 100e^{(\ln 4.2)t} = 100(4.2)^t$.

(b) $P(3) = 100(4.2)^3 = 7408.8 \approx 7409$ bacteria

(c) $dP/dt = kP \;\Rightarrow\; P'(3) = k\cdot P(3) = (\ln 4.2)\left(100(4.2)^3\right)$ [from part (a)] $\approx 10{,}632$ bacteria/h

(d) $P(t) = 100(4.2)^t = 10{,}000 \;\Rightarrow\; (4.2)^t = 100 \;\Rightarrow\; t = (\ln 100)/(\ln 4.2) \approx 3.2$ hours

5. (a) Let the population (in millions) in the year t be $P(t)$. Since the initial time is the year 1750, we substitute $t - 1750$ for t in

Theorem 2, so the exponential model gives $P(t) = P(1750)e^{k(t-1750)}$. Then $P(1800) = 980 = 790e^{k(1800-1750)}$ ⇒

$\frac{980}{790} = e^{k(50)}$ ⇒ $\ln \frac{980}{790} = 50k$ ⇒ $k = \frac{1}{50} \ln \frac{980}{790} \approx 0.0043104$. So with this model, we have

$P(1900) = 790e^{k(1900-1750)} \approx 1508$ million, and $P(1950) = 790e^{k(1950-1750)} \approx 1871$ million. Both of these

estimates are much too low.

(b) In this case, the exponential model gives $P(t) = P(1850)e^{k(t-1850)}$ ⇒ $P(1900) = 1650 = 1260e^{k(1900-1850)}$ ⇒

$\ln \frac{1650}{1260} = k(50)$ ⇒ $k = \frac{1}{50} \ln \frac{1650}{1260} \approx 0.005393$. So with this model, we estimate

$P(1950) = 1260e^{k(1950-1850)} \approx 2161$ million. This is still too low, but closer than the estimate of $P(1950)$ in part (a).

(c) The exponential model gives $P(t) = P(1900)e^{k(t-1900)}$ ⇒ $P(1950) = 2560 = 1650e^{k(1950-1900)}$ ⇒

$\ln \frac{2560}{1650} = k(50)$ ⇒ $k = \frac{1}{50} \ln \frac{2560}{1650} \approx 0.008785$. With this model, we estimate

$P(2000) = 1650e^{k(2000-1900)} \approx 3972$ million. This is much too low. The discrepancy is explained by the fact that the

world birth rate (average yearly number of births per person) is about the same as always, whereas the mortality rate

(especially the infant mortality rate) is much lower, owing mostly to advances in medical science and to the wars in the first

part of the twentieth century. The exponential model assumes, among other things, that the birth and mortality rates will

remain constant.

7. (a) If $y = [N_2O_5]$ then by Theorem 2, $\dfrac{dy}{dt} = -0.0005y$ ⇒ $y(t) = y(0)e^{-0.0005t} = Ce^{-0.0005t}$.

(b) $y(t) = Ce^{-0.0005t} = 0.9C$ ⇒ $e^{-0.0005t} = 0.9$ ⇒ $-0.0005t = \ln 0.9$ ⇒ $t = -2000 \ln 0.9 \approx 211$ s

9. (a) If $y(t)$ is the mass (in mg) remaining after t years, then $y(t) = y(0)e^{kt} = 100e^{kt}$.

$y(30) = 100e^{30k} = \frac{1}{2}(100)$ ⇒ $e^{30k} = \frac{1}{2}$ ⇒ $k = -(\ln 2)/30$ ⇒ $y(t) = 100e^{-(\ln 2)t/30} = 100 \cdot 2^{-t/30}$

(b) $y(100) = 100 \cdot 2^{-100/30} \approx 9.92$ mg

(c) $100e^{-(\ln 2)t/30} = 1$ ⇒ $-(\ln 2)t/30 = \ln \frac{1}{100}$ ⇒ $t = -30 \frac{\ln 0.01}{\ln 2} \approx 199.3$ years

11. Let $y(t)$ be the level of radioactivity. Thus, $y(t) = y(0)e^{-kt}$ and k is determined by using the half-life:

$y(5730) = \frac{1}{2}y(0)$ ⇒ $y(0)e^{-k(5730)} = \frac{1}{2}y(0)$ ⇒ $e^{-5730k} = \frac{1}{2}$ ⇒ $-5730k = \ln \frac{1}{2}$ ⇒ $k = -\dfrac{\ln \frac{1}{2}}{5730} = \dfrac{\ln 2}{5730}$.

If 74% of the ^{14}C remains, then we know that $y(t) = 0.74y(0)$ ⇒ $0.74 = e^{-t(\ln 2)/5730}$ ⇒ $\ln 0.74 = -\dfrac{t \ln 2}{5730}$ ⇒

$t = -\dfrac{5730(\ln 0.74)}{\ln 2} \approx 2489 \approx 2500$ years.

13. (a) Using Newton's Law of Cooling, $\frac{dT}{dt} = k(T - T_s)$, we have $\frac{dT}{dt} = k(T - 75)$. Now let $y = T - 75$, so

$y(0) = T(0) - 75 = 185 - 75 = 110$, so y is a solution of the initial-value problem $dy/dt = ky$ with $y(0) = 110$ and by

Theorem 2 we have $y(t) = y(0)e^{kt} = 110e^{kt}$.

$y(30) = 110e^{30k} = 150 - 75 \quad\Rightarrow\quad e^{30k} = \frac{75}{110} = \frac{15}{22} \quad\Rightarrow\quad k = \frac{1}{30}\ln\frac{15}{22}$, so $y(t) = 110e^{\frac{1}{30}t\ln\left(\frac{15}{22}\right)}$ and

$y(45) = 110e^{\frac{45}{30}\ln\left(\frac{15}{22}\right)} \approx 62°\text{F}$. Thus, $T(45) \approx 62 + 75 = 137°\text{F}$.

(b) $T(t) = 100 \quad\Rightarrow\quad y(t) = 25$. $y(t) = 110e^{\frac{1}{30}t\ln\left(\frac{15}{22}\right)} = 25 \quad\Rightarrow\quad e^{\frac{1}{30}t\ln\left(\frac{15}{22}\right)} = \frac{25}{110} \quad\Rightarrow\quad \frac{1}{30}t\ln\frac{15}{22} = \ln\frac{25}{110} \quad\Rightarrow$

$t = \dfrac{30\ln\frac{25}{110}}{\ln\frac{15}{22}} \approx 116$ min.

15. $\dfrac{dT}{dt} = k(T - 20)$. Letting $y = T - 20$, we get $\dfrac{dy}{dt} = ky$, so $y(t) = y(0)e^{kt}$. $\quad y(0) = T(0) - 20 = 5 - 20 = -15$, so

$y(25) = y(0)e^{25k} = -15e^{25k}$, and $y(25) = T(25) - 20 = 10 - 20 = -10$, so $-15e^{25k} = -10 \quad\Rightarrow\quad e^{25k} = \frac{2}{3}$. Thus,

$25k = \ln\left(\frac{2}{3}\right)$ and $k = \frac{1}{25}\ln\left(\frac{2}{3}\right)$, so $y(t) = y(0)e^{kt} = -15e^{(1/25)\ln(2/3)t}$. More simply, $e^{25k} = \frac{2}{3} \quad\Rightarrow\quad e^k = \left(\frac{2}{3}\right)^{1/25} \quad\Rightarrow$

$e^{kt} = \left(\frac{2}{3}\right)^{t/25} \quad\Rightarrow\quad y(t) = -15 \cdot \left(\frac{2}{3}\right)^{t/25}$.

(a) $T(50) = 20 + y(50) = 20 - 15 \cdot \left(\frac{2}{3}\right)^{50/25} = 20 - 15 \cdot \left(\frac{2}{3}\right)^2 = 20 - \frac{20}{3} = 13.\overline{3}\,°\text{C}$

(b) $15 = T(t) = 20 + y(t) = 20 - 15 \cdot \left(\frac{2}{3}\right)^{t/25} \quad\Rightarrow\quad 15 \cdot \left(\frac{2}{3}\right)^{t/25} = 5 \quad\Rightarrow\quad \left(\frac{2}{3}\right)^{t/25} = \frac{1}{3} \quad\Rightarrow$

$(t/25)\ln\left(\frac{2}{3}\right) = \ln\left(\frac{1}{3}\right) \quad\Rightarrow\quad t = 25\ln\left(\frac{1}{3}\right)/\ln\left(\frac{2}{3}\right) \approx 67.74$ min.

17. (a) Let $P(h)$ be the pressure at altitude h. Then $dP/dh = kP \quad\Rightarrow\quad P(h) = P(0)e^{kh} = 101.3e^{kh}$.

$P(1000) = 101.3e^{1000k} = 87.14 \quad\Rightarrow\quad 1000k = \ln\left(\frac{87.14}{101.3}\right) \quad\Rightarrow\quad k = \frac{1}{1000}\ln\left(\frac{87.14}{101.3}\right) \quad\Rightarrow$

$P(h) = 101.3\, e^{\frac{1}{1000}h\ln\left(\frac{87.14}{101.3}\right)}$, so $P(3000) = 101.3e^{3\ln\left(\frac{87.14}{101.3}\right)} \approx 64.5$ kPa.

(b) $P(6187) = 101.3\, e^{\frac{6187}{1000}\ln\left(\frac{87.14}{101.3}\right)} \approx 39.9$ kPa

19. (a) Using $A = A_0\left(1 + \dfrac{r}{n}\right)^{nt}$ with $A_0 = 3000$, $r = 0.05$, and $t = 5$, we have:

    (i) Annually: $n = 1$; $A = 3000\left(1 + \frac{0.05}{1}\right)^{1\cdot5} = \$3828.84$

    (ii) Semiannually: $n = 2$; $A = 3000\left(1 + \frac{0.05}{2}\right)^{2\cdot5} = \$3840.25$

    (iii) Monthly: $n = 12$; $A = 3000\left(1 + \frac{0.05}{12}\right)^{12\cdot5} = \$3850.08$

    (iv) Weekly: $n = 52$; $A = 3000\left(1 + \frac{0.05}{52}\right)^{52\cdot5} = \$3851.61$

    (v) Daily: $n = 365$; $A = 3000\left(1 + \frac{0.05}{365}\right)^{365\cdot5} = \$3852.01$

    (vi) Continuously: $A = 3000e^{(0.05)5} = \$3852.08$

(b) $dA/dt = 0.05A$ and $A(0) = 3000$.

21. (a) $\dfrac{dP}{dt} = kP - m = k\left(P - \dfrac{m}{k}\right)$. Let $y = P - \dfrac{m}{k}$, so $\dfrac{dy}{dt} = \dfrac{dP}{dt}$ and the differential equation becomes $\dfrac{dy}{dt} = ky$.

The solution is $y = y_0 e^{kt}$ $\Rightarrow$ $P - \dfrac{m}{k} = \left(P_0 - \dfrac{m}{k}\right)e^{kt}$ $\Rightarrow$ $P(t) = \dfrac{m}{k} + \left(P_0 - \dfrac{m}{k}\right)e^{kt}$.

(b) Since $k > 0$, there will be an exponential expansion $\Leftrightarrow$ $P_0 - \dfrac{m}{k} > 0$ $\Leftrightarrow$ $m < kP_0$.

(c) The population will be constant if $P_0 - \dfrac{m}{k} = 0$ $\Leftrightarrow$ $m = kP_0$. It will decline if $P_0 - \dfrac{m}{k} < 0$ $\Leftrightarrow$ $m > kP_0$.

(d) $P_0 = 8{,}000{,}000$, $k = \alpha - \beta = 0.016$, $m = 210{,}000$ $\Rightarrow$ $m > kP_0\ (= 128{,}000)$, so by part (c), the population was declining.

7.5   The Logistic Equation

1. (a) $dP/dt = 0.05P - 0.0005P^2 = 0.05P(1 - 0.01P) = 0.05P(1 - P/100)$. Comparing to Equation 4,

$dP/dt = kP(1 - P/M)$, we see that the carrying capacity is $M = 100$ and the value of k is 0.05.

(b) The slopes close to 0 occur where P is near 0 or 100. The largest slopes appear to be on the line $P = 50$. The solutions are increasing for $0 < P_0 < 100$ and decreasing for $P_0 > 100$.

(c) 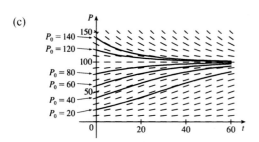 All of the solutions approach $P = 100$ as t increases. As in part (b), the solutions differ since for $0 < P_0 < 100$ they are increasing, and for $P_0 > 100$ they are decreasing. Also, some have an IP and some don't. It appears that the solutions which have $P_0 = 20$ and $P_0 = 40$ have inflection points at $P = 50$.

(d) The equilibrium solutions are $P = 0$ (trivial solution) and $P = 100$. The increasing solutions move away from $P = 0$ and all nonzero solutions approach $P = 100$ as $t \to \infty$.

3. (a) $\dfrac{dy}{dt} = ky\left(1 - \dfrac{y}{M}\right)$ $\Rightarrow$ $y(t) = \dfrac{M}{1 + Ae^{-kt}}$ with $A = \dfrac{M - y(0)}{y(0)}$. With $M = 8 \times 10^7$, $k = 0.71$, and

$y(0) = 2 \times 10^7$, we get the model $y(t) = \dfrac{8 \times 10^7}{1 + 3e^{-0.71t}}$, so $y(1) = \dfrac{8 \times 10^7}{1 + 3e^{-0.71}} \approx 3.23 \times 10^7$ kg.

(b) $y(t) = 4 \times 10^7$ $\Rightarrow$ $\dfrac{8 \times 10^7}{1 + 3e^{-0.71t}} = 4 \times 10^7$ $\Rightarrow$ $2 = 1 + 3e^{-0.71t}$ $\Rightarrow$ $e^{-0.71t} = \dfrac{1}{3}$ $\Rightarrow$

$-0.71t = \ln\tfrac{1}{3}$ $\Rightarrow$ $t = \dfrac{\ln 3}{0.71} \approx 1.55$ years

5. Using (4), $A = \dfrac{M - P_0}{P_0} = \dfrac{10{,}000 - 1000}{1000} = 9$, so $P(t) = \dfrac{10{,}000}{1 + 9e^{-kt}}$. $P(1) = 2500$ $\Rightarrow$ $2500 = \dfrac{10{,}000}{1 + 9e^{-k(1)}}$ $\Rightarrow$

$1 + 9e^{-k} = 4$ $\Rightarrow$ $9e^{-k} = 3$ $\Rightarrow$ $e^{-k} = \tfrac{1}{3}$ $\Rightarrow$ $-k = \ln\tfrac{1}{3}$ $\Rightarrow$ $k = \ln 3$. After another three years, $t = 4$,

and $P(4) = \dfrac{10{,}000}{1 + 9e^{-(\ln 3)4}} = \dfrac{10{,}000}{1 + 9\left(e^{\ln 3}\right)^{-4}} = \dfrac{10{,}000}{1 + 9(3)^{-4}} = \dfrac{10{,}000}{1 + \tfrac{1}{9}} = \dfrac{10{,}000}{\tfrac{10}{9}} = 9000$.

7. (a) We will assume that the difference in the birth and death rates is 20 million/year. Let $t = 0$ correspond to the year 1990 and use a unit of 1 billion for all calculations. $k \approx \frac{1}{P}\frac{dP}{dt} = \frac{1}{5.3}(0.02) = \frac{1}{265}$, so

$$\frac{dP}{dt} = kP\left(1 - \frac{P}{M}\right) = \frac{1}{265}P\left(1 - \frac{P}{100}\right), \qquad P \text{ in billions}$$

(b) $A = \frac{M - P_0}{P_0} = \frac{100 - 5.3}{5.3} = \frac{947}{53} \approx 17.8679.$ $P(t) = \frac{M}{1 + Ae^{-kt}} = \frac{100}{1 + \frac{947}{53}e^{-(1/265)t}}$, so $P(10) \approx 5.49$ billion.

(c) $P(110) \approx 7.81$, and $P(510) \approx 27.72$. The predictions are 7.81 billion in the year 2100 and 27.72 billion in 2500.

(d) If $M = 50$, then $P(t) = \frac{50}{1 + \frac{447}{53}e^{-(1/265)t}}$. So $P(10) \approx 5.48$, $P(110) \approx 7.61$, and $P(510) \approx 22.41$. The predictions become 5.48 billion in the year 2000, 7.61 billion in 2100, and 22.41 billion in the year 2500.

9. (a) Our assumption is that $\frac{dy}{dt} = ky(1 - y)$, where y is the fraction of the population that has heard the rumor.

(b) Using the logistic equation (1), $\frac{dP}{dt} = kP\left(1 - \frac{P}{M}\right)$, we substitute $y = \frac{P}{M}$, $P = My$, and $\frac{dP}{dt} = M\frac{dy}{dt}$,

to obtain $M\frac{dy}{dt} = k(My)(1 - y) \quad \Leftrightarrow \quad \frac{dy}{dt} = ky(1 - y)$, our equation in part (a).

Now the solution to (1) is $P(t) = \frac{M}{1 + Ae^{-kt}}$, where $A = \frac{M - P_0}{P_0}$.

We use the same substitution to obtain $My = \frac{M}{1 + \frac{M - My_0}{My_0}e^{-kt}} \quad \Rightarrow \quad y = \frac{y_0}{y_0 + (1 - y_0)e^{-kt}}$.

Alternatively, we could use the same steps as outlined in the solution of Equation 1.

(c) Let t be the number of hours since 8 AM. Then $y_0 = y(0) = \frac{80}{1000} = 0.08$ and $y(4) = \frac{1}{2}$, so

$\frac{1}{2} = y(4) = \frac{0.08}{0.08 + 0.92e^{-4k}}$. Thus, $0.08 + 0.92e^{-4k} = 0.16$, $e^{-4k} = \frac{0.08}{0.92} = \frac{2}{23}$, and $e^{-k} = \left(\frac{2}{23}\right)^{1/4}$,

so $y = \frac{0.08}{0.08 + 0.92(2/23)^{t/4}} = \frac{2}{2 + 23(2/23)^{t/4}}$. Solving this equation for t, we get

$2y + 23y\left(\frac{2}{23}\right)^{t/4} = 2 \quad \Rightarrow \quad \left(\frac{2}{23}\right)^{t/4} = \frac{2 - 2y}{23y} \quad \Rightarrow \quad \left(\frac{2}{23}\right)^{t/4} = \frac{2}{23}\cdot\frac{1 - y}{y} \quad \Rightarrow \quad \left(\frac{2}{23}\right)^{t/4-1} = \frac{1 - y}{y}$.

It follows that $\frac{t}{4} - 1 = \frac{\ln[(1 - y)/y]}{\ln\frac{2}{23}}$, so $t = 4\left[1 + \frac{\ln((1 - y)/y)}{\ln\frac{2}{23}}\right]$.

When $y = 0.9$, $\frac{1 - y}{y} = \frac{1}{9}$, so $t = 4\left(1 - \frac{\ln 9}{\ln\frac{2}{23}}\right) \approx 7.6$ h or 7 h 36 min. Thus, 90% of the population will have heard the rumor by 3:36 PM.

11. (a) $\frac{dP}{dt} = kP\left(1 - \frac{P}{M}\right) \quad \Rightarrow \quad \frac{d^2P}{dt^2} = k\left[P\left(-\frac{1}{M}\frac{dP}{dt}\right) + \left(1 - \frac{P}{M}\right)\frac{dP}{dt}\right] = k\frac{dP}{dt}\left(-\frac{P}{M} + 1 - \frac{P}{M}\right)$

$$= k\left[kP\left(1 - \frac{P}{M}\right)\right]\left(1 - \frac{2P}{M}\right) = k^2P\left(1 - \frac{P}{M}\right)\left(1 - \frac{2P}{M}\right)$$

(b) P grows fastest when P' has a maximum, that is, when $P'' = 0$. From part (a), $P'' = 0 \quad \Leftrightarrow \quad P = 0, P = M$, or $P = M/2$. Since $0 < P < M$, we see that $P'' = 0 \quad \Leftrightarrow \quad P = M/2$.

13. Following the hint, we choose $t = 0$ to correspond to 1960 and subtract

94,000 from each of the population figures. We then use a calculator to

obtain the models and add 94,000 to get the exponential function

$P_E(t) = 1578.3(1.0933)^t + 94,000$ and the logistic function

$P_L(t) = \dfrac{32,658.5}{1 + 12.75e^{-0.1706t}} + 94,000.$ P_L is a reasonably accurate

model, while P_E is not, since an exponential model would only be used

for the first few data points.

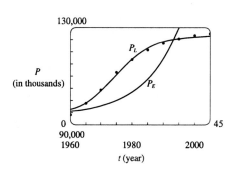

15. (a) The term -15 represents a harvesting of fish at a constant rate—in this case, 15 fish/week. This is the rate at which fish

are caught.

(b)

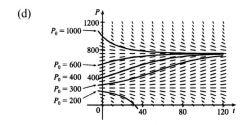

(c) From the graph in part (b), it appears that $P(t) = 250$ and $P(t) = 750$

are the equilibrium solutions. We confirm this analytically by solving the

equation $dP/dt = 0$ as follows: $0.08P(1 - P/1000) - 15 = 0 \Rightarrow$

$0.08P - 0.00008P^2 - 15 = 0 \Rightarrow$

$-0.00008(P^2 - 1000P + 187,500) = 0 \Rightarrow$

$(P - 250)(P - 750) = 0 \Rightarrow P = 250$ or 750.

(d)

For $0 < P_0 < 250$, $P(t)$ decreases to 0. For $P_0 = 250$, $P(t)$ remains

constant. For $250 < P_0 < 750$, $P(t)$ increases and approaches 750.

For $P_0 = 750$, $P(t)$ remains constant. For $P_0 > 750$, $P(t)$ decreases

and approaches 750.

(e) $\dfrac{dP}{dt} = 0.08P\left(1 - \dfrac{P}{1000}\right) - 15 \Leftrightarrow -\dfrac{100,000}{8} \cdot \dfrac{dP}{dt} = (0.08P - 0.00008P^2 - 15) \cdot \left(-\dfrac{100,000}{8}\right) \Leftrightarrow$

$-12,500\dfrac{dP}{dt} = P^2 - 1000P + 187,500 \Leftrightarrow \dfrac{dP}{(P - 250)(P - 750)} = -\dfrac{1}{12,500}dt \Leftrightarrow$

$\displaystyle\int\left(\dfrac{-1/500}{P - 250} + \dfrac{1/500}{P - 750}\right)dP = -\dfrac{1}{12,500}dt \Leftrightarrow \int\left(\dfrac{1}{P - 250} - \dfrac{1}{P - 750}\right)dP = \tfrac{1}{25}dt \Leftrightarrow$

$\ln|P - 250| - \ln|P - 750| = \tfrac{1}{25}t + C \Leftrightarrow \ln\left|\dfrac{P - 250}{P - 750}\right| = \tfrac{1}{25}t + C \Leftrightarrow \left|\dfrac{P - 250}{P - 750}\right| = e^{t/25+C} = ke^{t/25} \Leftrightarrow$

$\dfrac{P - 250}{P - 750} = ke^{t/25} \Leftrightarrow P - 250 = Pke^{t/25} - 750ke^{t/25} \Leftrightarrow P - Pke^{t/25} = 250 - 750ke^{t/25} \Leftrightarrow$

$P(t) = \dfrac{250 - 750ke^{t/25}}{1 - ke^{t/25}}$. If $t = 0$ and $P = 200$, then $200 = \dfrac{250 - 750k}{1 - k}$ ⇔ $200 - 200k = 250 - 750k$ ⇔

$550k = 50$ ⇔ $k = \frac{1}{11}$. Similarly, if $t = 0$ and $P = 300$, then

$k = -\frac{1}{9}$. Simplifying P with these two values of k gives us

$P(t) = \dfrac{250(3e^{t/25} - 11)}{e^{t/25} - 11}$ and $P(t) = \dfrac{750(e^{t/25} + 3)}{e^{t/25} + 9}$.

17. (a) $\dfrac{dP}{dt} = (kP)\left(1 - \dfrac{P}{M}\right)\left(1 - \dfrac{m}{P}\right)$. If $m < P < M$, then $dP/dt = (+)(+)(+) = +$ ⇒ P is increasing.

If $0 < P < m$, then $dP/dt = (+)(+)(-) = -$ ⇒ P is decreasing.

(b)

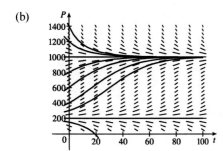

$k = 0.08$, $M = 1000$, and $m = 200$ ⇒

$$\dfrac{dP}{dt} = 0.08P\left(1 - \dfrac{P}{1000}\right)\left(1 - \dfrac{200}{P}\right)$$

For $0 < P_0 < 200$, the population dies out. For $P_0 = 200$, the population is steady. For $200 < P_0 < 1000$, the population increases and approaches 1000. For $P_0 > 1000$, the population decreases and approaches 1000.

The equilibrium solutions are $P(t) = 200$ and $P(t) = 1000$.

(c) $\dfrac{dP}{dt} = kP\left(1 - \dfrac{P}{M}\right)\left(1 - \dfrac{m}{P}\right) = kP\left(\dfrac{M - P}{M}\right)\left(\dfrac{P - m}{P}\right) = \dfrac{k}{M}(M - P)(P - m)$ ⇔

$\displaystyle\int \dfrac{dP}{(M - P)(P - m)} = \int \dfrac{k}{M}\,dt$. By partial fractions, $\dfrac{1}{(M - P)(P - m)} = \dfrac{A}{M - P} + \dfrac{B}{P - m}$, so

$A(P - m) + B(M - P) = 1$.

If $P = m$, $B = \dfrac{1}{M - m}$; if $P = M$, $A = \dfrac{1}{M - m}$, so $\dfrac{1}{M - m}\displaystyle\int\left(\dfrac{1}{M - P} + \dfrac{1}{P - m}\right)dP = \int \dfrac{k}{M}\,dt$ ⇒

$\dfrac{1}{M - m}\left(-\ln|M - P| + \ln|P - m|\right) = \dfrac{k}{M}t + C$ ⇒ $\dfrac{1}{M - m}\ln\left|\dfrac{P - m}{M - P}\right| = \dfrac{k}{M}t + C$ ⇒

$\ln\left|\dfrac{P - m}{M - P}\right| = (M - m)\dfrac{k}{M}t + C_1$ ⇔ $\dfrac{P - m}{M - P} = De^{(M-m)(k/M)t}$ $[D = \pm e^{C_1}]$.

Let $t = 0$: $\dfrac{P_0 - m}{M - P_0} = D$. So $\dfrac{P - m}{M - P} = \dfrac{P_0 - m}{M - P_0}e^{(M-m)(k/M)t}$.

Solving for P, we get $P(t) = \dfrac{m(M - P_0) + M(P_0 - m)e^{(M-m)(k/M)t}}{M - P_0 + (P_0 - m)e^{(M-m)(k/M)t}}$.

(d) If $P_0 < m$, then $P_0 - m < 0$. Let $N(t)$ be the numerator of the expression for $P(t)$ in part (c). Then

$N(0) = P_0(M - m) > 0$, and $P_0 - m < 0$ ⇔ $\displaystyle\lim_{t\to\infty} M(P_0 - m)e^{(M-m)(k/M)t} = -\infty$ ⇒ $\displaystyle\lim_{t\to\infty} N(t) = -\infty$.

Since N is continuous, there is a number t such that $N(t) = 0$ and thus $P(t) = 0$. So the species will become extinct.

19. (a) $dP/dt = kP\cos(rt - \phi)$ $\Rightarrow$ $(dP)/P = k\cos(rt - \phi)\,dt$ $\Rightarrow$ $\int(dP)/P = k\int\cos(rt - \phi)\,dt$ $\Rightarrow$

$\ln P = (k/r)\sin(rt - \phi) + C$. (Since this is a growth model, $P > 0$ and we can write $\ln P$ instead of $\ln|P|$.) Since

$P(0) = P_0$, we obtain $\ln P_0 = (k/r)\sin(-\phi) + C = -(k/r)\sin\phi + C$ $\Rightarrow$ $C = \ln P_0 + (k/r)\sin\phi$. Thus,

$\ln P = (k/r)\sin(rt - \phi) + \ln P_0 + (k/r)\sin\phi$, which we can rewrite as $\ln(P/P_0) = (k/r)[\sin(rt - \phi) + \sin\phi]$ or,

after exponentiation, $P(t) = P_0 e^{(k/r)[\sin(rt-\phi)+\sin\phi]}$.

(b) As k increases, the amplitude increases, but the minimum value stays the same.

As r increases, the amplitude and the period decrease.

A change in ϕ produces slight adjustments in the phase shift and amplitude.

Comparing values of k with
$P_0 = 1$, $r = 2$, and $\phi = \pi/2$

Comparing values of r with
$P_0 = 1$, $k = 1$, and $\phi = \pi/2$

Comparing values of ϕ with
$P_0 = 1$, $k = 1$, and $r = 2$

$P(t)$ oscillates between $P_0 e^{(k/r)(1+\sin\phi)}$ and $P_0 e^{(k/r)(-1+\sin\phi)}$ (the extreme values are attained when $rt - \phi$ is an odd

multiple of $\frac{\pi}{2}$), so $\lim_{t\to\infty} P(t)$ does not exist.

7.6   Predator-Prey Systems

1. (a) $dx/dt = -0.05x + 0.0001xy$. If $y = 0$, we have $dx/dt = -0.05x$, which indicates that in the absence of y, x declines at

a rate proportional to itself. So x represents the predator population and y represents the prey population. The growth of

the prey population, $0.1y$ (from $dy/dt = 0.1y - 0.005xy$), is restricted only by encounters with predators (the term

$-0.005xy$). The predator population increases only through the term $0.0001xy$; that is, by encounters with the prey and

not through additional food sources.

(b) $dy/dt = -0.015y + 0.00008xy$. If $x = 0$, we have $dy/dt = -0.015y$, which indicates that in the absence of x, y would

decline at a rate proportional to itself. So y represents the predator population and x represents the prey population. The

growth of the prey population, $0.2x$ (from $dx/dt = 0.2x - 0.0002x^2 - 0.006xy = 0.2x(1 - 0.001x) - 0.006xy$), is

restricted by a carrying capacity of 1000 [from the term $1 - 0.001x = 1 - x/1000$] and by encounters with predators (the

term $-0.006xy$). The predator population increases only through the term $0.00008xy$; that is, by encounters with the prey

and not through additional food sources.

3. (a) $dx/dt = 0.5x - 0.004x^2 - 0.001xy = 0.5x(1 - x/125) - 0.001xy$.

$dy/dt = 0.4y - 0.001y^2 - 0.002xy = 0.4y(1 - y/400) - 0.002xy$.

The system shows that x and y have carrying capacities of 125 and 400. An increase in x reduces the growth rate of y due to the negative term $-0.002xy$. An increase in y reduces the growth rate of x due to the negative term $-0.001xy$. Hence the system describes a competition model.

(b) $dx/dt = 0 \Rightarrow x(0.5 - 0.004x - 0.001y) = 0 \Rightarrow x(500 - 4x - y) = 0$ **(1)** and $dy/dt = 0 \Rightarrow$

$y(0.4 - 0.001y - 0.002x) = 0 \Rightarrow y(400 - y - 2x) = 0$ **(2)**.

From **(1)** and **(2)**, we get four equilibrium solutions.

(i) $x = 0$ and $y = 0$: If the populations are zero, there is no change.

(ii) $x = 0$ and $400 - y - 2x = 0 \Rightarrow x = 0$ and $y = 400$: In the absence of an x-population, the y-population stabilizes at 400.

(iii) $500 - 4x - y = 0$ and $y = 0 \Rightarrow x = 125$ and $y = 0$: In the absence of y-population, the x-population stabilizes at 125.

(iv) $500 - 4x - y = 0$ and $400 - y - 2x = 0 \Rightarrow y = 500 - 4x$ and $y = 400 - 2x \Rightarrow 500 - 4x = 400 - 2x \Rightarrow 100 = 2x \Rightarrow x = 50$ and $y = 300$: A y-population of 300 is just enough to support a constant x-population of 50.

5. (a) At $t = 0$, there are about 300 rabbits and 100 foxes. At $t = t_1$, the number of foxes reaches a minimum of about 20 while the number of rabbits is about 1000. At $t = t_2$, the number of rabbits reaches a maximum of about 2400, while the number of foxes rebounds to 100. At $t = t_3$, the number of rabbits decreases to about 1000 and the number of foxes reaches a maximum of about 315. As t increases, the number of foxes decreases greatly to 100, and the number of rabbits decreases to 300 (the initial populations), and the cycle starts again.

(b)

7.

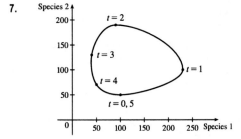

9. $\dfrac{dW}{dR} = \dfrac{-0.02W + 0.00002RW}{0.08R - 0.001RW}$　$\Leftrightarrow$　$(0.08 - 0.001W)R\,dW = (-0.02 + 0.00002R)W\,dR$　$\Leftrightarrow$

$\dfrac{0.08 - 0.001W}{W}\,dW = \dfrac{-0.02 + 0.00002R}{R}\,dR$　$\Leftrightarrow$　$\displaystyle\int\left(\dfrac{0.08}{W} - 0.001\right)dW = \int\left(-\dfrac{0.02}{R} + 0.00002\right)dR$　$\Leftrightarrow$

$0.08\ln|W| - 0.001W = -0.02\ln|R| + 0.00002R + K$　$\Leftrightarrow$　$0.08\ln W + 0.02\ln R = 0.001W + 0.00002R + K$　$\Leftrightarrow$

$\ln\left(W^{0.08}R^{0.02}\right) = 0.00002R + 0.001W + K$　$\Leftrightarrow$　$W^{0.08}R^{0.02} = e^{0.00002R + 0.001W + K}$　$\Leftrightarrow$

$R^{0.02}W^{0.08} = Ce^{0.00002R}e^{0.001W}$　$\Leftrightarrow$　$\dfrac{R^{0.02}W^{0.08}}{e^{0.00002R}e^{0.001W}} = C$. In general, if $\dfrac{dy}{dx} = \dfrac{-ry + bxy}{kx - axy}$, then $C = \dfrac{x^r y^k}{e^{bx}e^{ay}}$.

11. (a) Letting $W = 0$ gives us $dR/dt = 0.08R(1 - 0.0002R)$. $dR/dt = 0$　$\Leftrightarrow$　$R = 0$ or 5000. Since $dR/dt > 0$ for

$0 < R < 5000$, we would expect the rabbit population to *increase* to 5000 for these values of R. Since $dR/dt < 0$ for

$R > 5000$, we would expect the rabbit population to *decrease* to 5000 for these values of R. Hence, in the absence of

wolves, we would expect the rabbit population to stabilize at 5000.

(b) R and W are constant　$\Rightarrow$　$R' = 0$ and $W' = 0$　$\Rightarrow$

$$\begin{cases} 0 = 0.08R(1 - 0.0002R) - 0.001RW \\ 0 = -0.02W + 0.00002RW \end{cases} \Rightarrow \begin{cases} 0 = R[0.08(1 - 0.0002R) - 0.001W] \\ 0 = W(-0.02 + 0.00002R) \end{cases}$$

The second equation is true if $W = 0$ or $R = \frac{0.02}{0.00002} = 1000$. If $W = 0$ in the first equation, then either $R = 0$ or

$R = \frac{1}{0.0002} = 5000$　[as in part (a)]. If $R = 1000$, then $0 = 1000[0.08(1 - 0.0002 \cdot 1000) - 0.001W]$　$\Leftrightarrow$

$0 = 80(1 - 0.2) - W$　$\Leftrightarrow$　$W = 64$.

Case (i):　$W = 0$, $R = 0$: both populations are zero

Case (ii):　$W = 0$, $R = 5000$: see part (a)

Case (iii):　$R = 1000$, $W = 64$: the predator/prey interaction balances and the populations are stable.

(c) The populations of wolves and rabbits fluctuate around 64 and 1000, respectively, and eventually stabilize at those values.

(d)

7 Review

CONCEPT CHECK

1. (a) A differential equation is an equation that contains an unknown function and one or more of its derivatives.

(b) The order of a differential equation is the order of the highest derivative that occurs in the equation.

(c) An initial condition is a condition of the form $y(t_0) = y_0$.

2. $y' = x^2 + y^2 \geq 0$ for all x and y. $y' = 0$ only at the origin, so there is a horizontal tangent at $(0,0)$, but nowhere else. The graph of the solution is increasing on every interval.

3. See the paragraph preceding Example 1 in Section 7.2.

4. See the paragraph following Figure 14 in Section 7.2.

5. A separable equation is a first-order differential equation in which the expression for dy/dx can be factored as a function of x times a function of y, that is, $dy/dx = g(x)f(y)$. We can solve the equation by integrating both sides of the equation $dy/f(y) = g(x)dx$ and solving for y.

6. (a) $\dfrac{dy}{dt} = ky$; the relative growth rate, $\dfrac{1}{y}\dfrac{dy}{dt}$, is constant.

(b) The equation in part (a) is an appropriate model for population growth, assuming that there is enough room and nutrition to support the growth.

(c) If $y(0) = y_0$, then the solution is $y(t) = y_0 e^{kt}$.

7. (a) $dP/dt = kP(1 - P/M)$, where M is the carrying capacity.

(b) The equation in part (a) is an appropriate model for population growth, assuming that the population grows at a rate proportional to the size of the population in the beginning, but eventually levels off and approaches its carrying capacity because of limited resources.

8. (a) $dF/dt = kF - aFS$ and $dS/dt = -rS + bFS$.

(b) In the absence of sharks, an ample food supply would support exponential growth of the fish population, that is, $dF/dt = kF$, where k is a positive constant. In the absence of fish, we assume that the shark population would decline at a rate proportional to itself, that is, $dS/dt = -rS$, where r is a positive constant.

TRUE-FALSE QUIZ

1. True.   Since $y^4 \geq 0$, $y' = -1 - y^4 < 0$ and the solutions are decreasing functions.

3. False. $x + y$ cannot be written in the form $g(x)f(y)$.

5. True.   By comparing $\dfrac{dy}{dt} = 2y\left(1 - \dfrac{y}{5}\right)$ with the logistic differential equation (7.5.1), we see that the carrying capacity is 5; that is, $\lim\limits_{t \to \infty} y = 5$.

EXERCISES

1. (a)

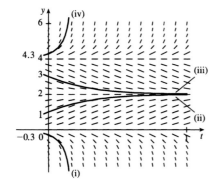

(b) $\lim\limits_{t \to \infty} y(t)$ appears to be finite for $0 \le c \le 4$. In fact

$\lim\limits_{t \to \infty} y(t) = 4$ for $c = 4$, $\lim\limits_{t \to \infty} y(t) = 2$ for $0 < c < 4$, and

$\lim\limits_{t \to \infty} y(t) = 0$ for $c = 0$. The equilibrium solutions are

$y(t) = 0$, $y(t) = 2$, and $y(t) = 4$.

3. (a)

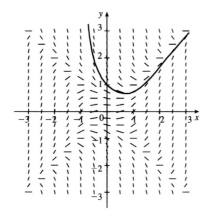

We estimate that when $x = 0.3$, $y = 0.8$, so $y(0.3) \approx 0.8$.

(b) $h = 0.1$, $x_0 = 0$, $y_0 = 1$ and $F(x, y) = x^2 - y^2$. So $y_n = y_{n-1} + 0.1\left(x_{n-1}^2 - y_{n-1}^2\right)$. Thus,

$y_1 = 1 + 0.1\left(0^2 - 1^2\right) = 0.9$, $y_2 = 0.9 + 0.1\left(0.1^2 - 0.9^2\right) = 0.82$, $y_3 = 0.82 + 0.1\left(0.2^2 - 0.82^2\right) = 0.75676$.

This is close to our graphical estimate of $y(0.3) \approx 0.8$.

(c) The centers of the horizontal line segments of the direction field are located on the lines $y = x$ and $y = -x$.

When a solution curve crosses one of these lines, it has a local maximum or minimum.

5. $2ye^{y^2} y' = 2x + 3\sqrt{x} \;\;\Rightarrow\;\; 2ye^{y^2} \dfrac{dy}{dx} = 2x + 3\sqrt{x} \;\;\Rightarrow\;\; 2ye^{y^2}\, dy = \left(2x + 3\sqrt{x}\right) dx \;\;\Rightarrow$

$\int 2ye^{y^2}\, dy = \int \left(2x + 3\sqrt{x}\right) dx \;\;\Rightarrow\;\; e^{y^2} = x^2 + 2x^{3/2} + C \;\;\Rightarrow\;\; y^2 = \ln(x^2 + 2x^{3/2} + C) \;\;\Rightarrow$

$y = \pm\sqrt{\ln(x^2 + 2x^{3/2} + C)}$

7. $\dfrac{dr}{dt} + 2tr = r \;\;\Rightarrow\;\; \dfrac{dr}{dt} = r - 2tr = r(1 - 2t) \;\;\Rightarrow\;\; \int \dfrac{dr}{r} = \int (1 - 2t)\, dt \;\;\Rightarrow\;\; \ln|r| = t - t^2 + C \;\;\Rightarrow$

$|r| = e^{t - t^2 + C} = ke^{t - t^2}$. Since $r(0) = 5$, $5 = ke^0 = k$. Thus, $r(t) = 5e^{t - t^2}$.

9. $\dfrac{d}{dx}(y) = \dfrac{d}{dx}(ke^x) \;\;\Rightarrow\;\; y' = ke^x = y$, so the orthogonal trajectories must have $y' = -\dfrac{1}{y} \;\;\Rightarrow\;\; \dfrac{dy}{dx} = -\dfrac{1}{y} \;\;\Rightarrow$

$y\, dy = -dx \;\;\Rightarrow\;\; \int y\, dy = -\int dx \;\;\Rightarrow\;\; \tfrac{1}{2}y^2 = -x + C \;\;\Rightarrow\;\; x = C - \tfrac{1}{2}y^2$, which are parabolas with a horizontal axis.

11. (a) $y(t) = y(0)e^{kt} = 200e^{kt}$ ⇒ $y(0.5) = 200e^{0.5k} = 360$ ⇒ $e^{0.5k} = 1.8$ ⇒ $0.5k = \ln 1.8$ ⇒

$k = 2\ln 1.8 = \ln(1.8)^2 = \ln 3.24$ ⇒ $y(t) = 200e^{(\ln 3.24)t} = 200(3.24)^t$

(b) $y(4) = 200(3.24)^4 \approx 22{,}040$ bacteria

(c) $y'(t) = 200(3.24)^t \cdot \ln 3.24$, so $y'(4) = 200(3.24)^4 \cdot \ln 3.24 \approx 25{,}910$ bacteria per hour

(d) $200(3.24)^t = 10{,}000$ ⇒ $(3.24)^t = 50$ ⇒ $t\ln 3.24 = \ln 50$ ⇒ $t = \ln 50/\ln 3.24 \approx 3.33$ hours

13. (a) $C'(t) = -kC(t)$ ⇒ $C(t) = C(0)e^{-kt}$ by Theorem 7.4.2. But $C(0) = C_0$, so $C(t) = C_0 e^{-kt}$.

(b) $C(30) = \frac{1}{2}C_0$ since the concentration is reduced by half. Thus, $\frac{1}{2}C_0 = C_0 e^{-30k}$ ⇒ $\ln\frac{1}{2} = -30k$ ⇒

$k = -\frac{1}{30}\ln\frac{1}{2} = \frac{1}{30}\ln 2$. Since 10% of the original concentration remains if 90% is eliminated, we want the value of t

such that $C(t) = \frac{1}{10}C_0$. Therefore, $\frac{1}{10}C_0 = C_0 e^{-t(\ln 2)/30}$ ⇒ $\ln 0.1 = -t(\ln 2)/30$ ⇒ $t = -\frac{30}{\ln 2}\ln 0.1 \approx 100$ h.

15. (a) Using (1) and (4) in Section 7.5, we see that for $\dfrac{dP}{dt} = 0.1P\left(1 - \dfrac{P}{2000}\right)$ with $P(0) = 100$, we have $k = 0.1$,

$M = 2000$, $P_0 = 100$, and $A = \dfrac{2000 - 100}{100} = 19$. Thus, the solution of the initial-value problem is

$P(t) = \dfrac{2000}{1 + 19e^{-0.1t}}$ and $P(20) = \dfrac{2000}{1 + 19e^{-2}} \approx 560$.

(b) $P = 1200$ ⇔ $1200 = \dfrac{2000}{1 + 19e^{-0.1t}}$ ⇔ $1 + 19e^{-0.1t} = \dfrac{2000}{1200}$ ⇔ $19e^{-0.1t} = \dfrac{5}{3} - 1$ ⇔

$e^{-0.1t} = \left(\frac{2}{3}\right)/19$ ⇔ $-0.1t = \ln\frac{2}{57}$ ⇔ $t = -10\ln\frac{2}{57} \approx 33.5$.

17. (a) $\dfrac{dL}{dt} \propto L_\infty - L$ ⇒ $\dfrac{dL}{dt} = k(L_\infty - L)$ ⇒ $\displaystyle\int \dfrac{dL}{L_\infty - L} = \int k\,dt$ ⇒ $-\ln|L_\infty - L| = kt + C$ ⇒

$\ln|L_\infty - L| = -kt - C$ ⇒ $|L_\infty - L| = e^{-kt-C}$ ⇒ $L_\infty - L = Ae^{-kt}$ ⇒ $L = L_\infty - Ae^{-kt}$.

At $t = 0$, $L = L(0) = L_\infty - A$ ⇒ $A = L_\infty - L(0)$ ⇒ $L(t) = L_\infty - [L_\infty - L(0)]e^{-kt}$.

(b) $L_\infty = 53$ cm, $L(0) = 10$ cm, and $k = 0.2$ ⇒ $L(t) = 53 - (53 - 10)e^{-0.2t} = 53 - 43e^{-0.2t}$.

19. Let P represent the population and I the number of infected people. The rate of spread dI/dt is jointly proportional to I and

to $P - I$, so for some constant k, $\dfrac{dI}{dt} = kI(P - I)$ ⇒ $I(t) = \dfrac{I_0 P}{I_0 + (P - I_0)e^{-kPt}}$ [from the discussion of logistic

growth in Section 7.5].

Now, measuring t in days, we substitute $t = 7$, $P = 5000$, $I_0 = 160$ and $I(7) = 1200$ to find k:

$1200 = \dfrac{160 \cdot 5000}{160 + (5000 - 160)e^{-5000\cdot 7\cdot k}}$ ⇔ $3 = \dfrac{2000}{160 + 4840e^{-35{,}000k}}$ ⇔ $480 + 14{,}520e^{-35{,}000k} = 2000$ ⇔

$e^{-35{,}000k} = \dfrac{2000 - 480}{14{,}520}$ ⇔ $-35{,}000k = \ln\dfrac{38}{363}$ ⇔ $k = \dfrac{-1}{35{,}000}\ln\dfrac{38}{363} \approx 0.00006448$. Next, let

$I = 5000 \times 80\% = 4000$, and solve for t: $4000 = \dfrac{160 \cdot 5000}{160 + (5000 - 160)e^{-k\cdot 5000\cdot t}}$ ⇔ $1 = \dfrac{200}{160 + 4840e^{-5000kt}}$ ⇔

$$160 + 4840e^{-5000kt} = 200 \quad \Leftrightarrow \quad e^{-5000kt} = \frac{200 - 160}{4840} \quad \Leftrightarrow \quad -5000kt = \ln\frac{1}{121} \quad \Leftrightarrow$$

$$t = \frac{-1}{5000k}\ln\frac{1}{121} = \frac{1}{\frac{1}{7}\ln\frac{38}{363}} \cdot \ln\frac{1}{121} = 7 \cdot \frac{\ln 121}{\ln\frac{363}{38}} \approx 14.875. \text{ So it takes about 15 days for 80\% of the population}$$

to be infected.

21. $\dfrac{dh}{dt} = -\dfrac{R}{V}\left(\dfrac{h}{k+h}\right) \;\Rightarrow\; \displaystyle\int \dfrac{k+h}{h}\,dh = \int\left(-\dfrac{R}{V}\right)dt \;\Rightarrow\; \int\left(1+\dfrac{k}{h}\right)dh = -\dfrac{R}{V}\int 1\,dt \;\Rightarrow$

$h + k\ln h = -\dfrac{R}{V}t + C.$ This equation gives a relationship between h and t, but it is not possible to isolate h and express it in

terms of t.

23. (a) $dx/dt = 0.4x(1 - 0.000005x) - 0.002xy$, $dy/dt = -0.2y + 0.000008xy$. If $y = 0$, then

$dx/dt = 0.4x(1 - 0.000005x)$, so $dx/dt = 0 \;\Leftrightarrow\; x = 0$ or $x = 200{,}000$, which shows that the insect population

increases logistically with a carrying capacity of 200,000. Since $dx/dt > 0$ for $0 < x < 200{,}000$ and $dx/dt < 0$ for

$x > 200{,}000$, we expect the insect population to stabilize at 200,000.

(b) x and y are constant $\;\Rightarrow\; x' = 0$ and $y' = 0 \;\Rightarrow$

$$\begin{cases} 0 = 0.4x(1 - 0.000005x) - 0.002xy \\ 0 = -0.2y + 0.000008xy \end{cases} \;\Rightarrow\; \begin{cases} 0 = 0.4x[(1 - 0.000005x) - 0.005y] \\ 0 = y(-0.2 + 0.000008x) \end{cases}$$

The second equation is true if $y = 0$ or $x = \frac{0.2}{0.000008} = 25{,}000$. If $y = 0$ in the first equation, then either $x = 0$

or $x = \frac{1}{0.000005} = 200{,}000$. If $x = 25{,}000$, then $0 = 0.4(25{,}000)[(1 - 0.000005 \cdot 25{,}000) - 0.005y] \;\Rightarrow$

$0 = 10{,}000[(1 - 0.125) - 0.005y] \;\Rightarrow\; 0 = 8750 - 50y \;\Rightarrow\; y = 175.$

Case (i): $\quad y = 0, x = 0$: Zero populations

Case (ii): $\quad y = 0, x = 200{,}000$: In the absence of birds, the insect population is always 200,000.

Case (iii): $\quad x = 25{,}000, y = 175$: The predator/prey interaction balances and the populations are stable.

(c) The populations of the birds and insects fluctuate

around 175 and 25,000, respectively, and

eventually stabilize at those values.

(d)

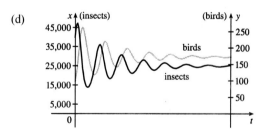

☐ FOCUS ON PROBLEM SOLVING

1. We use the Fundamental Theorem of Calculus to differentiate the given equation:

$$[f(x)]^2 = 100 + \int_0^x \left\{ [f(t)]^2 + [f'(t)]^2 \right\} dt \quad \Rightarrow \quad 2f(x)f'(x) = [f(x)]^2 + [f'(x)]^2 \quad \Rightarrow$$

$$[f(x)]^2 + [f'(x)]^2 - 2f(x)f'(x) = 0 \quad \Rightarrow \quad [f(x) - f'(x)]^2 = 0 \quad \Leftrightarrow \quad f(x) = f'(x). \text{ We can solve this as a separable}$$

equation, or else use Theorem 7.4.2 with $k = 1$, which says that the solutions are $f(x) = Ce^x$. Now $[f(0)]^2 = 100$, so

$f(0) = C = \pm 10$, and hence $f(x) = \pm 10e^x$ are the only functions satisfying the given equation.

3. $f'(x) = \lim\limits_{h \to 0} \dfrac{f(x+h) - f(x)}{h} = \lim\limits_{h \to 0} \dfrac{f(x)\,[f(h) - 1]}{h}$ [since $f(x+h) = f(x)f(h)$]

$$= f(x) \lim\limits_{h \to 0} \dfrac{f(h) - 1}{h} = f(x) \lim\limits_{h \to 0} \dfrac{f(h) - f(0)}{h - 0} = f(x)f'(0) = f(x)$$

Therefore, $f'(x) = f(x)$ for all x and from Theorem 7.4.2 we get $f(x) = Ae^x$.

Now $f(0) = 1 \quad \Rightarrow \quad A = 1 \quad \Rightarrow \quad f(x) = e^x$.

5. "The area under the graph of f from 0 to x is proportional to the $(n+1)$st power of $f(x)$" translates to

$$\int_0^x f(t)\,dt = k[f(x)]^{n+1} \text{ for some constant } k. \text{ By FTC1, } \frac{d}{dx} \int_0^x f(t)\,dt = \frac{d}{dx} \left\{ k[f(x)]^{n+1} \right\} \quad \Rightarrow$$

$$f(x) = k(n+1)[f(x)]^n f'(x) \quad \Rightarrow \quad 1 = k(n+1)[f(x)]^{n-1} f'(x) \quad \Rightarrow \quad 1 = k(n+1)y^{n-1} \frac{dy}{dx} \quad \Rightarrow$$

$$k(n+1)y^{n-1}\,dy = dx \quad \Rightarrow \quad \int k(n+1)y^{n-1}\,dy = \int dx \quad \Rightarrow \quad k(n+1)\frac{1}{n}y^n = x + C.$$

Now $f(0) = 0 \quad \Rightarrow \quad 0 = 0 + C \quad \Rightarrow \quad C = 0$ and then $f(1) = 1 \quad \Rightarrow \quad k(n+1)\dfrac{1}{n} = 1 \quad \Rightarrow \quad k = \dfrac{n}{n+1}$.

so $y^n = x$ and $y = f(x) = x^{1/n}$.

7. Let $y(t)$ denote the temperature of the peach pie t minutes after 5:00 PM and R the temperature of the room. Newton's Law of

Cooling gives us $dy/dt = k(y - R)$. Solving for y we get $\dfrac{dy}{y - R} = k\,dt \quad \Rightarrow \quad \ln|y - R| = kt + C \quad \Rightarrow$

$|y - R| = e^{kt+C} \quad \Rightarrow \quad y - R = \pm e^{kt} \cdot e^C \quad \Rightarrow \quad y = Me^{kt} + R$, where M is a nonzero constant. We are given

temperatures at three times.

$$y(0) = 100 \quad \Rightarrow \quad 100 = M + R \quad \Rightarrow \quad R = 100 - M$$

$$y(10) = 80 \quad \Rightarrow \quad 80 = Me^{10k} + R \tag{1}$$

$$y(20) = 65 \quad \Rightarrow \quad 65 = Me^{20k} + R \tag{2}$$

Substituting $100 - M$ for R in (1) and (2) gives us

$$-20 = Me^{10k} - M \quad \textbf{(3)} \quad \text{and} \quad -35 = Me^{20k} - M \quad \textbf{(4)}$$

Dividing (3) by (4) gives us $\dfrac{-20}{-35} = \dfrac{M\left(e^{10k} - 1\right)}{M\left(e^{20k} - 1\right)} \quad \Rightarrow \quad \dfrac{4}{7} = \dfrac{e^{10k} - 1}{e^{20k} - 1} \quad \Rightarrow \quad 4e^{20k} - 4 = 7e^{10k} - 7 \quad \Rightarrow$

$4e^{20k} - 7e^{10k} + 3 = 0$. This is a quadratic equation in e^{10k}. $\left(4e^{10k} - 3\right)\left(e^{10k} - 1\right) = 0 \;\Rightarrow\; e^{10k} = \frac{3}{4}$ or $1 \;\Rightarrow\;$

$10k = \ln\frac{3}{4}$ or $\ln 1 \;\Rightarrow\; k = \frac{1}{10}\ln\frac{3}{4}$ since k is a nonzero constant of proportionality. Substituting $\frac{3}{4}$ for e^{10k} in **(3)** gives us

$-20 = M\cdot\frac{3}{4} - M \;\Rightarrow\; -20 = -\frac{1}{4}M \;\Rightarrow\; M = 80$. Now $R = 100 - M$ so $R = 20°\mathrm{C}$.

9. (a) While running from $(L, 0)$ to (x, y), the dog travels a distance

$$s = \int_x^L \sqrt{1 + (dy/dx)^2}\, dx = -\int_L^x \sqrt{1 + (dy/dx)^2}\, dx, \text{ so}$$

$$\frac{ds}{dx} = -\sqrt{1 + (dy/dx)^2}.$$ The dog and rabbit run at the same speed, so the

rabbit's position when the dog has traveled a distance s is $(0, s)$. Since the

dog runs straight for the rabbit, $\dfrac{dy}{dx} = \dfrac{s - y}{0 - x}$ (see the figure).

Thus, $s = y - x\dfrac{dy}{dx} \;\Rightarrow\; \dfrac{ds}{dx} = \dfrac{dy}{dx} - \left(x\dfrac{d^2y}{dx^2} + 1\cdot\dfrac{dy}{dx}\right) = -x\dfrac{d^2y}{dx^2}$. Equating the two expressions for $\dfrac{ds}{dx}$

gives us $x\dfrac{d^2y}{dx^2} = \sqrt{1 + \left(\dfrac{dy}{dx}\right)^2}$, as claimed.

(b) Letting $z = \dfrac{dy}{dx}$, we obtain the differential equation $x\dfrac{dz}{dx} = \sqrt{1 + z^2}$, or $\dfrac{dz}{\sqrt{1 + z^2}} = \dfrac{dx}{x}$. Integrating:

$\ln x = \displaystyle\int \dfrac{dz}{\sqrt{1 + z^2}} \overset{25}{=} \ln\left(z + \sqrt{1 + z^2}\right) + C$. When $x = L$, $z = dy/dx = 0$, so $\ln L = \ln 1 + C$. Therefore, $C = \ln L$,

so $\ln x = \ln\left(\sqrt{1 + z^2} + z\right) + \ln L = \ln\left[L\left(\sqrt{1 + z^2} + z\right)\right] \;\Rightarrow\; x = L\left(\sqrt{1 + z^2} + z\right) \;\Rightarrow\; \sqrt{1 + z^2} = \dfrac{x}{L} - z \;\Rightarrow\;$

$1 + z^2 = \left(\dfrac{x}{L}\right)^2 - \dfrac{2xz}{L} + z^2 \;\Rightarrow\; \left(\dfrac{x}{L}\right)^2 - 2z\left(\dfrac{x}{L}\right) - 1 = 0 \;\Rightarrow\; z = \dfrac{(x/L)^2 - 1}{2(x/L)} = \dfrac{x^2 - L^2}{2Lx} = \dfrac{x}{2L} - \dfrac{L}{2}\dfrac{1}{x}$

[for $x > 0$]. Since $z = \dfrac{dy}{dx}$, $y = \dfrac{x^2}{4L} - \dfrac{L}{2}\ln x + C_1$. Since $y = 0$ when $x = L$, $0 = \dfrac{L}{4} - \dfrac{L}{2}\ln L + C_1 \;\Rightarrow\;$

$C_1 = \dfrac{L}{2}\ln L - \dfrac{L}{4}$. Thus, $y = \dfrac{x^2}{4L} - \dfrac{L}{2}\ln x + \dfrac{L}{2}\ln L - \dfrac{L}{4} = \dfrac{x^2 - L^2}{4L} - \dfrac{L}{2}\ln\left(\dfrac{x}{L}\right)$.

(c) As $x \to 0^+$, $y \to \infty$, so the dog never catches the rabbit.

11. (a) We are given that $V = \frac{1}{3}\pi r^2 h$, $dV/dt = 60{,}000\pi$ ft³/h, and $r = 1.5h = \frac{3}{2}h$. So $V = \frac{1}{3}\pi\left(\frac{3}{2}h\right)^2 h = \frac{3}{4}\pi h^3 \;\Rightarrow\;$

$\dfrac{dV}{dt} = \dfrac{3}{4}\pi\cdot 3h^2\dfrac{dh}{dt} = \dfrac{9}{4}\pi h^2\dfrac{dh}{dt}$. Therefore, $\dfrac{dh}{dt} = \dfrac{4(dV/dt)}{9\pi h^2} = \dfrac{240{,}000\pi}{9\pi h^2} = \dfrac{80{,}000}{3h^2}$ $(\star) \;\Rightarrow\;$

$\int 3h^2\, dh = \int 80{,}000\, dt \;\Rightarrow\; h^3 = 80{,}000t + C$. When $t = 0$, $h = 60$. Thus, $C = 60^3 = 216{,}000$, so

$h^3 = 80{,}000t + 216{,}000$. Let $h = 100$. Then $100^3 = 1{,}000{,}000 = 80{,}000t + 216{,}000 \;\Rightarrow\;$

$80{,}000t = 784{,}000 \;\Rightarrow\; t = 9.8$, so the time required is 9.8 hours.

(b) The floor area of the silo is $F = \pi \cdot 200^2 = 40{,}000\pi$ ft^2, and the area of the base of the pile is

$A = \pi r^2 = \pi\left(\frac{3}{2}h\right)^2 = \frac{9\pi}{4}h^2$. So the area of the floor which is not covered when $h = 60$ is

$F - A = 40{,}000\pi - 8100\pi = 31{,}900\pi \approx 100{,}217$ ft^2. Now $A = \frac{9\pi}{4}h^2 \;\Rightarrow\; dA/dt = \frac{9\pi}{4} \cdot 2h\,(dh/dt)$,

and from $(\star)$ in part (a) we know that when $h = 60$, $dh/dt = \frac{80{,}000}{3(60)^2} = \frac{200}{27}$ ft/h. Therefore,

$dA/dt = \frac{9\pi}{4}(2)(60)\left(\frac{200}{27}\right) = 2000\pi \approx 6283$ ft^2/h.

(c) At $h = 90$ ft, $dV/dt = 60{,}000\pi - 20{,}000\pi = 40{,}000\pi$ ft^3/h. From $(\star)$ in part (a),

$\dfrac{dh}{dt} = \dfrac{4(dV/dt)}{9\pi h^2} = \dfrac{4(40{,}000\pi)}{9\pi h^2} = \dfrac{160{,}000}{9h^2} \;\Rightarrow\; \int 9h^2\,dh = \int 160{,}000\,dt \;\Rightarrow\; 3h^3 = 160{,}000t + C$. When $t = 0$,

$h = 90$; therefore, $C = 3 \cdot 729{,}000 = 2{,}187{,}000$. So $3h^3 = 160{,}000t + 2{,}187{,}000$. At the top, $h = 100 \;\Rightarrow$

$3(100)^3 = 160{,}000t + 2{,}187{,}000 \;\Rightarrow\; t = \frac{813{,}000}{160{,}000} \approx 5.1$. The pile reaches the top after about 5.1 h.

13. Let $P(a, b)$ be any point on the curve. If m is the slope of the tangent line at P, then $m = y'(a)$, and an equation of the

normal line at P is $y - b = -\dfrac{1}{m}(x - a)$, or equivalently, $y = -\dfrac{1}{m}x + b + \dfrac{a}{m}$. The y-intercept is always 6, so

$b + \dfrac{a}{m} = 6 \;\Rightarrow\; \dfrac{a}{m} = 6 - b \;\Rightarrow\; m = \dfrac{a}{6 - b}$. We will solve the equivalent differential equation $\dfrac{dy}{dx} = \dfrac{x}{6 - y} \;\Rightarrow$

$(6 - y)\,dy = x\,dx \;\Rightarrow\; \displaystyle\int (6 - y)\,dy = \int x\,dx \;\Rightarrow\; 6y - \tfrac{1}{2}y^2 = \tfrac{1}{2}x^2 + C \;\Rightarrow\; 12y - y^2 = x^2 + K$.

Since $(3, 2)$ is on the curve, $12(2) - 2^2 = 3^2 + K \;\Rightarrow\; K = 11$. So the curve is given by $12y - y^2 = x^2 + 11 \;\Rightarrow$

$x^2 + y^2 - 12y + 36 = -11 + 36 \;\Rightarrow\; x^2 + (y - 6)^2 = 25$, a circle with center $(0, 6)$ and radius 5.

8 ☐ INFINITE SEQUENCES AND SERIES

8.1 Sequences

1. (a) A sequence is an ordered list of numbers. It can also be defined as a function whose domain is the set of positive integers.

   (b) The terms a_n approach 8 as n becomes large. In fact, we can make a_n as close to 8 as we like by taking n sufficiently large.

   (c) The terms a_n become large as n becomes large. In fact, we can make a_n as large as we like by taking n sufficiently large.

3. The first six terms of $a_n = \dfrac{n}{2n+1}$ are $\dfrac{1}{3}, \dfrac{2}{5}, \dfrac{3}{7}, \dfrac{4}{9}, \dfrac{5}{11}, \dfrac{6}{13}$. It appears that the sequence is approaching $\dfrac{1}{2}$.

$$\lim_{n \to \infty} \frac{n}{2n+1} = \lim_{n \to \infty} \frac{1}{2 + 1/n} = \frac{1}{2}$$

5. $\left\{1, \frac{1}{3}, \frac{1}{5}, \frac{1}{7}, \frac{1}{9}, \ldots\right\}$. The denominator of the nth term is the nth positive odd integer, so $a_n = \dfrac{1}{2n-1}$.

7. $\{2, 7, 12, 17, \ldots\}$. Each term is larger than the preceding one by 5, so $a_n = a_1 + d(n-1) = 2 + 5(n-1) = 5n - 3$.

9. $\left\{1, -\frac{2}{3}, \frac{4}{9}, -\frac{8}{27}, \ldots\right\}$. Each term is $-\frac{2}{3}$ times the preceding one, so $a_n = \left(-\frac{2}{3}\right)^{n-1}$.

11. $a_n = \dfrac{3 + 5n^2}{n + n^2} = \dfrac{(3 + 5n^2)/n^2}{(n + n^2)/n^2} = \dfrac{5 + 3/n^2}{1 + 1/n}$, so $a_n \to \dfrac{5 + 0}{1 + 0} = 5$ as $n \to \infty$. Converges

13. $a_n = 1 - (0.2)^n$, so $\lim\limits_{n \to \infty} a_n = 1 - 0 = 1$ by (7). Converges

15. Because the natural exponential function is continuous at 0, Theorem 5 enables us to write

$$\lim_{n \to \infty} a_n = \lim_{n \to \infty} e^{1/n} = e^{\lim_{n \to \infty}(1/n)} = e^0 = 1. \quad \text{Converges}$$

17. If $b_n = \dfrac{2n\pi}{1 + 8n}$, then $\lim\limits_{n \to \infty} b_n = \lim\limits_{n \to \infty} \dfrac{(2n\pi)/n}{(1 + 8n)/n} = \lim\limits_{n \to \infty} \dfrac{2\pi}{1/n + 8} = \dfrac{2\pi}{8} = \dfrac{\pi}{4}$. Since $\tan$ is continuous at $\frac{\pi}{4}$, by

Theorem 5, $\lim\limits_{n \to \infty} \tan\left(\dfrac{2n\pi}{1 + 8n}\right) = \tan\left(\lim\limits_{n \to \infty} \dfrac{2n\pi}{1 + 8n}\right) = \tan\dfrac{\pi}{4} = 1.$ Converges

19. $a_n = \dfrac{(-1)^{n-1} n}{n^2 + 1} = \dfrac{(-1)^{n-1}}{n + 1/n}$, so $0 \le |a_n| = \dfrac{1}{n + 1/n} \le \dfrac{1}{n} \to 0$ as $n \to \infty$, so $a_n \to 0$ by the Squeeze Theorem and

Theorem 4. Converges

21. $a_n = \dfrac{e^n + e^{-n}}{e^{2n} - 1} \cdot \dfrac{e^{-n}}{e^{-n}} = \dfrac{1 + e^{-2n}}{e^n - e^{-n}} \to 0$ as $n \to \infty$ because $1 + e^{-2n} \to 1$ and $e^n - e^{-n} \to \infty$. Converges

23. $a_n = n^2 e^{-n} = \dfrac{n^2}{e^n}$. Since $\lim\limits_{x \to \infty} \dfrac{x^2}{e^x} \overset{\mathrm{H}}{=} \lim\limits_{x \to \infty} \dfrac{2x}{e^x} \overset{\mathrm{H}}{=} \lim\limits_{x \to \infty} \dfrac{2}{e^x} = 0$, it follows from Theorem 2 that $\lim\limits_{n \to \infty} a_n = 0$. Converges

25. $0 \le \dfrac{\cos^2 n}{2^n} \le \dfrac{1}{2^n}$ [since $0 \le \cos^2 n \le 1$], so since $\lim\limits_{n \to \infty} \dfrac{1}{2^n} = 0$, $\left\{\dfrac{\cos^2 n}{2^n}\right\}$ converges to 0 by the Squeeze Theorem.

27. $y = \left(1 + \dfrac{2}{x}\right)^x \Rightarrow \ln y = x \ln\left(1 + \dfrac{2}{x}\right)$, so

$$\lim_{x \to \infty} \ln y = \lim_{x \to \infty} \frac{\ln(1 + 2/x)}{1/x} \overset{\text{H}}{=} \lim_{x \to \infty} \frac{\left(\dfrac{1}{1 + 2/x}\right)\left(-\dfrac{2}{x^2}\right)}{-1/x^2} = \lim_{x \to \infty} \frac{2}{1 + 2/x} = 2 \Rightarrow$$

$$\lim_{x \to \infty} \left(1 + \frac{2}{x}\right)^x = \lim_{x \to \infty} e^{\ln y} = e^2, \text{ so by Theorem 2, } \lim_{n \to \infty} \left(1 + \frac{2}{n}\right)^n = e^2. \quad \text{Convergent}$$

29. $a_n = \dfrac{(2n - 1)!}{(2n + 1)!} = \dfrac{(2n - 1)!}{(2n + 1)(2n)(2n - 1)!} = \dfrac{1}{(2n + 1)(2n)} \to 0$ as $n \to \infty$. Converges

31. $\{0, 1, 0, 0, 1, 0, 0, 0, 1, \ldots\}$ diverges since the sequence takes on only two values, 0 and 1, and never stays arbitrarily close to either one (or any other value) for n sufficiently large.

33. $a_n = \ln(2n^2 + 1) - \ln(n^2 + 1) = \ln\left(\dfrac{2n^2 + 1}{n^2 + 1}\right) = \ln\left(\dfrac{2 + 1/n^2}{1 + 1/n^2}\right) \to \ln 2$ as $n \to \infty$. Convergent

35.

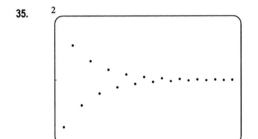

From the graph, it appears that the sequence converges to 1.

$\{(-2/e)^n\}$ converges to 0 by (7), and hence $\{1 + (-2/e)^n\}$

converges to $1 + 0 = 1$.

37.

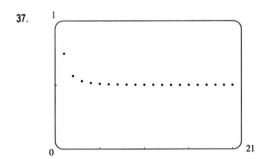

From the graph, it appears that the sequence converges to $\frac{1}{2}$.

As $n \to \infty$,

$$a_n = \sqrt{\frac{3 + 2n^2}{8n^2 + n}} = \sqrt{\frac{3/n^2 + 2}{8 + 1/n}} \Rightarrow \sqrt{\frac{0 + 2}{8 + 0}} = \sqrt{\frac{1}{4}} = \frac{1}{2},$$

so $\lim_{n \to \infty} a_n = \frac{1}{2}$.

39.

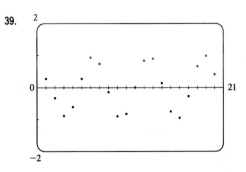

From the graph, it appears that the sequence $\{a_n\} = \left\{\dfrac{n^2 \cos n}{1 + n^2}\right\}$ is

divergent, since it oscillates between 1 and -1 (approximately). To

prove this, suppose that $\{a_n\}$ converges to L. If $b_n = \dfrac{n^2}{1 + n^2}$, then

$\{b_n\}$ converges to 1, and $\lim_{n \to \infty} \dfrac{a_n}{b_n} = \dfrac{L}{1} = L$. But $\dfrac{a_n}{b_n} = \cos n$, so

$\lim_{n \to \infty} \dfrac{a_n}{b_n}$ does not exist. This contradiction shows that $\{a_n\}$ diverges.

41. (a) $a_n = 1000(1.06)^n$ $\Rightarrow$ $a_1 = 1060, a_2 = 1123.60, a_3 = 1191.02, a_4 = 1262.48,$ and $a_5 = 1338.23.$

(b) $\lim\limits_{n \to \infty} a_n = 1000 \lim\limits_{n \to \infty} (1.06)^n,$ so the sequence diverges by (7) with $r = 1.06 > 1.$

43. (a) We are given that the initial population is 5000, so $P_0 = 5000.$ The number of catfish increases by 8% per month and is

decreased by 300 per month, so $P_1 = P_0 + 8\%P_0 - 300 = 1.08P_0 - 300, P_2 = 1.08P_1 - 300,$ and so on. Thus,

$P_n = 1.08P_{n-1} - 300.$

(b) Using the recursive formula with $P_0 = 5000,$ we get $P_1 = 5100, P_2 = 5208, P_3 = 5325$ (rounding any portion of a

catfish), $P_4 = 5451, P_5 = 5587,$ and $P_6 = 5734,$ which is the number of catfish in the pond after six months.

45. (a) $a_1 = 1, a_{n+1} = 4 - a_n$ for $n \geq 1.$ $a_1 = 1, a_2 = 4 - a_1 = 4 - 1 = 3, a_3 = 4 - a_2 = 4 - 3 = 1,$

$a_4 = 4 - a_3 = 4 - 1 = 3, a_5 = 4 - a_4 = 4 - 3 = 1.$ Since the terms of the sequence alternate between 1 and 3,

the sequence is divergent.

(b) $a_1 = 2, a_2 = 4 - a_1 = 4 - 2 = 2, a_3 = 4 - a_2 = 4 - 2 = 2.$ Since all of the terms are 2, $\lim\limits_{n \to \infty} a_n = 2$ and hence, the

sequence is convergent.

47. (a) Let a_n be the number of rabbit pairs in the nth month. Clearly $a_1 = 1 = a_2.$ In the nth month, each pair that is

2 or more months old (that is, a_{n-2} pairs) will produce a new pair to add to the a_{n-1} pairs already present. Thus,

$a_n = a_{n-1} + a_{n-2},$ so that $\{a_n\} = \{f_n\},$ the Fibonacci sequence.

(b) $a_n = \dfrac{f_{n+1}}{f_n}$ $\Rightarrow$ $a_{n-1} = \dfrac{f_n}{f_{n-1}} = \dfrac{f_{n-1} + f_{n-2}}{f_{n-1}} = 1 + \dfrac{f_{n-2}}{f_{n-1}} = 1 + \dfrac{1}{f_{n-1}/f_{n-2}} = 1 + \dfrac{1}{a_{n-2}}.$ If $L = \lim\limits_{n \to \infty} a_n,$

then $L = \lim\limits_{n \to \infty} a_{n-1}$ and $L = \lim\limits_{n \to \infty} a_{n-2},$ so L must satisfy $L = 1 + \dfrac{1}{L}$ $\Rightarrow$ $L^2 - L - 1 = 0$ $\Rightarrow$ $L = \frac{1+\sqrt{5}}{2}$

[since L must be positive].

49. $a_n = \dfrac{1}{2n + 3}$ is decreasing since $a_{n+1} = \dfrac{1}{2(n+1) + 3} = \dfrac{1}{2n + 5} < \dfrac{1}{2n + 3} = a_n$ for each $n \geq 1.$ The sequence is

bounded since $0 < a_n \leq \frac{1}{5}$ for all $n \geq 1.$ Note that $a_1 = \frac{1}{5}.$

51. The terms of $a_n = n(-1)^n$ alternate in sign, so the sequence is not monotonic. The first five terms are $-1, 2, -3, 4,$ and $-5.$

Since $\lim\limits_{n \to \infty} |a_n| = \lim\limits_{n \to \infty} n = \infty,$ the sequence is not bounded.

53. Since $\{a_n\}$ is a decreasing sequence, $a_n > a_{n+1}$ for all $n \geq 1.$ Because all of its terms lie between 5 and 8, $\{a_n\}$ is a

bounded sequence. By the Monotonic Sequence Theorem, $\{a_n\}$ is convergent; that is, $\{a_n\}$ has a limit $L.$ L must be less than

8 since $\{a_n\}$ is decreasing, so $5 \leq L < 8.$

55. $a_1 = 1, a_{n+1} = 3 - \dfrac{1}{a_n}.$ We show by induction that $\{a_n\}$ is increasing and bounded above by 3. Let P_n be the proposition

that $a_{n+1} > a_n$ and $0 < a_n < 3.$ Clearly P_1 is true. Assume that P_n is true. Then $a_{n+1} > a_n$ $\Rightarrow$ $\dfrac{1}{a_{n+1}} < \dfrac{1}{a_n}$ $\Rightarrow$

$-\dfrac{1}{a_{n+1}} > -\dfrac{1}{a_n}$. Now $a_{n+2} = 3 - \dfrac{1}{a_{n+1}} > 3 - \dfrac{1}{a_n} = a_{n+1}$ ⟺ P_{n+1}. This proves that $\{a_n\}$ is increasing and bounded

above by 3, so $1 = a_1 < a_n < 3$, that is, $\{a_n\}$ is bounded, and hence convergent by the Monotonic Sequence Theorem.

If $L = \lim\limits_{n\to\infty} a_n$, then $\lim\limits_{n\to\infty} a_{n+1} = L$ also, so L must satisfy $L = 3 - 1/L$ ⟹ $L^2 - 3L + 1 = 0$ ⟹ $L = \dfrac{3 \pm \sqrt{5}}{2}$.

But $L > 1$, so $L = \dfrac{3 + \sqrt{5}}{2}$.

57. $(0.8)^n < 0.000001$ ⟹ $\ln(0.8)^n < \ln(0.000001)$ ⟹ $n\ln(0.8) < \ln(0.000001)$ ⟹ $n > \dfrac{\ln(0.000001)}{\ln(0.8)}$ ⟹

$n > 61.9$, so n must be at least 62 to satisfy the given inequality.

59. (a) Suppose $\{p_n\}$ converges to p. Then $p_{n+1} = \dfrac{bp_n}{a + p_n}$ ⟹ $\lim\limits_{n\to\infty} p_{n+1} = \dfrac{b \lim\limits_{n\to\infty} p_n}{a + \lim\limits_{n\to\infty} p_n}$ ⟹ $p = \dfrac{bp}{a + p}$ ⟹

$p^2 + ap = bp$ ⟹ $p(p + a - b) = 0$ ⟹ $p = 0$ or $p = b - a$.

(b) $p_{n+1} = \dfrac{bp_n}{a + p_n} = \dfrac{\left(\dfrac{b}{a}\right)p_n}{1 + \dfrac{p_n}{a}} < \left(\dfrac{b}{a}\right)p_n$ since $1 + \dfrac{p_n}{a} > 1$.

(c) By part (b), $p_1 < \left(\dfrac{b}{a}\right)p_0$, $p_2 < \left(\dfrac{b}{a}\right)p_1 < \left(\dfrac{b}{a}\right)^2 p_0$, $p_3 < \left(\dfrac{b}{a}\right)p_2 < \left(\dfrac{b}{a}\right)^3 p_0$, etc. In general, $p_n < \left(\dfrac{b}{a}\right)^n p_0$,

so $\lim\limits_{n\to\infty} p_n \le \lim\limits_{n\to\infty} \left(\dfrac{b}{a}\right)^n \cdot p_0 = 0$ since $b < a$. $\left[\text{By (7), } \lim\limits_{n\to\infty} r^n = 0 \text{ if } -1 < r < 1. \text{ Here } r = \dfrac{b}{a} \in (0, 1).\right]$

(d) Let $a < b$. We first show, by induction, that if $p_0 < b - a$, then $p_n < b - a$ and $p_{n+1} > p_n$.

For $n = 0$, we have $p_1 - p_0 = \dfrac{bp_0}{a + p_0} - p_0 = \dfrac{p_0(b - a - p_0)}{a + p_0} > 0$ since $p_0 < b - a$. So $p_1 > p_0$.

Now we suppose the assertion is true for $n = k$, that is, $p_k < b - a$ and $p_{k+1} > p_k$. Then

$b - a - p_{k+1} = b - a - \dfrac{bp_k}{a + p_k} = \dfrac{a(b - a) + bp_k - ap_k - bp_k}{a + p_k} = \dfrac{a(b - a - p_k)}{a + p_k} > 0$ because $p_k < b - a$. So

$p_{k+1} < b - a$. And $p_{k+2} - p_{k+1} = \dfrac{bp_{k+1}}{a + p_{k+1}} - p_{k+1} = \dfrac{p_{k+1}(b - a - p_{k+1})}{a + p_{k+1}} > 0$ since $p_{k+1} < b - a$. Therefore,

$p_{k+2} > p_{k+1}$. Thus, the assertion is true for $n = k + 1$. It is therefore true for all n by mathematical induction.

A similar proof by induction shows that if $p_0 > b - a$, then $p_n > b - a$ and $\{p_n\}$ is decreasing.

In either case the sequence $\{p_n\}$ is bounded and monotonic, so it is convergent by the Monotonic Sequence Theorem.

It then follows from part (a) that $\lim\limits_{n\to\infty} p_n = b - a$.

8.2   Series

1. (a) A sequence is an ordered list of numbers whereas a series is the *sum* of a list of numbers.

   (b) A series is convergent if the sequence of partial sums is a convergent sequence. A series is divergent if it is not convergent.

3.

n	s_n
1	-2.40000
2	-1.92000
3	-2.01600
4	-1.99680
5	-2.00064
6	-1.99987
7	-2.00003
8	-1.99999
9	-2.00000
10	-2.00000

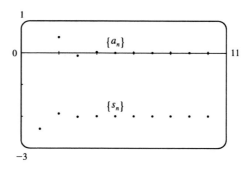

From the graph and the table, it seems that the series converges to -2. In fact, it is a geometric

series with $a = -2.4$ and $r = -\frac{1}{5}$, so its sum is $\displaystyle\sum_{n=1}^{\infty} \frac{12}{(-5)^n} = \frac{-2.4}{1 - \left(-\frac{1}{5}\right)} = \frac{-2.4}{1.2} = -2.$

Note that the dot corresponding to $n = 1$ is part of both $\{a_n\}$ and $\{s_n\}$.

TI-86 Note: To graph $\{a_n\}$ and $\{s_n\}$, set your calculator to Param mode and DrawDot mode. (DrawDot is under

GRAPH, MORE, FORMT (F3).) Now under E(t) = make the assignments: xt1=t, yt1=12/(-5)^t, xt2=t,

yt2=sum seq(yt1,t,1,t,1). (sum and seq are under LIST, OPS (F5), MORE.) Under WIND use

1,10,1,0,10,1,-3,1,1 to obtain a graph similar to the one above. Then use TRACE (F4) to see the values.

5.

n	s_n
1	0.44721
2	1.15432
3	1.98637
4	2.88080
5	3.80927
6	4.75796
7	5.71948
8	6.68962
9	7.66581
10	8.64639

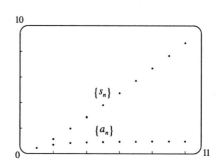

The series $\displaystyle\sum_{n=1}^{\infty} \frac{n}{\sqrt{n^2 + 4}}$ diverges, since its terms do not approach 0.

7.

n	s_n
1	0.29289
2	0.42265
3	0.50000
4	0.55279
5	0.59175
6	0.62204
7	0.64645
8	0.66667
9	0.68377
10	0.69849

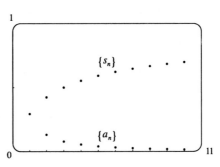

From the graph and the table, it seems that the series converges.

$$\sum_{n=1}^{k}\left(\frac{1}{\sqrt{n}}-\frac{1}{\sqrt{n+1}}\right)=\left(\frac{1}{\sqrt{1}}-\frac{1}{\sqrt{2}}\right)+\left(\frac{1}{\sqrt{2}}-\frac{1}{\sqrt{3}}\right)+\cdots+\left(\frac{1}{\sqrt{k}}-\frac{1}{\sqrt{k+1}}\right)$$

$$=1-\frac{1}{\sqrt{k+1}},$$

so $\displaystyle\sum_{n=1}^{\infty}\left(\frac{1}{\sqrt{n}}-\frac{1}{\sqrt{n+1}}\right)=\lim_{k\to\infty}\left(1-\frac{1}{\sqrt{k+1}}\right)=1.$

9. (a) $\displaystyle\lim_{n\to\infty}a_n=\lim_{n\to\infty}\frac{2n}{3n+1}=\frac{2}{3}$, so the *sequence* $\{a_n\}$ is convergent by (8.1.1).

(b) Since $\displaystyle\lim_{n\to\infty}a_n=\frac{2}{3}\neq 0$, the *series* $\displaystyle\sum_{n=1}^{\infty}a_n$ is divergent by the Test for Divergence.

11. $3-4+\frac{16}{3}-\frac{64}{9}+\cdots$ is a geometric series with ratio $r=-\frac{4}{3}$. Since $|r|=\frac{4}{3}>1$, the series diverges.

13. $10-2+0.4-0.08+\cdots$ is a geometric series with ratio $-\frac{2}{10}=-\frac{1}{5}$. Since $|r|=\frac{1}{5}<1$, the series converges to

$$\frac{a}{1-r}=\frac{10}{1-(-1/5)}=\frac{10}{6/5}=\frac{50}{6}=\frac{25}{3}.$$

15. $\displaystyle\sum_{n=1}^{\infty}6(0.9)^{n-1}$ is a geometric series with first term $a=6$ and ratio $r=0.9$. Since $|r|=0.9<1$, the series converges to

$$\frac{a}{1-r}=\frac{6}{1-0.9}=\frac{6}{0.1}=60.$$

17. $\displaystyle\sum_{n=0}^{\infty}\frac{\pi^n}{3^{n+1}}=\frac{1}{3}\sum_{n=0}^{\infty}\left(\frac{\pi}{3}\right)^n$ is a geometric series with ratio $r=\frac{\pi}{3}$. Since $|r|>1$, the series diverges.

19. $\displaystyle\sum_{n=1}^{\infty}\frac{n-1}{3n-1}$ diverges by the Test for Divergence since $\displaystyle\lim_{n\to\infty}a_n=\lim_{n\to\infty}\frac{n-1}{3n-1}=\frac{1}{3}\neq 0$.

21. $\displaystyle\sum_{k=2}^{\infty}\frac{k^2}{k^2-1}$ diverges by the Test for Divergence since $\displaystyle\lim_{k\to\infty}a_k=\lim_{k\to\infty}\frac{k^2}{k^2-1}=1\neq 0$.

23. Converges.

$$\sum_{n=1}^{\infty}\frac{1+2^n}{3^n}=\sum_{n=1}^{\infty}\left(\frac{1}{3^n}+\frac{2^n}{3^n}\right)=\sum_{n=1}^{\infty}\left[\left(\frac{1}{3}\right)^n+\left(\frac{2}{3}\right)^n\right]\qquad\text{[sum of two convergent geometric series]}$$

$$=\frac{1/3}{1-1/3}+\frac{2/3}{1-2/3}=\frac{1}{2}+2=\frac{5}{2}$$

25. $\displaystyle\sum_{n=1}^{\infty}\sqrt[n]{2}=2+\sqrt{2}+\sqrt[3]{2}+\sqrt[4]{2}+\cdots$ diverges by the Test for Divergence since

$$\lim_{n\to\infty}a_n=\lim_{n\to\infty}\sqrt[n]{2}=\lim_{n\to\infty}2^{1/n}=2^0=1\neq 0.$$

27. $\sum\limits_{n=1}^{\infty} \arctan n$ diverges by the Test for Divergence since $\lim\limits_{n\to\infty} a_n = \lim\limits_{n\to\infty} \arctan n = \frac{\pi}{2} \neq 0$.

29. $\sum\limits_{n=1}^{\infty} \dfrac{1}{e^n} = \sum\limits_{n=1}^{\infty} \left(\dfrac{1}{e}\right)^n$ is a geometric series with first term $a = \dfrac{1}{e}$ and ratio $r = \dfrac{1}{e}$. Since $|r| = \dfrac{1}{e} < 1$, the series converges

to $\dfrac{1/e}{1 - 1/e} = \dfrac{1/e}{1 - 1/e} \cdot \dfrac{e}{e} = \dfrac{1}{e-1}$. By Example 6, $\sum\limits_{n=1}^{\infty} \dfrac{1}{n(n+1)} = 1$. Thus, by Theorem 8(ii),

$$\sum_{n=1}^{\infty}\left(\frac{1}{e^n} + \frac{1}{n(n+1)}\right) = \sum_{n=1}^{\infty}\frac{1}{e^n} + \sum_{n=1}^{\infty}\frac{1}{n(n+1)} = \frac{1}{e-1} + 1 = \frac{1}{e-1} + \frac{e-1}{e-1} = \frac{e}{e-1}.$$

31. Using partial fractions, the partial sums of the series $\sum\limits_{n=2}^{\infty} \dfrac{2}{n^2 - 1}$ are

$$s_n = \sum_{i=2}^{n} \frac{2}{(i-1)(i+1)} = \sum_{i=2}^{n}\left(\frac{1}{i-1} - \frac{1}{i+1}\right)$$

$$= \left(1 - \frac{1}{3}\right) + \left(\frac{1}{2} - \frac{1}{4}\right) + \left(\frac{1}{3} - \frac{1}{5}\right) + \cdots + \left(\frac{1}{n-3} - \frac{1}{n-1}\right) + \left(\frac{1}{n-2} - \frac{1}{n}\right)$$

This sum is a telescoping series and $s_n = 1 + \dfrac{1}{2} - \dfrac{1}{n-1} - \dfrac{1}{n}$.

Thus, $\sum\limits_{n=2}^{\infty} \dfrac{2}{n^2 - 1} = \lim\limits_{n\to\infty} s_n = \lim\limits_{n\to\infty}\left(1 + \dfrac{1}{2} - \dfrac{1}{n-1} - \dfrac{1}{n}\right) = \dfrac{3}{2}$.

33. For the series $\sum\limits_{n=1}^{\infty} \dfrac{3}{n(n+3)}$, $s_n = \sum\limits_{i=1}^{n} \dfrac{3}{i(i+3)} = \sum\limits_{i=1}^{n}\left(\dfrac{1}{i} - \dfrac{1}{i+3}\right)$ [using partial fractions]. The latter sum is

$$\left(1 - \tfrac{1}{4}\right) + \left(\tfrac{1}{2} - \tfrac{1}{5}\right) + \left(\tfrac{1}{3} - \tfrac{1}{6}\right) + \left(\tfrac{1}{4} - \tfrac{1}{7}\right) + \cdots + \left(\tfrac{1}{n-3} - \tfrac{1}{n}\right) + \left(\tfrac{1}{n-2} - \tfrac{1}{n+1}\right) + \left(\tfrac{1}{n-1} - \tfrac{1}{n+2}\right) + \left(\tfrac{1}{n} - \tfrac{1}{n+3}\right)$$

$$= 1 + \tfrac{1}{2} + \tfrac{1}{3} - \tfrac{1}{n+1} - \tfrac{1}{n+2} - \tfrac{1}{n+3} \quad \text{[telescoping series]}$$

Thus, $\sum\limits_{n=1}^{\infty} \dfrac{3}{n(n+3)} = \lim\limits_{n\to\infty} s_n = \lim\limits_{n\to\infty}\left(1 + \tfrac{1}{2} + \tfrac{1}{3} - \tfrac{1}{n+1} - \tfrac{1}{n+2} - \tfrac{1}{n+3}\right) = 1 + \tfrac{1}{2} + \tfrac{1}{3} = \tfrac{11}{6}$. Converges

35. (a) Many people would guess that $x < 1$, but note that x consists of an infinite number of 9s.

(b) $x = 0.99999\ldots = \dfrac{9}{10} + \dfrac{9}{100} + \dfrac{9}{1000} + \dfrac{9}{10,000} + \cdots = \sum\limits_{n=1}^{\infty} \dfrac{9}{10^n}$, which is a geometric series with $a_1 = 0.9$ and

$r = 0.1$. Its sum is $\dfrac{0.9}{1 - 0.1} = \dfrac{0.9}{0.9} = 1$, that is, $x = 1$.

(c) The number 1 has two decimal representations, $1.00000\ldots$ and $0.99999\ldots$.

(d) Except for 0, all rational numbers that have a terminating decimal representation can be written in more than one way. For

example, 0.5 can be written as $0.49999\ldots$ as well as $0.50000\ldots$.

37. $0.\overline{2} = \dfrac{2}{10} + \dfrac{2}{10^2} + \cdots$ is a geometric series with $a = \dfrac{2}{10}$ and $r = \dfrac{1}{10}$. It converges to $\dfrac{a}{1-r} = \dfrac{2/10}{1 - 1/10} = \dfrac{2}{9}$.

39. $1.5\overline{342} = 1.53 + \dfrac{42}{10^4} + \dfrac{42}{10^6} + \cdots$. Now $\dfrac{42}{10^4} + \dfrac{42}{10^6} + \cdots$ is a geometric series with $a = \dfrac{42}{10^4}$ and $r = \dfrac{1}{10^2}$.

It converges to $\dfrac{a}{1-r} = \dfrac{42/10^4}{1 - 1/10^2} = \dfrac{42/10^4}{99/10^2} = \dfrac{42}{9900}$.

Thus, $1.5\overline{342} = 1.53 + \dfrac{42}{9900} = \dfrac{153}{100} + \dfrac{42}{9900} = \dfrac{15,147}{9900} + \dfrac{42}{9900} = \dfrac{15,189}{9900}$ or $\dfrac{5063}{3300}$.

41. $\sum_{n=1}^{\infty} \dfrac{x^n}{3^n} = \sum_{n=1}^{\infty} \left(\dfrac{x}{3}\right)^n$ is a geometric series with $r = \dfrac{x}{3}$, so the series converges $\Leftrightarrow |r| < 1 \Leftrightarrow \dfrac{|x|}{3} < 1 \Leftrightarrow |x| < 3$;

that is, $-3 < x < 3$. In that case, the sum of the series is $\dfrac{a}{1-r} = \dfrac{x/3}{1-x/3} = \dfrac{x/3}{1-x/3} \cdot \dfrac{3}{3} = \dfrac{x}{3-x}$.

43. $\sum_{n=0}^{\infty} \dfrac{\cos^n x}{2^n}$ is a geometric series with first term 1 and ratio $r = \dfrac{\cos x}{2}$, so it converges $\Leftrightarrow |r| < 1$. But $|r| = \dfrac{|\cos x|}{2} \leq \dfrac{1}{2}$

for all x. Thus, the series converges for all real values of x and the sum of the series is $\dfrac{1}{1-(\cos x)/2} = \dfrac{2}{2-\cos x}$.

45. After defining f, We use `convert (f,parfrac);` in Maple, `Apart` in Mathematica, or `Expand Rational` and

`Simplify` in Derive to find that the general term is $\dfrac{3n^2 + 3n + 1}{(n^2 + n)^3} = \dfrac{1}{n^3} - \dfrac{1}{(n+1)^3}$. So the nth partial sum is

$$s_n = \sum_{k=1}^{n} \left(\dfrac{1}{k^3} - \dfrac{1}{(k+1)^3}\right) = \left(1 - \dfrac{1}{2^3}\right) + \left(\dfrac{1}{2^3} - \dfrac{1}{3^3}\right) + \cdots + \left(\dfrac{1}{n^3} - \dfrac{1}{(n+1)^3}\right) = 1 - \dfrac{1}{(n+1)^3}$$

The series converges to $\lim_{n\to\infty} s_n = 1$. This can be confirmed by directly computing the sum using `sum (f,1..infinity);`

(in Maple), `Sum[f,{n,1,Infinity}]` (in Mathematica), or `Calculus Sum` (from 1 to ∞) and `Simplify` (in Derive).

47. For $n = 1$, $a_1 = 0$ since $s_1 = 0$. For $n > 1$,

$$a_n = s_n - s_{n-1} = \dfrac{n-1}{n+1} - \dfrac{(n-1)-1}{(n-1)+1} = \dfrac{(n-1)n - (n+1)(n-2)}{(n+1)n} = \dfrac{2}{n(n+1)}$$

Also, $\sum_{n=1}^{\infty} a_n = \lim_{n\to\infty} s_n = \lim_{n\to\infty} \dfrac{1-1/n}{1+1/n} = 1$.

49. (a) After the first pill is taken, 100 mg of the drug is in the body. After the second pill is taken, 100 mg plus 100(5%) mg

remains in the body. After the third pill is taken 100 mg plus 100(5%) mg plus $[100(5\%)](5\%)$ mg remains in the body.

This gives us $100 + 100(0.05) + 100(0.05)^2 = 105.25$ mg of the drug remaining in the body after the patient takes three

pills.

(b) Continuing the pattern established in part (a), we get $100 + 100(0.05) + 100(0.05)^2 + \cdots + 100(0.05)^{n-1}$ mg after n

pills are taken. By (3), this sum is $\dfrac{100(1 - 0.05^n)}{1 - 0.05}$.

(c) The amount of the drug remaining in the body in the long run can be approximated by summing the infinite series with

$a_1 = 100$ and $r = 0.05$. This sum is $\dfrac{100}{1 - 0.05} = \dfrac{100}{0.95} \approx 105.26$ mg.

51. (a) The first step in the chain occurs when the local government spends D dollars. The people who receive it spend a fraction c

of those D dollars, that is, Dc dollars. Those who receive the Dc dollars spend a fraction c of it, that is, Dc^2 dollars.

Continuing in this way, we see that the total spending after n transactions is

$$S_n = D + Dc + Dc^2 + \cdots + Dc^{n-1} = \dfrac{D(1 - c^n)}{1 - c} \text{ by (3).}$$

(b) $\lim_{n\to\infty} S_n = \lim_{n\to\infty} \dfrac{D(1 - c^n)}{1 - c} = \dfrac{D}{1 - c} \lim_{n\to\infty} (1 - c^n) = \dfrac{D}{1 - c} \left[\text{since } 0 < c < 1 \Rightarrow \lim_{n\to\infty} c^n = 0\right]$

$= \dfrac{D}{s} \quad [\text{since } c + s = 1] = kD \quad [\text{since } k = 1/s]$

If $c = 0.8$, then $s = 1 - c = 0.2$ and the multiplier is $k = 1/s = 5$.

53. $\sum_{n=2}^{\infty} (1+c)^{-n}$ is a geometric series with $a = (1+c)^{-2}$ and $r = (1+c)^{-1}$, so the series converges when

$$\left| (1+c)^{-1} \right| < 1 \quad \Leftrightarrow \quad |1+c| > 1 \quad \Leftrightarrow \quad 1+c > 1 \text{ or } 1+c < -1 \quad \Leftrightarrow \quad c > 0 \text{ or } c < -2. \text{ We calculate the sum of the}$$

series and set it equal to 2: $\dfrac{(1+c)^{-2}}{1-(1+c)^{-1}} = 2 \quad \Leftrightarrow \quad \left(\dfrac{1}{1+c}\right)^2 = 2 - 2\left(\dfrac{1}{1+c}\right) \quad \Leftrightarrow \quad 1 = 2(1+c)^2 - 2(1+c) \quad \Leftrightarrow$

$2c^2 + 2c - 1 = 0 \quad \Leftrightarrow \quad c = \dfrac{-2 \pm \sqrt{12}}{4} = \dfrac{\pm\sqrt{3}-1}{2}$. However, the negative root is inadmissible because $-2 < \dfrac{-\sqrt{3}-1}{2} < 0$.

So $c = \dfrac{\sqrt{3}-1}{2}$.

55. $e^{s_n} = e^{1 + \frac{1}{2} + \frac{1}{3} + \cdots + \frac{1}{n}} = e^1 e^{1/2} e^{1/3} \cdots e^{1/n} > (1+1)\left(1 + \tfrac{1}{2}\right)\left(1 + \tfrac{1}{3}\right) \cdots \left(1 + \tfrac{1}{n}\right)$ $[e^x > 1 + x]$

$$= \frac{2}{1} \frac{3}{2} \frac{4}{3} \cdots \frac{n+1}{n} = n + 1$$

Thus, $e^{s_n} > n + 1$ and $\lim_{n \to \infty} e^{s_n} = \infty$. Since $\{s_n\}$ is increasing, $\lim_{n \to \infty} s_n = \infty$, implying that the harmonic series is

divergent.

57. Let d_n be the diameter of C_n. We draw lines from the centers of the C_i to

the center of D (or C), and using the Pythagorean Theorem, we can write

$$1^2 + \left(1 - \tfrac{1}{2}d_1\right)^2 = \left(1 + \tfrac{1}{2}d_1\right)^2 \quad \Leftrightarrow$$

$1 = \left(1 + \tfrac{1}{2}d_1\right)^2 - \left(1 - \tfrac{1}{2}d_1\right)^2 = 2d_1$ [difference of squares] $\Rightarrow$ $d_1 = \tfrac{1}{2}$.

Similarly,

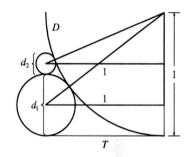

$$1 = \left(1 + \tfrac{1}{2}d_2\right)^2 - \left(1 - d_1 - \tfrac{1}{2}d_2\right)^2 = 2d_2 + 2d_1 - d_1^2 - d_1 d_2$$

$$= (2 - d_1)(d_1 + d_2) \quad \Leftrightarrow$$

$d_2 = \dfrac{1}{2 - d_1} - d_1 = \dfrac{(1 - d_1)^2}{2 - d_1}$, $1 = \left(1 + \tfrac{1}{2}d_3\right)^2 - \left(1 - d_1 - d_2 - \tfrac{1}{2}d_3\right)^2 \quad \Leftrightarrow \quad d_3 = \dfrac{[1 - (d_1 + d_2)]^2}{2 - (d_1 + d_2)}$, and in general,

$d_{n+1} = \dfrac{\left(1 - \sum_{i=1}^{n} d_i\right)^2}{2 - \sum_{i=1}^{n} d_i}$. If we actually calculate d_2 and d_3 from the formulas above, we find that they are $\dfrac{1}{6} = \dfrac{1}{2 \cdot 3}$ and

$\dfrac{1}{12} = \dfrac{1}{3 \cdot 4}$ respectively, so we suspect that in general, $d_n = \dfrac{1}{n(n+1)}$. To prove this, we use induction: Assume that for all

$k \le n$, $d_k = \dfrac{1}{k(k+1)} = \dfrac{1}{k} - \dfrac{1}{k+1}$. Then $\sum_{i=1}^{n} d_i = 1 - \dfrac{1}{n+1} = \dfrac{n}{n+1}$ [telescoping sum]. Substituting this into our

formula for d_{n+1}, we get $d_{n+1} = \dfrac{\left[1 - \dfrac{n}{n+1}\right]^2}{2 - \left(\dfrac{n}{n+1}\right)} = \dfrac{\dfrac{1}{(n+1)^2}}{\dfrac{n+2}{n+1}} = \dfrac{1}{(n+1)(n+2)}$, and the induction is complete.

Now, we observe that the partial sums $\sum_{i=1}^{n} d_i$ of the diameters of the circles approach 1 as $n \to \infty$; that is,

$$\sum_{n=1}^{\infty} a_n = \sum_{n=1}^{\infty} \frac{1}{n(n+1)} = 1, \text{ which is what we wanted to prove.}$$

59. The series $1 - 1 + 1 - 1 + 1 - 1 + \cdots$ diverges (geometric series with $r = -1$) so we cannot say that

$0 = 1 - 1 + 1 - 1 + 1 - 1 + \cdots$.

61. Suppose on the contrary that $\sum(a_n + b_n)$ converges. Then $\sum(a_n + b_n)$ and $\sum a_n$ are convergent series. So by Theorem 8(iii), $\sum[(a_n + b_n) - a_n]$ would also be convergent. But $\sum[(a_n + b_n) - a_n] = \sum b_n$, a contradiction, since $\sum b_n$ is given to be divergent.

63. The partial sums $\{s_n\}$ form an increasing sequence, since $s_n - s_{n-1} = a_n > 0$ for all n. Also, the sequence $\{s_n\}$ is bounded since $s_n \leq 1000$ for all n. So by the Monotonic Sequence Theorem, the sequence of partial sums converges, that is, the series $\sum a_n$ is convergent.

65. (a) At the first step, only the interval $\left(\frac{1}{3}, \frac{2}{3}\right)$ (length $\frac{1}{3}$) is removed. At the second step, we remove the intervals $\left(\frac{1}{9}, \frac{2}{9}\right)$ and $\left(\frac{7}{9}, \frac{8}{9}\right)$, which have a total length of $2 \cdot \left(\frac{1}{3}\right)^2$. At the third step, we remove 2^2 intervals, each of length $\left(\frac{1}{3}\right)^3$. In general, at the nth step we remove 2^{n-1} intervals, each of length $\left(\frac{1}{3}\right)^n$, for a length of $2^{n-1} \cdot \left(\frac{1}{3}\right)^n = \frac{1}{3}\left(\frac{2}{3}\right)^{n-1}$. Thus, the total length of all removed intervals is $\sum_{n=1}^{\infty} \frac{1}{3}\left(\frac{2}{3}\right)^{n-1} = \frac{1/3}{1 - 2/3} = 1$ [geometric series with $a = \frac{1}{3}$ and $r = \frac{2}{3}$]. Notice that at the nth step, the leftmost interval that is removed is $\left(\left(\frac{1}{3}\right)^n, \left(\frac{2}{3}\right)^n\right)$, so we never remove 0, and 0 is in the Cantor set. Also, the rightmost interval removed is $\left(1 - \left(\frac{2}{3}\right)^n, 1 - \left(\frac{1}{3}\right)^n\right)$, so 1 is never removed. Some other numbers in the Cantor set are $\frac{1}{3}, \frac{2}{3}, \frac{1}{9}, \frac{2}{9}, \frac{7}{9}$, and $\frac{8}{9}$.

(b) The area removed at the first step is $\frac{1}{9}$; at the second step, $8 \cdot \left(\frac{1}{9}\right)^2$; at the third step, $(8)^2 \cdot \left(\frac{1}{9}\right)^3$. In general, the area removed at the nth step is $(8)^{n-1}\left(\frac{1}{9}\right)^n = \frac{1}{9}\left(\frac{8}{9}\right)^{n-1}$, so the total area of all removed squares is

$$\sum_{n=1}^{\infty} \frac{1}{9}\left(\frac{8}{9}\right)^{n-1} = \frac{1/9}{1 - 8/9} = 1.$$

67. (a) For $\sum_{n=1}^{\infty} \frac{n}{(n+1)!}$, $s_1 = \frac{1}{1 \cdot 2} = \frac{1}{2}$, $s_2 = \frac{1}{2} + \frac{2}{1 \cdot 2 \cdot 3} = \frac{5}{6}$, $s_3 = \frac{5}{6} + \frac{3}{1 \cdot 2 \cdot 3 \cdot 4} = \frac{23}{24}$.

$s_4 = \frac{23}{24} + \frac{4}{1 \cdot 2 \cdot 3 \cdot 4 \cdot 5} = \frac{119}{120}$. The denominators are $(n+1)!$, so a guess would be $s_n = \frac{(n+1)! - 1}{(n+1)!}$.

(b) For $n = 1$, $s_1 = \frac{1}{2} = \frac{2! - 1}{2!}$, so the formula holds for $n = 1$. Assume $s_k = \frac{(k+1)! - 1}{(k+1)!}$. Then

$$s_{k+1} = \frac{(k+1)! - 1}{(k+1)!} + \frac{k+1}{(k+2)!} = \frac{(k+1)! - 1}{(k+1)!} + \frac{k+1}{(k+1)!(k+2)} = \frac{(k+2)! - (k+2) + k + 1}{(k+2)!}$$

$$= \frac{(k+2)! - 1}{(k+2)!}$$

Thus, the formula is true for $n = k + 1$. So by induction, the guess is correct.

(c) $\lim_{n \to \infty} s_n = \lim_{n \to \infty} \frac{(n+1)! - 1}{(n+1)!} = \lim_{n \to \infty} \left[1 - \frac{1}{(n+1)!}\right] = 1$ and so $\sum_{n=1}^{\infty} \frac{n}{(n+1)!} = 1$.

8.3   The Integral and Comparison Tests; Estimating Sums

1. The picture shows that $a_2 = \dfrac{1}{2^{1.3}} < \displaystyle\int_1^2 \dfrac{1}{x^{1.3}}\, dx,$

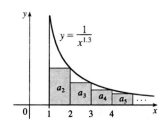

$a_3 = \dfrac{1}{3^{1.3}} < \displaystyle\int_2^3 \dfrac{1}{x^{1.3}}\, dx,$ and so on, so $\displaystyle\sum_{n=2}^{\infty} \dfrac{1}{n^{1.3}} < \int_1^{\infty} \dfrac{1}{x^{1.3}}\, dx.$ The

integral converges by (5.10.2) with $p = 1.3 > 1$, so the series converges.

3. (a) We cannot say anything about $\sum a_n$. If $a_n > b_n$ for all n and $\sum b_n$ is convergent, then $\sum a_n$ could be convergent or

divergent. (See the note after Example 2.)

(b) If $a_n < b_n$ for all n, then $\sum a_n$ is convergent. [This is part (i) of the Comparison Test.]

5. $\displaystyle\sum_{n=1}^{\infty} n^b$ is a p-series with $p = -b$. $\displaystyle\sum_{n=1}^{\infty} b^n$ is a geometric series. By (1), the p-series is convergent if $p > 1$. In this case,

$\displaystyle\sum_{n=1}^{\infty} n^b = \sum_{n=1}^{\infty} \left(1/n^{-b}\right),$ so $-b > 1 \;\Leftrightarrow\; b < -1$ are the values for which the series converge. A geometric series

$\displaystyle\sum_{n=1}^{\infty} ar^{n-1}$ converges if $|r| < 1$, so $\displaystyle\sum_{n=1}^{\infty} b^n$ converges if $|b| < 1 \;\Leftrightarrow\; -1 < b < 1.$

7. The function $f(x) = 1/\sqrt[5]{x} = x^{-1/5}$ is continuous, positive, and decreasing on $[1, \infty)$, so the Integral Test applies.

$\displaystyle\int_1^{\infty} x^{-1/5}\, dx = \lim_{t\to\infty} \int_1^t x^{-1/5}\, dx = \lim_{t\to\infty} \left[\tfrac{5}{4} x^{4/5}\right]_1^t = \lim_{t\to\infty} \left(\tfrac{5}{4} t^{4/5} - \tfrac{5}{4}\right) = \infty,$ so $\displaystyle\sum_{n=1}^{\infty} 1/\sqrt[5]{n}$ diverges.

9. $\dfrac{n}{2n^3 + 1} < \dfrac{n}{2n^3} = \dfrac{1}{2n^2} < \dfrac{1}{n^2}$ for all $n \geq 1$, so $\displaystyle\sum_{n=1}^{\infty} \dfrac{n}{2n^3 + 1}$ converges by comparison with $\displaystyle\sum_{n=1}^{\infty} \dfrac{1}{n^2},$ which converges

because it is a p-series with $p = 2 > 1$.

11. The series $\displaystyle\sum_{n=1}^{\infty} \dfrac{1}{n^{0.85}}$ is a p-series with $p = 0.85 \leq 1$, so it diverges by (1). Therefore, the series $\displaystyle\sum_{n=1}^{\infty} \dfrac{2}{n^{0.85}}$ must also diverge,

for if it converged, then $\displaystyle\sum_{n=1}^{\infty} \dfrac{1}{n^{0.85}}$ would have to converge [by Theorem 8(i) in Section 8.2].

13. $1 + \dfrac{1}{8} + \dfrac{1}{27} + \dfrac{1}{64} + \dfrac{1}{125} + \cdots = \displaystyle\sum_{n=1}^{\infty} \dfrac{1}{n^3}.$ This is a p-series with $p = 3 > 1$, so it converges by (1).

15. $f(x) = xe^{-x}$ is continuous and positive on $[1, \infty)$. $f'(x) = -xe^{-x} + e^{-x} = e^{-x}(1 - x) < 0$ for $x > 1$, so f is decreasing

on $[1, \infty)$. Thus, the Integral Test applies.

$\displaystyle\int_1^{\infty} xe^{-x}\, dx = \lim_{b\to\infty} \int_1^b xe^{-x}\, dx = \lim_{b\to\infty} \left[-xe^{-x} - e^{-x}\right]_1^b \quad \text{[by parts]} \quad = \lim_{b\to\infty} \left[-be^{-b} - e^{-b} + e^{-1} + e^{-1}\right] = 2/e$

since $\displaystyle\lim_{b\to\infty} be^{-b} = \lim_{b\to\infty} (b/e^b) \overset{\text{H}}{=} \lim_{b\to\infty} (1/e^b) = 0$ and $\displaystyle\lim_{b\to\infty} e^{-b} = 0.$ Thus, $\sum_{n=1}^{\infty} ne^{-n}$ converges.

17. $f(x) = \dfrac{1}{x \ln x}$ is continuous and positive on $[2, \infty)$, and also decreasing since $f'(x) = -\dfrac{1 + \ln x}{x^2(\ln x)^2} < 0$ for $x > 2$, so we can

use the Integral Test. $\displaystyle\int_2^\infty \dfrac{1}{x \ln x}\, dx = \lim_{t\to\infty} [\ln(\ln x)]_2^t = \lim_{t\to\infty} [\ln(\ln t) - \ln(\ln 2)] = \infty$, so the series $\displaystyle\sum_{n=2}^\infty \dfrac{1}{n \ln n}$ diverges.

19. $\dfrac{\cos^2 n}{n^2 + 1} \le \dfrac{1}{n^2 + 1} < \dfrac{1}{n^2}$, so the series $\displaystyle\sum_{n=1}^\infty \dfrac{\cos^2 n}{n^2 + 1}$ converges by comparison with the p-series $\displaystyle\sum_{n=1}^\infty \dfrac{1}{n^2}$ $[p = 2 > 1]$.

21. $\dfrac{n-1}{n\,4^n}$ is positive for $n > 1$ and $\dfrac{n-1}{n\,4^n} < \dfrac{n}{n\,4^n} = \dfrac{1}{4^n} = \left(\dfrac{1}{4}\right)^n$, so $\displaystyle\sum_{n=1}^\infty \dfrac{n-1}{n\,4^n}$ converges by comparison with the convergent

geometric series $\displaystyle\sum_{n=1}^\infty \left(\dfrac{1}{4}\right)^n$.

23. $1 + \dfrac{1}{3} + \dfrac{1}{5} + \dfrac{1}{7} + \dfrac{1}{9} + \cdots = \displaystyle\sum_{n=1}^\infty \dfrac{1}{2n-1}$. The function $f(x) = \dfrac{1}{2x-1}$ is

continuous, positive, and decreasing on $[1, \infty)$, so the Integral Test applies.

$\displaystyle\int_1^\infty \dfrac{1}{2x-1}\, dx = \lim_{t\to\infty}\int_1^t \dfrac{1}{2x-1}\, dx = \lim_{t\to\infty} \left[\tfrac{1}{2} \ln|2x-1|\right]_1^t = \tfrac{1}{2}\lim_{t\to\infty}(\ln(2t-1) - 0) = \infty$, so the series $\displaystyle\sum_{n=1}^\infty \dfrac{1}{2n-1}$

diverges.

25. Use the Limit Comparison Test with $a_n = \dfrac{1 + 4^n}{1 + 3^n}$ and $b_n = \dfrac{4^n}{3^n}$:

$$\lim_{n\to\infty} \dfrac{a_n}{b_n} = \lim_{n\to\infty} \dfrac{\frac{1+4^n}{1+3^n}}{\frac{4^n}{3^n}} = \lim_{n\to\infty} \dfrac{1+4^n}{1+3^n}\cdot\dfrac{3^n}{4^n} = \lim_{n\to\infty} \dfrac{1+4^n}{4^n}\cdot\dfrac{3^n}{1+3^n} = \lim_{n\to\infty} \left(\dfrac{1}{4^n} + 1\right)\cdot\dfrac{1}{\frac{1}{3^n} + 1} = 1 > 0$$

Since the geometric series $\sum b_n = \sum \left(\tfrac{4}{3}\right)^n$ diverges, so does $\displaystyle\sum_{n=1}^\infty \dfrac{1 + 4^n}{1 + 3^n}$. Alternatively, use the Comparison Test with

$\dfrac{1 + 4^n}{1 + 3^n} > \dfrac{1 + 4^n}{3^n + 3^n} > \dfrac{4^n}{2(3^n)} = \dfrac{1}{2}\left(\dfrac{4}{3}\right)^n$ or use the Test for Divergence.

27. $\dfrac{2 + (-1)^n}{n\sqrt{n}} \le \dfrac{3}{n\sqrt{n}}$, and $\displaystyle\sum_{n=1}^\infty \dfrac{3}{n\sqrt{n}}$ converges because it is a constant multiple of the convergent p-series $\displaystyle\sum_{n=1}^\infty \dfrac{1}{n\sqrt{n}}$

$\left[p = \tfrac{3}{2} > 1\right]$, so the given series converges by the Comparison Test.

29. Use the Limit Comparison Test with $a_n = \sin\left(\dfrac{1}{n}\right)$ and $b_n = \dfrac{1}{n}$. Then $\sum a_n$ and $\sum b_n$ are series with positive terms and

$\displaystyle\lim_{n\to\infty} \dfrac{a_n}{b_n} = \lim_{n\to\infty} \dfrac{\sin(1/n)}{1/n} = \lim_{\theta\to 0} \dfrac{\sin\theta}{\theta} = 1 > 0$. Since $\displaystyle\sum_{n=1}^\infty b_n$ is the divergent harmonic series,

$\displaystyle\sum_{n=1}^\infty \sin(1/n)$ also diverges. [Note that we could also use l'Hospital's Rule to evaluate the limit:

$\displaystyle\lim_{x\to\infty} \dfrac{\sin(1/x)}{1/x} \overset{H}{=} \lim_{x\to\infty} \dfrac{\cos(1/x)\cdot(-1/x^2)}{-1/x^2} = \lim_{x\to\infty} \cos\dfrac{1}{x} = \cos 0 = 1.$]

31. We have already shown (in Exercise 17) that when $p = 1$ the series $\sum\limits_{n=2}^{\infty} \dfrac{1}{n(\ln n)^p}$ diverges, so assume that $p \neq 1$.

$f(x) = \dfrac{1}{x(\ln x)^p}$ is continuous and positive on $[2, \infty)$, and $f'(x) = -\dfrac{p + \ln x}{x^2(\ln x)^{p+1}} < 0$ if $x > e^{-p}$, so that f is eventually decreasing and we can use the Integral Test.

$$\int_2^{\infty} \frac{1}{x(\ln x)^p}\,dx = \lim_{t \to \infty}\left[\frac{(\ln x)^{1-p}}{1-p}\right]_2^t \quad [\text{for } p \neq 1] = \lim_{t \to \infty}\left[\frac{(\ln t)^{1-p}}{1-p} - \frac{(\ln 2)^{1-p}}{1-p}\right]$$

This limit exists whenever $1 - p < 0 \iff p > 1$, so the series converges for $p > 1$.

33. (a) $f(x) = \dfrac{1}{x^2}$ is positive and continuous and $f'(x) = -\dfrac{2}{x^3}$ is negative for $x > 0$, and so the Integral Test applies.

$$\sum_{n=1}^{\infty}\frac{1}{n^2} \approx s_{10} = \frac{1}{1^2} + \frac{1}{2^2} + \frac{1}{3^2} + \cdots + \frac{1}{10^2} \approx 1.549768.$$

$R_{10} \leq \int_{10}^{\infty}\dfrac{1}{x^2}\,dx = \lim\limits_{t \to \infty}\left[\dfrac{-1}{x}\right]_{10}^{t} = \lim\limits_{t \to \infty}\left(-\dfrac{1}{t} + \dfrac{1}{10}\right) = \dfrac{1}{10}$, so the error is at most 0.1.

(b) $s_{10} + \displaystyle\int_{11}^{\infty}\frac{1}{x^2}\,dx \leq s \leq s_{10} + \int_{10}^{\infty}\frac{1}{x^2}\,dx \implies s_{10} + \frac{1}{11} \leq s \leq s_{10} + \frac{1}{10} \implies$

$1.549768 + 0.090909 = 1.640677 \leq s \leq 1.549768 + 0.1 = 1.649768$, so we get $s \approx 1.64522$ (the average of 1.640677 and 1.649768) with error ≤ 0.005 (the maximum of $1.649768 - 1.64522$ and $1.64522 - 1.640677$, rounded up).

(c) $R_n \leq \displaystyle\int_n^{\infty}\frac{1}{x^2}\,dx = \frac{1}{n}$. So $R_n < 0.001$ if $\dfrac{1}{n} < \dfrac{1}{1000} \iff n > 1000$.

35. $f(x) = 1/(2x+1)^6$ is continuous, positive, and decreasing on $[1, \infty)$, so the Integral Test applies. Using (3),

$R_n \leq \displaystyle\int_n^{\infty}(2x+1)^{-6}\,dx = \lim\limits_{t \to \infty}\left[\dfrac{-1}{10(2x+1)^5}\right]_n^t = \dfrac{1}{10(2n+1)^5}$. To be correct to five decimal places, we want

$\dfrac{1}{10(2n+1)^5} \leq \dfrac{5}{10^6} \iff (2n+1)^5 \geq 20{,}000 \iff n \geq \frac{1}{2}\left(\sqrt[5]{20{,}000} - 1\right) \approx 3.12$, so use $n = 4$.

$s_4 = \displaystyle\sum_{n=1}^{4}\frac{1}{(2n+1)^6} = \frac{1}{3^6} + \frac{1}{5^6} + \frac{1}{7^6} + \frac{1}{9^6} \approx 0.001\,446 \approx 0.00145.$

37. $\displaystyle\sum_{n=1}^{10}\frac{1}{\sqrt{n^4 + 1}} = \frac{1}{\sqrt{2}} + \frac{1}{\sqrt{17}} + \frac{1}{\sqrt{82}} + \cdots + \frac{1}{\sqrt{10{,}001}} \approx 1.24856$. Now $\dfrac{1}{\sqrt{n^4 + 1}} < \dfrac{1}{\sqrt{n^4}} = \dfrac{1}{n^2}$, so the error is

$R_{10} \leq T_{10} \leq \displaystyle\int_{10}^{\infty}\frac{1}{x^2}\,dx = \lim\limits_{t \to \infty}\left[-\frac{1}{x}\right]_{10}^{t} = \lim\limits_{t \to \infty}\left(-\frac{1}{t} + \frac{1}{10}\right) = \frac{1}{10} = 0.1.$

39. (a) From the figure, $a_2 + a_3 + \cdots + a_n \leq \int_1^n f(x)\,dx$, so with

$f(x) = \dfrac{1}{x}, \dfrac{1}{2} + \dfrac{1}{3} + \dfrac{1}{4} + \cdots + \dfrac{1}{n} \leq \displaystyle\int_1^n \frac{1}{x}\,dx = \ln n.$

Thus, $s_n = 1 + \dfrac{1}{2} + \dfrac{1}{3} + \dfrac{1}{4} + \cdots + \dfrac{1}{n} \leq 1 + \ln n.$

(b) By part (a), $s_{10^6} \leq 1 + \ln 10^6 \approx 14.82 < 15$ and

$s_{10^9} \leq 1 + \ln 10^9 \approx 21.72 < 22.$

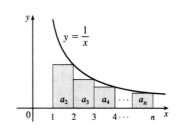

41. Since $\dfrac{d_n}{10^n} \le \dfrac{9}{10^n}$ for each n, and since $\displaystyle\sum_{n=1}^{\infty} \dfrac{9}{10^n}$ is a convergent geometric series $\left(|r| = \frac{1}{10} < 1\right)$, $0.d_1 d_2 d_3 \ldots = \displaystyle\sum_{n=1}^{\infty} \dfrac{d_n}{10^n}$

will always converge by the Comparison Test.

43. Yes. Since $\sum a_n$ is a convergent series with positive terms, $\displaystyle\lim_{n\to\infty} a_n = 0$ by Theorem 8.2.6, and $\sum b_n = \sum \sin(a_n)$ is a

series with positive terms (for large enough n). We have $\displaystyle\lim_{n\to\infty} \dfrac{b_n}{a_n} = \lim_{n\to\infty} \dfrac{\sin(a_n)}{a_n} = 1 > 0$ by Theorem 3.3.2. Thus, $\sum b_n$

is also convergent by the Limit Comparison Test.

45. $\displaystyle\lim_{n\to\infty} n a_n = \lim_{n\to\infty} \dfrac{a_n}{1/n}$, so we apply the Limit Comparison Test with $b_n = \dfrac{1}{n}$. Since $\displaystyle\lim_{n\to\infty} n a_n > 0$ we know that either both

series converge or both series diverge, and we also know that $\displaystyle\sum_{n=1}^{\infty} \dfrac{1}{n}$ diverges $[p\text{-series with } p = 1]$. Therefore, $\sum a_n$ must be

divergent.

8.4 Other Convergence Tests

1. (a) An alternating series is a series whose terms are alternately positive and negative.

(b) An alternating series $\displaystyle\sum_{n=1}^{\infty} a_n = \sum_{n=1}^{\infty} (-1)^{n-1} b_n$, where $b_n = |a_n|$, converges if $0 < b_{n+1} \le b_n$ for all n and $\displaystyle\lim_{n\to\infty} b_n = 0$.

(This is the Alternating Series Test.)

(c) The error involved in using the partial sum s_n as an approximation to the total sum s is the remainder $R_n = s - s_n$ and the

size of the error is smaller than b_{n+1}; that is, $|R_n| \le b_{n+1}$. (This is the Alternating Series Estimation Theorem.)

3. $\dfrac{4}{7} - \dfrac{4}{8} + \dfrac{4}{9} - \dfrac{4}{10} + \dfrac{4}{11} - \cdots = \displaystyle\sum_{n=1}^{\infty} (-1)^{n-1} \dfrac{4}{n+6}$. Now $b_n = \dfrac{4}{n+6} > 0$, $\{b_n\}$ is decreasing, and $\displaystyle\lim_{n\to\infty} b_n = 0$, so the

series converges by the Alternating Series Test.

5. $\displaystyle\sum_{n=1}^{\infty} a_n = \sum_{n=1}^{\infty} (-1)^{n-1} \dfrac{1}{2n+1} = \sum_{n=1}^{\infty} (-1)^{n-1} b_n$. Now $b_n = \dfrac{1}{2n+1} > 0$, $\{b_n\}$ is decreasing, and $\displaystyle\lim_{n\to\infty} b_n = 0$, so the

series converges by the Alternating Series Test.

7. $\displaystyle\sum_{n=1}^{\infty} a_n = \sum_{n=1}^{\infty} (-1)^n \dfrac{3n-1}{2n+1} = \sum_{n=1}^{\infty} (-1)^n b_n$. Now $\displaystyle\lim_{n\to\infty} b_n = \lim_{n\to\infty} \dfrac{3 - 1/n}{2 + 1/n} = \dfrac{3}{2} \ne 0$. Since $\displaystyle\lim_{n\to\infty} a_n \ne 0$

(in fact the limit does not exist), the series diverges by the Test for Divergence.

9. $b_n = \dfrac{n}{n^2 + 9} > 0$, and if $f(x) = \dfrac{x}{x^2 + 9}$, then $f'(x) = \dfrac{(x^2 + 9)(1) - x(2x)}{(x^2 + 9)^2} = \dfrac{9 - x^2}{(x^2 + 9)^2} < 0$ for $x > 3$, so $\{b_n\}$ is

eventually decreasing. Also, $\displaystyle\lim_{n\to\infty} b_n = \lim_{n\to\infty} \dfrac{n}{n^2 + 9} = \lim_{n\to\infty} \dfrac{1}{n + 9/n} = 0$, so the series $\displaystyle\sum_{n=1}^{\infty} (-1)^{n+1} \dfrac{n}{n^2 + 9}$ converges by

the Alternating Series Test.

11. $\displaystyle\sum_{n=1}^{\infty}\frac{(-1)^{n-1}}{n} = 1 - \frac{1}{2} + \frac{1}{3} - \frac{1}{4} + \cdots + \frac{1}{49} - \frac{1}{50} + \frac{1}{51} - \frac{1}{52} + \cdots$. The 50th partial sum of this series is an

underestimate, since $\displaystyle\sum_{n=1}^{\infty}\frac{(-1)^{n-1}}{n} = s_{50} + \left(\frac{1}{51} - \frac{1}{52}\right) + \left(\frac{1}{53} - \frac{1}{54}\right) + \cdots$, and the terms in parentheses are all positive.

The result can be seen geometrically in Figure 1.

13. If $p > 0$, $\dfrac{1}{(n+1)^p} \le \dfrac{1}{n^p}$ ($\{1/n^p\}$ is decreasing) and $\displaystyle\lim_{n\to\infty}\frac{1}{n^p} = 0$, so the series converges by the Alternating Series Test.

If $p \le 0$, $\displaystyle\lim_{n\to\infty}\frac{(-1)^{n-1}}{n^p}$ does not exist, so the series diverges by the Test for Divergence. Thus, $\displaystyle\sum_{n=1}^{\infty}\frac{(-1)^{n-1}}{n^p}$

converges $\Leftrightarrow$ $p > 0$.

15. The series $\displaystyle\sum_{n=1}^{\infty}\frac{(-1)^{n+1}}{n^6}$ satisfies (i) of the Alternating Series Test because $\dfrac{1}{(n+1)^6} < \dfrac{1}{n^6}$ and (ii) $\displaystyle\lim_{n\to\infty}\frac{1}{n^6} = 0$, so the

series is convergent. Now $b_5 = \dfrac{1}{5^6} = 0.000064 > 0.00005$ and $b_6 = \dfrac{1}{6^6} \approx 0.00002 < 0.00005$, so by the Alternating Series

Estimation Theorem, $n = 5$. (That is, since the 6th term is less than the desired error, we need to add the first 5 terms to get the

sum to the desired accuracy.)

17.

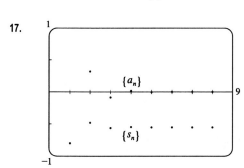

The graph gives us an estimate for the sum of the series

$$\sum_{n=1}^{\infty}\frac{(-0.8)^n}{n!} \text{ of } -0.55.$$

$b_8 = \dfrac{(0.8)^n}{8!} \approx 0.000004$, so

$$\sum_{n=1}^{\infty}\frac{(-0.8)^n}{n!} \approx s_7 = \sum_{n=1}^{7}\frac{(-0.8)^n}{n!}$$

$$\approx -0.8 + 0.32 - 0.085\overline{3} + 0.0170\overline{6} - 0.002\,731 + 0.000\,364 - 0.000\,042 \approx -0.5507$$

Adding b_8 to s_7 does not change the fourth decimal place of s_7, so the sum of the series, correct to four decimal places,

is -0.5507.

19. $b_7 = \dfrac{7^2}{10^7} = 0.000\,004\,9$, so

$$\sum_{n=1}^{\infty}\frac{(-1)^{n-1}n^2}{10^n} \approx s_6 = \sum_{n=1}^{6}\frac{(-1)^{n-1}n^2}{10^n} = \frac{1}{10} - \frac{4}{100} + \frac{9}{1000} - \frac{16}{10,000} + \frac{25}{100,000} - \frac{36}{1,000,000} = 0.067\,614$$

Adding b_7 to s_6 does not change the fourth decimal place of s_6, so the sum of the series, correct to four decimal places, is

0.0676.

21. Using the Ratio Test, $\lim\limits_{n\to\infty}\left|\dfrac{a_{n+1}}{a_n}\right| = \lim\limits_{n\to\infty}\left|\dfrac{(-3)^{n+1}/(n+1)^3}{(-3)^n/n^3}\right| = \lim\limits_{n\to\infty}\left|\dfrac{(-3)n^3}{(n+1)^3}\right| = 3\lim\limits_{n\to\infty}\left(\dfrac{n}{n+1}\right)^3 = 3 > 1,$

so the series $\sum\limits_{n=1}^{\infty}\dfrac{(-3)^n}{n^3}$ diverges.

23. Using the Ratio Test, $\lim\limits_{n\to\infty}\left|\dfrac{a_{n+1}}{a_n}\right| = \lim\limits_{n\to\infty}\left|\dfrac{(-10)^{n+1}}{(n+1)!}\cdot\dfrac{n!}{(-10)^n}\right| = \lim\limits_{n\to\infty}\left|\dfrac{-10}{n+1}\right| = 0 < 1$, so the series $\sum\limits_{n=0}^{\infty}\dfrac{(-10)^n}{n!}$ is

absolutely convergent.

25. $\lim\limits_{k\to\infty}\left|\dfrac{a_{k+1}}{a_k}\right| = \lim\limits_{k\to\infty}\left[\dfrac{(k+1)\left(\frac{2}{3}\right)^{k+1}}{k\left(\frac{2}{3}\right)^k}\right] = \lim\limits_{k\to\infty}\dfrac{k+1}{k}\left(\dfrac{2}{3}\right)^1 = \dfrac{2}{3}\lim\limits_{k\to\infty}\left(1+\dfrac{1}{k}\right) = \frac{2}{3}(1) = \frac{2}{3} < 1$, so the series

$\sum\limits_{n=1}^{\infty} k\left(\frac{2}{3}\right)^k$ is absolutely convergent by the Ratio Test. Since the terms of this series are positive, absolute convergence is the

same as convergence.

27. Consider the series whose terms are the absolute values of the terms of the given series. $\sum\limits_{n=1}^{\infty}\left|\dfrac{(-1)^{n-1}}{\sqrt{n}}\right| = \sum\limits_{n=1}^{\infty}\dfrac{1}{n^{1/2}}$, which is

a divergent p-series ($p = \frac{1}{2} \le 1$). Thus, $\sum\limits_{n=1}^{\infty}\dfrac{(-1)^{n-1}}{\sqrt{n}}$ is *not* absolutely convergent.

29. $\lim\limits_{n\to\infty}\left|\dfrac{a_{n+1}}{a_n}\right| = \lim\limits_{n\to\infty}\left[\dfrac{10^{n+1}}{(n+2)4^{2(n+1)+1}}\cdot\dfrac{(n+1)4^{2n+1}}{10^n}\right] = \lim\limits_{n\to\infty}\left[\dfrac{10^{n+1}}{(n+2)4^{2n+3}}\cdot\dfrac{(n+1)4^{2n+1}}{10^n}\right]$

$= \lim\limits_{n\to\infty}\left(\dfrac{10}{4^2}\cdot\dfrac{n+1}{n+2}\right) = \dfrac{5}{8} < 1,$

so the series $\sum\limits_{n=1}^{\infty}\dfrac{10^n}{(n+1)4^{2n+1}}$ is absolutely convergent by the Ratio Test. Since the terms of this series are positive, absolute

convergence is the same as convergence.

31. $\left|\dfrac{(-1)^n \arctan n}{n^2}\right| < \dfrac{\pi/2}{n^2}$, so since $\sum\limits_{n=1}^{\infty}\dfrac{\pi/2}{n^2} = \dfrac{\pi}{2}\sum\limits_{n=1}^{\infty}\dfrac{1}{n^2}$ converges ($p = 2 > 1$), the given series $\sum\limits_{n=1}^{\infty}\dfrac{(-1)^n \arctan n}{n^2}$

converges absolutely by the Comparison Test.

33. Use the Ratio Test with the series

$1 - \dfrac{1\cdot 3}{3!} + \dfrac{1\cdot 3\cdot 5}{5!} - \dfrac{1\cdot 3\cdot 5\cdot 7}{7!} + \cdots + (-1)^{n-1}\dfrac{1\cdot 3\cdot 5\cdots(2n-1)}{(2n-1)!} + \cdots = \sum\limits_{n=1}^{\infty}(-1)^{n-1}\dfrac{1\cdot 3\cdot 5\cdots(2n-1)}{(2n-1)!}.$

$\lim\limits_{n\to\infty}\left|\dfrac{a_{n+1}}{a_n}\right| = \lim\limits_{n\to\infty}\left|\dfrac{(-1)^n \cdot 1\cdot 3\cdot 5\cdots(2n-1)[2(n+1)-1]}{[2(n+1)-1]!}\cdot\dfrac{(2n-1)!}{(-1)^{n-1}\cdot 1\cdot 3\cdot 5\cdots(2n-1)}\right|$

$= \lim\limits_{n\to\infty}\left|\dfrac{(-1)(2n+1)(2n-1)!}{(2n+1)(2n)(2n-1)!}\right| = \lim\limits_{n\to\infty}\dfrac{1}{2n} = 0 < 1,$

so the given series is absolutely convergent and therefore convergent.

35. By the recursive definition, $\lim\limits_{n\to\infty}\left|\dfrac{a_{n+1}}{a_n}\right| = \lim\limits_{n\to\infty}\left|\dfrac{5n+1}{4n+3}\right| = \dfrac{5}{4} > 1$, so the series diverges by the Ratio Test.

37. (a) $\lim\limits_{n\to\infty}\left|\dfrac{1/(n+1)^3}{1/n^3}\right| = \lim\limits_{n\to\infty}\dfrac{n^3}{(n+1)^3} = \lim\limits_{n\to\infty}\dfrac{1}{(1+1/n)^3} = 1$. Inconclusive

(b) $\lim\limits_{n\to\infty}\left|\dfrac{(n+1)}{2^{n+1}}\cdot\dfrac{2^n}{n}\right| = \lim\limits_{n\to\infty}\dfrac{n+1}{2n} = \lim\limits_{n\to\infty}\left(\dfrac{1}{2}+\dfrac{1}{2n}\right) = \dfrac{1}{2}$. Conclusive (convergent)

(c) $\lim\limits_{n\to\infty}\left|\dfrac{(-3)^n}{\sqrt{n+1}}\cdot\dfrac{\sqrt{n}}{(-3)^{n-1}}\right| = 3\lim\limits_{n\to\infty}\sqrt{\dfrac{n}{n+1}} = 3\lim\limits_{n\to\infty}\sqrt{\dfrac{1}{1+1/n}} = 3$. Conclusive (divergent)

(d) $\lim\limits_{n\to\infty}\left|\dfrac{\sqrt{n+1}}{1+(n+1)^2}\cdot\dfrac{1+n^2}{\sqrt{n}}\right| = \lim\limits_{n\to\infty}\left[\sqrt{1+\dfrac{1}{n}}\cdot\dfrac{1/n^2+1}{1/n^2+(1+1/n)^2}\right] = 1$. Inconclusive

39. $\lim\limits_{n\to\infty}\sqrt[n]{|a_n|} = \lim\limits_{n\to\infty}\dfrac{n^2+1}{2n^2+1} = \lim\limits_{n\to\infty}\dfrac{1+1/n^2}{2+1/n^2} = \dfrac{1}{2} < 1$, so the series $\sum\limits_{n=1}^{\infty}\left(\dfrac{n^2+1}{2n^2+1}\right)^n$ is absolutely convergent by the

Root Test.

41. (a) $\lim\limits_{n\to\infty}\left|\dfrac{a_{n+1}}{a_n}\right| = \lim\limits_{n\to\infty}\left|\dfrac{x^{n+1}}{(n+1)!}\cdot\dfrac{n!}{x^n}\right| = \lim\limits_{n\to\infty}\left|\dfrac{x}{n+1}\right| = |x|\lim\limits_{n\to\infty}\dfrac{1}{n+1} = |x|\cdot 0 = 0 < 1$, so by the Ratio Test the

series $\sum\limits_{n=0}^{\infty}\dfrac{x^n}{n!}$ converges for all x.

(b) Since the series of part (a) always converges, we must have $\lim\limits_{n\to\infty}\dfrac{x^n}{n!} = 0$ by Theorem 8.2.6.

8.5 Power Series

1. A power series is a series of the form $\sum\limits_{n=0}^{\infty} c_n x^n = c_0 + c_1 x + c_2 x^2 + c_3 x^3 + \cdots$, where x is a variable and the c_n's are

constants called the coefficients of the series.

More generally, a series of the form $\sum\limits_{n=0}^{\infty} c_n (x-a)^n = c_0 + c_1(x-a) + c_2(x-a)^2 + \cdots$ is called a power series in

$(x-a)$ or a power series centered at a or a power series about a, where a is a constant.

3. If $a_n = \dfrac{x^n}{\sqrt{n}}$, then $\lim\limits_{n\to\infty}\left|\dfrac{a_{n+1}}{a_n}\right| = \lim\limits_{n\to\infty}\left|\dfrac{x^{n+1}}{\sqrt{n+1}}\cdot\dfrac{\sqrt{n}}{x^n}\right| = \lim\limits_{n\to\infty}\left|\dfrac{x}{\sqrt{n+1}/\sqrt{n}}\right| = \lim\limits_{n\to\infty}\dfrac{|x|}{\sqrt{1+1/n}} = |x|$.

By the Ratio Test, the series $\sum\limits_{n=1}^{\infty}\dfrac{x^n}{\sqrt{n}}$ converges when $|x| < 1$, so the radius of convergence $R = 1$. Now we'll check the

endpoints, that is, $x = \pm 1$. When $x = 1$, the series $\sum\limits_{n=1}^{\infty}\dfrac{1}{\sqrt{n}}$ diverges because it is a p-series with $p = \dfrac{1}{2} \le 1$. When $x = -1$,

the series $\sum\limits_{n=1}^{\infty}\dfrac{(-1)^n}{\sqrt{n}}$ converges by the Alternating Series Test. Thus, the interval of convergence is $I = [-1, 1)$.

5. If $a_n = \dfrac{(-1)^{n-1}x^n}{n^3}$, then

$$\lim_{n\to\infty}\left|\frac{a_{n+1}}{a_n}\right| = \lim_{n\to\infty}\left|\frac{(-1)^n x^{n+1}}{(n+1)^3} \cdot \frac{n^3}{(-1)^{n-1}x^n}\right| = \lim_{n\to\infty}\left|\frac{(-1)xn^3}{(n+1)^3}\right| = \lim_{n\to\infty}\left[\left(\frac{n}{n+1}\right)^3 |x|\right] = 1^3 \cdot |x| = |x|. \text{ By the}$$

Ratio Test, the series $\displaystyle\sum_{n=1}^{\infty} \frac{(-1)^{n-1}x^n}{n^3}$ converges when $|x| < 1$, so the radius of convergence $R = 1$. Now we'll check the

endpoints, that is, $x = \pm 1$. When $x = 1$, the series $\displaystyle\sum_{n=1}^{\infty} \frac{(-1)^{n-1}}{n^3}$ converges by the Alternating Series Test. When $x = -1$,

the series $\displaystyle\sum_{n=1}^{\infty} \frac{(-1)^{n-1}(-1)^n}{n^3} = -\sum_{n=1}^{\infty} \frac{1}{n^3}$ converges because it is a constant multiple of a convergent p-series $[p = 3 > 1]$.

Thus, the interval of convergence is $I = [-1, 1]$.

7. If $a_n = \dfrac{x^n}{n!}$, then $\displaystyle\lim_{n\to\infty}\left|\frac{a_{n+1}}{a_n}\right| = \lim_{n\to\infty}\left|\frac{x^{n+1}}{(n+1)!} \cdot \frac{n!}{x^n}\right| = \lim_{n\to\infty}\left|\frac{x}{n+1}\right| = |x| \lim_{n\to\infty}\frac{1}{n+1} = |x| \cdot 0 = 0 < 1$ for *all* real x.

So, by the Ratio Test, $R = \infty$ and $I = (-\infty, \infty)$.

9. If $a_n = (-1)^n \dfrac{n^2 x^n}{2^n}$, then

$$\lim_{n\to\infty}\left|\frac{a_{n+1}}{a_n}\right| = \lim_{n\to\infty}\left|\frac{(n+1)^2 x^{n+1}}{2^{n+1}} \cdot \frac{2^n}{n^2 x^n}\right| = \lim_{n\to\infty}\left|\frac{x(n+1)^2}{2n^2}\right| = \lim_{n\to\infty}\left[\frac{|x|}{2}\left(1 + \frac{1}{n}\right)^2\right] = \frac{|x|}{2}(1)^2 = \tfrac{1}{2}|x|. \text{ By the}$$

Ratio Test, the series $\displaystyle\sum_{n=1}^{\infty} (-1)^n \frac{n^2 x^n}{2^n}$ converges when $\tfrac{1}{2}|x| < 1 \iff |x| < 2$, so the radius of convergence is $R = 2$.

When $x = \pm 2$, both series $\displaystyle\sum_{n=1}^{\infty} (-1)^n \frac{n^2(\pm 2)^n}{2^n} = \sum_{n=1}^{\infty} (\mp 1)^n n^2$ diverge by the Test for Divergence since

$\displaystyle\lim_{n\to\infty}\left|(\mp 1)^n n^2\right| = \infty$. Thus, the interval of convergence is $I = (-2, 2)$.

11. $a_n = \dfrac{(-2)^n x^n}{\sqrt[4]{n}}$, so $\displaystyle\lim_{n\to\infty}\left|\frac{a_{n+1}}{a_n}\right| = \lim_{n\to\infty}\frac{2^{n+1}|x|^{n+1}}{\sqrt[4]{n+1}} \cdot \frac{\sqrt[4]{n}}{2^n |x|^n} = \lim_{n\to\infty} 2|x| \sqrt[4]{\frac{n}{n+1}} = 2|x|$, so by the Ratio Test, the

series converges when $2|x| < 1 \iff |x| < \tfrac{1}{2}$, so $R = \tfrac{1}{2}$. When $x = -\tfrac{1}{2}$, we get the divergent p-series $\displaystyle\sum_{n=1}^{\infty} \frac{1}{\sqrt[4]{n}}$

$\left[p = \tfrac{1}{4} \le 1\right]$. When $x = \tfrac{1}{2}$, we get the series $\displaystyle\sum_{n=1}^{\infty} \frac{(-1)^n}{\sqrt[4]{n}}$, which converges by the Alternating Series Test.

Thus, $I = \left(-\tfrac{1}{2}, \tfrac{1}{2}\right]$.

13. If $a_n = \dfrac{(x-2)^n}{n^2+1}$, then $\displaystyle\lim_{n\to\infty}\left|\frac{a_{n+1}}{a_n}\right| = \lim_{n\to\infty}\left|\frac{(x-2)^{n+1}}{(n+1)^2+1} \cdot \frac{n^2+1}{(x-2)^n}\right| = |x-2| \lim_{n\to\infty}\frac{n^2+1}{(n+1)^2+1} = |x-2|. \text{ By the}$

Ratio Test, the series $\displaystyle\sum_{n=0}^{\infty} \frac{(x-2)^n}{n^2+1}$ converges when $|x-2| < 1 \quad [R = 1] \iff -1 < x - 2 < 1 \iff 1 < x < 3$. When

$x = 1$, the series $\displaystyle\sum_{n=0}^{\infty} (-1)^n \frac{1}{n^2+1}$ converges by the Alternating Series Test; when $x = 3$, the series $\displaystyle\sum_{n=0}^{\infty} \frac{1}{n^2+1}$ converges by

comparison with the p-series $\displaystyle\sum_{n=1}^{\infty} \frac{1}{n^2}$ $[p = 2 > 1]$. Thus, the interval of convergence is $I = [1, 3]$.

15. If $a_n = \dfrac{3^n (x+4)^n}{\sqrt{n}}$, then $\lim\limits_{n\to\infty}\left|\dfrac{a_{n+1}}{a_n}\right| = \lim\limits_{n\to\infty}\left|\dfrac{3^{n+1}(x+4)^{n+1}}{\sqrt{n+1}}\cdot\dfrac{\sqrt{n}}{3^n(x+4)^n}\right| = 3\,|x+4|\,\lim\limits_{n\to\infty}\dfrac{\sqrt{n}}{\sqrt{n+1}} = 3\,|x+4|.$

By the Ratio Test, the series $\displaystyle\sum_{n=1}^{\infty}\dfrac{3^n(x+4)^n}{\sqrt{n}}$ converges when $3\,|x+4| < 1 \ \Leftrightarrow\ |x+4| < \tfrac{1}{3}\ \ \left[R = \tfrac{1}{3}\right]\ \Leftrightarrow$

$-\tfrac{1}{3} < x+4 < \tfrac{1}{3}\ \Leftrightarrow\ -\tfrac{13}{3} < x < -\tfrac{11}{3}$. When $x = -\tfrac{13}{3}$, the series $\displaystyle\sum_{n=1}^{\infty}(-1)^n\dfrac{1}{\sqrt{n}}$ converges by the Alternating Series

Test; when $x = -\tfrac{11}{3}$, the series $\displaystyle\sum_{n=1}^{\infty}\dfrac{1}{\sqrt{n}}$ diverges $\left[p = \tfrac{1}{2} \le 1\right]$. Thus, the interval of convergence is $I = \left[-\tfrac{13}{3}, -\tfrac{11}{3}\right)$.

17. $\lim\limits_{n\to\infty}\left|\dfrac{a_{n+1}}{a_n}\right| = \lim\limits_{n\to\infty}\left[\dfrac{|4x+1|^{n+1}}{(n+1)^2}\cdot\dfrac{n^2}{|4x+1|^n}\right] = \lim\limits_{n\to\infty}\dfrac{|4x+1|}{(1+1/n)^2} = |4x+1|$, so by the Ratio Test, the series

converges when $|4x+1| < 1\ \Leftrightarrow\ -1 < 4x+1 < 1\ \Leftrightarrow\ -2 < 4x < 0\ \Leftrightarrow\ -\tfrac{1}{2} < x < 0$, so $R = \tfrac{1}{4}$. When $x = -\tfrac{1}{2}$,

the series becomes $\displaystyle\sum_{n=1}^{\infty}\dfrac{(-1)^n}{n^2}$, which converges by the Alternating Series Test. When $x = 0$, the series becomes $\displaystyle\sum_{n=1}^{\infty}\dfrac{1}{n^2}$,

a convergent p-series $[p = 2 > 1]$. $\ \ I = \left[-\tfrac{1}{2}, 0\right]$.

19. If $a_n = n!\,(2x-1)^n$, then $\lim\limits_{n\to\infty}\left|\dfrac{a_{n+1}}{a_n}\right| = \lim\limits_{n\to\infty}\left|\dfrac{(n+1)!\,(2x-1)^{n+1}}{n!(2x-1)^n}\right| = \lim\limits_{n\to\infty}(n+1)\,|2x-1| \to \infty$ as $n \to \infty$

for all $x \neq \tfrac{1}{2}$. Since the series diverges for all $x \neq \tfrac{1}{2}$, $R = 0$ and $I = \left\{\tfrac{1}{2}\right\}$.

21. $a_n = \dfrac{n}{b^n}(x-a)^n$, where $b > 0$.

$\lim\limits_{n\to\infty}\left|\dfrac{a_{n+1}}{a_n}\right| = \lim\limits_{n\to\infty}\dfrac{(n+1)\,|x-a|^{n+1}}{b^{n+1}}\cdot\dfrac{b^n}{n\,|x-a|^n} = \lim\limits_{n\to\infty}\left(1+\dfrac{1}{n}\right)\dfrac{|x-a|}{b} = \dfrac{|x-a|}{b}.$

By the Ratio Test, the series converges when $\dfrac{|x-a|}{b} < 1\ \Leftrightarrow\ |x-a| < b\ \ [\text{so }R = b]\ \Leftrightarrow\ -b < x-a < b\ \Leftrightarrow$

$a - b < x < a + b$. When $|x-a| = b$, $\lim\limits_{n\to\infty}|a_n| = \lim\limits_{n\to\infty} n = \infty$, so the series diverges. Thus, $I = (a-b, a+b)$.

23. If $a_n = \dfrac{x^n}{1\cdot 3\cdot 5\cdot\ \cdots\ \cdot(2n-1)}$, then

$\lim\limits_{n\to\infty}\left|\dfrac{a_{n+1}}{a_n}\right| = \lim\limits_{n\to\infty}\left|\dfrac{x^{n+1}}{1\cdot 3\cdot 5\cdot\ \cdots\ \cdot(2n-1)(2n+1)}\cdot\dfrac{1\cdot 3\cdot 5\cdot\ \cdots\ \cdot(2n-1)}{x^n}\right| = \lim\limits_{n\to\infty}\dfrac{|x|}{2n+1} = 0 < 1.$ Thus, by

the Ratio Test, the series $\displaystyle\sum_{n=1}^{\infty}\dfrac{x^n}{1\cdot 3\cdot 5\cdot\ \cdots\ \cdot(2n-1)}$ converges for *all* real x and we have $R = \infty$ and $I = (-\infty, \infty)$.

25. (a) We are given that the power series $\sum_{n=0}^{\infty} c_n x^n$ is convergent for $x = 4$. So by Theorem 3, it must converge for at least

$-4 < x \le 4$. In particular, it converges when $x = -2$; that is, $\sum_{n=0}^{\infty} c_n(-2)^n$ is convergent.

(b) It does not follow that $\sum_{n=0}^{\infty} c_n(-4)^n$ is necessarily convergent. [See the comments after Theorem 3 about convergence at

the endpoint of an interval. An example is $c_n = (-1)^n/(n4^n)$.]

27. If $a_n = \dfrac{(n!)^k}{(kn)!} x^n$, then

$$\lim_{n\to\infty}\left|\frac{a_{n+1}}{a_n}\right| = \lim_{n\to\infty}\frac{[(n+1)!]^k (kn)!}{(n!)^k [k(n+1)]!}|x| = \lim_{n\to\infty}\frac{(n+1)^k}{(kn+k)(kn+k-1)\cdots(kn+2)(kn+1)}|x|$$

$$= \lim_{n\to\infty}\left[\frac{(n+1)}{(kn+1)}\frac{(n+1)}{(kn+2)}\cdots\frac{(n+1)}{(kn+k)}\right]|x|$$

$$= \lim_{n\to\infty}\left[\frac{n+1}{kn+1}\right]\lim_{n\to\infty}\left[\frac{n+1}{kn+2}\right]\cdots\lim_{n\to\infty}\left[\frac{n+1}{kn+k}\right]|x|$$

$$= \left(\frac{1}{k}\right)^k |x| < 1 \quad\Leftrightarrow\quad |x| < k^k \text{ for convergence, and the radius of convergence is } R = k^k.$$

29. (a) If $a_n = \dfrac{(-1)^n x^{2n+1}}{n!(n+1)!\,2^{2n+1}}$, then

$$\lim_{n\to\infty}\left|\frac{a_{n+1}}{a_n}\right| = \lim_{n\to\infty}\left|\frac{x^{2n+3}}{(n+1)!(n+2)!\,2^{2n+3}}\cdot\frac{n!(n+1)!\,2^{2n+1}}{x^{2n+1}}\right| = \left(\frac{x}{2}\right)^2\lim_{n\to\infty}\frac{1}{(n+1)(n+2)} = 0 \text{ for all } x.$$

So $J_1(x)$ converges for all x and its domain is $(-\infty, \infty)$.

(b), (c) The initial terms of $J_1(x)$ up to $n=5$ are $a_0 = \dfrac{x}{2}$,

$$a_1 = -\frac{x^3}{16},\ a_2 = \frac{x^5}{384},\ a_3 = -\frac{x^7}{18{,}432},\ a_4 = \frac{x^9}{1{,}474{,}560},$$

and $a_5 = -\dfrac{x^{11}}{176{,}947{,}200}$. The partial sums seem to

approximate $J_1(x)$ well near the origin, but as $|x|$ increases,

we need to take a large number of terms to get a good

approximation.

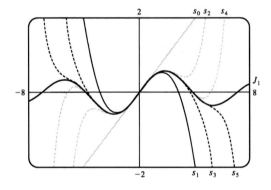

31. $s_{2n-1} = 1 + 2x + x^2 + 2x^3 + x^4 + 2x^5 + \cdots + x^{2n-2} + 2x^{2n-1}$

$= 1(1+2x) + x^2(1+2x) + x^4(1+2x) + \cdots + x^{2n-2}(1+2x) = (1+2x)(1 + x^2 + x^4 + \cdots + x^{2n-2})$

$= (1+2x)\dfrac{1-x^{2n}}{1-x^2}$ [by (8.2.3) with $r = x^2$] $\to \dfrac{1+2x}{1-x^2}$ as $n \to \infty$ by (8.2.4)], when $|x| < 1$.

Also $s_{2n} = s_{2n-1} + x^{2n} \to \dfrac{1+2x}{1-x^2}$ since $x^{2n} \to 0$ for $|x| < 1$. Therefore, $s_n \to \dfrac{1+2x}{1-x^2}$ since s_{2n} and s_{2n-1} both

approach $\dfrac{1+2x}{1-x^2}$ as $n \to \infty$. Thus, the interval of convergence is $(-1, 1)$ and $f(x) = \dfrac{1+2x}{1-x^2}$.

33. For $2 < x < 3$, $\sum c_n x^n$ diverges and $\sum d_n x^n$ converges. By Exercise 8.2.61, $\sum(c_n + d_n) x^n$ diverges. Since both series

converge for $|x| < 2$, the radius of convergence of $\sum(c_n + d_n) x^n$ is 2.

35. No. If a power series is centered at a, its interval of convergence is symmetric about a. If a power series has an infinite radius

of convergence, then its interval of convergence must be $(-\infty, \infty)$, not $[0, \infty)$.

8.6   Representations of Functions as Power Series

1. If $f(x) = \sum_{n=0}^{\infty} c_n x^n$ has radius of convergence 10, then $f'(x) = \sum_{n=1}^{\infty} nc_n x^{n-1}$ also has radius of convergence 10 by Theorem 2.

3. Our goal is to write the function in the form $\dfrac{1}{1-r}$, and then use Equation (1) to represent the function as a sum of a power series. $f(x) = \dfrac{1}{1+x} = \dfrac{1}{1-(-x)} = \sum_{n=0}^{\infty} (-x)^n = \sum_{n=0}^{\infty} (-1)^n x^n$ with $|-x| < 1$ $\Leftrightarrow$ $|x| < 1$, so $R = 1$ and $I = (-1, 1)$.

5. $f(x) = \dfrac{2}{3-x} = \dfrac{2}{3}\left(\dfrac{1}{1-x/3}\right) = \dfrac{2}{3}\sum_{n=0}^{\infty}\left(\dfrac{x}{3}\right)^n$ or, equivalently, $2\sum_{n=0}^{\infty}\dfrac{1}{3^{n+1}}x^n$. The series converges when $\left|\dfrac{x}{3}\right| < 1$, that is, when $|x| < 3$, so $R = 3$ and $I = (-3, 3)$.

7. $f(x) = \dfrac{x}{9+x^2} = \dfrac{x}{9}\left[\dfrac{1}{1+(x/3)^2}\right] = \dfrac{x}{9}\left[\dfrac{1}{1-\{-(x/3)^2\}}\right] = \dfrac{x}{9}\sum_{n=0}^{\infty}\left[-\left(\dfrac{x}{3}\right)^2\right]^n = \dfrac{x}{9}\sum_{n=0}^{\infty}(-1)^n\dfrac{x^{2n}}{9^n} = \sum_{n=0}^{\infty}(-1)^n\dfrac{x^{2n+1}}{9^{n+1}}$

The geometric series $\sum_{n=0}^{\infty}\left[-\left(\dfrac{x}{3}\right)^2\right]^n$ converges when $\left|-\left(\dfrac{x}{3}\right)^2\right| < 1$ $\Leftrightarrow$ $\dfrac{|x^2|}{9} < 1$ $\Leftrightarrow$ $|x|^2 < 9$ $\Leftrightarrow$ $|x| < 3$, so $R = 3$ and $I = (-3, 3)$.

9. $f(x) = \dfrac{1+x}{1-x} = (1+x)\left(\dfrac{1}{1-x}\right) = (1+x)\sum_{n=0}^{\infty} x^n = \sum_{n=0}^{\infty} x^n + \sum_{n=0}^{\infty} x^{n+1} = 1 + \sum_{n=1}^{\infty} x^n + \sum_{n=1}^{\infty} x^n = 1 + 2\sum_{n=1}^{\infty} x^n$.

The series converges when $|x| < 1$, so $R = 1$ and $I = (-1, 1)$.

A second approach: $f(x) = \dfrac{1+x}{1-x} = \dfrac{-(1-x)+2}{1-x} = -1 + 2\left(\dfrac{1}{1-x}\right) = -1 + 2\sum_{n=0}^{\infty} x^n = 1 + 2\sum_{n=1}^{\infty} x^n$.

A third approach:

$f(x) = \dfrac{1+x}{1-x} = (1+x)\left(\dfrac{1}{1-x}\right) = (1+x)(1+x+x^2+x^3+\cdots)$

$= (1+x+x^2+x^3+\cdots) + (x+x^2+x^3+x^4+\cdots) = 1 + 2x + 2x^2 + 2x^3 + \cdots = 1 + 2\sum_{n=1}^{\infty} x^n$.

11. (a) $f(x) = \dfrac{1}{(1+x)^2} = \dfrac{d}{dx}\left(\dfrac{-1}{1+x}\right) = -\dfrac{d}{dx}\left[\sum_{n=0}^{\infty}(-1)^n x^n\right]$ [from Exercise 3]

$= \sum_{n=1}^{\infty}(-1)^{n+1} nx^{n-1}$ [from Theorem 2(i)] $= \sum_{n=0}^{\infty}(-1)^n(n+1)x^n$ with $R = 1$.

In the last step, note that we *decreased* the initial value of the summation variable n by 1, and then *increased* each occurrence of n in the term by 1 [also note that $(-1)^{n+2} = (-1)^n$].

(b) $f(x) = \dfrac{1}{(1+x)^3} = -\dfrac{1}{2}\dfrac{d}{dx}\left[\dfrac{1}{(1+x)^2}\right] = -\dfrac{1}{2}\dfrac{d}{dx}\left[\sum_{n=0}^{\infty}(-1)^n(n+1)x^n\right]$ [from part (a)]

$= -\dfrac{1}{2}\sum_{n=1}^{\infty}(-1)^n(n+1)nx^{n-1} = \dfrac{1}{2}\sum_{n=0}^{\infty}(-1)^n(n+2)(n+1)x^n$ with $R = 1$.

(c) $f(x) = \dfrac{x^2}{(1+x)^3} = x^2\cdot\dfrac{1}{(1+x)^3} = x^2\cdot\dfrac{1}{2}\sum_{n=0}^{\infty}(-1)^n(n+2)(n+1)x^n$ [from part (b)]

$= \dfrac{1}{2}\sum_{n=0}^{\infty}(-1)^n(n+2)(n+1)x^{n+2}$

To write the power series with x^n rather than x^{n+2}, we will *decrease* each occurrence of n in the term by 2 and *increase* the initial value of the summation variable by 2. This gives us $\dfrac{1}{2}\sum_{n=2}^{\infty}(-1)^n(n)(n-1)x^n$ with $R = 1$.

13. $f(x) = \ln(5 - x) = -\int \dfrac{dx}{5 - x} = -\dfrac{1}{5}\int \dfrac{dx}{1 - x/5} = -\dfrac{1}{5}\int \left[\sum\limits_{n=0}^{\infty} \left(\dfrac{x}{5}\right)^n\right] dx = C - \dfrac{1}{5}\sum\limits_{n=0}^{\infty} \dfrac{x^{n+1}}{5^n (n + 1)} = C - \sum\limits_{n=1}^{\infty} \dfrac{x^n}{n \, 5^n}$

Putting $x = 0$, we get $C = \ln 5$. The series converges for $|x/5| < 1$ $\Leftrightarrow$ $|x| < 5$, so $R = 5$.

15. We know that $\dfrac{1}{1 + 4x} = \dfrac{1}{1 - (-4x)} = \sum\limits_{n=0}^{\infty} (-4x)^n$. Differentiating, we get

$\dfrac{-4}{(1 + 4x)^2} = \sum\limits_{n=1}^{\infty} (-4)^n n x^{n-1} = \sum\limits_{n=0}^{\infty} (-4)^{n+1}(n + 1)x^n$, so

$f(x) = \dfrac{x}{(1 + 4x)^2} = \dfrac{-x}{4} \cdot \dfrac{-4}{(1 + 4x)^2} = \dfrac{-x}{4}\sum\limits_{n=0}^{\infty} (-4)^{n+1}(n + 1)x^n = \sum\limits_{n=0}^{\infty} (-1)^n 4^n (n + 1)x^{n+1}$

for $|-4x| < 1$ $\Leftrightarrow$ $|x| < \frac{1}{4}$, so $R = \frac{1}{4}$.

17. By Example 5, $\dfrac{1}{(1 - x)^2} = \sum\limits_{n=0}^{\infty} (n + 1)x^n$. Thus,

$$f(x) = \dfrac{1 + x}{(1 - x)^2} = \dfrac{1}{(1 - x)^2} + \dfrac{x}{(1 - x)^2} = \sum\limits_{n=0}^{\infty} (n + 1)x^n + \sum\limits_{n=0}^{\infty} (n + 1)x^{n+1}$$

$$= \sum\limits_{n=0}^{\infty} (n + 1)x^n + \sum\limits_{n=1}^{\infty} n x^n \qquad \text{[make the starting values equal]}$$

$$= 1 + \sum\limits_{n=1}^{\infty} [(n + 1) + n]x^n = 1 + \sum\limits_{n=1}^{\infty} (2n + 1)x^n = \sum\limits_{n=0}^{\infty} (2n + 1)x^n \;\; \text{with } R = 1.$$

19. $f(x) = \dfrac{x}{x^2 + 16} = \dfrac{x}{16}\left(\dfrac{1}{1 - (-x^2/16)}\right) = \dfrac{x}{16}\sum\limits_{n=0}^{\infty} \left(-\dfrac{x^2}{16}\right)^n = \dfrac{x}{16}\sum\limits_{n=0}^{\infty} (-1)^n \dfrac{1}{16^n} x^{2n} = \sum\limits_{n=0}^{\infty} (-1)^n \dfrac{1}{16^{n+1}} x^{2n+1}$.

The series converges when $\left|-x^2/16\right| < 1$ $\Leftrightarrow$ $x^2 < 16$ $\Leftrightarrow$ $|x| < 4$, so $R = 4$. The partial sums are $s_1 = \dfrac{x}{16}$,

$s_2 = s_1 - \dfrac{x^3}{16^2}$, $s_3 = s_2 + \dfrac{x^5}{16^3}$, $s_4 = s_3 - \dfrac{x^7}{16^4}$, $s_5 = s_4 + \dfrac{x^9}{16^5}$, ... Note that s_1 corresponds to the first term of the infinite

sum, regardless of the value of the summation variable and the value of the exponent.

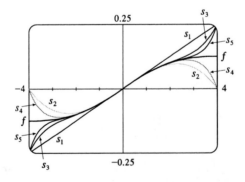

As n increases, $s_n(x)$ approximates f better on the interval of convergence, which is $(-4, 4)$.

21. $f(x) = \ln\left(\dfrac{1 + x}{1 - x}\right) = \ln(1 + x) - \ln(1 - x) = \int \dfrac{dx}{1 + x} + \int \dfrac{dx}{1 - x} = \int \dfrac{dx}{1 - (-x)} + \int \dfrac{dx}{1 - x}$

$= \int \left[\sum\limits_{n=0}^{\infty} (-1)^n x^n + \sum\limits_{n=0}^{\infty} x^n\right] dx = \int [(1 - x + x^2 - x^3 + x^4 - \cdots) + (1 + x + x^2 + x^3 + x^4 + \cdots)] \, dx$

$= \int (2 + 2x^2 + 2x^4 + \cdots) \, dx = \int \sum\limits_{n=0}^{\infty} 2x^{2n} \, dx = C + \sum\limits_{n=0}^{\infty} \dfrac{2x^{2n+1}}{2n + 1}$

But $f(0) = \ln\frac{1}{1} = 0$, so $C = 0$ and we have $f(x) = \sum\limits_{n=0}^{\infty} \dfrac{2x^{2n+1}}{2n+1}$ with $R = 1$. If $x = \pm 1$, then $f(x) = \pm 2 \sum\limits_{n=0}^{\infty} \dfrac{1}{2n+1}$,

which both diverge by the Limit Comparison Test with $b_n = \dfrac{1}{n}$. The partial sums are $s_1 = \dfrac{2x}{1}$, $s_2 = s_1 + \dfrac{2x^3}{3}$,

$s_3 = s_2 + \dfrac{2x^5}{5}, \ldots$.

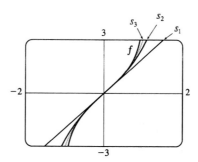

As n increases, $s_n(x)$ approximates f better on the interval of convergence, which is $(-1, 1)$.

23. $\dfrac{t}{1-t^8} = t \cdot \dfrac{1}{1-t^8} = t \sum\limits_{n=0}^{\infty} (t^8)^n = \sum\limits_{n=0}^{\infty} t^{8n+1} \quad\Rightarrow\quad \displaystyle\int \dfrac{t}{1-t^8}\,dt = C + \sum\limits_{n=0}^{\infty} \dfrac{t^{8n+2}}{8n+2}$. The series for $\dfrac{1}{1-t^8}$ converges

when $|t^8| < 1 \quad\Leftrightarrow\quad |t| < 1$, so $R = 1$ for that series and also the series for $t/(1-t^8)$. By Theorem 2, the series for

$\displaystyle\int \dfrac{t}{1-t^8}\,dt$ also has $R = 1$.

25. By Example 7, $\tan^{-1} x = \sum\limits_{n=0}^{\infty} (-1)^n \dfrac{x^{2n+1}}{2n+1}$ with $R = 1$, so

$x - \tan^{-1} x = x - \left(x - \dfrac{x^3}{3} + \dfrac{x^5}{5} - \dfrac{x^7}{7} + \cdots \right) = \dfrac{x^3}{3} - \dfrac{x^5}{5} + \dfrac{x^7}{7} - \cdots = \sum\limits_{n=1}^{\infty} (-1)^{n+1} \dfrac{x^{2n+1}}{2n+1}$ and

$\dfrac{x - \tan^{-1} x}{x^3} = \sum\limits_{n=1}^{\infty} (-1)^{n+1} \dfrac{x^{2n-2}}{2n+1}$, so

$\displaystyle\int \dfrac{x - \tan^{-1} x}{x^3}\,dx = C + \sum\limits_{n=1}^{\infty} (-1)^{n+1} \dfrac{x^{2n-1}}{(2n+1)(2n-1)} = C + \sum\limits_{n=1}^{\infty} (-1)^{n+1} \dfrac{x^{2n-1}}{4n^2 - 1}$. By Theorem 2, $R = 1$.

27. $\dfrac{1}{1+x^5} = \dfrac{1}{1-(-x^5)} = \sum\limits_{n=0}^{\infty} (-x^5)^n = \sum\limits_{n=0}^{\infty} (-1)^n x^{5n} \quad\Rightarrow$

$\displaystyle\int \dfrac{1}{1+x^5}\,dx = \int \sum\limits_{n=0}^{\infty} (-1)^n x^{5n}\,dx = C + \sum\limits_{n=0}^{\infty} (-1)^n \dfrac{x^{5n+1}}{5n+1}$. Thus,

$I = \displaystyle\int_0^{0.2} \dfrac{1}{1+x^5}\,dx = \left[x - \dfrac{x^6}{6} + \dfrac{x^{11}}{11} - \cdots \right]_0^{0.2} = 0.2 - \dfrac{(0.2)^6}{6} + \dfrac{(0.2)^{11}}{11} - \cdots$. The series is alternating, so if we use

the first two terms, the error is at most $(0.2)^{11}/11 \approx 1.9 \times 10^{-9}$. So $I \approx 0.2 - (0.2)^6/6 \approx 0.199\,989$ to six decimal places.

29. We substitute $3x$ for x in Example 7, and find that

$$\int x \arctan(3x)\, dx = \int x \sum_{n=0}^{\infty} (-1)^n \frac{(3x)^{2n+1}}{2n+1}\, dx = \int \sum_{n=0}^{\infty} (-1)^n \frac{3^{2n+1}\, x^{2n+2}}{2n+1}\, dx = C + \sum_{n=0}^{\infty} (-1)^n \frac{3^{2n+1}\, x^{2n+3}}{(2n+1)(2n+3)}$$

So
$$\int_0^{0.1} x \arctan(3x)\, dx = \left[\frac{3x^3}{1\cdot 3} - \frac{3^3 x^5}{3\cdot 5} + \frac{3^5 x^7}{5\cdot 7} - \frac{3^7 x^9}{7\cdot 9} + \cdots \right]_0^{0.1}$$

$$= \frac{1}{10^3} - \frac{9}{5\times 10^5} + \frac{243}{35\times 10^7} - \frac{2187}{63\times 10^9} + \cdots .$$

The series is alternating, so if we use three terms, the error is at most $\dfrac{2187}{63\times 10^9} \approx 3.5\times 10^{-8}$. So

$$\int_0^{0.1} x \arctan(3x)\, dx \approx \frac{1}{10^3} - \frac{9}{5\times 10^5} + \frac{243}{35\times 10^7} \approx 0.000\,983 \text{ to six decimal places.}$$

31. By Example 7, $\arctan x = x - \dfrac{x^3}{3} + \dfrac{x^5}{5} - \dfrac{x^7}{7} + \cdots$, so $\arctan 0.2 = 0.2 - \dfrac{(0.2)^3}{3} + \dfrac{(0.2)^5}{5} - \dfrac{(0.2)^7}{7} + \cdots$.

The series is alternating, so if we use three terms, the error is at most $\dfrac{(0.2)^7}{7} \approx 0.000\,002$.

Thus, to five decimal places, $\arctan 0.2 \approx 0.2 - \dfrac{(0.2)^3}{3} + \dfrac{(0.2)^5}{5} \approx 0.197\,40$.

33. (a) $J_0(x) = \displaystyle\sum_{n=0}^{\infty} \frac{(-1)^n\, x^{2n}}{2^{2n}(n!)^2}$, $J_0'(x) = \displaystyle\sum_{n=1}^{\infty} \frac{(-1)^n\, 2n x^{2n-1}}{2^{2n}(n!)^2}$, and $J_0''(x) = \displaystyle\sum_{n=1}^{\infty} \frac{(-1)^n\, 2n(2n-1)x^{2n-2}}{2^{2n}(n!)^2}$, so

$$x^2 J_0''(x) + x J_0'(x) + x^2 J_0(x) = \sum_{n=1}^{\infty} \frac{(-1)^n\, 2n(2n-1)x^{2n}}{2^{2n}(n!)^2} + \sum_{n=1}^{\infty} \frac{(-1)^n\, 2n x^{2n}}{2^{2n}(n!)^2} + \sum_{n=0}^{\infty} \frac{(-1)^n\, x^{2n+2}}{2^{2n}(n!)^2}$$

$$= \sum_{n=1}^{\infty} \frac{(-1)^n\, 2n(2n-1)x^{2n}}{2^{2n}(n!)^2} + \sum_{n=1}^{\infty} \frac{(-1)^n\, 2n x^{2n}}{2^{2n}(n!)^2} + \sum_{n=1}^{\infty} \frac{(-1)^{n-1}\, x^{2n}}{2^{2n-2}\,[(n-1)!]^2}$$

$$= \sum_{n=1}^{\infty} \frac{(-1)^n\, 2n(2n-1)x^{2n}}{2^{2n}(n!)^2} + \sum_{n=1}^{\infty} \frac{(-1)^n\, 2n x^{2n}}{2^{2n}(n!)^2} + \sum_{n=1}^{\infty} \frac{(-1)^n(-1)^{-1}2^2 n^2 x^{2n}}{2^{2n}(n!)^2}$$

$$= \sum_{n=1}^{\infty} (-1)^n \left[\frac{2n(2n-1) + 2n - 2^2 n^2}{2^{2n}(n!)^2} \right] x^{2n}$$

$$= \sum_{n=1}^{\infty} (-1)^n \left[\frac{4n^2 - 2n + 2n - 4n^2}{2^{2n}(n!)^2} \right] x^{2n} = 0$$

(b) $\displaystyle\int_0^1 J_0(x)\, dx = \int_0^1 \left[\sum_{n=0}^{\infty} \frac{(-1)^n\, x^{2n}}{2^{2n}(n!)^2} \right] dx = \int_0^1 \left(1 - \frac{x^2}{4} + \frac{x^4}{64} - \frac{x^6}{2304} + \cdots \right) dx$

$$= \left[x - \frac{x^3}{3\cdot 4} + \frac{x^5}{5\cdot 64} - \frac{x^7}{7\cdot 2304} + \cdots \right]_0^1 = 1 - \frac{1}{12} + \frac{1}{320} - \frac{1}{16{,}128} + \cdots$$

Since $\frac{1}{16{,}128} \approx 0.000062$, it follows from The Alternating Series Estimation Theorem that, correct to three decimal places,

$\int_0^1 J_0(x)\, dx \approx 1 - \frac{1}{12} + \frac{1}{320} \approx 0.920$.

35. (a) $f(x) = \sum_{n=0}^{\infty} \frac{x^n}{n!}$ $\Rightarrow$ $f'(x) = \sum_{n=1}^{\infty} \frac{nx^{n-1}}{n!} = \sum_{n=1}^{\infty} \frac{x^{n-1}}{(n-1)!} = \sum_{n=0}^{\infty} \frac{x^n}{n!} = f(x)$

(b) By Theorem 7.4.2, the only solution to the differential equation $df(x)/dx = f(x)$ is $f(x) = Ke^x$, but $f(0) = 1$, so

$K = 1$ and $f(x) = e^x$.

Or: We could solve the equation $df(x)/dx = f(x)$ as a separable differential equation.

37. If $a_n = \frac{x^n}{n^2}$, then by the Ratio Test, $\lim_{n \to \infty} \left| \frac{a_{n+1}}{a_n} \right| = \lim_{n \to \infty} \left| \frac{x^{n+1}}{(n+1)^2} \cdot \frac{n^2}{x^n} \right| = |x| \lim_{n \to \infty} \left(\frac{n}{n+1} \right)^2 = |x| < 1$ for

convergence, so $R = 1$. When $x = \pm 1$, $\sum_{n=1}^{\infty} \left| \frac{x^n}{n^2} \right| = \sum_{n=1}^{\infty} \frac{1}{n^2}$ which is a convergent p-series ($p = 2 > 1$), so the interval of

convergence for f is $[-1, 1]$. By Theorem 2, the radii of convergence of f' and f'' are both 1, so we need only check the

endpoints. $f(x) = \sum_{n=1}^{\infty} \frac{x^n}{n^2}$ $\Rightarrow$ $f'(x) = \sum_{n=1}^{\infty} \frac{nx^{n-1}}{n^2} = \sum_{n=0}^{\infty} \frac{x^n}{n+1}$, and this series diverges for $x = 1$ (harmonic series)

and converges for $x = -1$ (Alternating Series Test), so the interval of convergence is $[-1, 1)$. $f''(x) = \sum_{n=1}^{\infty} \frac{nx^{n-1}}{n+1}$ diverges

at both 1 and -1 (Test for Divergence) since $\lim_{n \to \infty} \frac{n}{n+1} = 1 \neq 0$, so its interval of convergence is $(-1, 1)$.

39. By Example 7, $\tan^{-1} x = \sum_{n=0}^{\infty} (-1)^n \frac{x^{2n+1}}{2n+1}$ for $|x| < 1$. In particular, for $x = \frac{1}{\sqrt{3}}$, we

have $\frac{\pi}{6} = \tan^{-1}\left(\frac{1}{\sqrt{3}} \right) = \sum_{n=0}^{\infty} (-1)^n \frac{(1/\sqrt{3})^{2n+1}}{2n+1} = \sum_{n=0}^{\infty} (-1)^n \left(\frac{1}{3} \right)^n \frac{1}{\sqrt{3}} \frac{1}{2n+1}$, so

$\pi = \frac{6}{\sqrt{3}} \sum_{n=0}^{\infty} \frac{(-1)^n}{(2n+1)3^n} = 2\sqrt{3} \sum_{n=0}^{\infty} \frac{(-1)^n}{(2n+1)3^n}$.

8.7 Taylor and Maclaurin Series

1. Using Theorem 5 with $\sum_{n=0}^{\infty} b_n(x-5)^n$, $b_n = \frac{f^{(n)}(a)}{n!}$, so $b_8 = \frac{f^{(8)}(5)}{8!}$.

3. Since $f^{(n)}(0) = (n+1)!$, Equation 7 gives the Maclaurin series

$\sum_{n=0}^{\infty} \frac{f^{(n)}(0)}{n!} x^n = \sum_{n=0}^{\infty} \frac{(n+1)!}{n!} x^n = \sum_{n=0}^{\infty} (n+1)x^n$. Applying the Ratio Test with $a_n = (n+1)x^n$ gives us

$\lim_{n \to \infty} \left| \frac{a_{n+1}}{a_n} \right| = \lim_{n \to \infty} \left| \frac{(n+2)x^{n+1}}{(n+1)x^n} \right| = |x| \lim_{n \to \infty} \frac{n+2}{n+1} = |x| \cdot 1 = |x|$. For convergence, we must have $|x| < 1$, so the

radius of convergence $R = 1$.

5.

n	$f^{(n)}(x)$	$f^{(n)}(0)$
0	$(1-x)^{-2}$	1
1	$2(1-x)^{-3}$	2
2	$6(1-x)^{-4}$	6
3	$24(1-x)^{-5}$	24
4	$120(1-x)^{-6}$	120
$\vdots$	$\vdots$	$\vdots$

$$(1-x)^{-2} = f(0) + f'(0)x + \frac{f''(0)}{2!}x^2 + \frac{f'''(0)}{3!}x^3 + \frac{f^{(4)}(0)}{4!}x^4 + \cdots$$

$$= 1 + 2x + \tfrac{6}{2}x^2 + \tfrac{24}{6}x^3 + \tfrac{120}{24}x^4 + \cdots$$

$$= 1 + 2x + 3x^2 + 4x^3 + 5x^4 + \cdots = \sum_{n=0}^{\infty} (n+1)x^n$$

$$\lim_{n\to\infty} \left| \frac{a_{n+1}}{a_n} \right| = \lim_{n\to\infty} \left| \frac{(n+2)x^{n+1}}{(n+1)x^n} \right| = |x| \lim_{n\to\infty} \frac{n+2}{n+1} = |x|(1) = |x| < 1$$

for convergence, so $R = 1$.

7.

n	$f^{(n)}(x)$	$f^{(n)}(0)$
0	$\sin \pi x$	0
1	$\pi \cos \pi x$	π
2	$-\pi^2 \sin \pi x$	0
3	$-\pi^3 \cos \pi x$	$-\pi^3$
4	$\pi^4 \sin \pi x$	0
5	$\pi^5 \cos \pi x$	π^5
$\vdots$	$\vdots$	$\vdots$

$$\sin \pi x = f(0) + f'(0)x + \frac{f''(0)}{2!}x^2 + \frac{f'''(0)}{3!}x^3$$

$$+ \frac{f^{(4)}(0)}{4!}x^4 + \frac{f^{(5)}(0)}{5!}x^5 + \cdots$$

$$= 0 + \pi x + 0 - \frac{\pi^3}{3!}x^3 + 0 + \frac{\pi^5}{5!}x^5 + \cdots$$

$$= \pi x - \frac{\pi^3}{3!}x^3 + \frac{\pi^5}{5!}x^5 - \frac{\pi^7}{7!}x^7 + \cdots$$

$$= \sum_{n=0}^{\infty} (-1)^n \frac{\pi^{2n+1}}{(2n+1)!} x^{2n+1}$$

$$\lim_{n\to\infty} \left| \frac{a_{n+1}}{a_n} \right| = \lim_{n\to\infty} \left| \frac{\pi^{2n+3} x^{2n+3}}{(2n+3)!} \cdot \frac{(2n+1)!}{\pi^{2n+1} x^{2n+1}} \right| = \lim_{n\to\infty} \frac{\pi^2 x^2}{(2n+3)(2n+2)}$$

$$= 0 < 1 \quad \text{for all } x, \text{ so } R = \infty.$$

9.

n	$f^{(n)}(x)$	$f^{(n)}(0)$
0	e^{5x}	1
1	$5e^{5x}$	5
2	$5^2 e^{5x}$	25
3	$5^3 e^{5x}$	125
4	$5^4 e^{5x}$	625
$\vdots$	$\vdots$	$\vdots$

$$e^{5x} = \sum_{n=0}^{\infty} \frac{f^{(n)}(0)}{n!} x^n = \sum_{n=0}^{\infty} \frac{5^n}{n!} x^n.$$

$$\lim_{n\to\infty} \left| \frac{a_{n+1}}{a_n} \right| = \lim_{n\to\infty} \left[\frac{5^{n+1} |x|^{n+1}}{(n+1)!} \cdot \frac{n!}{5^n |x|^n} \right]$$

$$= \lim_{n\to\infty} \frac{5|x|}{n+1} = 0 < 1 \quad \text{for all } x, \text{ so } R = \infty.$$

11.

n	$f^{(n)}(x)$	$f^{(n)}(1)$
0	$x^4 - 3x^2 + 1$	-1
1	$4x^3 - 6x$	-2
2	$12x^2 - 6$	6
3	$24x$	24
4	24	24
5	0	0
6	0	0
$\vdots$	$\vdots$	$\vdots$

$f^{(n)}(x) = 0$ for $n \geq 5$, so f has a finite series expansion about $a = 1$.

$$f(x) = x^4 - 3x^2 + 1 = \sum_{n=0}^{4} \frac{f^{(n)}(1)}{n!}(x-1)^n$$

$$= \frac{-1}{0!}(x-1)^0 + \frac{-2}{1!}(x-1)^1 + \frac{6}{2!}(x-1)^2 + \frac{24}{3!}(x-1)^3 + \frac{24}{4!}(x-1)^4$$

$$= -1 - 2(x-1) + 3(x-1)^2 + 4(x-1)^3 + (x-1)^4$$

A finite series converges for all x, so $R = \infty$.

13. $f(x) = e^x$ $\Rightarrow$ $f^{(n)}(x) = e^x$, so $f^{(n)}(3) = e^3$ and $e^x = \sum\limits_{n=0}^{\infty} \dfrac{e^3}{n!}(x-3)^n$. If $a_n = \dfrac{e^3}{n!}(x-3)^n$, then

$$\lim_{n \to \infty} \left| \frac{a_{n+1}}{a_n} \right| = \lim_{n \to \infty} \left| \frac{e^3(x-3)^{n+1}}{(n+1)!} \cdot \frac{n!}{e^3(x-3)^n} \right| = \lim_{n \to \infty} \frac{|x-3|}{n+1} = 0 < 1 \text{ for all } x \text{, so } R = \infty.$$

15.

n	$f^{(n)}(x)$	$f^{(n)}(\pi)$
0	$\cos x$	-1
1	$-\sin x$	0
2	$-\cos x$	1
3	$\sin x$	0
4	$\cos x$	-1
$\vdots$	$\vdots$	$\vdots$

$$\cos x = \sum_{k=0}^{\infty} \frac{f^{(k)}(\pi)}{k!}(x-\pi)^k = -1 + \frac{(x-\pi)^2}{2!} - \frac{(x-\pi)^4}{4!} + \frac{(x-\pi)^6}{6!} - \cdots$$

$$= \sum_{n=0}^{\infty} (-1)^{n+1} \frac{(x-\pi)^{2n}}{(2n)!}$$

$$\lim_{n \to \infty} \left| \frac{a_{n+1}}{a_n} \right| = \lim_{n \to \infty} \left[\frac{|x-\pi|^{2n+2}}{(2n+2)!} \cdot \frac{(2n)!}{|x-\pi|^{2n}} \right]$$

$$= \lim_{n \to \infty} \frac{|x-\pi|^2}{(2n+2)(2n+1)} = 0 < 1 \quad \text{for all } x \text{, so } R = \infty.$$

17.

n	$f^{(n)}(x)$	$f^{(n)}(9)$
0	$x^{-1/2}$	$\frac{1}{3}$
1	$-\frac{1}{2}x^{-3/2}$	$-\frac{1}{2} \cdot \frac{1}{3^3}$
2	$\frac{3}{4}x^{-5/2}$	$-\frac{1}{2} \cdot \left(-\frac{3}{2}\right) \cdot \frac{1}{3^5}$
3	$-\frac{15}{8}x^{-7/2}$	$-\frac{1}{2} \cdot \left(-\frac{3}{2}\right) \cdot \left(-\frac{5}{2}\right) \cdot \frac{1}{3^7}$
$\vdots$	$\vdots$	$\vdots$

$$\frac{1}{\sqrt{x}} = \frac{1}{3} - \frac{1}{2 \cdot 3^3}(x-9) + \frac{3}{2^2 \cdot 3^5}\frac{(x-9)^2}{2!}$$

$$- \frac{3 \cdot 5}{2^3 \cdot 3^7}\frac{(x-9)^3}{3!} + \cdots$$

$$= \frac{1}{3} + \sum_{n=1}^{\infty} (-1)^n \frac{1 \cdot 3 \cdot 5 \cdot \cdots \cdot (2n-1)}{2^n \cdot 3^{2n+1} \cdot n!}(x-9)^n.$$

$$\lim_{n \to \infty} \left| \frac{a_{n+1}}{a_n} \right| = \lim_{n \to \infty} \left[\frac{1 \cdot 3 \cdot 5 \cdot \cdots \cdot (2n-1)[2(n+1)-1]\,|x-9|^{n+1}}{2^{n+1} \cdot 3^{[2(n+1)+1]} \cdot (n+1)!} \cdot \frac{2^n \cdot 3^{2n+1} \cdot n!}{1 \cdot 3 \cdot 5 \cdot \cdots \cdot (2n-1)\,|x-9|^n} \right]$$

$$= \lim_{n \to \infty} \left[\frac{(2n+1)\,|x-9|}{2 \cdot 3^2(n+1)} \right] = \frac{1}{9}|x-9| < 1 \quad \text{for convergence, so } |x-9| < 9 \text{ and } R = 9.$$

19. If $f(x) = \sin \pi x$, then $f^{(n+1)}(x) = \pm\pi^{n+1} \sin \pi x$ or $\pm\pi^{n+1}\cos \pi x$. In each case, $\left| f^{(n+1)}(x) \right| \leq \pi^{n+1}$, so by Formula 9

with $a = 0$ and $M = \pi^{n+1}$, $|R_n(x)| \leq \dfrac{\pi^{n+1}}{(n+1)!}\,|x|^{n+1} = \dfrac{|\pi x|^{n+1}}{(n+1)!}$. Thus, $|R_n(x)| \to 0$ as $n \to \infty$ by Equation 10.

So $\lim\limits_{n \to \infty} R_n(x) = 0$ and, by Theorem 8, the series in Exercise 7 represents $\sin \pi x$ for all x.

21. The general binomial series in (17) is

$$(1+x)^k = \sum_{n=0}^{\infty} \binom{k}{n} x^n = 1 + kx + \frac{k(k-1)}{2!}x^2 + \frac{k(k-1)(k-2)}{3!}x^3 + \cdots.$$

$$(1+x)^{1/2} = \sum_{n=0}^{\infty} \binom{\frac{1}{2}}{n} x^n = 1 + \left(\tfrac{1}{2}\right)x + \frac{\left(\frac{1}{2}\right)\left(-\frac{1}{2}\right)}{2!}x^2 + \frac{\left(\frac{1}{2}\right)\left(-\frac{1}{2}\right)\left(-\frac{3}{2}\right)}{3!}x^3 + \cdots$$

$$= 1 + \frac{x}{2} - \frac{x^2}{2^2 \cdot 2!} + \frac{1 \cdot 3 \cdot x^3}{2^3 \cdot 3!} - \frac{1 \cdot 3 \cdot 5 \cdot x^4}{2^4 \cdot 4!} + \cdots$$

$$= 1 + \frac{x}{2} + \sum_{n=2}^{\infty} \frac{(-1)^{n-1} 1 \cdot 3 \cdot 5 \cdot \cdots \cdot (2n-3)x^n}{2^n \cdot n!} \quad \text{for } |x| < 1, \quad \text{so } R = 1.$$

23. $\dfrac{1}{(2+x)^3} = \dfrac{1}{[2(1+x/2)]^3} = \dfrac{1}{8}\left(1+\dfrac{x}{2}\right)^{-3} = \dfrac{1}{8}\displaystyle\sum_{n=0}^{\infty}\binom{-3}{n}\left(\dfrac{x}{2}\right)^n$. The binomial coefficient is

$$\binom{-3}{n} = \dfrac{(-3)(-4)(-5)\cdots\cdots(-3-n+1)}{n!} = \dfrac{(-3)(-4)(-5)\cdots\cdots[-(n+2)]}{n!}$$

$$= \dfrac{(-1)^n \cdot 2\cdot3\cdot4\cdot5\cdots\cdots(n+1)(n+2)}{2\cdot n!} = \dfrac{(-1)^n(n+1)(n+2)}{2}$$

Thus, $\dfrac{1}{(2+x)^3} = \dfrac{1}{8}\displaystyle\sum_{n=0}^{\infty}\dfrac{(-1)^n(n+1)(n+2)}{2}\dfrac{x^n}{2^n} = \sum_{n=0}^{\infty}\dfrac{(-1)^n(n+1)(n+2)x^n}{2^{n+4}}$ for $\left|\dfrac{x}{2}\right| < 1 \iff |x| < 2$, so $R = 2$.

25. $\sin x = \displaystyle\sum_{n=0}^{\infty}(-1)^n\dfrac{x^{2n+1}}{(2n+1)!} \implies f(x) = \sin(\pi x) = \sum_{n=0}^{\infty}(-1)^n\dfrac{(\pi x)^{2n+1}}{(2n+1)!} = \sum_{n=0}^{\infty}(-1)^n\dfrac{\pi^{2n+1}}{(2n+1)!}x^{2n+1},\ R = \infty$.

27. $e^x = \displaystyle\sum_{n=0}^{\infty}\dfrac{x^n}{n!} \implies e^{2x} = \sum_{n=0}^{\infty}\dfrac{(2x)^n}{n!} = \sum_{n=0}^{\infty}\dfrac{2^n x^n}{n!}$, so $f(x) = e^x + e^{2x} = \sum_{n=0}^{\infty}\dfrac{1}{n!}x^n + \sum_{n=0}^{\infty}\dfrac{2^n}{n!}x^n = \sum_{n=0}^{\infty}\dfrac{2^n+1}{n!}x^n$.

$R = \infty$.

29. $\cos x = \displaystyle\sum_{n=0}^{\infty}(-1)^n\dfrac{x^{2n}}{(2n)!} \implies \cos\left(\tfrac{1}{2}x^2\right) = \sum_{n=0}^{\infty}(-1)^n\dfrac{\left(\tfrac{1}{2}x^2\right)^{2n}}{(2n)!} = \sum_{n=0}^{\infty}(-1)^n\dfrac{x^{4n}}{2^{2n}(2n)!}$, so

$f(x) = x\cos\left(\tfrac{1}{2}x^2\right) = \displaystyle\sum_{n=0}^{\infty}(-1)^n\dfrac{1}{2^{2n}(2n)!}x^{4n+1},\ R = \infty$.

31. We must write the binomial in the form $(1+$ expression$)$, so we'll factor out a 4.

$$\dfrac{x}{\sqrt{4+x^2}} = \dfrac{x}{\sqrt{4(1+x^2/4)}} = \dfrac{x}{2\sqrt{1+x^2/4}} = \dfrac{x}{2}\left(1+\dfrac{x^2}{4}\right)^{-1/2} = \dfrac{x}{2}\sum_{n=0}^{\infty}\binom{-\tfrac{1}{2}}{n}\left(\dfrac{x^2}{4}\right)^n$$

$$= \dfrac{x}{2}\left[1 + \left(-\tfrac{1}{2}\right)\dfrac{x^2}{4} + \dfrac{\left(-\tfrac{1}{2}\right)\left(-\tfrac{3}{2}\right)}{2!}\left(\dfrac{x^2}{4}\right)^2 + \dfrac{\left(-\tfrac{1}{2}\right)\left(-\tfrac{3}{2}\right)\left(-\tfrac{5}{2}\right)}{3!}\left(\dfrac{x^2}{4}\right)^3 + \cdots\right]$$

$$= \dfrac{x}{2} + \dfrac{x}{2}\sum_{n=1}^{\infty}(-1)^n\dfrac{1\cdot3\cdot5\cdots\cdots(2n-1)}{2^n\cdot4^n\cdot n!}x^{2n}$$

$$= \dfrac{x}{2} + \sum_{n=1}^{\infty}(-1)^n\dfrac{1\cdot3\cdot5\cdots\cdots(2n-1)}{n!\,2^{3n+1}}x^{2n+1}\ \text{and}\ \dfrac{x^2}{4} < 1 \iff \dfrac{|x|}{2} < 1 \iff |x| < 2,\ \text{so}\ R = 2.$$

33. $\sin^2 x = \dfrac{1}{2}(1-\cos 2x) = \dfrac{1}{2}\left[1 - \displaystyle\sum_{n=0}^{\infty}\dfrac{(-1)^n(2x)^{2n}}{(2n)!}\right] = \dfrac{1}{2}\left[1 - 1 - \sum_{n=1}^{\infty}\dfrac{(-1)^n(2x)^{2n}}{(2n)!}\right] = \sum_{n=1}^{\infty}\dfrac{(-1)^{n+1}2^{2n-1}x^{2n}}{(2n)!}$,

$R = \infty$

35. $\cos x = \displaystyle\sum_{n=0}^{\infty}(-1)^n\dfrac{x^{2n}}{(2n)!} \implies f(x) = \cos(x^2) = \sum_{n=0}^{\infty}\dfrac{(-1)^n\left(x^2\right)^{2n}}{(2n)!} = \sum_{n=0}^{\infty}\dfrac{(-1)^n x^{4n}}{(2n)!},\ R = \infty$

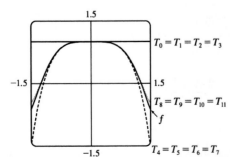

Notice that, as n increases, $T_n(x)$

becomes a better approximation to $f(x)$.

37. $e^x \overset{(11)}{=} \displaystyle\sum_{n=0}^{\infty} \frac{x^n}{n!}$, so $e^{-x} = \displaystyle\sum_{n=0}^{\infty} \frac{(-x)^n}{n!} = \sum_{n=0}^{\infty} (-1)^n \frac{x^n}{n!}$, so

$$f(x) = xe^{-x} = \sum_{n=0}^{\infty} (-1)^n \frac{1}{n!} x^{n+1}$$

$$= x - x^2 + \tfrac{1}{2}x^3 - \tfrac{1}{6}x^4 + \tfrac{1}{24}x^5 - \tfrac{1}{120}x^6 + \cdots$$

$$= \sum_{n=1}^{\infty} (-1)^{n-1} \frac{x^n}{(n-1)!}$$

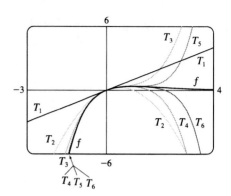

The series for e^x converges for all x, so the same is true of the series
for $f(x)$; that is, $R = \infty$. From the graphs of f and the first few Taylor

polynomials, we see that $T_n(x)$ provides a closer fit to $f(x)$ near 0 as n increases.

39. $e^x = \displaystyle\sum_{n=0}^{\infty} \frac{x^n}{n!}$, so $e^{-0.2} = \displaystyle\sum_{n=0}^{\infty} \frac{(-0.2)^n}{n!} = 1 - 0.2 + \frac{1}{2!}(0.2)^2 - \frac{1}{3!}(0.2)^3 + \frac{1}{4!}(0.2)^4 - \frac{1}{5!}(0.2)^5 + \frac{1}{6!}(0.2)^6 - \cdots$.

But $\dfrac{1}{6!}(0.2)^6 = 8.\overline{8} \times 10^{-8}$, so by the Alternating Series Estimation Theorem, $e^{-0.2} \approx \displaystyle\sum_{n=0}^{5} \frac{(-0.2)^n}{n!} \approx 0.81873$, correct to

five decimal places.

41. (a) $1/\sqrt{1-x^2} = \left[1 + \left(-x^2\right)\right]^{-1/2} = 1 + \left(-\tfrac{1}{2}\right)\left(-x^2\right) + \dfrac{\left(-\frac{1}{2}\right)\left(-\frac{3}{2}\right)}{2!}\left(-x^2\right)^2 + \dfrac{\left(-\frac{1}{2}\right)\left(-\frac{3}{2}\right)\left(-\frac{5}{2}\right)}{3!}\left(-x^2\right)^3 + \cdots$

$$= 1 + \sum_{n=1}^{\infty} \frac{1 \cdot 3 \cdot 5 \cdot \cdots \cdot (2n-1)}{2^n \cdot n!} x^{2n}$$

(b) $\sin^{-1} x = \displaystyle\int \frac{1}{\sqrt{1-x^2}}\, dx = C + x + \sum_{n=1}^{\infty} \frac{1 \cdot 3 \cdot 5 \cdot \cdots \cdot (2n-1)}{(2n+1)2^n \cdot n!} x^{2n+1}$

$$= x + \sum_{n=1}^{\infty} \frac{1 \cdot 3 \cdot 5 \cdot \cdots \cdot (2n-1)}{(2n+1)2^n \cdot n!} x^{2n+1} \quad \text{since } 0 = \sin^{-1} 0 = C.$$

43. $\cos x \overset{(16)}{=} \displaystyle\sum_{n=0}^{\infty} (-1)^n \frac{x^{2n}}{(2n)!} \quad \Rightarrow \quad \cos(x^3) = \sum_{n=0}^{\infty} (-1)^n \frac{(x^3)^{2n}}{(2n)!} = \sum_{n=0}^{\infty} (-1)^n \frac{x^{6n}}{(2n)!} \quad \Rightarrow$

$x\cos(x^3) = \displaystyle\sum_{n=0}^{\infty} (-1)^n \frac{x^{6n+1}}{(2n)!} \quad \Rightarrow \quad \int x\cos(x^3)\, dx = C + \sum_{n=0}^{\infty} (-1)^n \frac{x^{6n+2}}{(6n+2)(2n)!}$, with $R = \infty$.

45. $\cos x \overset{(16)}{=} \displaystyle\sum_{n=0}^{\infty} (-1)^n \frac{x^{2n}}{(2n)!} \quad \Rightarrow \quad \cos x - 1 = \sum_{n=1}^{\infty} (-1)^n \frac{x^{2n}}{(2n)!} \quad \Rightarrow \quad \frac{\cos x - 1}{x} = \sum_{n=1}^{\infty} (-1)^n \frac{x^{2n-1}}{(2n)!} \quad \Rightarrow$

$\displaystyle\int \frac{\cos x - 1}{x}\, dx = C + \sum_{n=1}^{\infty} (-1)^n \frac{x^{2n}}{2n \cdot (2n)!}$, with $R = \infty$.

47. By Exercise 43, $\displaystyle\int x\cos(x^3)\, dx = C + \sum_{n=0}^{\infty} (-1)^n \frac{x^{6n+2}}{(6n+2)(2n)!}$, so

$\displaystyle\int_0^1 x\cos(x^3)\, dx = \left[\sum_{n=0}^{\infty} (-1)^n \frac{x^{6n+2}}{(6n+2)(2n)!}\right]_0^1 = \sum_{n=0}^{\infty} \frac{(-1)^n}{(6n+2)(2n)!} = \frac{1}{2} - \frac{1}{8 \cdot 2!} + \frac{1}{14 \cdot 4!} - \frac{1}{20 \cdot 6!} + \cdots$, but

$\dfrac{1}{20 \cdot 6!} = \dfrac{1}{14{,}400} \approx 0.000\,069$, so $\displaystyle\int_0^1 x\cos(x^3)\, dx \approx \frac{1}{2} - \frac{1}{16} + \frac{1}{336} \approx 0.440$ (correct to three decimal places) by the

Alternating Series Estimation Theorem.

49. $\sqrt{1+x^4} = (1+x^4)^{1/2} = \sum\limits_{n=0}^{\infty} \binom{1/2}{n}(x^4)^n$, so $\int \sqrt{1+x^4}\,dx = C + \sum\limits_{n=0}^{\infty} \binom{1/2}{n}\dfrac{x^{4n+1}}{4n+1}$ and hence, since $0.4 < 1$,

we have

$$I = \int_0^{0.4} \sqrt{1+x^4}\,dx = \sum_{n=0}^{\infty} \binom{1/2}{n}\frac{(0.4)^{4n+1}}{4n+1}$$

$$= (1)\frac{(0.4)^1}{0!} + \frac{\frac{1}{2}}{1!}\frac{(0.4)^5}{5} + \frac{\frac{1}{2}\left(-\frac{1}{2}\right)}{2!}\frac{(0.4)^9}{9} + \frac{\frac{1}{2}\left(-\frac{1}{2}\right)\left(-\frac{3}{2}\right)}{3!}\frac{(0.4)^{13}}{13} + \frac{\frac{1}{2}\left(-\frac{1}{2}\right)\left(-\frac{3}{2}\right)\left(-\frac{5}{2}\right)}{4!}\frac{(0.4)^{17}}{17} + \cdots$$

$$= 0.4 + \frac{(0.4)^5}{10} - \frac{(0.4)^9}{72} + \frac{(0.4)^{13}}{208} - \frac{5(0.4)^{17}}{2176} + \cdots$$

Now $\dfrac{(0.4)^9}{72} \approx 3.6 \times 10^{-6} < 5 \times 10^{-6}$, so by the Alternating Series Estimation Theorem, $I \approx 0.4 + \dfrac{(0.4)^5}{10} \approx 0.40102$

(correct to five decimal places).

51. $\lim\limits_{x\to 0} \dfrac{x - \ln(1+x)}{x^2} = \lim\limits_{x\to 0}\dfrac{x - \left(x - \frac{1}{2}x^2 + \frac{1}{3}x^3 - \frac{1}{4}x^4 + \frac{1}{5}x^5 - \cdots\right)}{x^2} = \lim\limits_{x\to 0}\dfrac{\frac{1}{2}x^2 - \frac{1}{3}x^3 + \frac{1}{4}x^4 - \frac{1}{5}x^5 + \cdots}{x^2}$

$$= \lim_{x\to 0}\left(\tfrac{1}{2} - \tfrac{1}{3}x + \tfrac{1}{4}x^2 - \tfrac{1}{5}x^3 + \cdots\right) = \tfrac{1}{2}$$

since power series are continuous functions.

53. $\lim\limits_{x\to 0}\dfrac{\sin x - x + \frac{1}{6}x^3}{x^5} = \lim\limits_{x\to 0}\dfrac{\left(x - \frac{1}{3!}x^3 + \frac{1}{5!}x^5 - \frac{1}{7!}x^7 + \cdots\right) - x + \frac{1}{6}x^3}{x^5}$

$$= \lim_{x\to 0}\frac{\frac{1}{5!}x^5 - \frac{1}{7!}x^7 + \cdots}{x^5} = \lim_{x\to 0}\left(\frac{1}{5!} - \frac{x^2}{7!} + \frac{x^4}{9!} - \cdots\right) = \frac{1}{5!} = \frac{1}{120}$$

since power series are continuous functions.

55. From Equation 11, we have $e^{-x^2} = 1 - \dfrac{x^2}{1!} + \dfrac{x^4}{2!} - \dfrac{x^6}{3!} + \cdots$ and we know that $\cos x = 1 - \dfrac{x^2}{2!} + \dfrac{x^4}{4!} - \cdots$ from

Equation 16. Therefore, $e^{-x^2}\cos x = \left(1 - x^2 + \frac{1}{2}x^4 - \cdots\right)\left(1 - \frac{1}{2}x^2 + \frac{1}{24}x^4 - \cdots\right)$. Writing only the terms with

degree ≤ 4, we get $e^{-x^2}\cos x = 1 - \frac{1}{2}x^2 + \frac{1}{24}x^4 - x^2 + \frac{1}{2}x^4 + \frac{1}{2}x^4 + \cdots = 1 - \frac{3}{2}x^2 + \frac{25}{24}x^4 + \cdots$.

57. $\dfrac{x}{\sin x} \overset{(15)}{=} \dfrac{x}{x - \frac{1}{6}x^3 + \frac{1}{120}x^5 - \cdots}$.

$$
\begin{array}{r}
1 + \frac{1}{6}x^2 + \frac{7}{360}x^4 + \cdots \\
x - \frac{1}{6}x^3 + \frac{1}{120}x^5 - \cdots \enclose{longdiv}{} \\
\end{array}
$$

$$x - \tfrac{1}{6}x^3 + \tfrac{1}{120}x^5 - \cdots \overline{\big)\, x}$$

$$\underline{x - \tfrac{1}{6}x^3 + \tfrac{1}{120}x^5 - \cdots}$$

$$\tfrac{1}{6}x^3 - \tfrac{1}{120}x^5 + \cdots$$

$$\underline{\tfrac{1}{6}x^3 - \tfrac{1}{36}x^5 + \cdots}$$

$$\tfrac{7}{360}x^5 + \cdots$$

$$\underline{\tfrac{7}{360}x^5 + \cdots}$$

$$\cdots$$

From the long division above, $\dfrac{x}{\sin x} = 1 + \frac{1}{6}x^2 + \frac{7}{360}x^4 + \cdots$.

59. $\sum\limits_{n=0}^{\infty}(-1)^n\dfrac{x^{4n}}{n!} = \sum\limits_{n=0}^{\infty}\dfrac{(-x^4)^n}{n!} = e^{-x^4}$, by (11).

61. $\displaystyle\sum_{n=1}^{\infty}(-1)^{n-1}\frac{3^n}{n5^n} = \sum_{n=1}^{\infty}(-1)^{n-1}\frac{(3/5)^n}{n} = \ln\left(1+\frac{3}{5}\right)$ [from Table 1] $= \ln\frac{8}{5}$

63. $\displaystyle\sum_{n=0}^{\infty}\frac{(-1)^n\,\pi^{2n+1}}{4^{2n+1}(2n+1)!} = \sum_{n=0}^{\infty}\frac{(-1)^n\left(\frac{\pi}{4}\right)^{2n+1}}{(2n+1)!} = \sin\frac{\pi}{4} = \frac{1}{\sqrt{2}}$, by (15).

65. $3 + \dfrac{9}{2!} + \dfrac{27}{3!} + \dfrac{81}{4!} + \cdots = \dfrac{3^1}{1!} + \dfrac{3^2}{2!} + \dfrac{3^3}{3!} + \dfrac{3^4}{4!} + \cdots = \displaystyle\sum_{n=1}^{\infty}\frac{3^n}{n!} = \sum_{n=0}^{\infty}\frac{3^n}{n!} - 1 = e^3 - 1$, by (11).

67. Assume that $|f'''(x)| \le M$, so $f'''(x) \le M$ for $a \le x \le a + d$. Now $\int_a^x f'''(t)\,dt \le \int_a^x M\,dt \Rightarrow$

$f''(x) - f''(a) \le M(x - a) \Rightarrow f''(x) \le f''(a) + M(x - a)$. Thus, $\int_a^x f''(t)\,dt \le \int_a^x[f''(a) + M(t - a)]\,dt \Rightarrow$

$f'(x) - f'(a) \le f''(a)(x - a) + \frac{1}{2}M(x - a)^2 \Rightarrow f'(x) \le f'(a) + f''(a)(x - a) + \frac{1}{2}M(x - a)^2 \Rightarrow$

$\int_a^x f'(t)\,dt \le \int_a^x\left[f'(a) + f''(a)(t - a) + \frac{1}{2}M(t - a)^2\right]dt \Rightarrow$

$f(x) - f(a) \le f'(a)(x - a) + \frac{1}{2}f''(a)(x - a)^2 + \frac{1}{6}M(x - a)^3$. So

$f(x) - f(a) - f'(a)(x - a) - \frac{1}{2}f''(a)(x - a)^2 \le \frac{1}{6}M(x - a)^3$. But

$R_2(x) = f(x) - T_2(x) = f(x) - f(a) - f'(a)(x - a) - \frac{1}{2}f''(a)(x - a)^2$, so $R_2(x) \le \frac{1}{6}M(x - a)^3$.

A similar argument using $f'''(x) \ge -M$ shows that $R_2(x) \ge -\frac{1}{6}M(x - a)^3$. So $|R_2(x_2)| \le \frac{1}{6}M\,|x - a|^3$.

Although we have assumed that $x > a$, a similar calculation shows that this inequality is also true if $x < a$.

69. (a) $g(x) = \displaystyle\sum_{n=0}^{\infty}\binom{k}{n}x^n \Rightarrow g'(x) = \sum_{n=1}^{\infty}\binom{k}{n}nx^{n-1}$, so

$(1 + x)g'(x) = (1 + x)\displaystyle\sum_{n=1}^{\infty}\binom{k}{n}nx^{n-1} = \sum_{n=1}^{\infty}\binom{k}{n}nx^{n-1} + \sum_{n=1}^{\infty}\binom{k}{n}nx^n$

$= \displaystyle\sum_{n=0}^{\infty}\binom{k}{n+1}(n+1)x^n + \sum_{n=0}^{\infty}\binom{k}{n}nx^n \qquad \begin{bmatrix}\text{Replace }n\text{ with }n+1\\ \text{in the first series}\end{bmatrix}$

$= \displaystyle\sum_{n=0}^{\infty}(n+1)\frac{k(k-1)(k-2)\cdots(k-n+1)(k-n)}{(n+1)!}x^n + \sum_{n=0}^{\infty}\left[(n)\frac{k(k-1)(k-2)\cdots(k-n+1)}{n!}\right]x^n$

$= \displaystyle\sum_{n=0}^{\infty}\frac{(n+1)k(k-1)(k-2)\cdots(k-n+1)}{(n+1)!}[(k-n)+n]\,x^n$

$= k\displaystyle\sum_{n=0}^{\infty}\frac{k(k-1)(k-2)\cdots(k-n+1)}{n!}x^n = k\sum_{n=0}^{\infty}\binom{k}{n}x^n = kg(x)$

Thus, $g'(x) = \dfrac{kg(x)}{1 + x}$.

(b) $h(x) = (1 + x)^{-k}g(x) \Rightarrow$

$h'(x) = -k(1 + x)^{-k-1}g(x) + (1 + x)^{-k}g'(x)$ [Product Rule]

$= -k(1 + x)^{-k-1}g(x) + (1 + x)^{-k}\dfrac{kg(x)}{1 + x}$ [from part (a)]

$= -k(1 + x)^{-k-1}g(x) + k(1 + x)^{-k-1}g(x) = 0$

(c) From part (b) we see that $h(x)$ must be constant for $x \in (-1, 1)$, so $h(x) = h(0) = 1$ for $x \in (-1, 1)$.

Thus, $h(x) = 1 = (1 + x)^{-k}g(x) \Leftrightarrow g(x) = (1 + x)^k$ for $x \in (-1, 1)$.

8.8 Applications of Taylor Polynomials

1. (a)

n	$f^{(n)}(x)$	$f^{(n)}(0)$	$T_n(x)$
0	$\cos x$	1	1
1	$-\sin x$	0	1
2	$-\cos x$	-1	$1 - \frac{1}{2}x^2$
3	$\sin x$	0	$1 - \frac{1}{2}x^2$
4	$\cos x$	1	$1 - \frac{1}{2}x^2 + \frac{1}{24}x^4$
5	$-\sin x$	0	$1 - \frac{1}{2}x^2 + \frac{1}{24}x^4$
6	$-\cos x$	-1	$1 - \frac{1}{2}x^2 + \frac{1}{24}x^4 - \frac{1}{720}x^6$

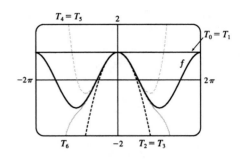

(b)

x	f	$T_0 = T_1$	$T_2 = T_3$	$T_4 = T_5$	T_6
$\frac{\pi}{4}$	0.7071	1	0.6916	0.7074	0.7071
$\frac{\pi}{2}$	0	1	-0.2337	0.0200	-0.0009
π	-1	1	-3.9348	0.1239	-1.2114

(c) As n increases, $T_n(x)$ is a good approximation to $f(x)$ on a larger and larger interval.

3.

n	$f^{(n)}(x)$	$f^{(n)}(2)$
0	$1/x$	$\frac{1}{2}$
1	$-1/x^2$	$-\frac{1}{4}$
2	$2/x^3$	$\frac{1}{4}$
3	$-6/x^4$	$-\frac{3}{8}$

$$T_3(x) = \sum_{n=0}^{3} \frac{f^{(n)}(2)}{n!}(x-2)^n$$

$$= \frac{\frac{1}{2}}{0!} - \frac{\frac{1}{4}}{1!}(x-2) + \frac{\frac{1}{4}}{2!}(x-2)^2 - \frac{\frac{3}{8}}{3!}(x-2)^3$$

$$= \frac{1}{2} - \frac{1}{4}(x-2) + \frac{1}{8}(x-2)^2 - \frac{1}{16}(x-2)^3$$

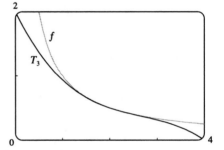

5.

n	$f^{(n)}(x)$	$f^{(n)}(\pi/2)$
0	$\cos x$	0
1	$-\sin x$	-1
2	$-\cos x$	0
3	$\sin x$	1

$$T_3(x) = \sum_{n=0}^{3} \frac{f^{(n)}(\pi/2)}{n!}\left(x - \frac{\pi}{2}\right)^n$$

$$= -\left(x - \frac{\pi}{2}\right) + \frac{1}{6}\left(x - \frac{\pi}{2}\right)^3$$

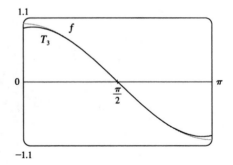

7.

n	$f^{(n)}(x)$	$f^{(n)}(0)$
0	xe^{-2x}	0
1	$(1-2x)e^{-2x}$	1
2	$4(x-1)e^{-2x}$	-4
3	$4(3-2x)e^{-2x}$	12

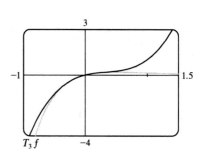

$$T_3(x) = \sum_{n=0}^{3} \frac{f^{(n)}(0)}{n!} x^n = \frac{0}{1} \cdot 1 + \frac{1}{1} x^1 + \frac{-4}{2} x^2 + \frac{12}{6} x^3 = x - 2x^2 + 2x^3$$

9. You may be able to simply find the Taylor polynomials for

$f(x) = \cot x$ using your CAS. We will list the values of $f^{(n)}(\pi/4)$

for $n = 0$ to $n = 5$.

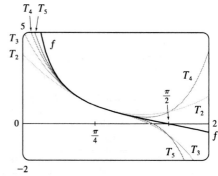

n	0	1	2	3	4	5
$f^{(n)}(\pi/4)$	1	-2	4	-16	80	-512

$$T_5(x) = \sum_{n=0}^{5} \frac{f^{(n)}(\pi/4)}{n!} \left(x - \tfrac{\pi}{4}\right)^n$$

$$= 1 - 2\left(x - \tfrac{\pi}{4}\right) + 2\left(x - \tfrac{\pi}{4}\right)^2 - \tfrac{8}{3}\left(x - \tfrac{\pi}{4}\right)^3 + \tfrac{10}{3}\left(x - \tfrac{\pi}{4}\right)^4 - \tfrac{64}{15}\left(x - \tfrac{\pi}{4}\right)^5$$

For $n = 2$ to $n = 5$, $T_n(x)$ is the polynomial consisting of all the terms up to and including the $\left(x - \tfrac{\pi}{4}\right)^n$ term.

11.

n	$f^{(n)}(x)$	$f^{(n)}(4)$
0	$\sqrt{x}$	2
1	$\frac{1}{2}x^{-1/2}$	$\frac{1}{4}$
2	$-\frac{1}{4}x^{-3/2}$	$-\frac{1}{32}$
3	$\frac{3}{8}x^{-5/2}$	

(a) $f(x) = \sqrt{x} \approx T_2(x) = 2 + \dfrac{1}{4}(x-4) - \dfrac{1/32}{2!}(x-4)^2$

$$= 2 + \tfrac{1}{4}(x-4) - \tfrac{1}{64}(x-4)^2$$

(b) $|R_2(x)| \le \dfrac{M}{3!}|x-4|^3$, where $|f'''(x)| \le M$. Now $4 \le x \le 4.2 \;\;\Rightarrow$

$|x-4| \le 0.2 \;\;\Rightarrow\;\; |x-4|^3 \le 0.008$. Since $f'''(x)$ is decreasing

on $[4, 4.2]$, we can take $M = |f'''(4)| = \tfrac{3}{8}4^{-5/2} = \tfrac{3}{256}$, so

$$|R_2(x)| \le \dfrac{3/256}{6}(0.008) = \dfrac{0.008}{512} = 0.000\,015\,625.$$

(c)

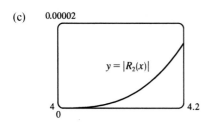

From the graph of $|R_2(x)| = |\sqrt{x} - T_2(x)|$, it seems that the

error is less than 1.52×10^{-5} on $[4, 4.2]$.

13.

n	$f^{(n)}(x)$	$f^{(n)}(1)$
0	$x^{2/3}$	1
1	$\frac{2}{3}x^{-1/3}$	$\frac{2}{3}$
2	$-\frac{2}{9}x^{-4/3}$	$-\frac{2}{9}$
3	$\frac{8}{27}x^{-7/3}$	$\frac{8}{27}$
4	$-\frac{56}{81}x^{-10/3}$	

(a) $f(x) = x^{2/3} \approx T_3(x) = 1 + \frac{2}{3}(x-1) - \frac{2/9}{2!}(x-1)^2 + \frac{8/27}{3!}(x-1)^3$

$$= 1 + \frac{2}{3}(x-1) - \frac{1}{9}(x-1)^2 + \frac{4}{81}(x-1)^3$$

(b) $|R_3(x)| \le \dfrac{M}{4!}|x-1|^4$, where $\left|f^{(4)}(x)\right| \le M$. Now $0.8 \le x \le 1.2$ $\Rightarrow$

$|x-1| \le 0.2$ $\Rightarrow$ $|x-1|^4 \le 0.0016$. Since $\left|f^{(4)}(x)\right|$ is decreasing

on $[0.8, 1.2]$, we can take $M = \left|f^{(4)}(0.8)\right| = \frac{56}{81}(0.8)^{-10/3}$, so

$$|R_3(x)| \le \frac{\frac{56}{81}(0.8)^{-10/3}}{24}(0.0016) \approx 0.000\,096\,97.$$

(c) 0.00006

$y = |R_3(x)|$

0.8 1 1.2
0

From the graph of $|R_3(x)| = \left|x^{2/3} - T_3(x)\right|$, it seems that the

error is less than $0.000\,053\,3$ on $[0.8, 1.2]$.

15.

n	$f^{(n)}(x)$	$f^{(n)}(0)$
0	e^{x^2}	1
1	$e^{x^2}(2x)$	0
2	$e^{x^2}(2+4x^2)$	2
3	$e^{x^2}(12x+8x^3)$	0
4	$e^{x^2}(12+48x^2+16x^4)$	

(a) $f(x) = e^{x^2} \approx T_3(x) = 1 + \frac{2}{2!}x^2 = 1 + x^2$

(b) $|R_3(x)| \le \dfrac{M}{4!}|x|^4$, where $\left|f^{(4)}(x)\right| \le M$. Now $0 \le x \le 0.1$ $\Rightarrow$

$x^4 \le (0.1)^4$, and letting $x = 0.1$ gives

$$|R_3(x)| \le \frac{e^{0.01}(12+0.48+0.0016)}{24}(0.1)^4 \approx 0.00006.$$

(c) 0.00008

$y = |R_3(x)|$

0 0.1

From the graph of $|R_3(x)| = \left|e^{x^2} - T_3(x)\right|$, it appears that the

error is less than $0.000\,051$ on $[0, 0.1]$.

17.

n	$f^{(n)}(x)$	$f^{(n)}(0)$
0	$x\sin x$	0
1	$\sin x + x\cos x$	0
2	$2\cos x - x\sin x$	2
3	$-3\sin x - x\cos x$	0
4	$-4\cos x + x\sin x$	-4
5	$5\sin x + x\cos x$	

(a) $f(x) = x\sin x \approx T_4(x) = \frac{2}{2!}(x-0)^2 + \frac{-4}{4!}(x-0)^4 = x^2 - \frac{1}{6}x^4$

(b) $|R_4(x)| \le \dfrac{M}{5!}|x|^5$, where $\left|f^{(5)}(x)\right| \le M$. Now $-1 \le x \le 1$ $\Rightarrow$

$|x| \le 1$, and a graph of $f^{(5)}(x)$ shows that $\left|f^{(5)}(x)\right| \le 5$ for $-1 \le x \le 1$.

Thus, we can take $M = 5$ and get $|R_4(x)| \le \dfrac{5}{5!}\cdot 1^5 = \dfrac{1}{24} = 0.041\overline{6}$.

(c)

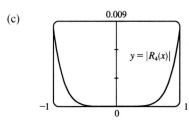

From the graph of $|R_4(x)| = |x\sin x - T_4(x)|$, it seems that the error is less than 0.0082 on $[-1, 1]$.

19. From Exercise 5, $\cos x = -\left(x - \frac{\pi}{2}\right) + \frac{1}{6}\left(x - \frac{\pi}{2}\right)^3 + R_3(x)$, where $|R_3(x)| \le \dfrac{M}{4!}\left|x - \frac{\pi}{2}\right|^4$ with

$\left|f^{(4)}(x)\right| = |\cos x| \le M = 1$. Now $x = 80° = (90° - 10°) = \left(\frac{\pi}{2} - \frac{\pi}{18}\right) = \frac{4\pi}{9}$ radians, so the error is

$\left|R_3\left(\frac{4\pi}{9}\right)\right| \le \frac{1}{24}\left(\frac{\pi}{18}\right)^4 \approx 0.000\,039$, which means our estimate would *not* be accurate to five decimal places. However,

$T_3 = T_4$, so we can use $\left|R_4\left(\frac{4\pi}{9}\right)\right| \le \frac{1}{120}\left(\frac{\pi}{18}\right)^5 \approx 0.000\,001$. Therefore, to five decimal places,

$\cos 80° \approx -\left(-\frac{\pi}{18}\right) + \frac{1}{6}\left(-\frac{\pi}{18}\right)^3 \approx 0.17365$.

21. All derivatives of e^x are e^x, so $|R_n(x)| \le \dfrac{e^x}{(n+1)!}|x|^{n+1}$, where $0 < x < 0.1$. Letting $x = 0.1$,

$R_n(0.1) \le \dfrac{e^{0.1}}{(n+1)!}(0.1)^{n+1} < 0.00001$, and by trial and error we find that $n = 3$ satisfies this inequality since

$R_3(0.1) < 0.0000046$. Thus, by adding the four terms of the Maclaurin series for e^x corresponding to $n = 0, 1, 2$, and 3,

we can estimate $e^{0.1}$ to within 0.00001. (In fact, this sum is $1.1051\overline{6}$ and $e^{0.1} \approx 1.10517$.)

23. $\sin x = x - \dfrac{1}{3!}x^3 + \dfrac{1}{5!}x^5 - \cdots$. By the Alternating Series

Estimation Theorem, the error in the approximation

$\sin x = x - \dfrac{1}{3!}x^3$ is less than $\left|\dfrac{1}{5!}x^5\right| < 0.01 \iff$

$|x^5| < 120(0.01) \iff |x| < (1.2)^{1/5} \approx 1.037$. The curves

$y = x - \frac{1}{6}x^3$ and $y = \sin x - 0.01$ intersect at $x \approx 1.043$, so

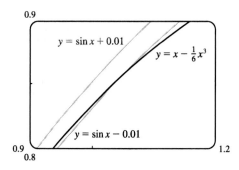

the graph confirms our estimate. Since both the sine function

and the given approximation are odd functions, we need to check the estimate only for $x > 0$. Thus, the desired range of

values for x is $-1.037 < x < 1.037$.

25. $\arctan x = x - \dfrac{x^3}{3} + \dfrac{x^5}{5} - \dfrac{x^7}{7} + \cdots$. By the Alternating Series

Estimation Theorem, the error is less than $\left|-\frac{1}{7}x^7\right| < 0.05 \iff$

$|x^7| < 0.35 \iff |x| < (0.35)^{1/7} \approx 0.8607$. The curves

$y = x - \frac{1}{3}x^3 + \frac{1}{5}x^5$ and $y = \arctan x + 0.05$ intersect at

$x \approx 0.9245$, so the graph confirms our estimate. Since both the

arctangent function and the given approximation are odd functions,

we need to check the estimate only for $x > 0$. Thus, the desired

range of values for x is $-0.86 < x < 0.86$.

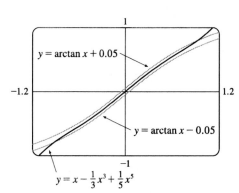

27. Let $s(t)$ be the position function of the car, and for convenience set $s(0) = 0$. The velocity of the car is $v(t) = s'(t)$ and the acceleration is $a(t) = s''(t)$, so the second degree Taylor polynomial is $T_2(t) = s(0) + v(0)t + \dfrac{a(0)}{2}t^2 = 20t + t^2$. We estimate the distance traveled during the next second to be $s(1) \approx T_2(1) = 20 + 1 = 21$ m. The function $T_2(t)$ would not be accurate over a full minute, since the car could not possibly maintain an acceleration of 2 m/s^2 for that long (if it did, its final speed would be 140 m/s ≈ 313 mi/h!).

29. $E = \dfrac{q}{D^2} - \dfrac{q}{(D+d)^2} = \dfrac{q}{D^2} - \dfrac{q}{D^2(1+d/D)^2} = \dfrac{q}{D^2}\left[1 - \left(1 + \dfrac{d}{D}\right)^{-2}\right].$

We use the Binomial Series to expand $(1 + d/D)^{-2}$:

$$E = \dfrac{q}{D^2}\left[1 - \left(1 - 2\left(\dfrac{d}{D}\right) + \dfrac{2\cdot 3}{2!}\left(\dfrac{d}{D}\right)^2 - \dfrac{2\cdot 3\cdot 4}{3!}\left(\dfrac{d}{D}\right)^3 + \cdots\right)\right] = \dfrac{q}{D^2}\left[2\left(\dfrac{d}{D}\right) - 3\left(\dfrac{d}{D}\right)^2 + 4\left(\dfrac{d}{D}\right)^3 - \cdots\right]$$

$$\approx \dfrac{q}{D^2}\cdot 2\left(\dfrac{d}{D}\right) = 2qd\cdot\dfrac{1}{D^3}$$

when D is much larger than d; that is, when P is far away from the dipole.

31. (a) L is the length of the arc subtended by the angle θ, so $L = R\theta \implies$

$\theta = L/R$. Now $\sec\theta = (R+C)/R \implies R\sec\theta = R + C \implies$

$C = R\sec\theta - R = R\sec(L/R) - R.$

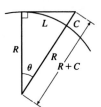

(b) First we'll find a Taylor polynomial $T_4(x)$ for $f(x) = \sec x$ at $x = 0$.

n	$f^{(n)}(x)$	$f^{(n)}(0)$
0	$\sec x$	1
1	$\sec x \tan x$	0
2	$\sec x(2\tan^2 x + 1)$	1
3	$\sec x \tan x(6\tan^2 x + 5)$	0
4	$\sec x(24\tan^4 x + 28\tan^2 x + 5)$	5

Thus, $f(x) = \sec x \approx T_4(x) = 1 + \frac{1}{2!}(x-0)^2 + \frac{5}{4!}(x-0)^4 = 1 + \frac{1}{2}x^2 + \frac{5}{24}x^4$. By part (a),

$$C \approx R\left[1 + \dfrac{1}{2}\left(\dfrac{L}{R}\right)^2 + \dfrac{5}{24}\left(\dfrac{L}{R}\right)^4\right] - R = R + \dfrac{1}{2}R\cdot\dfrac{L^2}{R^2} + \dfrac{5}{24}R\cdot\dfrac{L^4}{R^4} - R = \dfrac{L^2}{2R} + \dfrac{5L^4}{24R^3}.$$

(c) Taking $L = 100$ km and $R = 6370$ km, the formula in part (a) says that

$C = R\sec(L/R) - R = 6370\sec(100/6370) - 6370 \approx 0.785\,009\,965\,44$ km.

The formula in part (b) says that $C \approx \dfrac{L^2}{2R} + \dfrac{5L^4}{24R^3} = \dfrac{100^2}{2\cdot 6370} + \dfrac{5\cdot 100^4}{24\cdot 6370^3} \approx 0.785\,009\,957\,36$ km.

The difference between these two results is only $0.000\,000\,008\,08$ km, or $0.000\,008\,08$ m!

33. Using $f(x) = T_n(x) + R_n(x)$ with $n = 1$ and $x = r$, we have $f(r) = T_1(r) + R_1(r)$, where T_1 is the first-degree Taylor polynomial of f at a. Because $a = x_n$, $f(r) = f(x_n) + f'(x_n)(r - x_n) + R_1(r)$. But r is a root of f, so $f(r) = 0$

and we have $0 = f(x_n) + f'(x_n)(r - x_n) + R_1(r)$. Taking the first two terms to the left side gives us

$f'(x_n)(x_n - r) - f(x_n) = R_1(r)$. Dividing by $f'(x_n)$, we get $x_n - r - \dfrac{f(x_n)}{f'(x_n)} = \dfrac{R_1(r)}{f'(x_n)}$. By the formula for Newton's

method, the left side of the preceding equation is $x_{n+1} - r$, so $|x_{n+1} - r| = \left| \dfrac{R_1(r)}{f'(x_n)} \right|$. Taylor's Inequality gives us

$|R_1(r)| \leq \dfrac{|f''(r)|}{2!} |r - x_n|^2$. Combining this inequality with the facts $|f''(x)| \leq M$ and $|f'(x)| \geq K$ gives us

$|x_{n+1} - r| \leq \dfrac{M}{2K} |x_n - r|^2$.

8 Review

CONCEPT CHECK

1. (a) See Definition 8.1.1.

   (b) See Definition 8.2.2.

   (c) The terms of the sequence $\{a_n\}$ approach 3 as n becomes large.

   (d) By adding sufficiently many terms of the series, we can make the partial sums as close to 3 as we like.

2. (a) See the definition on page 561.

   (b) A sequence is monotonic if it is either increasing or decreasing.

   (c) By Theorem 8.1.8, every bounded, monotonic sequence is convergent.

3. (a) See (4) in Section 8.2.

   (b) The p-series $\sum\limits_{n=1}^{\infty} \dfrac{1}{n^p}$ is convergent if $p > 1$.

4. If $\sum a_n = 3$, then $\lim\limits_{n \to \infty} a_n = 0$ and $\lim\limits_{n \to \infty} s_n = 3$.

5. (a) *Test for Divergence:* If $\lim\limits_{n \to \infty} a_n$ does not exist or if $\lim\limits_{n \to \infty} a_n \neq 0$, then the series $\sum_{n=1}^{\infty} a_n$ is divergent.

   (b) *Integral Test:* Suppose f is a continuous, positive, decreasing function on $[1, \infty)$ and let $a_n = f(n)$. Then the series

   $\sum_{n=1}^{\infty} a_n$ is convergent if and only if the improper integral $\int_1^{\infty} f(x)\,dx$ is convergent. In other words:

   (i) If $\int_1^{\infty} f(x)\,dx$ is convergent, then $\sum_{n=1}^{\infty} a_n$ is convergent.

   (ii) If $\int_1^{\infty} f(x)\,dx$ is divergent, then $\sum_{n=1}^{\infty} a_n$ is divergent.

   (c) *Comparison Test:* Suppose that $\sum a_n$ and $\sum b_n$ are series with positive terms.

   (i) If $\sum b_n$ is convergent and $a_n \leq b_n$ for all n, then $\sum a_n$ is also convergent.

   (ii) If $\sum b_n$ is divergent and $a_n \geq b_n$ for all n, then $\sum a_n$ is also divergent.

   (d) *Limit Comparison Test:* Suppose that $\sum a_n$ and $\sum b_n$ are series with positive terms. If $\lim\limits_{n \to \infty} (a_n / b_n) = c$, where c is a

   finite number and $c > 0$, then either both series converge or both diverge.

(e) *Alternating Series Test:* If the alternating series $\sum_{n=1}^{\infty}(-1)^{n-1}b_n = b_1 - b_2 + b_3 - b_4 + b_5 - b_6 + \cdots$ $[b_n > 0]$

satisfies (i) $b_{n+1} \le b_n$ for all n and (ii) $\lim_{n \to \infty} b_n = 0$, then the series is convergent.

(f) *Ratio Test:*

(i) If $\lim_{n \to \infty} \left| \dfrac{a_{n+1}}{a_n} \right| = L < 1$, then the series $\sum_{n=1}^{\infty} a_n$ is absolutely convergent (and therefore convergent).

(ii) If $\lim_{n \to \infty} \left| \dfrac{a_{n+1}}{a_n} \right| = L > 1$ or $\lim_{n \to \infty} \left| \dfrac{a_{n+1}}{a_n} \right| = \infty$, then the series $\sum_{n=1}^{\infty} a_n$ is divergent.

(iii) If $\lim_{n \to \infty} \left| \dfrac{a_{n+1}}{a_n} \right| = 1$, the Ratio Test is inconclusive; that is, no conclusion can be drawn about the convergence or

divergence of $\sum a_n$.

6. (a) A series $\sum a_n$ is called *absolutely convergent* if the series of absolute values $\sum |a_n|$ is convergent.

(b) If a series $\sum a_n$ is absolutely convergent, then it is convergent.

7. (a) Use (4) in Section 8.3.

(b) See Example 8 in Section 8.3.

(c) By adding terms until you reach the desired accuracy given by the Alternating Series Estimation Theorem.

8. (a) $\sum_{n=0}^{\infty} c_n(x-a)^n$

(b) Given the power series $\sum_{n=0}^{\infty} c_n(x-a)^n$, the radius of convergence is:

(i) 0 if the series converges only when $x = a$

(ii) ∞ if the series converges for all x, or

(iii) a positive number R such that the series converges if $|x - a| < R$ and diverges if $|x - a| > R$.

(c) The interval of convergence of a power series is the interval that consists of all values of x for which the series converges.

Corresponding to the cases in part (b), the interval of convergence is: (i) the single point $\{a\}$, (ii) all real numbers, that is,

the real number line $(-\infty, \infty)$, or (iii) an interval with endpoints $a - R$ and $a + R$ which can contain neither, either, or

both of the endpoints. In this case, we must test the series for convergence at each endpoint to determine the interval of

convergence.

9. (a), (b) See Theorem 8.6.2.

10. (a) $T_n(x) = \sum_{i=0}^{n} \dfrac{f^{(i)}(a)}{i!}(x-a)^i$

(b) $\sum_{n=0}^{\infty} \dfrac{f^{(n)}(a)}{n!}(x-a)^n$

(c) $\displaystyle\sum_{n=0}^{\infty} \frac{f^{(n)}(0)}{n!} x^n$ [$a = 0$ in part (b)]

(d) See Theorem 8.7.8.

(e) See Taylor's Inequality (8.7.9).

11. (a)–(f) See the table on page 613.

12. See the binomial series (8.7.17) for the expansion. The radius of convergence for the binomial series is 1.

TRUE-FALSE QUIZ

1. False.   See Note 2 after Theorem 8.2.6.

3. True.   If $\lim_{n\to\infty} a_n = L$, then as $n \to \infty$, $2n + 1 \to \infty$, so $a_{2n+1} \to L$.

5. False.   For example, take $c_n = (-1)^n/(n6^n)$.

7. False, since $\displaystyle\lim_{n\to\infty} \left|\frac{a_{n+1}}{a_n}\right| = \lim_{n\to\infty} \left|\frac{1}{(n+1)^3} \cdot \frac{n^3}{1}\right| = \lim_{n\to\infty} \left|\frac{n^3}{(n+1)^3} \cdot \frac{1/n^3}{1/n^3}\right| = \lim_{n\to\infty} \frac{1}{(1+1/n)^3} = 1.$

9. False.   See the note after Example 4 in Section 8.3.

11. True.   See (7) in Section 8.1.

13. True.   By Theorem 8.7.5 the coefficient of x^3 is $\dfrac{f'''(0)}{3!} = \dfrac{1}{3} \;\Rightarrow\; f'''(0) = 2.$

Or: Use Theorem 8.6.2 to differentiate f three times.

15. False.   For example, let $a_n = b_n = (-1)^n$. Then $\{a_n\}$ and $\{b_n\}$ are divergent, but $a_n b_n = 1$, so $\{a_n b_n\}$ is convergent.

17. True by Theorem 8.4.1. $\left[\sum (-1)^n a_n \text{ is absolutely convergent and hence convergent.}\right]$

19. True. $0.99999\ldots = 0.9 + 0.9(0.1)^1 + 0.9(0.1)^2 + 0.9(0.1)^3 + \cdots = \displaystyle\sum_{n=1}^{\infty} (0.9)(0.1)^{n-1} = \frac{0.9}{1 - 0.1} = 1$ by the formula

for the sum of a geometric series $[S = a_1/(1 - r)]$ with ratio r satisfying $|r| < 1$.

EXERCISES

1. $\left\{\dfrac{2 + n^3}{1 + 2n^3}\right\}$ converges since $\displaystyle\lim_{n\to\infty} \frac{2 + n^3}{1 + 2n^3} = \lim_{n\to\infty} \frac{2/n^3 + 1}{1/n^3 + 2} = \frac{1}{2}.$

3. $\displaystyle\lim_{n\to\infty} a_n = \lim_{n\to\infty} \frac{n^3}{1 + n^2} = \lim_{n\to\infty} \frac{n}{1/n^2 + 1} = \infty$, so the sequence diverges.

5. $|a_n| = \left|\dfrac{n \sin n}{n^2 + 1}\right| \leq \dfrac{n}{n^2 + 1} < \dfrac{1}{n}$, so $|a_n| \to 0$ as $n \to \infty$. Thus, $\displaystyle\lim_{n\to\infty} a_n = 0$. The sequence $\{a_n\}$ is convergent.

7. $\left\{\left(1+\dfrac{3}{n}\right)^{4n}\right\}$ is convergent. Let $y = \left(1+\dfrac{3}{x}\right)^{4x}$. Then

$$\lim_{x\to\infty} \ln y = \lim_{x\to\infty} 4x \ln(1+3/x) = \lim_{x\to\infty} \frac{\ln(1+3/x)}{1/(4x)} \stackrel{\text{H}}{=} \lim_{x\to\infty} \frac{\frac{1}{1+3/x}\left(-\frac{3}{x^2}\right)}{-1/(4x^2)} = \lim_{x\to\infty} \frac{12}{1+3/x} = 12, \text{ so}$$

$$\lim_{x\to\infty} y = \lim_{n\to\infty}\left(1+\frac{3}{n}\right)^{4n} = e^{12}.$$

9. $\dfrac{n}{n^3+1} < \dfrac{n}{n^3} = \dfrac{1}{n^2}$, so $\displaystyle\sum_{n=1}^{\infty} \dfrac{n}{n^3+1}$ converges by the Comparison Test with the convergent p-series $\displaystyle\sum_{n=1}^{\infty} \dfrac{1}{n^2}$ $[p=2>1]$.

11. $\displaystyle\lim_{n\to\infty}\left|\dfrac{a_{n+1}}{a_n}\right| = \lim_{n\to\infty}\left[\dfrac{(n+1)^3}{5^{n+1}} \cdot \dfrac{5^n}{n^3}\right] = \lim_{n\to\infty}\left(1+\dfrac{1}{n}\right)^3 \cdot \dfrac{1}{5} = \dfrac{1}{5} < 1$, so $\displaystyle\sum_{n=1}^{\infty} \dfrac{n^3}{5^n}$ converges by the Ratio Test.

13. Let $f(x) = \dfrac{1}{x\sqrt{\ln x}}$. Then f is continuous, positive, and decreasing on $[2,\infty)$, so the Integral Test applies.

$$\int_2^{\infty} f(x)\,dx = \lim_{t\to\infty}\int_2^t \frac{1}{x\sqrt{\ln x}}\,dx \quad \left[u=\ln x,\, du=\frac{1}{x}\,dx\right] = \lim_{t\to\infty}\int_{\ln 2}^{\ln t} u^{-1/2}\,du = \lim_{t\to\infty}\left[2\sqrt{u}\right]_{\ln 2}^{\ln t}$$

$$= \lim_{t\to\infty}\left(2\sqrt{\ln t} - 2\sqrt{\ln 2}\right) = \infty,$$

so the series $\displaystyle\sum_{n=2}^{\infty} \dfrac{1}{n\sqrt{\ln n}}$ diverges.

15. $b_n = \dfrac{\sqrt{n}}{n+1} > 0$, $\{b_n\}$ is decreasing, and $\displaystyle\lim_{n\to\infty} b_n = 0$, so the series $\displaystyle\sum_{n=1}^{\infty}(-1)^{n-1}\dfrac{\sqrt{n}}{n+1}$ converges by the Alternating

Series Test.

17. $\displaystyle\lim_{n\to\infty}\left|\dfrac{a_{n+1}}{a_n}\right| = \lim_{n\to\infty}\dfrac{1\cdot 3\cdot 5\cdot\cdots\cdot(2n-1)(2n+1)}{5^{n+1}(n+1)!} \cdot \dfrac{5^n n!}{1\cdot 3\cdot 5\cdot\cdots\cdot(2n-1)} = \lim_{n\to\infty}\dfrac{2n+1}{5(n+1)} = \dfrac{2}{5} < 1$, so the series

converges by the Ratio Test.

19. $\displaystyle\sum_{n=1}^{\infty} \dfrac{(-3)^{n-1}}{2^{3n}} = \sum_{n=1}^{\infty} \dfrac{(-3)^{n-1}}{(2^3)^n} = \sum_{n=1}^{\infty} \dfrac{(-3)^{n-1}}{8^n} = \dfrac{1}{8}\sum_{n=1}^{\infty} \dfrac{(-3)^{n-1}}{8^{n-1}} = \dfrac{1}{8}\sum_{n=1}^{\infty}\left(-\dfrac{3}{8}\right)^{n-1} = \dfrac{1}{8}\left(\dfrac{1}{1-(-3/8)}\right)$

$$= \dfrac{1}{8}\cdot\dfrac{8}{11} = \dfrac{1}{11}$$

21. $\displaystyle\sum_{n=1}^{\infty}[\tan^{-1}(n+1) - \tan^{-1} n] = \lim_{n\to\infty} s_n$

$$= \lim_{n\to\infty}[(\tan^{-1}2 - \tan^{-1}1) + (\tan^{-1}3 - \tan^{-1}2) + \cdots + (\tan^{-1}(n+1) - \tan^{-1}n)]$$

$$= \lim_{n\to\infty}[\tan^{-1}(n+1) - \tan^{-1}1] = \dfrac{\pi}{2} - \dfrac{\pi}{4} = \dfrac{\pi}{4}$$

23. $1.2345345345\ldots = 1.2 + 0.0\overline{345} = \dfrac{12}{10} + \dfrac{345/10{,}000}{1 - 1/1000} = \dfrac{12}{10} + \dfrac{345}{9990} = \dfrac{4111}{3330}$

25. $\displaystyle\sum_{n=1}^{\infty} \dfrac{(-1)^{n+1}}{n^5} = 1 - \dfrac{1}{32} + \dfrac{1}{243} - \dfrac{1}{1024} + \dfrac{1}{3125} - \dfrac{1}{7776} + \dfrac{1}{16{,}807} - \dfrac{1}{32{,}768} + \cdots.$

Since $b_8 = \dfrac{1}{8^5} = \dfrac{1}{32{,}768} < 0.000031$, $\displaystyle\sum_{n=1}^{\infty} \dfrac{(-1)^{n+1}}{n^5} \approx \sum_{n=1}^{7} \dfrac{(-1)^{n+1}}{n^5} \approx 0.9721.$

27. $\displaystyle\sum_{n=1}^{\infty} \dfrac{1}{2 + 5^n} \approx \sum_{n=1}^{8} \dfrac{1}{2 + 5^n} \approx 0.18976224.$ To estimate the error, note that $\dfrac{1}{2 + 5^n} < \dfrac{1}{5^n}$, so the remainder term is

$R_8 = \displaystyle\sum_{n=9}^{\infty} \dfrac{1}{2 + 5^n} < \sum_{n=9}^{\infty} \dfrac{1}{5^n} = \dfrac{1/5^9}{1 - 1/5} = 6.4 \times 10^{-7}$ [geometric series with $a = \frac{1}{5^9}$ and $r = \frac{1}{5}$].

29. Use the Limit Comparison Test. $\displaystyle\lim_{n\to\infty} \left| \dfrac{\left(\frac{n+1}{n}\right) a_n}{a_n} \right| = \lim_{n\to\infty} \dfrac{n+1}{n} = \lim_{n\to\infty} \left(1 + \dfrac{1}{n}\right) = 1 > 0.$

Since $\sum |a_n|$ is convergent, so is $\sum \left| \left(\dfrac{n+1}{n}\right) a_n \right|$, by the Limit Comparison Test.

31. $\displaystyle\lim_{n\to\infty} \left| \dfrac{a_{n+1}}{a_n} \right| = \lim_{n\to\infty} \left[\dfrac{|x+2|^{n+1}}{(n+1)\,4^{n+1}} \cdot \dfrac{n\,4^n}{|x+2|^n} \right] = \lim_{n\to\infty} \left[\dfrac{n}{n+1}\,\dfrac{|x+2|}{4} \right] = \dfrac{|x+2|}{4} < 1 \iff |x+2| < 4,$ so $R = 4.$

$|x+2| < 4 \iff -4 < x + 2 < 4 \iff -6 < x < 2.$ If $x = -6$, then the series $\displaystyle\sum_{n=1}^{\infty} \dfrac{(x+2)^n}{n\,4^n}$ becomes

$\displaystyle\sum_{n=1}^{\infty} \dfrac{(-4)^n}{n4^n} = \sum_{n=1}^{\infty} \dfrac{(-1)^n}{n}$, the alternating harmonic series, which converges by the Alternating Series Test. When $x = 2$, the

series becomes the harmonic series $\displaystyle\sum_{n=1}^{\infty} \dfrac{1}{n}$, which diverges. Thus, $I = [-6, 2).$

33. $\displaystyle\lim_{n\to\infty} \left| \dfrac{a_{n+1}}{a_n} \right| = \lim_{n\to\infty} \left| \dfrac{2^{n+1}(x-3)^{n+1}}{\sqrt{n+4}} \cdot \dfrac{\sqrt{n+3}}{2^n (x-3)^n} \right| = 2\,|x-3| \lim_{n\to\infty} \sqrt{\dfrac{n+3}{n+4}} = 2\,|x-3| < 1 \iff |x-3| < \frac{1}{2},$

so $R = \frac{1}{2}.$ $|x-3| < \frac{1}{2} \iff -\frac{1}{2} < x - 3 < \frac{1}{2} \iff \frac{5}{2} < x < \frac{7}{2}.$ For $x = \frac{7}{2}$, the series $\displaystyle\sum_{n=1}^{\infty} \dfrac{2^n (x-3)^n}{\sqrt{n+3}}$ becomes

$\displaystyle\sum_{n=0}^{\infty} \dfrac{1}{\sqrt{n+3}} = \sum_{n=3}^{\infty} \dfrac{1}{n^{1/2}}$, which diverges $\left[p = \frac{1}{2} \leq 1 \right]$, but for $x = \frac{5}{2}$, we get $\displaystyle\sum_{n=0}^{\infty} \dfrac{(-1)^n}{\sqrt{n+3}}$, which is a convergent

alternating series, so $I = \left[\frac{5}{2}, \frac{7}{2} \right).$

35.

n	$f^{(n)}(x)$	$f^{(n)}\left(\frac{\pi}{6}\right)$
0	$\sin x$	$\frac{1}{2}$
1	$\cos x$	$\frac{\sqrt{3}}{2}$
2	$-\sin x$	$-\frac{1}{2}$
3	$-\cos x$	$-\frac{\sqrt{3}}{2}$
4	$\sin x$	$\frac{1}{2}$
$\vdots$	$\vdots$	$\vdots$

$$\sin x = f\left(\frac{\pi}{6}\right) + f'\left(\frac{\pi}{6}\right)\left(x - \frac{\pi}{6}\right) + \frac{f''\left(\frac{\pi}{6}\right)}{2!}\left(x - \frac{\pi}{6}\right)^2 + \frac{f^{(3)}\left(\frac{\pi}{6}\right)}{3!}\left(x - \frac{\pi}{6}\right)^3 + \frac{f^{(4)}\left(\frac{\pi}{6}\right)}{4!}\left(x - \frac{\pi}{6}\right)^4 + \cdots$$

$$= \frac{1}{2}\left[1 - \frac{1}{2!}\left(x - \frac{\pi}{6}\right)^2 + \frac{1}{4!}\left(x - \frac{\pi}{6}\right)^4 - \cdots\right] + \frac{\sqrt{3}}{2}\left[\left(x - \frac{\pi}{6}\right) - \frac{1}{3!}\left(x - \frac{\pi}{6}\right)^3 + \cdots\right]$$

$$= \frac{1}{2}\sum_{n=0}^{\infty}(-1)^n\frac{1}{(2n)!}\left(x - \frac{\pi}{6}\right)^{2n} + \frac{\sqrt{3}}{2}\sum_{n=0}^{\infty}(-1)^n\frac{1}{(2n+1)!}\left(x - \frac{\pi}{6}\right)^{2n+1}$$

37. $\dfrac{1}{1+x} = \dfrac{1}{1-(-x)} = \sum_{n=0}^{\infty}(-x)^n = \sum_{n=0}^{\infty}(-1)^n\,x^n$ for $|x| < 1 \quad \Rightarrow \quad \dfrac{x^2}{1+x} = \sum_{n=0}^{\infty}(-1)^n\,x^{n+2}$ with $R = 1$.

39. $\displaystyle\int \frac{1}{4-x}\,dx = -\ln(4-x) + C$ and

$$\int \frac{1}{4-x}\,dx = \frac{1}{4}\int \frac{1}{1-x/4}\,dx = \frac{1}{4}\int \sum_{n=0}^{\infty}\left(\frac{x}{4}\right)^n dx = \frac{1}{4}\int \sum_{n=0}^{\infty}\frac{x^n}{4^n}\,dx = \frac{1}{4}\sum_{n=0}^{\infty}\frac{x^{n+1}}{4^n(n+1)} + C. \text{ So}$$

$$\ln(4-x) = -\frac{1}{4}\sum_{n=0}^{\infty}\frac{x^{n+1}}{4^n(n+1)} + C = -\sum_{n=0}^{\infty}\frac{x^{n+1}}{4^{n+1}(n+1)} + C = -\sum_{n=1}^{\infty}\frac{x^n}{n4^n} + C. \text{ Putting } x = 0, \text{ we get } C = \ln 4.$$

Thus, $f(x) = \ln(4-x) = \ln 4 - \displaystyle\sum_{n=1}^{\infty}\frac{x^n}{n4^n}$. The series converges for $|x/4| < 1 \quad \Leftrightarrow \quad |x| < 4$, so $R = 4$.

Another solution:

$$\ln(4-x) = \ln[4(1 - x/4)] = \ln 4 + \ln(1 - x/4) = \ln 4 + \ln[1 + (-x/4)]$$

$$= \ln 4 + \sum_{n=1}^{\infty}(-1)^{n+1}\frac{(-x/4)^n}{n} \text{ [from Table 1]} = \ln 4 + \sum_{n=1}^{\infty}(-1)^{2n+1}\frac{x^n}{n4^n} = \ln 4 - \sum_{n=1}^{\infty}\frac{x^n}{n4^n}.$$

41. $\sin x = \displaystyle\sum_{n=0}^{\infty}\frac{(-1)^n\,x^{2n+1}}{(2n+1)!} \quad \Rightarrow \quad \sin(x^4) = \sum_{n=0}^{\infty}\frac{(-1)^n\,(x^4)^{2n+1}}{(2n+1)!} = \sum_{n=0}^{\infty}\frac{(-1)^n\,x^{8n+4}}{(2n+1)!}$ for all x, so the radius of

convergence is ∞.

43. $f(x) = \dfrac{1}{\sqrt[4]{16-x}} = \dfrac{1}{\sqrt[4]{16(1-x/16)}} = \dfrac{1}{\sqrt[4]{16}\left(1-\frac{1}{16}x\right)^{1/4}} = \frac{1}{2}\left(1-\frac{1}{16}x\right)^{-1/4}$

$\quad = \dfrac{1}{2}\left[1 + \left(-\dfrac{1}{4}\right)\left(-\dfrac{x}{16}\right) + \dfrac{\left(-\frac{1}{4}\right)\left(-\frac{5}{4}\right)}{2!}\left(-\dfrac{x}{16}\right)^2 + \dfrac{\left(-\frac{1}{4}\right)\left(-\frac{5}{4}\right)\left(-\frac{9}{4}\right)}{3!}\left(-\dfrac{x}{16}\right)^3 + \cdots\right]$

$\quad = \dfrac{1}{2} + \displaystyle\sum_{n=1}^{\infty} \dfrac{1\cdot 5\cdot 9\cdot\cdots\cdot(4n-3)}{2\cdot 4^n\cdot n!\cdot 16^n}x^n = \dfrac{1}{2} + \displaystyle\sum_{n=1}^{\infty}\dfrac{1\cdot 5\cdot 9\cdot\cdots\cdot(4n-3)}{2^{6n+1}\,n!}x^n$

for $\left|-\dfrac{x}{16}\right| < 1 \quad\Leftrightarrow\quad |x| < 16$, so $R = 16$.

45. $e^x = \displaystyle\sum_{n=0}^{\infty}\dfrac{x^n}{n!}$, so $\dfrac{e^x}{x} = \dfrac{1}{x}\displaystyle\sum_{n=0}^{\infty}\dfrac{x^n}{n!} = \displaystyle\sum_{n=0}^{\infty}\dfrac{x^{n-1}}{n!} = x^{-1} + \displaystyle\sum_{n=1}^{\infty}\dfrac{x^{n-1}}{n!} = \dfrac{1}{x} + \displaystyle\sum_{n=1}^{\infty}\dfrac{x^{n-1}}{n!}$ and

$\displaystyle\int\dfrac{e^x}{x}\,dx = C + \ln|x| + \displaystyle\sum_{n=1}^{\infty}\dfrac{x^n}{n\cdot n!}$.

47. (a)

n	$f^{(n)}(x)$	$f^{(n)}(1)$
0	$x^{1/2}$	1
1	$\frac{1}{2}x^{-1/2}$	$\frac{1}{2}$
2	$-\frac{1}{4}x^{-3/2}$	$-\frac{1}{4}$
3	$\frac{3}{8}x^{-5/2}$	$\frac{3}{8}$
4	$-\frac{15}{16}x^{-7/2}$	$-\frac{15}{16}$
$\vdots$	$\vdots$	$\vdots$

$\sqrt{x} \approx T_3(x) = 1 + \dfrac{1/2}{1!}(x-1) - \dfrac{1/4}{2!}(x-1)^2 + \dfrac{3/8}{3!}(x-1)^3$

$\qquad\qquad = 1 + \frac{1}{2}(x-1) - \frac{1}{8}(x-1)^2 + \frac{1}{16}(x-1)^3$

(b)

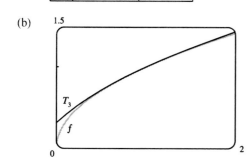

(c) $|R_3(x)| \le \dfrac{M}{4!}|x-1|^4$, where $\left|f^{(4)}(x)\right| \le M$ with

$\quad f^{(4)}(x) = -\frac{15}{16}x^{-7/2}$. Now $0.9 \le x \le 1.1 \Rightarrow$

$\quad -0.1 \le x-1 \le 0.1 \Rightarrow (x-1)^4 \le (0.1)^4$,

$\quad$ and letting $x = 0.9$ gives $M = \dfrac{15}{16(0.9)^{7/2}}$, so

$\quad |R_3(x)| \le \dfrac{15}{16(0.9)^{7/2}\,4!}(0.1)^4 \approx 0.000\,005\,648$

$\qquad\qquad \approx 0.000\,006 = 6\times 10^{-6}$

(d)

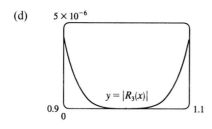

From the graph of $|R_3(x)| = |\sqrt{x} - T_3(x)|$, it appears that

the error is less than 5×10^{-6} on $[0.9, 1.1]$.

49. $\sin x = \displaystyle\sum_{n=0}^{\infty}(-1)^n\dfrac{x^{2n+1}}{(2n+1)!} = x - \dfrac{x^3}{3!} + \dfrac{x^5}{5!} - \dfrac{x^7}{7!} + \cdots$, so $\sin x - x = -\dfrac{x^3}{3!} + \dfrac{x^5}{5!} - \dfrac{x^7}{7!} + \cdots$ and

$\dfrac{\sin x - x}{x^3} = -\dfrac{1}{3!} + \dfrac{x^2}{5!} - \dfrac{x^4}{7!} + \cdots$. Thus, $\displaystyle\lim_{x\to 0}\dfrac{\sin x - x}{x^3} = \lim_{x\to 0}\left(-\dfrac{1}{6} + \dfrac{x^2}{120} - \dfrac{x^4}{5040} + \cdots\right) = -\dfrac{1}{6}$.

☐ FOCUS ON PROBLEM SOLVING

1. It would be far too much work to compute 15 derivatives of f. The key idea is to remember that $f^{(n)}(0)$ occurs in the coefficient of x^n in the Maclaurin series of f. We start with the Maclaurin series for sin: $\sin x = x - \dfrac{x^3}{3!} + \dfrac{x^5}{5!} - \cdots$.

Then $\sin(x^3) = x^3 - \dfrac{x^9}{3!} + \dfrac{x^{15}}{5!} - \cdots$, and so the coefficient of x^{15} is $\dfrac{f^{(15)}(0)}{15!} = \dfrac{1}{5!}$. Therefore,

$f^{(15)}(0) = \dfrac{15!}{5!} = 6 \cdot 7 \cdot 8 \cdot 9 \cdot 10 \cdot 11 \cdot 12 \cdot 13 \cdot 14 \cdot 15 = 10{,}897{,}286{,}400.$

3. (a) At each stage, each side is replaced by four shorter sides, each of length $\frac{1}{3}$ of the side length at the preceding stage. Writing s_0 and ℓ_0 for the number of sides and the length of the side of the initial triangle, we generate the table at right. In general, we have $s_n = 3 \cdot 4^n$ and

$s_0 = 3$	$\ell_0 = 1$
$s_1 = 3 \cdot 4$	$\ell_1 = 1/3$
$s_2 = 3 \cdot 4^2$	$\ell_2 = 1/3^2$
$s_3 = 3 \cdot 4^3$	$\ell_3 = 1/3^3$
$\vdots$	$\vdots$

$\ell_n = \left(\frac{1}{3}\right)^n$, so the length of the perimeter at the nth stage of construction is $p_n = s_n \ell_n = 3 \cdot 4^n \cdot \left(\frac{1}{3}\right)^n = 3 \cdot \left(\frac{4}{3}\right)^n$.

(b) $p_n = \dfrac{4^n}{3^{n-1}} = 4\left(\dfrac{4}{3}\right)^{n-1}$. Since $\frac{4}{3} > 1$, $p_n \to \infty$ as $n \to \infty$.

(c) The area of each of the small triangles added at a given stage is one-ninth of the area of the triangle added at the preceding stage. Let a be the area of the original triangle. Then the area a_n of each of the small triangles added at stage n is

$a_n = a \cdot \dfrac{1}{9^n} = \dfrac{a}{9^n}$. Since a small triangle is added to each side at every stage, it follows that the total area A_n added to the

figure at the nth stage is $A_n = s_{n-1} \cdot a_n = 3 \cdot 4^{n-1} \cdot \dfrac{a}{9^n} = a \cdot \dfrac{4^{n-1}}{3^{2n-1}}$. Then the total area enclosed by the snowflake

curve is $A = a + A_1 + A_2 + A_3 + \cdots = a + a \cdot \dfrac{1}{3} + a \cdot \dfrac{4}{3^3} + a \cdot \dfrac{4^2}{3^5} + a \cdot \dfrac{4^3}{3^7} + \cdots$. After the first term, this is a

geometric series with common ratio $\dfrac{4}{9}$, so $A = a + \dfrac{a/3}{1 - \frac{4}{9}} = a + \dfrac{a}{3} \cdot \dfrac{9}{5} = \dfrac{8a}{5}$. But the area of the original equilateral

triangle with side 1 is $a = \dfrac{1}{2} \cdot 1 \cdot \sin\dfrac{\pi}{3} = \dfrac{\sqrt{3}}{4}$. So the area enclosed by the snowflake curve is $\dfrac{8}{5} \cdot \dfrac{\sqrt{3}}{4} = \dfrac{2\sqrt{3}}{5}$.

5. $\ln\left(1 - \dfrac{1}{n^2}\right) = \ln\left(\dfrac{n^2 - 1}{n^2}\right) = \ln\dfrac{(n+1)(n-1)}{n^2} = \ln[(n+1)(n-1)] - \ln n^2$

$= \ln(n+1) + \ln(n-1) - 2\ln n = \ln(n-1) - \ln n - \ln n + \ln(n+1)$

$= \ln\dfrac{n-1}{n} - [\ln n - \ln(n+1)] = \ln\dfrac{n-1}{n} - \ln\dfrac{n}{n+1}$.

Let $s_k = \sum\limits_{n=2}^{k} \ln\left(1 - \dfrac{1}{n^2}\right) = \sum\limits_{n=2}^{k} \left(\ln\dfrac{n-1}{n} - \ln\dfrac{n}{n+1}\right)$ for $k \geq 2$. Then

$s_k = \left(\ln\dfrac{1}{2} - \ln\dfrac{2}{3}\right) + \left(\ln\dfrac{2}{3} - \ln\dfrac{3}{4}\right) + \cdots + \left(\ln\dfrac{k-1}{k} - \ln\dfrac{k}{k+1}\right) = \ln\dfrac{1}{2} - \ln\dfrac{k}{k+1}$, so

$\sum\limits_{n=2}^{\infty} \ln\left(1 - \dfrac{1}{n^2}\right) = \lim\limits_{k\to\infty} s_k = \lim\limits_{k\to\infty}\left(\ln\dfrac{1}{2} - \ln\dfrac{k}{k+1}\right) = \ln\dfrac{1}{2} - \ln 1 = \ln 1 - \ln 2 - \ln 1 = -\ln 2.$

7. $u = 1 + \dfrac{x^3}{3!} + \dfrac{x^6}{6!} + \dfrac{x^9}{9!} + \cdots, \; v = x + \dfrac{x^4}{4!} + \dfrac{x^7}{7!} + \dfrac{x^{10}}{10!} + \cdots, \; w = \dfrac{x^2}{2!} + \dfrac{x^5}{5!} + \dfrac{x^8}{8!} + \cdots.$

Use the Ratio Test to show that the series for u, v, and w have positive radii of convergence (∞ in each case), so

Theorem 8.6.2 applies, and hence, we may differentiate each of these series:

$$\frac{du}{dx} = \frac{3x^2}{3!} + \frac{6x^5}{6!} + \frac{9x^8}{9!} + \cdots = \frac{x^2}{2!} + \frac{x^5}{5!} + \frac{x^8}{8!} + \cdots = w$$

Similarly, $\dfrac{dv}{dx} = 1 + \dfrac{x^3}{3!} + \dfrac{x^6}{6!} + \dfrac{x^9}{9!} + \cdots = u$, and $\dfrac{dw}{dx} = x + \dfrac{x^4}{4!} + \dfrac{x^7}{7!} + \dfrac{x^{10}}{10!} + \cdots = v$.

So $u' = w$, $v' = u$, and $w' = v$. Now differentiate the left hand side of the desired equation:

$$\frac{d}{dx}(u^3 + v^3 + w^3 - 3uvw) = 3u^2u' + 3v^2v' + 3w^2w' - 3(u'vw + uv'w + uvw')$$

$$= 3u^2w + 3v^2u + 3w^2v - 3(vw^2 + u^2w + uv^2) = 0 \quad \Rightarrow$$

$u^3 + v^3 + w^3 - 3uvw = C$. To find the value of the constant C, we put $x = 0$ in the last equation and get

$1^3 + 0^3 + 0^3 - 3(1 \cdot 0 \cdot 0) = C \; \Rightarrow \; C = 1$, so $u^3 + v^3 + w^3 - 3uvw = 1$.

9. If L is the length of a side of the equilateral triangle, then the area is $A = \frac{1}{2}L \cdot \frac{\sqrt{3}}{2}L = \frac{\sqrt{3}}{4}L^2$ and so $L^2 = \frac{4}{\sqrt{3}}A$.

Let r be the radius of one of the circles. When there are n rows of circles, the figure shows that

$$L = \sqrt{3}r + r + (n-2)(2r) + r + \sqrt{3}r = r(2n - 2 + 2\sqrt{3}), \text{ so } r = \frac{L}{2(n + \sqrt{3} - 1)}.$$

The number of circles is $1 + 2 + \cdots + n = \dfrac{n(n+1)}{2}$, and so the total area of the circles is

$$A_n = \frac{n(n+1)}{2}\pi r^2 = \frac{n(n+1)}{2}\pi \frac{L^2}{4(n + \sqrt{3} - 1)^2}$$

$$= \frac{n(n+1)}{2}\pi \frac{4A/\sqrt{3}}{4(n + \sqrt{3} - 1)^2} = \frac{n(n+1)}{(n + \sqrt{3} - 1)^2}\frac{\pi A}{2\sqrt{3}} \quad \Rightarrow$$

$$\frac{A_n}{A} = \frac{n(n+1)}{(n + \sqrt{3} - 1)^2}\frac{\pi}{2\sqrt{3}}$$

$$= \frac{1 + 1/n}{[1 + (\sqrt{3} - 1)/n]^2}\frac{\pi}{2\sqrt{3}} \to \frac{\pi}{2\sqrt{3}} \text{ as } n \to \infty$$

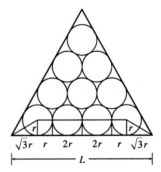

$\sqrt{3}r \quad r \quad 2r \quad 2r \quad r \quad \sqrt{3}r$

L

11. By Table 1 in Section 8.7, $\tan^{-1} x = \displaystyle\sum_{n=0}^{\infty}(-1)^n \frac{x^{2n+1}}{2n+1}$ for $|x| < 1$. In particular, for $x = \dfrac{1}{\sqrt{3}}$, we

have $\dfrac{\pi}{6} = \tan^{-1}\left(\dfrac{1}{\sqrt{3}}\right) = \displaystyle\sum_{n=0}^{\infty}(-1)^n \frac{(1/\sqrt{3})^{2n+1}}{2n+1} = \displaystyle\sum_{n=0}^{\infty}(-1)^n \left(\dfrac{1}{3}\right)^n \dfrac{1}{\sqrt{3}}\dfrac{1}{2n+1}$, so

$$\pi = \frac{6}{\sqrt{3}}\sum_{n=0}^{\infty}\frac{(-1)^n}{(2n+1)3^n} = 2\sqrt{3}\sum_{n=0}^{\infty}\frac{(-1)^n}{(2n+1)3^n} = 2\sqrt{3}\left(1 + \sum_{n=1}^{\infty}\frac{(-1)^n}{(2n+1)3^n}\right) \; \Rightarrow \; \sum_{n=1}^{\infty}\frac{(-1)^n}{(2n+1)3^n} = \frac{\pi}{2\sqrt{3}} - 1.$$

13. Let $f(x)$ denote the left-hand side of the equation $1 + \dfrac{x}{2!} + \dfrac{x^2}{4!} + \dfrac{x^3}{6!} + \dfrac{x^4}{8!} + \cdots = 0$. If $x \geq 0$, then $f(x) \geq 1$ and there are

no solutions of the equation. Note that $f(-x^2) = 1 - \dfrac{x^2}{2!} + \dfrac{x^4}{4!} - \dfrac{x^6}{6!} + \dfrac{x^8}{8!} - \cdots = \cos x$. The solutions of $\cos x = 0$ for

$x < 0$ are given by $x = \dfrac{\pi}{2} - \pi k$, where k is a positive integer. Thus, the solutions of $f(x) = 0$ are $x = -\left(\dfrac{\pi}{2} - \pi k\right)^2$, where

k is a positive integer.

15. Call the series S. We group the terms according to the number of digits in their denominators:

$$S = \underbrace{\left(\tfrac{1}{1} + \tfrac{1}{2} + \cdots + \tfrac{1}{8} + \tfrac{1}{9}\right)}_{g_1} + \underbrace{\left(\tfrac{1}{11} + \cdots + \tfrac{1}{99}\right)}_{g_2} + \underbrace{\left(\tfrac{1}{111} + \cdots + \tfrac{1}{999}\right)}_{g_3} + \cdots$$

Now in the group g_n, since we have 9 choices for each of the n digits in the denominator, there are 9^n terms.

Furthermore, each term in g_n is less than $\dfrac{1}{10^{n-1}}$ [except for the first term in g_1]. So $g_n < 9^n \cdot \dfrac{1}{10^{n-1}} = 9\left(\tfrac{9}{10}\right)^{n-1}$.

Now $\displaystyle\sum_{n=1}^{\infty} 9\left(\tfrac{9}{10}\right)^{n-1}$ is a geometric series with $a = 9$ and $r = \tfrac{9}{10} < 1$. Therefore, by the Comparison Test,

$$S = \sum_{n=1}^{\infty} g_n < \sum_{n=1}^{\infty} 9\left(\tfrac{9}{10}\right)^{n-1} = \dfrac{9}{1 - 9/10} = 90.$$

☐ APPENDIXES

A Intervals, Inequalities, and Absolute Values

1. $|5 - 23| = |-18| = 18$

3. $\left|\sqrt{5} - 5\right| = -\left(\sqrt{5} - 5\right) = 5 - \sqrt{5}$ because $\sqrt{5} - 5 < 0$.

5. If $x < 2$, $x - 2 < 0$, so $|x - 2| = -(x - 2) = 2 - x$.

7. $|x + 1| = \begin{cases} x + 1 & \text{if } x + 1 \geq 0 \\ -(x + 1) & \text{if } x + 1 < 0 \end{cases} = \begin{cases} x + 1 & \text{if } x \geq -1 \\ -x - 1 & \text{if } x < -1 \end{cases}$

9. $\left|x^2 + 1\right| = x^2 + 1$ [since $x^2 + 1 \geq 0$ for all x].

11. $2x + 7 > 3 \quad \Leftrightarrow \quad 2x > -4 \quad \Leftrightarrow \quad x > -2$, so $x \in (-2, \infty)$.

13. $1 - x \leq 2 \quad \Leftrightarrow \quad -x \leq 1 \quad \Leftrightarrow \quad x \geq -1$, so $x \in [-1, \infty)$.

15. $0 \leq 1 - x < 1 \quad \Leftrightarrow \quad -1 \leq -x < 0 \quad \Leftrightarrow \quad 1 \geq x > 0$, so $x \in (0, 1]$.

17. $(x - 1)(x - 2) > 0$.

Case 1: (both factors are positive, so their product is positive) $x - 1 > 0 \quad \Leftrightarrow \quad x > 1$,

and $x - 2 > 0 \quad \Leftrightarrow \quad x > 2$, so $x \in (2, \infty)$.

Case 2: (both factors are negative, so their product is positive) $x - 1 < 0 \quad \Leftrightarrow \quad x < 1$,

and $x - 2 < 0 \quad \Leftrightarrow \quad x < 2$, so $x \in (-\infty, 1)$.

Thus, the solution set is $(-\infty, 1) \cup (2, \infty)$.

19. $x^2 < 3 \quad \Leftrightarrow \quad x^2 - 3 < 0 \quad \Leftrightarrow \quad \left(x - \sqrt{3}\right)\left(x + \sqrt{3}\right) < 0$.

Case 1: $x > \sqrt{3}$ and $x < -\sqrt{3}$, which is impossible.

Case 2: $x < \sqrt{3}$ and $x > -\sqrt{3}$.

Thus, the solution set is $\left(-\sqrt{3}, \sqrt{3}\right)$.

Another method: $x^2 < 3 \quad \Leftrightarrow \quad |x| < \sqrt{3} \quad \Leftrightarrow \quad -\sqrt{3} < x < \sqrt{3}$.

21. $x^3 - x^2 \leq 0 \quad \Leftrightarrow \quad x^2(x - 1) \leq 0$. Since $x^2 \geq 0$ for all x, the inequality is satisfied when $x - 1 \leq 0 \quad \Leftrightarrow \quad x \leq 1$.

Thus, the solution set is $(-\infty, 1]$.

23. $x^3 > x \quad \Leftrightarrow \quad x^3 - x > 0 \quad \Leftrightarrow \quad x(x^2 - 1) > 0 \quad \Leftrightarrow \quad x(x - 1)(x + 1) > 0$. Construct a chart:

Interval	x	$x - 1$	$x + 1$	$x(x - 1)(x + 1)$
$x < -1$	$-$	$-$	$-$	$-$
$-1 < x < 0$	$-$	$-$	$+$	$+$
$0 < x < 1$	$+$	$-$	$+$	$-$
$x > 1$	$+$	$+$	$+$	$+$

Since $x^3 > x$ when the last column is positive, the solution set is $(-1, 0) \cup (1, \infty)$.

25. $1/x < 4$. This is clearly true for $x < 0$. So suppose $x > 0$. then $1/x < 4$ $\Leftrightarrow$
$1 < 4x$ $\Leftrightarrow$ $\frac{1}{4} < x$. Thus, the solution set is $(-\infty, 0) \cup \left(\frac{1}{4}, \infty\right)$.

27. $C = \frac{5}{9}(F - 32)$ $\Rightarrow$ $F = \frac{9}{5}C + 32$. So $50 \le F \le 95$ $\Rightarrow$ $50 \le \frac{9}{5}C + 32 \le 95$ $\Rightarrow$ $18 \le \frac{9}{5}C \le 63$ $\Rightarrow$
$10 \le C \le 35$. So the interval is $[10, 35]$.

29. (a) Let T represent the temperature in degrees Celsius and h the height in km. $T = 20$ when $h = 0$ and T decreases by $10°C$
for every km ($1°C$ for each 100-m rise). Thus, $T = 20 - 10h$ when $0 \le h \le 12$.

(b) From part (a), $T = 20 - 10h$ $\Rightarrow$ $10h = 20 - T$ $\Rightarrow$ $h = 2 - T/10$. So $0 \le h \le 5$ $\Rightarrow$ $0 \le 2 - T/10 \le 5$ $\Rightarrow$
$-2 \le -T/10 \le 3$ $\Rightarrow$ $-20 \le -T \le 30$ $\Rightarrow$ $20 \ge T \ge -30$ $\Rightarrow$ $-30 \le T \le 20$. Thus, the range of
temperatures (in $°C$) to be expected is $[-30, 20]$.

31. $|x + 3| = |2x + 1|$ $\Leftrightarrow$ either $x + 3 = 2x + 1$ or $x + 3 = -(2x + 1)$. In the first case, $x = 2$, and in the second case,
$x + 3 = -2x - 1$ $\Leftrightarrow$ $3x = -4$ $\Leftrightarrow$ $x = -\frac{4}{3}$. So the solutions are $-\frac{4}{3}$ and 2.

33. By Property 5 of absolute values, $|x| < 3$ $\Leftrightarrow$ $-3 < x < 3$, so $x \in (-3, 3)$.

35. $|x - 4| < 1$ $\Leftrightarrow$ $-1 < x - 4 < 1$ $\Leftrightarrow$ $3 < x < 5$, so $x \in (3, 5)$.

37. $|x + 5| \ge 2$ $\Leftrightarrow$ $x + 5 \ge 2$ or $x + 5 \le -2$ $\Leftrightarrow$ $x \ge -3$ or $x \le -7$, so $x \in (-\infty, -7] \cup [-3, \infty)$.

39. $|2x - 3| \le 0.4$ $\Leftrightarrow$ $-0.4 \le 2x - 3 \le 0.4$ $\Leftrightarrow$ $2.6 \le 2x \le 3.4$ $\Leftrightarrow$ $1.3 \le x \le 1.7$, so $x \in [1.3, 1.7]$.

41. $a(bx - c) \ge bc$ $\Leftrightarrow$ $bx - c \ge \dfrac{bc}{a}$ $\Leftrightarrow$ $bx \ge \dfrac{bc}{a} + c = \dfrac{bc + ac}{a}$ $\Leftrightarrow$ $x \ge \dfrac{bc + ac}{ab}$

43. $|ab| = \sqrt{(ab)^2} = \sqrt{a^2 b^2} = \sqrt{a^2}\,\sqrt{b^2} = |a|\,|b|$

B Coordinate Geometry

1. Use the distance formula with $P_1(x_1, y_1) = (1, 1)$ and $P_2(x_2, y_2) = (4, 5)$ to get
$$|P_1 P_2| = \sqrt{(4 - 1)^2 + (5 - 1)^2} = \sqrt{3^2 + 4^2} = \sqrt{25} = 5$$

3. The slope m of the line through $P(-3, 3)$ and $Q(-1, -6)$ is $m = \dfrac{-6 - 3}{-1 - (-3)} = -\dfrac{9}{2}$.

5. Using $A(-2, 9)$, $B(4, 6)$, $C(1, 0)$, and $D(-5, 3)$, we have
$$|AB| = \sqrt{[4 - (-2)]^2 + (6 - 9)^2} = \sqrt{6^2 + (-3)^2} = \sqrt{45} = \sqrt{9}\,\sqrt{5} = 3\sqrt{5},$$
$$|BC| = \sqrt{(1 - 4)^2 + (0 - 6)^2} = \sqrt{(-3)^2 + (-6)^2} = \sqrt{45} = \sqrt{9}\,\sqrt{5} = 3\sqrt{5},$$
$$|CD| = \sqrt{(-5 - 1)^2 + (3 - 0)^2} = \sqrt{(-6)^2 + 3^2} = \sqrt{45} = \sqrt{9}\,\sqrt{5} = 3\sqrt{5}, \text{ and}$$
$$|DA| = \sqrt{[-2 - (-5)]^2 + (9 - 3)^2} = \sqrt{3^2 + 6^2} = \sqrt{45} = \sqrt{9}\,\sqrt{5} = 3\sqrt{5}. \text{ So all sides are of equal length}$$

and we have a rhombus. Moreover, $m_{AB} = \dfrac{6-9}{4-(-2)} = -\dfrac{1}{2}$, $m_{BC} = \dfrac{0-6}{1-4} = 2$, $m_{CD} = \dfrac{3-0}{-5-1} = -\dfrac{1}{2}$, and

$m_{DA} = \dfrac{9-3}{-2-(-5)} = 2$, so the sides are perpendicular. Thus, A, B, C, and D are vertices of a square.

7. The graph of the equation $x = 3$ is a vertical line with x-intercept 3. The line does not have a slope.

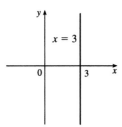

9. $xy = 0 \iff x = 0$ or $y = 0$. The graph consists of the coordinate axes.

11. By the point-slope form of the equation of a line, an equation of the line through $(2, -3)$ with slope 6 is

$y - (-3) = 6(x - 2)$ or $y = 6x - 15$.

13. The slope of the line through $(2, 1)$ and $(1, 6)$ is $m = \dfrac{6-1}{1-2} = -5$, so an equation of the line is

$y - 1 = -5(x - 2)$ or $y = -5x + 11$.

15. By the slope-intercept form of the equation of a line, an equation of the line is $y = 3x - 2$.

17. Since the line passes through $(1, 0)$ and $(0, -3)$, its slope is $m = \dfrac{-3 - 0}{0 - 1} = 3$, so an equation is $y = 3x - 3$.

Another method: From Exercise 46, $\dfrac{x}{1} + \dfrac{y}{-3} = 1 \;\Rightarrow\; -3x + y = -3 \;\Rightarrow\; y = 3x - 3$.

19. The line is parallel to the x-axis, so it is horizontal and must have the form $y = k$. Since it goes through the point $(x, y) = (4, 5)$, the equation is $y = 5$.

21. Putting the line $x + 2y = 6$ into its slope-intercept form gives us $y = -\dfrac{1}{2}x + 3$, so we see that this line has slope $-\dfrac{1}{2}$. Thus, we want the line of slope $-\dfrac{1}{2}$ that passes through the point $(1, -6)$: $y - (-6) = -\dfrac{1}{2}(x - 1) \iff y = -\dfrac{1}{2}x - \dfrac{11}{2}$.

23. $2x + 5y + 8 = 0 \iff y = -\dfrac{2}{5}x - \dfrac{8}{5}$. Since this line has slope $-\dfrac{2}{5}$, a line perpendicular to it would have slope $\dfrac{5}{2}$, so the required line is $y - (-2) = \dfrac{5}{2}[x - (-1)] \iff y = \dfrac{5}{2}x + \dfrac{1}{2}$.

25. $x + 3y = 0 \iff y = -\dfrac{1}{3}x$, so the slope is $-\dfrac{1}{3}$ and the y-intercept is 0.

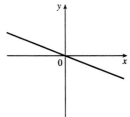

27. $3x - 4y = 12 \iff y = \dfrac{3}{4}x - 3$, so the slope is $\dfrac{3}{4}$ and the y-intercept is -3.

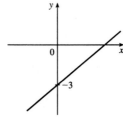

29. $\{(x, y) \mid x < 0\}$

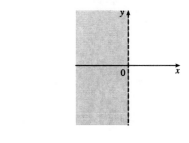

31. $\left\{(x, y) \,\middle|\, |x| \leq 2\right\} =$

$\{(x, y) \mid -2 \leq x \leq 2\}$

33. $\{(x, y) \mid 0 \leq y \leq 4, x \leq 2\}$

35. $\{(x, y) \mid 1 + x \leq y \leq 1 - 2x\}$

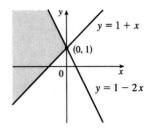

37. An equation of the circle with center $(3, -1)$ and radius 5 is $(x - 3)^2 + (y + 1)^2 = 5^2 = 25$.

39. $x^2 + y^2 - 4x + 10y + 13 = 0 \iff x^2 - 4x + y^2 + 10y = -13 \iff$

$(x^2 - 4x + 4) + (y^2 + 10y + 25) = -13 + 4 + 25 = 16 \iff (x - 2)^2 + (y + 5)^2 = 4^2$. Thus, we have a circle with

center $(2, -5)$ and radius 4.

41. $2x - y = 4 \iff y = 2x - 4 \implies m_1 = 2$ and $6x - 2y = 10 \iff 2y = 6x - 10 \iff y = 3x - 5 \implies m_2 = 3$.

Since $m_1 \neq m_2$, the two lines are not parallel. To find the point of intersection: $2x - 4 = 3x - 5 \iff x = 1 \implies$

$y = -2$. Thus, the point of intersection is $(1, -2)$.

43. Let M be the point $\left(\dfrac{x_1 + x_2}{2}, \dfrac{y_1 + y_2}{2}\right)$. Then

$$|MP_1|^2 = \left(x_1 - \frac{x_1 + x_2}{2}\right)^2 + \left(y_1 - \frac{y_1 + y_2}{2}\right)^2 = \left(\frac{x_1 - x_2}{2}\right)^2 + \left(\frac{y_1 - y_2}{2}\right)^2$$

$$|MP_2|^2 = \left(x_2 - \frac{x_1 + x_2}{2}\right)^2 + \left(y_2 - \frac{y_1 + y_2}{2}\right)^2 = \left(\frac{x_2 - x_1}{2}\right)^2 + \left(\frac{y_2 - y_1}{2}\right)^2$$

Hence, $|MP_1| = |MP_2|$; that is, M is equidistant from P_1 and P_2.

45. With $A(1, 4)$ and $B(7, -2)$, the slope of segment AB is $\frac{-2-4}{7-1} = -1$, so its perpendicular bisector has slope 1. The midpoint

of AB is $\left(\frac{1+7}{2}, \frac{4+(-2)}{2}\right) = (4, 1)$, so an equation of the perpendicular bisector is $y - 1 = 1(x - 4)$ or $y = x - 3$.

47. If $P(x, y)$ is any point on the parabola, then the distance from P to the focus is $|PF| = \sqrt{x^2 + (y - p)^2}$ and the distance

from P to the directrix is $|y + p|$. (Figure 14 in the text illustrates the case where $p > 0$.) The defining property of a parabola

is that these distances are equal: $\sqrt{x^2 + (y - p)^2} = |y + p|$. We get an equivalent equation by squaring and simplifying:

$x^2 + (y - p)^2 = |y + p|^2 = (y + p)^2 \iff x^2 + y^2 - 2py + p^2 = y^2 + 2py + p^2 \iff x^2 = 4py$. Thus, an equation of a

parabola with focus $(0, p)$ and directrix $y = -p$ is $x^2 = 4py$.

49. See Figure 20 in the text. $P(x, y)$ is a point on the ellipse when $|PF_1| + |PF_2| = 2a$; that is,

$\sqrt{(x + c)^2 + y^2} + \sqrt{(x - c)^2 + y^2} = 2a$ or $\sqrt{(x + c)^2 + y^2} = 2a - \sqrt{(x + c)^2 + y^2}$. Squaring both sides, we have

$x^2 - 2cx + c^2 + y^2 = 4a^2 - 4a\sqrt{(x + c)^2 + y^2} + x^2 + 2cx + c^2 + y^2$, which simplifies to $a\sqrt{(x + c)^2 + y^2} = a^2 + cx$.

We square again: $a^2\left(x^2 + 2cx + c^2 + y^2\right) = a^4 + 2a^2cx + c^2x^2$, which becomes $\left(a^2 - c^2\right)x^2 + a^2y^2 = a^2\left(a^2 - c^2\right)$.

From triangle F_1F_2P in Figure 20, we see that $2c < 2a$, so $c < a$ and, therefore, $a^2 - c^2 > 0$. For convenience, let $b^2 = a^2 - c^2$. Then the equation of the ellipse becomes $b^2x^2 + a^2y^2 = a^2b^2$ or, if both sides are divided by a^2b^2,

$$\frac{x^2}{a^2} + \frac{y^2}{b^2} = 1.$$

51. From Figure 23 in the text, $|PF_1| - |PF_2| = \pm 2a \iff \sqrt{(x + c)^2 + y^2} - \sqrt{(x - c)^2 + y^2} = \pm 2a \iff$

$\sqrt{(x + c)^2 + y^2} = \sqrt{(x - c)^2 + y^2} \pm 2a \iff$

$(x + c)^2 + y^2 = (x - c)^2 + y^2 + 4a^2 \pm 4a\sqrt{(x - c)^2 + y^2} \iff 4cx - 4a^2 = \pm 4a\sqrt{(x - c)^2 + y^2} \iff$

$c^2x^2 - 2a^2cx + a^4 = a^2\left(x^2 - 2cx + c^2 + y^2\right) \iff \left(c^2 - a^2\right)x^2 - a^2y^2 = a^2\left(c^2 - a^2\right) \iff$

$b^2x^2 - a^2y^2 = a^2b^2$ [where $b^2 = c^2 - a^2$] $\iff \dfrac{x^2}{a^2} - \dfrac{y^2}{b^2} = 1.$

53.

$x + 4y = 8$, $x = 2y^2 - 8$. Substitute x from the second equation into the first: $(2y^2 - 8) + 4y = 8 \iff 2y^2 + 4y - 16 = 0 \iff$

$y^2 + 2y - 8 = 0 \iff (y + 4)(y - 2) = 0 \iff y = -4$ or 2. So the points of intersection are $(24, -4)$ and $(0, 2)$.

55. Differentiating implicitly, $\dfrac{x^2}{a^2} + \dfrac{y^2}{b^2} = 1 \implies \dfrac{2x}{a^2} + \dfrac{2yy'}{b^2} = 0 \implies y' = -\dfrac{b^2x}{a^2y}$ $[y \neq 0]$. Thus, the slope of the tangent

line at P is $-\dfrac{b^2x_1}{a^2y_1}$. The slope of F_1P is $\dfrac{y_1}{x_1 + c}$ and of F_2P is $\dfrac{y_1}{x_1 - c}$. By the formula in Problem 17 on text page 253,

we have

$$\tan\alpha = \frac{\dfrac{y_1}{x_1 + c} + \dfrac{b^2x_1}{a^2y_1}}{1 - \dfrac{b^2x_1y_1}{a^2y_1(x_1 + c)}} = \frac{a^2y_1^2 + b^2x_1(x_1 + c)}{a^2y_1(x_1 + c) - b^2x_1y_1} = \frac{a^2b^2 + b^2cx_1}{c^2x_1y_1 + a^2cy_1} \quad \left[\begin{array}{l}\text{using } b^2x_1^2 + a^2y_1^2 = a^2b^2, \\ \text{and } a^2 - b^2 = c^2\end{array}\right]$$

$$= \frac{b^2\left(cx_1 + a^2\right)}{cy_1\left(cx_1 + a^2\right)} = \frac{b^2}{cy_1}$$

and

$$\tan\beta = \frac{-\dfrac{b^2x_1}{a^2y_1} - \dfrac{y_1}{x_1 - c}}{1 - \dfrac{b^2x_1y_1}{a^2y_1(x_1 - c)}} = \frac{-a^2y_1^2 - b^2x_1(x_1 - c)}{a^2y_1(x_1 - c) - b^2x_1y_1} = \frac{-a^2b^2 + b^2cx_1}{c^2x_1y_1 - a^2cy_1} = \frac{b^2\left(cx_1 - a^2\right)}{cy_1\left(cx_1 - a^2\right)} = \frac{b^2}{cy_1}$$

Thus, $\alpha = \beta$.

C Trigonometry

1. (a) $210° = 210° \left(\frac{\pi}{180°}\right) = \frac{7\pi}{6}$ rad

(b) $9° = 9° \left(\frac{\pi}{180°}\right) = \frac{\pi}{20}$ rad

3. (a) 4π rad $= 4\pi \left(\frac{180°}{\pi}\right) = 720°$

(b) $-\frac{3\pi}{8}$ rad $= -\frac{3\pi}{8} \left(\frac{180°}{\pi}\right) = -67.5°$

5. Using Formula 3, $a = r\theta = 36 \cdot \frac{\pi}{12} = 3\pi$ cm.

7. Using Formula 3, $\theta = a/r = \frac{1}{1.5} = \frac{2}{3}$ rad $= \frac{2}{3}\left(\frac{180°}{\pi}\right) = \left(\frac{120}{\pi}\right)° \approx 38.2°$.

9. (a)

(b)

11.

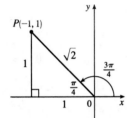

From the diagram we see that a point on the terminal side is $P(-1, 1)$. Therefore, taking $x = -1$, $y = 1$, $r = \sqrt{2}$ in the definitions of the trigonometric ratios, we have $\sin\frac{3\pi}{4} = \frac{1}{\sqrt{2}}$, $\cos\frac{3\pi}{4} = -\frac{1}{\sqrt{2}}$, $\tan\frac{3\pi}{4} = -1$, $\csc\frac{3\pi}{4} = \sqrt{2}$, $\sec\frac{3\pi}{4} = -\sqrt{2}$, and $\cot\frac{3\pi}{4} = -1$.

13. $\sin\theta = y/r = \frac{3}{5}$ $\Rightarrow$ $y = 3$, $r = 5$, and $x = \sqrt{r^2 - y^2} = 4$ (since $0 < \theta < \frac{\pi}{2}$). Therefore taking $x = 4$, $y = 3$, $r = 5$ in the definitions of the trigonometric ratios, we have $\cos\theta = \frac{4}{5}$, $\tan\theta = \frac{3}{4}$, $\csc\theta = \frac{5}{3}$, $\sec\theta = \frac{5}{4}$, and $\cot\theta = \frac{4}{3}$.

15. $\sin 35° = \frac{x}{10}$ $\Rightarrow$ $x = 10\sin 35° \approx 5.73576$ cm

17. $\tan\frac{2\pi}{5} = \frac{x}{8}$ $\Rightarrow$ $x = 8\tan\frac{2\pi}{5} \approx 24.62147$ cm

19.

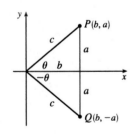

(a) From the diagram we see that $\sin\theta = \frac{y}{r} = \frac{a}{c}$, and $\sin(-\theta) = \frac{-a}{c} = -\frac{a}{c} = -\sin\theta$.

(b) Again from the diagram we see that $\cos\theta = \frac{x}{r} = \frac{b}{c} = \cos(-\theta)$.

21. Using (12a), we have $\sin\left(\frac{\pi}{2} + x\right) = \sin\frac{\pi}{2}\cos x + \cos\frac{\pi}{2}\sin x = 1 \cdot \cos x + 0 \cdot \sin x = \cos x$.

23. Using (6), we have $\sin\theta\cot\theta = \sin\theta \cdot \frac{\cos\theta}{\sin\theta} = \cos\theta$.

25. Using (14a), we have $\tan 2\theta = \tan(\theta + \theta) = \frac{\tan\theta + \tan\theta}{1 - \tan\theta\tan\theta} = \frac{2\tan\theta}{1 - \tan^2\theta}$.

27. Since $\sin x = \frac{1}{3}$ we can label the opposite side as having length 1, the hypotenuse as having length 3, and use the Pythagorean Theorem to get that the adjacent side has length $\sqrt{8}$. Then, from the diagram,

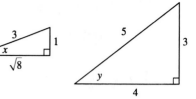

$\cos x = \frac{\sqrt{8}}{3}$. Similarly we have that $\sin y = \frac{3}{5}$. Now use (12a):

$\sin(x + y) = \sin x \cos y + \cos x \sin y = \frac{1}{3} \cdot \frac{4}{5} + \frac{\sqrt{8}}{3} \cdot \frac{3}{5} = \frac{4}{15} + \frac{3\sqrt{8}}{15} = \frac{4 + 6\sqrt{2}}{15}$.

29. $2\cos x - 1 = 0 \iff \cos x = \frac{1}{2} \implies x = \frac{\pi}{3}, \frac{5\pi}{3}$ for $x \in [0, 2\pi]$.

31. Using (15a), we have $\sin 2x = \cos x \iff 2\sin x \cos x - \cos x = 0 \iff \cos x(2\sin x - 1) = 0 \iff \cos x = 0$ or $2\sin x - 1 = 0 \implies x = \frac{\pi}{2}, \frac{3\pi}{2}$ or $\sin x = \frac{1}{2} \implies x = \frac{\pi}{6}$ or $\frac{5\pi}{6}$. Therefore, the solutions are $x = \frac{\pi}{6}, \frac{\pi}{2}, \frac{5\pi}{6}, \frac{3\pi}{2}$.

33. We know that $\sin x = \frac{1}{2}$ when $x = \frac{\pi}{6}$ or $\frac{5\pi}{6}$, and from Figure 13(a), we see that $\sin x \leq \frac{1}{2} \implies 0 \leq x \leq \frac{\pi}{6}$ or $\frac{5\pi}{6} \leq x \leq 2\pi$ for $x \in [0, 2\pi]$.

35. $\tan x = -1$ when $x = \frac{3\pi}{4}, \frac{7\pi}{4}$, and $\tan x = 1$ when $x = \frac{\pi}{4}$ or $\frac{5\pi}{4}$. From Figure 14(a) we see that $-1 < \tan x < 1 \implies 0 \leq x < \frac{\pi}{4}, \frac{3\pi}{4} < x < \frac{5\pi}{4}$, and $\frac{7\pi}{4} < x \leq 2\pi$.

37. $y = \cos\left(x - \frac{\pi}{3}\right)$. We start with the graph of $y = \cos x$ and shift it $\frac{\pi}{3}$ units to the right.

39. $y = \frac{1}{3}\tan\left(x - \frac{\pi}{2}\right)$. We start with the graph of $y = \tan x$, shift it $\frac{\pi}{2}$ units to the right and compress it to $\frac{1}{3}$ of its original vertical size.

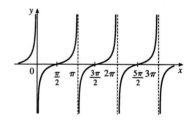

41. From the figure in the text, we see that $x = b\cos\theta$, $y = b\sin\theta$, and from the distance formula we have that the distance c from (x, y) to $(a, 0)$ is $c = \sqrt{(x - a)^2 + (y - 0)^2} \implies$

$$c^2 = (b\cos\theta - a)^2 + (b\sin\theta)^2 = b^2\cos^2\theta - 2ab\cos\theta + a^2 + b^2\sin^2\theta$$
$$= a^2 + b^2(\cos^2\theta + \sin^2\theta) - 2ab\cos\theta = a^2 + b^2 - 2ab\cos\theta \quad \text{[by (7)]}$$

43. Using the Law of Cosines, we have $c^2 = 1^2 + 1^2 - 2(1)(1)\cos(\alpha - \beta) = 2[1 - \cos(\alpha - \beta)]$. Now, using the distance formula, $c^2 = |AB|^2 = (\cos\alpha - \cos\beta)^2 + (\sin\alpha - \sin\beta)^2$. Equating these two expressions for c^2, we get

$2[1 - \cos(\alpha - \beta)] = \cos^2\alpha + \sin^2\alpha + \cos^2\beta + \sin^2\beta - 2\cos\alpha\cos\beta - 2\sin\alpha\sin\beta \implies$

$1 - \cos(\alpha - \beta) = 1 - \cos\alpha\cos\beta - \sin\alpha\sin\beta \implies \cos(\alpha - \beta) = \cos\alpha\cos\beta + \sin\alpha\sin\beta$.

45. In Exercise 44 we used the subtraction formula for cosine to prove the addition formula for cosine. Using that formula with $x = \frac{\pi}{2} - \alpha$, $y = \beta$, we get $\cos\left[\left(\frac{\pi}{2} - \alpha\right) + \beta\right] = \cos\left(\frac{\pi}{2} - \alpha\right)\cos\beta - \sin\left(\frac{\pi}{2} - \alpha\right)\sin\beta \implies$

$\cos\left[\frac{\pi}{2} - (\alpha - \beta)\right] = \cos\left(\frac{\pi}{2} - \alpha\right)\cos\beta - \sin\left(\frac{\pi}{2} - \alpha\right)\sin\beta$. Now we use the identities given in the problem,

$\cos\left(\frac{\pi}{2} - \theta\right) = \sin\theta$ and $\sin\left(\frac{\pi}{2} - \theta\right) = \cos\theta$, to get $\sin(\alpha - \beta) = \sin\alpha\cos\beta - \cos\alpha\sin\beta$.

D   PRECISE DEFINITIONS OF LIMITS

1. On the left side of $x = 2$, we need $|x - 2| < \left|\frac{10}{7} - 2\right| = \frac{4}{7}$. On the right side, we need $|x - 2| < \left|\frac{10}{3} - 2\right| = \frac{4}{3}$. For both of

these conditions to be satisfied at once, we need the more restrictive of the two to hold, that is, $|x - 2| < \frac{4}{7}$. So we can choose

$\delta = \frac{4}{7}$, or any smaller positive number.

3. The leftmost question mark is the solution of $\sqrt{x} = 1.6$ and the rightmost, $\sqrt{x} = 2.4$. So the values are $1.6^2 = 2.56$ and

$2.4^2 = 5.76$. On the left side, we need $|x - 4| < |2.56 - 4| = 1.44$. On the right side, we need $|x - 4| < |5.76 - 4| = 1.76$.

To satisfy both conditions, we need the more restrictive condition to hold—namely, $|x - 4| < 1.44$. Thus, we can choose

$\delta = 1.44$, or any smaller positive number.

5.

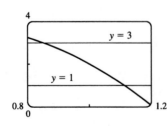

From the graph, we find that $\tan x = 0.8$ when $x \approx 0.675$, so

$\frac{\pi}{4} - \delta_1 \approx 0.675 \quad \Rightarrow \quad \delta_1 \approx \frac{\pi}{4} - 0.675 \approx 0.1106$. Also, $\tan x = 1.2$

when $x \approx 0.876$, so $\frac{\pi}{4} + \delta_2 \approx 0.876 \quad \Rightarrow \quad \delta_2 = 0.876 - \frac{\pi}{4} \approx 0.0906$.

Thus, we choose $\delta = 0.0906$ (or any smaller positive number) since this is

the smaller of δ_1 and δ_2.

7. For $\varepsilon = 1$, the definition of a limit requires that we find δ such that $\left|(4 + x - 3x^3) - 2\right| < 1 \quad \Leftrightarrow \quad 1 < 4 + x - 3x^3 < 3$

whenever $0 < |x - 1| < \delta$. If we plot the graphs of $y = 1$, $y = 4 + x - 3x^3$ and $y = 3$ on the same screen, we see that we

need $0.86 \le x \le 1.11$. So since $|1 - 0.86| = 0.14$ and $|1 - 1.11| = 0.11$, we choose $\delta = 0.11$ (or any smaller positive

number). For $\varepsilon = 0.1$, we must find δ such that $\left|(4 + x - 3x^3) - 2\right| < 0.1 \quad \Leftrightarrow \quad 1.9 < 4 + x - 3x^3 < 2.1$ whenever

$0 < |x - 1| < \delta$. From the graph, we see that we need $0.988 \le x \le 1.012$. So since $|1 - 0.988| = 0.012$ and

$|1 - 1.012| = 0.012$, we choose $\delta = 0.012$ (or any smaller positive number) for the inequality to hold.

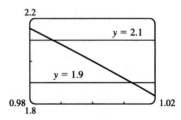

9. Given $\varepsilon > 0$, we need $\delta > 0$ such that if $|x| < \delta$ then $\left|x^3 - 0\right| < \varepsilon \quad \Leftrightarrow \quad |x|^3 < \varepsilon \quad \Leftrightarrow \quad |x| < \sqrt[3]{\varepsilon}$. Take $\delta = \sqrt[3]{\varepsilon}$.

Then $|x - 0| < \delta \quad \Rightarrow \quad \left|x^3 - 0\right| < \delta^3 = \varepsilon$. Thus, $\lim\limits_{x \to 0} x^3 = 0$ by the definition of a limit.

11. (a) $A = \pi r^2$ and $A = 1000 \text{ cm}^2 \quad \Rightarrow \quad \pi r^2 = 1000 \quad \Rightarrow \quad r^2 = \frac{1000}{\pi} \quad \Rightarrow \quad r = \sqrt{\frac{1000}{\pi}} \quad (r > 0) \quad \approx 17.8412 \text{ cm}$.

(b) $|A - 1000| \le 5 \implies -5 \le \pi r^2 - 1000 \le 5 \implies 1000 - 5 \le \pi r^2 \le 1000 + 5 \implies$

$\sqrt{\frac{995}{\pi}} \le r \le \sqrt{\frac{1005}{\pi}} \implies 17.7966 \le r \le 17.8858.$ $\sqrt{\frac{1000}{\pi}} - \sqrt{\frac{995}{\pi}} \approx 0.04466$ and $\sqrt{\frac{1005}{\pi}} - \sqrt{\frac{1000}{\pi}} \approx 0.04455.$ So

if the machinist gets the radius within 0.0445 cm of 17.8412, the area will be within 5 cm^2 of 1000.

(c) x is the radius, $f(x)$ is the area, a is the target radius given in part (a), L is the target area (1000), ε is the tolerance in the area (5), and δ is the tolerance in the radius given in part (b).

13. (a) $|4x - 8| = 4|x - 2| < 0.1 \iff |x - 2| < \dfrac{0.1}{4}$, so $\delta = \dfrac{0.1}{4} = 0.025.$

(b) $|4x - 8| = 4|x - 2| < 0.01 \iff |x - 2| < \dfrac{0.01}{4}$, so $\delta = \dfrac{0.01}{4} = 0.0025.$

15. Given $\varepsilon > 0$, we need $\delta > 0$ such that if $0 < |x - (-3)| < \delta$, then

$|(1 - 4x) - 13| < \varepsilon.$ But $|(1 - 4x) - 13| < \varepsilon \iff$

$|-4x - 12| < \varepsilon \iff |-4|\,|x + 3| < \varepsilon \iff |x - (-3)| < \varepsilon/4.$

So if we choose $\delta = \varepsilon/4$, then $0 < |x - (-3)| < \delta \implies$

$|(1 - 4x) - 13| < \varepsilon.$ Thus, $\lim\limits_{x \to -3}(1 - 4x) = 13$ by the definition of

a limit.

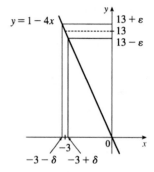

17. $\left| \dfrac{6x^2 + 5x - 3}{2x^2 - 1} - 3 \right| < 0.2 \iff 2.8 < \dfrac{6x^2 + 5x - 3}{2x^2 - 1} < 3.2.$ So

we graph the three parts of this inequality on the same screen, and

find that the curve $y = \dfrac{6x^2 + 5x - 3}{2x^2 - 1}$ seems to lie between the lines

$y = 2.8$ and $y = 3.2$ whenever $x > 12.8.$ So we can choose $N = 13$

(or any larger number) so that the inequality holds whenever $x \ge N.$

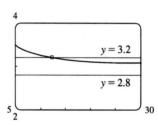

19. (a) $1/x^2 < 0.0001 \iff x^2 > 1/0.0001 = 10{,}000 \iff x > 100 \quad (x > 0)$

(b) If $\varepsilon > 0$ is given, then $1/x^2 < \varepsilon \iff x^2 > 1/\varepsilon \iff x > 1/\sqrt{\varepsilon}.$ Let $N = 1/\sqrt{\varepsilon}.$ Then $x > N \implies$

$x > \dfrac{1}{\sqrt{\varepsilon}} \implies \left| \dfrac{1}{x^2} - 0 \right| = \dfrac{1}{x^2} < \varepsilon,$ so $\lim\limits_{x \to \infty} \dfrac{1}{x^2} = 0.$

21. (a)

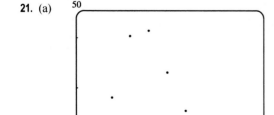

From the graph, it appears that the sequence $\left\{ \dfrac{n^5}{n!} \right\}$

converges to 0, that is, $\lim\limits_{n \to \infty} \dfrac{n^5}{n!} = 0.$

(b)

From the first graph, it seems that the smallest possible value of N corresponding to $\varepsilon = 0.1$ is 9, since $n^5/n! < 0.1$

whenever $n \geq 10$, but $9^5/9! > 0.1$. From the second graph, it seems that for $\varepsilon = 0.001$, the smallest possible value for N

is 11 since $n^5/n! < 0.001$ whenever $n \geq 12$.

23. If $\lim\limits_{n\to\infty} |a_n| = 0$, then $\lim\limits_{n\to\infty} (-|a_n|) = 0$, and since $-|a_n| \leq a_n \leq |a_n|$, we have that $\lim\limits_{n\to\infty} a_n = 0$ by the Squeeze

Theorem.

F Sigma Notation

1. $\sum\limits_{i=1}^{5} \sqrt{i} = \sqrt{1} + \sqrt{2} + \sqrt{3} + \sqrt{4} + \sqrt{5}$

3. $\sum\limits_{i=4}^{6} 3^i = 3^4 + 3^5 + 3^6$

5. $\sum\limits_{k=0}^{4} \dfrac{2k-1}{2k+1} = -1 + \dfrac{1}{3} + \dfrac{3}{5} + \dfrac{5}{7} + \dfrac{7}{9}$

7. $\sum\limits_{i=1}^{n} i^{10} = 1^{10} + 2^{10} + 3^{10} + \cdots + n^{10}$

9. $\sum\limits_{j=0}^{n-1} (-1)^j = 1 - 1 + 1 - 1 + \cdots + (-1)^{n-1}$

11. $1 + 2 + 3 + 4 + \cdots + 10 = \sum\limits_{i=1}^{10} i$

13. $\dfrac{1}{2} + \dfrac{2}{3} + \dfrac{3}{4} + \dfrac{4}{5} + \cdots + \dfrac{19}{20} = \sum\limits_{i=1}^{19} \dfrac{i}{i+1}$

15. $2 + 4 + 6 + 8 + \cdots + 2n = \sum\limits_{i=1}^{n} 2i$

17. $1 + 2 + 4 + 8 + 16 + 32 = \sum\limits_{i=0}^{5} 2^i$

19. $x + x^2 + x^3 + \cdots + x^n = \sum\limits_{i=1}^{n} x^i$

21. $\sum\limits_{i=4}^{8} (3i - 2) = [3(4) - 2] + [3(5) - 2] + [3(6) - 2] + [3(7) - 2] + [3(8) - 2] = 10 + 13 + 16 + 19 + 22 = 80$

23. $\sum\limits_{j=1}^{6} 3^{j+1} = 3^2 + 3^3 + 3^4 + 3^5 + 3^6 + 3^7 = 9 + 27 + 81 + 243 + 729 + 2187 = 3276$
(For a more general method, see Exercise 47.)

25. $\sum\limits_{n=1}^{20} (-1)^n = -1 + 1 - 1 + 1 - 1 + 1 - 1 + 1 - 1 + 1 - 1 + 1 - 1 + 1 - 1 + 1 - 1 + 1 - 1 + 1 = 0$

27. $\sum\limits_{i=0}^{4} (2^i + i^2) = (1 + 0) + (2 + 1) + (4 + 4) + (8 + 9) + (16 + 16) = 61$

29. $\displaystyle\sum_{i=1}^{n} 2i = 2 \sum_{i=1}^{n} i = 2 \cdot \frac{n(n+1)}{2}$ [by Theorem 3(c)] $= n(n+1)$

31. $\displaystyle\sum_{i=1}^{n} (i^2 + 3i + 4) = \sum_{i=1}^{n} i^2 + 3 \sum_{i=1}^{n} i + \sum_{i=1}^{n} 4 = \frac{n(n+1)(2n+1)}{6} + \frac{3n(n+1)}{2} + 4n$

$$= \tfrac{1}{6}[(2n^3 + 3n^2 + n) + (9n^2 + 9n) + 24n] = \tfrac{1}{6}(2n^3 + 12n^2 + 34n) = \tfrac{1}{3}n(n^2 + 6n + 17)$$

33. $\displaystyle\sum_{i=1}^{n} (i+1)(i+2) = \sum_{i=1}^{n} (i^2 + 3i + 2) = \sum_{i=1}^{n} i^2 + 3 \sum_{i=1}^{n} i + \sum_{i=1}^{n} 2 = \frac{n(n+1)(2n+1)}{6} + \frac{3n(n+1)}{2} + 2n$

$$= \frac{n(n+1)}{6}[(2n+1) + 9] + 2n = \frac{n(n+1)}{3}(n+5) + 2n$$

$$= \frac{n}{3}[(n+1)(n+5) + 6] = \frac{n}{3}(n^2 + 6n + 11)$$

35. $\displaystyle\sum_{i=1}^{n} (i^3 - i - 2) = \sum_{i=1}^{n} i^3 - \sum_{i=1}^{n} i - \sum_{i=1}^{n} 2 = \left[\frac{n(n+1)}{2}\right]^2 - \frac{n(n+1)}{2} - 2n$

$$= \tfrac{1}{4}n(n+1)[n(n+1) - 2] - 2n = \tfrac{1}{4}n(n+1)(n+2)(n-1) - 2n$$

$$= \tfrac{1}{4}n[(n+1)(n-1)(n+2) - 8] = \tfrac{1}{4}n[(n^2-1)(n+2) - 8] = \tfrac{1}{4}n(n^3 + 2n^2 - n - 10)$$

37. By Theorem 2(a) and Example 3, $\displaystyle\sum_{i=1}^{n} c = c \sum_{i=1}^{n} 1 = cn.$

39. $\displaystyle\sum_{i=1}^{n} [(i+1)^4 - i^4] = (2^4 - 1^4) + (3^4 - 2^4) + (4^4 - 3^4) + \cdots + [(n+1)^4 - n^4]$

$$= (n+1)^4 - 1^4 = n^4 + 4n^3 + 6n^2 + 4n$$

On the other hand,

$$\sum_{i=1}^{n} [(i+1)^4 - i^4] = \sum_{i=1}^{n} (4i^3 + 6i^2 + 4i + 1) = 4 \sum_{i=1}^{n} i^3 + 6 \sum_{i=1}^{n} i^2 + 4 \sum_{i=1}^{n} i + \sum_{i=1}^{n} 1$$

$$= 4S + n(n+1)(2n+1) + 2n(n+1) + n \qquad \left[\text{where } S = \sum_{i=1}^{n} i^3\right]$$

$$= 4S + 2n^3 + 3n^2 + n + 2n^2 + 2n + n = 4S + 2n^3 + 5n^2 + 4n$$

Thus, $n^4 + 4n^3 + 6n^2 + 4n = 4S + 2n^3 + 5n^2 + 4n$, from which it follows that

$$4S = n^4 + 2n^3 + n^2 = n^2(n^2 + 2n + 1) = n^2(n+1)^2 \text{ and } S = \left[\frac{n(n+1)}{2}\right]^2.$$

41. (a) $\displaystyle\sum_{i=1}^{n} \left[i^4 - (i-1)^4\right] = (1^4 - 0^4) + (2^4 - 1^4) + (3^4 - 2^4) + \cdots + [n^4 - (n-1)^4] = n^4 - 0 = n^4$

(b) $\displaystyle\sum_{i=1}^{100} (5^i - 5^{i-1}) = (5^1 - 5^0) + (5^2 - 5^1) + (5^3 - 5^2) + \cdots + (5^{100} - 5^{99}) = 5^{100} - 5^0 = 5^{100} - 1$

(c) $\displaystyle\sum_{i=3}^{99} \left(\frac{1}{i} - \frac{1}{i+1}\right) = \left(\frac{1}{3} - \frac{1}{4}\right) + \left(\frac{1}{4} - \frac{1}{5}\right) + \left(\frac{1}{5} - \frac{1}{6}\right) + \cdots + \left(\frac{1}{99} - \frac{1}{100}\right) = \frac{1}{3} - \frac{1}{100} = \frac{97}{300}$

(d) $\displaystyle\sum_{i=1}^{n} (a_i - a_{i-1}) = (a_1 - a_0) + (a_2 - a_1) + (a_3 - a_2) + \cdots + (a_n - a_{n-1}) = a_n - a_0$

43. $\lim\limits_{n\to\infty}\sum\limits_{i=1}^{n}\dfrac{1}{n}\left(\dfrac{i}{n}\right)^2 = \lim\limits_{n\to\infty}\dfrac{1}{n^3}\sum\limits_{i=1}^{n}i^2 = \lim\limits_{n\to\infty}\dfrac{1}{n^3}\dfrac{n(n+1)(2n+1)}{6} = \lim\limits_{n\to\infty}\dfrac{1}{6}\left(1+\dfrac{1}{n}\right)\left(2+\dfrac{1}{n}\right) = \frac{1}{6}(1)(2) = \frac{1}{3}$

45. $\lim\limits_{n\to\infty}\sum\limits_{i=1}^{n}\dfrac{2}{n}\left[\left(\dfrac{2i}{n}\right)^3 + 5\left(\dfrac{2i}{n}\right)\right] = \lim\limits_{n\to\infty}\sum\limits_{i=1}^{n}\left[\dfrac{16}{n^4}i^3 + \dfrac{20}{n^2}i\right] = \lim\limits_{n\to\infty}\left[\dfrac{16}{n^4}\sum\limits_{i=1}^{n}i^3 + \dfrac{20}{n^2}\sum\limits_{i=1}^{n}i\right]$

$$= \lim\limits_{n\to\infty}\left[\dfrac{16}{n^4}\dfrac{n^2(n+1)^2}{4} + \dfrac{20}{n^2}\dfrac{n(n+1)}{2}\right] = \lim\limits_{n\to\infty}\left[\dfrac{4(n+1)^2}{n^2} + \dfrac{10n(n+1)}{n^2}\right]$$

$$= \lim\limits_{n\to\infty}\left[4\left(1+\dfrac{1}{n}\right)^2 + 10\left(1+\dfrac{1}{n}\right)\right] = 4\cdot 1 + 10\cdot 1 = 14$$

47. Let $S = \sum\limits_{i=1}^{n}ar^{i-1} = a + ar + ar^2 + \cdots + ar^{n-1}$. Multiplying both sides by r gives us

$rS = ar + ar^2 + \cdots + ar^{n-1} + ar^n$. Subtracting the first equation from the second, we find

$(r-1)S = ar^n - a = a(r^n - 1)$, so $S = \dfrac{a(r^n - 1)}{r - 1}$ [since $r \neq 1$].

49. $\sum\limits_{i=1}^{n}(2i + 2^i) = 2\sum\limits_{i=1}^{n}i + \sum\limits_{i=1}^{n}2\cdot 2^{i-1} = 2\dfrac{n(n+1)}{2} + \dfrac{2(2^n - 1)}{2 - 1} = 2^{n+1} + n^2 + n - 2.$

For the first sum we have used Theorems 2(a) and 3(c), and for the second, Exercise 47 with $a = r = 2$.

G   INTEGRATION OF RATIONAL FUNCTIONS BY PARTIAL FRACTIONS

1. (a) $\dfrac{2x}{(x+3)(3x+1)} = \dfrac{A}{x+3} + \dfrac{B}{3x+1}$

(b) $\dfrac{1}{x^3 + 2x^2 + x} = \dfrac{1}{x(x^2 + 2x + 1)} = \dfrac{1}{x(x+1)^2} = \dfrac{A}{x} + \dfrac{B}{x+1} + \dfrac{C}{(x+1)^2}$

3. (a) $\dfrac{x^4 + 1}{x^5 + 4x^3} = \dfrac{x^4 + 1}{x^3(x^2 + 4)} = \dfrac{A}{x} + \dfrac{B}{x^2} + \dfrac{C}{x^3} + \dfrac{Dx + E}{x^2 + 4}$

(b) $\dfrac{1}{(x^2 - 9)^2} = \dfrac{1}{[(x+3)(x-3)]^2} = \dfrac{1}{(x+3)^2(x-3)^2} = \dfrac{A}{x+3} + \dfrac{B}{(x+3)^2} + \dfrac{C}{x-3} + \dfrac{D}{(x-3)^2}$

5. (a) $\dfrac{x^4}{x^4 - 1} = \dfrac{(x^4 - 1) + 1}{x^4 - 1} = 1 + \dfrac{1}{x^4 - 1}$ [or use long division] $= 1 + \dfrac{1}{(x^2 - 1)(x^2 + 1)}$

$$= 1 + \dfrac{1}{(x-1)(x+1)(x^2 + 1)} = 1 + \dfrac{A}{x-1} + \dfrac{B}{x+1} + \dfrac{Cx + D}{x^2 + 1}$$

(b) $\dfrac{t^4 + t^2 + 1}{(t^2 + 1)(t^2 + 4)^2} = \dfrac{At + B}{t^2 + 1} + \dfrac{Ct + D}{t^2 + 4} + \dfrac{Et + F}{(t^2 + 4)^2}$

7. $\displaystyle\int\dfrac{x}{x-6}\,dx = \int\dfrac{(x-6)+6}{x-6}\,dx = \int\left(1 + \dfrac{6}{x-6}\right)dx = x + 6\ln|x-6| + C$

9. $\dfrac{x-9}{(x+5)(x-2)} = \dfrac{A}{x+5} + \dfrac{B}{x-2}$. Multiply both sides by $(x+5)(x-2)$ to get $x - 9 = A(x-2) + B(x+5)(*)$, or

equivalently, $x - 9 = (A+B)x - 2A + 5B$. Equating coefficients of x on each side of the equation gives us $1 = A + B$ **(1)**

and equating constants gives us $-9 = -2A + 5B$ **(2)**. Adding two times **(1)** to **(2)** gives us $-7 = 7B$ $\Leftrightarrow$ $B = -1$ and

hence, $A = 2$. [Alternatively, to find the coefficients A and B, we may use substitution as follows: substitute 2 for x in $(*)$ to

get $-7 = 7B$ $\Leftrightarrow$ $B = -1$, then substitute -5 for x in $(*)$ to get $-14 = -7A$ $\Leftrightarrow$ $A = 2$.] Thus,

$$\int \frac{x-9}{(x+5)(x-2)}\,dx = \int \left(\frac{2}{x+5} + \frac{-1}{x-2} \right)dx = 2\ln|x+5| - \ln|x-2| + C.$$

11. $\dfrac{1}{x^2-1} = \dfrac{1}{(x+1)(x-1)} = \dfrac{A}{x+1} + \dfrac{B}{x-1}$. Multiply both sides by $(x+1)(x-1)$ to get $1 = A(x-1) + B(x+1)$.

Substituting 1 for x gives $1 = 2B$ $\Leftrightarrow$ $B = \frac{1}{2}$. Substituting -1 for x gives $1 = -2A$ $\Leftrightarrow$ $A = -\frac{1}{2}$. Thus,

$$\int_2^3 \frac{1}{x^2-1}\,dx = \int_2^3 \left(\frac{-1/2}{x+1} + \frac{1/2}{x-1} \right)dx = \left[-\tfrac{1}{2}\ln|x+1| + \tfrac{1}{2}\ln|x-1| \right]_2^3$$

$$= \left(-\tfrac{1}{2}\ln 4 + \tfrac{1}{2}\ln 2 \right) - \left(-\tfrac{1}{2}\ln 3 + \tfrac{1}{2}\ln 1 \right) = \tfrac{1}{2}(\ln 2 + \ln 3 - \ln 4) \quad \left[\text{or } \tfrac{1}{2}\ln\tfrac{3}{2}\right]$$

13. $\displaystyle\int \frac{ax}{x^2-bx}\,dx = \int \frac{ax}{x(x-b)}\,dx = \int \frac{a}{x-b}\,dx = a\ln|x-b| + C$

15. $\dfrac{x^3-2x^2-4}{x^3-2x^2} = 1 + \dfrac{-4}{x^2(x-2)}$. Write $\dfrac{-4}{x^2(x-2)} = \dfrac{A}{x} + \dfrac{B}{x^2} + \dfrac{C}{x-2}$. Multiplying both sides by $x^2(x-2)$ gives

$-4 = Ax(x-2) + B(x-2) + Cx^2$. Substituting 0 for x gives $-4 = -2B$ $\Leftrightarrow$ $B = 2$. Substituting 2 for x gives

$-4 = 4C$ $\Leftrightarrow$ $C = -1$. Equating coefficients of x^2, we get $0 = A + C$, so $A = 1$. Thus,

$$\int_3^4 \frac{x^3-2x^2-4}{x^3-2x^2}\,dx = \int_3^4 \left(1 + \frac{1}{x} + \frac{2}{x^2} - \frac{1}{x-2} \right)dx = \left[x + \ln|x| - \frac{2}{x} - \ln|x-2| \right]_3^4$$

$$= \left[\left(4 + \ln 4 - \tfrac{1}{2} - \ln 2 \right) - \left(3 + \ln 3 - \tfrac{2}{3} - 0 \right) \right] = \tfrac{7}{6} + \ln\tfrac{2}{3}$$

17. $\dfrac{4y^2-7y-12}{y(y+2)(y-3)} = \dfrac{A}{y} + \dfrac{B}{y+2} + \dfrac{C}{y-3}$ $\Rightarrow$ $4y^2 - 7y - 12 = A(y+2)(y-3) + By(y-3) + Cy(y+2)$. Setting

$y = 0$ gives $-12 = -6A$, so $A = 2$. Setting $y = -2$ gives $18 = 10B$, so $B = \frac{9}{5}$. Setting $y = 3$ gives $3 = 15C$,

so $C = \frac{1}{5}$. Now

$$\int_1^2 \frac{4y^2-7y-12}{y(y+2)(y-3)}\,dy = \int_1^2 \left(\frac{2}{y} + \frac{9/5}{y+2} + \frac{1/5}{y-3} \right)dy = \left[2\ln|y| + \tfrac{9}{5}\ln|y+2| + \tfrac{1}{5}\ln|y-3| \right]_1^2$$

$$= 2\ln 2 + \tfrac{9}{5}\ln 4 + \tfrac{1}{5}\ln 1 - 2\ln 1 - \tfrac{9}{5}\ln 3 - \tfrac{1}{5}\ln 2$$

$$= 2\ln 2 + \tfrac{18}{5}\ln 2 - \tfrac{1}{5}\ln 2 - \tfrac{9}{5}\ln 3 = \tfrac{27}{5}\ln 2 - \tfrac{9}{5}\ln 3 = \tfrac{9}{5}(3\ln 2 - \ln 3) = \tfrac{9}{5}\ln\tfrac{8}{3}$$

19. $\dfrac{1}{(x+5)^2(x-1)} = \dfrac{A}{x+5} + \dfrac{B}{(x+5)^2} + \dfrac{C}{x-1}$ $\Rightarrow$ $1 = A(x+5)(x-1) + B(x-1) + C(x+5)^2$.

Setting $x = -5$ gives $1 = -6B$, so $B = -\frac{1}{6}$. Setting $x = 1$ gives $1 = 36C$, so $C = \frac{1}{36}$. Setting $x = -2$ gives

$$1 = A(3)(-3) + B(-3) + C(3^2) = -9A - 3B + 9C = -9A + \tfrac{1}{2} + \tfrac{1}{4} = -9A + \tfrac{3}{4}, \text{ so } 9A = -\tfrac{1}{4} \text{ and } A = -\tfrac{1}{36}. \text{ Now}$$

$$\int \frac{1}{(x+5)^2(x-1)} \, dx = \int \left[\frac{-1/36}{x+5} - \frac{1/6}{(x+5)^2} + \frac{1/36}{x-1} \right] dx = -\frac{1}{36} \ln|x+5| + \frac{1}{6(x+5)} + \frac{1}{36} \ln|x-1| + C.$$

21. $\dfrac{5x^2 + 3x - 2}{x^3 + 2x^2} = \dfrac{5x^2 + 3x - 2}{x^2(x+2)} = \dfrac{A}{x} + \dfrac{B}{x^2} + \dfrac{C}{x+2}$. Multiply by $x^2(x+2)$ to get

$5x^2 + 3x - 2 = Ax(x+2) + B(x+2) + Cx^2$. Set $x = -2$ to get $C = 3$, and take

$x = 0$ to get $B = -1$. Equating the coefficients of x^2 gives $5 = A + C \;\Rightarrow\; A = 2$. So

$$\int \frac{5x^2 + 3x - 2}{x^3 + 2x^2} \, dx = \int \left(\frac{2}{x} - \frac{1}{x^2} + \frac{3}{x+2} \right) dx = 2\ln|x| + \frac{1}{x} + 3\ln|x+2| + C.$$

23. $\dfrac{10}{(x-1)(x^2+9)} = \dfrac{A}{x-1} + \dfrac{Bx+C}{x^2+9}$. Multiply both sides by $(x-1)(x^2+9)$ to get

$10 = A(x^2 + 9) + (Bx + C)(x - 1)$ (∗). Substituting 1 for x gives $10 = 10A \;\Leftrightarrow\; A = 1$.

Substituting 0 for x gives $10 = 9A - C \;\Rightarrow\; C = 9(1) - 10 = -1$.

The coefficients of the x^2-terms in (∗) must be equal, so $0 = A + B \;\Rightarrow\; B = -1$. Thus,

$$\int \frac{10}{(x-1)(x^2+9)} \, dx = \int \left(\frac{1}{x-1} + \frac{-x-1}{x^2+9} \right) dx = \int \left(\frac{1}{x-1} - \frac{x}{x^2+9} - \frac{1}{x^2+9} \right) dx$$

$$= \ln|x-1| - \tfrac{1}{2}\ln(x^2+9) \ [\text{let } u = x^2 + 9] - \tfrac{1}{3}\tan^{-1}\left(\tfrac{x}{3}\right) \ [\text{Formula 10}] + C$$

25. $\dfrac{x^3 + x^2 + 2x + 1}{(x^2+1)(x^2+2)} = \dfrac{Ax+B}{x^2+1} + \dfrac{Cx+D}{x^2+2}$. Multiply both sides by $(x^2+1)(x^2+2)$ to get

$x^3 + x^2 + 2x + 1 = (Ax + B)(x^2 + 2) + (Cx + D)(x^2 + 1) \;\Leftrightarrow\;$

$x^3 + x^2 + 2x + 1 = (Ax^3 + Bx^2 + 2Ax + 2B) + (Cx^3 + Dx^2 + Cx + D) \;\Leftrightarrow\;$

$x^3 + x^2 + 2x + 1 = (A + C)x^3 + (B + D)x^2 + (2A + C)x + (2B + D)$. Comparing coefficients gives us the following

system of equations:

$$A + C = 1 \quad \textbf{(1)} \qquad\qquad B + D = 1 \quad \textbf{(2)}$$
$$2A + C = 2 \quad \textbf{(3)} \qquad\qquad 2B + D = 1 \quad \textbf{(4)}$$

Subtracting equation **(1)** from equation **(3)** gives us $A = 1$, so $C = 0$. Subtracting equation **(2)** from equation **(4)** gives us

$B = 0$, so $D = 1$. Thus, $I = \displaystyle\int \frac{x^3 + x^2 + 2x + 1}{(x^2+1)(x^2+2)} \, dx = \int \left(\frac{x}{x^2+1} + \frac{1}{x^2+2} \right) dx$. For $\displaystyle\int \frac{x}{x^2+1} \, dx$, let $u = x^2 + 1$

so $du = 2x\,dx$ and then $\displaystyle\int \frac{x}{x^2+1} \, dx = \frac{1}{2}\int \frac{1}{u} \, du = \frac{1}{2}\ln|u| + C = \frac{1}{2}\ln(x^2+1) + C$. For $\displaystyle\int \frac{1}{x^2+2} \, dx$, use

Formula 10 with $a = \sqrt{2}$. So $\displaystyle\int \frac{1}{x^2+2} \, dx = \int \frac{1}{x^2+(\sqrt{2})^2} \, dx = \frac{1}{\sqrt{2}}\tan^{-1}\frac{x}{\sqrt{2}} + C$.

Thus, $I = \dfrac{1}{2}\ln(x^2+1) + \dfrac{1}{\sqrt{2}}\tan^{-1}\dfrac{x}{\sqrt{2}} + C$.

27. $\displaystyle \int \frac{x+4}{x^2+2x+5}\,dx = \int \frac{x+1}{x^2+2x+5}\,dx + \int \frac{3}{x^2+2x+5}\,dx = \frac{1}{2}\int \frac{(2x+2)\,dx}{x^2+2x+5} + \int \frac{3\,dx}{(x+1)^2+4}$

$$= \frac{1}{2}\ln\left|x^2+2x+5\right| + 3\int \frac{2\,du}{4(u^2+1)} \qquad \begin{bmatrix} \text{where } x+1=2u, \\ \text{and } dx = 2\,du \end{bmatrix}$$

$$= \frac{1}{2}\ln(x^2+2x+5) + \frac{3}{2}\tan^{-1}u + C = \frac{1}{2}\ln(x^2+2x+5) + \frac{3}{2}\tan^{-1}\left(\frac{x+1}{2}\right) + C$$

29. $\displaystyle \frac{1}{x^3-1} = \frac{1}{(x-1)(x^2+x+1)} = \frac{A}{x-1} + \frac{Bx+C}{x^2+x+1} \quad \Rightarrow \quad 1 = A(x^2+x+1) + (Bx+C)(x-1).$

Take $x=1$ to get $A = \frac{1}{3}$. Equating coefficients of x^2 and then comparing the constant terms, we get

$0 = \frac{1}{3} + B, 1 = \frac{1}{3} - C$, so $B = -\frac{1}{3}, C = -\frac{2}{3} \quad \Rightarrow$

$$\int \frac{1}{x^3-1}\,dx = \int \frac{\frac{1}{3}}{x-1}\,dx + \int \frac{-\frac{1}{3}x - \frac{2}{3}}{x^2+x+1}\,dx = \frac{1}{3}\ln|x-1| - \frac{1}{3}\int \frac{x+2}{x^2+x+1}\,dx$$

$$= \frac{1}{3}\ln|x-1| - \frac{1}{3}\int \frac{x+1/2}{x^2+x+1}\,dx - \frac{1}{3}\int \frac{(3/2)\,dx}{(x+1/2)^2+3/4}$$

$$= \frac{1}{3}\ln|x-1| - \frac{1}{6}\ln\left(x^2+x+1\right) - \frac{1}{2}\left(\frac{2}{\sqrt{3}}\right)\tan^{-1}\left(\frac{x+\frac{1}{2}}{\sqrt{3}/2}\right) + K$$

$$= \frac{1}{3}\ln|x-1| - \frac{1}{6}\ln(x^2+x+1) - \frac{1}{\sqrt{3}}\tan^{-1}\left(\frac{1}{\sqrt{3}}(2x+1)\right) + K$$

31. $\displaystyle \frac{1}{x(x^2+4)^2} = \frac{A}{x} + \frac{Bx+C}{x^2+4} + \frac{Dx+E}{(x^2+4)^2} \quad \Rightarrow \quad 1 = A(x^2+4)^2 + (Bx+C)x(x^2+4) + (Dx+E)x.$ Setting $x=0$

gives $1 = 16A$, so $A = \frac{1}{16}$. Now compare coefficients.

$$1 = \frac{1}{16}(x^4+8x^2+16) + (Bx^2+Cx)(x^2+4) + Dx^2 + Ex$$

$$1 = \frac{1}{16}x^4 + \frac{1}{2}x^2 + 1 + Bx^4 + Cx^3 + 4Bx^2 + 4Cx + Dx^2 + Ex$$

$$1 = \left(\frac{1}{16} + B\right)x^4 + Cx^3 + \left(\frac{1}{2} + 4B + D\right)x^2 + (4C+E)x + 1$$

So $B + \frac{1}{16} = 0 \quad \Rightarrow \quad B = -\frac{1}{16}, C = 0, \frac{1}{2} + 4B + D = 0 \quad \Rightarrow \quad D = -\frac{1}{4}$, and $4C + E = 0 \quad \Rightarrow \quad E = 0$. Thus,

$$\int \frac{dx}{x(x^2+4)^2} = \int \left(\frac{\frac{1}{16}}{x} + \frac{-\frac{1}{16}x}{x^2+4} + \frac{-\frac{1}{4}x}{(x^2+4)^2}\right)dx = \frac{1}{16}\ln|x| - \frac{1}{16}\cdot\frac{1}{2}\ln|x^2+4| - \frac{1}{4}\left(-\frac{1}{2}\right)\frac{1}{x^2+4} + C$$

$$= \frac{1}{16}\ln|x| - \frac{1}{32}\ln(x^2+4) + \frac{1}{8(x^2+4)} + C$$

33. $\displaystyle \int \frac{x-3}{(x^2+2x+4)^2}\,dx = \int \frac{x-3}{(x^2+2x+4)^2}\,dx = \int \frac{x-3}{[(x+1)^2+3]^2}\,dx = \int \frac{u-4}{(u^2+3)^2}\,du \qquad \text{[with } u = x+1]$

$$= \int \frac{u\,du}{(u^2+3)^2} - 4\int \frac{du}{(u^2+3)^2} = \frac{1}{2}\int \frac{dv}{v^2} - 4\int \frac{\sqrt{3}\sec^2\theta\,d\theta}{9\sec^4\theta} \qquad \begin{bmatrix} v = u^2+3 \text{ in the first integral;} \\ u = \sqrt{3}\tan\theta \text{ in the second} \end{bmatrix}$$

$$= \frac{-1}{(2v)} - \frac{4\sqrt{3}}{9}\int \cos^2\theta\,d\theta = \frac{-1}{2(u^2+3)} - \frac{2\sqrt{3}}{9}(\theta + \sin\theta\cos\theta) + C$$

$$= \frac{-1}{2(x^2+2x+4)} - \frac{2\sqrt{3}}{9}\left[\tan^{-1}\left(\frac{x+1}{\sqrt{3}}\right) + \frac{\sqrt{3}(x+1)}{x^2+2x+4}\right] + C$$

$$= \frac{-1}{2(x^2+2x+4)} - \frac{2\sqrt{3}}{9}\tan^{-1}\left(\frac{x+1}{\sqrt{3}}\right) - \frac{2(x+1)}{3(x^2+2x+4)} + C$$

35. Let $u = \sqrt{x}$, so $u^2 = x$ and $dx = 2u\,du$. Thus,

$$\int_9^{16} \frac{\sqrt{x}}{x-4}\,dx = \int_3^4 \frac{u}{u^2-4} 2u\,du = 2\int_3^4 \frac{u^2}{u^2-4}\,du = 2\int_3^4 \left(1 + \frac{4}{u^2-4}\right)du \quad \text{[by long division]}$$

$$= 2 + 8\int_3^4 \frac{du}{(u+2)(u-2)} \quad (\ast)$$

Multiply $\dfrac{1}{(u+2)(u-2)} = \dfrac{A}{u+2} + \dfrac{B}{u-2}$ by $(u+2)(u-2)$ to get $1 = A(u-2) + B(u+2)$. Equating coefficients we

get $A + B = 0$ and $-2A + 2B = 1$. Solving gives us $B = \frac{1}{4}$ and $A = -\frac{1}{4}$, so $\dfrac{1}{(u+2)(u-2)} = \dfrac{-1/4}{u+2} + \dfrac{1/4}{u-2}$ and $(\ast)$ is

$$2 + 8\int_3^4 \left(\frac{-1/4}{u+2} + \frac{1/4}{u-2}\right)du = 2 + 8\left[-\tfrac{1}{4}\ln|u+2| + \tfrac{1}{4}\ln|u-2|\right]_3^4$$

$$= 2 + \left[2\ln|u-2| - 2\ln|u+2|\right]_3^4 = 2 + 2\left[\ln\left|\frac{u-2}{u+2}\right|\right]_3^4$$

$$= 2 + 2\left(\ln\tfrac{2}{6} - \ln\tfrac{1}{5}\right) = 2 + 2\ln\tfrac{2/6}{1/5}$$

$$= 2 + 2\ln\tfrac{5}{3} \text{ or } 2 + \ln\left(\tfrac{5}{3}\right)^2 = 2 + \ln\tfrac{25}{9}$$

37. Let $u = e^x$. Then $x = \ln u$, $dx = \dfrac{du}{u}$ $\Rightarrow$

$$\int \frac{e^{2x}\,dx}{e^{2x} + 3e^x + 2} = \int \frac{u^2\,(du/u)}{u^2 + 3u + 2} = \int \frac{u\,du}{(u+1)(u+2)} = \int \left[\frac{-1}{u+1} + \frac{2}{u+2}\right]du$$

$$= 2\ln|u+2| - \ln|u+1| + C = \ln\left[\frac{(e^x+2)^2}{e^x+1}\right] + C$$

39.

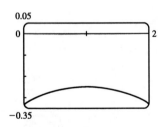

From the graph, we see that the integral will be negative, and we guess that the area is about the same as that of a rectangle with width 2 and height 0.3, so we estimate the integral to be $-(2 \cdot 0.3) = -0.6$. Now

$$\frac{1}{x^2 - 2x - 3} = \frac{1}{(x-3)(x+1)} = \frac{A}{x-3} + \frac{B}{x+1} \quad \Leftrightarrow$$

$$1 = (A+B)x + A - 3B, \text{ so } A = -B \text{ and } A - 3B = 1 \quad \Leftrightarrow \quad A = \tfrac{1}{4}$$

and $B = -\frac{1}{4}$, so the integral becomes

$$\int_0^2 \frac{dx}{x^2 - 2x - 3} = \frac{1}{4}\int_0^2 \frac{dx}{x-3} - \frac{1}{4}\int_0^2 \frac{dx}{x+1} = \frac{1}{4}\left[\ln|x-3| - \ln|x+1|\right]_0^2 = \frac{1}{4}\left[\ln\left|\frac{x-3}{x+1}\right|\right]_0^2$$

$$= \tfrac{1}{4}\left(\ln\tfrac{1}{3} - \ln 3\right) = -\tfrac{1}{2}\ln 3 \approx -0.55$$

41. $\dfrac{P+S}{P[(r-1)P - S]} = \dfrac{A}{P} + \dfrac{B}{(r-1)P - S}$ $\Rightarrow$ $P + S = A[(r-1)P - S] + BP = [(r-1)A + B]P - AS$ $\Rightarrow$

$(r-1)A + B = 1, -A = 1$ $\Rightarrow$ $A = -1, B = r$. Now

$$t = \int \frac{P+S}{P[(r-1)P - S]}\,dP = \int \left[\frac{-1}{P} + \frac{r}{(r-1)P - S}\right]dP = -\int \frac{dP}{P} + \frac{r}{r-1}\int \frac{r-1}{(r-1)P - S}\,dP$$

so $t = -\ln P + \dfrac{r}{r-1}\ln|(r-1)P - S| + C$. Here $r = 0.10$ and $S = 900$, so

$$t = -\ln P + \tfrac{0.1}{-0.9}\ln|-0.9P - 900| + C = -\ln P - \tfrac{1}{9}\ln\big(|-1|\,|0.9P + 900|\,\big) = -\ln P - \tfrac{1}{9}\ln(0.9P + 900) + C.$$

When $t = 0$, $P = 10{,}000$, so $0 = -\ln 10{,}000 - \tfrac{1}{9}\ln(9900) + C$. Thus, $C = \ln 10{,}000 + \tfrac{1}{9}\ln 9900$ [≈ 10.2326],

so our equation becomes

$$t = \ln 10{,}000 - \ln P + \tfrac{1}{9}\ln 9900 - \tfrac{1}{9}\ln(0.9P + 900) = \ln\frac{10{,}000}{P} + \frac{1}{9}\ln\frac{9900}{0.9P + 900}$$

$$= \ln\frac{10{,}000}{P} + \frac{1}{9}\ln\frac{1100}{0.1P + 100} = \ln\frac{10{,}000}{P} + \frac{1}{9}\ln\frac{11{,}000}{P + 1000}$$

43. (a) In Maple, we define $f(x)$, and then use `convert(f,parfrac,x);` to obtain

$$f(x) = \frac{24{,}110/4879}{5x + 2} - \frac{668/323}{2x + 1} - \frac{9438/80{,}155}{3x - 7} + \frac{(22{,}098x + 48{,}935)/260{,}015}{x^2 + x + 5}$$

In Mathematica, we use the command `Apart`, and in Derive, we use `Expand`.

(b) $\displaystyle\int f(x)\,dx = \tfrac{24{,}110}{4879}\cdot\tfrac{1}{5}\ln|5x + 2| - \tfrac{668}{323}\cdot\tfrac{1}{2}\ln|2x + 1| - \tfrac{9438}{80{,}155}\cdot\tfrac{1}{3}\ln|3x - 7|$

$$+ \frac{1}{260{,}015}\int\frac{22{,}098\left(x + \tfrac{1}{2}\right) + 37{,}886}{\left(x + \tfrac{1}{2}\right)^2 + \tfrac{19}{4}}\,dx + C$$

$$= \tfrac{24{,}110}{4879}\cdot\tfrac{1}{5}\ln|5x + 2| - \tfrac{668}{323}\cdot\tfrac{1}{2}\ln|2x + 1| - \tfrac{9438}{80{,}155}\cdot\tfrac{1}{3}\ln|3x - 7|$$

$$+ \tfrac{1}{260{,}015}\left[22{,}098\cdot\tfrac{1}{2}\ln\left(x^2 + x + 5\right) + 37{,}886\cdot\sqrt{\tfrac{4}{19}}\,\tan^{-1}\left(\tfrac{1}{\sqrt{19/4}}\left(x + \tfrac{1}{2}\right)\right)\right] + C$$

$$= \tfrac{4822}{4879}\ln|5x + 2| - \tfrac{334}{323}\ln|2x + 1| - \tfrac{3146}{80{,}155}\ln|3x - 7| + \tfrac{11{,}049}{260{,}015}\ln\left(x^2 + x + 5\right)$$

$$+ \tfrac{75{,}772}{260{,}015\sqrt{19}}\,\tan^{-1}\left[\tfrac{1}{\sqrt{19}}(2x + 1)\right] + C$$

Using a CAS, we get

$$\frac{4822\ln(5x + 2)}{4879} - \frac{334\ln(2x + 1)}{323} - \frac{3146\ln(3x - 7)}{80{,}155}$$

$$+ \frac{11{,}049\ln(x^2 + x + 5)}{260{,}015} + \frac{3988\sqrt{19}}{260{,}015}\,\tan^{-1}\left[\frac{\sqrt{19}}{19}(2x + 1)\right]$$

The main difference in this answer is that the absolute value signs and the constant of integration have been omitted. Also, the fractions have been reduced and the denominators rationalized.

45. There are only finitely many values of x where $Q(x) = 0$ (assuming that Q is not the zero polynomial). At all other values of x, $F(x)/Q(x) = G(x)/Q(x)$, so $F(x) = G(x)$. In other words, the values of F and G agree at all except perhaps finitely many values of x. By continuity of F and G, the polynomials F and G must agree at those values of x too.

More explicitly: If a is a value of x such that $Q(a) = 0$, then $Q(x) \neq 0$ for all x sufficiently close to a. Thus,

$$\begin{aligned}
F(a) &= \lim_{x\to a} F(x) && \text{[by continuity of } F] \\
&= \lim_{x\to a} G(x) && \text{[whenever } Q(x) \neq 0] \\
&= G(a) && \text{[by continuity of } G]
\end{aligned}$$

H POLAR COORDINATES

H.1 Curves in Polar Coordinates

1. (a) $\left(2, \frac{\pi}{3}\right)$

By adding 2π to $\frac{\pi}{3}$, we obtain the point $\left(2, \frac{7\pi}{3}\right)$. The direction opposite $\frac{\pi}{3}$ is $\frac{4\pi}{3}$, so $\left(-2, \frac{4\pi}{3}\right)$ is a point that satisfies the $r < 0$ requirement.

(b) $\left(1, -\frac{3\pi}{4}\right)$

$r > 0$: $\left(1, -\frac{3\pi}{4} + 2\pi\right) = \left(1, \frac{5\pi}{4}\right)$

$r < 0$: $\left(-1, -\frac{3\pi}{4} + \pi\right) = \left(-1, \frac{\pi}{4}\right)$

(c) $\left(-1, \frac{\pi}{2}\right)$

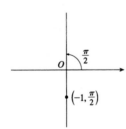

$r > 0$: $\left(-(-1), \frac{\pi}{2} + \pi\right) = \left(1, \frac{3\pi}{2}\right)$

$r < 0$: $\left(-1, \frac{\pi}{2} + 2\pi\right) = \left(-1, \frac{5\pi}{2}\right)$

3. (a)

$x = 1\cos\pi = 1(-1) = -1$ and
$y = 1\sin\pi = 1(0) = 0$ give us
the Cartesian coordinates $(-1, 0)$.

(b)

$x = 2\cos\left(-\frac{2\pi}{3}\right) = 2\left(-\frac{1}{2}\right) = -1$ and

$y = 2\sin\left(-\frac{2\pi}{3}\right) = 2\left(-\frac{\sqrt{3}}{2}\right) = -\sqrt{3}$

give us $\left(-1, -\sqrt{3}\right)$.

(c)

$x = -2\cos\frac{3\pi}{4} = -2\left(-\frac{\sqrt{2}}{2}\right) = \sqrt{2}$ and

$y = -2\sin\frac{3\pi}{4} = -2\left(\frac{\sqrt{2}}{2}\right) = -\sqrt{2}$

gives us $\left(\sqrt{2}, -\sqrt{2}\right)$.

$\left(-2, \frac{3\pi}{4}\right)$

5. (a) $x = 2$ and $y = -2$ $\Rightarrow$ $r = \sqrt{2^2 + (-2)^2} = 2\sqrt{2}$ and $\theta = \tan^{-1}\left(\frac{-2}{2}\right) = -\frac{\pi}{4}$. Since $(2, -2)$ is in the fourth

quadrant, the polar coordinates are (i) $\left(2\sqrt{2}, \frac{7\pi}{4}\right)$ and (ii) $\left(-2\sqrt{2}, \frac{3\pi}{4}\right)$.

(b) $x = -1$ and $y = \sqrt{3}$ $\Rightarrow$ $r = \sqrt{(-1)^2 + \left(\sqrt{3}\right)^2} = 2$ and $\theta = \tan^{-1}\left(\frac{\sqrt{3}}{-1}\right) = \frac{2\pi}{3}$. Since $\left(-1, \sqrt{3}\right)$ is in the second

quadrant, the polar coordinates are (i) $\left(2, \frac{2\pi}{3}\right)$ and (ii) $\left(-2, \frac{5\pi}{3}\right)$.

7. The curves $r = 1$ and $r = 2$ represent circles with center O and radii 1 and 2. The region in the plane

satisfying $1 \leq r \leq 2$ consists of both circles and the shaded region between them in the figure.

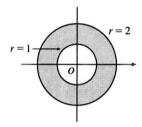

9. The region satisfying $0 \leq r < 4$ and $-\pi/2 \leq \theta < \pi/6$ does

not include the circle $r = 4$ nor the line $\theta = \frac{\pi}{6}$.

11. $2 < r < 3$, $\frac{5\pi}{3} \leq \theta \leq \frac{7\pi}{3}$

13. $r = 3\sin\theta$ $\Rightarrow$ $r^2 = 3r\sin\theta$ $\Leftrightarrow$ $x^2 + y^2 = 3y$ $\Leftrightarrow$ $x^2 + \left(y - \frac{3}{2}\right)^2 = \left(\frac{3}{2}\right)^2$, a circle of radius $\frac{3}{2}$ centered at $\left(0, \frac{3}{2}\right)$.

The first two equations are actually equivalent since $r^2 = 3r\sin\theta$ $\Rightarrow$ $r(r - 3\sin\theta) = 0$ $\Rightarrow$ $r = 0$ or $r = 3\sin\theta$. But

$r = 3\sin\theta$ gives the point $r = 0$ (the pole) when $\theta = 0$. Thus, the single equation $r = 3\sin\theta$ is equivalent to the compound

condition $(r = 0$ or $r = 3\sin\theta)$.

15. $r = \csc\theta$ $\Leftrightarrow$ $r = \dfrac{1}{\sin\theta}$ $\Leftrightarrow$ $r\sin\theta = 1$ $\Leftrightarrow$ $y = 1$, a horizontal line 1 unit above the x-axis.

17. $x = -y^2$ $\Leftrightarrow$ $r\cos\theta = -r^2\sin^2\theta$ $\Leftrightarrow$ $\cos\theta = -r\sin^2\theta$ $\Leftrightarrow$ $r = -\dfrac{\cos\theta}{\sin^2\theta} = -\cot\theta\csc\theta$.

19. $x^2 + y^2 = 2cx \iff r^2 = 2cr\cos\theta \iff r^2 - 2cr\cos\theta = 0 \iff r(r - 2c\cos\theta) = 0 \iff r = 0$ or $r = 2c\cos\theta$.

$r = 0$ is included in $r = 2c\cos\theta$ when $\theta = \frac{\pi}{2} + n\pi$, so the curve is represented by the single equation $r = 2c\cos\theta$.

21. (a) The description leads immediately to the polar equation $\theta = \frac{\pi}{6}$, and the Cartesian equation $y = \tan\left(\frac{\pi}{6}\right)x = \frac{1}{\sqrt{3}}x$ is

slightly more difficult to derive.

(b) The easier description here is the Cartesian equation $x = 3$.

23. $\theta = -\pi/6$

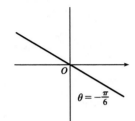

25. $r = \sin\theta \iff r^2 = r\sin\theta \iff x^2 + y^2 = y \iff$

$x^2 + \left(y - \frac{1}{2}\right)^2 = \left(\frac{1}{2}\right)^2$. The reasoning here is the same

as in Exercise 13. This is a circle of radius $\frac{1}{2}$ centered at $\left(0, \frac{1}{2}\right)$.

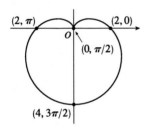

27. $r = 2(1 - \sin\theta)$. This curve is a cardioid.

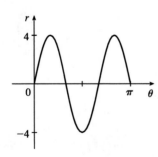

29. $r = \theta, \quad \theta \geq 0$

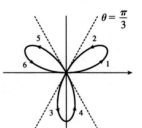

31. $r = 4\sin 3\theta$

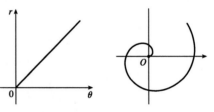

33. $r = 2\cos 4\theta$

35. $r = 1 - 2\sin\theta$

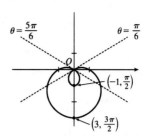

37. $r^2 = 9\sin 2\theta$

39. $r = 2\cos\left(\frac{3}{2}\theta\right)$

41. $r = 1 + 2\cos 2\theta$

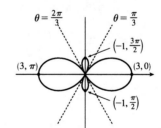

43. For $\theta = 0$, π, and 2π, r has its minimum value of about 0.5. For $\theta = \frac{\pi}{2}$ and $\frac{3\pi}{2}$, r attains its maximum value of 2.

We see that the graph has a similar shape for $0 \le \theta \le \pi$ and $\pi \le \theta \le 2\pi$.

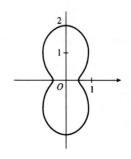

45. $x = (r)\cos\theta = (4 + 2\sec\theta)\cos\theta = 4\cos\theta + 2$. Now, $r \to \infty \Rightarrow$

$(4 + 2\sec\theta) \to \infty \Rightarrow \theta \to \left(\frac{\pi}{2}\right)^{-}$ or $\theta \to \left(\frac{3\pi}{2}\right)^{+}$ [since we need only

consider $0 \leq \theta < 2\pi$], so $\lim\limits_{r \to \infty} x = \lim\limits_{\theta \to \pi/2^{-}} (4\cos\theta + 2) = 2$. Also,

$r \to -\infty \Rightarrow (4 + 2\sec\theta) \to -\infty \Rightarrow \theta \to \left(\frac{\pi}{2}\right)^{+}$ or $\theta \to \left(\frac{3\pi}{2}\right)^{-}$, so

$\lim\limits_{r \to -\infty} x = \lim\limits_{\theta \to \pi/2^{+}} (4\cos\theta + 2) = 2$. Therefore, $\lim\limits_{r \to \pm\infty} x = 2 \Rightarrow x = 2$ is a vertical asymptote.

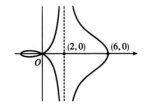

47. (a) We see that the curve $r = 1 + c\sin\theta$ crosses itself at the origin, where $r = 0$ (in fact the inner loop corresponds to

negative r-values,) so we solve the equation of the limaçon for $r = 0 \Leftrightarrow c\sin\theta = -1 \Leftrightarrow \sin\theta = -1/c$. Now if

$|c| < 1$, then this equation has no solution and hence there is no inner loop. But if $c < -1$, then on the interval $(0, 2\pi)$

the equation has the two solutions $\theta = \sin^{-1}(-1/c)$ and $\theta = \pi - \sin^{-1}(-1/c)$, and if $c > 1$, the solutions are

$\theta = \pi + \sin^{-1}(1/c)$ and $\theta = 2\pi - \sin^{-1}(1/c)$. In each case, $r < 0$ for θ between the two solutions, indicating a loop.

(b) For $0 < c < 1$, the dimple (if it exists) is characterized by the fact that y has a local maximum at $\theta = \frac{3\pi}{2}$. So we

determine for what c-values $\dfrac{d^2y}{d\theta^2}$ is negative at $\theta = \frac{3\pi}{2}$, since by the Second Derivative Test this indicates a maximum:

$y = r\sin\theta = \sin\theta + c\sin^2\theta \Rightarrow \dfrac{dy}{d\theta} = \cos\theta + 2c\sin\theta\cos\theta = \cos\theta + c\sin 2\theta \Rightarrow \dfrac{d^2y}{d\theta^2} = -\sin\theta + 2c\cos 2\theta$.

At $\theta = \frac{3\pi}{2}$, this is equal to $-(-1) + 2c(-1) = 1 - 2c$, which is negative only for $c > \frac{1}{2}$. A similar argument shows that

for $-1 < c < 0$, y only has a local minimum at $\theta = \frac{\pi}{2}$ (indicating a dimple) for $c < -\frac{1}{2}$.

49. $r = 1/\theta \Rightarrow x = r\cos\theta = (\cos\theta)/\theta$, $y = r\sin\theta = (\sin\theta)/\theta \Rightarrow$

$$\dfrac{dy}{dx} = \dfrac{dy/d\theta}{dx/d\theta} = \dfrac{\sin\theta(-1/\theta^2) + (1/\theta)\cos\theta}{\cos\theta(-1/\theta^2) - (1/\theta)\sin\theta} \cdot \dfrac{\theta^2}{\theta^2} = \dfrac{-\sin\theta + \theta\cos\theta}{-\cos\theta - \theta\sin\theta}$$

When $\theta = \pi$, $\dfrac{dy}{dx} = \dfrac{-0 + \pi(-1)}{-(-1) - \pi(0)} = \dfrac{-\pi}{1} = -\pi$.

51. $r = \cos 2\theta \Rightarrow x = r\cos\theta = \cos 2\theta\cos\theta$, $y = r\sin\theta = \cos 2\theta\sin\theta \Rightarrow$

$$\dfrac{dy}{dx} = \dfrac{dy/d\theta}{dx/d\theta} = \dfrac{\cos 2\theta\cos\theta + \sin\theta(-2\sin 2\theta)}{\cos 2\theta(-\sin\theta) + \cos\theta(-2\sin 2\theta)}$$

When $\theta = \dfrac{\pi}{4}$, $\dfrac{dy}{dx} = \dfrac{0(\sqrt{2}/2) + (\sqrt{2}/2)(-2)}{0(-\sqrt{2}/2) + (\sqrt{2}/2)(-2)} = \dfrac{-\sqrt{2}}{-\sqrt{2}} = 1$.

53. $r = 3\cos\theta \Rightarrow x = r\cos\theta = 3\cos\theta\cos\theta$, $y = r\sin\theta = 3\cos\theta\sin\theta \Rightarrow$

$\frac{dy}{d\theta} = -3\sin^2\theta + 3\cos^2\theta = 3\cos 2\theta = 0 \Rightarrow 2\theta = \frac{\pi}{2}$ or $\frac{3\pi}{2} \Leftrightarrow \theta = \frac{\pi}{4}$ or $\frac{3\pi}{4}$.

So the tangent is horizontal at $\left(\frac{3}{\sqrt{2}}, \frac{\pi}{4}\right)$ and $\left(-\frac{3}{\sqrt{2}}, \frac{3\pi}{4}\right)$ $\left[\text{same as } \left(\frac{3}{\sqrt{2}}, -\frac{\pi}{4}\right)\right]$.

$\frac{dx}{d\theta} = -6\sin\theta\cos\theta = -3\sin 2\theta = 0 \Rightarrow 2\theta = 0$ or $\pi \Leftrightarrow \theta = 0$ or $\frac{\pi}{2}$. So the tangent is vertical at $(3, 0)$ and $\left(0, \frac{\pi}{2}\right)$.

55. $r = 1 + \cos\theta$ $\Rightarrow$ $x = r\cos\theta = \cos\theta\,(1 + \cos\theta)$, $y = r\sin\theta = \sin\theta\,(1 + \cos\theta)$ $\Rightarrow$

$\frac{dy}{d\theta} = (1 + \cos\theta)\cos\theta - \sin^2\theta = 2\cos^2\theta + \cos\theta - 1 = (2\cos\theta - 1)(\cos\theta + 1) = 0$ $\Rightarrow$ $\cos\theta = \frac{1}{2}$ or -1 $\Rightarrow$

$\theta = \frac{\pi}{3}$, π, or $\frac{5\pi}{3}$ $\Rightarrow$ horizontal tangent at $\left(\frac{3}{2}, \frac{\pi}{3}\right)$, $(0, \pi)$, and $\left(\frac{3}{2}, \frac{5\pi}{3}\right)$.

$\frac{dx}{d\theta} = -(1 + \cos\theta)\sin\theta - \cos\theta\sin\theta = -\sin\theta\,(1 + 2\cos\theta) = 0$ $\Rightarrow$ $\sin\theta = 0$ or $\cos\theta = -\frac{1}{2}$ $\Rightarrow$

$\theta = 0$, π, $\frac{2\pi}{3}$, or $\frac{4\pi}{3}$ $\Rightarrow$ vertical tangent at $(2, 0)$, $\left(\frac{1}{2}, \frac{2\pi}{3}\right)$, and $\left(\frac{1}{2}, \frac{4\pi}{3}\right)$.

Note that the tangent is horizontal, not vertical when $\theta = \pi$, since $\lim\limits_{\theta \to \pi} \dfrac{dy/d\theta}{dx/d\theta} = 0$.

57. $r = a\sin\theta + b\cos\theta$ $\Rightarrow$ $r^2 = ar\sin\theta + br\cos\theta$ $\Rightarrow$ $x^2 + y^2 = ay + bx$ $\Rightarrow$

$x^2 - bx + \left(\frac{1}{2}b\right)^2 + y^2 - ay + \left(\frac{1}{2}a\right)^2 = \left(\frac{1}{2}b\right)^2 + \left(\frac{1}{2}a\right)^2$ $\Rightarrow$ $\left(x - \frac{1}{2}b\right)^2 + \left(y - \frac{1}{2}a\right)^2 = \frac{1}{4}(a^2 + b^2)$, and this is a circle

with center $\left(\frac{1}{2}b, \frac{1}{2}a\right)$ and radius $\frac{1}{2}\sqrt{a^2 + b^2}$.

Note for Exercises 59–62: Maple is able to plot polar curves using the `polarplot` command, or using the `coords=polar` option in a regular `plot` command. In Mathematica, use `PolarPlot`. In Derive, change to `Polar` under `Options State`. If your graphing device cannot plot polar equations, you must convert to parametric equations. For example, in Exercise 59, $x = r\cos\theta = [1 + 2\sin(\theta/2)]\cos\theta$, $y = r\sin\theta = [1 + 2\sin(\theta/2)]\sin\theta$.

59. $r = e^{\sin\theta} - 2\cos(4\theta)$.

The parameter interval is $[0, 2\pi]$.

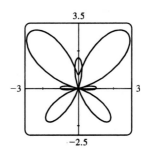

61. $r = 2 - 5\sin(\theta/6)$.

The parameter interval is $[-6\pi, 6\pi]$.

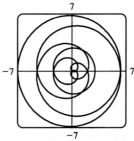

63. It appears that the graph of $r = 1 + \sin\left(\theta - \frac{\pi}{6}\right)$ is the same shape as the graph of $r = 1 + \sin\theta$, but rotated counterclockwise about the origin by $\frac{\pi}{6}$. Similarly, the graph of $r = 1 + \sin\left(\theta - \frac{\pi}{3}\right)$ is rotated by $\frac{\pi}{3}$. In general, the graph of $r = f(\theta - \alpha)$ is the same shape as that of $r = f(\theta)$, but rotated counterclockwise through α about the origin. That is, for any point (r_0, θ_0) on the curve $r = f(\theta)$, the point $(r_0, \theta_0 + \alpha)$ is on the curve $r = f(\theta - \alpha)$, since $r_0 = f(\theta_0) = f((\theta_0 + \alpha) - \alpha)$.

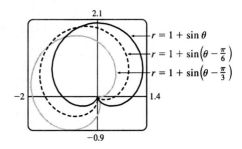

65. (a) $r = \sin n\theta$.

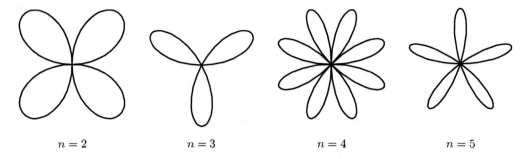

$$n = 2 \qquad n = 3 \qquad n = 4 \qquad n = 5$$

From the graphs, it seems that when n is even, the number of loops in the curve (called a rose) is $2n$, and when n is odd, the number of loops is simply n. This is because in the case of n odd, every point on the graph is traversed twice, due to the fact that

$$r(\theta + \pi) = \sin[n(\theta + \pi)] = \sin n\theta \cos n\pi + \cos n\theta \sin n\pi = \begin{cases} \sin n\theta & \text{if } n \text{ is even} \\ -\sin n\theta & \text{if } n \text{ is odd} \end{cases}$$

(b) The graph of $r = |\sin n\theta|$ has $2n$ loops whether n is odd or even, since $r(\theta + \pi) = r(\theta)$.

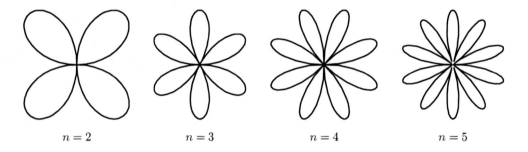

$$n = 2 \qquad n = 3 \qquad n = 4 \qquad n = 5$$

67. $r = \dfrac{1 - a\cos\theta}{1 + a\cos\theta}$. We start with $a = 0$, since in this case the curve is simply the circle $r = 1$.

As a increases, the graph moves to the left, and its right side becomes flattened. As a increases through about 0.4, the right side seems to grow a dimple, which upon closer investigation (with narrower θ-ranges) seems to appear at $a \approx 0.42$ [the actual value is $\sqrt{2} - 1$]. As $a \to 1$, this dimple becomes more pronounced, and the curve begins to stretch out horizontally, until at $a = 1$ the denominator vanishes at $\theta = \pi$, and the dimple becomes an actual cusp. For $a > 1$ we must choose our parameter interval carefully, since $r \to \infty$ as $1 + a\cos\theta \to 0 \iff \theta \to \pm\cos^{-1}(-1/a)$. As a increases from 1, the curve splits into two parts. The left part has a loop, which grows larger as a increases, and the right part grows broader vertically, and its left tip develops a dimple when $a \approx 2.42$ [actually, $\sqrt{2} + 1$]. As a increases, the dimple grows more and more pronounced. If $a < 0$, we get the same graph as we do for the corresponding positive a-value, but with a rotation through π about the pole, as happened when c was replaced with $-c$ in Exercise 66.

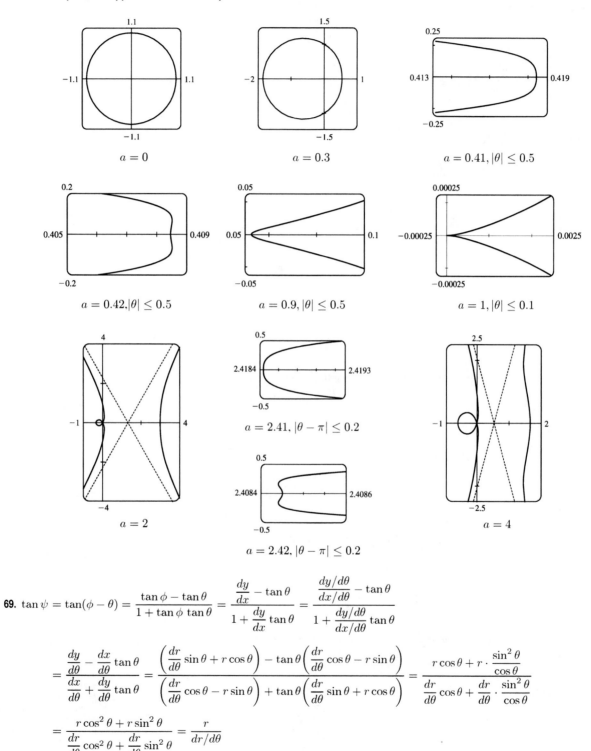

$a = 0$

$a = 0.3$

$a = 0.41, |\theta| \leq 0.5$

$a = 0.42, |\theta| \leq 0.5$

$a = 0.9, |\theta| \leq 0.5$

$a = 1, |\theta| \leq 0.1$

$a = 2$

$a = 2.41, |\theta - \pi| \leq 0.2$

$a = 2.42, |\theta - \pi| \leq 0.2$

$a = 4$

69. $\tan \psi = \tan(\phi - \theta) = \dfrac{\tan \phi - \tan \theta}{1 + \tan \phi \tan \theta} = \dfrac{\dfrac{dy}{dx} - \tan \theta}{1 + \dfrac{dy}{dx} \tan \theta} = \dfrac{\dfrac{dy/d\theta}{dx/d\theta} - \tan \theta}{1 + \dfrac{dy/d\theta}{dx/d\theta} \tan \theta}$

$= \dfrac{\dfrac{dy}{d\theta} - \dfrac{dx}{d\theta} \tan \theta}{\dfrac{dx}{d\theta} + \dfrac{dy}{d\theta} \tan \theta} = \dfrac{\left(\dfrac{dr}{d\theta} \sin \theta + r \cos \theta \right) - \tan \theta \left(\dfrac{dr}{d\theta} \cos \theta - r \sin \theta \right)}{\left(\dfrac{dr}{d\theta} \cos \theta - r \sin \theta \right) + \tan \theta \left(\dfrac{dr}{d\theta} \sin \theta + r \cos \theta \right)} = \dfrac{r \cos \theta + r \cdot \dfrac{\sin^2 \theta}{\cos \theta}}{\dfrac{dr}{d\theta} \cos \theta + \dfrac{dr}{d\theta} \cdot \dfrac{\sin^2 \theta}{\cos \theta}}$

$= \dfrac{r \cos^2 \theta + r \sin^2 \theta}{\dfrac{dr}{d\theta} \cos^2 \theta + \dfrac{dr}{d\theta} \sin^2 \theta} = \dfrac{r}{dr/d\theta}$

H.2 Areas and Lengths in Polar Coordinates

1. $r = \theta^2$, $0 \le \theta \le \frac{\pi}{4}$. $A = \int_0^{\pi/4} \frac{1}{2} r^2\, d\theta = \int_0^{\pi/4} \frac{1}{2} (\theta^2)^2\, d\theta = \int_0^{\pi/4} \frac{1}{2}\theta^4\, d\theta = \left[\frac{1}{10}\theta^5\right]_0^{\pi/4} = \frac{1}{10}\left(\frac{\pi}{4}\right)^5 = \frac{1}{10,240}\pi^5$

3. $r = \sin\theta$, $\frac{\pi}{3} \le \theta \le \frac{2\pi}{3}$.

$$A = \int_{\pi/3}^{2\pi/3} \frac{1}{2}\sin^2\theta\, d\theta = \frac{1}{4}\int_{\pi/3}^{2\pi/3}(1 - \cos 2\theta)\, d\theta = \frac{1}{4}\left[\theta - \frac{1}{2}\sin 2\theta\right]_{\pi/3}^{2\pi/3} = \frac{1}{4}\left[\frac{2\pi}{3} - \frac{1}{2}\sin\frac{4\pi}{3} - \frac{\pi}{3} + \frac{1}{2}\sin\frac{2\pi}{3}\right]$$

$$= \frac{1}{4}\left[\frac{2\pi}{3} - \frac{1}{2}\left(-\frac{\sqrt{3}}{2}\right) - \frac{\pi}{3} + \frac{1}{2}\left(\frac{\sqrt{3}}{2}\right)\right] = \frac{1}{4}\left(\frac{\pi}{3} + \frac{\sqrt{3}}{2}\right) = \frac{\pi}{12} + \frac{\sqrt{3}}{8}$$

5. $r = \sqrt{\theta}$, $0 \le \theta \le 2\pi$. $A = \int_0^{2\pi} \frac{1}{2} r^2\, d\theta = \int_0^{2\pi} \frac{1}{2}\left(\sqrt{\theta}\right)^2 d\theta = \int_0^{2\pi} \frac{1}{2}\theta\, d\theta = \left[\frac{1}{4}\theta^2\right]_0^{2\pi} = \pi^2$

7. $r = 4 + 3\sin\theta$, $-\frac{\pi}{2} \le \theta \le \frac{\pi}{2}$.

$$A = \int_{-\pi/2}^{\pi/2} \frac{1}{2}((4 + 3\sin\theta)^2\, d\theta = \frac{1}{2}\int_{-\pi/2}^{\pi/2}(16 + 24\sin\theta + 9\sin^2\theta)\, d\theta$$

$$= \frac{1}{2}\int_{-\pi/2}^{\pi/2}(16 + 9\sin^2\theta)\, d\theta \qquad \text{[by Theorem 5.5.6(a)]}$$

$$= \frac{1}{2}\cdot 2\int_0^{\pi/2}\left[16 + 9\cdot\frac{1}{2}(1 - \cos 2\theta)\right]d\theta \qquad \text{[by Theorem 5.5.6(a)]}$$

$$= \int_0^{\pi/2}\left(\frac{41}{2} - \frac{9}{2}\cos 2\theta\right)d\theta = \left[\frac{41}{2}\theta - \frac{9}{4}\sin 2\theta\right]_0^{\pi/2} = \left(\frac{41\pi}{4} - 0\right) - (0 - 0) = \frac{41\pi}{4}$$

9. The curve goes through the pole when $\theta = \pi/4$, so we'll find the area for $0 \le \theta \le \pi/4$ and multiply it by 4.

$A = 4\int_0^{\pi/4} \frac{1}{2} r^2\, d\theta = 2\int_0^{\pi/4}(4\cos 2\theta)\, d\theta$

$= 8\int_0^{\pi/4}\cos 2\theta\, d\theta = 4\left[\sin 2\theta\right]_0^{\pi/4} = 4$

11. One-sixth of the area lies above the polar axis and is bounded by the curve $r = 2\cos 3\theta$ for $\theta = 0$ to $\theta = \pi/6$.

$$A = 6\int_0^{\pi/6} \frac{1}{2}(2\cos 3\theta)^2\, d\theta = 12\int_0^{\pi/6}\cos^2 3\theta\, d\theta$$

$$= \frac{12}{2}\int_0^{\pi/6}(1 + \cos 6\theta)\, d\theta$$

$$= 6\left[\theta + \frac{1}{6}\sin 6\theta\right]_0^{\pi/6} = 6\left(\frac{\pi}{6}\right) = \pi$$

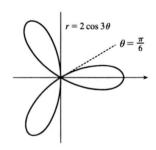

$r = 2\cos 3\theta$

$\theta = \frac{\pi}{6}$

13. $A = \int_0^{2\pi} \frac{1}{2}(1 + 2\sin 6\theta)^2\, d\theta = \frac{1}{2}\int_0^{2\pi}(1 + 4\sin 6\theta + 4\sin^2 6\theta)\, d\theta$

$= \frac{1}{2}\int_0^{2\pi}\left[1 + 4\sin 6\theta + 4\cdot\frac{1}{2}(1 - \cos 12\theta)\right]d\theta$

$= \frac{1}{2}\int_0^{2\pi}(3 + 4\sin 6\theta - 2\cos 12\theta)\, d\theta$

$= \frac{1}{2}\left[3\theta - \frac{2}{3}\cos 6\theta - \frac{1}{6}\sin 12\theta\right]_0^{2\pi}$

$= \frac{1}{2}\left[(6\pi - \frac{2}{3} - 0) - (0 - \frac{2}{3} - 0)\right] = 3\pi$

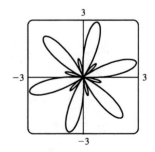

15. The shaded loop is traced out from $\theta = 0$ to $\theta = \pi/2$.

$$A = \int_0^{\pi/2} \tfrac{1}{2} r^2\, d\theta = \tfrac{1}{2} \int_0^{\pi/2} \sin^2 2\theta\, d\theta$$

$$= \tfrac{1}{2} \int_0^{\pi/2} \tfrac{1}{2}(1 - \cos 4\theta)\, d\theta = \tfrac{1}{4}\left[\theta - \tfrac{1}{4}\sin 4\theta\right]_0^{\pi/2}$$

$$= \tfrac{1}{4}\left(\tfrac{\pi}{2}\right) = \tfrac{\pi}{8}$$

17.

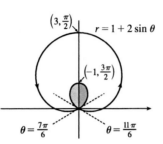

This is a limaçon, with inner loop traced out between $\theta = \frac{7\pi}{6}$ and $\frac{11\pi}{6}$ [found by solving $r = 0$].

$$A = 2 \int_{7\pi/6}^{3\pi/2} \tfrac{1}{2}(1 + 2\sin\theta)^2\, d\theta = \int_{7\pi/6}^{3\pi/2} \left(1 + 4\sin\theta + 4\sin^2\theta\right) d\theta = \int_{7\pi/6}^{3\pi/2} \left[1 + 4\sin\theta + 4 \cdot \tfrac{1}{2}(1 - \cos 2\theta)\right] d\theta$$

$$= \left[\theta - 4\cos\theta + 2\theta - \sin 2\theta\right]_{7\pi/6}^{3\pi/2} = \left(\tfrac{9\pi}{2}\right) - \left(\tfrac{7\pi}{2} + 2\sqrt{3} - \tfrac{\sqrt{3}}{2}\right) = \pi - \tfrac{3\sqrt{3}}{2}$$

19. $2\cos\theta = 1 \;\Rightarrow\; \cos\theta = \tfrac{1}{2} \;\Rightarrow\; \theta = \tfrac{\pi}{3}$ or $\tfrac{5\pi}{3}$.

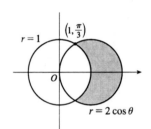

$$A = 2 \int_0^{\pi/3} \tfrac{1}{2}\left[(2\cos\theta)^2 - 1^2\right] d\theta = \int_0^{\pi/3} (4\cos^2\theta - 1)\, d\theta$$

$$= \int_0^{\pi/3} \left\{4\left[\tfrac{1}{2}(1 + \cos 2\theta)\right] - 1\right\} d\theta = \int_0^{\pi/3} (1 + 2\cos 2\theta)\, d\theta$$

$$= \left[\theta + \sin 2\theta\right]_0^{\pi/3} = \tfrac{\pi}{3} + \tfrac{\sqrt{3}}{2}$$

21. $3\cos\theta = 1 + \cos\theta \;\Leftrightarrow\; \cos\theta = \tfrac{1}{2} \;\Rightarrow\; \theta = \tfrac{\pi}{3}$ or $-\tfrac{\pi}{3}$.

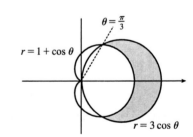

$$A = 2 \int_0^{\pi/3} \tfrac{1}{2}\left[(3\cos\theta)^2 - (1 + \cos\theta)^2\right] d\theta$$

$$= \int_0^{\pi/3} (8\cos^2\theta - 2\cos\theta - 1)\, d\theta = \int_0^{\pi/3} [4(1 + \cos 2\theta) - 2\cos\theta - 1]\, d\theta$$

$$= \int_0^{\pi/3} (3 + 4\cos 2\theta - 2\cos\theta)\, d\theta = \left[3\theta + 2\sin 2\theta - 2\sin\theta\right]_0^{\pi/3}$$

$$= \pi + \sqrt{3} - \sqrt{3} = \pi$$

23. $\sqrt{3}\cos\theta = \sin\theta \;\Rightarrow\; \sqrt{3} = \dfrac{\sin\theta}{\cos\theta} \;\Rightarrow\; \tan\theta = \sqrt{3} \;\Rightarrow\; \theta = \tfrac{\pi}{3}$.

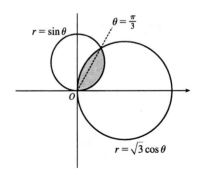

$$A = \int_0^{\pi/3} \tfrac{1}{2}(\sin\theta)^2\, d\theta + \int_{\pi/3}^{\pi/2} \tfrac{1}{2}\left(\sqrt{3}\cos\theta\right)^2 d\theta$$

$$= \int_0^{\pi/3} \tfrac{1}{2} \cdot \tfrac{1}{2}(1 - \cos 2\theta)\, d\theta + \int_{\pi/3}^{\pi/2} \tfrac{1}{2} \cdot 3 \cdot \tfrac{1}{2}(1 + \cos 2\theta)\, d\theta$$

$$= \tfrac{1}{4}\left[\theta - \tfrac{1}{2}\sin 2\theta\right]_0^{\pi/3} + \tfrac{3}{4}\left[\theta + \tfrac{1}{2}\sin 2\theta\right]_{\pi/3}^{\pi/2}$$

$$= \tfrac{1}{4}\left[\left(\tfrac{\pi}{3} - \tfrac{\sqrt{3}}{4}\right) - 0\right] + \tfrac{3}{4}\left[\left(\tfrac{\pi}{2} + 0\right) - \left(\tfrac{\pi}{3} + \tfrac{\sqrt{3}}{4}\right)\right]$$

$$= \tfrac{\pi}{12} - \tfrac{\sqrt{3}}{16} + \tfrac{\pi}{8} - \tfrac{3\sqrt{3}}{16} = \tfrac{5\pi}{24} - \tfrac{\sqrt{3}}{4}$$

25. $\sin 2\theta = \cos 2\theta \quad \Rightarrow \quad \dfrac{\sin 2\theta}{\cos 2\theta} = 1 \quad \Rightarrow \quad \tan 2\theta = 1 \quad \Rightarrow \quad 2\theta = \frac{\pi}{4} \quad \Rightarrow$

$\theta = \frac{\pi}{8} \quad \Rightarrow$

$\qquad A = 8 \cdot 2 \int_0^{\pi/8} \frac{1}{2} \sin^2 2\theta \, d\theta = 8 \int_0^{\pi/8} \frac{1}{2}(1 - \cos 4\theta) \, d\theta$

$\qquad = 4\left[\theta - \frac{1}{4}\sin 4\theta\right]_0^{\pi/8} = 4\left(\frac{\pi}{8} - \frac{1}{4} \cdot 1\right) = \frac{\pi}{2} - 1$

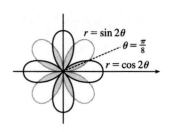

27. The darker shaded region (from $\theta = 0$ to $\theta = 2\pi/3$) represents $\frac{1}{2}$ of the desired area plus $\frac{1}{2}$ of the area of the inner loop.

From this area, we'll subtract $\frac{1}{2}$ of the area of the inner loop (the lighter shaded region from $\theta = 2\pi/3$ to $\theta = \pi$), and then

double that difference to obtain the desired area.

$A = 2\left[\int_0^{2\pi/3} \frac{1}{2}\left(\frac{1}{2} + \cos\theta\right)^2 d\theta - \int_{2\pi/3}^{\pi} \frac{1}{2}\left(\frac{1}{2} + \cos\theta\right)^2 d\theta\right]$

$\quad = \int_0^{2\pi/3} \left(\frac{1}{4} + \cos\theta + \cos^2\theta\right) d\theta - \int_{2\pi/3}^{\pi} \left(\frac{1}{4} + \cos\theta + \cos^2\theta\right) d\theta$

$\quad = \int_0^{2\pi/3} \left[\frac{1}{4} + \cos\theta + \frac{1}{2}(1 + \cos 2\theta)\right] d\theta$

$\qquad\qquad - \int_{2\pi/3}^{\pi} \left[\frac{1}{4} + \cos\theta + \frac{1}{2}(1 + \cos 2\theta)\right] d\theta$

$\quad = \left[\dfrac{\theta}{4} + \sin\theta + \dfrac{\theta}{2} + \dfrac{\sin 2\theta}{4}\right]_0^{2\pi/3} - \left[\dfrac{\theta}{4} + \sin\theta + \dfrac{\theta}{2} + \dfrac{\sin 2\theta}{4}\right]_{2\pi/3}^{\pi}$

$\quad = \left(\frac{\pi}{6} + \frac{\sqrt{3}}{2} + \frac{\pi}{3} - \frac{\sqrt{3}}{8}\right) - \left(\frac{\pi}{4} + \frac{\pi}{2}\right) + \left(\frac{\pi}{6} + \frac{\sqrt{3}}{2} + \frac{\pi}{3} - \frac{\sqrt{3}}{8}\right)$

$\quad = \frac{\pi}{4} + \frac{3}{4}\sqrt{3} = \frac{1}{4}\left(\pi + 3\sqrt{3}\right)$

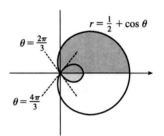

29. $2\sin 2\theta = 1 \quad \Rightarrow \quad \sin 2\theta = \frac{1}{2} \quad \Rightarrow \quad 2\theta = \frac{\pi}{6}, \frac{5\pi}{6}, \frac{13\pi}{6}, \text{ or } \frac{17\pi}{6}.$

By symmetry, the eight points of intersection are given by

$(1, \theta)$, where $\theta = \frac{\pi}{12}, \frac{5\pi}{12}, \frac{13\pi}{12}, \text{ and } \frac{17\pi}{12}, \text{ and}$

$(-1, \theta)$, where $\theta = \frac{7\pi}{12}, \frac{11\pi}{12}, \frac{19\pi}{12}, \text{ and } \frac{23\pi}{12}.$

[There are many ways to describe these points.]

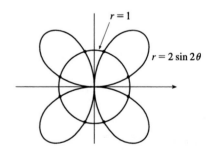

31. The pole is a point of intersection. $\sin\theta = \sin 2\theta = 2\sin\theta \cos\theta \quad \Leftrightarrow$

$\sin\theta \, (1 - 2\cos\theta) = 0 \quad \Leftrightarrow \quad \sin\theta = 0 \text{ or } \cos\theta = \frac{1}{2} \quad \Rightarrow$

$\theta = 0, \pi, \frac{\pi}{3}, \text{ or } -\frac{\pi}{3} \quad \Rightarrow \quad$ the other intersection points are $\left(\frac{\sqrt{3}}{2}, \frac{\pi}{3}\right)$

and $\left(\frac{\sqrt{3}}{2}, \frac{2\pi}{3}\right)$ [by symmetry].

33.

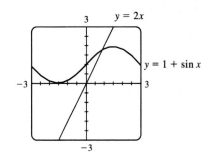

From the first graph, we see that the pole is one point of intersection. By zooming in or using the cursor, we find the θ-values of the intersection points to be $\alpha \approx 0.88786 \approx 0.89$ and $\pi - \alpha \approx 2.25$. (The first of these values may be more easily estimated by plotting $y = 1 + \sin x$ and $y = 2x$ in rectangular coordinates; see the second graph.) By symmetry, the total area contained is twice the area contained in the first quadrant, that is,

$$A = 2 \int_0^\alpha \tfrac{1}{2}(2\theta)^2 \, d\theta + 2 \int_\alpha^{\pi/2} \tfrac{1}{2}(1 + \sin\theta)^2 \, d\theta = \int_0^\alpha 4\theta^2 \, d\theta + \int_\alpha^{\pi/2} \left[1 + 2\sin\theta + \tfrac{1}{2}(1 - \cos 2\theta)\right] d\theta$$

$$= \left[\tfrac{4}{3}\theta^3\right]_0^\alpha + \left[\theta - 2\cos\theta + \left(\tfrac{1}{2}\theta - \tfrac{1}{4}\sin 2\theta\right)\right]_\alpha^{\pi/2} = \tfrac{4}{3}\alpha^3 + \left[\left(\tfrac{\pi}{2} + \tfrac{\pi}{4}\right) - \left(\alpha - 2\cos\alpha + \tfrac{1}{2}\alpha - \tfrac{1}{4}\sin 2\alpha\right)\right] \approx 3.4645$$

35. $L = \displaystyle\int_a^b \sqrt{r^2 + (dr/d\theta)^2} \, d\theta = \int_0^{\pi/3} \sqrt{(3\sin\theta)^2 + (3\cos\theta)^2} \, d\theta = \int_0^{\pi/3} \sqrt{9(\sin^2\theta + \cos^2\theta)} \, d\theta$

$\quad = 3 \displaystyle\int_0^{\pi/3} d\theta = 3\left[\theta\right]_0^{\pi/3} = 3\left(\tfrac{\pi}{3}\right) = \pi.$

As a check, note that the circumference of a circle with radius $\tfrac{3}{2}$ is $2\pi\left(\tfrac{3}{2}\right) = 3\pi$, and since $\theta = 0$ to $\pi = \tfrac{\pi}{3}$ traces out $\tfrac{1}{3}$ of the circle (from $\theta = 0$ to $\theta = \pi$), $\tfrac{1}{3}(3\pi) = \pi.$

37. $L = \displaystyle\int_a^b \sqrt{r^2 + (dr/d\theta)^2} \, d\theta = \int_0^{2\pi} \sqrt{(\theta^2)^2 + (2\theta)^2} \, d\theta = \int_0^{2\pi} \sqrt{\theta^4 + 4\theta^2} \, d\theta$

$\quad = \displaystyle\int_0^{2\pi} \sqrt{\theta^2(\theta^2 + 4)} \, d\theta = \int_0^{2\pi} \theta\sqrt{\theta^2 + 4} \, d\theta$

Now let $u = \theta^2 + 4$, so that $du = 2\theta \, d\theta \quad \left[\theta \, d\theta = \tfrac{1}{2} \, du\right]$ and

$$\int_0^{2\pi} \theta\sqrt{\theta^2 + 4} \, d\theta = \int_4^{4\pi^2+4} \tfrac{1}{2}\sqrt{u} \, du = \tfrac{1}{2} \cdot \tfrac{2}{3}\left[u^{3/2}\right]_4^{4(\pi^2+1)} = \tfrac{1}{3}\left[4^{3/2}(\pi^2 + 1)^{3/2} - 4^{3/2}\right] = \tfrac{8}{3}\left[(\pi^2 + 1)^{3/2} - 1\right]$$

39. The curve $r = 3\sin 2\theta$ is completely traced with $0 \le \theta \le 2\pi$.

$\quad r^2 + \left(\tfrac{dr}{d\theta}\right)^2 = (3\sin 2\theta)^2 + (6\cos 2\theta)^2 \quad \Rightarrow$

$\quad L = \int_0^{2\pi} \sqrt{9\sin^2 2\theta + 36\cos^2 2\theta} \, d\theta \approx 29.0653$

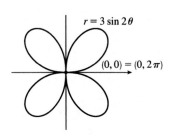

I Complex Numbers

---

1. $(5 - 6i) + (3 + 2i) = (5 + 3) + (-6 + 2)i = 8 + (-4)i = 8 - 4i$

3. $(2 + 5i)(4 - i) = 2(4) + 2(-i) + (5i)(4) + (5i)(-i) = 8 - 2i + 20i - 5i^2 = 8 + 18i - 5(-1)$

$$= 8 + 18i + 5 = 13 + 18i$$

5. $\overline{12 + 7i} = 12 - 7i$

7. $\dfrac{1 + 4i}{3 + 2i} = \dfrac{1 + 4i}{3 + 2i} \cdot \dfrac{3 - 2i}{3 - 2i} = \dfrac{3 - 2i + 12i - 8(-1)}{3^2 + 2^2} = \dfrac{11 + 10i}{13} = \dfrac{11}{13} + \dfrac{10}{13}i$

9. $\dfrac{1}{1 + i} = \dfrac{1}{1 + i} \cdot \dfrac{1 - i}{1 - i} = \dfrac{1 - i}{1 - (-1)} = \dfrac{1 - i}{2} = \dfrac{1}{2} - \dfrac{1}{2}i$

11. $i^3 = i^2 \cdot i = (-1)i = -i$

13. $\sqrt{-25} = \sqrt{25}\,i = 5i$

15. $\overline{12 - 5i} = 12 + 15i$ and $|12 - 15i| = \sqrt{12^2 + (-5)^2} = \sqrt{144 + 25} = \sqrt{169} = 13$

17. $\overline{-4i} = \overline{0 - 4i} = 0 + 4i = 4i$ and $|-4i| = \sqrt{0^2 + (-4)^2} = \sqrt{16} = 4$

19. $4x^2 + 9 = 0 \iff 4x^2 = -9 \iff x^2 = -\frac{9}{4} \iff x = \pm\sqrt{-\frac{9}{4}} = \pm\sqrt{\frac{9}{4}}\,i = \pm\frac{3}{2}i.$

21. By the quadratic formula, $x^2 + 2x + 5 = 0 \iff x = \dfrac{-2 \pm \sqrt{2^2 - 4(1)(5)}}{2(1)} = \dfrac{-2 \pm \sqrt{-16}}{2} = \dfrac{-2 \pm 4i}{2} = -1 \pm 2i.$

23. By the quadratic formula, $z^2 + z + 2 = 0 \iff z = \dfrac{-1 \pm \sqrt{1^2 - 4(1)(2)}}{2(1)} = \dfrac{-1 \pm \sqrt{-7}}{2} = -\dfrac{1}{2} \pm \dfrac{\sqrt{7}}{2}i.$

25. For $z = -3 + 3i$, $r = \sqrt{(-3)^2 + 3^2} = 3\sqrt{2}$ and $\tan\theta = \frac{3}{-3} = -1 \Rightarrow \theta = \frac{3\pi}{4}$ (since z lies in the second quadrant).

Therefore, $-3 + 3i = 3\sqrt{2}\left(\cos\frac{3\pi}{4} + i\sin\frac{3\pi}{4}\right).$

27. For $z = 3 + 4i$, $r = \sqrt{3^2 + 4^2} = 5$ and $\tan\theta = \frac{4}{3} \Rightarrow \theta = \tan^{-1}\left(\frac{4}{3}\right)$ (since z lies in the first quadrant). Therefore,

$3 + 4i = 5\left\{\cos\left[\tan^{-1}\left(\frac{4}{3}\right)\right] + i\sin\left[\tan^{-1}\left(\frac{4}{3}\right)\right]\right\}.$

29. For $z = \sqrt{3} + i$, $r = \sqrt{\left(\sqrt{3}\right)^2 + 1^2} = 2$ and $\tan\theta = \frac{1}{\sqrt{3}} \Rightarrow \theta = \frac{\pi}{6} \Rightarrow z = 2\left(\cos\frac{\pi}{6} + i\sin\frac{\pi}{6}\right).$

For $w = 1 + \sqrt{3}\,i$, $r = 2$ and $\tan\theta = \sqrt{3} \Rightarrow \theta = \frac{\pi}{3} \Rightarrow w = 2\left(\cos\frac{\pi}{3} + i\sin\frac{\pi}{3}\right).$

Therefore, $zw = 2 \cdot 2\left[\cos\left(\frac{\pi}{6} + \frac{\pi}{3}\right) + i\sin\left(\frac{\pi}{6} + \frac{\pi}{3}\right)\right] = 4\left(\cos\frac{\pi}{2} + i\sin\frac{\pi}{2}\right),$

$z/w = \frac{2}{2}\left[\cos\left(\frac{\pi}{6} - \frac{\pi}{3}\right) + i\sin\left(\frac{\pi}{6} - \frac{\pi}{3}\right)\right] = \cos\left(-\frac{\pi}{6}\right) + i\sin\left(-\frac{\pi}{6}\right),$ and $1 = 1 + 0i = 1(\cos 0 + i\sin 0) \Rightarrow$

$1/z = \frac{1}{2}\left[\cos\left(0 - \frac{\pi}{6}\right) + i\sin\left(0 - \frac{\pi}{6}\right)\right] = \frac{1}{2}\left[\cos\left(-\frac{\pi}{6}\right) + i\sin\left(-\frac{\pi}{6}\right)\right].$ For $1/z$, we could also use the formula that precedes

Example 5 to obtain $1/z = \frac{1}{2}\left(\cos\frac{\pi}{6} - i\sin\frac{\pi}{6}\right).$

31. For $z = 2\sqrt{3} - 2i$, $r = \sqrt{\left(2\sqrt{3}\right)^2 + (-2)^2} = 4$ and $\tan\theta = \frac{-2}{2\sqrt{3}} = -\frac{1}{\sqrt{3}}$ $\Rightarrow$ $\theta = -\frac{\pi}{6}$ $\Rightarrow$

$z = 4\left[\cos\left(-\frac{\pi}{6}\right) + i\sin\left(-\frac{\pi}{6}\right)\right]$. For $w = -1 + i$, $r = \sqrt{2}$, $\tan\theta = \frac{1}{-1} = -1$ $\Rightarrow$ $\theta = \frac{3\pi}{4}$ $\Rightarrow$

$w = \sqrt{2}\left(\cos\frac{3\pi}{4} + i\sin\frac{3\pi}{4}\right)$. Therefore, $zw = 4\sqrt{2}\left[\cos\left(-\frac{\pi}{6} + \frac{3\pi}{4}\right) + i\sin\left(-\frac{\pi}{6} + \frac{3\pi}{4}\right)\right] = 4\sqrt{2}\left(\cos\frac{7\pi}{12} + i\sin\frac{7\pi}{12}\right)$,

$z/w = \frac{4}{\sqrt{2}}\left[\cos\left(-\frac{\pi}{6} - \frac{3\pi}{4}\right) + i\sin\left(-\frac{\pi}{6} - \frac{3\pi}{4}\right)\right] = \frac{4}{\sqrt{2}}\left[\cos\left(-\frac{11\pi}{12}\right) + i\sin\left(-\frac{11\pi}{12}\right)\right] = 2\sqrt{2}\left(\cos\frac{13\pi}{12} + i\sin\frac{13\pi}{12}\right)$, and

$1/z = \frac{1}{4}\left[\cos\left(-\frac{\pi}{6}\right) - i\sin\left(-\frac{\pi}{6}\right)\right] = \frac{1}{4}\left(\cos\frac{\pi}{6} + i\sin\frac{\pi}{6}\right)$.

33. For $z = 1 + i$, $r = \sqrt{2}$ and $\tan\theta = \frac{1}{1} = 1$ $\Rightarrow$ $\theta = \frac{\pi}{4}$ $\Rightarrow$ $z = \sqrt{2}\left(\cos\frac{\pi}{4} + i\sin\frac{\pi}{4}\right)$. So by De Moivre's Theorem,

$$(1+i)^{20} = \left[\sqrt{2}\left(\cos\frac{\pi}{4} + i\sin\frac{\pi}{4}\right)\right]^{20} = \left(2^{1/2}\right)^{20}\left(\cos\frac{20\cdot\pi}{4} + i\sin\frac{20\cdot\pi}{4}\right) = 2^{10}(\cos 5\pi + i\sin 5\pi)$$

$$= 2^{10}[-1 + i(0)] = -2^{10} = -1024$$

35. For $z = 2\sqrt{3} + 2i$, $r = \sqrt{\left(2\sqrt{3}\right)^2 + 2^2} = \sqrt{16} = 4$ and $\tan\theta = \frac{2}{2\sqrt{3}} = \frac{1}{\sqrt{3}}$ $\Rightarrow$ $\theta = \frac{\pi}{6}$ $\Rightarrow$ $z = 4\left(\cos\frac{\pi}{6} + i\sin\frac{\pi}{6}\right)$.

So by De Moivre's Theorem,

$$\left(2\sqrt{3} + 2i\right)^5 = \left[4\left(\cos\frac{\pi}{6} + i\sin\frac{\pi}{6}\right)\right]^5 = 4^5\left(\cos\frac{5\pi}{6} + i\sin\frac{5\pi}{6}\right) = 1024\left[-\frac{\sqrt{3}}{2} + \frac{1}{2}i\right] = -512\sqrt{3} + 512i.$$

37. $1 = 1 + 0i = 1(\cos 0 + i\sin 0)$. Using Equation 3 with $r = 1$, $n = 8$, and $\theta = 0$, we have

$$w_k = 1^{1/8}\left[\cos\left(\frac{0 + 2k\pi}{8}\right) + i\sin\left(\frac{0 + 2k\pi}{8}\right)\right] = \cos\frac{k\pi}{4} + i\sin\frac{k\pi}{4}, \text{ where } k = 0, 1, 2, \ldots, 7.$$

$w_0 = 1(\cos 0 + i\sin 0) = 1$, $w_1 = 1\left(\cos\frac{\pi}{4} + i\sin\frac{\pi}{4}\right) = \frac{1}{\sqrt{2}} + \frac{1}{\sqrt{2}}i$,

$w_2 = 1\left(\cos\frac{\pi}{2} + i\sin\frac{\pi}{2}\right) = i$, $w_3 = 1\left(\cos\frac{3\pi}{4} + i\sin\frac{3\pi}{4}\right) = -\frac{1}{\sqrt{2}} + \frac{1}{\sqrt{2}}i$,

$w_4 = 1(\cos\pi + i\sin\pi) = -1$, $w_5 = 1\left(\cos\frac{5\pi}{4} + i\sin\frac{5\pi}{4}\right) = -\frac{1}{\sqrt{2}} - \frac{1}{\sqrt{2}}i$,

$w_6 = 1\left(\cos\frac{3\pi}{2} + i\sin\frac{3\pi}{2}\right) = -i$, $w_7 = 1\left(\cos\frac{7\pi}{4} + i\sin\frac{7\pi}{4}\right) = \frac{1}{\sqrt{2}} - \frac{1}{\sqrt{2}}i$

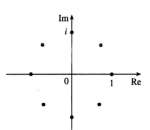

39. $i = 0 + i = 1\left(\cos\frac{\pi}{2} + i\sin\frac{\pi}{2}\right)$. Using Equation 3 with $r = 1$, $n = 3$, and $\theta = \frac{\pi}{2}$, we have

$$w_k = 1^{1/3}\left[\cos\left(\frac{\frac{\pi}{2} + 2k\pi}{3}\right) + i\sin\left(\frac{\frac{\pi}{2} + 2k\pi}{3}\right)\right], \text{ where } k = 0, 1, 2.$$

$w_0 = \left(\cos\frac{\pi}{6} + i\sin\frac{\pi}{6}\right) = \frac{\sqrt{3}}{2} + \frac{1}{2}i$

$w_1 = \left(\cos\frac{5\pi}{6} + i\sin\frac{5\pi}{6}\right) = -\frac{\sqrt{3}}{2} + \frac{1}{2}i$

$w_2 = \left(\cos\frac{9\pi}{6} + i\sin\frac{9\pi}{6}\right) = -i$

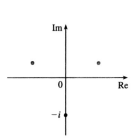

41. Using Euler's formula (6) with $y = \frac{\pi}{2}$, we have $e^{i\pi/2} = \cos\frac{\pi}{2} + i\sin\frac{\pi}{2} = 0 + 1i = i$.

43. Using Euler's formula (6) with $y = \frac{\pi}{3}$, we have $e^{i\pi/3} = \cos\frac{\pi}{3} + i\sin\frac{\pi}{3} = \frac{1}{2} + \frac{\sqrt{3}}{2}i$.

45. Using Equation 7 with $x = 2$ and $y = \pi$, we have $e^{2+i\pi} = e^2 e^{i\pi} = e^2(\cos\pi + i\sin\pi) = e^2(-1 + 0) = -e^2$.

47. Take $r = 1$ and $n = 3$ in De Moivre's Theorem to get

$$[1(\cos\theta + i\sin\theta)]^3 = 1^3(\cos 3\theta + i\sin 3\theta)$$

$$(\cos\theta + i\sin\theta)^3 = \cos 3\theta + i\sin 3\theta$$

$$\cos^3\theta + 3(\cos^2\theta)(i\sin\theta) + 3(\cos\theta)(i\sin\theta)^2 + (i\sin\theta)^3 = \cos 3\theta + i\sin 3\theta$$

$$\cos^3\theta + (3\cos^2\theta\,\sin\theta)i - 3\cos\theta\,\sin^2\theta - (\sin^3\theta)i = \cos 3\theta + i\sin 3\theta$$

$$(\cos^3\theta - 3\sin^2\theta\,\cos\theta) + (3\sin\theta\,\cos^2\theta - \sin^3\theta)i = \cos 3\theta + i\sin 3\theta$$

Equating real and imaginary parts gives $\cos 3\theta = \cos^3\theta - 3\sin^2\theta\,\cos\theta$ and $\sin 3\theta = 3\sin\theta\,\cos^2\theta - \sin^3\theta$.

49. $F(x) = e^{rx} = e^{(a+bi)x} = e^{ax+bxi} = e^{ax}(\cos bx + i\sin bx) = e^{ax}\cos bx + i(e^{ax}\sin bx)$ $\Rightarrow$

$$F'(x) = (e^{ax}\cos bx)' + i(e^{ax}\sin bx)'$$

$$= (ae^{ax}\cos bx - be^{ax}\sin bx) + i(ae^{ax}\sin bx + be^{ax}\cos bx)$$

$$= a[e^{ax}(\cos bx + i\sin bx)] + b[e^{ax}(-\sin bx + i\cos bx)]$$

$$= ae^{rx} + b[e^{ax}(i^2\sin bx + i\cos bx)]$$

$$= ae^{rx} + bi[e^{ax}(\cos bx + i\sin bx)] = ae^{rx} + bie^{rx} = (a+bi)e^{rx} = re^{rx}$$